AUTOMATIC
CONTROL
SYSTEMS

BENJAMIN C. KUO

Professor Emeritus
Department of Electrical and Computer Engineering
University of Illinois at Urbana-Champaign

SIXTH EDITION

AUTOMATIC
CONTROL
SYSTEMS

Prentice-Hall International, Inc.

© 1991, 1987, 1982, 1975, 1967, 1962 by Prentice-Hall, Inc.
A Division of Simon & Schuster
Englewood Cliffs, New Jersey 07632

Printed in the United States of America
10   9   8   7   6   5   4   3   2   1

ISBN 0-13-053505-2

Prentice-Hall International (UK) Limited, *London*
Prentice-Hall of Australia Pty. Limited, *Sydney*
Prentice-Hall Canada Inc., *Toronto*
Prentice-Hall Hispanoamericana, S.A., *Mexico*
Prentice-Hall of India Private Limited, *New Delhi*
Prentice-Hall of Japan, Inc., *Tokyo*
Simon & Schuster Asia Pte. Ltd., *Singapore*
Editora Prentice-Hall do Brasil, Ltda., *Rio de Janeiro*
Prentice-Hall, Inc., *Englewood Cliffs, New Jersey*

To my wife
who has endured through six editions of this book.

To my children
whom I am proud to have.

CONTENTS

PREFACE   xiii

PREFACE TO SOFTWARE   xvii

1   INTRODUCTION   1

1-1   CONTROL SYSTEMS   1
1-2   WHAT IS FEEDBACK AND WHAT ARE ITS EFFECTS?   6
1-3   TYPES OF FEEDBACK CONTROL SYSTEMS   10
1-4   SUMMARY   14

2   MATHEMATICAL FOUNDATION   16

2-1   INTRODUCTION   16
2-2   COMPLEX-VARIABLE CONCEPT   17
2-3   DIFFERENTIAL EQUATIONS   19
2-4   LAPLACE TRANSFORM   22
2-5   INVERSE LAPLACE TRANSFORM BY PARTIAL-FRACTION EXPANSION   28
2-6   APPLICATION OF LAPLACE TRANSFORM TO THE SOLUTION OF LINEAR ORDINARY DIFFERENTIAL EQUATIONS   33
2-7   ELEMENTARY MATRIX THEORY   36
2-8   MATRIX ALGEBRA   41
2-9   VECTOR-MATRIX FORM OF STATE EQUATIONS   47
2-10   DIFFERENCE EQUATIONS   48
2-11   THE z-TRANSFORM   50

2-12   APPLICATION OF THE *z*-TRANSFORM TO THE SOLUTION
        OF LINEAR DIFFERENCE EQUATIONS   59
2-13   SUMMARY   60

3   TRANSFER FUNCTIONS, BLOCK DIAGRAMS, AND SIGNAL-FLOW GRAPHS   67

3-1   INTRODUCTION   67
3-2   IMPULSE RESPONSE AND TRANSFER FUNCTIONS OF LINEAR
       SYSTEMS   68
3-3   BLOCK DIAGRAMS   71
3-4   SIGNAL-FLOW GRAPHS   77
3-5   SUMMARY OF THE BASIC PROPERTIES OF SIGNAL-FLOW
       GRAPHS   79
3-6   DEfiNITIONS OF SIGNAL-FLOW GRAPHS   79
3-7   SIGNAL-FLOW-GRAPH ALGEBRA   82
3-8   EXAMPLES OF THE CONSTRUCTION OF SIGNAL-FLOW GRAPHS   84
3-9   GENERAL GAIN FORMULA FOR SIGNAL-FLOW GRAPHS   87
3-10   APPLICATION OF THE GENERAL GAIN FORMULA TO BLOCK
       DIAGRAMS   93
3-11   STATE DIAGRAM   94
3-12   TRANSFER FUNCTIONS OF DISCRETE-DATA SYSTEMS   100
3-13   SUMMARY   108

4   MATHEMATICAL MODELING OF PHYSICAL SYSTEMS   123

4-1   INTRODUCTION   123
4-2   EQUATIONS OF ELECTRIC NETWORKS   124
4-3   MODELING OF MECHANICAL SYSTEM ELEMENTS   127
4-4   EQUATIONS OF MECHANICAL SYSTEMS   142
4-5   SENSORS AND ENCODERS IN CONTROL SYSTEMS   150
4-6   DC MOTORS IN CONTROL SYSTEMS   164
4-7   LINEARIZATION OF NONLINEAR SYSTEMS   176
4-8   SYSTEMS WITH TRANSPORTATION LAGS   181
4-9   SUMMARY   183

5   STATE-VARIABLE ANALYSIS OF LINEAR DYNAMIC SYSTEMS   203

5-1   INTRODUCTION   203
5-2   MATRIX REPRESENTATION OF STATE EQUATIONS   204
5-3   STATE-TRANSITION MATRIX   206
5-4   STATE-TRANSITION EQUATION   211

5-5   RELATIONSHIP BETWEEN STATE EQUATIONS AND HIGH-ORDER
       DIFFERENTIAL EQUATIONS   217
5-6   TRANSFORMATION TO PHASE-VARIABLE CANONICAL FORM   222
5-7   RELATIONSHIP BETWEEN STATE EQUATIONS AND TRANSFER
       FUNCTIONS   226
5-8   CHARACTERISTIC EQUATION AND EIGENVALUES   229
5-9   DECOMPOSITION OF TRANSFER FUNCTIONS   231
5-10   CONTROLLABILITY OF LINEAR SYSTEMS   236
5-11   OBSERVABILITY OF LINEAR SYSTEMS   241
5-12   STATE EQUATIONS OF LINEAR DISCRETE-DATA SYSTEMS   243
5-13 z-TRANSFORM SOLUTION OF DISCRETE STATE EQUATIONS   247
5-14   STATE DIAGRAMS OF DISCRETE-DATA SYSTEMS   250
5-15   A FINAL ILLUSTRATIVE EXAMPLE: MAGNETIC-BALL-SUSPENSION
       SYSTEM   255
5-16   SUMMARY   258

6 STABILITY OF LINEAR CONTROL SYSTEMS 279

6-1   INTRODUCTION   279
6-2   BOUNDED-INPUT BOUNDED-OUTPUT (BIBO) STABILITY
       CONTINUOUS-DATA SYSTEMS   280
6-3   ZERO-INPUT STABILITY CONTINUOUS-DATA SYSTEMS   282
6-4   METHODS OF DETERMINING STABILITY OF LINEAR CONTROL
       SYSTEMS CONTINUOUS-DATA SYSTEMS   284
6-5   ROUTH–HURWITZ CRITERION   285
6-6   STABILITY OF DISCRETE-DATA SYSTEMS   293
6-7   STABILITY TESTS OF DISCRETE-DATA SYSTEMS   295
6-8   SUMMARY   298

7 TIME-DOMAIN ANALYSIS OF CONTROL SYSTEMS 307

7-1   TIME RESPONSE OF CONTINUOUS-DATA SYSTEMS:
       INTRODUCTION   307
7-2   TYPICAL TEST SIGNALS FOR THE TIME RESPONSE OF CONTROL
       SYSTEMS   308
7-3   TIME-DOMAIN PERFORMANCE OF CONTINUOUS-DATA CONTROL
       SYSTEMS:
       THE STEADY-STATE ERROR   310
7-4   TIME-DOMAIN PERFORMANCE OF CONTROL SYSTEMS:
       TRANSIENT RESPONSE   325
7-5   TRANSIENT RESPONSE OF A PROTOTYPE SECOND-ORDER
       SYSTEM   327
7-6   TIME-DOMAIN ANALYSIS OF A POSITION-CONTROL SYSTEM   338
7-7   EFFECTS OF ADDING POLES AND ZEROS TO TRANSFER
       FUNCTIONS   349

7-8   DOMINANT POLES OF TRANSFER FUNCTIONS   355
7-9   APPROXIMATION OF HIGH-ORDER SYSTEMS BY LOW-ORDER
       SYSTEMS
       CONTINUOUS-DATA SYSTEMS   357
7-10   TIME RESPONSE OF DISCRETE-DATA SYSTEMS: INTRODUCTION   369
7-11   MAPPING BETWEEN s-PLANE AND z-PLANE TRAJECTORIES   373
7-12   RELATION BETWEEN CHARACTERISTIC-EQUATION ROOTS
       AND TRANSIENT RESPONSE   378
7-13   STEADY-STATE-ERROR ANALYSIS OF DISCRETE-DATA SYSTEMS   380
7-14   SUMMARY   385

8 ROOT-LOCUS TECHNIQUE 398

8-1   INTRODUCTION   398
8-2   BASIC PROPERTIES OF THE ROOT LOCI   399
8-3   PROPERTIES AND CONSTRUCTION OF THE COMPLETE ROOT
       LOCI   405
8-4   SOME IMPORTANT ASPECTS OF THE CONSTRUCTION OF THE
       ROOT LOCI   433
8-5   ROOT CONTOUR: MULTIPLE-PARAMETER VARIATION   440
8-6   ROOT LOCI OF DISCRETE-DATA CONTROL SYSTEMS   447
8-7   SUMMARY   451

9 TIME-DOMAIN DESIGN OF CONTROL SYSTEMS 460

9-1   INTRODUCTION   460
9-2   TIME-DOMAIN DESIGN WITH THE PID CONTROLLER   463
9-3   TIME-DOMAIN DESIGN OF THE PHASE-LEAD CONTROLLER   479
9-4   TIME-DOMAIN DESIGN OF THE PHASE-LAG CONTROLLER   488
9-5   LEAD-LAG CONTROLLER   496
9-6   POLE-ZERO-CANCELLATION COMPENSATION   500
9-7   FORWARD AND FEEDFORWARD COMPENSATION   508
9-8   DESIGN OF ROBUST CONTROL SYSTEMS   514
9-9   MINOR-LOOP FEEDBACK CONTROL   521
9-10   STATE-FEEDBACK CONTROL   525
9-11   POLE-PLACEMENT DESIGN THROUGH STATE FEEDBACK   527
9-12   STATE FEEDBACK WITH INTEGRAL CONTROL   531
9-13   DIGITAL IMPLEMENTATION OF CONTROLLERS   538
9-14   SUMMARY   543

10 FREQUENCY-DOMAIN ANALYSIS OF CONTROL SYSTEMS 562

10-1   INTRODUCTION   562
10-2 M_p, ω_p, AND BANDWIDTH OF THE PROTOTYPE SECOND-ORDER
       SYSTEM   566

10-3   EFFECTS OF ADDING A ZERO TO THE OPEN-LOOP TRANSFER FUNCTION   571

10-4   EFFECTS OF ADDING A POLE TO THE OPEN-LOOP TRANSFER FUNCTION   575

10-5   NYQUIST STABILITY CRITERION   577

10-6   RELATION BETWEEN THE ROOT LOCI AND THE NYQUIST PLOT   592

10-7   ILLUSTRATIVE EXAMPLES OF THE APPLICATIONS OF THE NYQUIST CRITERION   595

10-8   EFFECTS OF ADDITION OF POLES AND ZEROS TO $G(s)H(s)$ ON THE SHAPE OF THE NYQUIST LOCUS   606

10-9   STABILITY ANALYSIS OF MULTILOOP SYSTEMS   611

10-10 STABILITY OF LINEAR CONTROL SYSTEMS WITH TIME DELAYS   613

10-11 RELATIVE STABILITY: GAIN MARGIN AND PHASE MARGIN   618

10-12 STABILITY ANALYSIS WITH THE BODE PLOT   623

10-13 RELATIVE STABILITY FROM THE SLOPE OF THE MAGNITUDE CURVE OF THE BODE PLOT   627

10-14 STABILITY ANALYSIS WITH THE GAIN-PHASE PLOT   630

10-15 CONSTANT-M LOCI IN THE $G(j\omega)$-PLANE   631

10-16 CONSTANT-PHASE LOCI IN THE $G(j\omega)$-PLANE   635

10-17 CONSTANT-M AND CONSTANT-N LOCI IN THE MAGNITUDE-VERSUS-PHASE PLANE: THE NICHOLS CHART   638

10-18 NICHOLS CHART APPLICATIONS FOR NONUNITY-FEEDBACK SYSTEMS   642

10-19 SENSITIVITY STUDIES IN THE FREQUENCY DOMAIN   642

10-20 FREQUENCY-DOMAIN ANALYSIS OF DISCRETE-DATA CONTROL SYSTEMS   646

10-21 SUMMARY   650

11 FREQUENCY-DOMAIN DESIGN OF CONTROL SYSTEMS   664

11-1   INTRODUCTION   664

11-2   DESIGN WITH THE PD CONTROLLER   669

11-3   DESIGN WITH THE PI CONTROLLER   672

11-4   DESIGN WITH THE PID CONTROLLER   677

11-5   DESIGN WITH THE PHASE-LEAD CONTROLLER   679

11-6   DESIGN WITH THE PHASE-LAG CONTROLLER   692

11-7   DESIGN WITH LEAD-LAG (LAG-LEAD) CONTROLLERS   704

11-8   DESIGN WITH BRIDGED-T (NOTCH) CONTROLLER   706

11-9   SUMMARY   713

APPENDIX A
FREQUENCY-DOMAIN PLOTS   721

A-1   POLAR PLOTS   722
A-2   BODE PLOT (CORNER PLOT OR ASYMPTOTIC PLOT)   727
A-3   MAGNITUDE-VERSUS-PHASE PLOT   738
A-4   GAIN AND PHASE CROSSOVER POINTS   739
A-5   MINIMUM-PHASE AND NONMINIMUM-PHASE FUNCTIONS   740

APPENDIX B
LAPLACE TRANSFORM TABLE   742

ANSWERS AND HINTS FOR SELECTED PROBLEMS   745

INDEX   753

PREFACE (Readme)

I hope that your first impression of the sixth edition is favorable. A considerable amount of thought, feedback, and effort have gone into the writing and production of this revision. The two-color format of this text should enhance its readability and effectiveness, as much of the material on control systems is heavily dependent on graphical techniques.

The fifth edition of *Automatic Control Systems,* published in 1987, was one of the first texts on control systems accompanied by computer software disks. The idea apears to have taken hold, as today, many similar texts have accompanying software. The sixth edition follows the same philosophy of computer-aided learning as the previous edition, with the computer software, Automatic Control Systems Programs (ACSP), expanded and improved. Although there are many commercially available control-systems software packages that are more elaborate and more powerful, the ACSP software contains more than just a set of problem-solving computer programs. It is designed to enhance the learning experience from this book. The reader should note that all the numerical examples in the text are solved and that the graphical data are generated from the computer programs in the ACSP. Furthermore, all the computational aspects of the problems in the text can be obtained with the aid of one or more of these programs. Several programs contain drill problems that should be valuable for self-study. The program disks for the ACSP software are available for either the IBM PC (or compatibles) or the Macintosh computer. More details on the programs are found in the *Preface to ACSP Computer Disks And Software Manual* following this preface.

A significant portion of the text has been revised and expanded. Highlights of the major revisions are as follows:

1. A chapter is devoted to the stability of linear systems, Chapter 6.
2. The material on state-variable analysis in Chapter 5 is revamped and slightly condensed.
3. There is a complete rewrite of the Nyquist criterion, Chapter 10, resulting in a simplified procedure for minimum-phase and nonminimum-phase transfer functions.
4. The chapters on design, Chapters 9 and 11, have been extensively revised and expanded. The sections on PID control, phase-lead and phase-lag control, lead-lag control, and pole-zero cancellation control have all been expanded. In addition, discussions on disturbance-rejection and design sensitivity specifications have been included. Another new addition is the discussion and demonstration of robust control design at the entry level.
5. The majority of the problems at the end of each chapter have been revised and expanded. New values have been assigned to old problems, and many new problems have been added.
6. Summary and review sections are added at the end of each chapter.
7. Correlation between the Nyquist plot and the root-locus plot is illustrated extensively.
8. Additional computer problems are devised for applicable chapters. More computer problems are found in the ASCP Software Manual.

The following paragraphs are aimed at three groups: the professors who have adopted the text, or who we hope will select this as their text; the practicing engineers who hope to find answers and solutions in the text to solve their design problems, and, finally, the students, who have no choice since most likely the reason they are reading it is because the professor has adopted the book.

To the Professors: The material assembled in this book is an outgrowth of a senior-level control-system course taught by the author at the University of Illinois at Urbana-Champaign for over 30 years. The first five editions have been adopted by hundreds of universities in the United States and around the world, and have been translated into six different languages. The sixth edition has been expanded to a total of eleven chapters and two appendices. Great emphasis has been placed on the design of control systems. Many systems, such as the **sun-seeker system**, **the liquid-level control system**, **magnetic-ball-suspension control**, **broom-balancing**, **aircraft-attitude control**, **dc-motor control**, **space-vehicle payload control**, and others, are carried out as numerical examples throughout the text from analysis to design. There is more material in the text than can be covered in one semester. The subject on discrete-data control systems follows the treatment of continuous-data control systems in selected chapters. Personally, I find it easier to introduce discrete-data-system methods as extensions to their analog counterparts, although, realistically, it is difficult to cover all the subjects on digital control in a one-semester course. Typically, in a sixteen-week semester, I could assign and cover most of the major topics offered in the text with the exception of discrete-data systems beyond Chapter 5. Mathematical modeling of mechanical systems and dc motors in Chapter 4 are covered along with the examples and discussions starting from Chapter 6. The rest of the material in Chapter 4 is left to the students to read on their own. A sample section-by-section selection for a one-semester course is given in the Instructor's Manual, which is available from the publisher to qualified instructors. The *Manual* also contains detailed solutions to all the problems in the text, as well as additional problems for future assignments and possibly for quizzes and exams. Upon request, the publisher can also provide to the qualified instructor a set of transparency masters for selected illustrations. Contemporary professors and students are becoming busier and busier, and the objective of this text and all its supplemental materials is to make their teaching and learning efforts more efficient and enjoyable.

To Practicing Engineers: This book is written with the readers in mind, and is very suitable for self-study. The objective is to treat the subjects clearly and thoroughly. The book does not follow the theorm–proof–Q.E.D. format and is without heavy mathematics. The author has consulted extensively for the industry over his 35-year teaching career, and has participated in solving numerous control-systems problems, ranging from aerospace systems, industrial controls, automotive controls, and control of computer peripherals. Although it is difficult to adopt all the details and realisms of practical problems in a textbook at this level, some examples and problems reflect simplified versions of real-life systems.

The ACSP software should be a useful tool since even most companies are equipped with mainframe computers and commercially available software. As mentioned before, the programs in the ACSP are integrated with the material in the text, and are very convenient to apply. Utility programs such as the ones for root solving, partial-fraction expansion, matrix multiplication and inversion are handy for day-to-day applications. I hope that you will find this book useful for solving your engineering problems.

To the Students: You have had it now that you have signed up for this course, and your professor has adopted this book! You had no say about the choice. Worse yet, one of the reasons that your professor made the selection is because he or she intends to give you a great deal of work. Please don't misunderstand me. What I really mean is that although this is an easy book to study from, it is a no-nonsense book. It does not have any cartoons or nice-looking photographs to amuse you. From here on it is all business and hard work. You should have the prerequisites on subjects found in a typical linear-systems course, such as how to solve linear ordinary differential equations, Laplace transforms and their applications, time response and frequency-domain analysis of linear systems. In this text you should find no new mathematics that you have not been exposed to. What is interesting and challenging is that you are going to learn how to apply some of the mathematics that you have acquired during the last two or three years of study in college.

This book has a total of 171 illustrative examples. Some of these are simple for the purpose of illustrating new ideas and subject matters. Some examples are quite elaborate, in order to bring the practical world closer to you. Furthermore, the objective of this book is to present the complex subjects in a clear and thorough way. The professor already knows the subjects well, so the book is written with the students in mind. One of the important learning strategies for the student is not to rely strictly on the assigned textbook. When studying a certain subject, go to the library and check out a few similar texts to see how the other authors treat the same problem. You may gain new perspectives on the subject and discover how one author may treat the material with more care and thoroughness than the others.

The ACSP software package should be a great learning aid to accompany this text. It is possible that your professor has made available other programs that can be run on a mainframe computer or PCs, but the dedicated feature of the programs and the convenience of having your personal copy of the software should be of great advantage to you.

Most students in college do not believe that the material they learn in the classroom is ever going to be applied directly in industry. Some of my students came back from field and interview trips totally surprised to find that the material they learned in the course on control systems is being used in the industry today. They are surprised to find that this text is also a popular reference to the practicing engineer. Unfortunately, these fact-finding and self-motivating trips usually occur near the end of the semester, which is often too late for students to get motivated. I am telling you this at an early stage with the hope that the book and your professor will motivate you to learn something about control sytems. Your professor probably will be telling you the same thing in the lecture, but you may not believe him or her. The bottom line is that the subjects treated in this text are not difficult, but you have to do the exercises to gain full comprehension. There are many learning aids available to you: the ACSP contains programs for problem solving; the book contains Review Questions and a Summary at the end of each chapter. There is also a section on Answers and Hints for Selected Problems at the end of the text. The *Computer Disks and Software Manual of the ACSP* also contains additional computer and drill problems for you to advance at your own pace.

I hope that you enjoy the book and will be able to learn from it. This book represents another major textbook acquisition in your college career. My advice to you is not to sell it back to the bookstore at the end of the semester. If you do so, not only will you be violating the cardinal law of trading, of buying low and selling high, but most likely when you get into the industry, your first assignment may involve the design of a control system, regardless of your specialization.

B.C. Kuo
Champaign, Illinois

PREFACE TO ACSP COMPUTER DISKS AND SOFTWARE MANUAL

In recent years, the analysis and design of control systems have been dramatically affected by the widespread use of personal computers. With the mainframe computer, the user may not have the convenience of getting access at any time or at any place. Personal computers have become so powerful and advanced that they can be used to solve sophisticated and complex control systems problems with ease. Today, most schools and companies are equipped with personal computers that are easily accessible and inexpensive. The ACSP software package is designed to complement the material in this text, and is made available at a very reasonable cost. The programs found in the ACSP are capable of solving complex linear control systems with reasonable speed. All the illustrative examples as well as all the problems found in the text are solved using the programs in the ACSP.

The ACSP software is contained in two $5\frac{1}{4}$-in. floppy disks for use on the IBM PC, XT, AT, PS/2®, or any compatible computer. The ACSP software is also available for the Apple Macintosh® computers. You should make the proper selection when ordering with the Order Form found at the end of the book.

The programs contained in the ACSP disks are as follows:

LINSYS: linear system analysis in the state-variable domain. The system parameters are entered in the state-equation form. The following computations can be conducted: inverse of the **A** matrix, determinant of **A**, matrix coefficients of the numerator of $(s\mathbf{I} - \mathbf{A})^{-1}$, characteristic equation of **A**, eigenvalues of **A**, state-transition matrix, exp $(\mathbf{A}t)$, phase-variable canonical form, check controllability and observability, time response with step input and/or initial states, and pole-placement design with state feedback.

POLYROOT: polynomial root solving.

MATRIXM: multiplication of two matrices.

PFE: partial-fraction expansion of s-domain transfer functions.

CLRSP: time response (impulse, step, and ramp response) of continuous-data control systems (open-loop and closed-loop).

ROOTLOCI: root locus diagram of s-diagram equations (can be applied to z-domain equations). Computes K for prescribed damping and stability.

FREQRP: frequency-domain analysis of linear continuous-data systems. Bode plot, Nyquist plot, gain-phase plot, and Nichols chart (requires HP plotter). Computes gain and phase margins, M_p, bandwidth, sensitivity, and resonant frequency.

FREQCAD: computer-aided design of linear continuous-data control system in the frequency domain. The user specifies the phase margin or the gain margin, selects the controller type, such as phase-lead, phase-lag, etc., and the program computes the controller parameters and performance data.

SDCS: analysis of sampled-data control system with/without a zero-order hold. A digital controller can be included.

INVZ: time response of a digital control system by inputting the z-transfer function.

One of the advantages of the ACSP is that the programs are easy to use, and they can be run on a moderately configured PC with two floppy disk drives or one hard disk, 512K of installed memory, and a monochrome or color monitor. An EPSON or IBM or any other compatible dot matrix printer is needed for printouts and graphs. Plots can be made at the end of each run, and need not be made separately. The computer does not have to be equipped with hardware for graphics. The ouputs of some of the programs can be better plotted on the HP-7470-series or the HP ColorPro or compatible digital plotter with two or more pens, but the plotter is not necessary, except for the case of the Nichols chart.

For those who are studying from this text, either in an orgnaized course or self-study, the ACSP software should make the learning experience more rewarding and enjoyable.

The *Computer Disks and Software Manual* of the ACSP also contains additional computer problems on selected text material.

B.C. Kuo
Champaign, Illinois

AUTOMATIC
CONTROL
SYSTEMS

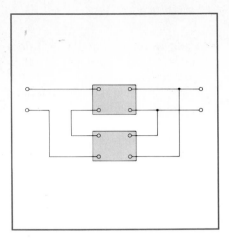

INTRODUCTION

CONTROL SYSTEMS

In this introductory chapter, we attempt to familiarize the reader with the following subjects:

1.   What a control system is.
2.   Why control systems are important.
3.   What the basic components of a control system are.
4.   Why feedback is incorporated into most control systems.
5.   Types of control systems.

One of the most commonly asked questions by a novice to the subject of control systems is: What is a control system? To answer the question, we can cite that in our daily life there are numerous "objectives" that need to be accomplished. For instance, in the domestic domain, we need to regulate the temperature and humidity of homes and buildings for comfortable living. For transportation, we need to control the automobile and airplane to go from one point to another accurately and safely. Industrially, manufacturing processes contain numerous objectives for products that will satisfy the precision and cost-effectiveness requirements. An average human being is capable of performing a wide range of tasks, including decision making. Some of these tasks, such as picking up objects and walking from one point to another, are commonly carried out in a routine fashion. Under certain conditions, some of these tasks are to be performed in the best possible way. For instance, an athlete running a 100-yard dash has the objective of running that distance in the shortest possible time. A marathon runner, on the other hand, not only must run the distance as quickly as possible, but in doing so, he or she must control the consumption of energy and devise the best strategy for the race. The means of achieving these "objectives" usually involve the use of control systems.

In recent years, control systems have assumed an increasingly important role in the development and advancement of modern civilization and technology. Practically every aspect of our day-to-day activities is affected by some type of control system. Control systems are found in abundance in all sectors of industry, such as quality control of manufactured products, automatic assembly line, machine-tool control, space technology and weapon systems, computer control, transportation systems, power systems, robotics, and many others. Even such problems as inventory control and social and economic-systems control may be approached from the theory of automatic control.

The basic ingredients of a control system can be described by:

1. objectives of control,
2. control-system components,
3. results or outputs.

actuating signals

controlled variables

In block diagram form, the basic relationship between these three components is illustrated in Fig. 1-1. In more technical terms, the **objectives** can be identified with **inputs**, or **actuating signals**, *u*, and the results are also called **outputs**, or the **controlled variables**, *c*. In general, the objective of the control system is to control the outputs in some prescribed manner by the inputs through the elements of the control system.

As a simple example of the control system fashioned in Fig. 1-1, consider the steering control of an automobile. The direction of the two front wheels can be regarded as the controlled variable, or the output, *c*; the direction of the steering wheel is the actuating signal, or the input, *u*. The control system, or process in this case, is composed of the steering mechanism and the dynamics of the entire automobile. However, if the objective is to control the speed of the automobile, then the amount of pressure exerted on the accelerator is the actuating signal, and the vehicle speed is the controlled variable. As a whole, we can regard the automobile control system as one with two inputs (steering and accelerator) and two outputs (heading and speed). In this case, the two controls and two outputs are independent of each other, but, in general, there are systems for which the controls are coupled. Systems with more than one input and one output are called **multivariable systems**.

As another example of a control system, we consider the idle-speed control of an automobile engine. The objective of such a control system is to maintain the engine

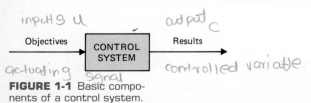

FIGURE 1-1 Basic components of a control system.

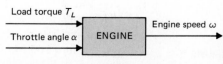

FIGURE 1-2 Idle-speed control system.

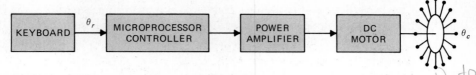

FIGURE 1-3 Open-loop printwheel control system.

idle speed at a relatively low value (for fuel economy) regardless of the applied engine loads (e.g., transmission, power steering, air conditioning, etc.). Without the idle-speed control, any sudden engine-load application would cause a drop in engine speed that might cause the engine to stall. Thus, the main objectives of the idle-speed control system are (1) to eliminate or minimize the speed droop when engine loading is applied, and (2) to maintain the engine idle speed at a desired value. Figure 1-2 shows the block diagram of the idle-speed control system from the standpoint of inputs–system–outputs. In this case, the throttle angle α and the load torque T_L (due to the application of air conditioning, power steering, transmission, or power brakes, etc.) are the inputs, and the engine speed ω is the output. The engine is the controlled process, or system.

printwheel control

microprocessor

      Figure 1-3 shows an example of the printwheel control system of a word processor or electronic typewriter. The printwheel, which typically has 96 or 100 characters, is to be rotated to position the desired character in front of the hammer for impact printing. The character selection is done in the usual manner from a keyboard. Once a certain key on the keyboard is depressed, a command for the printwheel to rotate from the present position to the next position is initiated. The microprocessor computes the direction and the distance to be traveled, and sends out a control logic signal to the power amplifier, which in turn controls the motor that drives the printwheel. In practice, the control signals generated by the microprocessor controller should be able to drive the printwheel from one position to another sufficiently fast and with high print quality, which means that the position of the printwheel should be controlled accurately. Figure 1-4 shows a typical set of input and output of the system. When a reference command input is given, the signal is represented as a step function. Since the electric circuit of the motor has inductance and the mechanical load has inertia, the printwheel cannot respond to the input instantaneously. Typically, it will follow the response as shown, and settle at the new position after some time t_1. Printing should not

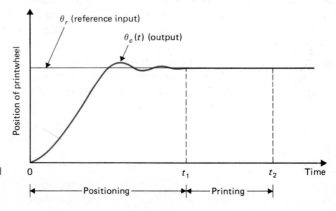

FIGURE 1-4 Typical input and output of the printwheel control system.

begin until the printwheel has come to a complete stop; otherwise, the character will be smeared. Figure 1-4 shows that after the printwheel has settled, the period from t_1 to t_2 is reserved for printing, so that the system is ready to receive a new command after time t_2.

Open-Loop Control Systems (Nonfeedback Systems)

open-loop systems

The idle-speed control system and the printwheel control system illustrated in Figs. 1-2 and 1-3, respectively, are rather unsophisticated and are called **open-loop control systems**. It is not difficult to see that the systems as shown would not satisfactorily fulfill critical performance requirements. For instance, if the throttle angle α is set at a certain initial value, which corresponds to a certain engine speed, when a load torque T_L is then applied, there is no way to prevent a drop in the engine speed. The only way to make the system work is to have a means of adjusting α in response to a change in the load torque in order to maintain ω at the desired level. Similarly, there is no guarantee that the printwheel will stop at the desired position once the command is given.

The conventional electric washing machine is another example of an open-loop control system because, typically, the amount of machine wash time is entirely determined by the judgment and estimation of the human operator.

controller

controlled process

The elements of an open-loop control system can usually be divided into two parts: the **controller** and the **controlled process**, as shown by the block diagram of Fig. 1-5. An input signal or command r is applied to the controller, whose output acts as the actuating signal u; the actuating signal then controls the controlled process so that the controlled variable c will perform according to some prescribed standards. In simple cases, the controller can be an amplifier, mechanical linkages, filter, or other control elements, depending on the nature of the system. In more sophisticated cases, the controller can be a computer such as a microprocessor.

Because of the simplicity and economy of open-loop control systems, we find this type of system in many noncritical applications.

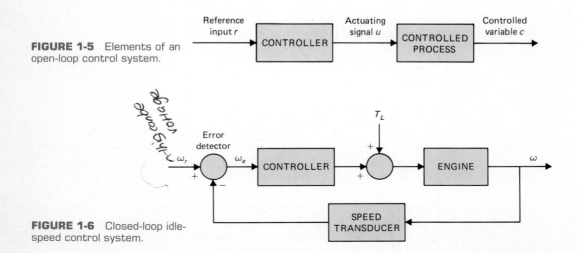

FIGURE 1-5   Elements of an open-loop control system.

FIGURE 1-6   Closed-loop idle-speed control system.

Closed-Loop Control Systems (Feedback Control Systems)

What is missing in the open-loop control system for more accurate and more adaptive control is a link or feedback from the output to the input of the system. To obtain more accurate control, the controlled signal $c(t)$ should be fed back and compared with the reference input, and an actuating signal proportional to the difference of the input and the output must be sent through the system to correct the error. A system with one or more feedback paths such as that just described is called a **closed-loop system**.

closed-loop systems

The block diagram of a closed-loop idle-speed control system is shown in Fig. 1-6. The reference input ω_r sets the desired idling speed. The engine speed at idle should agree with the reference value ω_r, and any difference between the actual speed and the desired speed caused by any disturbance such as the load torque T_L is sensed by the speed transducer and the error detector. The controller will operate on the difference and provide a signal to adjust the throttle angle α to correct the error. Figure 1-7 illustrates a comparison of the typical performances of the open-loop and closed-loop

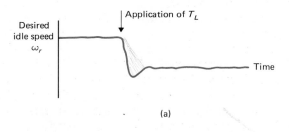

(a)

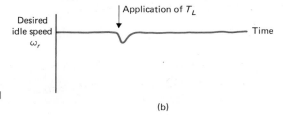

(b)

FIGURE 1-7 (a) Typical response of the open-loop idle-speed control system. (b) Typical response of the closed-loop idle-speed control system.

idle-speed control systems. In Fig. 1-7(a), the idle speed of the open-loop system will drop and settle at a lower value after a load torque is applied. In Fig. 1-7(b), the idle speed of the closed-loop system is shown to recover quickly to the preset value after the application of T_L.

regulator system

The idle-speed control system illustrated is also known as a **regulator system**, whose objective is to maintain the system output at some prescribed level.

Figure 1-8 shows the block diagram of the printwheel control system with feedback. In this case, the position of the printwheel is detected by a position sensor whose

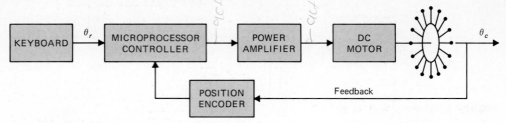

FIGURE 1-8 Closed-loop printwheel control system.

output is compared with the desired position entered from the keyboard and processed in the microprocessor. The motor is thus controlled in such a way as to drive the printwheel to the desired position accurately. Information on the speed of the printwheel can also be processed in the microprocessor from the position data so that the profile of the printwheel motion can be better controlled.

1-2

WHAT IS FEEDBACK AND WHAT ARE ITS EFFECTS?

stability

bandwidth

impedance

sensitivity

cause and effect

The motivation for using feedback illustrated by the examples in Section 1-1 is somewhat oversimplified. In these examples, the use of feedback is shown to be for the purpose of reducing the error between the reference input and the system output. However, the significance of the effects of feedback in control systems is more complex than is demonstrated by these examples. The reduction of system error is merely one of the many important effects that feedback may have upon a system. We show in the following sections that feedback also has effects on such system performance characteristics as **stability**, **bandwidth**, **overall gain**, **impedance**, and **sensitivity**.

To understand the effects of feedback on a control system, it is essential that we examine this phenomenon in a broad sense. When feedback is deliberately introduced for the purpose of control, its existence is easily identified. However, there are numerous situations wherein a physical system that we normally recognize as an inherently nonfeedback system may turn out to have feedback when it is observed in a certain manner. In general, we can state that whenever a closed sequence of **cause-and-effect relationships** exists among the variables of a system, feedback is said to exist. This viewpoint will inevitably admit feedback in a large number of systems that ordinarily would be identified as nonfeedback systems. However, with the availability of the feedback and control-system theory, this general definition of feedback enables numerous systems, with or without physical feedback, to be studied in a systematic way once the existence of feedback in the previously mentioned sense is established.

We shall now investigate the effects of feedback on the various aspects of system performance. Without the necessary mathematical foundation of linear-system theory, at this point we can only rely on simple static-system notation for our discussion. Let us consider the simple feedback system configuration shown in Fig. 1-9, where r is the input signal, c the output signal, e the error, and b the feedback signal. The parameters G and H may be considered as constant gains. By simple algebraic manipulations, it is simple to show that the input–output relation of the system is

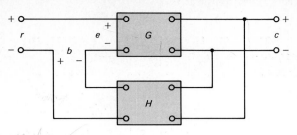

FIGURE 1-9 Feedback system.

$$M = \frac{c}{r} = \frac{G}{1 + GH} \qquad (1\text{-}1)$$

Using this basic relationship of the feedback system structure, we can uncover some of the significant effects of feedback.

Effect of Feedback on Overall Gain

feedback

As seen from Eq. (1-1), feedback affects the gain G of a nonfeedback system by a factor $1 + GH$. The system of Fig. 1-9 is said to have **negative feedback**, since a minus sign is assigned to the feedback signal. The quantity GH may itself include a minus sign, so *the general effect of feedback is that it may increase or decrease the gain G*. In a practical control system, G and H are functions of frequency, so the magnitude of $1 + GH$ may be greater than 1 in one frequency range but less than 1 in another. Therefore, *feedback could increase the gain of the system in one frequency range but decrease it in another*.

Effect of Feedback on Stability

Stability is a notion that describes whether the system will be able to follow the input command. In a nonrigorous manner, a system is said to be unstable if its output is out of control or increases without bound.

To investigate the effect of feedback on stability, we can again refer to the expression in Eq. (1-1). If $GH = -1$, the output of the system is infinite for any finite input, and the system is said to be unstable. Therefore, we may state that *feedback can cause a system that is originally stable to become unstable*. Certainly, feedback is a two-edged sword; when it is improperly used, it can be harmful. It should be pointed out, however, that we are only dealing with the static case here, and, in general, $GH = -1$ is not the only condition for instability.

It can be demonstrated that one of the advantages of incorporating feedback is that it can stabilize an unstable system. Let us assume that the feedback system in Fig. 1-9 is unstable because $GH = -1$. If we introduce another feedback loop through a negative feedback gain of F, as shown in Fig. 1-10, the input–output relation of the overall system is

$$\frac{c}{r} = \frac{G}{1 + GH + GF} \qquad (1\text{-}2)$$

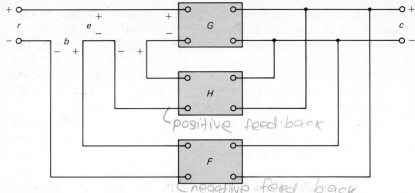

FIGURE 1-10 Feedback system with two feedback loops.

It is apparent that although the properties of G and H are such that the inner-loop feedback system is unstable, because $GH = -1$, the overall system can be stable by properly selecting the outer-loop feedback gain F.

In practice, GH is a function of frequency, and the stability condition of the closed-loop system depends on the **magnitude** and **phase** of GH. The bottom line is that *feedback can improve stability or be harmful to stability if it is not properly applied.*

stability

Effect of Feedback on Sensitivity

Sensitivity considerations often are important in the design of control systems. Since all physical elements have properties that change with environment and age, we cannot always consider that the parameters of a control system to be completely stationary over the entire operating life of the system. For instance, the winding resistance of an electric motor changes as the temperature of the motor rises during operation. In general, a good control system should be very insensitive to parameter variations but sensitive to the input commands. We shall investigate what effect feedback has on the sensitivity to parameter variations.

Referring to the system in Fig. 1-9, we consider G as a gain parameter that may vary. The sensitivity of the gain of the overall system, M, to the variation in G is defined as

$$S_G^M = \frac{\partial M/M}{\partial G/G} = \frac{\text{percentage change in } M}{\text{percentage change in } G} \qquad (1\text{-}3)$$

sensitivity

where ∂M denotes the incremental change in M due to the incremental change in G, ∂G. By using Eq. (1-1), the sensitivity function is written

$$S_G^M = \frac{\partial M}{\partial G}\frac{G}{M} = \frac{1}{1 + GH} \qquad (1\text{-}4)$$

This relation shows that if GH is a positive constant, the magnitude of the sensitivity function can be made arbitrarily small by increasing GH, provided that the system remains stable. It is apparent that in an open-loop system, the gain of the system will respond in a one-to-one fashion to the variation in G; i.e., $S_G^M = 1$. It should again be reminded that in practice, GH is a function of frequency; the magnitude of $1 + GH$ may be less than unity over some frequency ranges, so that feedback could be harmful to the sensitivity to parameter variations in certain cases.

In general, the sensitivity of the system gain of a feedback system to parameter variations depends on where the parameter is located. The reader can derive the sensitivity of the system in Fig. 1-9 due to the variation of H.

Effect of Feedback on External Disturbance or Noise

All physical systems are subject to some types of extraneous signals or noise during operation. Examples of these signals are thermal-noise voltage in electronic circuits and brush or commutator noise in electric motors. External disturbance, such as a wind gust acting on an antenna, is also quite common in control systems. Therefore, in the design of a control system, considerations should be given so that the system is insensitive to noise and disturbances and sensitive to input commands.

disturbance

The effect of feedback on noise and disturbance depends greatly on where these extraneous signals occur in the system. No general conclusions can be made; but in many situations, *feedback can reduce the effect of noise and disturbance on system performance*.

Let us refer to the system shown in Fig. 1-11, in which r denotes the command signal and n is the noise signal. In the absence of feedback, $H = 0$, the output c is

$$c = G_1 G_2 e + G_2 n \tag{1-5}$$

where $e = r$. The signal-to-noise ratio of the output is defined as

signal-to-noise ratio

$$\frac{\text{Output due to signal}}{\text{Output due to noise}} = \frac{G_1 G_2 e}{G_2 n} = G_1 \frac{e}{n} \tag{1-6}$$

To increase the signal-to-noise ratio, evidently, we should either increase the magnitude of G_1 or e relative to n. Varying the magnitude of G_2 would have no effect whatsoever on the ratio.

With the presence of feedback, the system output due to r and n acting simultaneously is

$$c = \frac{G_1 G_2}{1 + G_1 G_2 H} r + \frac{G_2}{1 + G_1 G_2 H} n \tag{1-7}$$

Comparing Eq. (1-7) with Eq. (1-5) shows that the noise component in the output of Eq. (1-7) is reduced by the factor $1 + G_1 G_2 H$ if the latter is greater than unity, but the signal component is also changed by the same amount. The signal-to-noise ratio is

$$\frac{\text{Output due to signal}}{\text{Output due to noise}} = \frac{G_1 G_2 r / (1 + G_1 G_2 H)}{G_2 n / (1 + G_1 G_2 H)} = G_1 \frac{r}{n} \tag{1-8}$$

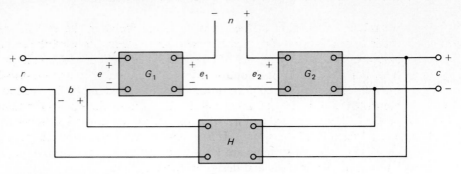

FIGURE 1-11 Feedback system with a noise signal.

and is the same as that without feedback. In this case, feedback is shown to have no direct effect on the output signal-to-noise ratio of the system in Fig. 1-11. However, the application of feedback suggests a possibility of improving the signal-to-noise ratio under certain conditions. Let us assume that in the system of Fig. 1-11, if the magnitude of G_1 is increased to G_1' and that of the input r to r', with all other parameters unchanged, the output due to the input signal acting alone is at the same level as that when feedback is absent. In other words, we let

$$c\big|_{n=0} = \frac{G_1' G_2 r'}{1 + G_1' G_2 H} = G_1 G_2 r \tag{1-9}$$

With G_1 increased to G_1', the output due to noise acting alone becomes

$$c\big|_{r=0} = \frac{G_2 n}{1 + G_1' G_2 H} \tag{1-10}$$

which is smaller than the output due to n when G_1 is not increased. The signal-to-noise ratio is now

$$\frac{G_1 G_2 r}{G_2 n / (1 + G_1' G_2 H)} = \frac{G_1 r}{n}(1 + G_1' G_2 H) \tag{1-11}$$

which is greater than that of the system without feedback by a factor of $(1 + G_1' G_2 H)$.

In Chapter 9, the feedforward and forward controller configurations are used along with feedback to reduce the effects of disturbance and noise inputs.

In general, feedback also has effects on such performance characteristics as bandwidth, impedance, transient response, and frequency response. These effects will become known as we continue with the text.

1-3

TYPES OF FEEDBACK CONTROL SYSTEMS

analysis
and design

Feedback control systems may be classified in a number of ways, depending upon the purpose of the classification. For instance, according to the method of analysis and design, control systems are classified as **linear** or **nonlinear, time-varying** or **time-**

modulated system

unmodulated system

invariant. According to the types of signal found in the system, reference is often made to **continuous-data** and **discrete-data** systems, or **modulated** and **unmodulated** systems. Control systems are often classified according to the main purpose of the system. For instance, a **positional-control system** and a **velocity-control system** control the output variables according to the way the names imply. In Chapter 7, the **type** of a control system is defined according to the form of the open-loop transfer function. In general, there are many other ways of identifying control systems according to some special features of the system. It is important that some of these more common ways of classifying control systems are known so that proper perspective is gained before embarking on the analysis and design of these systems.

Linear versus Nonlinear Control Systems

nonlinear system

This classification is made according to the methods of analysis and design. Strictly speaking, linear systems do not exist in practice, since all physical systems are nonlinear to some extent. Linear feedback control systems are idealized models fabricated by the analyst purely for the simplicity of analysis and design. When the magnitudes of signals in a control systems are limited to ranges in which system components exhibit linear characteristics (i.e., the principle of superposition applies), the system is essentially linear. But when the magnitudes of signals are extended beyond the range of the linear operation, depending on the severity of the nonlinearity, the system should no longer be considered linear. For instance, amplifiers used in control systems often exhibit a saturation effect when their input signals become large; the magnetic field of a motor usually has saturation properties. Other common nonlinear effects found in control systems are the backlash or dead play between coupled gear members, nonlinear spring characteristics, nonlinear frictional force or torque between moving members, and so on. Quite often, nonlinear characteristics are intentionally introduced in a control system to improve its performance or provide more effective control. For instance, to achieve minimum-time control, an on–off (bang-bang or relay) type controller is used in many missile or spacecraft control systems. Typically in these systems, jets are mounted on the sides of the vehicle to provide reaction torque for attitude control.

attitude control

These jets are often controlled in a full-on or full-off fashion, so a fixed amount of air is applied from a given jet for a certain time duration to control the attitude of the space vehicle.

For linear systems, there exists a wealth of analytical and graphical techniques for design and analysis purposes. A majority of the material in this text is devoted to the analysis and design of linear systems. Nonlinear systems, on the other hand, are usually difficult to treat mathematically, and there are no general methods available for solving a wide class of nonlinear systems. In the design of control systems, it is practical to first design the controller based on the linear-system model by neglecting the nonlinearities of the system. The designed controller is then applied to the nonlinear system model for evaluation or redesign by computer simulation.

Time-Invariant versus Time-Varying Systems

time-varying system

When the parameters of a control system are stationary with respect to time during the operation of the system, the system is called a time-invariant system. In practice, most physical systems contain elements that drift or vary with time. For example, the winding resistance of an electric motor will vary when the motor is being first excited and its temperature is rising. Another example of a time-varying system is a guided-missile control system in which the mass of the missile decreases as the fuel on board is being consumed during flight. Although a time-varying system without nonlinearity is still a linear system, the analysis and design of this class of systems are usually much more complex than that of the linear time-invariant systems.

Continuous-Data Control Systems

A continuous-data system is one in which the signals at various parts of the system are all functions of the continuous time variable t. Among all continuous-data control systems, the signals may be further classified as ac or dc. Unlike the general definitions of ac and dc signals used in electrical engineering, ac and dc control systems carry special significances. When one refers to an ac control system, it usually means that the signals in the system are *modulated* by some kind of modulation scheme. On the other hand, when a dc control system is referred to, it does not mean that all the signals in the system are unidirectional; then there would be no corrective control movement. A **dc**

*dc control
system*

control system simply implies that the signals are unmodulated, but they are still ac signals according to the conventional definition. The schematic diagram of a closed-loop dc control system is shown in Fig. 1-12. Typical waveforms of the signals in response to a step-function input are shown in the figure. Typical components of a dc control system are potentiometers, dc amplifiers, dc motors, dc tachometers, etc.

*ac control
system*

The schematic diagram of a typical **ac control system** that performs essentially the same task as that in Fig. 1-12 is shown in Fig. 1-13. In this case, the signals in the system are modulated; that is, the information is transmitted by an ac carrier signal. Notice that the output-controlled variable still behaves similar to that of the dc system. In this case, the modulated signals are demodulated by the low-pass characteristics of the ac motor. Ac control systems are used extensively in aircraft and missile control

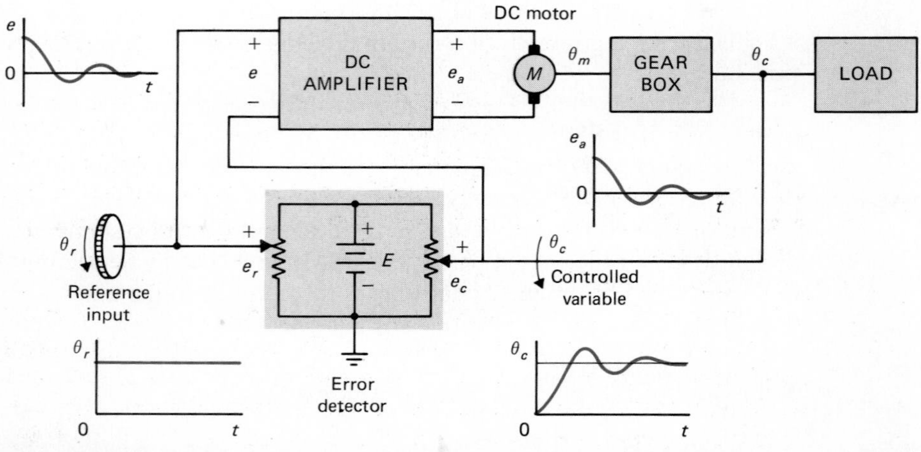

FIGURE 1-12 Schematic diagram of a typical dc closed-loop control system.

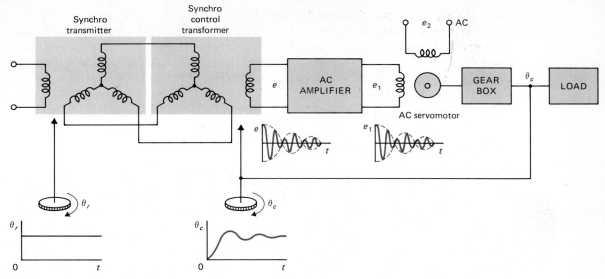

FIGURE 1-13 Schematic diagram of a typical ac closed-loop control system.

systems in which noise and disturbance often create problems. By using modulated ac control systems with carrier frequencies of 400 Hz or higher frequencies, the system will be less susceptible to low-frequency noise. Typical components of an ac control system are synchros, resolvers, ac amplifiers, ac motors, gyroscopes, accelerometers, etc.

In practice, not all control systems are strictly of the ac or the dc type. A system may incorporate a mixture of ac and dc components, using modulators and demodulators to match the signals at various points in the system.

Discrete-Data Control Systems

digital control system

microprocessor

Discrete-data control systems differ from the continuous-data systems in that the signals at one or more points of the system are in the form of either a pulse train or a digital code. Usually, discrete-data control systems are subdivided into **sampled-data** and **digital control systems**. Sampled-data control systems refer to a more general class of discrete-data systems in which the signals are in the form of pulse data. A digital control system refers to the use of a digital computer or controller in the system, so that the signals are digitally coded, such as in binary code. For example, the printwheel control system shown in Fig. 1-8 is a typical digital control system, since the microprocessor receives and outputs digital data.

In general, a sampled-data system receives data or information only intermittently at specific instants of time. For instance, the error signal in a control system can be supplied only in the form of pulses, in which case the control system receives no information about the error signal during the periods between two consecutive pulses. Strictly, a sampled-data system can also be classified as an ac system, since the signal of the system is pulse modulated.

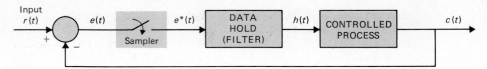

FIGURE 1-14 Block diagram of a sampled-data control system.

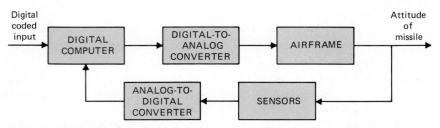

FIGURE 1-15 Digital autopilot system for a guided missile.

Figure 1-14 illustrates how a typical sampled-data system operates. A continuous input signal $r(t)$ is applied to the system. The error signal $e(t)$ is sampled by a sampling device, the sampler, and the output of the sampler is a sequence of pulses. The sampling rate of the sampler may or may not be uniform. There are many advantages of incorporating sampling into a control system. One important advantage of the sampling operation is that expensive equipment used in the system may be time-shared among several control channels. Another advantage is that pulse data are usually less susceptible to noise.

Because digital computers provide many advantages in size and flexibility, computer control has become increasingly popular in recent years. Many airborne systems contain digital controllers that can pack thousands of discrete elements in a space no larger than the size of this book. Figure 1-15 shows the basic elements of digital autopilot for a guided missile.

1-4

SUMMARY

In this chapter, we have introduced some of the basic concepts of what a control system is and what it is supposed to accomplish. The basic components of a control system are described. By demonstrating the effects of feedback in a rudimentary way, the question on why most control systems are closed-loop systems is also clarified. Most important, it is pointed out that feedback is a double-edged sword, so that it can benefit as well as harm a system to be controlled. This points out the challenging task in the design of a control system, which involves considerations on such performance criteria as stability, sensitivity, bandwidth, and accuracy. Finally, the various types of control systems are categorized according to the system signals, linearity, and control objectives. Several typical control-system examples are given to illustrate the points of emphasis in the analysis and design of control systems. Points are also made on the fact that most systems encountered in real life are nonlinear to a varying extent. The concentration on the studies of linear systems is mainly due to the availability of unified and simple-to-understand analytical methods in the analysis and design of linear systems.

REVIEW QUESTIONS

1. List the advantages and the disadvantages of an open-loop system.
2. List the advantages and the disadvantages of a closed-loop system.
3. Give the definitions of ac and dc control systems.
4. Give the advantages of a digital control system over a continuous-data control system.
5. A closed-loop control system is usually more accurate than an open-loop system.   (T)   (F)
6. Feedback is sometimes used to improve the sensitivity of a control system.   (T)   (F)
7. If an open-loop system is unstable, then applying feedback will always improve its stability.   (T)   (F)
8. Feedback can increase the gain of a system in one frequency range but decrease it in another.   (T)   (F)
9. Nonlinear elements are sometimes intentionally introduced to a control system to improve its performance.   (T)   (F)

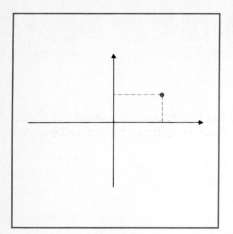

MATHEMATICAL FOUNDATION

INTRODUCTION

The studies of control systems rely to a great extent on the use of applied mathematics. Since the analysis and design of practical control systems have to deal with real problems, we cannot completely ignore the hardware and laboratory aspects of the problems. However, one of the major purposes of control-systems studies is to develop a set of analytical tools so that the designer can arrive at reasonably predictable and reliable designs without depending completely on the drudgery of experimentation or computer simulation.

For the study of classical control theory, which represents a good portion of this text, the required mathematical background includes such subjects as **complex-variable theory**, **differential and difference equations**, **Laplace transformation and z-transformation**, and so on. Modern control theory, on the other hand, requires considerably more intensive mathematical background. In addition to the above-mentioned subjects, modern control theory is based on the foundation of matrix theory, set theory, linear algebra and transformation, variational calculus, mathematical programming, probability theory, and other advanced mathematics.

In this chapter, we present the background material that is needed for the topics on control systems discussed in this text. Because of space limitations, the treatment of these mathematical subjects cannot be exhaustive. The reader who wishes to conduct an in-depth study of any of these subjects should refer to specialized books.

2-2
COMPLEX-VARIABLE CONCEPT

Complex Variable

complex variables

The classical control-systems theory is based on the application of complex variables and their functions, since both the Laplace transform variable s and the z-transform variable z are complex variables.

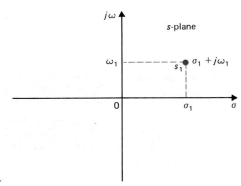

FIGURE 2-1 Complex s-plane.

A complex variable s has two components: a real component σ and an imaginary component ω. Graphically, the real component of s is represented by a σ axis in the horizontal direction, and the imaginary component is measured along the vertical $j\omega$ axis, in the complex s-plane. Figure 2-1 illustrates the complex s-plane, in which any arbitrary point $s = s_1$ is defined by the coordinates $\sigma = \sigma_1$ and $\omega = \omega_1$, or simply $s_1 = \sigma_1 + j\omega_1$.

Functions of a Complex Variable

The function $G(s)$ is said to be a function of the complex variable s if for every value of s, there is one or more corresponding values of $G(s)$. Since s is defined to have real and imaginary parts, the function $G(s)$ is also represented by its real and imaginary parts; that is,

$$G(s) = \text{Re } G(s) + j \text{ Im } G(s) \qquad (2\text{-}1)$$

where Re $G(s)$ denotes the real part of $G(s)$, and Im $G(s)$ represents the imaginary part of $G(s)$. Thus, the function $G(s)$ can also be represented by the complex $G(s)$-plane, whose horizontal axis represents Re $G(s)$ and whose vertical axis measures the imaginary component of $G(s)$. If for every value of s (every point in the s-plane), there is only one corresponding value for $G(s)$ [one corresponding point in the $G(s)$-plane], $G(s)$ is said to be a **single-valued function**, and the mapping from points in the s-plane

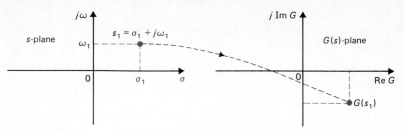

FIGURE 2-2 Single-valued mapping from the s-plane to the $G(s)$-plane.

onto points in the $G(s)$-plane is described as **single-valued** (Fig. 2-2). If the mapping from the $G(s)$-plane to the s-plane is also single-valued, the mapping is called **one-to-one**. However, there are many functions for which the mapping from the function plane to the complex-variable plane is not single-valued. For instance, given the function

$$G(s) = \frac{1}{s(s+1)} \tag{2-2}$$

it is apparent that for each value of s, there is only one unique corresponding value for $G(s)$. However, the reverse mapping is not true; for instance, the point $G(s) = \infty$ is mapped onto two points, $s = 0$ and $s = -1$, in the s-plane.

Analytic Function

analytic function

A function $G(s)$ of the complex variable s is called an analytic function in a region of the s-plane if the function and all its derivatives exist in the region. For instance, the function given in Eq. (2-2) is analytic at every point in the s-plane except at the point $s = 0$ and $s = -1$. At these two points, the value of the function is infinite. As another example, the function $G(s) = s + 2$ is analytic at every point in the finite s-plane.

Singularities and Poles of a Function

singularities poles

The **singularities** of a function are the points in the s-plane at which the function or its derivatives does not exist. A **pole** is the most common type of singularity and plays a very important role in the studies of classical control theory.

The definition of a pole can be stated as: *If a function $G(s)$ is analytic and single-valued in the neighborhood of s_i, it is said to have a pole of order r at $s = s_i$ if the limit*

$$\lim_{s \to s_i}[(s - s_i)^r G(s)]$$

has a finite, nonzero value. In other words, the denominator of $G(s)$ must include the factor $(s - s_i)^r$, so when $s = s_i$, the function becomes infinite. If $r = 1$, the pole at $s = s_i$ is called a **simple pole**. As an example, the function

$$G(s) = \frac{10(s + 2)}{s(s + 1)(s + 3)^2} \tag{2-3}$$

has a pole of order 2 at $s = -3$ and simple poles at $s = 0$ and $s = -1$. It can also be said that the function $G(s)$ is analytic in the s-plane except at these poles.

Zeros of a Function

zeros

The definition of a **zero** of a function can be stated as: *If the function $G(s)$ is analytic at $s = s_i$, it is said to have a zero of order r at $s = s_i$ if the limit*

$$\lim_{s \to s_i}[s - s_i)^{-r}G(s)] \tag{2-4}$$

has a finite, nonzero value. Or, simply, $G(s)$ has a zero of order r at $s = s_i$ if $1/G(s)$ has an rth-order pole at $s = s_i$. For example, the function in Eq. (2-3) has a simple zero at $s = -2$.

If the function under consideration is a rational function of s, that is, a quotient of two polynomials of s, the total number of poles equals the total number of zeros, counting the multiple-order poles and zeros, and taking into account of the poles and zeros at infinity. The function in Eq. (2-3) has four finite poles at $s = 0$, -1, -3, and -3; there is one finite zero at $s = -2$, but there are three zeros at infinity, since

$$\lim_{s \to \infty} G(s) = \lim \frac{10}{s^3} = 0 \tag{2-5}$$

Therefore, the function has a total of four poles and four zeros in the entire s-plane, including infinity.

2-3

DIFFERENTIAL EQUATIONS

Linear Ordinary Differential Equations

A large class of systems and phenomena in engineering are most conveniently formulated in terms of differential equations. These equations generally involve derivatives and integrals of the dependent variables with respect to the independent variable. For instance, a series electric RLC (resistance–inductance–capacitance) network can be represented by the differential equation

$$Ri(t) + L\frac{di(t)}{dt} + \frac{1}{C}\int i(t)\, dt = v(t) \tag{2-6}$$

where R is the resistance, L the inductance, C the capacitance, $i(t)$ the current in the network, and $v(t)$ the applied voltage. In this case, $v(t)$ is the forcing function, t the independent variable, and $i(t)$ the dependent variable or the unknown that is to be determined by solving the differential equation.

Similarly, for a series mechanical mass–spring–damper system, the differential equation of the system can be written as

$$M\frac{d^2 y(t)}{dt^2} + B\frac{dy(t)}{dt} + Ky(t) = f(t) \tag{2-7}$$

where $f(t)$ is the applied force, M the mass, B the damping coefficient, K the linear spring constant, and $y(t)$ the displacement.

The last two equations are defined as second-order differential equations, and we refer to the systems as second-order systems. Strictly, Eq. (2-6) should be referred to as an integrodifferential equation, since an integral is involved.

In general, the differential equation of an nth-order system is written

$$a_{n+1}\frac{dy^n(t)}{dt^n} + a_n\frac{dy^{n-1}(t)}{dt^{n-1}} + \cdots + a_2\frac{dy(t)}{dt} + a_1 y(t) = f(t) \tag{2-8}$$

The differential equations in Eqs. (2-6) through (2-8) are also known as **linear ordinary differential equations** if the coefficients $a_1, a_2, \ldots, a_{n+1}$ are not functions of $y(t)$, and since $y(t)$ and its derivatives are all of the first power.

In this book, since we treat only systems that contain lumped parameters, the differential equations encountered are all of the ordinary type. For systems with distributed parameters, such as in heat-transfer systems, partial differential equations are used.

Nonlinear Differential Equations

Many physical systems are nonlinear and are described by nonlinear differential equations. For instance, the differential equation that describes the motion of the pendulum shown in Fig. 2-3 is

$$ML\frac{d^2 \theta(t)}{dt^2} + Mg \sin \theta(t) = 0 \tag{2-9}$$

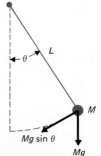

FIGURE 2-3 A simple pendulum.

nonlinear
system

Since $\theta(t)$ appears as a sine function, Eq. (2-9) is nonlinear, and the system is called a **nonlinear system**.

First-Order Differential Equations: State Equations

In general, an nth-order differential equation can be decomposed into n first-order differential equations. Since, in principle, first-order differential equations are simpler to solve than higher-order ones, there are reasons why first-order differential equations are used in the analytical studies of control systems.

For the second-order differential equation in Eq. (2-7), if we let

$$x_1(t) = y(t) \tag{2-10}$$

and

$$x_2(t) = \frac{dy(t)}{dt} = \frac{dx_1(t)}{dt} \tag{2-11}$$

the equation is written as

$$M\frac{dx_2(t)}{dt} = -Kx_1(t) - Bx_2(t) + f(t) \tag{2-12}$$

Rearranging Eqs. (2-11) and (2-12), we get two first-order differential equations:

$$\frac{dx_1(t)}{dt} = x_2(t) \tag{2-13}$$

$$\frac{dx_2(t)}{dt} = -\frac{K}{M}x_1(t) - \frac{B}{M}x_2(t) + \frac{1}{M}f(t) \tag{2-14}$$

For Eq. (2-6), we let

$$x_1(t) = \int i(t)\, dt \tag{2-15}$$

and

$$x_2(t) = \frac{dx_1(t)}{dt} = i(t) \tag{2-16}$$

Then Eq. (2-6) is decomposed into the following two first-order differential equations:

$$\frac{dx_1(t)}{dt} = x_2(t) \tag{2-17}$$

$$\frac{dx_2(t)}{dt} = -\frac{1}{LC}x_1(t) - \frac{R}{L}x_2(t) + \frac{1}{L}v(t) \tag{2-18}$$

In a similar fashion, we can decompose Eq. (2-8) into n first-order differential equations.

state equations
state variables

In control-systems theory, the set of first-order differential equations is called the **state equations**, and $x_1, x_2, \ldots, x_n$ are called the **state variables**. Later in this chapter, we show that the state equations can be written in a compact vector-matrix form. The reader should refer to Chapter 5 for a more in-depth discussion on the state variables and state equations.

2-4

LAPLACE TRANSFORM

The Laplace transform is one of the mathematical tools used for the solution of linear ordinary differential equations. In comparison with the classical method of solving linear differential equations, the Laplace transform method has the following two attractive features:

> 1. The homogeneous equation and the particular integral of the solution are obtained in one operation.
> 2. The Laplace transform converts the differential equation into an algebraic equation in s. It is then possible to manipulate the algebraic equation by simple algebraic rules to obtain the solution in the s-domain. The final solution is obtained by taking the inverse Laplace transform.

Definition of the Laplace Transform

Given the real function $f(t)$ that satisfies the condition

$$\int_0^\infty \left| f(t)e^{-\sigma t} \right| dt < \infty \tag{2-19}$$

for some finite real σ, the Laplace transform of $f(t)$ is defined as

$$F(s) = \int_0^\infty f(t)e^{-st} \, dt \tag{2-20}$$

or

$$F(s) = \text{Laplace transform of } f(t) = \mathcal{L}[f(t)] \tag{2-21}$$

Laplace operator   The variable s is referred to as the **Laplace operator**, which is a complex variable; that is, $s = \sigma + j\omega$. The defining equation of Eq. (2-20) is also known as the **one-sided Laplace transform**, as the integration is evaluated from $t = 0$ to ∞. This simply means that all information contained in $f(t)$ prior to $t = 0$ is ignored or considered to be zero. This assumption does not impose any limitation on the applications of the Laplace transform to linear-system problems, since in the usual time-domain studies, time reference is often chosen at the instant $t = 0$. Furthermore, for a physical system when an input is applied at $t = 0$, the response of the system does not start sooner than $t = 0$; that is, response does not precede excitation. Such a system is also known *causal system*   as being **causal**.

Strictly, the one-sided Laplace transform should be defined from $t = 0^-$ to $t = \infty$. The symbol 0^- implies that the limit of $t \to 0$ is taken from the left side of $t = 0$. This limit process will take care of situations under which the function $f(t)$ has a jump discontinuity or an impulse at $t = 0$. For the subjects treated in this text, the defining equation of the Laplace transform in Eq. (2-20) is almost never used in problem solving, since the transform expressions are either given or can be found from the Laplace transform table. Thus, the fine point of using 0^- or 0^+ never needs to be addressed. For simplicity, we shall, in general, use $t = 0$ or $t = t_0$ as the initial time in all subsequent chapters.

The following examples illustrate how Eq. (2-20) is used for the evaluation of the Laplace transform of $f(t)$.

EXAMPLE 2-1   Let $f(t)$ be a unit-step function that is defined as

unit-step function

$$f(t) = u_s(t) = 1 \qquad t > 0$$
$$= 0 \qquad t < 0 \qquad (2\text{-}22)$$

The Laplace transform of $f(t)$ is obtained as

$$F(s) = \mathscr{L}[u_s(t)] = \int_0^\infty u_s(t)e^{-st}\,dt = -\frac{1}{s}e^{-st}\bigg|_0^\infty = \frac{1}{s} \qquad (2\text{-}23)$$

The last equation is valid if

$$\int_0^\infty |u_s(t)e^{-\sigma t}|\,dt = \int_0^\infty |e^{-\sigma t}|\,dt < \infty \qquad (2\text{-}24)$$

which means that the real part of s, σ, must be greater than zero. In practice, we simply refer to the Laplace transform of the unit-step function as $1/s$, and rarely do we have to be concerned with the region in the s-plane in which the transform integral converges absolutely.

EXAMPLE 2-2   Consider the exponential function

$$f(t) = e^{-\alpha t} \qquad t \geq 0 \qquad (2\text{-}25)$$

where α is a constant. The Laplace transform of $f(t)$ is written

$$F(s) = \int_0^\infty e^{-\alpha t} e^{-st}\, dt = -\frac{e^{-(s+\alpha)t}}{s+\alpha}\bigg|_0^\infty = \frac{1}{s+\alpha} \tag{2-26}$$

Inverse Laplace Transformation

inverse Laplace transform

The operation of obtaining $f(t)$ from the Laplace transform $F(s)$ is termed the **inverse Laplace transformation**, and is denoted by

$$f(t) = \mathscr{L}^{-1}[F(s)] \tag{2-27}$$

The inverse Laplace transform integral is given as

$$f(t) = \frac{1}{2\pi j}\int_{c-j\infty}^{c+j\infty} F(s)e^{st}\, ds \tag{2-28}$$

where c is a real constant that is greater than the real parts of all the singularities of $F(s)$. Equation (2-28) represents a line integral that is to be evaluated in the s-plane. However, for most engineering purposes, the inverse Laplace transform operation can be carried out simply by referring to the Laplace transform table, such as the one given in Appendix B.

Important Theorems of the Laplace Transform

The applications of the Laplace transform in many instances are simplified by the utilization of the properties of the transform. These properties are presented by the following theorems, and no proofs are given.

THEOREM 1.   *Multiplication by a Constant*
        Let k be a constant, and F(s) be the Laplace transform of f(t). Then

$$\mathscr{L}[kf(t)] = kF(s) \tag{2-29}$$

THEOREM 2.   *Sum and Difference*
        Let $F_1(s)$ and $F_2(s)$ be the Laplace transforms of $f_1(t)$ and $f_2(t)$, respectively. Then

$$\mathscr{L}[\,f_1(t) \pm f_2(t)] = F_1(s) \pm F_2(s) \qquad (2\text{-}30)$$

■

THEOREM 3.   *Differentiation*
Let $F(s)$ be the Laplace transform of $f(t)$, and $f(0)$ is the
limit of $f(t)$ as t approaches 0. The Laplace transform of the time
derivative of $f(t)$ is

$$\mathscr{L}\left(\frac{df(t)}{dt}\right) = sF(s) - \lim_{t \to 0} f(t) = sF(s) - f(0) \qquad (2\text{-}31)$$

In general, for higher-order derivatives of $f(t)$,

$$\mathscr{L}\left(\frac{d^n f(t)}{dt^n}\right) = s^n F(s) - \lim_{t \to 0}\left(s^{n-1}f(t) + s^{n-2}\frac{df(t)}{dt} + \cdots + \frac{d^{n-1}f(t)}{dt^{n-1}}\right)$$
$$= s^n F(s) - s^{n-1}f(0) - s^{n-2}f^{(1)}(0) - \cdots - f^{(n-1)}(0) \qquad (2\text{-}32)$$

where $f^{(i)}(0)$ denotes the ith-order derivative of $f(t)$ with respect
to t, evaluated at $t = 0$.   ■

THEOREM 4.   *Integration*
The Laplace transform of the first integral of $f(t)$ with re-
spect to time is the Laplace transform of $f(t)$ divided by s; that
is,

$$\mathscr{L}\left(\int_0^t f(\tau)\,d\tau\right) = \frac{F(s)}{s} \qquad (2\text{-}33)$$

In general, for nth-order integration,

$$\mathscr{L}\left(\int_0^{t_1}\int_0^{t_2}\cdots\int_0^{t_n} f(\tau)\,d\tau\,dt_1\,dt_2\cdots dt_{n-1}\right) = \frac{F(s)}{s^n} \qquad (2\text{-}34)$$

■

time shift   **THEOREM 5.**   *Shift in Time*
The Laplace transform of $f(t)$ delayed by time T is equal to
the Laplace transform $f(t)$ multiplied by e^{-Ts}; that is,

$$\mathcal{L}[f(t - T)u_s(t - T)] = e^{-Ts}F(s) \qquad (2\text{-}35)$$

where $u_s(t - T)$ denotes the unit-step function that is shifted in time to the right by T. ■

THEOREM 6. *Initial-Value Theorem*
If the Laplace transform of f(t) is F(s), then

$$\lim_{t \to 0} f(t) = \lim_{s \to \infty} sF(s) \qquad (2\text{-}36)$$

if the time limit exists. ■

final-value theorem

THEOREM 7. *Final-Value Theorem*
If the Laplace transform of f(t) is F(s), and if sF(s) is analytic on the imaginary axis and in the right half of the s-plane, then

$$\lim_{t \to \infty} f(t) = \lim_{s \to 0} sF(s) \qquad (2\text{-}37)$$

■

The final-value theorem is very useful for the analysis and design of control systems, since it gives the final value of a time function by knowing the behavior of its Laplace transform at $s = 0$. The final-value theorem is *not* valid if $sF(s)$ contains any pole whose real part is zero or positive, which is equivalent to the analytic requirement of $sF(s)$ stated in the theorem. The following examples illustrate the care that must be taken in applying the theorem.

EXAMPLE 2-3

Consider the function

$$F(s) = \frac{5}{s(s^2 + s + 2)} \qquad (2\text{-}38)$$

Since $sF(s)$ is analytic on the imaginary axis and in the right-half s-plane, the final-value theorem may be applied. Using Eq. (2-37),

$$\lim_{t \to \infty} f(t) = \lim_{s \to 0} sF(s) = \lim_{s \to 0} \frac{5}{s^2 + s + 2} = \frac{5}{2} \qquad (2\text{-}39)$$

EXAMPLE 2-4

Consider the function

$$F(s) = \frac{\omega}{s^2 + \omega^2} \qquad (2\text{-}40)$$

which is the Laplace transform of $f(t) = \sin \omega t$. Since the function $sF(s)$ has two poles on the imaginary axis of the s-plane, the final-value theorem *cannot* be applied in this case. In other words, although the final-value theorem would yield a value of zero as the final value of $f(t)$, the result is erroneous.

THEOREM 8. *Complex Shifting*
The Laplace transform of $f(t)$ multiplied by $e^{\mp \alpha t}$, where α is a constant, is equal to the Laplace transform $F(s)$ with s replaced by $s \pm \alpha$; that is,

$$\mathcal{L}[e^{\mp \alpha t}f(t)] = F(s \pm \alpha) \qquad (2\text{-}41)$$

THEOREM 9. *Real Convolution (Complex Multiplication)*
Let $F_1(s)$ and $F_2(s)$ be the Laplace transforms of $f_1(t)$ and $f_2(t)$, respectively, and $f_1(t) = 0$, $f_2(t) = 0$, for $t < 0$, then

$$F_1(s)F_2(s) = \mathcal{L}[f_1(t) * f_2(t)]$$
$$= \mathcal{L}\left[\int_0^t f_1(\tau)f_2(t - \tau)\, d\tau\right] = \mathcal{L}\left[\int_0^t f_2(\tau)f_1(t - \tau)\, d\tau\right]$$
$$(2\text{-}42)$$

convolution

where the symbol " $*$ " denotes **convolution** *in the time domain.*

Equation (2-42) shows that multiplication of two transformed functions in the complex s-domain is equivalent to the convolution of two corresponding real functions of t in the t-domain. An important fact to remember is that the inverse Laplace transform of the product of two functions in the s-domain is *not* equal to the product of the two corresponding real functions in the t-domain; that is, in general,

$$\mathcal{L}^{-1}[F_1(s)F_2(s)] \neq f_1(t)f_2(t) \qquad (2\text{-}43)$$

There is also a dual relation to the real convolution theorem, called the **complex convolution**, or **real multiplication**. Essentially, the theorem states that multiplication in the real t-domain is equivalent to convolution in the complex s-domain that is,

$$\mathcal{L}[f_1(t)f_2(t)] = F_1(s) * F_2(s) \qquad (2\text{-}44)$$

where $*$ denotes complex convolution in this case. Details of the complex convolution formula are not given here.

TABLE 2-1   Theorems of Laplace Transforms

Multiplication by a constant	$\mathscr{L}[kf(t)] = kF(s)$
Sum and difference	$\mathscr{L}[f_1(t) \pm f_2(t)] = F_1(s) \pm F_2(s)$
Differentiation	$\mathscr{L}\left[\dfrac{df(t)}{dt}\right] = sF(s) - f(0)$
	$\mathscr{L}\left[\dfrac{d^n f(t)}{dt^n}\right] = s^n F(s) - s^{n-1}f(0) - s^{n-2}f^{(1)}(0)$
	$\qquad\qquad - \cdots - sf^{(n-2)}(0) - f^{(n-1)}(0)$
	where
	$f^{(k)}(0) = \dfrac{d^k f(t)}{dt^k}\bigg|_{t=0}$
Integration	$\mathscr{L}\left[\displaystyle\int_0^t f(\tau)d\tau\right] = \dfrac{F(s)}{s}$
	$\mathscr{L}\left[\displaystyle\int_0^{t_1}\int_0^{t_2}\cdots\int_0^{t_n} f(\tau)\, d\tau\, dt_1\, dt_2 \cdots dt_{n-1}\right] = \dfrac{F(s)}{s^n}$
Shift in time	$\mathscr{L}[f(t-T)u_s(t-T)] = e^{-Ts}F(s)$
Initial-value theorem	$\lim\limits_{t\to 0} f(t) = \lim\limits_{s\to\infty} sF(s)$
Final-value theorem	$\lim\limits_{t\to\infty} f(t) = \lim\limits_{s\to 0} sF(s)$ if $sF(s)$ does not have poles on or to the right of the imaginary axis in the s-plane
Complex shifting	$\mathscr{L}[e^{\mp \alpha t}f(t)] = F(s \pm \alpha)$
Real convolution	$F_1(s)F_2(s) = \mathscr{L}\left[\displaystyle\int_0^t f_1(\tau)f_2(t-\tau)\, d\tau\right]$
	$\qquad\qquad = \mathscr{L}\left[\displaystyle\int_0^t f_2(\tau)f_1(t-\tau)\, d\tau\right]$
	$\qquad\qquad = \mathscr{L}[f_1(t) * f_2(t)]$

Table 2-1 summarizes the theorems of the Laplace transforms presented.

2-5

INVERSE LAPLACE TRANSFORM
BY PARTIAL-FRACTION EXPANSION

In a majority of problems in control systems, the evaluation of the inverse Laplace transform does not rely on the use of the inversion integral of Eq. (2-28). Rather, the inverse Laplace transform operation involving rational functions can be carried out using a Laplace transform table and partial-fraction expansion. When the Laplace transform solution of a differential equation is a rational function in s, it can be written as

$$X(s) = \frac{Q(s)}{P(s)} \tag{2-45}$$

where $P(s)$ and $Q(s)$ are polynomials of s. It is assumed that *the order of $P(s)$ in s is greater than that of $Q(s)$*. The polynomial $P(s)$ may be written

$$P(s) = s^n + a_1 s^{n-1} + \cdots + a_{n-1}s + a_n \tag{2-46}$$

where $a_1, a_2, \ldots, a_n$ are real coefficients. The zeros of $Q(s)$ are either real or in complex-conjugate pairs, simple or in multiple order. The methods of partial-fraction expansion will now be given for the cases of simple poles, multiple-order poles, and complex-conjugate poles of $X(s)$.

Partial-Fraction Expansion When All the Poles of X(s) Are Simple and Real

If all the poles of $X(s)$ are simple and real, Eq. (2-45) can be written

$$X(s) = \frac{Q(s)}{P(s)} = \frac{Q(s)}{(s + s_1)(s + s_2) \cdots (s + s_n)} \tag{2-47}$$

where $s_1 \neq s_2 \neq \cdots \neq s_n$. Applying the partial-fraction expansion, Eq. (2-47) is written

$$X(s) = \frac{K_{s1}}{s + s_1} + \frac{K_{s2}}{s + s_2} + \cdots + \frac{K_{sn}}{s + s_n} \tag{2-48}$$

The coefficient K_{si} $(i = 1, 2, \ldots, n)$ is determined by multiplying both sides of Eq. (2-47) or (2-48) by the factor $(s + s_i)$ and then setting s equal to $-s_i$. To find the coefficient K_{s1}, for instance, we multiply both sides of Eq. (2-47) by $(s + s_1)$ and let $s = -s_1$. Thus,

$$K_{s1} = \left[(s + s_1) \frac{Q(s)}{P(s)} \right]\Bigg|_{s=-s_1} = \frac{Q(-s_1)}{(s_2 - s_1)(s_3 - s_1) \cdots (s_n - s_1)} \tag{2-49}$$

EXAMPLE 2-5   Consider the function

$$X(s) = \frac{5s + 3}{(s + 1)(s + 2)(s + 3)} \tag{2-50}$$

which is written in the partial-fraction expanded form:

$$X(s) = \frac{K_{-1}}{s + 1} + \frac{K_{-2}}{s + 2} + \frac{K_{-3}}{s + 3} \tag{2-51}$$

The coefficients K_{-1}, K_{-2}, and K_{-3} are determined as follows:

$$K_{-1} = [(s + 1)X(s)]\big|_{s=-1} = \frac{5(-1) + 3}{(2 - 1)(3 - 1)} = -1 \tag{2-52}$$

$$K_{-2} = [(s + 2)X(s)]\big|_{s=-2} = \frac{5(-2) + 3}{(1 - 2)(3 - 2)} = 7 \tag{2-53}$$

$$K_{-3} = [(s + 3)X(s)]\big|_{s=-3} = \frac{5(-3) + 3}{(1 - 3)(2 - 3)} = -6 \tag{2-54}$$

Thus eq. (2-51) becomes

$$X(s) = \frac{-1}{s + 1} + \frac{7}{s + 2} - \frac{6}{s + 3} \tag{2-55}$$

Partial-Fraction Expansion When Some Poles of X(s) Are of Multiple Order

repeated poles   If r of the n poles of $X(s)$ are identical, or, say, the pole at $s = -s_i$ is of multiplicity r, $X(s)$ is written

$$X(s) = \frac{Q(s)}{P(s)} = \frac{Q(s)}{(s + s_1)(s + s_2) \cdots (s + s_{n-r})(s + s_i)^r} \tag{2-56}$$

Then $X(s)$ can be expanded as

$$X(s) = \frac{K_{s1}}{s + s_1} + \frac{K_{s2}}{s + s_2} + \cdots + \frac{K_{s(n-r)}}{s + s_{n-r}}$$
$$|\leftarrow (n - r) \text{ terms of simple poles} \rightarrow|$$
$$+ \frac{A_1}{s + s_i} + \frac{A_2}{(s + s_i)^2} + \cdots + \frac{A_r}{(s + s_i)^r} \tag{2-57}$$
$$|\longleftarrow r \text{ terms of repeated poles} \longrightarrow|$$

The $(n - r)$ coefficients, which correspond to simple poles, $K_{s1}, K_{s2}, \ldots, K_{s(n-r)}$, may be evaluated by the method described by Eq. (2-49). The determination of the coefficients that correspond to the multiple-order poles is described in what follows.

$$A_r = [(s + s_i)^r X(s)]\big|_{s=-s_i} \tag{2-58}$$

$$A_{r-1} = \frac{d}{ds}[(s + s_i)^r X(s)]\bigg|_{s=-s_i} \tag{2-59}$$

$$A_{r-2} = \frac{1}{2!}\frac{d^2}{ds^2}[(s + s_i)^r X(s)]\bigg|_{s=-s_i} \tag{2-60}$$

$$\vdots$$

$$A_1 = \frac{1}{(r-1)!} \frac{d^{r-1}}{ds^{r-1}} [(s + s_i)^r X(s)] \Bigg|_{s=-s_i} \qquad (2\text{-}61)$$

EXAMPLE 2-6   Consider the function

$$X(s) = \frac{1}{s(s+1)^3(s+2)} \qquad (2\text{-}62)$$

By using the format of Eq. (2-57), $X(s)$ is written

$$X(s) = \frac{K_0}{s} + \frac{K_{-2}}{s+2} + \frac{A_1}{s+1} + \frac{A_2}{(s+1)^2} + \frac{A_3}{(s+1)^3} \qquad (2\text{-}63)$$

The coefficients corresponding to the simple poles are

$$K_0 = [sX(s)]|_{s=0} = \tfrac{1}{2} \qquad (2\text{-}64)$$
$$K_{-2} = [(s+2)X(s)]|_{s=-2} = \tfrac{1}{2} \qquad (2\text{-}65)$$

and those of the third-order pole are

$$A_3 = [(s+1)^3 X(s)]|_{s=-1} = -1 \qquad (2\text{-}66)$$
$$A_2 = \frac{d}{ds}[(s+1)^3 X(s)]|_{s=-1} = \frac{d}{ds}\left[\frac{1}{s(s+2)}\right]\Bigg|_{s=-1} = 0 \qquad (2\text{-}67)$$
$$A_1 = \frac{1}{2!}\frac{d^2}{ds^2}[(s+1)^3 X(s)]\Bigg|_{s=-1} = \frac{1}{2}\frac{d^2}{ds^2}\left[\frac{1}{s(s+2)}\right]\Bigg|_{s=-1} = -1 \qquad (2\text{-}68)$$

The completed partial-fraction expansion is

$$X(s) = \frac{1}{2s} + \frac{1}{2(s+2)} - \frac{1}{s+1} - \frac{1}{(s+1)^3} \qquad (2\text{-}69)$$

Partial-Fraction Expansion of Simple Complex-Conjugate Poles

complex poles   The partial-fraction expansion of Eq. (2-48) is valid also for simple complex-conjugate poles. Since complex-conjugate poles are more difficult to handle and are of special interest in control-systems studies, they deserve special treatment here.

Suppose that $X(s)$ of Eq. (2-45) contains a pair of complex poles:

$$s = -\alpha + j\omega \qquad \text{and} \qquad s = -\alpha - j\omega$$

The corresponding coefficients of these poles are

$$K_{-\alpha+j\omega} = (s + \alpha - j\omega)X(s)\big|_{s=-\alpha+j\omega} \tag{2-70}$$

$$K_{-\alpha-j\omega} = (s + \alpha + j\omega)X(s)\big|_{s=-\alpha-j\omega} \tag{2-71}$$

EXAMPLE 2-7   Consider the function

$$X(s) = \frac{\omega_n^2}{s^2 + 2\zeta\omega_n s + \omega_n^2} \tag{2-72}$$

Let us assume that the values of ζ and ω_n are such that the poles of $X(s)$ are complex. Then, $X(s)$ is expanded as follows:

$$X(s) = \frac{K_{-\alpha+j\omega}}{s + \alpha - j\omega} + \frac{K_{-\alpha-j\omega}}{s + \alpha + j\omega} \tag{2-73}$$

where

$$\alpha = \zeta\omega_n \tag{2-74}$$

and

$$\omega = \omega_n\sqrt{1 - \zeta^2} \tag{2-75}$$

The coefficients in Eq. (2-73) are determined as

$$K_{-\alpha+j\omega} = (s + \alpha - j\omega)X(s)\big|_{s=-\alpha+j\omega} = \frac{\omega_n^2}{2j\omega} \tag{2-76}$$

$$K_{-\alpha-j\omega} = (s + \alpha + j\omega)X(s)\big|_{s=-\alpha-j\omega} = -\frac{\omega_n^2}{2j\omega} \tag{2-77}$$

The complete partial-fraction expansion of Eq. (2-72) is

$$X(s) = \frac{\omega_n^2}{2j\omega}\left[\frac{1}{s + \alpha - j\omega} - \frac{1}{s + \alpha + j\omega}\right] \tag{2-78}$$

Taking the inverse Laplace transform on both sides of the last equation gives

$$x(t) = \frac{\omega_n^2}{2j\omega}e^{-\alpha t}(e^{j\omega t} - e^{-j\omega t})$$

$$= \frac{\omega_n}{\sqrt{1 - \zeta^2}}e^{-\zeta\omega_n t}\sin \omega_n\sqrt{1 - \zeta^2}\, t \qquad t \geq 0 \tag{2-79}$$

Computer Solution of the Partial-Fraction Expansion

computer solution   The partial-fraction expansion can be carried out by a digital computer program. The PFE program contained in the ACSP software accompanying this book can handle real multiple-order poles of $X(s)$ up to the third order. Complex-conjugate poles must be simple.

2-6

APPLICATION OF LAPLACE TRANSFORM TO THE SOLUTION OF LINEAR ORDINARY DIFFERENTIAL EQUATIONS

Linear ordinary differential equations can be solved by the Laplace transform method with the aid of the theorems on Laplace transform given in Section 2-4 and a table of Laplace transforms. The procedure is outlined as follows:

> 1. Transform the differential equation to the s-domain by Laplace transform using the Laplace transform table.
> 2. Manipulate the transformed algebraic equation and solve for the output variable.
> 3. Perform partial-fraction expansion so that the inverse Laplace transform can be obtained from the Laplace transform table.
> 4. Perform the inverse Laplace transform.

Let us illustrate the method by several examples.

EXAMPLE 2-8   Consider the differential equation

$$\frac{d^2x(t)}{dt^2} + 3\frac{dx(t)}{dt} + 2x(t) = 5u_s(t) \tag{2-80}$$

where $u_s(t)$ is the unit-step function. The initial conditions are $x(0) = -1$ and $x^{(1)}(0) = dx(t)/dt|_{t=0} = 2$. To solve the differential equation, we first take the Laplace transform on both sides of Eq. (2-80):

$$s^2X(s) - sx(0) - x^{(1)}(0) + 3sX(s) - 3x(0) + 2X(s) = 5/s \tag{2-81}$$

Substituting the values of the initial conditions into the last equation and solving for $X(s)$, we get

$$X(s) = \frac{-s^2 - s + 5}{s(s^2 + 3s + 2)} = \frac{-s^2 - s + 5}{s(s + 1)(s + 2)} \tag{2-82}$$

Equation (2-82) is expanded by partial-fraction expansion to give

$$X(s) = \frac{5}{2s} - \frac{5}{s + 1} + \frac{3}{2(s + 2)} \tag{2-83}$$

Taking the inverse Laplace transform of Eq. (2-83), we get the complete solution as

$$x(t) = \frac{5}{2} - 5e^{-t} + \frac{3}{2}e^{-2t} \qquad t \ge 0 \tag{2-84}$$

The first term in Eq. (2-84) is the steady-state solution; the last two terms represent the transient solution. Unlike the classical method, which requires separate steps to give the transient and the steady-state solutions, the Laplace transform method gives the entire solution in one operation.

If only the magnitude of the steady-state solution is of interest, the final-value theorem may be applied. Thus,

$$\lim_{t \to \infty} x(t) = \lim_{s \to 0} sX(s) = \lim_{s \to 0} \frac{-s^2 - s + 5}{s^2 + 3s + 2} = \frac{5}{2} \tag{2-85}$$

where we have first checked and found that the function $sX(s)$ has poles only in the left-half s-plane, so that the final-value theorem is valid.

EXAMPLE 2-9   Consider the linear differential equation

$$\frac{d^2x(t)}{dt^2} + 34.5\frac{dx(t)}{dt} + 1000x(t) = 1000u_s(t) \tag{2-86}$$

The initial values of $x(t)$ and $dx(t)/dt$ are zero. Taking the Laplace transform on both sides of Eq. (2-86) and solving for $X(s)$, we have

$$X(s) = \frac{1000}{s(s^2 + 34.5s + 1000)} = \frac{\omega_n^2}{s(s^2 + 2\zeta\omega_n s + \omega_n^2)} \tag{2-87}$$

where $\zeta = 0.5455$, and $\omega_n = 31.62$. The inverse Laplace transform of the last equation can be executed in a number of ways. The Laplace transform table in Appendix B gives the transform pair of the expression in Eq. (2-87) directly. The result is

$$x(t) = 1 - \frac{e^{-\zeta\omega_n t}}{\sqrt{1 - \zeta^2}} \sin\left(\omega_n\sqrt{1 - \zeta^2}\, t + \theta\right) \qquad t \ge 0 \tag{2-88}$$

where

$$\theta = \cos^{-1}\zeta = 56.94° \tag{2-89}$$

Thus,

$$x(t) = 1 - 1.193e^{-17.25t} \sin(26.5t + 56.94°) \qquad t \ge 0 \tag{2-90}$$

Equation (2-88) can be derived by performing the partial-fraction expansion of Eq. (2-87) knowing that the poles are at $s = 0$, $-\alpha + j\omega$, and $-\alpha - j\omega$, where

$$\alpha = \zeta\omega_n = 17.25 \tag{2-91}$$

$$\omega = \omega_n\sqrt{1 - \zeta^2} = 26.5 \tag{2-92}$$

The partial-fraction expansion of Eq. (2-87) is written

$$X(s) = \frac{K_0}{s} + \frac{K_{-\alpha+j\omega}}{s + \alpha - j\omega} + \frac{K_{-\alpha-j\omega}}{s + \alpha + j\omega} \tag{2-93}$$

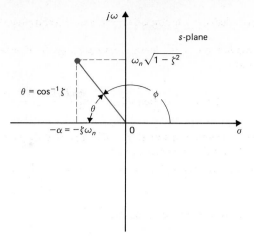

FIGURE 2-4 Root location in the s-plane.

where

$$K_0 = sX(s)\big|_{s=0} = 1 \tag{2-94}$$

$$K_{-\alpha+j\omega} = (s + \alpha - j\omega)X(s)\big|_{s=-\alpha+j\omega} = \frac{e^{-j\phi}}{2j\sqrt{1 - \zeta^2}} \tag{2-95}$$

$$K_{-\alpha-k\omega} = (s + \alpha + j\omega)X(s)\big|_{s=-\alpha-j\omega} = \frac{-e^{j\phi}}{2j\sqrt{1 - \zeta^2}} \tag{2-96}$$

The angle ϕ is given by

$$\phi = 180° - \cos^{-1}\zeta \tag{2-97}$$

and is illustrated in Fig. 2-4.

The inverse Laplace transform of Eq. (2-93) is now written

$$x(t) = 1 + \frac{1}{2j\sqrt{1 - \zeta^2}}e^{-\zeta\omega_n t}[e^{j(\omega t - \phi)} - e^{-j(\omega t - \phi)}]$$

$$= 1 + \frac{1}{\sqrt{1 - \zeta^2}}e^{-\zeta\omega_n t}\sin[\omega_n\sqrt{1 - \zeta^2}\,t - \phi] \qquad t \geq 0 \tag{2-98}$$

Substituting Eq. (2-97) into the last equation for ϕ, we have

$$x(t) = 1 - \frac{1}{\sqrt{1 - \zeta^2}}e^{-\zeta\omega_n t}\sin[\omega_n\sqrt{1 - \zeta^2}\,t + \cos^{-1}\zeta] \qquad t \geq 0 \tag{2-99}$$

or

$$x(t) = 1 - 1.193e^{-17.25t}\sin(26.5t + 56.94°) \qquad t \geq 0 \tag{2-100}$$

2-7

ELEMENTARY MATRIX THEORY

In the study of modern control theory, it is often desirable to use matrix notation to simplify complex mathematical expressions. The matrix notation usually makes the equations much easier to handle and manipulate.

As a motivation to the reason of using matrix notation, let us consider the following set of n simultaneous algebraic equations:

$$
\begin{aligned}
a_{11}x_1 + a_{12}x_2 + \cdots + a_{1n}x_n &= y_1 \\
a_{21}x_1 + a_{22}x_2 + \cdots + a_{2n}x_n &= y_2 \\
\cdots\cdots\cdots\cdots\cdots\cdots\cdots\cdots\cdots\cdots \\
a_{n1}x_1 + a_{n2}x_2 + \cdots + a_{nn}x_n &= y_n
\end{aligned}
\tag{2-101}
$$

We may use the matrix equation

$$
\mathbf{Ax} = \mathbf{y} \tag{2-102}
$$

as a simplified representation of Eq. (2-101). The symbols $\mathbf{A}$, $\mathbf{x}$, and $\mathbf{y}$ are defined as matrices, which contain the coefficients and variables of the original equations as their elements. In terms of matrix algebra, which will be discussed later, Eq. (2-102) can be stated as: *The product of the matrices $\mathbf{A}$ and $\mathbf{x}$ is equal to the matrix $\mathbf{y}$.* The three matrices involved are defined as

$$
\mathbf{A} = \begin{bmatrix}
a_{11} & a_{12} & \cdots & a_{1n} \\
a_{21} & a_{22} & \cdots & a_{2n} \\
\vdots & \vdots & \ddots & \vdots \\
a_{n1} & a_{n2} & \cdots & a_{nn}
\end{bmatrix}
\tag{2-103}
$$

$$
\mathbf{x} = \begin{bmatrix}
x_1 \\
x_2 \\
\vdots \\
x_n
\end{bmatrix}
\tag{2-104}
$$

$$
\mathbf{y} = \begin{bmatrix}
y_1 \\
y_2 \\
\vdots \\
y_n
\end{bmatrix}
\tag{2-105}
$$

which are simply bracketed arrays of coefficients and variables.

Definition of a Matrix

A matrix is a collection of elements arranged in a rectangular or square array. There are several ways of bracketing a matrix. In this text, the square brackets, such as in Eq. (2-105), are used to represent matrices. It is important to distinguish between a matrix and a determinant:

Matrix	**Determinant**
An array of numbers or elements with n rows and m columns.	An array of numbers or elements with n rows and n columns (always square).
Does not have a value, although a square matrix ($n = m$) has a determinant.	Has a value.

Some important definitions of matrices are given in the following.

MATRIX ELEMENTS.   When a matrix is written

$$\mathbf{A} = \begin{bmatrix} a_{11} & a_{12} & a_{13} \\ a_{21} & a_{22} & a_{23} \\ a_{31} & a_{32} & a_{33} \end{bmatrix} \tag{2-106}$$

where a_{ij} is identified as the element in the ith **row** and the jth **column** of the matrix. As a rule, we always refer to the row first and the column last.

ORDER OF A MATRIX.   *The order of a matrix refers to the total number of rows and columns of the matrix*. For example, the matrix in Eq. (2-106) has three rows and three columns and is called a 3×3 (three-by-three) matrix. In general, a matrix with n rows and m columns is termed $n \times m$ or **n by m**.

SQUARE MATRIX.   *A square matrix is one that has the same number of rows as columns*.

COLUMN MATRIX.   *A column matrix is one that has one column and more than one row, that is, an $m \times 1$ matrix, $m > 1$.* Quite often, a column matrix is referred to as a **column vector** or simply an ***m*-vector** if there are m rows and one column. The matrix in Eq. (2-104) is a typical n-vector.

ROW MATRIX.   *A row matrix is one that has one row and more than one column, that is, a $1 \times n$ matrix, where $n > 1$.* A row matrix can also be referred to as a **row-vector**.

DIAGONAL MATRIX.   *A diagonal matrix is a square matrix with $a_{ij} = 0$ for all $i \neq j$.* Examples of a diagonal matrix are

$$\begin{bmatrix} a_{11} & 0 & 0 \\ 0 & a_{22} & 0 \\ 0 & 0 & a_{33} \end{bmatrix} \qquad \begin{bmatrix} 5 & 0 \\ 0 & 3 \end{bmatrix} \tag{2-107}$$

UNITY MATRIX (IDENTITY MATRIX).   *A unity matrix is a diagonal matrix with all the elements on the main diagonal ($i = j$) equal to 1.* A unity matrix is often designated by **I** or **U**. An example of a unity matrix is

$$\mathbf{I} = \begin{bmatrix} 1 & 0 & 0 \\ 0 & 1 & 0 \\ 0 & 0 & 1 \end{bmatrix} \tag{2-108}$$

NULL MATRIX.   *A null matrix is one whose elements are all equal to zero.*

SYMMETRIC MATRIX.   *A symmetric matrix is a square matrix that satisfies the condition*

$$a_{ij} = a_{ji} \tag{2-109}$$

for all i and j. A symmetric matrix has the property that if its rows are interchanged with its columns, the same matrix is preserved. Two examples of the symmetric matrix are

$$\mathbf{A} = \begin{bmatrix} 6 & 5 & 1 \\ 5 & 0 & 10 \\ 1 & 10 & -1 \end{bmatrix} \qquad \mathbf{B} = \begin{bmatrix} 1 & -4 \\ -4 & 1 \end{bmatrix} \tag{2-110}$$

DETERMINANT OF A MATRIX.   *With each square matrix, a determinant having the same elements and order as the matrix may be defined. The determinant of a square matrix* **A** *is designated by*

$$\det \mathbf{A} = \Delta_A = |\mathbf{A}| \tag{2-111}$$

For example, the determinant of the matrix **A** in Eq. (2-106) is written

$$|\mathbf{A}| = \begin{vmatrix} a_{11} & a_{12} & a_{13} \\ a_{21} & a_{22} & a_{23} \\ a_{31} & a_{32} & a_{33} \end{vmatrix} \tag{2-112}$$

COFACTOR OF A DETERMINANT ELEMENT.   *Given any nth-order determinant* $|\mathbf{A}|$*, the cofactor of any element a_{ij}, A_{ij}, is the determinant obtained by eliminating all elements of the ith row and jth column and then multiplied by* $(-1)^{i+j}$. For example, the cofactor of the element a_{11} of $|\mathbf{A}|$ in Eq. (2-112) is written

$$A_{11} = (-1)^{1+1} \begin{vmatrix} a_{22} & a_{23} \\ a_{32} & a_{33} \end{vmatrix} = a_{22}a_{23} - a_{23}a_{32} \qquad (2\text{-}113)$$

In general, the value of a determinant can be written in terms of the cofactors. Let **A** be an $n \times n$ matrix, then the determinant of **A** can be written in terms of the cofactors of any row, or the cofactors of any column. That is,

$$\det \mathbf{A} = \sum_{j=1}^{n} a_{ij} A_{ij} \qquad (i = 1, \text{ or } 2, \dots, \text{ or } n) \qquad (2\text{-}114)$$

or

$$\det \mathbf{A} = \sum_{i=1}^{n} a_{ij} A_{ij} \qquad (j = 1, \text{ or } 2, \dots, \text{ or } n) \qquad (2\text{-}115)$$

EXAMPLE 2-10   The value of the determinant in Eq. (2-112) is

$$\det \mathbf{A} = |\mathbf{A}| = a_{11}A_{11} + a_{12}A_{12} + a_{13}A_{13}$$
$$= a_{11}(a_{22}a_{33} - a_{23}a_{32}) - a_{12}(a_{21}a_{33} - a_{23}a_{31}) \qquad (2\text{-}116)$$
$$+ a_{13}(a_{21}a_{32} - a_{22}a_{31})$$

singular matrix   **SINGULAR MATRIX.**   *A square matrix is said to be singular if the value of its determinant is zero.* If a square matrix has a nonzero determinant, it is called a **nonsingular matrix**. When a matrix is singular, it usually means that not all the rows or not all the columns of the matrix are independent of each other. When the matrix is used to represent a set of algebraic equations, singularity of the matrix means that these equations are not independent of each other.

EXAMPLE 2-11   Consider the following set of equations:

$$\begin{aligned} 2x_1 - 3x_2 + x_3 &= 0 \\ -x_1 + x_2 + x_3 &= 0 \\ x_1 - 2x_2 + 2x_3 &= 0 \end{aligned} \qquad (2\text{-}117)$$

The third equation is equal to the sum of the first two. Thus, these three equations are not completely independent. In matrix form, these equations may be written as

$$\mathbf{Ax} = \mathbf{0} \qquad (2\text{-}118)$$

where

$$\mathbf{A} = \begin{bmatrix} 2 & -3 & 1 \\ -1 & 1 & 1 \\ 1 & -2 & 2 \end{bmatrix} \qquad \mathbf{x} = \begin{bmatrix} x_1 \\ x_2 \\ x_3 \end{bmatrix} \qquad (2\text{-}119)$$

and $\mathbf{0}$ is a 3×1 null matrix. The determinant of $\mathbf{A}$ is 0, and, thus, the matrix $\mathbf{A}$ is singular. In this case, the rows of $\mathbf{A}$ are dependent.

transpose

TRANSPOSE OF A MATRIX. *The transpose of a matrix $\mathbf{A}$ is defined as the matrix that is obtained by interchanging the corresponding rows and columns in $\mathbf{A}$. Let $\mathbf{A}$ be an $n \times m$ matrix that is represented by*

$$\mathbf{A} = [a_{ij}]_{n, m} \tag{2-120}$$

The transpose of $\mathbf{A}$, denoted by $\mathbf{A}'$, is given by

$$\mathbf{A}' = \text{transpose of } \mathbf{A} = [a_{ij}]_{m,n} \tag{2-121}$$

Notice that the order of $\mathbf{A}$ is $n \times m$, but the order of $\mathbf{A}'$ is $m \times n$.

EXAMPLE 2-12

Consider the 2×3 matrix

$$\mathbf{A} = \begin{bmatrix} 3 & 2 & 1 \\ 0 & -1 & 5 \end{bmatrix} \tag{2-122}$$

The transpose of $\mathbf{A}$ is obtained by interchanging the rows and the columns.

$$\mathbf{A}' = \begin{bmatrix} 3 & 0 \\ 2 & -1 \\ 1 & 5 \end{bmatrix} \tag{2-123}$$

SOME PROPERTIES OF MATRIX TRANSPOSE

1. $(\mathbf{A}')' = \mathbf{A}$ (2-124)
2. $(k\mathbf{A})' = k\mathbf{A}'$, where k is a scalar   (2-125)
3. $(\mathbf{A} + \mathbf{B})' = \mathbf{A}' + \mathbf{B}'$ (2-126)
4. $(\mathbf{AB})' = \mathbf{B}'\mathbf{A}'$ (2-127)

adjoint

ADJOINT OF A MATRIX. *Let $\mathbf{A}$ be a square matrix of order n. The adjoint matrix of $\mathbf{A}$, denoted by adj $\mathbf{A}$, is defined as*

$$\text{adj } \mathbf{A} = [A_{ij} \text{ of det } \mathbf{A}]'_{n, n} \tag{2-128}$$

where A_{ij} denotes the cofactor of a_{ij}.

EXAMPLE 2-13   Consider the 2×2 matrix

$$\mathbf{A} = \begin{bmatrix} a_{11} & a_{12} \\ a_{21} & a_{22} \end{bmatrix} \tag{2-129}$$

The cofactors are $A_{11} = a_{22}$, $A_{12} = -a_{21}$, $A_{21} = -a_{12}$, and $A_{22} = a_{11}$. Thus, the adjoint matrix of $\mathbf{A}$ is

$$\text{adj } \mathbf{A} = \begin{bmatrix} A_{11} & A_{12} \\ A_{21} & A_{22} \end{bmatrix}' = \begin{bmatrix} a_{22} & -a_{21} \\ -a_{12} & a_{11} \end{bmatrix}' = \begin{bmatrix} a_{22} & -a_{12} \\ -a_{21} & a_{11} \end{bmatrix} \tag{2-130}$$

2-8

MATRIX ALGEBRA

When carrying out matrix operations, it is necessary to define matrix algebra in the form of addition, subtraction, multiplication, and division.

Equality of Matrices

Two matrices $\mathbf{A}$ and $\mathbf{B}$ are said to be equal to each other if they satisfy the following conditions:

1. They are of the same order.
2. The corresponding elements are equal; that is,

$$a_{ij} = b_{ij} \qquad \text{for every } i \text{ and } j$$

EXAMPLE 2-14

$$\mathbf{A} = \begin{bmatrix} a_{11} & a_{12} \\ a_{21} & a_{22} \end{bmatrix} = \mathbf{B} = \begin{bmatrix} b_{11} & b_{12} \\ b_{21} & b_{22} \end{bmatrix} \tag{2-131}$$

implies that $a_{11} = b_{11}$, $a_{12} = b_{12}$, $a_{21} = b_{21}$, and $a_{22} = b_{22}$.

Addition and Subtraction of Matrices

A and B of same Order

Two matrices $\mathbf{A}$ and $\mathbf{B}$ can be added or subtracted to form $\mathbf{A} \pm \mathbf{B}$ if they are of the same order. That is,

$$\mathbf{A} \pm \mathbf{B} = [a_{ij}]_{n, m} \pm [b_{ij}]_{n, m} = \mathbf{C} = [c_{ij}]_{n, m} \tag{2-132}$$

where

$$c_{ij} = a_{ij} \pm b_{ij} \tag{2-133}$$

for all i and j.

The order of the matrices is preserved after addition or subtraction.

EXAMPLE 2-15

Consider the matrices

$$\mathbf{A} = \begin{bmatrix} 3 & 2 \\ -1 & 4 \\ 0 & -1 \end{bmatrix} \quad \mathbf{B} = \begin{bmatrix} 0 & 3 \\ -1 & 2 \\ 1 & 0 \end{bmatrix} \tag{2-134}$$

which are of the same order. Then the sum of $\mathbf{A}$ and $\mathbf{B}$ is

$$\mathbf{C} = \mathbf{A} + \mathbf{B} = \begin{bmatrix} 3+0 & 2+3 \\ -1-1 & 4+2 \\ 0+1 & -1+0 \end{bmatrix} = \begin{bmatrix} 3 & 5 \\ -2 & 6 \\ 1 & -1 \end{bmatrix} \tag{2-135}$$

Associative Law of Matrix (Addition and Subtraction)

The associative law of scalar algebra still holds for matrix addition and subtraction. That is,

$$(\mathbf{A} + \mathbf{B}) + \mathbf{C} = \mathbf{A} + (\mathbf{B} + \mathbf{C}) \tag{2-136}$$

Commutative Law of Matrix (Addition and Subtraction)

The commutative law for matrix addition and subtraction states that the following matrix relationship is true:

$$\mathbf{A} + \mathbf{B} + \mathbf{C} = \mathbf{B} + \mathbf{C} + \mathbf{A} = \mathbf{A} + \mathbf{C} + \mathbf{B} \tag{2-137}$$

Matrix Multiplication

conformable   The matrices $\mathbf{A}$ and $\mathbf{B}$ may be multiplied together to form the product $\mathbf{AB}$ if they are **conformable.** This means that the number of columns of $\mathbf{A}$ must equal the number of rows of $\mathbf{B}$. In other words, let

$$\mathbf{A} = [a_{ij}]_{n,p} \quad \mathbf{B} = [b_{ij}]_{q,m} \tag{2-138}$$

Then $\mathbf{A}$ and $\mathbf{B}$ are conformable to form the product

$$\mathbf{C} = \mathbf{AB} = [a_{ij}]_{n,p}[b_{ij}]_{q,m} = [c_{ij}]_{n,m} \tag{2-139}$$

if and only if $p = q$. The matrix $\mathbf{C}$ will have the same number of rows as $\mathbf{A}$ and the same number of columns as $\mathbf{B}$.

It is important to note that $\mathbf{A}$ and $\mathbf{B}$ may be conformable to form $\mathbf{AB}$, but they may not be conformable for the product $\mathbf{BA}$, unless in Eq. (2-139), n also equals m. This points out an important fact that *the commutative law is not generally valid for matrix multiplication*. It is also noteworthy that even though $\mathbf{A}$ and $\mathbf{B}$ are conformable for both $\mathbf{AB}$ and $\mathbf{BA}$, usually $\mathbf{AB} \neq \mathbf{BA}$. In general, the following references are made with respect to matrix multiplication whenever they exist:

$$\mathbf{AB} = \mathbf{A} \text{ postmultiplied by } \mathbf{B} \quad \text{or} \quad \mathbf{B} \text{ premultiplied by } \mathbf{A}$$

Rules of Matrix Multiplication

When the matrices $\mathbf{A}$ ($n \times p$) and $\mathbf{B}$ ($p \times m$) are conformable to form the matrix $\mathbf{C} = \mathbf{AB}$, the ijth element of $\mathbf{C}$, c_{ij}, is given by

$$c_{ij} = \sum_{k=1}^{p} a_{ik} b_{kj} \tag{2-140}$$

for $i = 1, 2, \ldots, n$, and $j = 1, 2, \ldots, m$.

EXAMPLE 2-16   Given the matrices

$$\mathbf{A} = [a_{ij}]_{2,3} \qquad \mathbf{B} = [b_{ij}]_{3,1} \tag{2-141}$$

The two matrices are conformable for the product $\mathbf{AB}$ but not for $\mathbf{BA}$. Thus,

$$\mathbf{AB} = \begin{bmatrix} a_{11} & a_{12} & a_{13} \\ a_{21} & a_{22} & a_{23} \end{bmatrix} \begin{bmatrix} b_{11} \\ b_{21} \\ b_{31} \end{bmatrix} = \begin{bmatrix} a_{11}b_{11} + a_{12}b_{21} + a_{13}b_{31} \\ a_{21}b_{11} + a_{22}b_{21} + a_{23}b_{31} \end{bmatrix} \tag{2-142}$$

EXAMPLE 2-17   Given the matrices

$$\mathbf{A} = \begin{bmatrix} 3 & -1 \\ 0 & 1 \\ 2 & 0 \end{bmatrix} \qquad \mathbf{B} = \begin{bmatrix} 1 & 0 & -1 \\ 2 & 1 & 0 \end{bmatrix} \tag{2-143}$$

The two matrices are conformable for $\mathbf{AB}$ and $\mathbf{BA}$.

$$\mathbf{AB} = \begin{bmatrix} 3 & -1 \\ 0 & 1 \\ 2 & 0 \end{bmatrix} \begin{bmatrix} 1 & 0 & -1 \\ 2 & 1 & 0 \end{bmatrix}$$

$$= \begin{bmatrix} (3)(1) + (-1)(2) & (3)(0) + (-1)(1) & (3)(-1) + (-1)(0) \\ (0)(1) + (1)(2) & (0)(0) + (1)(1) & (0)(-1) + (1)(0) \\ (2)(1) + (0)(2) & (2)(0) + (0)(1) & (2)(-1) + (0)(0) \end{bmatrix} \quad (2\text{-}144)$$

$$= \begin{bmatrix} 1 & -1 & -3 \\ 2 & 1 & 0 \\ 2 & 0 & -2 \end{bmatrix}$$

$$\mathbf{BA} = \begin{bmatrix} 1 & 0 & -1 \\ 2 & 1 & 0 \end{bmatrix} \begin{bmatrix} 3 & -1 \\ 0 & 1 \\ 2 & 0 \end{bmatrix}$$

$$= \begin{bmatrix} (1)(3) + (0)(0) + (-1)(2) & (1)(-1) + (0)(1) + (-1)(0) \\ (2)(3) + (1)(0) + (0)(2) & (2)(-1) + (1)(1) + (0)(0) \end{bmatrix} \quad (2\text{-}145)$$

$$= \begin{bmatrix} 1 & -1 \\ 6 & -1 \end{bmatrix}$$

Therefore, even though **AB** and **BA** both exist, they are not equal. In fact, in this case, the products are not of the same order.

Although the commutative law does not hold in general for matrix multiplication, the **associate** and **distributive** laws are valid. For the distributive law, we state that

$$\mathbf{A(B + C)} = \mathbf{AB} + \mathbf{AC} \quad (2\text{-}146)$$

if the products are conformable. For the associative law,

$$\mathbf{(AB)C} = \mathbf{A(BC)} \quad (2\text{-}147)$$

if the product is conformable.

Multiplication by a Scalar *k*

Multiplying a matrix **A** *by any scalar k is equivalent to multiplying each element of* **A** *by k. If* $\mathbf{A} = [a_{ij}]_{n,m}$,

$$k\mathbf{A} = [ka_{ij}]_{n,m} \quad (2\text{-}148)$$

Inverse of a Matrix (Matrix Division)

In the algebra for scalar quantities, when we write $y = ax$, it implies that $x = y/a$ is also true. In matrix algebra, if

$$\mathbf{A}\mathbf{x} = \mathbf{y} \tag{2-149}$$

then it *may be possible* to write

$$\mathbf{x} = \mathbf{A}^{-1}\mathbf{y} \tag{2-150}$$

where $\mathbf{A}^{-1}$ denotes the **matrix inverse of A**. The conditions that $\mathbf{A}^{-1}$ exists are

1. **A** *is a square matrix.*
2. **A** *must be nonsingular.*
3. *If* $\mathbf{A}^{-1}$ *exists, it is given by*

$$\mathbf{A}^{-1} = \frac{\text{adj } \mathbf{A}}{|\mathbf{A}|} \tag{2-151}$$

EXAMPLE 2-18

Given the matrix

$$\mathbf{A} = \begin{bmatrix} a_{11} & a_{12} \\ a_{21} & a_{22} \end{bmatrix} \tag{2-152}$$

matrix inverse

the inverse of **A** is given by

$$\mathbf{A}^{-1} = \frac{\text{adj } \mathbf{A}}{|\mathbf{A}|} = \frac{\begin{bmatrix} a_{22} & -a_{12} \\ -a_{21} & a_{11} \end{bmatrix}}{a_{11}a_{22} - a_{12}a_{21}} \tag{2-153}$$

where for **A** to be nonsingular, $|\mathbf{A}| \neq 0$, or $a_{11}a_{22} - a_{12}a_{21} \neq 0$.

Equation (2-153) shows that adj **A** of a 2×2 matrix is obtained by *interchanging the two elements on the main diagonal and changing the signs of the elements off the diagonal of* **A**.

EXAMPLE 2-19

Given the matrix

$$\mathbf{A} = \begin{bmatrix} a_{11} & a_{12} & a_{13} \\ a_{21} & a_{22} & a_{23} \\ a_{31} & a_{32} & a_{33} \end{bmatrix} \tag{2-154}$$

To find the inverse of **A**, the adjoint matrix of **A** is

$$\text{adj } \mathbf{A} = \begin{bmatrix} a_{22}a_{33} - a_{23}a_{32} & -(a_{12}a_{33} - a_{13}a_{32}) & a_{12}a_{23} - a_{13}a_{22} \\ -(a_{21}a_{33} - a_{23}a_{31}) & a_{11}a_{33} - a_{13}a_{31} & -(a_{11}a_{23} - a_{21}a_{13}) \\ a_{21}a_{32} - a_{22}a_{31} & -(a_{11}a_{32} - a_{12}a_{31}) & a_{11}a_{22} - a_{12}a_{21} \end{bmatrix} \quad (2\text{-}155)$$

The determinant of $\mathbf{A}$ is

$$|\mathbf{A}| = a_{11}a_{22}a_{33} + a_{12}a_{23}a_{31} + a_{13}a_{32}a_{21} - a_{13}a_{22}a_{31} - a_{12}a_{21}a_{33} - a_{11}a_{23}a_{32} \quad (2\text{-}156)$$

SOME PROPERTIES OF MATRIX INVERSE

1. $\mathbf{AA}^{-1} = \mathbf{A}^{-1}\mathbf{A} = \mathbf{I}$ (2-157)
2. $(\mathbf{A}^{-1})^{-1} = \mathbf{A}$ (2-158)
3. In matrix algebra, in general,

if A is not a square matrix   B ≠ C

$$\mathbf{AB} = \mathbf{AC} \quad (2\text{-}159)$$

does not necessarily imply that $\mathbf{B} = \mathbf{C}$. The reader can easily construct an example to illustrate this property. However, if $\mathbf{A}$ is a square matrix, and is nonsingular, we can premultiply both sides of Eq. (2-159) by $\mathbf{A}^{-1}$. Then,

$$\mathbf{A}^{-1}\mathbf{AB} = \mathbf{A}^{-1}\mathbf{AC} \quad (2\text{-}160)$$

or

$$\mathbf{IB} = \mathbf{IC} \quad (2\text{-}161)$$

which leads to

$$\mathbf{B} = \mathbf{C}$$

4. If $\mathbf{A}$ and $\mathbf{B}$ are square matrices and are nonsingular, then

$$(\mathbf{AB})^{-1} = \mathbf{B}^{-1}\mathbf{A}^{-1} \quad (2\text{-}162)$$

Rank of a Matrix

rank   The rank of a matrix $\mathbf{A}$ is the maximum number of linearly independent columns of $\mathbf{A}$; or, it is the order of the largest nonsingular matrix contained in $\mathbf{A}$.

EXAMPLE 2-20   Several examples on the rank of a matrix are as follows:

$$\begin{bmatrix} 0 & 1 \\ 0 & 0 \end{bmatrix} \quad \text{rank} = 1 \qquad \begin{bmatrix} 0 & 5 & 1 & 4 \\ 3 & 0 & 3 & 2 \end{bmatrix} \quad \text{rank} = 2$$

$$\begin{bmatrix} 3 & 9 & 2 \\ 1 & 3 & 0 \\ 2 & 6 & 1 \end{bmatrix} \quad \text{rank} = 2 \qquad \begin{bmatrix} 3 & 0 & 0 \\ 1 & 2 & 0 \\ 0 & 0 & 1 \end{bmatrix} \quad \text{rank} = 3$$

The following properties are useful in the determination of the rank of a matrix. Given an $n \times m$ matrix $\mathbf{A}$.

1.   Rank of $\mathbf{A}$ = Rank of $\mathbf{A}'$
2.   Rank of $\mathbf{A}$ = Rank of $\mathbf{A}'\mathbf{A}$.
3.   Rank of $\mathbf{A}$ = Rank of $\mathbf{AA}'$.

Properties 2 and 3 are useful in the determination of rank; since $\mathbf{A}'\mathbf{A}$ and $\mathbf{AA}'$ are always square, the rank condition can be checked by evaluating the determinant of these matrices.

Computer-Aided Solutions of Matrices

The MATRIXM program in the ACSP software that accompanies this text gives the results of the multiplication of two matrices. The inverse and the determinant of a matrix can be obtained using the LINSYS program in ACSP.

2-9

VECTOR-MATRIX FORM OF STATE EQUATIONS

state vector

The state equations defined in Section 2-3 can be conveniently expressed in vector-matrix form. Consider that a linear second-order system is described by the following state equations:

$$\frac{dx_1(t)}{dt} = a_{11}x_1(t) + a_{12}x_2(t) + b_{11}u(t) \tag{2-163}$$

$$\frac{dx_2(t)}{dt} = a_{21}x_1(t) + a_{22}x_2(t) + b_{21}u(t) \tag{2-164}$$

where $x_1(t)$ and $x_2(t)$ are the state variables; $u(t)$ is the input; a_{11}, a_{12}, a_{21}, and a_{22} are constant coefficients. Let the **state vector** be defined as the 2×1 matrix,

$$\mathbf{x}(t) = \begin{bmatrix} x_1(t) \\ x_2(t) \end{bmatrix} \tag{2-165}$$

The state equations in vector-matrix form are expressed as

$$\frac{d\mathbf{x}(t)}{dt} = \dot{\mathbf{x}}(t) = \mathbf{A}\mathbf{x}(t) + \mathbf{B}\mathbf{u}(t) \qquad (2\text{-}166)$$

where

$$\mathbf{A} = \begin{bmatrix} a_{11} & a_{12} \\ a_{21} & a_{22} \end{bmatrix} \qquad \mathbf{B} = \begin{bmatrix} b_{11} \\ b_{21} \end{bmatrix} \qquad (2\text{-}167)$$

In general, Eq. (2-166) can be used to represent a linear nth-order system with n state variables, n state equations, and p inputs. In this case, $\mathbf{x}(t)$ and $d\mathbf{x}(t)/dt$ are $n \times 1$ vectors, $\mathbf{A}$ is $n \times n$, $\mathbf{B}$ is $n \times p$, and the input $\mathbf{u}(t)$ is a $p \times 1$ vector.

The vector-matrix equation form can also be applied to nonlinear state equations. The general expression is

$$\frac{d\mathbf{x}(t)}{dt} = \mathbf{f}[\mathbf{x}(t), \mathbf{u}(t), t] \qquad (2\text{-}168)$$

where $\mathbf{f}[\,\cdot\,]$ represents a $n \times 1$ function matrix.

2-10

DIFFERENCE EQUATIONS

Because digital controllers are frequently used in control systems, it is necessary to establish equations that relate digital and discrete-time signals. Just as differential equations are used to represent systems with analog signals, difference equations are used for systems with discrete or digital data. Difference equations are also used to approximate differential equations, since the former are more easily programmed on a digital computer, or are generally easier to solve.

In general, a linear nth-order difference equation with constant coefficients can be written as

$$a_{n+1}y(k + n) + a_n y(k + n - 1) + \cdots + a_2 y(k + 1) + a_1 y(k) = f(k) \qquad (2\text{-}169)$$

where $y(i)$, $i = k, k + 1, \ldots, k + n$, denotes the discrete dependent variable y at the ith instant if the independent variable is time. In general, the independent variable can represent any real quantity.

Similar to the case of the analog systems, it is convenient to use a set of first-order difference equations (state equations) to represent a high-order difference equation. For the difference equation in Eq. (2-169), if we let

$$x_1(k) = y(k)$$
$$x_2(k) = x_1(k + 1) = y(k + 1)$$
$$\vdots$$
$$x_{n-1}(k) = x_{n-2}(k + 1) = y(k + n - 2) \tag{2-170}$$

then the equation is written as

$$x_n(k + 1) = -\frac{a_1}{a_{n+1}} x_1(k) - \frac{a_2}{a_{n+1}} x_2(k) - \cdots - \frac{a_n}{a_{n+1}} x_n(k) + \frac{1}{a_{n+1}} f(k) \tag{2-171}$$

The first $(n - 1)$ state equations are taken directly from Eq. (2-170), and the last one is given by Eq. (2-171). Writing these n first-order difference state equations in vector-matrix form, we have

$$\mathbf{x}(k + 1) = \mathbf{A}\mathbf{x}(k) + \mathbf{B}u(k) \tag{2-172}$$

where

$$\mathbf{x}(k) = \begin{bmatrix} x_1(k) \\ x_2(k) \\ \vdots \\ x_n(k) \end{bmatrix} \tag{2-173}$$

is the $n \times 1$ state vector, and

$$\mathbf{A} = \begin{bmatrix} 0 & 1 & 0 & \cdots & 0 \\ 0 & 0 & 1 & \cdots & 0 \\ \cdots & \cdots & \cdots & \cdots & \cdots \\ 0 & 0 & 0 & \cdots & 1 \\ -\dfrac{a_1}{a_{n+1}} & -\dfrac{a_2}{a_{n+1}} & -\dfrac{a_3}{a_{n+1}} & \cdots & -\dfrac{a_n}{a_{n+1}} \end{bmatrix} \tag{2-174}$$

$$\mathbf{B} = \begin{bmatrix} 0 \\ 0 \\ \vdots \\ 1 \\ \dfrac{1}{a_{n+1}} \end{bmatrix} \tag{2-175}$$

and $u(k) = f(k)$.

2-11

THE z-TRANSFORM

In Section 2-6, we have shown that the Laplace transform can be used to solve linear ordinary differential equations. The z-transform represents an operational method for solving linear difference equations and linear systems with discrete or digital data. Let us first consider the analysis of a discrete-data system that is represented by the block diagram of Fig. 2-5. One way of describing the discrete nature of the signals is to consider that the input and the output of the system are sequences of numbers. These numbers appear at uniform time intervals T. Thus, the input and the output sequences may be represented by $u(kT)$ and $c(kT)$, respectively, $k = 0, 1, 2, \ldots$.

In order to represent the input and the output sequences by time-domain expressions, we introduce an impulse train such that the numbers are represented by the strengths of the impulses at the corresponding time instants. Thus, the input sequence is expressed as the impulse train:

$$u^*(t) = \sum_{k=0}^{\infty} u(kT)\, \delta(t - kT) \tag{2-176}$$

A similar expression can be written for the output sequence $c(kT)$.

Another important type of discrete-data system is the sampled-data system in which some of the signals are continuous but are converted to discrete data by devices such as the analog-to-digital (A/D) converter. The formation of discrete data from continuous data can be represented by samplers. Figure 2-6 shows the schematic diagram of a uniform-rate sampler that converts the continuous-data signal $u(t)$ into a discrete-data signal $u^*(t)$, which is described by Eq. (2-176). The time duration between the closings of the sampler is T and is called the **sampling period** (s). Since the impulse $\delta(t - kT)$ in Eq. (2-176) has a zero pulse width, the sampler so defined closes only for an infinitesimally small time duration. Therefore, such a sampler is not real, and is called an **ideal sampler**.

sampling period

ideal sampler

Now we are ready to investigate the application of transform methods to discrete-data systems. Taking the Laplace transform on both sides of Eq. (2-176), we have

$$U^*(s) = \sum_{k=0}^{\infty} u(kT)e^{-kTs} \tag{2-177}$$

FIGURE 2-5 Block diagram of a discrete-data system.

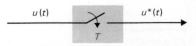

FIGURE 2-6 Ideal sampler.

The fact that Eq. (2-177) contains the exponential term e^{-kTs} reveals the difficulty of using the Laplace transform for the general treatment of discrete-data systems, since the transfer function relations will no longer be algebraic as in the continuous-data case. One simple fact is that the commonly used Laplace transform tables do not have entries with transcendental functions in s. This leads to the motivation of introducing the z-transform, which simply converts transcendental functions in s into algebraic functions in z.

Definition of the z-Transform

The z-transform is defined as

$$z = e^{Ts} \qquad\qquad ((2\text{-}178))$$

where s is the Laplace-transform variable, and T is the sampling period. Equation (2-178) also leads to

$$s = \frac{1}{T} \ln z \qquad\qquad (2\text{-}179)$$

By using Eq. (2-178), the expression in Eq. (2-177) is written

$$U^*\left[s = \frac{1}{T} \ln z\right] = U(z) = \sum_{k=0}^{\infty} u(kT)z^{-k} \qquad\qquad (2\text{-}180)$$

or

$$U(z) = z\text{-transform of } u^*(t) = \mathfrak{z}[u^*(t)]$$
$$= [\text{Laplace transform of } u^*(t)]|_{s=\ln z/T} \qquad (2\text{-}181)$$

The following examples illustrate the derivation of the z-transforms of two simple functions.

EXAMPLE 2-21   Consider the sequence

$$r(kT) = e^{-\alpha kT} \qquad k = 0, 1, 2, \ldots \qquad\qquad (2\text{-}182)$$

where α is a real constant. From Eq. (2-176),

$$r^*(t) = \sum_{k=0}^{\infty} e^{-\alpha kT} \delta(t - kT) \tag{2-183}$$

Then, the Laplace transform of $r^*(t)$ is written

$$R^*(s) = \sum_{k=0}^{\infty} e^{-\alpha kT} e^{-kTs} \tag{2-184}$$

By multiplying both sides of the last equation by $e^{-(s+\alpha)T}$ and subtracting the resulting equation from Eq. (2-184), $R^*(s)$ is expressed in a closed form:

$$R^*(s) = \frac{1}{1 - e^{-(s+\alpha)T}} \tag{2-185}$$

for

$$\left| e^{-(s+\alpha)T} \right| < 1 \tag{2-186}$$

The z-transform of $r^*(t)$ is obtained by setting $z = e^{Ts}$ in Eq. (2-185).

$$R(z) = \frac{1}{1 - e^{-\alpha T} z^{-1}} = \frac{z}{z - e^{-\alpha T}} \tag{2-187}$$

for $\left| e^{-\alpha T} z^{-1} \right| < 1$.

EXAMPLE 2-22   In example 2-21, if $\alpha = 0$, we have

$$r(kT) = 1 \qquad k = 0, 1, 2, \ldots \tag{2-188}$$

which represents a sequence of numbers all equal to unity. Then,

$$R(z) = 1 + z^{-1} + z^{-2} + \cdots = \frac{z}{z - 1} \tag{2-189}$$

for $|z| > 1$. Equation (2-189) is known as the z-transform of the unit-step function $u_s(t)$.

In general, the z-transforms of more complex functions may be obtained with the help of some of the z-transform theorems.

If a time function $r(t)$ is given, the procedure of finding its z-transform is to first form the sequence $r(kT)$ and then use Eq. (2-180) to get $R(z)$. An equivalent interpretation of this step is to send the signal $r(t)$ through the ideal sampler whose output is $r^*(t)$. $R^*(s)$ is obtained by taking the Laplace transform of $r^*(t)$, and $R(z)$ is finally determined by substituting z for e^{Ts} in $R^*(s)$.

TABLE 2-2   Table of z-Transforms

LAPLACE TRANSFORM	TIME FUNCTION	z-TRANSFORM
1	Unit impulse $\delta(t)$	1
$\dfrac{1}{s}$	Unit step $u_s(t)$	$\dfrac{z}{z-1}$
$\dfrac{1}{1-e^{-Ts}}$	$\delta_T(t) = \displaystyle\sum_{n=0}^{\infty} \delta(t-nT)$	$\dfrac{z}{z-1}$
$\dfrac{1}{s^2}$	t	$\dfrac{Tz}{(z-1)^2}$
$\dfrac{1}{s^3}$	$\dfrac{t^2}{2}$	$\dfrac{T^2 z(z+1)}{2(z-1)^3}$
$\dfrac{1}{s^{n+1}}$	$\dfrac{t^n}{n!}$	$\displaystyle\lim_{\alpha\to 0} \dfrac{(-1)^n}{n!}\dfrac{\partial^n}{\partial\alpha^n}\left[\dfrac{z}{z-e^{-\alpha T}}\right]$
$\dfrac{1}{s+\alpha}$	$e^{-\alpha t}$	$\dfrac{z}{z-e^{-\alpha T}}$
$\dfrac{1}{(s+\alpha)^2}$	$te^{-\alpha t}$	$\dfrac{Tze^{-\alpha T}}{(z-e^{-\alpha T})^2}$
$\dfrac{\alpha}{s(s+\alpha)}$	$1-e^{-\alpha t}$	$\dfrac{(1-e^{-\alpha T})z}{(z-1)(z-e^{-\alpha T})}$
$\dfrac{\omega}{s^2+\omega^2}$	$\sin \omega t$	$\dfrac{z\sin\omega T}{z^2 - 2z\cos\omega T + 1}$
$\dfrac{\omega}{(s+\alpha)^2+\omega^2}$	$e^{-\alpha t}\sin\omega t$	$\dfrac{ze^{-\alpha T}\sin\omega T}{z^2 - 2ze^{-\alpha T}\cos\omega T + e^{-2\alpha T}}$
$\dfrac{s}{s^2+\omega^2}$	$\cos \omega t$	$\dfrac{z(z-\cos\omega T)}{z^2 - 2z\cos\omega T + 1}$
$\dfrac{s+\alpha}{(s+\alpha)^2+\omega^2}$	$e^{-\alpha t}\cos\omega t$	$\dfrac{z^2 - ze^{-\alpha T}\cos\omega T}{z^2 - 2ze^{-\alpha T}\cos\omega T + e^{-2\alpha T}}$

Table 2-2 gives the z-transforms of some of the functions commonly used in systems analysis. A more extensive table can be found in [8].

Some Important Theorems of the z-Transformation

Some of the commonly used theorems of the z-transform are stated in the following without proof. Just as in the case of the Laplace transform, these theorems are useful in many aspects of the z-transform analysis.

THEOREM 1.   *Addition and Subtraction*
  If $r_1(kT)$ and $r_2(kT)$ have z-transforms $R_1(z)$ and $R_2(z)$, respectively, then

$$\mathscr{z}[r_1(kT) \pm r_2(kT)] = R_1(z) \pm R_2(z) \qquad (2\text{-}190)$$

THEOREM 2.   *Multiplication by a Constant*

$$\mathfrak{z}[\alpha r(kT)] = \alpha \, \mathfrak{z}[r(kT)] = \alpha R(z) \qquad (2\text{-}191)$$

where α is a constant. ∎

time delay

THEOREM 3.   *Real Translation (Time Delay)*

$$\mathfrak{z}[r(kT - nT)] = z^{-n}R(z) \qquad (2\text{-}192)$$

and

$$\mathfrak{z}[r(kT + nT)] = z^n\left[R(z) - \sum_{k=0}^{n-1} r(kT)z^{-k}\right] \qquad (2\text{-}193)$$

where n is a positive integer. Equation (2-192) represents the z-transform of a time sequence that is shifted to the right by nT, and Eq. (2-193) denotes that of a time sequence shifted to the left by nT. The reason that the right-hand side of Eq. (2-193) is not just $z^n R(z)$ is because the z-transform, similar to the Laplace transform, is defined only for $k \geq 0$. Thus, the second term on the right-hand side of Eq. (2-193) simply represents the sequence that is lost after it is shifted to the left by nT. ∎

THEOREM 4.   *Complex Translation*

$$\mathfrak{z}[e^{\mp\alpha kT}r(kT)] = R(ze^{\pm\alpha T}) \qquad (2\text{-}194)$$

∎

THEOREM 5.   *Initial-Value Theorem*

$$\lim_{k\to 0} r(kT) = \lim_{z\to\infty} R(z) \qquad (2\text{-}195)$$

if the limit exists. ∎

*final-value
theorem*

THEOREM 6.   *Final-Value Theorem*

$$\lim_{k\to\infty} r(kT) = \lim_{z\to 1} (1 - z^{-1})R(z) \qquad (2\text{-}196)$$

TABLE 2-3   Theorems of z-Transforms

Addition and subtraction	$\mathfrak{z}[r_1(kT) \pm r_2(kT)] = R_1(z) \pm R_2(z)$
Multiplication by a constant	$\mathfrak{z}[\alpha r(kT)] = \alpha\,\mathfrak{z}[r(kT)] = \alpha R(z)$
Real translation	$\mathfrak{z}[r(kT - nT)] = z^{-n}R(z)$
	$\mathfrak{z}[r(kT + nT)] = z^{n}\left[R(z) - \sum_{k=0}^{n-1} r(kT)z^{-k}\right]$
	where n = positive integer
Complex translation	$\mathfrak{z}[e^{\mp \alpha kT}r(kT)] = R(ze^{\pm \alpha T})$
Initial-Value theorem	$\lim_{k \to 0} r(kT) = \lim_{z \to \infty} R(z)$
Final-Value theorem	$\lim_{k \to \infty} r(kT) = \lim_{z \to 1}(1 - z^{-1})R(z)$
	if $(1 - z^{-1})R(z)$ has no poles on or outside the unit circle $|z| = 1$.
Real convolution	$F_1(\mathfrak{z})F_2(\mathfrak{z}) = \mathfrak{z}\left[\sum_{k=0}^{n} f_1(k)f_2(N - k)\right]$
	$= \mathfrak{z}\left[\sum_{k=0}^{N} f_2(k)f_1(N - k)\right]$
	$= \mathfrak{z}[f_1(k) * f_2(k)]$

if the function $(1 - z^{-1})R(z)$ has no poles on or outside the unit circle $|z| = 1$. ■

THEOREM 7.   *Real Convolution*

$$F_1(z)F_2(z) = \mathfrak{z}\left[\sum_{k=0}^{N} f_1(k)f_2(N - k)\right] = \mathfrak{z}\left[\sum_{k=0}^{N} f_2(k)f_1(N - k)\right]$$

$$= \mathfrak{z}[\, f_1(k) * f_2(k)] \qquad (2\text{-}197)$$

where "∗" denotes real convolution in the discrete-time domain. Thus, we see that just as in the Laplace transformation, the z-transform of the product of two real functions $f_1(k)$ and $f_2(k)$ is not equal to the product of the z-transform $F_1(z)$ and $F_2(z)$. One exception to this is if one of the two functions is the integral delay e^{-NTs}; then

$$\mathfrak{z}[e^{-NTs}F(s)] = \mathfrak{z}[e^{-NTs}]\mathfrak{z}[F(s)] = z^{-N}F(z) \qquad (2\text{-}198)$$

■

Table 2-3 summarizes the theorems on z-transforms that were given. The following examples illustrate the usefulness of some of these theorems.

EXAMPLE 2-23

(Complex Translation Theorem)

The complex tanslation theorem is used to find the z-transform of $f(t) = te^{-\alpha t}$, $t \geq 0$. Let

$r(t) = t, t \geq 0$; then

$$R(z) = \mathfrak{z}[tu_s(t)] = \mathfrak{z}(kT) = \frac{Tz}{(z - 1)^2} \tag{2-199}$$

Using the complex translation theorem in Eq. (2-194), we obtain

$$F(z) = \mathfrak{z}[te^{-\alpha t}u_s(t)] = R(ze^{\alpha T}) = \frac{Tze^{-\alpha T}}{(z - e^{-\alpha T})^2} \tag{2-200}$$

EXAMPLE 2-24   *(Final-Value Theorem)*
Given the function

$$R(z) = \frac{0.792z^2}{(z - 1)(z^2 - 0.416z + 0.208)} \tag{2-201}$$

determine the value of $r(kT)$ as k approaches infinity.

Since the function $(1 - z^{-1})R(z)$ does not have any pole on or outside the unit circle $|z| = 1$ in the z-plane, the final-value theorem in Eq. (2-196) can be applied. Thus,

$$\lim_{k \to \infty} r(kT) = \lim_{z \to 1} \frac{0.792z}{z^2 - 0.416z + 0.208} = 1 \tag{2-202}$$

Inverse *z*-Transformation

inverse
z-transform

Just as in the Laplace transformation, one of the major objectives of the z-transformation is that algebraic manipulations can be made first in the z-domain, and then the final time response is determined by the inverse z-transformation. *In general, the inverse z-transform of $R(z)$ can yield information on $r(kT)$ only, not on $r(t)$. In other words, the z-transform carries information only at the sampling instants.* When the time signal $r(t)$ is sampled by the ideal sampler, only information on the signal at the sampling instants, $t = kT$, is retained. With this in mind, the inverse z-transformation can be carried out by one of the following three methods:

1. The partial-fraction expansion method.
2. The power-series method.
3. The inversion formula.

partial-
fraction
expansion

PARTIAL-FRACTION EXPANSION METHOD.   The z-transform function $R(z)$ is expanded by partial-fraction expansion into a sum of simple recognizable terms, and the z-transform table is used to determine the corresponding $r(kT)$. In carrying out the partial-fraction expansion, there is a slight difference between the z-transform and the Laplace-transform procedures. With reference to the z-transform table, we note that practically all the z-transform functions have the term z in the numerator. Therefore, we should expand $R(z)$ into the form of

$$R(z) = \frac{K_1 z}{z - e^{-\alpha T}} + \frac{K_2 z}{z - e^{-\beta T}} + \cdots \qquad (2\text{-}203)$$

For this, we should first expand $R(z)/z$ into fractions and then multiply z across to obtain the final expression. The following example will illustrate this recommended procedure.

EXAMPLE 2-25   Given the z-transform function

$$R(z) = \frac{(1 - e^{-\alpha T})z}{(z - 1)(z - e^{-\alpha T})} \qquad (2\text{-}204)$$

it is desired to find the inverse z-transform. Expanding $R(z)/z$ by partial-fraction expansion, we have

$$\frac{R(z)}{z} = \frac{1}{z - 1} - \frac{1}{z - e^{-\alpha T}} \qquad (2\text{-}205)$$

The final expanded expression for $R(z)$ is

$$R(z) = \frac{z}{z - 1} - \frac{z}{z - e^{-\alpha T}} \qquad (2\text{-}206)$$

From the z-transform table of Table 2-2, the corresponding inverse z-transform of $R(z)$ is found to be

$$r(kT) = 1 - e^{-\alpha kT} \qquad k = 0, 1, 2, \ldots \qquad (2\text{-}207)$$

It should be pointed out that if $R(z)$ does not contain any factors of z in the numerator, this usually means that the time sequence has a time delay, and the partial-fraction expansion of $R(z)$ may be carried out without first dividing $R(z)$ by z. The following example illustrates this situation.

EXAMPLE 2-26   Consider the function

$$R(z) = \frac{(1 - e^{-\alpha T})}{(z - 1)(z - e^{-\alpha T})} \qquad (2\text{-}208)$$

which does not contain any powers of z as a factor in the numerator. In this case, the partial-fraction expansion of $R(z)$ may be carried out in the usual manner. We have

$$R(z) = \frac{1}{z - 1} - \frac{1}{z - e^{-\alpha T}} \qquad (2\text{-}209)$$

Although the z-transform table does not contain exact matches for the component functions in the last equation, we recognize that the inverse z-transform of the first term on the right-hand side can be written as

$$\mathfrak{z}^{-1}\left[\frac{1}{z-1}\right] = \mathfrak{z}^{-1}\left[z^{-1}\frac{z}{z-1}\right] = u_s[(k-1)T] \tag{2-210}$$

for $k = 1, 2, \ldots$. Similarly, the second term on the right-hand side of Eq. (2-209) can be identified with a time delay of T seconds. Thus, the inverse z-transform of $R(z)$ is written

$$r(kT) = (1 - e^{-\alpha(k-1)T})u_s[(k-1)T] \tag{2-211}$$

for $k = 1, 2, \ldots$.

Computer Solution of the Partial-Fraction Expansion of R(z)/z

Whether the function to be expanded by the partial fraction is in the form of $R(z)/z$ or $R(z)$, the PFE program in the ACSP software in the s form can still be applied.

computer solution POWER-SERIES METHOD. The definition of the z-transform in Eq. (2-180) gives a straightforward method of carrying out the inverse z-transform. From Eq. (2-180), we can clearly see that the coefficient of z^{-k} in $U(z)$ is simply $u(kT)$. Thus, if we expand $U(z)$ into a power series in powers of z^{-1}, we can find the values of $u(kT)$ for $k = 0, 1, 2, \ldots$.

EXAMPLE 2-27 Consider the $R(z)$ given in Eq. (2-204), which can be expanded into a power series of z^{-1} by dividing the numerator polynomial by the denominator polynomial by long division. The result is

$$R(z) = (1 - e^{-\alpha T})z^{-1} + (1 - e^{-2\alpha T})z^{-2} + \cdots + (1 - e^{-k\alpha T})z^{-k} + \cdots \tag{2-212}$$

Thus,

$$r(kT) = 1 - e^{-\alpha kT} \quad k = 0, 1, 2, \ldots \tag{2-213}$$

which is the same result as in Eq. (2-207).

INVERSION FORMULA. The time sequence $r(kT)$ can be determined from $R(z)$ by use of the inversion formula:

$$r(kT) = \frac{1}{2\pi j}\oint_\Gamma R(z)z^{k-1}\,dz \tag{2-214}$$

which is a contour integration along the path Γ, where Γ is a circle of radius $|z| = e^{cT}$ centered at the origin in the z-plane, and c is of such a value that all the poles of $R(z)z^{k-1}$ are inside the circle. The reader should notice that the inversion formula of the z-transform is similar to that of the inverse Laplace-transform integral given in Eq. (2-28). One way of evaluating the contour integration of Eq. (2-214) is by use of the residue theorem of complex-variable theory, and the details are not covered here.

2-12
APPLICATION OF THE z-TRANSFORM TO THE SOLUTION OF LINEAR DIFFERENCE EQUATIONS

difference equation

The z-transform can be used for the purpose of solving linear difference equations. Let us consider the first-order unforced difference equation

$$x(k + 1) + x(k) = 0 \qquad (2\text{-}215)$$

To solve this equation, we take the z-transform on both sides of the equation. By this we mean that we multiply both sides of the equation by z^{-k} and take the sum from $k = 0$ to $k = \infty$. We have

$$\sum_{k=0}^{\infty} x(k + 1)z^{-k} + \sum_{k=0}^{\infty} x(k)z^{-k} = 0 \qquad (2\text{-}216)$$

By using the definition of $X(z)$ and the real-translation theorem of Eq. (2-193), the last equation is written

$$z[X(z) - x(0)] + X(z) = 0 \qquad (2\text{-}217)$$

Solving for $X(z)$, we get

$$X(z) = \frac{z}{z + 1}x(0) \qquad (2\text{-}218)$$

The inverse z-transform of the last equation can be obtained by expanding $X(z)$ into a power series in z^{-1} by long division. We have

$$X(z) = (1 - z^{-1} + z^{-2} - z^{-3} + \cdots)x(0) \qquad (2\text{-}219)$$

Thus, $x(k)$ is written

$$x(k) = (-1)^k x(0) \qquad k = 0, 1, 2, \ldots \qquad (2\text{-}220)$$

The z-transform can be applied to solve difference state equations of high-order systems that are presented in vector-matrix form. The subject is formally treated in Chapter 5, Section 5-13.

EXAMPLE 2-28   Consider the second-order difference equation

$$x(k + 2) + 0.5x(k + 1) + 0.2x(k) = u(k) \tag{2-221}$$

where

$$u(k) = u_s(k) = 1 \qquad \text{for } k = 0, 1, 2, \ldots \tag{2-222}$$

The initial conditions of $x(k)$ are $x(0) = 0$ and $x(1) = 0$.

    Taking the z-transform on both sides of Eq. (2-221), we get

$$[z^2 X(z) - z^2 x(0) - zx(1)] + 0.5[zX(z) - zx(0)] + 0.2X(z) = U(z) \tag{2-223}$$

The z-transform of $u(k)$ is $U(z) = z/(z - 1)$. Substituting the initial conditions of $x(k)$ and $U(z)$ into Eq. (2-223) and solving for $X(z)$, we have

$$X(z) = \frac{z}{(z - 1)(z^2 + 0.5z + 0.2)} \tag{2-224}$$

The partial-fraction expansion of $X(z)/z$ is

$$\frac{X(z)}{z} = \frac{0.588}{z - 1} - \frac{1.036e^{j1.283}}{z + 0.25 + j0.37} - \frac{1.036e^{-j1.283}}{z + 0.25 - j0.37} \tag{2-225}$$

where the exponents in the numerator coefficients are in radians.

    Taking the inverse z-transform of $X(z)$, we get

$$x(k) = 0.588 - 1.036(0.447)^k [e^{-j(2.165k - 1.283)} + e^{j(2.165k - 1.283)}]$$

$$= 0.588 - 2.072(0.447)^k \cos(2.165k - 1.283) \qquad k \geq 0 \tag{2-226}$$

2-13

SUMMARY

In this chapter, we have introduced some of the mathematical fundamentals required for the studies of linear control systems. Specifically, the Laplace transform is used for the solution of linear ordinary differential equations, and the z-transform is devised for the solution of linear difference equations. Both transform methods are characterized by first transforming the time-domain equations into algebraic equations in the transform domain. The solutions are first obtained in the transform domain using the familiar methods of solving algebraic equations. The final solution in the time domain is obtained by taking the inverse transform. For engineering problems, the transform tables and the partial-fraction expansion method are recommended for the inverse transformation.

    Matrix algebra is introduced to simplify the modeling and manipulations of systems that involve multiple numbers of variables and equations. The definitions and properties of the matrix algebra are given.

REVIEW QUESTIONS

1. Give the definitions of the pole and zero of a function of the complex variable s.
2. What are the attractive features of the Laplace-transform method of solving linear ordinary differential equations over the classical method?
3. What are state equations?
4. What is a causal system?
5. Give the defining equation of the one-sided Laplace transform.
6. Give the defining equation of the inverse Laplace transform.
7. Write the expression of the final-value theorem of the Laplace transform. What is the condition under which the theorem is valid?
8. What is the Laplace transform of the unit-step function, $u_s(t)$?
9. What is the Laplace transform of the unit-ramp function, $tu_s(t)$?
10. Write the Laplace transform of $f(t)$ shifted to the right (delayed) by T_d in terms of the Laplace transform of $f(t)$, $F(s)$.
11. If $\mathcal{L}[f_1(t)] = F_1(s)$ and $\mathcal{L}[f_2(t)] = F_2(s)$, then $\mathcal{L}[f_1(t)f_2(t)] = F_1(s)F_2(s)$.   **(T)**   **(F)**
12. Do you know how to handle the exponential term in performing the partial-fraction expansion of

$$F(s) = \frac{10}{(s+1)(s+2)} e^{-2s}$$

13. How would you handle the situation of performing the partial-fraction expansion of $F(s)$ with the final objective of finding $f(t)$?

$$F(s) = \frac{1}{(s+5)^3}$$

14. Do you know how to perform the following matrix operations: transpose, determinant, adjoint, and inverse of a square matrix?
15. What is the definition of the rank of a matrix?
16. Give the definition of the z-transform.
17. What is an ideal sampler?
18. A signal $f(t)$ is sent through an ideal sampler with the sampling period T. Write the expression of the output of the sampler as a function of $f(t)$.
19. Give the z-transform the unit-step function $u_s(t)$.
20. Give the z-transform of the unit-ramp function $tu_s(t)$.
21. Once the z-transform of $f(t)$ is taken to get $F(z)$, the inverse z-transform of $F(z)$ will give $f(t)$.   **(T)**   **(F)**
22. List all the methods you know on performing the inverse z-transform.
23. If the function $F(z)$ has at least one zero at $z = 0$, what step must be taken first before performing the partial-fraction expansion?
24. What is noticeably different between the time sequences that correspond to the following two functions?

$$F_1(z) = \frac{z}{z-1} \qquad F_2(z) = \frac{z}{z+1}$$

REFERENCES

Complex Variables, Laplace Transforms, and Matrix Algebra

1. F. B. HILDEBRAND, *Methods of Applied Mathematics,* 2nd ed., Prentice Hall, Englewood Cliffs, NJ, 1965.
2. R. BELLMAN, *Introduction to Matrix Analysis,* McGraw-Hill Book Company, New York, 1960.
3. F. AYRES, JR., *Theory and Problems of MATRICES,* Schaum's Outline Series, McGraw-Hill Book Company, New York, 1962.
4. B. C. KUO, *Linear Networks and Systems,* McGraw-Hill Book Company, New York, 1967.
5. C. R. WYLIE, JR., *Advanced Engineering Mathematics,* 2nd ed., McGraw-Hill Book Company, New York, 1960.

Partial-Fraction Expansion

6. C. POTTLE, "On the Partial Fraction Expansion of a Rational Function with Multiple Poles by Digital Computer," *IEEE Trans. Circuit Theory,* Vol. CT-11, pp. 161–162, Mar. 1964.
7. B. O. WATKINS, "A Partial Fraction Algorithm," *IEEE Trans. Automatic Control,* Vol. AC-16, pp. 489–491, Oct. 1971.

z-Transforms

8. B. C. KUO, *Digital Control Systems,* Holt, Rinehart and Winston, New York, 1980.
9. E. I. JURY, *Theory and Application of the z-Transform Method,* Robert E. Krieger Publishing Co., Huntington, NY, 1973.
10. R. J. MAYHAN, *Discrete-Time and Continuous-Time Linear Systems,* Addison-Wesley, Reading, MA, 1985.

PROBLEMS

2-1. Find the poles and zeros of the following functions (include the ones at infinity, if any). Mark the finite poles with x and the finite zeros with o in the *s*-plane.

(a) $G(s) = \dfrac{10(s+2)}{s^2(s+1)(s+10)}$ (b) $G(s) = \dfrac{10s(s+1)}{(s+2)(s^2+3s+2)}$

(c) $G(s) = \dfrac{10(s+2)}{s(s^2+2s+2)}$ (d) $G(s) = \dfrac{e^{-2s}}{10s(s+1)(s+2)}$

2-2. Find the Laplace transforms of the following functions. Use the theorems on Laplace transforms if applicable.

(a) $g(t) = 5te^{-5t} \quad t \geq 0$
(b) $g(t) = t\sin 2t + e^{-2t} \quad t \geq 0$
(c) $g(t) = 2e^{-2t}\sin 2t \quad t \geq 0$
(d) $g(t) = \sin 2t \cos 2t u_s(t) \quad$ where $u_s(t) =$ unit-step function
(e) $g(t) = \sum_{k=0}^{\infty} e^{-5kT}\delta(t-kT) \quad$ where $\delta(t) =$ unit-impulse function

2-3. Find the Laplace transforms of the functions shown in Fig. 2P-3. First, write a complete expression for $g(t)$, and then take the Laplace transform. Let $g_T(t)$ be the description of

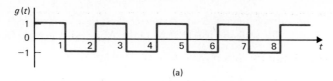

(a)

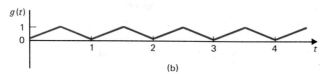

(b)

FIGURE 2P-3

the function over the basic period and then delay $g_T(t)$ appropriately to get $g(t)$. Take the Laplace transform of $g(t)$ to get $G(s)$.

2-4. Find the Laplace transform of the following function.

$$g(t) = \begin{cases} t + 1 & 0 \le t < 1 \\ 0 & 1 \le t < 2 \\ 2 - t & 2 \le t < 3 \\ 0 & t \ge 3 \end{cases}$$

2-5. Solve the following differential equations by means of the Laplace transform.

(a) $\dfrac{d^2f(t)}{dt^2} + 5\dfrac{df(t)}{dt} + 4f(t) = e^{-2t}u_s(t)$

Assume zero initial conditions.

(b) $\dfrac{dx_1(t)}{dt} = x_2(t)$

$\dfrac{dx_2(t)}{dt} = -2x_1(t) - 3x_2(t) + u_s(t)$

$x_1(0) = 1, \quad x_2(0) = 0$

2-6. Find the inverse Laplace transforms of the following functions. Perform partial-fraction expansion on $G(s)$ first, and then use the Laplace transform table. The program PFE in the ACSP software can be used for partial-fraction expansion.

(a) $G(s) = \dfrac{1}{s(s + 2)(s + 3)}$

(b) $G(s) = \dfrac{10}{(s + 1)^2(s + 3)}$

(c) $G(s) = \dfrac{100(s + 2)}{s(s^2 + 4)(s + 1)}e^{-s}$

(d) $G(s) = \dfrac{2(s + 1)}{s(s^2 + s + 2)}$

(e) $G(s) = \dfrac{1}{(s + 1)^3}$

2-7. Carry out the following matrix sums and differences.

(a) $\begin{bmatrix} 5 & -6 \\ 0 & 3 \end{bmatrix} + \begin{bmatrix} 10 & -3 \\ 2 & -3 \end{bmatrix}$

(b) $\begin{bmatrix} 2 & -2 \\ 0 & 10 \\ 3 & 0 \end{bmatrix} - \begin{bmatrix} 10 & 0 \\ 3 & 4 \\ 0 & -4 \end{bmatrix}$

(c) $\begin{bmatrix} \dfrac{1}{s+1} & 3 & \dfrac{1}{s} \\ \dfrac{2}{s} & 0 & \dfrac{1}{s-3} \end{bmatrix} + \begin{bmatrix} 1 & 0 & -10 \\ s & \dfrac{1}{s} & 1 \end{bmatrix}$

2-8. Determine if the following matrices are conformable for the products $\mathbf{AB}$ and $\mathbf{BA}$. Find the valid products. The program MATRIXM in the ACSP software can be used for matrix multiplication.

(a)
$$\mathbf{A} = \begin{bmatrix} 3 \\ 2 \\ 1 \end{bmatrix} \qquad \mathbf{B} = \begin{bmatrix} 6 & 1 & 0 \end{bmatrix}$$

(b)
$$\mathbf{A} = \begin{bmatrix} -1 & 2 \\ 0 & 3 \end{bmatrix} \qquad \mathbf{B} = \begin{bmatrix} 0 & 10 & -10 \\ -1 & 1 & 1 \end{bmatrix}$$

2-9. Express the following set of algebraic equations in matrix form, $\mathbf{Ax} = \mathbf{B}$.

(a)
$$\begin{aligned} x_1 + x_2 - x_3 &= 1 \\ -x_1 + 3x_2 - x_3 &= 1 \\ 3x_1 - 5x_2 - 2x_3 &= 0 \end{aligned}$$

(b)
$$\begin{aligned} x_1 + x_2 - x_3 &= 1 \\ -x_1 + 3x_2 - x_3 &= 1 \\ 2x_1 - 2x_2 &= 0 \end{aligned}$$

Find the inverse of the matrix $\mathbf{A}$ to see if these equations are linearly independent. The program LINSYS of the ACSP software package can be used to find the determinant and the inverse of $\mathbf{A}$. If the equations are not linearly independent, can you still solve for x_1, x_2, and x_3?

2-10. Express the following set of differential equations in the form of

$$\begin{aligned} \dot{\mathbf{x}}(t) &= \mathbf{Ax}(t) + \mathbf{Bu}(t) \\ \dot{x}_1(t) &= -x_1(t) + 2x_2(t) \\ \dot{x}_2(t) &= -2x_2(t) + 3x_3(t) + u_1(t) \\ \dot{x}_3(t) &= -x_1(t) - 3x_2(t) - x_3(t) + u_2(t) \end{aligned}$$

2-11. Find the inverse of the following matrices. Do the problems by hand, and then if the ACSP software or any other computer program for matrix inversion is available, check the answers with the computer.

(a)
$$\mathbf{A} = \begin{bmatrix} 2 & 5 \\ 10 & -1 \end{bmatrix}$$

(b)
$$\mathbf{A} = \begin{bmatrix} 3 & 0 & -1 \\ -2 & 1 & 2 \\ 0 & 1 & -1 \end{bmatrix}$$

(c)
$$\mathbf{A} = \begin{bmatrix} 1 & 3 & 4 \\ -1 & 1 & 0 \\ -1 & 0 & -1 \end{bmatrix}$$

(d)
$$\mathbf{A} = \begin{bmatrix} 0 & 1 & 0 \\ 2 & -2 & 3 \\ 0 & 1 & 5 \end{bmatrix}$$

2-12. Determine the ranks of the following matrices.

(a)
$$\begin{bmatrix} 3 & 2 \\ 6 & 1 \\ 3 & 0 \end{bmatrix}$$

(b)
$$\begin{bmatrix} 2 & 4 & 0 & 8 \\ 1 & 2 & 6 & 3 \end{bmatrix}$$

(c)
$$\begin{bmatrix} 1 & 0 & 0 \\ 5 & 0 & 0 \end{bmatrix}$$

(d)
$$\begin{bmatrix} 1 & 0 & 0 & 1 \\ 0 & 0 & 5 & 0 \end{bmatrix}$$

2-13. The following signals are sampled by an ideal sampler with a sampling period of T seconds. Express the output of the sampler, $f^*(t)$, as an impulse train. Find the Laplace transform of $f^*(t)$, $F^*(s)$. Express $F^*(s)$ in closed form.

(a) $f(t) = te^{-3t}$ (b) $f(t) = t \sin 2t$
(c) $f(t) = e^{-2t} \sin \omega t$ (d) $f(t) = t^2 e^{-2t}$

2-14. Determine the z-transform of the functions in Problem 2-13.

2-15.   Determine the z-transform of the following sequences.
   (a) $f(kT) = kT \sin 2kT$
   (b) $f(k) = \begin{cases} 1 & k = 0, 2, 4, 6, \ldots, \text{ even integers} \\ -1 & k = 1, 3, 5, 7, \ldots, \text{ odd integers} \end{cases}$

2-16.   Determine the z-transform of the following functions. Strictly, one has to take the inverse Laplace transform of $F(s)$, resulting in $f(t)$. Then, follow the steps outlined in Problems 2-13 and 2-14 to arrive at $F(z)$. The simplest way to determine $F(z)$ is to perform a partial-fraction expansion of $F(s)$, and then use the z-transform table.
   (a) $F(s) = \dfrac{1}{(s + 5)^n}$ $n = $ positive integer   (b) $F(s) = \dfrac{1}{s^3(s + 1)}$

   (c) $F(s) = \dfrac{10}{s(s + 5)^2}$ (d) $F(s) = \dfrac{5}{s(s^2 + 2)}$

2-17.   Find the inverse z-transform $f(kT)$ of the following functions. Apply partial-fraction expansion to $F(z)$, and then use the z-transform table.
   (a) $F(z) = \dfrac{10z}{(z - 1)(z - 0.2)}$ (b) $F(z) = \dfrac{z}{(z - 1)(z^2 + z + 1)}$

   (c) $F(z) = \dfrac{z}{(z - 1)(z + 0.85)}$ (d) $F(z) = \dfrac{10}{(z - 1)(z - 0.5)}$

2-18.   Given that $\mathscr{z}[f(t)] = F(z)$, find the value of $f(kT)$ as k approachs infinity without applying the inverse z-transform to $F(z)$. Use the final-value theorem if it is applicable.
   (a) $F(z) = \dfrac{0.368z}{(z - 1)(z^2 - 1.364z + 0.732)}$

   (b) $F(z) = \dfrac{10z}{(z - 1)(z + 1)}$ (c) $F(z) = \dfrac{z^2}{(z - 1)(z - 0.5)^2}$
   Check the answers by carrying out the long division of $F(z)$ and express it in a power series of z^{-1}.

2-19.   Solve the following difference equations by means of the z-transform.
   (a) $x(k + 2) - x(k + 1) + 0.16x(k) = u_s(k)$ $x(0) = x(1) = 0$
   (b) $x(k + 2) - x(k) = 0$ $x(0) = 1,\ x(1) = 0$

2-20.   This problem deals with the application of the difference equations and the z-transform to a loan-amortization problem. Consider that a new car is purchased with a loan of P_0 dollars over a period of N months, at a monthly interest rate of r percent. The principal and interest are to be paid back in N equal payments of u dollars each.
   (a)   Show that the difference equation that describes the load process can be written as

$$P(k + 1) = (1 + r)P(k) - u$$

   where $P(k) = $ amount owed after the kth period, $k = 0, 1, 2, \ldots, N$
   $P(0) = P_0 = $ initial amount borrowed
   $P(N) = 0$ (after N periods, owe nothing)
   The last two conditions are also known as the boundary conditions.
   (b)   Solve the difference equation in part (a) by the recursive method, starting with $k = 0$, and then $k = 1, 2, \ldots$, and substituting successively. Show that the solution of the equation is

$$u = \dfrac{(1 + r)^N P_0 r}{(1 + r)^N - 1}$$

(c) Solve the difference equation in part (a) by means of the z-transform method.

(d) Consider that $P_0 = \$15{,}000$, $r = 0.01$ (1 percent per month), and $N = 48$ months. Find u, the monthly payment.

ADDITIONAL COMPUTER PROBLEMS

The following problems can be solved by computer programs contained in the ACSP software package or other computer programs available to the reader. The names of the suitable programs in the ACSP are indicated.

2-21. Perform a partial-fraction expansion of the following functions (PFE).

(a) $G(s) = \dfrac{10(s + 1)}{s^2(s + 4)(s + 6)}$

(b) $G(s) = \dfrac{s + 1}{s(s + 2)(s^2 + 2s + 2)}$

(c) $G(s) = \dfrac{5(s + 2)}{s^2(s + 1)(s + 5)}$

(d) $G(s) = \dfrac{5e^{-2s}}{(s + 1)(s^2 + s + 1)}$

2-22. Find $g(k)$ for $k = 0$ through $k = 40$ for the following discrete-data function (INVZ).

(a) $G(z) = \dfrac{5z}{(z - 1)(z - 0.1)}$

(b) $G(z) = \dfrac{10z(z - 0.2)}{(z - 1)(z - 0.5)(z - 0.8)}$

(c) $G(z) = \dfrac{z}{(z - 1)(z + 1)}$

(d) $G(z) = \dfrac{2z}{(z - 1)(z^2 - z + 1)}$

2-23. Find the inverses of the following matrices (if the inverse exists) (LINSYS).

(a) $\mathbf{A} = \begin{bmatrix} 10 & 5 \\ -2 & -1 \end{bmatrix}$

(b) $\mathbf{A} = \begin{bmatrix} 2.5 & 1 \\ 3 & 10 \end{bmatrix}$

(c) $\mathbf{A} = \begin{bmatrix} 1 & 0 & -1 \\ 3 & 2 & 0 \\ 4 & 0 & 3 \end{bmatrix}$

(d) $\mathbf{A} = \begin{bmatrix} 0 & 1 & 0 \\ 0 & 0 & 1 \\ -1 & -2 & -3 \end{bmatrix}$

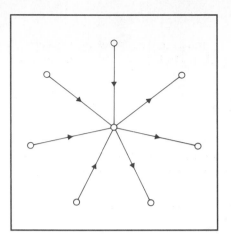

3

TRANSFER FUNCTIONS, BLOCK DIAGRAMS, AND SIGNAL-FLOW GRAPHS

3-1

INTRODUCTION

An important first step in the analysis and design of control systems is the mathematical modeling of the controlled process. In general, given a controlled process, the set of variables that identify the dynamic characteristics of the process should first be defined. For instance, consider a motor used for control purposes. We may identify the applied voltage, current in the armature windings, developed torque on the rotor shaft, and angular displacement and velocity of the rotor as the system variables. These variables are interrelated through established physical laws that lead to mathematical equations that describe the dynamics of the motor. Depending on the operating condition of the motor, as well as on the emphasis of the modeling, the system equations may be linear or nonlinear, time-varying or time-invariant.

The physical laws that govern the principles of operation of systems in real life can often be quite complex, and realistic characterization of the systems may require nonlinear and/or time-varying equations that are very difficult to solve. For practical reasons, in order to establish a class of applicable analysis and design tools for control systems, assumptions and approximations are made to the physical systems whenever possible so that these systems can be studied using linear systems theory. There are two ways of justifying the linear-systems approach. One is that the system is basically linear, or the system is operated in the linear region so that the conditions of linearity are mostly satisfied. The second is that the system is basically nonlinear or operated in a nonlinear region, but in order to apply the linear analysis and design tools, we linearize

operating condition emphasis of modeling

⇓

linear or non linear
time varying or time invariant

67

the system about a nominal operating point, such as that described in Section 4-7. It should be kept in mind that the analysis is applicable only for the range of the variables in which the linearization is valid.

3-2

IMPULSE RESPONSE AND TRANSFER FUNCTIONS OF LINEAR SYSTEMS

Impulse Response

Consider that a linear time-invariant system has the input $r(t)$ and the output $c(t)$. The system can be characterized by its **impulse response** $g(t)$, which is defined as the output when the input is a unit-impulse function $\delta(t)$. Once the impulse response of a linear system is known, the output of the system, $c(t)$, with *any* input, $r(t)$, can be found by using the **transfer function.**

impulse response

transfer function

Transfer Function (Single-Input Single-Output Systems)

The transfer function of a linear time-invariant system is defined as the Laplace transform of the impulse response, with all the initial conditions set to zero.

Let $G(s)$ denote the transfer function of a single-input single-output system with input $r(t)$ and output $c(t)$ and impulse response $g(t)$. Then, the transfer function $G(s)$ is defined as

$$G(s) = \mathscr{L}[g(t)] \tag{3-1}$$

The transfer function $G(s)$ is related to the Laplace transform of the input and the output through the following relation:

$$G(s) = \frac{C(s)}{R(s)} \tag{3-2}$$

with all the initial conditions set to zero, and $C(s)$ and $R(s)$ are the Laplace transforms of $c(t)$ and $r(t)$, respectively.

Although the transfer function of a linear system is defined in terms of the impulse response, in practice, the input–output relation of a linear time-invariant system with continuous-data input is often described by a differential equation, so that it is more convenient to derive the transfer function directly from the differential equation. Let us consider that the input–output relation of a linear time-invariant system is described by the following nth-order differential equation with constant real coefficients:

$$\frac{d^n c(t)}{dt^n} + a_n \frac{d^{n-1} c(t)}{dt^{n-1}} + \cdots + a_2 \frac{dc(t)}{dt} + a_1 c(t)$$

$$= b_{m+1} \frac{d^m r(t)}{dt^m} + b_m \frac{d^{m-1} r(t)}{dt^{m-1}} + \cdots + b_2 \frac{dr(t)}{dt} + b_1 r(t) \qquad (3\text{-}3)$$

The coefficients $a_1, a_2, \ldots, a_n$ and $b_1, b_2, \ldots, b_{m+1}$ are real constants, and $n \geq m$. Once the input $r(t)$ for $t \geq t_0$ and the initial conditions of $c(t)$ and the derivatives of $c(t)$ are specified at the initial time $t = t_0$, the output response $c(t)$ for $t \geq t_0$ is determined by solving Eq. (3-3). However, from the standpoint of linear-system analysis and design, the method of using differential equations exclusively is quite cumbersome. Thus, differential equations of the form of Eq. (3-3) are seldom used in their original form for the analysis and design of control systems. It should be pointed out that although efficient subroutines are available on digital computers for the solution of high-order differential equations, the basic philosophy of linear control theory is that of developing analysis and design tools that would avoid the exact solution of the system differential equations, except when computer-simulation solutions are desired for final presentation or verification.

initial
conditions

To obtain the transfer function of the linear system that is represented by Eq. (3-3), we simply take the Laplace transform on both sides of the equation, and assume **zero initial conditions**. The result is

$$(s^n + a_n s^{n-1} + \cdots + a_2 s + a_1) C(s)$$

$$= (b_{m+1} s^m + b_m s^{m-1} + \cdots + b_2 s + b_1) R(s) \qquad (3\text{-}4)$$

The transfer function between $r(t)$ and $c(t)$ is given by

$$G(s) = \frac{C(s)}{R(s)} = \frac{b_{m+1} s^m + b_m s^{m-1} + \cdots + b_2 s + b_1}{s^n + a_n s^{n-1} + \cdots + a_2 s + a_1} \qquad (3\text{-}5)$$

The properties of the transfer function are summarized as follows:

1. The transfer function is defined only for a linear time-invariant system. It is not defined for nonlinear systems.
2. The transfer function between an input variable and an output variable of a system is defined as the Laplace transform of the impulse response. Alternately, the transfer function between a pair of input and output variables is the ratio of the Laplace transform of the output to the Laplace transform of the input.
3. All initial conditions of the system are set to zero.
4. The transfer function is independent of the input of the system.
5. The transfer function of a continuous-data system is expressed only as a function of the complex variable s. It is not a function of the real variable, time, or any other variable that is used as the independent variable. For discrete-data systems modeled by difference equations, the transfer function is a function of z when the z-transform is used.

characteristic equation

THE CHARACTERISTIC EQUATION. *The* **characteristic equation** *of a linear system is defined as the equation obtained by setting the denominator polynomial of the transfer function to zero.* Thus, from Eq. (3-5), the characteristic equation of the system described by Eq. (3-3) is

$$s^n + a_n s^{n-1} + \cdots + a_2 s + a_1 = 0 \qquad (3\text{-}6)$$

Later we shall show that the stability of linear single-input single-output systems is completely governed by the roots of the characteristic equation.

Transfer Function (Multivariable Systems)

multivariable system

The definition of transfer function is easily extended to a system with a multiple number of inputs and outputs. A system of this type is often referred to as a **multivariable system**. In a multivariable system, a differential equation of the form of Eq. (3-3) may be used to describe the relationship between a pair of input and output. When dealing with the relationship between one input and one output, it is assumed that all other inputs are set to zero. Since the principle of superposition is valid for linear systems, the total effect on any output due to all the inputs acting simultaneously is obtained by adding up the outputs due to each input acting alone.

idle-speed control

Examples on multivariable systems are plentiful in practice. For example, in the control of the speed $\omega(t)$ of a motor subjecting to an external disturbance torque, $T_d(t)$, by controlling the input voltage, $v(t)$, the system is considered to have two inputs in $v(t)$ and $T_d(t)$, and one output in $\omega(t)$. Another example is the idle-speed control system of an automobile engine. In this case, the two inputs are the amounts of fuel and air intake to the engine, and the output is the idle speed of the engine. In the control of turbo-propeller engine, the input variables are the fuel rate and the propeller blade angle. The output variables are the speed of rotation of the engine and the turbine-inlet temperature. In general, either one of the outputs is affected by the changes in both inputs. For instance, when the blade angle of the propeller is increased, the speed of the rotation of the engine will decrease and the temperature usually increases. The following transfer function relations may be determined from tests performed on the system:

turboprop control

$$C_1(s) = G_{11}(s)R_1(s) + G_{12}(s)R_2(s) \qquad (3\text{-}7)$$

$$C_2(s) = G_{21}(s)R_1(s) + G_{22}(s)R_2(s) \qquad (3\text{-}8)$$

where $C_1(s)$ = speed of rotation of engine
$C_2(s)$ = turbine-inlet temperature
$R_1(s)$ = fuel rate
$R_2(s)$ = propeller blade angle

all in Laplace-transform variables measured from some reference level. The transfer function $G_{11}(s)$ represents the transfer function between the fuel rate and the engine speed with the propeller blade angle held at the reference value; that is, $R_2(s) = 0$. Similar definitions can be given to the other transfer functions, $G_{12}(s)$, $G_{21}(s)$, and $G_{22}(s)$.

In general, if a linear system has p inputs and q outputs, the transfer function between the jth input and the ith output is defined as

$$G_{ij}(s) = \frac{C_i(s)}{R_j(s)} \tag{3-9}$$

with $R_k(s) = 0$, $k = 1, 2, \ldots, p$, $k \neq j$. Note that Eq. (3-9) is defined with only the jth input in effect, whereas the other inputs are set to zero. When all the inputs are in action, the ith output transform is written

$$C_i(s) = G_{i1}(s)R_1(s) + G_{i2}(s)R_2(s) + \cdots + G_{ip}(s)R_p(s)$$

$$= \sum_{j=1}^{p} G_{ij}(s)R_j(s) \quad i = 1, 2, \ldots, q \tag{3-10}$$

where $G_{ij}(s)$ is defined in Eq. (3-9).

It is convenient to express Eq. (3-10) in matrix form:

$$\mathbf{C}(s) = \mathbf{G}(s)\mathbf{R}(s) \tag{3-11}$$

where

$$\mathbf{C}(s) = \begin{bmatrix} C_1(s) \\ C_2(s) \\ \vdots \\ C_q(s) \end{bmatrix} \tag{3-12}$$

is the $q \times 1$ **transformed output vector;**

$$\mathbf{R}(s) = \begin{bmatrix} R_1(s) \\ R_2(s) \\ \vdots \\ R_p(s) \end{bmatrix} \tag{3-13}$$

is the $p \times 1$ **transformed input vector;** and

$$\mathbf{G}(s) = \begin{bmatrix} G_{11}(s) & G_{12}(s) & \cdots & G_{1p}(s) \\ G_{21}(s) & G_{22}(s) & \cdots & G_{2p}(s) \\ \cdots\cdots\cdots\cdots\cdots\cdots\cdots\cdots \\ G_{q1}(s) & G_{q2}(s) & \cdots & G_{qp}(s) \end{bmatrix} \tag{3-14}$$

is the $q \times p$ **transfer-function matrix.**

3-3

BLOCK DIAGRAMS

block diagrams   Because of their simplicity and versatility, **block diagrams** are often used by control engineers to model all types of systems. A block diagram can be used simply to describe the composition and interconnection of a system. Or it can be used, together

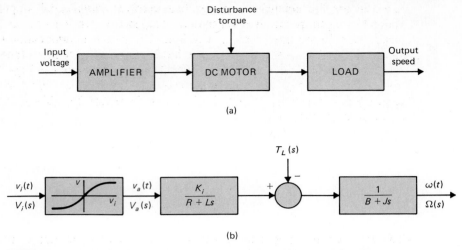

FIGURE 3-1 (a) Block diagram of a dc-motor control system. (b) Block diagram with transfer functions and amplifier characteristics.

with transfer functions, to describe the cause-and-effect relationships throughout the system. For instance, the block diagram of Fig. 3-1(a) models an open-loop dc-motor speed-control system. The block diagram in this case simply shows how the system components are interconnected, and no mathematical details are given. If the mathematical and functional relationships of all the system elements are known, the block diagram can be used as a tool for the analytical or computer solution of the system. In general, block diagrams can be used to model linear as well as nonlinear systems. For example, the input–output relations of the dc-motor control system may be represented by the block diagram shown in Fig. 3-1(b). In the figure, the input voltage to the motor is the output of the power amplifier, which realistically has a nonlinear characteristic. If the motor is linear, or, more appropriately, we should say that if it is operated in the linear region of its characteristics, its dynamics can be represented by transfer functions. The nonlinear amplifier gain can only be described between the time variables $v_i(t)$ and $v_a(t)$, and no transfer function exists between the Laplace transform variables $V_i(s)$ and $V_a(s)$. However, if the magnitude of $v_i(t)$ is limited to the linear range of the amplifier, then the amplifier can be regarded as linear, and the transfer function between $V_i(s)$ and $V_a(s)$ is

[margin note: transfer function is only associated with linear regions of application]

$$\frac{V_a(s)}{V_i(s)} = K \qquad (3\text{-}15)$$

where K is the slope of the linear region of the amplifier characteristics.

Block Diagrams of Control Systems

We shall now define the block-diagram elements used frequently in control systems and the related algebra. One of the important components of a control system is the *sensing device* that acts as a function point for signal comparisons. The physical components

[margin note: sensing device]

involved are the potentiometer, synchro, resolver, differential amplifier, multiplier, and other signal-processing transducers. In general, sensing devices perform simple mathematical operations such as addition, subtraction, multiplication (nonlinear), and sometimes combinations of these. The block-diagram elements of these operations are illustrated in Fig. 3-2. The addition and subtraction operations in Figs. 3-2(a), (b), and (c) are linear, so that the input and output variables of these block-diagram elements can be time-domain variables or Laplace-transform variables. Thus, in Fig. 3-2(a), the block diagram implies

$$e(t) = r(t) - c(t) \tag{3-16}$$

or

$$E(s) = R(s) - C(s) \tag{3-17}$$

The multiplication operation depicted in Fig. 3-2(d) is nonlinear, so that the input–output relation has meaning only in the real (time) domain:

$$e(t) = r(t)c(t) \tag{3-18}$$

It is important to keep in mind that multiplication in the real domain *does not* carry over to the Laplace-transform domain, so that $E(s) \neq R(s)C(s)$. Rather, Eq. (3-18), when transformed, should read $E(s) = R(s) * C(s)$, where the symbol $*$ in this case represents **complex convolution** of $R(s)$ and $C(s)$.

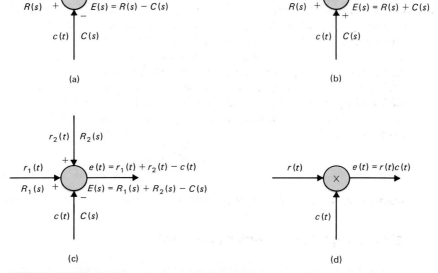

FIGURE 3-2 Block-diagram elements of typical sensing devices of control systems. (a) Subtraction. (b) Addition. (c) Addition and subtraction. (d) Multiplication.

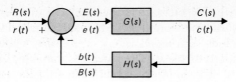

FIGURE 3-3 Basic block diagram of a feedback control
system.

Figure 3-3 shows the block diagram of a linear feedback control system. The following terminology is defined with reference to the diagram.

$$r(t), R(s) = \text{reference input (command)}$$
$$c(t), C(s) = \text{output (controlled variable)}$$
$$b(t), B(s) = \text{feedback signal}$$
$$e(t), E(s) = \text{error signal } [e(t) = r(t) - b(t)]$$
$$H(s) = \text{feedback transfer function}$$
$$G(s)H(s) = \text{loop or open-loop transfer function}$$
$$G(s) = \text{forward-path transfer function}$$
$$M(s) = C(s)/R(s) = \text{closed-loop transfer function}$$

The closed-loop transfer function $M(s)$ can be expressed as a function of $G(s)$ and $H(s)$. From Fig. 3-3, we write

$$C(s) = G(s)E(s) \tag{3-19}$$

and

$$B(s) = H(s)C(s) \tag{3-20}$$

The error signal is written

$$E(s) = R(s) - B(s) \tag{3-21}$$

Substituting Eq. (3-21) into Eq. (3-19) yields

$$C(s) = G(s)R(s) - G(s)B(s) \tag{3-22}$$

Substituting Eq. (3-20) into Eq. (3-22) and then solving for $C(s)/R(s)$ gives the closed-loop transfer function:

$$M(s) = \frac{C(s)}{R(s)} = \frac{G(s)}{1 + G(s)H(s)} \tag{3-23}$$

In general, a control system may contain more than one feedback loop, and the evaluation of the transfer function from the block diagram by means of the algebraic method just described may be tedious. Although, in principle, the block diagram of a system with one input and one output can always be reduced to the basic single-loop form of Fig. 3-3, the algebraic steps involved in the reduction process may again be quite tedious. We shall show in Section 3-4 that the transfer function of any linear system can be obtained directly from its block diagram by use of the signal-flow-graph gain formula.

Block Diagrams and Transfer Functions of Multivariable Systems

multivariable systems

In this section, we shall illustrate the block-diagram and matrix representations of multivariable systems. Two block-diagram representations of a multivariable system with p inputs and q outputs are shown in Figs. 3-4(a) and (b). In Fig. 3-4(a), the individual input and output signals are designated, whereas in the block diagram of Fig. 3-4(b), the multiplicity of the inputs and outputs is denoted by vectors. The case of Fig. 3-4(b) is preferable in practice because of its simplicity.

Figure 3-5 shows the block diagram of a multivariable feedback control system. The transfer-function relationships of the system are expressed in matrix form:

$$\mathbf{C}(s) = \mathbf{G}(s)\mathbf{E}(s) \tag{3-24}$$

$$\mathbf{E}(s) = \mathbf{R}(s) - \mathbf{B}(s) \tag{3-25}$$

$$\mathbf{B}(s) = \mathbf{H}(s)\mathbf{C}(s) \tag{3-26}$$

where $\mathbf{C}(s)$ is the $q \times 1$ output vector; $\mathbf{E}(s)$, $\mathbf{R}(s)$, and $\mathbf{B}(s)$ are all $p \times 1$ vectors; and $\mathbf{G}(s)$ and $\mathbf{H}(s)$ are $q \times p$ and $p \times q$ transfer-function matrices, respectively. Substituting Eq. (3-26) into Eq. (3-25) and then from Eq. (3-25) to Eq. (3-24), we get

$$\mathbf{C}(s) = \mathbf{G}(s)\mathbf{R}(s) - \mathbf{G}(s)\mathbf{H}(s)\mathbf{C}(s) \tag{3-27}$$

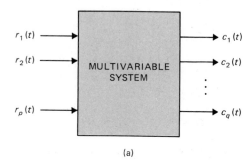

(a)

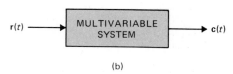

(b)

FIGURE 3-4 Block-diagram representations of a multivariable system.

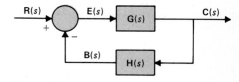

FIGURE 3-5 Block diagram of a multivariable feedback control system.

Solving for $\mathbf{C}(s)$ from Eq. (3-27) gives

$$\mathbf{C}(s) = [\mathbf{I} + \mathbf{G}(s)\mathbf{H}(s)]^{-1}\mathbf{G}(s)\mathbf{R}(s) \tag{3-28}$$

provided that $\mathbf{I} + \mathbf{G}(s)\mathbf{H}(s)$ is nonsingular. The closed-loop transfer matrix is defined as

$$\mathbf{M}(s) = [\mathbf{I} + \mathbf{G}(s)\mathbf{H}(s)]^{-1}\mathbf{G}(s) \tag{3-29}$$

Then Eq. (3-28) is written

$$\mathbf{C}(s) = \mathbf{M}(s)\mathbf{R}(s) \tag{3-30}$$

EXAMPLE 3-1

Consider that the forward-path transfer-function matrix and the feedback-path transfer function matrix of the system shown in Fig. 3-5 are

$$\mathbf{G}(s) = \begin{bmatrix} \dfrac{1}{s+1} & -\dfrac{1}{s} \\ 2 & \dfrac{1}{s+2} \end{bmatrix} \qquad \mathbf{H}(s) = \begin{bmatrix} 1 & 0 \\ 0 & 1 \end{bmatrix} \tag{3-31}$$

respectively. The closed-loop transfer matrix of the system is given by Eq. (3-29), and is evaluated as follows:

$$\mathbf{I} + \mathbf{G}(s)\mathbf{H}(s) = \begin{bmatrix} 1 + \dfrac{1}{s+1} & -\dfrac{1}{s} \\ 2 & 1 + \dfrac{1}{s+2} \end{bmatrix} = \begin{bmatrix} \dfrac{s+2}{s+1} & -\dfrac{1}{s} \\ 2 & \dfrac{s+3}{s+2} \end{bmatrix} \tag{3-32}$$

The closed-loop transfer matrix is

$$\mathbf{M}(s) = [\mathbf{I} + \mathbf{G}(s)\mathbf{H}(s)]^{-1}\mathbf{G}(s) = \frac{1}{\Delta} \begin{bmatrix} \dfrac{s+3}{s+2} & \dfrac{1}{s} \\ -2 & \dfrac{s+2}{s+1} \end{bmatrix} \begin{bmatrix} \dfrac{1}{s+1} & -\dfrac{1}{s} \\ 2 & \dfrac{1}{s+2} \end{bmatrix} \tag{3-33}$$

where

$$\Delta = \frac{s+2}{s+1}\frac{s+3}{s+2} + \frac{2}{s} = \frac{s^2 + 5s + 2}{s(s+1)} \tag{3-34}$$

Thus,

$$\mathbf{M}(s) = \frac{s(s+1)}{s^2 + 5s + 2} \begin{bmatrix} \dfrac{3s^2 + 9s + 4}{s(s+1)(s+2)} & -\dfrac{1}{s} \\ 2 & \dfrac{3s+2}{s(s+1)} \end{bmatrix} \tag{3-35}$$

simplified version of the block diagram. **77**

3-4

SIGNAL-FLOW GRAPHS

A signal-flow graph may be regarded as a simplified version of a block diagram. Originally, the signal-flow graph was introduced by S. J. Mason [2] for the cause-and-effect representation of linear systems that are modeled by algebraic equations. Besides the difference in the physical appearances of the signal-flow graph and the block diagram, we may regard the signal-flow graph to be constrained by more rigid mathematical rules, whereas the usage of the block-diagram notation is less stringent.

more rigid mathematical rules for flow graph

A signal-flow graph may be defined as a graphical means of portraying the input–output relationships between the variables of a set of linear algebraic equations.

Consider that a linear system is described by the set of N algebraic equations:

$$y_j = \sum_{k=1}^{N} a_{kj} y_k \qquad j = 1, 2, \ldots, N \qquad (3\text{-}36)$$

cause and effect

It should be pointed out that these N equations are written in the form of cause-and-effect relations:

$$j\text{th effect} = \sum_{k=1}^{N} (\text{gain from } k \text{ to } j) \times (k\text{th cause}) \qquad (3\text{-}37)$$

or simply

$$\text{Output} = \sum (\text{gain}) \times (\text{input}) \qquad (3\text{-}38)$$

This is the single most important axiom in the construction of the set of algebraic equations from which a signal-flow graph is drawn.

In the case when a system is represented by a set of integrodifferential equations, we must first transform these into Laplace-transform equations and then rearrange the latter into the form of Eq. (3-36), or

$$Y_j(s) = \sum_{k=1}^{N} G_{kj}(s) Y_k(s) \qquad j = 1, 2, \ldots, N \qquad (3\text{-}39)$$

Basic Elements of a Signal-Flow Graph

When constructing a signal-flow graph, junction points, or **nodes**, are used to represent the variables y_j and y_k. The nodes are connected together by line segments called **branches**, according to the cause-and-effect equations. The branches have associated branch gains and directions. *A signal can transmit through a branch only in the direction of the arrow.* In general, given a set of equations such as those of Eq. (3-36) or Eq. (3-39), the construction of the signal-flow graph is basically a matter of following through the cause-and-effect relations relating each variable in terms of itself and the other variables. For instance, consider that a linear system is represented by the simple equation

nodes

branches

$$y_2 = a_{12} y_1 \qquad (3\text{-}40)$$

FIGURE 3-6 Signal-flow graph of $y_2 = a_{12}y_1$.

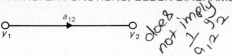

where y_1 is the input, y_2 is the output, and a_{12} the gain, or transmittance, between the two variables. The signal-flow-graph representation of Eq. (3-40) is shown in Fig. 3-6. Notice that the branch directing from node y_1 (input) to node y_2 (output) expresses the dependence of y_2 upon y_1, but not the reverse. An important consideration in the application of signal-flow graphs is that the branch between the two nodes y_1 and y_2 should be interpreted as a unilateral amplifier with gain a_{12}, so that when a signal of one unit is applied to the input y_1, the signal is multiplied by a_{12} and a signal of strength $a_{12}y_1$ is delivered at node y_2. Although algebraically Eq. (3-40) can be written as

$$y_1 = \frac{1}{a_{12}}y_2 \qquad (3\text{-}41)$$

the signal-flow graph of Fig. 3-6 does not imply this relationship. If Eq. (3-41) is valid as a cause-and-effect equation, a new signal-flow graph must be drawn with y_2 as the input and y_1 as the output.

EXAMPLE 3-2

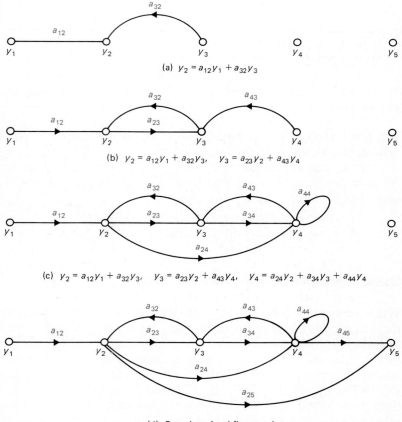

(a) $y_2 = a_{12}y_1 + a_{32}y_3$

(b) $y_2 = a_{12}y_1 + a_{32}y_3, \quad y_3 = a_{23}y_2 + a_{43}y_4$

(c) $y_2 = a_{12}y_1 + a_{32}y_3, \quad y_3 = a_{23}y_2 + a_{43}y_4, \quad y_4 = a_{24}y_2 + a_{34}y_3 + a_{44}y_4$

(d) Complete signal-flow graph

FIGURE 3-7 Step-by-step construction of the signal flow graph for Eq. (3-42).

As an example on the construction of a signal-flow graph, consider the following set of algebraic equations:

$$y_2 = a_{12}y_1 + a_{32}y_3$$

$$y_3 = a_{23}y_2 + a_{43}y_4$$

$$y_4 = a_{24}y_2 + a_{34}y_3 + a_{44}y_4 \qquad\qquad (3\text{-}42)$$

$$y_5 = a_{25}y_2 + a_{45}y_4$$

The signal-flow graph for these equations is constructed step by step, as shown in Fig. 3-7, although the indicated sequence of steps is not essential. The nodes representing the variables y_1, y_2, y_3, y_4, and y_5 are located in order from left to right. The first equation states that y_2 depends upon two signals, $a_{12}y_1$ and $a_{32}y_3$; the signal-flow graph representing this equation is shown in Fig. 3-7(a). The second equation states that y_3 depends upon the sum of $a_{23}y_2$ and $a_{43}y_4$; therefore, on the signal-flow graph of Fig. 3-7(a), a branch of gain a_{23} is drawn from node y_2 to y_3, and a branch of gain a_{43} is drawn from y_4 to y_3, with the directions of the branches indicated by the arrows, as shown in Fig. 3-7(b). Similarly, with the consideration of the third equation, Fig. 3-7(c) is obtained. Finally, when the last equation of Eq. (3-42) is modeled, the complete signal-flow graph of Fig. 3-7(d) is obtained.

3-5

SUMMARY OF THE BASIC PROPERTIES OF SIGNAL-FLOW GRAPHS

At this point, it is best to summarize some of the important properties of the signal-flow graph.

properties

1. A signal-flow graph applies only to linear systems. block diagram to di. an N. 2 9%
2. The equations for which a signal-flow graph is drawn must be algebraic equations in the form of effects as functions of causes.
3. Nodes are used to represent variables. Normally, the nodes are arranged from left to right, following a succession of causes and effects through the system.
4. Signals travel along branches only in the direction described by the arrows of the branches.
5. The branch directing from node y_k to y_j represents the dependence of the variable y_j upon y_k, but not the reverse.
6. A signal y_k traveling along a branch between nodes y_k and y_j is multiplied by the gain of the branch, a_{kj}, so that a signal $a_{kj}y_k$ is delivered at node y_j.

3-6

DEFINITIONS OF SIGNAL-FLOW GRAPHS

In addition to the branches and nodes defined earlier for the signal-flow graph, the following terms are useful for the purpose of identification and reference.

input node   **INPUT NODE (SOURCE).**   *An **input node** is a node that has only outgoing branches.* (Example: node y_1 in Fig. 3-7.)

output node   **OUTPUT NODE (SINK).**   *An **output node** is a node that has only incoming branches.* (Example: node y_5 in Fig. 3-7.) However, this condition is not always readily met by an output node. For instance, the signal-flow graph shown in Fig. 3-8(a) does not have any node that satisfies the condition of an output node. It may be necessary to regard nodes y_2 and/or y_3 as output nodes. To make y_2 an output node, we simply connect a branch with unity gain from the existing node y_2 to a new node y_2, as shown in Fig. 3-8(b). The same procedure can be applied to y_3. Notice that in the modified signal-flow graph of Fig. 3-8(b), it is equivalent that the equations $y_2 = y_2$ and $y_3 = y_3$ are added to the original equations. In general, we can state: *any noninput node of a signal-flow graph can be made an output node by the procedure just illustrated.* However, we cannot convert a noninput node into an input node by using a similar operation. For instance, node y_2 of the signal-flow graph of Fig. 3-8(a) does not satisfy the definition of an input node. If we attempt to convert it into an input node by adding an incoming branch with unity gain from another identical node y_2, the signal-flow graph of Fig. 3-9 would result. The equation that portrays the relationship at node y_2 now reads

$$y_2 = y_2 + a_{12} y_1 + a_{32} y_3 \tag{3-43}$$

which is different from the original equation, as written from Fig. 3-8(a),

$$y_2 = a_{12} y_1 + a_{32} y_3 \tag{3-44}$$

Since the only proper way that a signal-flow graph can be drawn is from a set of cause-and-effect equations, that is, with the causes on the right side of the equations and the effects on the left side of the equations, we must transfer y_2 to the right side of Eq. (3-44) if it were to be an input. By rearranging Eq. (3-44), the two equations orig-

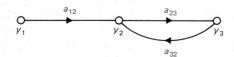

(a) Original signal-flow graph

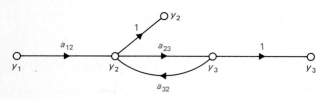

(b) Modified signal-flow graph

FIGURE 3-8 Modification of a signal-flow graph so that y_2 and y_3 satisfy the requirement as output nodes.

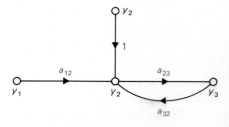

FIGURE 3-9 Erroneous way to make node y_2 an input node.

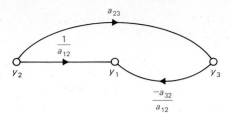

FIGURE 3-10 Signal-flow graph with y_2 as an input node.

inally for the signal-flow graph of Fig. 3-8 now become

$$y_1 = \frac{1}{a_{12}} y_2 - \frac{a_{32}}{a_{12}} y_3 \qquad (3\text{-}45)$$

$$y_3 = a_{23} y_2 \qquad (3\text{-}46)$$

The signal-flow graph for these two equations is shown in Fig. 3-10, with y_2 as an input node.

path

PATH.   *A **path** is any collection of a continuous succession of branches traversed in the same direction.* The definition of a path is entirely general, since it does not prevent any node from being traversed more than once. Therefore, as simple as the signal flow graph of Fig. 3-8(a) is, it may have numerous paths just by traversing the branches a_{23} and a_{32} continuously.

node can be traversed more than once

forward path

FORWARD PATH.   *A **forward path** is a path that starts at an input node and ends at an output node, and along which no node is traversed more than once.* For example, in the signal-flow graph of Fig. 3-7(d), y_1 is the input node, and there are four possible output nodes in y_2, y_3, y_4, and y_5. The forward path between y_1 and y_2 is simply the connecting branch between y_1 and y_2. There are two forward paths between y_1 and y_3; one contains the branches from y_1 to y_2 to y_3, and the other one contains the branches from y_1 to y_2 to y_4 (through the branch with gain a_{24}) and then back to y_3 (through the branch with gain a_{43}). The reader should try to determine the *two* forward paths between y_1 and y_4. Similarly, there are *three* forward paths between y_1 and y_5.

not more than once

loop

LOOP.   *A **loop** is a path that originates and terminates on the same node, and along which no other node is encountered more than once.* For example, there are four loops in the signal-flow graph of Fig. 3-7(d). These are shown in Fig. 3-11.

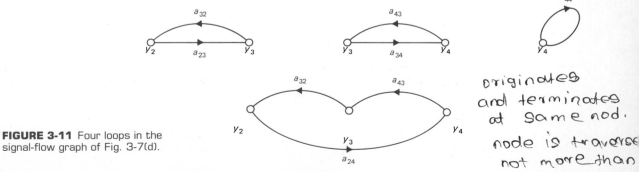

FIGURE 3-11 Four loops in the signal-flow graph of Fig. 3-7(d).

originates and terminates at same nod.

node is traverse not more than once.

path gain

PATH GAIN.   *The product of the branch gains encountered in traversing a path is called the* **path gain**. For example, the path gain for the path $y_1 - y_2 - y_3 - y_4$ in Fig. 3-7(d) is $a_{12}a_{23}a_{34}$.

forward-path gain

FORWARD-PATH GAIN.   *The* **forward-path gain** *is the path gain of a forward path*.

LOOP GAIN.   *The* **loop gain** *is the path gain of a loop*. For example, the loop gain of the loop $y_2 - y_4 - y_3 - y_2$ in Fig. 3-11 is $a_{24}a_{43}a_{32}$.

3-7

SIGNAL-FLOW-GRAPH ALGEBRA

algebra

Based on the properties of the signal-flow graph, we can outline the following manipulation rules and algebra of the signal-flow graph.

1. The value of the variable represented by a node is equal to the sum of all the signals entering the node. For the signal-flow graph of Fig. 3-12, the value of y_1 is equal to the sum of the signals transmitted through all the incoming branches; that is,

$$y_1 = a_{21}y_2 + a_{31}y_3 + a_{41}y_4 + a_{51}y_5 \qquad (3\text{-}47)$$

2. The value of the variable represented by a node is transmitted through all branches leaving the node. In the signal-flow graph of Fig. 3-12, we have

$$y_6 = a_{16}y_1$$
$$y_7 = a_{17}y_1 \qquad (3\text{-}48)$$
$$y_8 = a_{18}y_1$$

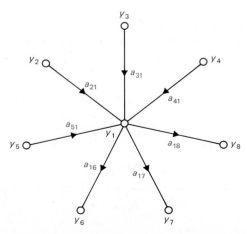

FIGURE 3-12 Node as a summing point and as a transmitting point.

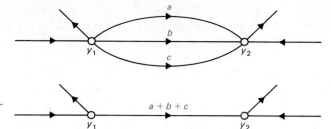

FIGURE 3-13 Signal-flow graph with parallel paths replaced by one with a single branch.

3. Parallel branches in the same direction connecting two nodes can be replaced by a single branch with gain equal to the sum of the gains of the parallel branches. An example of this case is illustrated in Fig. 3-13.

4. A series connection of unidirectional branches, as shown in Fig. 3-14, can be replaced by a single branch with gain equal to the product of the branch gains.

5. *Signal-flow graph of a feedback control system.* Figure 3-15 shows the signal-flow graph of a feedback control system whose block diagram is given in Fig. 3-3. Writing the equations for the signals at the nodes $E(s)$ and $C(s)$, we have

$$E(s) = R(s) - H(s)C(s) \qquad (3\text{-}49)$$

and

$$C(s) = G(s)E(s) \qquad (3\text{-}50)$$

The closed-loop transfer function in Eq. (3-23) can be obtained from these two equations:

$$\frac{C(s)}{R(s)} = \frac{G(s)}{1 + G(s)H(s)} \qquad (3\text{-}51)$$

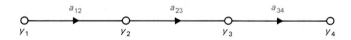

FIGURE 3-14 Signal-flow graph with cascaded unidirectional branches replaced by a single branch.

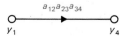

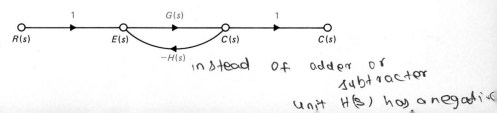

FIGURE 3-15 Signal-flow graph of a feedback control system.

instead of adder or subtractor

unit H(s) has a negative

In general, we do not have to rely on algebraic manipulation to determine the input–output relation of a signal-flow graph. In Section 3-9, a general gain formula is introduced that allows the determination of the gain between an input node and an output node by inspection.

3-8

EXAMPLES OF THE CONSTRUCTION OF SIGNAL-FLOW GRAPHS

examples

In this section, we give two simple illustrative examples on how the cause-and-effect equations are written, and how the signal-flow graph is drawn based on these equations. Owing to the lack of background on system modeling at this point, we shall use two electric networks as examples. More elaborate systems will be discussed in Chapter 4, where the modeling of systems are formally discussed.

EXAMPLE 3-3

passive network

The passive network shown in Fig. 3-16(a) is considered to consist of R, L, and C elements so that the network elements can be represented by impedance functions, $Z(s)$, and admittance functions, $Y(s)$. The Laplace transform of the input voltage is denoted by $E_i(s)$ and that of the output voltage is $E_o(s)$. In this case, it is more convenient to use the branch currents and node voltages designated, as shown in Fig. 3-16. One set of independent cause-and-effect equations is written

$$I_1(s) = [E_i(s) - E_2(s)]Y_1(s) \tag{3-52}$$

$$E_2(s) = [I_1(s) - I_3(s)]Z_2(s) \tag{3-53}$$

$$I_3(s) = [E_2(s) - E_o(s)]Y_3(s) \tag{3-54}$$

$$E_o(s) = Z_4(s)I_3(s) \tag{3-55}$$

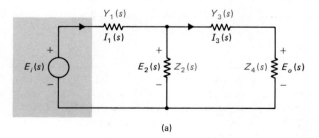

(a)

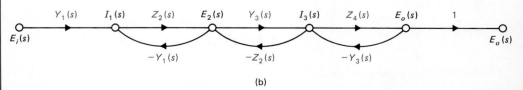

(b)

FIGURE 3-16 (a) Passive ladder network. (b) A signal-flow graph for the network.

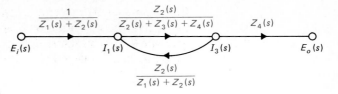

FIGURE 3-17 Signal-flow graph of the network in Fig. 3-16(a) using the loop equations as a starting point.

With the variables $E_i(s)$, $I_1(s)$, $E_2(s)$, $I_3(s)$, and $E_o(s)$ arranged from left to right in order, the signal-flow graph of the network using Eqs. (3-52) through (3-55) is constructed as shown in Fig. 3-16(b).

In the case of network analysis, the cause-and-effect equations that are most suitable for signal-flow graphs are neither the loop equations nor the node equations, but often a mixture of both. Of course, this does not mean that we cannot construct a signal-flow graph starting from the loop or the node equations. For instance, in Fig. 3-16(a), if we let $I_1(s)$ and $I_3(s)$ be the loop currents of the two loops, the loop equations are

$$E_i(s) = [Z_1(s) + Z_2(s)]I_1(s) - Z_2(s)I_3(s) \tag{3-56}$$

$$0 = -Z_2(s)I_1(s) + [Z_2(s) + Z_3(s) + Z_4(s)]I_3(s) \tag{3-57}$$

$$E_o(s) = Z_4(s)I_3(s) \tag{3-58}$$

Equations (3-56) and (3-57) should be rearranged, since only effect variables can appear on the left-hand sides of the equations. Therefore, solving for $I_1(s)$ from Eq. (3-56) and $I_3(s)$ from Eq. (3-57), we get

$$I_1(s) = \frac{1}{Z_1(s) + Z_2(s)}E_i(s) + \frac{Z_2(s)}{Z_1(s) + Z_2(s)}I_3(s) \tag{3-59}$$

$$I_3(s) = \frac{Z_2(s)}{Z_2(s) + Z_3(s) + Z_4(s)}I_1(s) \tag{3-60}$$

Thus, Eqs. (3-58), (3-59), and (3-60) are all in the form of cause-and-effect equations. The signal-flow graph portraying these equations is shown in Fig. 3-17. This exercise also illustrates that the signal-flow graph of a system is *not unique*.

EXAMPLE 3-4
RLC network

Let us now illustrate how a signal-flow graph can be drawn for a system that is originally described by differential equations. An *RLC* network is shown in Fig. 3-18(a). We define the current $i(t)$ and the voltage $e_c(t)$ as the dependent variables of the network. Writing the voltage across the inductance and the current in the capacitor in terms of the input $e_1(t)$, $i(t)$, and $e_c(t)$, we have the following.

Voltage across $i(t)$:

$$L\frac{di(t)}{dt} = e_1(t) - Ri(t) - e_c(t) \tag{3-61}$$

Current in $e_c(t)$:

$$C\frac{de_c(t)}{dt} = i(t) \tag{3-62}$$

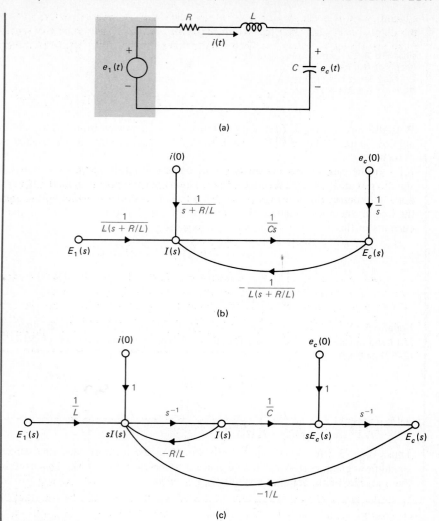

FIGURE 3-18 (a) RLC network. (b) Signal-flow graph. (c) Alternative signal-
flow graph.

We cannot construct a signal-flow graph using these two equations since they are differential
equations. To convert these equations to algebraic equations, we first divide Eqs. (3-61) and
(3-62) by L and C, respectively. Then we take the Laplace transform on both sides of these
equations, which gives

$$sI(s) = i(0) + \frac{1}{L}E_1(s) - \frac{R}{L}I(s) - \frac{1}{L}E_c(s) \qquad (3\text{-}63)$$

$$sE_c(s) = e_c(0) + \frac{1}{C}I(s) \qquad (3\text{-}64)$$

where $i(0)$ and $e_c(0)$ are the initial values of $i(t)$ and $e_c(t)$, respectively, at $t = 0$. From the signal-flow-graph standpoint, $e_c(0)$, $i(0)$, and $E_1(s)$ are the input variables, and $I(s)$ and $E_c(s)$ are the output variables. There are several possible ways of constructing the signal-flow graph for these transformed equations. One way is to solve for $I(s)$ from Eq. (3-63) and $E_c(s)$ from Eq. (3-64); we get

$$I(s) = \frac{1}{s + R/L}i(0) + \frac{1}{L(s + R/L)}E_1(s) - \frac{1}{L(s + R/L)}E_c(s) \qquad (3\text{-}65)$$

$$E_c(s) = \frac{1}{s}e_c(0) + \frac{1}{Cs}I(s) \qquad (3\text{-}66)$$

The signal-flow graph for these equations is shown in Fig. 3-18(b).

As an alternative, we can use Eqs. (3-63) and (3-64) directly, and define $I(s)$, $E_c(s)$, $sI(s)$, and $sE_c(s)$ as the noninput variables. These four variables are related by the equations

$$I(s) = s^{-1}[sI(s)] \qquad (3\text{-}67)$$

$$E_c(s) = s^{-1}[sE_c(s)] \qquad (3\text{-}68)$$

The significance of using s^{-1} as a branch gain is that it represents pure integration in the time domain. The signal-flow graph using Eqs. (3-63), (3-64), (3-67), and (3-68) is shown in Fig. 3-18(c). Notice that in this signal-flow graph, the Laplace-transform variable appears only in the form of s^{-1}. Therefore, this signal-flow graph may be used as a basis for an analog or digital

state diagrams   computer solution. Signal-flow graphs in this form are defined as the **state diagrams**.

3-9

GENERAL GAIN FORMULA FOR SIGNAL-FLOW GRAPHS [3]

gain formula   Given a signal-flow graph or a block diagram, the task of solving for the input–output relations by algebraic manipulation could be quite tedious. Fortunately, there is a general gain formula available that allows the determination of the input–output relations of a signal-flow graph by inspection. The general gain formula is

$$M = \frac{y_{out}}{y_{in}} = \sum_{k=1}^{N} \frac{M_k \Delta_k}{\Delta} \qquad (3\text{-}69)$$

where y_{in} = input-node variable

$\quad y_{out}$ = output-node variable

$\quad M$ = gain between y_{in} and y_{out}

$\quad N$ = total number of forward paths between y_{in} and y_{out}

$\quad M_k$ = gain of the kth forward path between y_{in} and y_{out}

$$\Delta = 1 - \sum_m P_{m1} + \sum_m P_{m2} - \sum_m P_{m3} + \cdots \qquad (3\text{-}70)$$

nontouching

P_{mr} = gain product of the mth possible combination of r nontouching loops $(1 \le r \le N)$. (Two parts of a signal-flow graph are **nontouching** if they do not share a common node).

or

$\Delta = 1 - $ (sum of the gains of all individual loops) + (sum of products of gains of all possible combinations of two nontouching loops) − (sum of products of gains of all possible combinations of three nontouching loops) + · · ·   (3-71)

Δ_k = the Δ for that part of the signal-flow graph that is nontouching with the kth forward path.

The gain formula in Eq. (3-69) may seem formidable to use at first glance. However, Δ is the only term in the formula that could be complicated if the signal-flow graph has a large number of nontouching loops.

Care must be taken when applying the gain formula to ensure that *the gain formula can be applied only between an input node and an output node; i.e., y_{in} must be an input node, and y_{out} must be an output node.*

EXAMPLE 3-5   Consider that the closed-loop transfer function $C(s)/R(s)$ of the signal-flow graph in Fig. 3-15 is to be determined by use of the gain formula, Eq. (3-69). The following results are obtained by inspection of the signal-flow graph:

1.   There is only one forward path between $R(s)$ and $C(s)$, and the forward-path gain is

$$M_1 = G(s) \qquad (3\text{-}72)$$

2.   There is only one loop; the loop gain is

$$P_{11} = -G(s)H(s) \qquad (3\text{-}73)$$

3.   There are no nontouching loops since there is only one loop. Furthermore, the forward path is in touch with the only loop. Thus, $\Delta_1 = 1$, and $\Delta = 1 - P_{11} = 1 + G(s)H(s)$.

By using Eq. (3-69), the closed-loop transfer function is written

$$\frac{C(s)}{R(s)} = \frac{M_1 \Delta_1}{\Delta} = \frac{G(s)}{1 + G(s)H(s)} \qquad (3\text{-}74)$$

which agrees with Eq. (3-51).

EXAMPLE 3-6   Consider that the functional relation between E_i and E_o is to be determined for the signal-flow graph in Fig. 3-16(b). The signal-flow graph is redrawn in Fig. 3-19(a). The following results are obtained by inspection of the signal-flow graph:

1.   There is only one forward path between E_i and E_o, as shown in Fig. 3-19(b). The forward-path gain is

$$M_1 = Y_1 Z_2 Y_3 Z_4 \qquad (3\text{-}75)$$

2.   There are three individual loops, as shown in Fig. 3-19(c); the loop gains are

$$P_{11} = -Z_2 Y_1 \qquad (3\text{-}76)$$

$$P_{21} = -Z_2 Y_3 \qquad (3\text{-}77)$$

$$P_{31} = -Z_4 Y_3 \qquad (3\text{-}78)$$

nontouching loops

3.   There is one pair of nontouching loops, as shown in Fig. 3-19(d); the loop gains of these two loops are $-Z_2 Y_1$ and $-Z_4 Y_3$. Thus,

P_{12} = product of gains of the first (and only) possible
combination of two nontouching loops = $Z_2 Z_4 Y_1 Y_3$ $\qquad (3\text{-}79)$

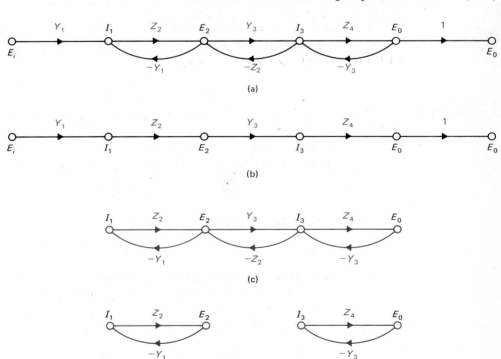

FIGURE 3-19 (a) Signal-flow graph of the passive network in Fig. 3-16(a).
(b) Forward path between E_i and E_o. (c) Three individual loops.
(d) Two nontouching loops.

4. There are no combinations of nontouching loops more than *two*, so $P_{m3} = 0$, $P_{m4} = 0, \ldots$.

From Eq. (3-70),

$$\Delta = 1 - (P_{11} + P_{21} + P_{31}) + P_{12}$$
$$= 1 + Z_2 Y_1 + Z_2 Y_3 + Z_4 Y_3 + Z_2 Z_4 Y_1 Y_3 \tag{3-80}$$

5. All three loops are in touch with the forward path; thus, $\Delta_1 = 1$. Substituting all the above components in Eq. (3-69), we have

$$\frac{E_o}{E_i} = \frac{M_1 \Delta_1}{\Delta} = \frac{Y_1 Y_3 Z_2 Z_4}{1 + Z_2 Y_1 + Z_2 Y_3 + Z_4 Y_3 + Z_2 Z_4 Y_1 Y_3} \tag{3-81}$$

EXAMPLE 3-7   Consider the determination of the relations between I and the three inputs, E_1, $i(0)$, and $e_c(0)$, for the signal-flow graph of Fig. 3-18(c). Since the system is linear, the principle of superposition applies. The gain between one input and one output is determined by applying the gain formula to the two variables and setting the rest of the inputs to zero.

The signal-flow graph is redrawn, as shown in Fig. 3-20(a). The forward paths between each input and I are shown in Figs. 3-20(b), (c), and (d), respectively.

The signal-flow graph has two loops, and there are no nontouching loops; Δ is given by

$$\Delta = 1 + \frac{R}{L}s^{-1} + \frac{1}{LC}s^{-2} \tag{3-82}$$

All the forward paths are in touch with the two loops; thus $\Delta_k = 1$ for $k = 1, 2,$ and 3. Considering each input separately, we have

$$\left.\frac{I}{E_1}\right|_{i(0)=0,\,e_c(0)=0} = \frac{(1/L)s^{-1}}{\Delta} \tag{3-83}$$

$$\left.\frac{I}{i(0)}\right|_{E_1=0,\,e_c(0)=0} = \frac{s^{-1}}{\Delta} \tag{3-84}$$

$$\left.\frac{I}{e_c(0)}\right|_{i(0)=0,\,E_1=0} = \frac{-(1/L)s^{-2}}{\Delta} \tag{3-85}$$

When all three inputs are applied simultaneously, the output I is

$$I = \frac{1}{\Delta}\left[\frac{1}{L}s^{-1}E_1 + s^{-1}i(0) - \frac{1}{L}s^{-2}e_c(0)\right] \tag{3-86}$$

In a similar fashion, the reader should verify that when E_c is considered as the output variable, we have

$$E_c = \frac{1}{\Delta}\left\{\frac{1}{LC}s^{-2}E_1 + \frac{1}{C}s^{-2}i(0) + s^{-1}\left[1 + \frac{R}{L}s^{-1}\right]e_c(0)\right\} \tag{3-87}$$

In this case, the loop between the nodes sI and I is not in touch with the forward path between $e_c(0)$ and E_c.

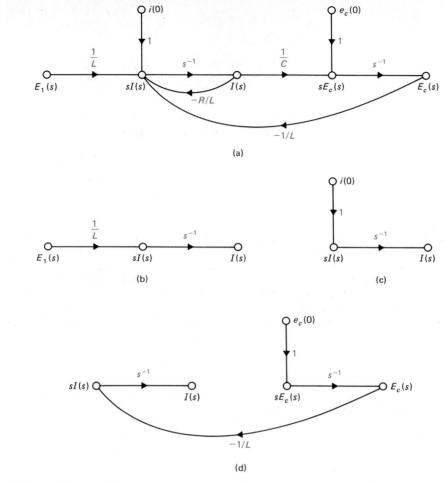

FIGURE 3-20 (a) Signal-flow graph of the *RLC* network in Fig. 3-18(a). (b) Forward path between E_1 and *I*. (c) Forward path between $i(0)$ and *I*. (d) Forward path between $e_c(0)$ and *I*.

EXAMPLE 3-8   Consider the signal-flow graph in Fig. 3-21. The following input–output relations are obtained by use of the general gain formula:

$$\frac{y_2}{y_1} = \frac{1 + G_3 H_2 + H_4 + G_3 H_2 H_4}{\Delta} \tag{3-88}$$

$$\frac{y_4}{y_1} = \frac{G_1 G_2 (1 + H_4)}{\Delta} \tag{3-89}$$

$$\frac{y_6}{y_1} = \frac{y_7}{y_1} = \frac{G_1 G_2 G_3 G_4 + G_1 G_5 (1 + G_3 H_2)}{\Delta} \tag{3-90}$$

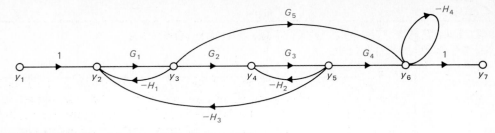

FIGURE 3-21 Signal-flow graph for Example 3-8.

where

$$\Delta = 1 + G_1H_1 + G_3H_2 + G_1G_2G_3H_3 + H_4 + G_1H_1G_3H_2 + G_1H_1H_4 + G_3H_2H_4$$
$$+ G_1G_2G_3H_3H_4 + G_1H_1G_3H_2H_4 \qquad (3-91)$$

for different outputs different Δ.

Application of the General Gain Formula Between Output Nodes and Noninput Nodes

It was emphasized earlier that the gain formula can only be applied between a pair of input and output nodes. Often, it is of interest to find the relation between an output-node variable and a noninput-node variable. For example, in the signal-flow graph of Fig. 3-21, it may be of interest to find the relation y_7/y_2, which represents the dependence of y_7 upon y_2, which is not an input.

Let y_{in} be an input and y_{out} be an output of a signal-flow graph. The quotient y_{out}/y_2, where y_2 is not an input, may be written as

$$\frac{y_{out}}{y_2} = \frac{\dfrac{y_{out}}{y_{in}}}{\dfrac{y_2}{y_{in}}} = \frac{\dfrac{\sum M_k \Delta_k|_{\text{from } y_{in} \text{ to } y_{out}}}{\Delta}}{\dfrac{\sum M_k \Delta_k|_{\text{from } y_{in} \text{ to } y_2}}{\Delta}} \qquad (3-92)$$

Since Δ is independent of the inputs and the outputs, the last equation is written

$$\frac{y_{out}}{y_2} = \frac{\sum M_k \Delta_k|_{\text{from } y_{in} \text{ to } y_{out}}}{\sum M_k \Delta_k|_{\text{from } y_{in} \text{ to } y_2}} \qquad (3-93)$$

Notice that Δ does not appear in the last equation.

EXAMPLE 3-9   From the signal-flow graph in Fig. 3-21, the gain relation between y_2 and y_7 is written

$$\frac{y_7}{y_2} = \frac{y_7/y_1}{y_2/y_1} = \frac{G_1G_2G_3G_4 + G_1G_5(1 + G_3H_2)}{1 + G_3H_2 + H_4 + G_3H_2H_4} \qquad (3-94)$$

APPLICATION OF THE GENERAL GAIN FORMULA TO BLOCK DIAGRAMS

application examples

Because of the similarity between the block diagram and the signal-flow graph, the gain formula in Eq. (3-69) can be applied to determine the input–output relationships of either. In general, given a block diagram of a linear system, we apply the gain formula directly to it. In order to be able to identify all the loops and nontouching parts clearly, sometimes it may be helpful if an equivalent signal-flow graph is drawn for a block diagram first before applying the gain formula.

EXAMPLE 3-10

To illustrate how the signal-flow graph and the block diagram are related, the two equivalent representations of a control system are shown in Fig. 3-22. Note that since a node on the signal-flow graph is interpreted as a summing point of all incoming signals to the node, the negative feedbacks on the block diagram are represented by assigning negative gains to the feedback paths on the signal-flow graph.

The closed-loop transfer function of the system is obtained by applying Eq. (3-69) to either the block diagram or the signal-flow graph:

$$\frac{C(s)}{R(s)} = \frac{G_1 G_2 G_3 + G_1 G_4}{\Delta} \tag{3-95}$$

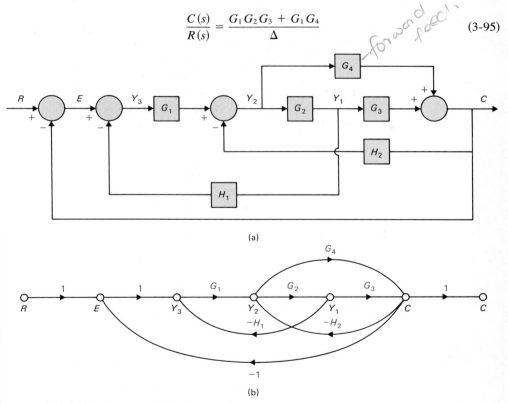

FIGURE 3-22 (a) Block diagram of a control system. (b) Equivalent signal-flow graph.

where

$$\Delta = 1 + G_1 G_2 H_1 + G_2 G_3 H_2 + G_1 G_2 G_3 + G_4 H_2 + G_1 G_4 \tag{3-96}$$

Similarly,

$$\frac{E(s)}{R(s)} = \frac{1 + G_1 G_2 H_1 + G_2 G_3 H_2 + G_4 H_2}{\Delta} \tag{3-97}$$

$$\frac{C(s)}{E(s)} = \frac{G_1 G_2 G_3 + G_1 G_4}{1 + G_1 G_2 H_1 + G_2 G_3 H_2 + G_4 H_2} \tag{3-98}$$

3-11

STATE DIAGRAM

state diagram

In this section, we introduce the methods of the **state diagram**, which represents an extension of the signal-flow graph to portray state equations and differential equations. The significance of the state diagram is that it forms a close relationship among the state equations, computer simulation, and transfer functions. A state diagram is constructed following all the rules of the signal-flow graph using the transformed state equations.

integration

The basic elements of a state diagram are similar to the conventional signal-flow graph, except for the **integration** operation. Let the variables $x_1(t)$ and $x_2(t)$ be related by the first-order differentiation:

$$\frac{dx_1(t)}{dt} = x_2(t) \tag{3-99}$$

Integrating both sides of the last equation with respect to t from the initial time t_0, we get

$$x_1(t) = \int_{t_0}^{t} x_2(\tau) \, d\tau + x_1(t_0) \tag{3-100}$$

Since the signal-flow-graph algebra does not handle integration in the time domain, we must take the Laplace transform on both sides of Eq. (3-100). We have

$$
\begin{aligned}
X_1(s) &= \mathscr{L}\left[\int_{t_0}^{t} x_2(\tau) \, d\tau\right] + \frac{x_1(t_0)}{s} \\
&= \mathscr{L}\left[\int_{0}^{t} x_2(\tau) \, d\tau - \int_{0}^{t_0} x_2(\tau) \, d\tau\right] + \frac{x_1(t_0)}{s} \\
&= \frac{X_2(s)}{s} - \mathscr{L}\left[\int_{0}^{t_0} x_2(\tau) \, d\tau\right] + \frac{x_1(t_0)}{s}
\end{aligned}
\tag{3-101}
$$

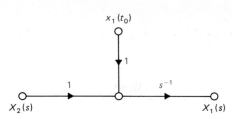

FIGURE 3-23 Signal-flow-graph representation of $X_1(s) = [X_2(s)/s] + [x_1(t_0)/s]$.

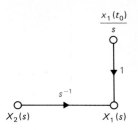

FIGURE 3-24 Signal-flow-graph representation of $X_1(s) = [X_2(s)/s] + [x_1(t_0)/s]$.

Since the past history of the integrator is represented by $x_1(t_0)$, and the state transition is assumed to start at $\tau = t_0$, $x_2(\tau) = 0$ for $0 < \tau < t_0$. Thus, Eq. (3-101) becomes

$$X_1(s) = \frac{X_2(s)}{s} + \frac{x_1(t_0)}{s} \qquad \tau \geq t_0 \qquad (3\text{-}102)$$

Equation (3-102) is now algebraic and can be represented by a single-flow graph, as shown in Fig. 3-23. An alternative signal-flow graph with fewer elements for Eq. (3-102) is shown in Fig. 3-24. *Figure 3-24 shows that the output of the integrator is equal to s^{-1} times the input, plus the initial condition $x_1(t_0)/s$.*

Before embarking on several illustrative examples on the construction of state diagrams, let us point out the important usages of the state diagrams.

decomposition

1. A state diagram can be constructed directly from the system's differential equation. This allows the determination of the state variables and the state equations.

2. A state diagram can be constructed from the system's transfer function. This step is defined as the **decomposition** of transfer functions (Section 5-11).

3. The state diagram can be used for the programming of the system on an analog computer.

4. The state diagram can be used for the simulation of the system on a digital computer.

5. The state-transition equation in the Laplace transform domain may be obtained from the state diagram by means of the signal-flow-graph gain formula.

6. The transfer functions of a system can be determined from the state diagram.

7. The state equations and the output equations can be determined from the state diagram.

The details of these techniques follow.

From Differential Equation to State Diagram

*differential
equation*

When a linear system is described by a high-order differential equation, a state diagram can be constructed from these equations, although a direct approach is not always the most convenient. Consider the following differential equation:

$$\frac{d^n c(t)}{dt^n} + a_n \frac{d^{n-1} c(t)}{d^{n-1}} + \cdots + a_2 \frac{dc(t)}{dt} + a_1 c(t) = r(t) \qquad (3\text{-}103)$$

To construct a state diagram using this equation, we rearrange the equation as

$$\frac{d^n c(t)}{dt^n} = -a_n \frac{d^{n-1} c(t)}{dt^{n-1}} - \cdots - a_2 \frac{dc(t)}{dt} - a_1 c(t) + r(t) \qquad (3\text{-}104)$$

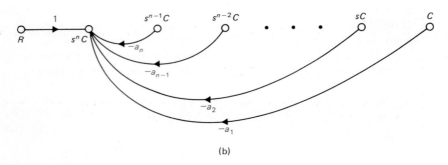

(a)

(b)

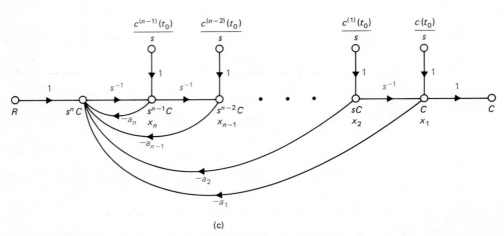

(c)

FIGURE 3-25 State-diagram representation of the differential equation of Eq. (3-103).

As a first step, the nodes representing $R(s)$, $s^n C(s)$, $s^{n-1} C(s)$, . . . , $sC(s)$, and $C(s)$ are arranged from left to right, as shown in Fig. 3-25(a). Since $s^i C(s)$ corresponds to $d^i c(t)/dt^i$, $i = 0, 1, 2, \ldots, n$, in the Laplace domain, as the next step, the nodes in Fig. 3-25(a) are connected by branches to portray Eq. (3-104), resulting in Fig. 3-25(b). Finally, the integrator branches with gains of s^{-1} are inserted, and the initial conditions are added to the outputs of the integrators, according to the basic scheme in Fig. 3-24. The complete state diagram is drawn as shown in Fig. 3-25(c). *The outputs of the integrators are defined as the state variables,* $x_1, x_2, \ldots, x_n$. This is usually the natural choice of state variables once the state diagram is drawn.

state variables

When the differential equation has derivatives of the input on the right side, the problem of drawing the state diagram directly is not as straightforward as just illustrated. We will show that, in general, it is more convenient to obtain the transfer function from the differential equation first and then arrive at the state diagram through decomposition (Section 5-9).

EXAMPLE 3-11   Consider the differential equation

$$\frac{d^2 c(t)}{dt^2} + 3\frac{dc(t)}{dt} + 2c(t) = r(t) \qquad (3\text{-}105)$$

Equating the highest-ordered term of the last equation to the rest of the terms, we have

$$\frac{d^2 c(t)}{dt^2} = -3\frac{dc(t)}{dt} - 2c(t) + r(t) \qquad (3\text{-}106)$$

Following the procedure just outlined, the state diagram of the system is shown in Fig. 3-26. The state variables x_1 and x_2 are assigned as shown.

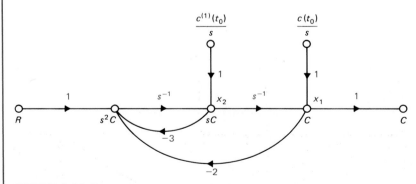

FIGURE 3-26 State diagram for Eq. (3-105).

From State Diagram to Transfer Function

The transfer function between an input and an output is obtained from the state diagram by use of the gain formula, and setting all other inputs and the initial states to zero.

EXAMPLE 3-12   Consider the state diagram of Fig. 3-26. The transfer function between $R(s)$ and $C(s)$ is obtained by applying the gain formula between these two nodes and setting $x_1(t_0) = 0$ and $x_2(t_0) = 0$. We have

$$\frac{C(s)}{R(s)} = \frac{1}{s^2 + 3s + 2} \tag{3-107}$$

From State Diagram to State Equations

state equations   The state equations and the output equations can be obtained directly from the state
output   equation by use of the gain formula. With reference to the general form of the state
equations   equation, the following procedure is outlined:

> 1. Delete the initial states and the integrator branches with gains s^{-1} from the state diagram, since the state equations do not contain the Laplace operator s or the initial states.
> 2. Regard the nodes that represent the derivatives of the state variables as output nodes, since these variables appear on the left-hand side of the state equations.
> 3. Regard the state variables and the inputs as input variables on the state diagram, since these variables are found on the right-hand sides of the state equation.
> 4. Apply the gain formula.

EXAMPLE 3-13   Figure 3-27 shows the state diagram of Fig. 3-26 with the integrator branches and the initial states eliminated. Using $dx_1(t)/dt$ and $dx_2(t)$ as the output nodes and $x_1(t)$, $x_2(t)$, and $r(t)$ as input nodes, and applying the gain formula between these nodes, the state equations are obtained as

$$\frac{dx_1(t)}{dt} = x_2(t)$$

$$\frac{dx_2(t)}{dt} = -2x_1(t) - 3x_2(t) + r(t) \tag{3-108}$$

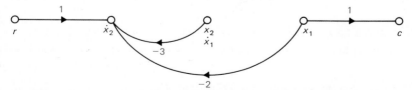

FIGURE 3-27 State diagram of Fig. 3-26 with the initial states and the integrator branches eliminated.

EXAMPLE 3-14   As another example on the determination of the state equations from the state diagram, consider the state diagram shown in Fig. 3-28(a). This example will also emphasize the importance of applying the gain formula to the problem. Figure 3-28(b) shows the state diagram with the initial states and the integrator branches eliminated. Notice that in this case, the state diagram in Fig. 3-28(b) still contains a loop. By applying the gain formula to the state diagram in Fig. 3-28(b) with $\dot{x}_1(t)$, $\dot{x}_2(t)$, and $\dot{x}_3(t)$ as the output-node variables and $r(t)$, $x_1(t)$, $x_2(t)$, and $x_3(t)$ as input nodes, the state equations are obtained as follows:

$$
\begin{bmatrix}
\dfrac{dx_1(t)}{dt} \\[2ex]
\dfrac{dx_2(t)}{dt} \\[2ex]
\dfrac{dx_3(t)}{dt}
\end{bmatrix}
=
\begin{bmatrix}
0 & 1 & 0 \\[2ex]
\dfrac{-(a_2 + a_3)}{1 + a_0 a_3} & -a_1 & \dfrac{1 - a_0 a_2}{1 + a_0 a_3} \\[2ex]
0 & 0 & 0
\end{bmatrix}
\begin{bmatrix}
x_1(t) \\[2ex]
x_2(t) \\[2ex]
x_3(t)
\end{bmatrix}
+
\begin{bmatrix}
0 \\[2ex]
0 \\[2ex]
1
\end{bmatrix}
r(t)
\qquad (3\text{-}109)
$$

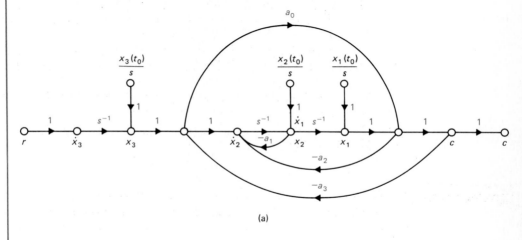

(a)

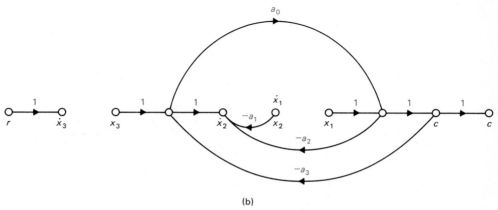

(b)

FIGURE 3-28 (a) State diagram. (b) State diagram in part (a) with all initial states and integrators eliminated.

3-12

TRANSFER FUNCTIONS OF DISCRETE-DATA SYSTEMS

ideal sampler

z.o.h.

impulses

Discrete-data and digital control systems have two unique features in that the signals in these systems are in the form of pulse trains, and the controlled processes are often analog. For instance, the dc motor can be controlled either by a controller that puts out analog signals or by a digital controller that sends out digital data. In the latter case, an interface such as a digital-to-analog (D/A) converter is necessary to couple the digital component to the analog devices. Figure 3-29 shows the block diagram of a discrete-data system with the digital operation modeled by an **ideal sampler** with sampling period T. The output of the ideal sampler is a train of impulses. The **data hold** acts as an interface, or filter, that converts the impulses into an analog signal. One of the most commonly used data holds in practice is the **zero-order hold (z.o.h.)**. Functionally, the z.o.h. simply holds the magnitude of the signal carried by the impulse at a given time instant, say, kT, for the entire sampling period T, until the next impulse arrives at $t = (k + 1)T$. Analytically, the z.o.h. is also used to model the operations of a D/A. Figure 3-30 illustrates a set of typical signals represented by $r(t)$, $r^*(t)$, and $h(t)$ of Fig. 3-29 when the data hold is a z.o.h. In Fig. 3-30(b), the impulses are represented by arrows, since by definition, an impulse has zero pulse width and infinite height. The lengths of the arrows represent the *strengths* of, or areas under, the impulses. The out-

FIGURE 3-29 Block diagram of a discrete-data system.

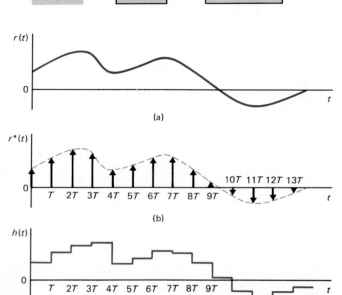

FIGURE 3-30 (a) Input signal to ideal sampler. (b) Output signal of ideal sampler. (c) Output signal of z.o.h.

FIGURE 3-31 Block diagram of a digital process.

put of the z.o.h. is a staircase approximation of the input signal $r(t)$. It is easy to see that $h(t)$ approaches $r(t)$ as the sampling period T approaches zero; that is,

$$\lim_{T \to 0} h(t) = r(t) \tag{3-110}$$

It should be pointed out that since $r*(t)$ is an impulse train, its limit as T approaches zero *does not* have any physical meaning.

Another situation often encountered in a discrete-data or digital control system is that the process receives discrete or digital data and sends out signals in the same form, such as in the case of a microcomputer. Figure 3-31 shows the block diagram representation of such a system.

There are several ways of deriving the transfer-function representation of the system of Fig. 3-29. The following derivation is based on the Fourier-series representation of the signal $r*(t)$. We begin by writing

$$r*(t) = r(t)\delta_T(t) \tag{3-111}$$

where $\delta_T(t)$ is the unit-impulse train,

$$\delta_T(t) = \sum_{k=-\infty}^{\infty} \delta(t - kT) \tag{3-112}$$

Since $\delta_T(t)$ is a periodic function with period T, it can be expressed as a Fourier series:

$$\delta_T(t) = \sum_{n=-\infty}^{\infty} C_n e^{j2\pi nt/T} \tag{3-113}$$

where C_n is the Fourier coefficient, and is given by

$$C_n = \frac{1}{T} \int_0^T \delta_T(t) e^{-jn\omega_s t} \, dt \tag{3-114}$$

sampling frequency   where $\omega_s = 2\pi/T$ is the sampling frequency in rad/s.

Since the unit impulse is defined as a pulse with a width of δ and a height of $1/\delta$, and $\delta \to 0$, C_n is written

$$C_n = \lim_{\delta \to 0} \frac{1}{T\delta} \int_0^{\delta} e^{-jn\omega_s t} \, dt = \lim_{\delta \to 0} \frac{1 - e^{-jn\omega_s \delta}}{jn\omega_s T\delta} = \frac{1}{T} \tag{3-115}$$

Substituting Eq. (3-115) in Eq. (3-113), and then the latter in Eq. (3-111), we get

$$r*(t) = \frac{1}{T} \sum_{n=-\infty}^{\infty} r(t) e^{jn\omega_s t} \tag{3-116}$$

*complex
shifting*

Taking the Laplace transform on both sides of the last equation, and using the complex shifting property of Eq. (2-41), we get

$$R^*(s) = \frac{1}{T} \sum_{n=-\infty}^{\infty} R(s - jn\omega_s) = \frac{1}{T} \sum_{n=-\infty}^{\infty} R(s + jn\omega_s) \qquad (3\text{-}117)$$

Equation (3-117) represents an alternative expression of $R^*(s)$ to that of Eq. (2-177) for $U^*(s)$. Since the summing limits of $R^*(s)$ is from $-\infty$ to ∞, if s is replaced by $s + jm\omega_s$ in Eq. (3-117), where m is any integer, we have

$$R^*(s + jm\omega_s) = R^*(s) \qquad (3\text{-}118)$$

Now we are ready to derive the transfer function for the discrete-data system in Fig. 3-29. The Laplace transform of the system output $c(t)$ is written

$$C(s) = G(s)R^*(s) \qquad (3\text{-}119)$$

Although in principle the output $c(t)$ is obtained by taking the inverse Laplace transform on both sides of Eq. (3-119), in reality this step is difficult to execute because $G(s)$ and $R^*(s)$ represent different types of signals. To overcome this problem, we apply a fictitious sampler at the output of the system, as shown in Fig. 3-32. The fictitious sampler S_2 has the same sampling period T and is synchronized to the original sampler S_1. The sampled form of $c(t)$ is $c^*(t)$. Applying Eq. (3-117) to $c^*(t)$, and using Eq. (3-119), we get

$$C^*(s) = \frac{1}{T} \sum_{n=-\infty}^{\infty} G(s + jn\omega_s)R^*(s + jn\omega_s) \qquad (3\text{-}120)$$

In view of the relationship in Eq. (3-118), the last equation is written

$$C^*(s) = R^*(s)\frac{1}{T} \sum_{n=-\infty}^{\infty} G(s + jn\omega_s) = R^*(s)G^*(s) \qquad (3\text{-}121)$$

*pulse-transfer
function*

where $G^*(s)$ is defined the same way as $R^*(s)$ in Eq. (3-117), and is called the **pulse-transfer function**.

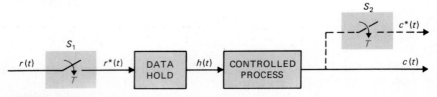

FIGURE 3-32 Discrete-data system with fictitious sampler.

Now that all the functions in Eq. (3-121) are in sampled form, we can take the z-transform on both sides of the equation by substituting $z = e^{Ts}$. We have

$$C(z) = G(z)R(z) \qquad (3\text{-}122)$$

z-transfer function

where $G(z)$ is defined as the **z-transfer function** of the linear process, and is given by

$$G(z) = \sum_{k=\infty}^{\infty} g(kT)z^{-k} \qquad (3\text{-}123)$$

Thus, for the discrete-data system of Figs. 3-29 and 3-32, the z-transform of the output is equal to the product of the z-transfer function of the process and the z-transform of the input. The transfer-function relation in Eq. (3-122) is directly applicable to the all-digital system shown in Fig. 3-31.

Transfer Functions of Discrete-Data Systems with Cascaded Elements

discrete-data systems

The transfer-function representation of discrete-data systems with cascaded elements is slightly more involved than that for continuous-data systems, because of the variation of having or not having any samplers in between the elements. Figure 3-33 shows two different situations of a discrete-data system that contains two cascaded elements. In the system of Fig. 3-33(a), the two elements are separated by a sampler S_2, which is

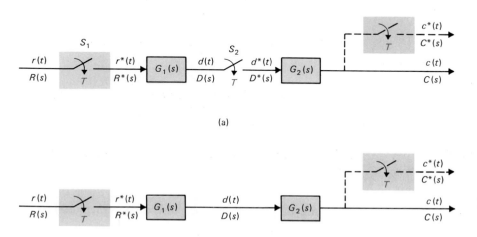

(a)

(b)

FIGURE 3-33 (a) Discrete-data system with cascaded elements and sampler separates the two elements. (b) Discrete-data system with cascaded elements and no sampler in between.

synchronized to, and has the same period as, the sampler S_1. The two elements in the system of Fig. 3-33(b) are connected directly together. It is important to distinguish these two cases when deriving the pulse-transfer function and the z-transfer function. For the system in Fig. 3-33(a), the output of $G_1(s)$ is written

$$D(s) = G_1(s)R*(s) \qquad (3\text{-}124)$$

and the system output is

$$C(s) = G_2(s)D*(s) \qquad (3\text{-}125)$$

Taking the pulse transform on both sides of Eq. (3-124) and substituting the result in Eq. (3-125) yields

$$C(s) = G_2(s)G_1^*(s)R*(s) \qquad (3\text{-}126)$$

Taking the pulse transform on both sides of the last equation gives

$$C*(s) = G_2^*(s)G_1^*(s)R*(s) \qquad (3\text{-}127)$$

where we have made use of the relations in Eqs. (3-117) and (3-118). The corresponding z-transform expression of Eq. (3-127) is

$$C(z) = G_2(z)G_1(z)R(z) \qquad (3\text{-}128)$$

We conclude that the z-transform of two linear elements separated by a sampler is equal to the product of the z-transforms of the two individual transfer functions.

The Laplace transform of the output of the system in Fig. 3-33(b) is

$$C(s) = G_1(s)G_2(s)R*(s) \qquad (3\text{-}129)$$

The pulse transform of Eq. (3-129) is

$$C*(s) = [G_1(s)G_2(s)]*R*(s) \qquad (3\text{-}130)$$

where

$$[G_1(s)G_2(s)]* = \frac{1}{T}\sum_{n=-\infty}^{\infty} G_1(s + jn\omega_s)G_2(s + jn\omega_s) \qquad (3\text{-}131)$$

Notice that since $G_1(s)$ and $G_2(s)$ are not separated by a sampler, they have to be treated as one element when taking the pulse transform. For simplicity, we define the following notation:

$$[G_1(s)G_2(s)]* = G_1 G_2^*(s) = G_2 G_1^*(s) \qquad (3\text{-}132)$$

Then Eq. (3-130) becomes

$$C^*(s) = G_1 G_2^*(s) R^*(s) \tag{3-133}$$

Taking the z-transform on both sides of Eq. (3-133) gives

$$C(z) = G_1 G_2(z) R(z) \tag{3-134}$$

where $G_1 G_2(z)$ is defined as the z-transform of $G_1(s)G_2(s)$. It is important to note that, in general,

$$G_1 G_2^*(s) \neq G_1^*(s) G_2^*(s) \tag{3-135}$$

and

$$G_1 G_2(z) \neq G_1(z) G_2(z) \tag{3-136}$$

Transfer Function of the Zero-Order Hold

Based on the description of the z.o.h. given earlier, its impulse response is shown in Fig. 3-34. The transfer function of the z.o.h. is

$$G_h(s) = \mathscr{L}[g_h(t)] = \frac{1 - e^{-Ts}}{s} \tag{3-137}$$

Thus, if the z.o.h. is connected in cascade with a linear process with transfer function $G_p(s)$, as shown in Fig. 3-35, the z-transform of the combination is written

$$G(z) = \mathscr{z}[G_h(s)G_p(s)] = \mathscr{z}\left[\frac{1 - e^{-Ts}}{s} G_p(s)\right] \tag{3-138}$$

By using the time-delay property of z-transforms, Eq. (3-138) is simplified to

$$G(z) = (1 - z^{-1})\mathscr{z}\left[\frac{G_p(s)}{s}\right] \tag{3-139}$$

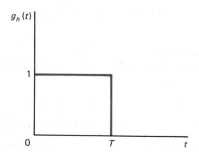

FIGURE 3-34 Impulse response of the z.o.h.

EXAMPLE 3-15   Consider that for the system shown in Fig. 3-35,

$$G_p(s) = \frac{1}{s(s + 0.5)} \tag{3-140}$$

The z-transfer function of the system between the input and the output is determined using Eq. (3-139).

$$G(z) = (1 - z^{-1})\mathscr{Z}\left[\frac{1}{s^2(s + 0.5)}\right] = \frac{0.426z + 0.361}{z^2 - 1.606z + 0.606} \tag{3-141}$$

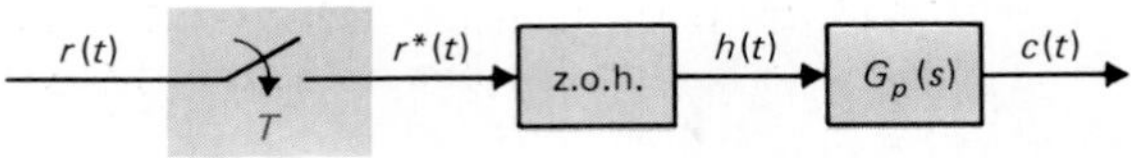

FIGURE 3-35 Linear process cascaded with a z.o.h.

$r(t)$ ⟶ $\underset{T}{\times}$ ⟶ $r^*(t)$ ⟶ [z.o.h.] ⟶ $h(t)$ ⟶ [$G_p(s)$] ⟶ $c(t)$

Transfer Functions of Closed-Loop Discrete-Data Systems

The transfer functions of closed-loop discrete-data systems are derived using the following procedures:

1. Regard the outputs of samplers as "inputs" to the system.
2. All other noninputs of the system are treated as outputs.
3. Write cause-and-effect equations between the inputs and the outputs of the system using the signal-flow-graph gain formula.
4. Manipulate the cause-and-effect equations to give the transfer functions.

EXAMPLE 3-16   Consider the discrete-data system shown in Fig. 3-36 that has a sampler in the forward path. The output of the sampler is regarded as an input to the system. Thus, the system has inputs $R(s)$ and $E^*(s)$. The signals $E(s)$ and $C(s)$ are regarded as outputs of the system.

Writing the cause-and-effect equations for $E(s)$ and $C(s)$ using the gain formula, we get,

$$E(s) = R(s) - G(s)H(s)E^*(s) \tag{3-142}$$

$$C(s) = G(s)E^*(s) \tag{3-143}$$

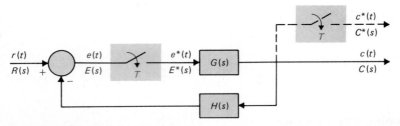

FIGURE 3-36 Closed-loop discrete-data system.

Notice that the right-hand side of the last two equations contains only the inputs $R(s)$ and $E*(s)$ and the transfer functions. Taking the pulse transform on both sides of Eq. (3-142) and solving for $E*(s)$, we get

$$E*(s) = \frac{R*(s)}{1 + GH*(s)} \tag{3-144}$$

Substituting $E*(s)$ from the last equation into Eq. (3-143), we get

$$C(s) = \frac{G(s)}{1 + GH*(s)} R*(s) \tag{3-145}$$

Taking the pulse transform on both sides of Eq. (3-143) and then using Eq. (3-144), we arrive at the pulse-transfer function of the closed-loop system:

$$\frac{C*(s)}{R*(s)} = \frac{G*(s)}{1 + GH*(s)} \tag{3-146}$$

The z-transfer function of the system is

$$\frac{C(z)}{R(z)} = \frac{G(z)}{1 + GH(z)} \tag{3-147}$$

EXAMPLE 3-17   We shall show in this example that although it is possible to define a transfer function for the system in Fig. 3-36, in general, this may not be possible for all discrete-data systems. Let us consider the system shown in Fig. 3-37, which has a sampler in the feedback path. In this case, the outputs of the sampler, $C*(s)$ and $R(s)$, are the inputs of the system; $C(s)$ and $E(s)$ are regarded as the outputs. Writing $E(s)$ and $C(s)$ in terms of the inputs and using the gain formula, we get

$$C(s) = G(s)E(s) \tag{3-148}$$

$$E(s) = R(s) - H(s)C*(s) \tag{3-149}$$

By taking the pulse transform on both sides of the last two equations, after simple algebraic manipulations, the pulse transform of the output is written

$$C*(s) = \frac{GR*(s)}{1 + GH*(s)} \tag{3-150}$$

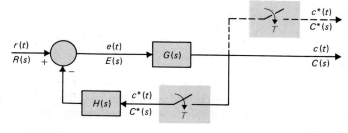

FIGURE 3-37 Closed-loop discrete-data system.

Note that the input $R(s)$ and the transfer function $G(s)$ are now combined as one function, $GR^*(s)$, and we cannot define a transfer function in the form of $C^*(s)/R^*(s)$. The z-transform of the output is written

$$C(z) = \frac{GR(z)}{1 + GH(z)} \qquad (3\text{-}151)$$

Although we have been able to arrive at the input–output transfer function and transfer relation of the systems in Figs. 3-36 and 3-37 by algebraic means without difficulty, for more complex system configurations, the algebraic method may become tedious. The signal-flow-graph method may be extended to the analysis of discrete-data systems; the reader may refer to the literature [6] for more details.

Computer Solution of System Transfer Functions

*computer
solutions*

The SDCS program of the ACSP software package or some other commercially available programs for microcomputers or mainframe computers can be used to determine the open-loop and the closed-loop transfer functions of the discrete-data system shown in Fig. 3-38 when $D(z)$ and $G_p(s)$ and the sampling period T are given.

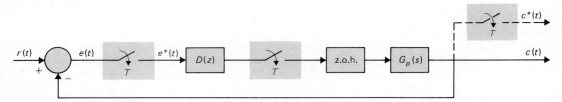

FIGURE 3-38 A discrete-data system with a digital controller.

3-13

SUMMARY

In this chapter, we have introduced the mathematical modeling of linear systems using transfer functions, block diagrams, and signal-flow graphs. The transfer function of a linear system is defined in terms of the impulse response as well as the differential equation. Multivariable as well as single-variable systems are treated.

The block-diagram representation is shown to be a versatile method of portraying linear and nonlinear systems.

A powerful means of representing the interrelationships between signals of a linear system is the signal-flow graph. When properly applied, the signal-flow graph allows the derivation of the gains between input and output variables of a linear system using the gain formula. The state diagram is a signal-flow graph that is applied to dynamic systems that are represented by differential equations.

Finally, the transfer function of a discrete-data system is defined, and the rules governing the manipulations of transfer functions between interconnected discrete-data systems are described.

REVIEW QUESTIONS

1. Give the definition of the transfer function of a linear time-invariant system in terms of its impulse response.

2. When defining the transfer function, what happens to the initial conditions of the system?

3. Give the definition of the characteristic equation of a linear system in terms of the transfer function.

4. What is referred to as a multivariable system?

5. Can signal-flow graphs be applied to nonlinear systems?

6. Can signal-flow graphs be applied to systems that are described by differential equations?

7. Give the definition of the input node of a signal-flow graph.

8. Give the definition of the output node of a signal-flow graph.

9. State the form to which the equations must first be conditioned before drawing the signal-flow graph.

10. What does the arrow on the branch of a signal-flow graph represent?

11. All the noninput nodes of a signal-flow graph can be regarded as output nodes.   **(T)   (F)**

12. The gain formula can be applied between any two nodes of a signal-flow graph.   **(T)   (F)**

13. Two loops of a signal-flow graph are said to be nontouching if they do not share a common node.   **(T)   (F)**

14. The Δ of a signal-flow graph depends only on the loop configuration of the signal-flow graph.

15. List the advantages and utilities of the state diagram.

16. Given the state diagram of a linear dynamic system, how should you define the state variables?

17. Given the state diagram of a linear dynamic system, how do you find the transfer function between a pair of input and output variables?

18. Given the state diagram of a linear dynamic system, how do you write the state equations of the system from the state diagram?

19. What is a zero-order hold and what is it used for?

20. Is the zero-order hold a linear device?

REFERENCES

Block Diagrams and Signal-Flow Graphs

1. T. D. GRAYBEAL, "Block Diagram Network Transformation," *Elec. Eng.,* Vol. 70, pp. 985–990, 1951.

2. S. J. MASON, "Feedback Theory—Some Properties of Signal Flow Graphs," *Proc. IRE,* Vol. 41, No. 9, pp. 1144–1156, Sept. 1953.

3. S. J. MASON, "Feedback Theory—Further Properties of Signal Flow Graphs," *Proc. IRE,* Vol. 44, No. 7, pp. 920–926, July 1956.

4. L. P. A. ROBICHAUD, M. BOISVERT, and J. ROBERT, *Signal Flow Graphs and Applications,* Prentice Hall, Englewood Cliffs, NJ, 1962.

5. B. C. KUO, *Linear Networks and Systems,* McGraw-Hill Book Company, New York, 1967.

Discrete-Data Control Systems

6. B. C. KUO, *Digital Control Systems,* Holt, Rinehart and Winston, Inc., New York, 1980.

7. G. F. FRANKLIN, J. D. POWELL, and M. L. WORKMAN, *Digital Control of Dynamic Systems,* Second Edition, Addison-Wesley, Reading, MA, 1990.

8. C. L. PHILLIPS and H. T. NAGLE, Jr., *Digital Control System Analysis and Design,* Prentice Hall, Englewood Cliffs, NJ, 1984.

9. K. OGATA, *Discrete-Time Control Systems,* Prentice Hall, Englewood Cliffs, NJ, 1987.

10. G. H. HOSTETTER, *Digital Control System Design,* Holt, Rinehart and Winston, Inc., New York, 1988.

PROBLEMS

3-1. The following differential equations represent linear time-invariant systems, where $r(t)$ denotes the input, and $c(t)$ the output. Find the transfer function $C(s)/R(s)$ for each of the systems.

(a) $\dfrac{d^3c(t)}{dt^3} + 2\dfrac{d^2c(t)}{dt^2} + 5\dfrac{dc(t)}{dt} + 6c(t) = 3\dfrac{dr(t)}{dt} + r(t)$

(b) $\dfrac{d^4c(t)}{dt^4} + 10\dfrac{d^2c(t)}{dt^2} + \dfrac{dc(t)}{dt} + 5c(t) = 5r(t)$

(c) $\dfrac{d^3c(t)}{dt^3} + 10\dfrac{d^2c(t)}{dt^2} + 2\dfrac{dc(t)}{dt} + c(t) + 2\displaystyle\int_0^t c(\tau)\, d\tau = \dfrac{dr(t)}{dt} + 2r(t)$

(d) $2\dfrac{d^2c(t)}{dt^2} + \dfrac{dc(t)}{dt} + 5c(t) = r(t) + 2r(t-1)$

3-2. A linear time-invariant multivariable system with inputs $r_1(t)$ and $r_2(t)$ and outputs $c_1(t)$ and $c_2(t)$ is described by the following set of differential equations.

$$\frac{d^2c_1(t)}{dt^2} + 2\frac{dc_1(t)}{dt} + 3c_2(t) = r_1(t) + r_2(t)$$

$$\frac{d^2c_2(t)}{dt^2} + 3\frac{dc_1(t)}{dt} + c_1(t) - c_2(t) = r_2(t) + \frac{dr_1(t)}{dt}$$

Find the following transfer functions:

$$\left.\frac{C_1(s)}{R_1(s)}\right|_{R_2=0} \qquad \left.\frac{C_2(s)}{R_1(s)}\right|_{R_2=0} \qquad \left.\frac{C_1(s)}{R_2(s)}\right|_{R_1=0} \qquad \left.\frac{C_2(s)}{R_2(s)}\right|_{R_1=0}$$

3-3. The block diagram of an electric train control system is shown in Fig. 3P-3. The system parameters and variables are

$e_r(t)$ = voltage representing the desired train speed, V
$v(t)$ = speed of train, ft/s
M = mass of train = 30,000 lb/ft/s^2
K = amplifier gain
K_t = gain of speed indicator = 0.15 V/ft/s

To determine the transfer function of the controller, we apply a step function of 1 volt to the input of the controller; i.e., $e_c(t) = u_s(t)$. The output of the controller is measured

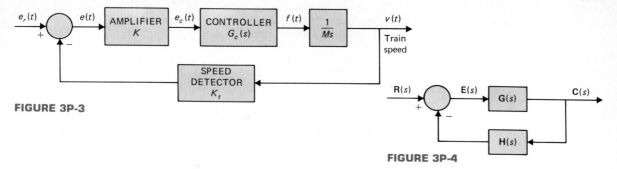

FIGURE 3P-3

FIGURE 3P-4

and is described by the following expression:

$$f(t) = 100(1 - 0.3e^{-6t} - 0.7e^{-10t}) \qquad t \geq 0$$

(a)   Find the transfer function $G_c(s)$ of the controller.
(b)   Derive the open-loop transfer function $V(s)/E(s)$ of the system. The feedback path is opened in this case.
(c)   Derive the closed-loop transfer function $V(s)/E_r(s)$ of the system.
(d)   Assuming that K is set at a value so that the train will not run away (unstable), find the steady-state speed of the train in ft/s when the input is $e_r(t) = u_s(t)$ V.

3-4.   The block diagram of a multivariable feedback control system is shown in Fig. 3P-4. The transfer function matrices are

$$\mathbf{G}(s) = \begin{bmatrix} \dfrac{1}{s+1} & \dfrac{2}{s(s+2)} \\ \dfrac{5}{s} & 10 \end{bmatrix} \qquad \mathbf{H}(s) = \begin{bmatrix} 1 & 0 \\ 0 & 1 \end{bmatrix}$$

Find the closed-loop transfer-function matrix of the system.

3-5.   Draw a signal-flow graph for the following set of algebraic equations.
(a) $x_1 = \qquad -x_2 - 3x_3 + 5$ (b) $2x_1 + x_2 + 5x_3 = -1$
        $x_2 = 5x_1 - x_2 - x_3$ $x_1 - 2x_2 + x_3 = 1$
        $x_3 = 4x_1 + x_2 - 5x_3 + 1$ $x_2 + 2x_3 = 0$

These equations should be in the form of cause-and-effect before a signal-flow graph can be drawn. Show that there are many possible signal-flow graphs for each set of equations.

3-6.   The block diagram of a control system is shown in Fig. 3P-6. Draw an equivalent signal-flow graph for the system. Find the following transfer functions by applying Mason's

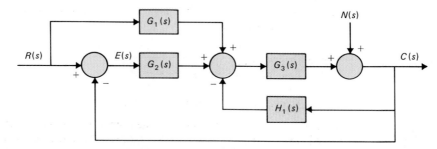

FIGURE 3P-6

gain formula directly to the block diagram. Compare the answers by applying the gain formula to the equivalent signal-flow graph.

$$\frac{C(s)}{R(s)}\bigg|_{N=0} \qquad \frac{C(s)}{N(s)}\bigg|_{R=0} \qquad \frac{E(s)}{R(s)}\bigg|_{N=0} \qquad \frac{E(s)}{N(s)}\bigg|_{R=0}$$

3-7. Find the gains Y_5/Y_1, Y_2/Y_1, and Y_5/Y_2 for the signal-flow graphs shown in Fig. 3P-7.

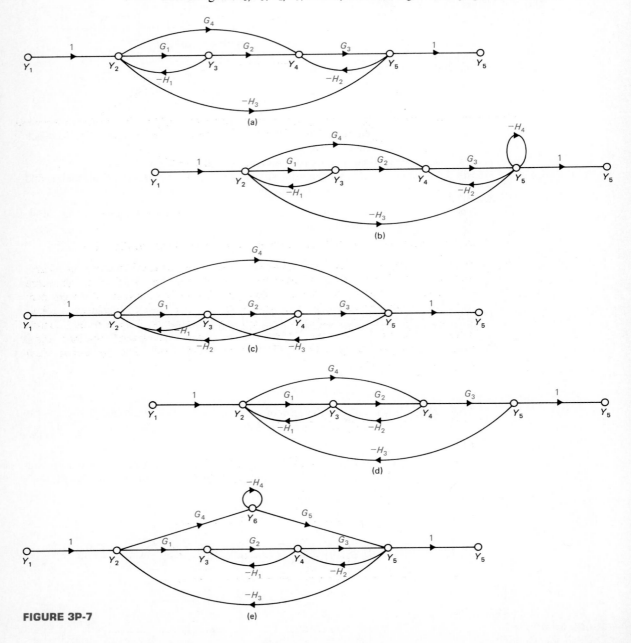

FIGURE 3P-7

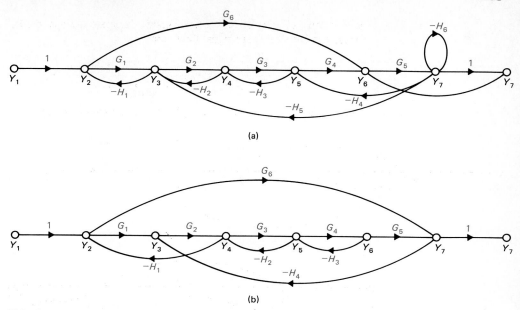

FIGURE 3P-8

3-8.   Find the gains Y_7/Y_1 and Y_2/Y_1 for the signal-flow graphs shown in Fig. 3P-8.

3-9.   Signal-flow graphs may be used to solve a variety of electrical network problems. Shown in Fig. 3P-9 is the equivalent circuit of an electronic circuit. The voltage source $e_d(t)$ represents a disturbance voltage. The objective is to find the value of the constant k so that the output voltage $e_o(t)$ is not affected by $e_d(t)$. To solve the problem, it is best to first write a set of cause-and-effect equations for the network. This involves a combination of node and loop equations. Then construct a signal-flow graph using these equations. Find the gain $e_o(t)/e_d(t)$ with all other inputs set to zero. For $e_d(t)$ not to affect $e_o(t)$, set $e_o(t)/e_d(t)$ to zero.

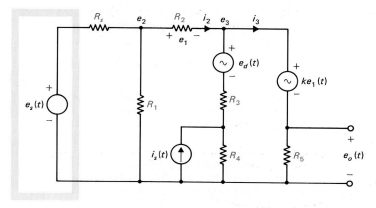

FIGURE 3P-9

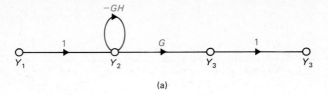

(a)

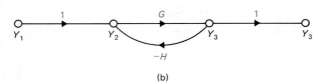

(b)

FIGURE 3P-10

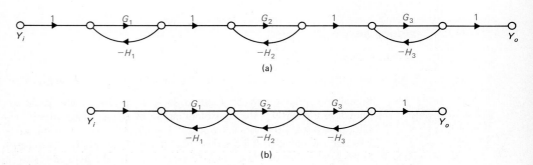

(a)

(b)

FIGURE 3P-11

3-10. Show that the two systems shown in Figs. 3P-10(a) and (b) are equivalent.

3-11. Show that the two systems shown in Figs. 3P-11(a) and (b) are *not* equivalent.

3-12. Find the following gain relations for the signal-flow graphs shown in Fig. 3P-12.

$$\left.\frac{Y_6}{Y_1}\right|_{Y_7=0} \qquad \left.\frac{Y_6}{Y_7}\right|_{Y_1=0}$$

3-13. Find the following gain relations for the signal-flow graph shown in Fig. 3P-13. Comment on why the results for parts (c) and (d) are not the same.

(a) $\left.\dfrac{Y_7}{Y_1}\right|_{Y_8=0}$ (b) $\left.\dfrac{Y_7}{Y_8}\right|_{Y_1=0}$ (c) $\left.\dfrac{Y_7}{Y_4}\right|_{Y_8=0}$ (d) $\left.\dfrac{Y_7}{Y_4}\right|_{Y_1=0}$

3-14. The block diagram of a feedback control system is shown in Fig. 3P-14.
   (a) Find the transfer function $C(s)/E(s)$, $N(s) = 0$.
   (b) Find the transfer function $C(s)/R(s)$, $N(s) = 0$.
   (c) Find the transfer function $C(s)/N(s)$, $R(s) = 0$.
   (d) Find the output $C(s)$ when $R(s)$ and $N(s)$ are both applied.

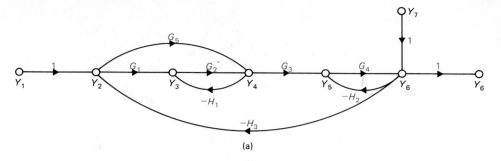

(a)

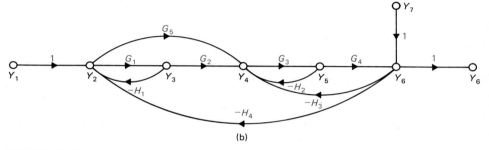

(b)

FIGURE 3P-12

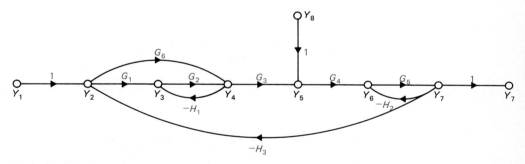

FIGURE 3P-13

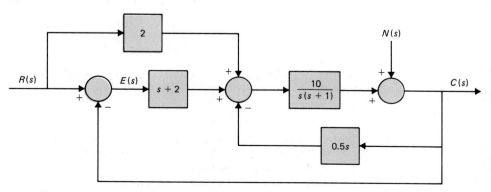

FIGURE 3P-14

3-15. The block diagram of a feedback control system is shown in Fig. 3P-15.
   (a) Apply Mason's gain formula directly to the block diagram to find the transfer functions

$$\left.\frac{C(s)}{R(s)}\right|_{N=0} \qquad \left.\frac{C(s)}{N(s)}\right|_{R=0}$$

   Express $C(s)$ in terms of $R(s)$ and $N(s)$ when both inputs are applied simultaneously.
   (b) Find the desired relation among the transfer functions $G_1(s)$, $G_2(s)$, $G_3(s)$, $G_4(s)$, $H_1(s)$, and $H_2(s)$ so that the output $C(s)$ is not affected by the disturbance signal $N(s)$ at all.

3-16. Figure 3P-16 shows the block diagram of a control system that has a disturbance input $N(s)$. The feedforward transfer function $G_d(s)$ is used to eliminate the effect of $N(s)$ on the output $C(s)$. Find the transfer function $C(s)/N(s)|_{R=0}$. Determine the expression of $G_d(s)$ so that the previously mentioned condition is achieved.

3-17. A linear feedback control system has the block diagram shown in Fig. 3P-17.
   (a) Find the transfer function $H(s)$ so that the output $C(s)$ is not affected by the noise $N(s)$; i.e., $C(s)/N(s)|_{R=0} = 0$.
   (b) With $H(s)$ as determined in part (a), find the value of K so that the steady-state value of $e(t)$ is equal to 0.1 when the input is a unit-ramp function, $r(t) = tu_s(t)$, $R(s) = 1/s^2$, and $N(s) = 0$. Apply the final-value theorem.

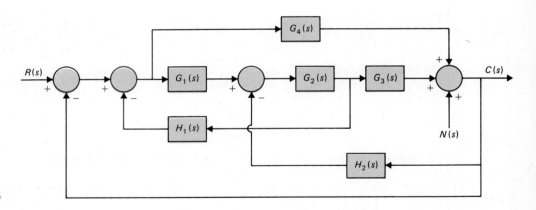

FIGURE 3P-15

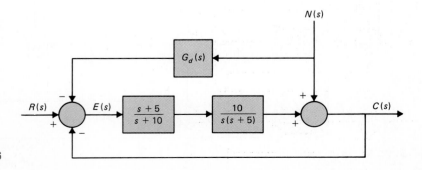

FIGURE 3P-16

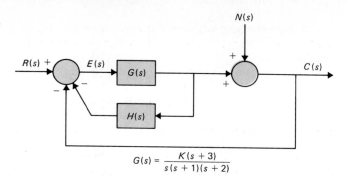

FIGURE 3P-17

$$G(s) = \frac{K(s+3)}{s(s+1)(s+2)}$$

3-18. The block diagram of a position-control system utilizing a dc motor is shown in Fig. 3P-18.
   (a) Find the open-loop transfer function $\Theta_o(s)/\Theta_e(s)$ (the outer feedback path is open).
   (b) Find the closed-loop transfer function $\Theta_o(s)/\Theta_r(s)$.

3-19. The block diagram of a feedback control system is shown in Fig. 3P-19.
   (a) Draw an equivalent signal-flow graph for the system.
   (b) Find the Δ of the system by means of Mason's gain formula.

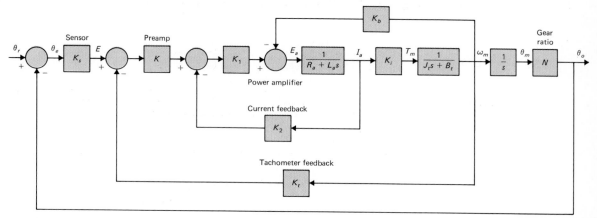

FIGURE 3P-18

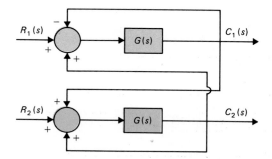

FIGURE 3P-19

(c)   Find the following transfer functions:

$$\left.\frac{C_1(s)}{R_1(s)}\right|_{R_2=0} \qquad \left.\frac{C_1(s)}{R_2(s)}\right|_{R_1=0} \qquad \left.\frac{C_2(s)}{R_1(s)}\right|_{R_2=0} \qquad \left.\frac{C_2(s)}{R_2(s)}\right|_{R_1=0}$$

(d)   Express the transfer-function relations in matrix form, $\mathbf{C}(s) = \mathbf{G}(s)\mathbf{R}(s)$, where $\mathbf{C}(s)$ is the 2×1 output vector, $\mathbf{R}(s)$ is the 2×1 input vector, and $\mathbf{G}(s)$ is the 2×2 transfer-function matrix.

3-20.   The block diagram of a linear control system is shown in Fig. 3P-20, where $G_p(s)$ is the transfer function of the controlled process, and $G_c(s)$ and $H(s)$ are controller transfer functions.

(a)   Derive   the   transfer   functions $C(s)/R(s)|_{N=0}$ and $C(s)/N(s)|_{R=0}$.   Find $C(s)/R(s)|_{N=0}$ when $G_c(s) = G_p(s)$.

(b)   Let

$$G_p(s) = G_c(s) = \frac{100}{(s + 1)(s + 5)}$$

Find the output response $c(t)$ when $N(s) = 0$ and $r(t) = u_s(t)$.

(c)   With $G_p(s)$ and $G_c(s)$ as given in part (b), select $H(s)$ among the following choices such that when $n(t) = u_s(t)$ and $r(t) = 0$, the steady-state value of $c(t)$ is equal to zero. (There may be more than one answer.)

$$H(s) = \frac{10}{s(s + 1)} \qquad H(s) = \frac{10}{(s + 1)(s + 2)}$$

$$H(s) = \frac{10(s + 1)}{s + 2} \qquad H(s) = \frac{K}{s^n} \quad (n = \text{positive integer} \geq 1)$$

3-21.   (a)   Draw a state diagram for the following state equations.

$$\frac{dx_1(t)}{dt} = -2x_1(t) + 3x_2(t)$$

$$\frac{dx_2(t)}{dt} = -5x_1(t) - 5x_2(t) + 2r(t)$$

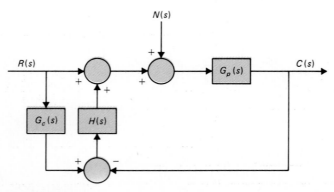

FIGURE 3P-20

(b) Find the characteristic equation of the system.

(c) Find the transfer functions $X_1(s)/R(s)$ and $X_2(s)/R(s)$.

3-22. The differential equation of a linear system is given by

$$\frac{d^3c(t)}{dt^3} + 5\frac{d^2c(t)}{dt^2} + 6\frac{dc(t)}{dt} + 10c(t) = r(t)$$

where $c(t)$ is the output, and $r(t)$ is the input.

(a) Draw a state diagram for the system.

(b) Write the state equations from the state diagram. Define the state variables from right to left in ascending order.

(c) Find the characteristic equation and its roots. (The program POLYROOT in the ACSP software package or any root-finding program may be used.)

(d) Find the transfer function $C(s)/R(s)$.

3-23. Repeat Problem 3-22 for the following differential equation:

$$\frac{d^4c(t)}{dt^4} + 5\frac{d^3c(t)}{dt^3} + 3\frac{d^2c(t)}{dt^2} + 10\frac{dc(t)}{dt} + c(t) = 2r(t)$$

3-24. The block diagram of a feedback control system is shown in Fig. 3P-24.

(a) Derive the following transfer functions:

$$\left.\frac{C(s)}{R(s)}\right|_{n=0} \qquad \left.\frac{C(s)}{N(s)}\right|_{R=0} \qquad \left.\frac{E(s)}{R(s)}\right|_{N=0}$$

(b) Find the transfer function $G_4(s)$ so that the output $C(s)$ is totally independent of $N(s)$.

(c) Find the steady-state value of $e(t)$ when the input $r(t)$ is a unit-step function. Set $n(t) = 0$. Assume that all the poles of $sE(s)$ are in the left-half s-plane.

3-25. A linear time-invariant digital control system has an output that is described by the time sequence

$$e(kT) = 1 - e^{-2kT} \qquad k = 0, 1, 2, \ldots$$

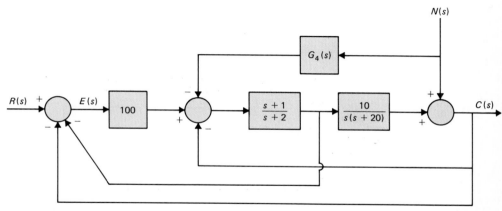

FIGURE 3P-24

when the system is subject to an input sequence described by $r(kT) = 1$ for all $k \geq 0$. Find the transfer function $G(z) = C(z)/R(z)$ of the system.

3-26. Find the transfer function $C(z)/R(z)$ of the discrete-data systems shown in Fig. 3P-26. The sampling period is 0.5 s. Do the problems analytically, and then check the answers using the SDCS program of the ACSP software, or any other computer program, if available.

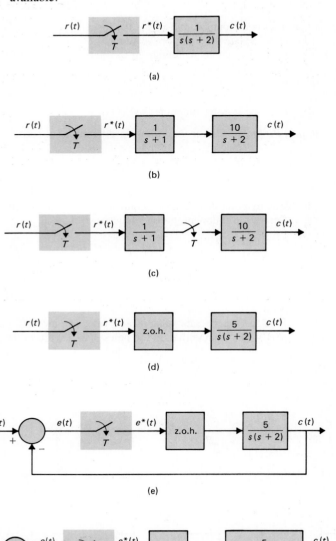

FIGURE 3P-26

3-27.   It is well known that the transfer function of an analog integrator is

$$G(s) = \frac{Y(s)}{X(s)} = \frac{1}{s}$$

where $X(s)$ and $Y(s)$ are the Laplace transforms of the input and the output of the integrator, respectively. There are many ways of implementing integration digitally. In a basic computer course, the rectangular integration is described by the schemes shown in Fig. 3P-27. The continuous signal $x(t)$ is approximated by a staircase signal; T is the sampling period. The integral of $x(t)$, which is the area under $x(t)$, is approximated by the area under the rectangular approximation signal.

(a)   Let $y(kT)$ denote the digital approximation of the integral of $x(t)$ from $t = 0$ to $t = kT$. Then $y(kT)$ can be written as

$$y(kT) = y[(k-1)T] + Tx(kT) \qquad (1)$$

where $y[(k-1)T]$ denotes the area under $x(t)$ from $t = 0$ to $t = (k-1)T$. Take the z-transform on both sides of Eq. (1) and show that the transfer function of the

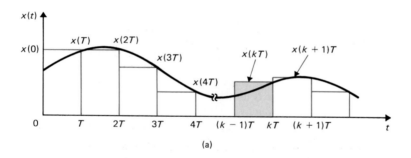

(a)

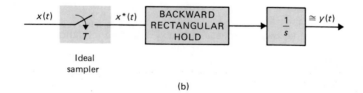

(b)

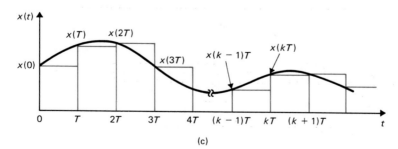

(c)

FIGURE 3P-27

digital integrator is

$$G(z) = \frac{Y(z)}{X(z)} = \frac{Tz}{z-1}$$

(b) The rectangular integration described in Fig. 3P-27(a) can be interpreted as a sample-and-hold operation, as shown in Fig. 3P-27(b). The signal $x(t)$ is first sent through an ideal sampler with sampling period T. The output of the sampler is the sequence $x(0), x(T), \ldots, x(kT), \ldots$. These numbers are then sent through a "backward" hold device to give the rectangle of height $x(kT)$ during the time interval from $(k-1)T$ to kT. Verify the result obtained in part (a) for $G(z)$ using the "backward" sample-and-hold interpretation.

(c) As an alternative, we can use a "forward" rectangular hold, as shown in Fig. 3P-27(c). Find the transfer function $G(z)$ for such a rectangular integrator.

ADDITIONAL COMPUTER PROBLEM

The following problem can be solved using the program SDCS of the ACSP software or any other suitable computer program that the reader may have.

3-28. The block diagram of a digital control system is shown in Fig. 3P-28.

(a) Find the transfer function $C(z)/R(z)$ of the system. First, do the problem by hand and then check the answers using the computer. Let $D(z) = 1$ and $T = 1$ s. Show that the system is unstable.

(b) For $D(z) = 1$ and $T = 0.1$ s, repeat part (a) and find the output response $c(kT)$ when the input is a unit-step function. Show that the system is now stable. Use zero initial conditions. Plot the output for 200 samples using the computer program. Now you should have learned how the system stability is dependent on the sampling period. Find the final value of the output with the unit step analytically using the final-value theorem of the z-transforms. Check the answer with the computer results.

(c) The unit-step response obtained in part (b) is quite oscillatory. Now let the transfer function of the controller be

$$D(z) = \frac{z - 0.8}{0.25(z - 0.2)}$$

$T = 0.1$ s. Repeat part (b).

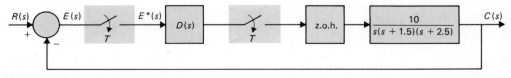

FIGURE 3P-28

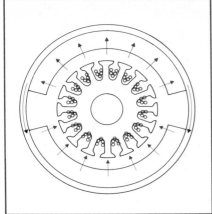

4

MATHEMATICAL MODELING OF PHYSICAL SYSTEMS

INTRODUCTION

One of the most important tasks in the analysis and design of control systems is mathematical modeling of the systems. In the preceding chapters, we have introduced a number of methods of modeling linear systems. The two most common are the transfer-function approach and the state-equation approach. The transfer-function method is valid only for linear time-invariant systems, whereas the state equations can be applied to model linear as well as nonlinear systems. In reality, since all physical systems are nonlinear to some extent, in order to use transfer functions and linear state equations, the system must first be linearized, or its range of operation be confined to a linear range.

Although the analysis and design of linear control systems have been well developed, their counterparts for nonlinear systems are usually quite complex. Therefore, the control-systems engineer often has the task of determining not only how to accurately describe a system mathematically, but, more importantly, how to make proper assumptions and approximations, whenever necessary, so that the system may be adequately characterized by a linear mathematical model.

It is important to point out that the modern control engineer should place special emphasis on the mathematical modeling of the system so that the analysis and design problems can be adaptable for computer solutions. Therefore, the main objectives of this chapter are as follows:

mathematical modelling
linearizing the

1.   To demonstrate the mathematical modeling of control systems and components.
2.   To demonstrate how the modeling will lead to computer solutions.
3.   To linearize nonlinear systems.

The modeling of many system components and systems will be illustrated with the emphasis placed on the approach to the problem, and no attempt is made to cover all possible types of systems.

Since state equations will be used for system modeling, we shall first introduce the basic concept of state.

state

To begin with the state-variable approach, we should first define the **state** of a system. As the word implies, the **state** refers to the **past**, **present**, and **future** conditions of the system. In general, the state can be described by a set of numbers, a curve, an equation, or something that is more abstract in nature. From a mathematical sense,

state variables
state equations

it is convenient to define a set of **state variables** and **state equations** to model dynamic systems. In Chapter 2, we have shown that the state equations are first-order differential equations. In general, there are some basic rules regarding the definition of a state variable and what constitutes a state equation. Consider that $x_1(t)$, $x_2(t)$, . . . , $x_n(t)$ are chosen to be the state variables of a dynamic system. Then, these variables must satisfy the following conditions:

1.   At any initial time $t = t_0$, the state variables $x_1(t_0)$, $x_2(t_0)$, . . . , $x_n(t_0)$

initial states

define the **initial states** of the system.
2.   Once the inputs of the system for $t \geq t_0$ and the initial states defined above are specified, the state variables should completely define the future behavior of the system.

Formally, the state variables are defined as follows:

DEFINITION OF STATE VARIABLES.   *The state variables of a system are defined as a minimal set of variables, $x_1(t)$, $x_2(t)$, . . . , $x_n(t)$, such that knowledge of these variables at any time t_0, and information on the input excitation subsequently applied, are sufficient to determine the state of the system at any time $t > t_0$.*

output

One should not confuse the state variables with the outputs of a system. An output of a system is a variable that can be **measured**, but a state variable does not always satisfy this requirement. In general, an output variable can be expressed as a linear combination of the state variables.

The concept and definition of state just given are applied to the formulation of state equations of electrical and mechanical systems in the following sections.

4-2

EQUATIONS OF ELECTRIC NETWORKS

The classical way of writing equations of electric networks is based on the loop method or the node method, which are formulated from the two laws of Kirchhoff. However, the loop and node equations are not natural for computer solutions. A modern method of writing network equations is the state-variable method, which is briefly illustrated in Section 2-3. Since the networks encountered in most control systems are rather simple, we shall present the subject here only at the introductory level. More detailed discussions on the state equations of electric networks can be found in texts on network theory.

EXAMPLE 4-1

Let us consider the *RLC* network shown in Fig. 4-1(a). A practical way is to assign the current in the inductor *L*, $i(t)$, and the voltage across the capacitor *C*, $e_c(t)$, as the state variables. The reason for this choice is because the state variables are directly related to the energy-storage elements of a system. The inductor is a storage for kinetic energy, and the capacitor is a storage of electric potential energy. By assigning $i(t)$ and $e_c(t)$ as state variables, we have a complete description of the past history (via the initial states), the present, and future states of the network.

The state equations for the network in Fig. 4-1(a) are written by first equating the current in *C* and the voltage across *L* in terms of the state variables and the applied voltage $e(t)$. We have:

Current in *C*:

$$C\frac{de_c(t)}{dt} = i(t) \tag{4-1}$$

Voltage across *L*:

$$L\frac{di(t)}{dt} = -e_c(t) - Ri(t) + e(t) \tag{4-2}$$

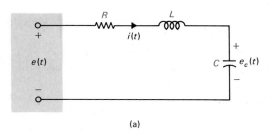

(a)

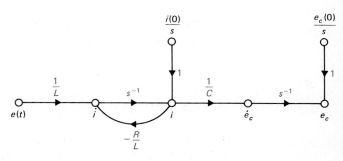

(b)

FIGURE 4-1 (a) *RLC* network.
(b) State diagram.

The state equations in vector-matrix form are then written as

$$
\begin{bmatrix} \dfrac{de_c(t)}{dt} \\[2ex] \dfrac{di(t)}{dt} \end{bmatrix}
=
\begin{bmatrix} 0 & \dfrac{1}{C} \\[2ex] -\dfrac{1}{L} & -\dfrac{R}{L} \end{bmatrix}
\begin{bmatrix} e_c(t) \\[1ex] i(t) \end{bmatrix}
+
\begin{bmatrix} 0 \\[1ex] \dfrac{1}{L} \end{bmatrix} e(t)
\tag{4-3}
$$

The state diagram of the network is shown in Fig. 4-1(b). Notice that the outputs of the integrators are defined as the state variables. The transfer functions of the system are obtained by applying Mason's gain formula to the state diagram when all the initial states are set to zero.

gain formula

$$
\frac{E_c(s)}{E(s)} = \frac{\dfrac{1}{LC}s^{-2}}{1 + \dfrac{R}{L}s^{-1}} = \frac{1}{Cs(Ls + R)}
\tag{4-4}
$$

$$
\frac{I(s)}{E(s)} = \frac{\dfrac{1}{L}s^{-1}}{1 + \dfrac{R}{L}s^{-1}} = \frac{1}{Ls + R}
\tag{4-5}
$$

EXAMPLE 4-2

As another example of writing the state equations of an electric network, consider the network shown in Fig. 4-2(a). According to the foregoing discussion, the voltage across the capacitor, $e_c(t)$, and the currents in the inductors, $i_1(t)$ and $i_2(t)$, are assigned as state variables, as

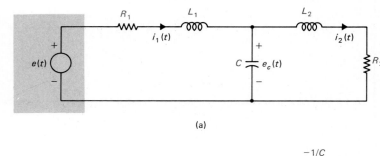

(a)

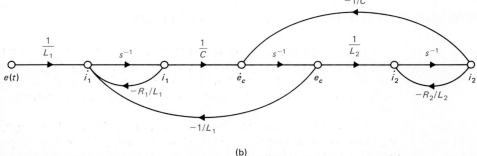

(b)

FIGURE 4-2 (a) Network of Example 4-2. (b) State diagram.

shown in Fig. 4-2(a). The state equations of the network are obtained by writing the voltages across the inductors and the currents in the capacitor in terms of the three state variables. The state equations are

$$L_1 \frac{di_1(t)}{dt} = -R_1 i_1(t) - e_c(t) + e(t) \tag{4-6}$$

$$L_2 \frac{di_2(t)}{dt} = -R_2 i_2(t) + e_c(t) \tag{4-7}$$

$$C \frac{de_c(t)}{dt} = i_1(t) - i_2(t) \tag{4-8}$$

In vector-matrix form, the state equations are

$$\begin{bmatrix} \dfrac{di_1(t)}{dt} \\[2ex] \dfrac{di_2(t)}{dt} \\[2ex] \dfrac{de_c(t)}{dt} \end{bmatrix} = \begin{bmatrix} -\dfrac{R_1}{L_1} & 0 & -\dfrac{1}{L_1} \\[2ex] 0 & -\dfrac{R_2}{L_2} & \dfrac{1}{L_2} \\[2ex] \dfrac{1}{C} & -\dfrac{1}{C} & 0 \end{bmatrix} \begin{bmatrix} i_1(t) \\[2ex] i_2(t) \\[2ex] e_c(t) \end{bmatrix} + \begin{bmatrix} \dfrac{1}{L_1} \\[2ex] 0 \\[2ex] 0 \end{bmatrix} e(t) \tag{4-9}$$

The state diagram of the network is shown in Fig. 4-2(b), where the initial states are omitted. The transfer functions between $I_1(s)$, $I_2(s)$ and $E_c(s)$ and $E(s)$, respectively, are written from the state diagram.

$$\frac{I_1(s)}{E(s)} = \frac{L_2 C s^2 + R_2 C s + 1}{\Delta} \tag{4-10}$$

$$\frac{I_2(s)}{E(s)} = \frac{1}{\Delta} \tag{4-11}$$

$$\frac{E_c(s)}{E(s)} = \frac{L_2 s + R_2}{\Delta} \tag{4-12}$$

where

$$\Delta = L_1 L_2 C s^3 + (R_1 L_2 + R_2 L_1) C s^2 + (L_1 + L_2 + R_1 R_2 C) s + R_1 + R_2 \tag{4-13}$$

4-3

MODELING OF MECHANICAL SYSTEM ELEMENTS

Most control systems contain mechanical as well as electrical components. From a mathematical viewpoint, the description of electrical and mechanical elements are analogous. In fact, we can show that given an electrical device, there is usually an analogous mechanical counterpart, and vice versa, that is described by similar equations.

translational motion   The motion of mechanical elements can be described in various dimensions as **translational**, **rotational**, or combinations of both. The equations governing the motion of mechanical systems are often directly or indirectly formulated from Newton's law of motion.

Translational Motion

acceleration
velocity
displacement

The motion of translation is defined as a motion that takes place along a straight line. The variables that are used to describe translational motion are **acceleration**, **velocity**, and **displacement**.

Newton's law of motion states that the *algebraic sum of forces acting on a rigid body in a given direction is equal to the product of the mass of the body and its acceleration in the same direction.* The law can be expressed as

$$\sum \text{forces} = Ma \qquad (4\text{-}14)$$

where M denotes the mass, and a is the acceleration in the direction considered. For translational motion, the following elements are usually involved:

mass

1. *Mass.* *Mass is considered as a property of an element that stores the kinetic energy of translational motion.* Mass is analogous to inductance of electric networks. If W denotes the weight of a body, then M is given by

    $$M = \frac{W}{g} \qquad (4\text{-}15)$$

    where g is the acceleration of free fall of the body due to gravity. ($g = 32.174$ ft/s^2 in British units, and $g = 9.8066$ m/s^2 in SI units.)

    The consistent sets of basic units in the British and SI units are as follows:

UNITS	MASS M	ACCELERATION	FORCE
SI	kilogram (kg)	m/s^2	Newton (N)
British	slug	ft/s^2	pound (lb force)

Conversion factors between these and other secondary units are as follows:

Force:

$$1 \text{ N} = 0.2248 \text{ lb (force)} = 3.5969 \text{ oz (force)}$$

Mass:

$$1 \text{ kg} = 1000 \text{ g} = 2.2046 \text{ lb (mass)}$$
$$= 35.274 \text{ oz (mass)}$$
$$= 0.06852 \text{ slug}$$

Distance:

$$1 \text{ m} = 3.2808 \text{ ft} = 39.37 \text{ in.}$$
$$1 \text{ in} = 25.4 \text{ mm}$$
$$1 \text{ ft} = 0.3048 \text{ m}$$

FIGURE 4-3 Force–mass system.

Figure 4-3 illustrates the situation where a force is acting on a body with mass M. The force equation is written

$$f(t) = Ma(t) = M\frac{d^2y(t)}{dt^2} = M\frac{dv(t)}{dt} \tag{4-16}$$

linear spring

2. *Linear spring.*   In practice, a linear spring may be an actual spring or the compliance of a cable or a belt. In general, *a spring is considered to be an element that stores potential energy*. It is analogous to a capacitor in electric networks. In practice, all springs are nonlinear to some extent. However, if the deformation of the spring is small, its behavior can be approximated by a linear relationship:

$$f(t) = Ky(t) \tag{4-17}$$

where K is the **spring constant**, or simply **stiffness**.
The two basic unit systems for the spring constant are as follows:

UNITS	SPRING CONSTANT K
SI	N/m
British	lb/ft

Equation (4-17) implies that the force acting on the spring is directly proportional to the displacement (deformation) of the spring. The model representing a linear spring element is shown in Fig. 4-4.

preload tension

If the spring is preloaded with a preload tension of T, then Eq. (4-17) should be modified to

$$f(t) - T = Ky(t) \tag{4-18}$$

friction

FRICTION FOR TRANSLATION MOTION.   Whenever there is motion or tendency of motion between two elements, frictional forces exist. The frictional forces encoun-

FIGURE 4-4 Force–spring system.

tered in physical systems are usually of a nonlinear nature. The characteristics of the frictional forces between two contacting surfaces often depend on such factors as the composition of the surfaces, the pressure between the surfaces, their relative velocity, and others, so that an exact mathematical description of the frictional force is difficult. Three different types of friction are commonly used in practical systems: **viscous friction**, **static friction**, and **Coulomb friction**. These are discussed separately in the following.

viscous friction

1. *Viscous friction.   Viscous friction represents a retarding force that is a linear relationship between the applied force and velocity.* The schematic diagram element for viscous friction is often represented by a dashpot, such as that shown in Fig. 4-5. The mathematical expression of viscous friction is

$$f(t) = B\frac{dy(t)}{dt} \tag{4-19}$$

viscous frictional coefficient

where B is the **viscous frictional coefficient**.
    The units of B are as follows:

UNITS	VISCOUS FRICTIONAL COEFFICIENT B
SI	N/m/s
British	lb/ft/s

Figure 4-6(a) shows the functional relation between the viscous frictional force and velocity.

static friction

2. *Static friction.   Static friction represents a retarding force that tends to prevent motion from beginning.* The static frictional force can be represented by the expression

$$f(t) = \pm(F_s)|_{\dot{y}=0} \tag{4-20}$$

which is defined as a frictional force that exists only when the body is stationary but has a tendency of moving. The sign of the friction depends on the direction of motion or the initial direction of velocity. The force-to-velocity relation of static friction is illustrated in Fig. 4-6(b). Notice that

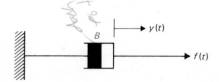

FIGURE 4-5 Dashpot for viscous friction.

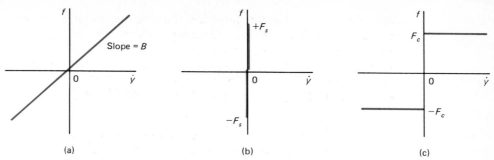

FIGURE 4-6 Functional relationships of linear and nonlinear frictional forces. (a) Viscous friction. (b) Static friction. (c) Coulomb friction.

once motion begins, the static frictional force vanishes and other frictions take over.

Coulomb friction

3. *Coulomb friction.   Coulomb friction is a retarding force that has a constant amplitude with respect to the change in velocity, but the sign of the frictional force changes with the reversal of the direction of velocity.* The mathematical relation for the Coulomb friction is given by

$$f(t) = F_c \frac{\left[\dfrac{dy(t)}{dt}\right]}{\left|\left[\dfrac{dy(t)}{dt}\right]\right|} \tag{4-21}$$

Coulomb friction coefficient

where F_c is the **Coulomb friction coefficient**. The functional description of the friction-to-velocity relation is shown in Fig. 4-6(c).

It should be pointed out that the three types of frictions cited are merely practical models that have been devised to portray frictional phenomena found in physical systems. They are by no means exhaustive. In many unusual situations, we have to use other frictional models to represent the actual phenomenon accurately. One such example is rolling dry friction [5, 6], which turns out to have nonlinear hysteresis properties.

Rotational Motion

The rotational motion of a body can be defined as motion about a fixed axis. The extension of Newton's law of motion for rotational motion states that the *algebraic sum of moments or torques about a fixed axis is equal to the product of the inertia and the angular acceleration about the axis. Or,*

$$\sum \text{torques} = J\alpha \tag{4-22}$$

where J denotes the inertia, and α is the angular acceleration. The other variables gen-

torque

angular velocity

angular displacement

inertia

erally used to describe the motion of rotation are **torque** T, **angular velocity** ω, and **angular displacement** θ. The elements involved with the rotation motion are as follows.

1. *Inertia.   Inertia, J, is considered as a property of an element that stores the kinetic energy of rotational motion.* The inertia of a given element depends on the geometric composition about the axis of rotation and its density. For instance, the inertia of a circular disk or shaft about its geometric axis is given by

$$J = \tfrac{1}{2}Mr^2 \tag{4-23}$$

where M is the mass of the disk or shaft, and r is its radius.

EXAMPLE 4-3

Given a disk that is 1 in. in radius, 0.25 in. thick, and weighs 5 oz, its inertia is

$$J = \frac{1}{2}\frac{Wr^2}{g} = \frac{1}{2}\frac{(5 \text{ oz})(1 \text{ in.})^2}{386 \text{ in.}/s^2} = 0.00647 \text{ oz-in.-}s^2 \tag{4-24}$$

Usually, the density of the material is given in weight per unit volume. *Then, for a circular disk or shaft, it can be shown that the inertia is proportional to the fourth power of the radius and the first power of the thickness or length.* If the weight W is expressed as

$$W = \rho(\pi r^2 h) \tag{4-25}$$

where ρ is the density in weight per unit volume, r the radius, and h the thickness or length, Eq. (4-24) is written as

$$J = \frac{1}{2}\frac{\rho\pi h r^4}{g} = 0.00406\rho h r^4 \tag{4-26}$$

where h and r are in inches.

steel

For steel, $\rho = 4.53$ oz/in.^{3}; Eq. (4-26) becomes

$$J = 0.0184hr^4 \tag{4-27}$$

aluminum

For aluminum, ρ is 1.56 oz/in.^{3}; Eq. (4-26) becomes

$$J = 0.00636hr^4 \tag{4-28}$$

When a torque is applied to a body with inertia J, as shown in Fig. 4-7, the torque equation is written

$$T(t) = J\alpha(t) = J\frac{d\omega(t)}{dt} = J\frac{d^2\theta(t)}{dt^2} \tag{4-29}$$

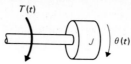

$T(t)$

$\theta(t)$

FIGURE 4-7 Torque-inertia system.

The SI and British units for the quantities in Eq. (4-29) are tabulated as follows:

UNITS	INERTIA	TORQUE	ANGULAR DISPLACEMENT
SI	kg-m^2	N-m	radian
		dyne-cm	radian
British	slug-ft^2	lb-ft	
	(lb-ft-s^2)		
	oz-in.-s^2	oz-in.	radian

The following conversion factors are often found useful:

ANGULAR DISPLACEMENT

$$1 \text{ rad} = \frac{180}{\pi} = 57.3°$$

ANGULAR VELOCITY

$$1 \text{ rpm} = \frac{2\pi}{60} = 0.1047 \text{ rad/s}$$

$$1 \text{ rpm} = 6 \text{ deg/s}$$

TORQUE

$$1 \text{ g-cm} = 0.0139 \text{ oz-in.}$$

$$1 \text{ lb-ft} = 192 \text{ oz-in.}$$

$$1 \text{ oz-in.} = 0.00521 \text{ lb-ft}$$

INERTIA

$$1 \text{ g-cm} = 1.417 \times 10^{-5} \text{ oz-in.-s}^2$$

$$1 \text{ lb-ft-s}^2 = 192 \text{ oz-in.-s}^2 = 32.2 \text{ lb-ft}^2$$

$$1 \text{ oz-in-s}^2 = 386 \text{ oz-in.}^2$$

$$1 \text{ g-cm-s}^2 = 980 \text{ g-cm}^2$$

torsional spring

TORSIONAL SPRING.   As with the linear spring for translational motion, a **torsional spring constant K**, in torque per unit angular displacement, can be devised to represent the compliance of a rod or a shaft when it is subject to an applied torque. Fig-

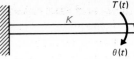

FIGURE 4-8 Torque–torsional spring system.

ure 4-8 illustrates a simple torque–spring system that can be represented by the equation

$$T(t) = K\theta(t) \tag{4-30}$$

The units of the spring constant K in the SI and British systems are as follows:

UNITS	SPRING CONSTANT K
SI	N-m/rad
British	ft-lb/rad

If the torsional spring is preloaded by a preload torque of TP, Eq. (4-30) is modified to

$$T(t) - TP = K\theta(t) \tag{4-31}$$

friction   **FRICTION FOR ROTATIONAL MOTION.**   The three types of friction described for translational motion can be carried over to the motion of rotation. Therefore, Eqs. (4-19), (4-20), and (4-21) can be replaced, respectively, by their counterparts:

Viscous friction:

$$T(t) = B\frac{d\theta(t)}{dt} \tag{4-32}$$

Static friction:

$$T(t) = \pm(F_s)\big|_{\dot{\theta}=0} \tag{4-33}$$

Coulomb friction:

$$T(t) = F_c\frac{\left[\dfrac{d\theta(t)}{dt}\right]}{\left|\left[\dfrac{d\theta(t)}{dt}\right]\right|} \tag{4-34}$$

Relation Between Translational and Rotational Motions

In motion-control problems, it is often necessary to convert rotational motion into translation, For instance, a load may be controlled to move along a straight line through a rotary motor and screw assembly, such as that shown in Fig. 4-9. Figure 4-10 shows a similar situation in which a rack and pinion is used as the mechanical linkage. Another common system in motion control is the control of a mass through a pulley by a rotary prime mover, such as that shown in Fig. 4-11. The systems shown in Figs. 4-9, 4-10, and 4-11 can all be represented by a simple system with an equivalent inertia connected directly to the drive motor. For instance, the mass in Fig. 4-11 can be regarded as a point mass that moves about the pulley, which has a radius r. By disregarding the inertia of the pulley, the equivalent inertia that the motor sees is

$$J = Mr^2 = \frac{W}{g}r^2 \qquad (4\text{-}35)$$

If the radius of the pinion in Fig. 4-10 is r, the equivalent inertia that the motor sees is also given by Eq. (4-35).

Now consider the system of Fig. 4-9. The lead of the screw L is defined as the linear distance that the mass travels per revolution of the screw. In principle, the two systems in Figs. 4-10 and 4-11 are equivalent. In Fig. 4-10, the distance traveled by the mass per revolution of the pinion is $2\pi r$. By using Eq. (4-35) as the equivalent inertia for the system of Fig. 4-9,

$$J = \frac{W}{g}\left(\frac{L}{2\pi}\right)^2 \qquad (4\text{-}36)$$

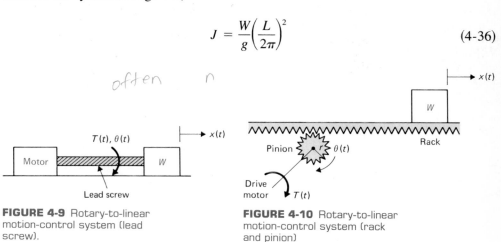

FIGURE 4-9 Rotary-to-linear motion-control system (lead screw).

FIGURE 4-10 Rotary-to-linear motion-control system (rack and pinion)

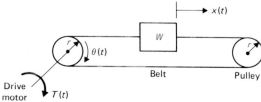

FIGURE 4-11 Rotary-to-linear motion-control system (belt and pulley).

where, in the British units,

$$J = \text{inertia (oz-in.-s}^2)$$

$$W = \text{weight (oz)}$$

$$L = \text{screw lead (in.)}$$

$$g = \text{gravitational force (386.4 in./s}^2)$$

Mechanical Energy and Power

Energy and power play an important role in the design of electromechanical systems. Stored energy in the form of kinetic and potential energy controls the dynamics of the system, whereas dissipative energy usually is spent in the form of heat, which must be closely controlled.

The kinetic energy of a moving mass with a velocity v is

$$W_k = \tfrac{1}{2}Mv^2 \tag{4-37}$$

kinetic energy   The following consistent sets of units are given for the kinetic-energy relations:

UNITS	ENERGY	MASS	VELOCITY
SI	joules or N-m	N/m/s^2	m/s
British	ft-lb	lb/ft/s^2 (slug)	ft/s

For a rotational system, the kinetic energy is written

$$W_k = \tfrac{1}{2}J\omega^2 \tag{4-38}$$

where J is the moment of inertia, and ω is the angular velocity. The following units are given for the rotational kinetic energy:

UNITS	ENERGY	INERTIA	ANGULAR VELOCITY
SI	joules or N-m	kg-m^2	rad/s
British	oz-in.	oz-in.-s^2	rad/s

potential
energy
Potential energy stored in a mechanical element represents the amount of work required to change the configuration. For a linear spring that is deformed in length by y, the potential energy stored in the spring is

$$W_p = \tfrac{1}{2}Ky^2 \tag{4-39}$$

where K is the spring constant. For a torsional spring that is subject to an angular deformation of θ, the potential energy stored is

$$W_p = \tfrac{1}{2} K\theta^2 \qquad (4\text{-}40)$$

When dealing with a frictional element, the form of energy differs from the previous two cases in that the energy represents a loss of dissipation by the system in overcoming the frictional force. Power is the time rate of doing work. Therefore, the power dissipated in a frictional element is the product of force $f(t)$ and velocity $v(t)$; that is,

$$P = f(t)v(t) \qquad (4\text{-}41)$$

Since $f(t) = Bv(t)$, where B is the viscous-friction coefficient, Eq. (4-41) becomes

$$P = B[v(t)]^2 \qquad (4\text{-}42)$$

horsepower

The SI unit for power is in N-m/s, or watts (W). In British units, power is represented in ft-lb/s, or horsepower (hp). Furthermore,

$$1 \text{ hp} = 746 \text{ W} = 550 \text{ ft-lb/s} \qquad (4\text{-}43)$$

Since power is the rate at which energy is being dissipated, the energy dissipated in a frictional element is

$$W_d = B \int [v(t)]^2 \, dt \qquad (4\text{-}44)$$

Gear Trains, Levers, and Timing Belts

gear train

A gear train, a lever, or a timing belt over pulleys is a mechanical device that transmits energy from one part of a system to another in such a way that force, torque, speed, and displacement may be altered. These devices can also be regarded as matching devices used to attain maximum power transfer. Two gears are shown coupled together in Fig. 4-12. The inertia and friction of the gears are neglected in the ideal case considered.

The relationships between the torques T_1 and T_2, angular displacements θ_1 and θ_2,

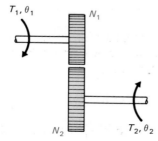

FIGURE 4-12 Gear train.

and the teeth numbers N_1 and N_2 of the gear train are derived from the following facts:

1. *The number of teeth on the surface of the gears is proportional to the radii r_1 and r_2 of the gears*; that is,

$$r_1 N_2 = r_2 N_1 \qquad (4\text{-}45)$$

2. *The distance traveled along the surface of each gear is the same.* Thus,

$$\theta_1 r_1 = \theta_2 r_2 \qquad (4\text{-}46)$$

3. *The work done by one gear is equal to that of the other* since there is assumed to be no loss. Thus,

$$T_1 \theta_1 = T_2 \theta_2 \qquad (4\text{-}47)$$

If the angular velocities of the two gears, ω_1 and ω_2, are brought into the picture, Eqs. (4-45) through (4-47) lead to

$$\frac{T_1}{T_2} = \frac{\theta_2}{\theta_1} = \frac{N_1}{N_2} = \frac{\omega_2}{\omega_1} = \frac{r_1}{r_2} \qquad (4\text{-}48)$$

In practice, gears do have inertia and friction between the coupled gear teeth that often cannot be neglected. An equivalent representation of a gear train with viscous friction, Coulomb friction, and inertia, considered as lumped parameters, is shown in Fig. 4-13, and the variables and parameters are defined as follows:

$$T = \text{applied torque}$$
$$\theta_1, \theta_2 = \text{angular displacements}$$
$$T_1, T_2 = \text{torque transmitted to gears}$$
$$J_1, J_2 = \text{inertia of gears}$$
$$N_1, N_2 = \text{number of teeth}$$
$$F_{c1}, F_{c2} = \text{Coulomb friction coefficients}$$
$$B_1, B_2 = \text{viscous frictional coefficients}$$

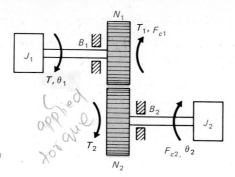

FIGURE 4-13 Gear train with friction and inertia.

The torque equation for gear 2 is

$$T_2(t) = J_2 \frac{d^2\theta_2(t)}{dt^2} + B_2 \frac{d\theta_2(t)}{dt} + F_{c2} \frac{\dot{\theta}_2}{|\dot{\theta}_2|} \tag{4-49}$$

The torque equation on the side of gear 1 is

$$T(t) = J_1 \frac{d^2\theta_1(t)}{dt^2} + B_1 \frac{d\theta_1(t)}{dt} + F_{c1} \frac{\dot{\theta}_1}{|\dot{\theta}_1|} + T_1(t) \tag{4-50}$$

By use of Eq. (4-48), Eq. (4-49) is converted to

$$T_1(t) = \frac{N_1}{N_2} T_2(t) = \left[\frac{N_1}{N_2}\right]^2 J_2 \frac{d^2\theta_1(t)}{dt^2} + \left[\frac{N_1}{N_2}\right]^2 B_2 \frac{d\theta_1(t)}{dt} + \frac{N_1}{N_2} F_{c2} \frac{\dot{\theta}_2}{|\dot{\theta}_2|} \tag{4-51}$$

Equation (4-51) indicates that *it is possible to reflect inertia, friction, compliance, torque, speed, and displacement from one side of a gear train to the other.* The following quantities are obtained when reflecting from gear 2 to gear 1:

Inertia: $\left[\dfrac{N_1}{N_2}\right]^2 J_2$

Viscous frictional coefficient: $\left[\dfrac{N_1}{N_2}\right]^2 B_2$

Torque: $\dfrac{N_1}{N_2} T_2$

Angular displacement: $\dfrac{N_2}{N_1} \omega_2$

Angular velocity: $\dfrac{N_2}{N_1} \omega_2$

Coulomb friction torque: $\dfrac{N_1}{N_2} F_{c2} \dfrac{\omega_2}{|\omega_2|}$

If torsional spring effect were present, the spring constant is also multiplied by $(N_1/N_2)^2$ in reflecting from gear 2 to gear 1. Now substituting Eq. (4-51) into Eq. (4-50), we get

$$T(t) = J_{1e}\frac{d^2\theta_1(t)}{dt^2} + B_{1e}\frac{d\theta_1(t)}{dt} + T_F \qquad (4\text{-}52)$$

where

$$J_{1e} = J_1 + \left[\frac{N_1}{N_2}\right]^2 J_2 \qquad (4\text{-}53)$$

$$B_{1e} = B_1 + \left[\frac{N_1}{N_2}\right]^2 B_2 \qquad (4\text{-}54)$$

$$T_F = F_{c1}\frac{\dot\theta_1}{|\dot\theta_1|} + \frac{N_1}{N_2}F_{c2}\frac{\dot\theta_2}{|\dot\theta_2|} \qquad (4\text{-}55)$$

EXAMPLE 4-4   Given a load that has inertia of 0.05 oz-in.-s² and a Coulomb friction torque of 2 oz-in., find the inertia and frictional torque reflected through a 1:5 gear train ($N_1/N_2 = 1/5$, with N_2 on the load side). The reflected inertia on the side of N_1 is $(1/5)^2 \times 0.05 = 0.002$ oz-in.-s². The reflected Coulomb friction is $(1/5) \times 2 = 0.4$ oz-in.

timing belts   Timing belts and chain drives serve the same purpose as the gear train except that they allow the transfer of energy over a longer distance without using an excessive number of gears. Figure 4-14 shows the diagram of a belt or chain drive between two pulleys. Assuming that there is no slippage between the belt and the pulleys, it is easy

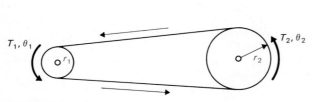

FIGURE 4-14 Belt or chain drive.

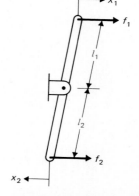

FIGURE 4-15 Lever system.

to see that Eq. (4-48) still applies to this case. In fact, the reflection or transmittance of torque, inertia, friction, and so on is similar to that of a gear train.

lever system   The lever system shown in Fig. 4-15 transmits translational motion and forces in the same way that gear trains transmit rotational motion. The relation between the forces and distances is

$$\frac{f_1}{f_2} = \frac{l_2}{l_1} = \frac{x_2}{x_1} \tag{4-56}$$

Backlash and Dead Zone

backlash   Backlash and dead zone are commonly found in gear trains and similar mechanical linkages where the coupling is not perfect. In a majority of situations, backlash may give rise to undesirable inaccuracy, oscillations, and instability in control systems. In addition, it has a tendency to wear out the mechanical elements. Regardless of the ac-
dead zone   tual mechanical elements, a physical model of backlash or dead zone between an input and an output member is shown in Fig. 4-16. The model can be used for a rotational system as well as for a translational system. The amount of backlash is $b/2$ on either side of the reference position.

In general, the dynamics of the mechanical linkage with backlash depend upon the relative inertia-to-friction ratio of the output member. If the inertia of the output member is very small compared with that of the input member, the motion is controlled predominantly by friction. This means that the output member will not coast whenever there is no contact between the two members. When the output is driven by the input, the two members will travel together until the input member reverses its direction; then the output member will be at a standstill until the backlash is taken up on the other side, at which time it is assumed that the output member instantaneously takes on the velocity of the input menber. The transfer characteristic between the input and output displacements of a system with backlash with negligible output inertia is shown in Fig. 4-17.

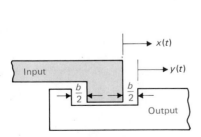

FIGURE 4-16 Physical model of backlash between two mechanical elements.

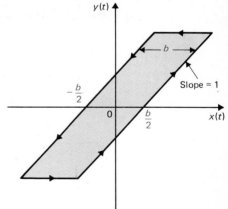

FIGURE 4-17 Input–output characteristic of backlash.

4-4

EQUATIONS OF MECHANICAL SYSTEMS

free-body diagram

The equations of a linear mechanical system are written by first constructing a model of the system containing interconnected linear elements, and then by applying Newton's law of motion to the free-body diagram. For translational motion, the equation of motion is Eq. (4-14), and for rotational motion, Eq. (4-22) is used. The following examples illustrate how equations of mechanical systems are written.

EXAMPLE 4-5

Consider the mass–spring–friction system shown in Fig. 4-18(a). The linear motion concerned is in the horizonal direction. The free-body diagram of the system is shown in Fig. 4-18(b). The force equation of the system is

$$f(t) = M\frac{d^2y(t)}{dt^2} + B\frac{dy(t)}{dt} + Ky(t) \tag{4-57}$$

The last equation is rearranged by equating the highest-order derivative term to the rest of the terms:

$$\frac{d^2y(t)}{dt^2} = -\frac{B}{M}\frac{dy(t)}{dt} - \frac{K}{M}y(t) + \frac{1}{M}f(t) \tag{4-58}$$

The state diagram of the system is constructed as shown in Fig. 4-18(c). By defining the outputs of the integrators on the state diagram as state variables x_1 and x_2, the state equations are

$$\frac{dx_1(t)}{dt} = x_2(t) \tag{4-59}$$

$$\frac{dx_2(t)}{dt} = -\frac{K}{M}x_1(t) - \frac{B}{M}x_2(t) + \frac{1}{M}f(t) \tag{4-60}$$

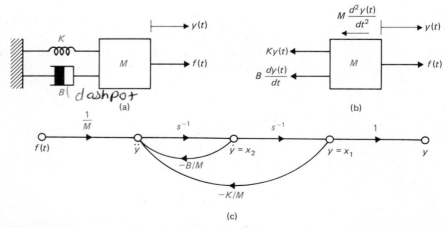

FIGURE 4-18 (a) Mass–spring–friction system. (b) Free-body diagram. (c) State diagram.

It is not difficult to see that this mechanical system is analogous to a series *RLC* electric network. With this analogy, the state equations can be written directly using a different set of state variables. Consider that mass M is analogous to inductance L, the spring constant K is analogous to the inverse of capacitance $1/C$, and the viscous-frictional coefficient B is analogous to resistance R. It is logical to assign $v(t)$, the velocity, and $f_k(t)$, the force acting on the spring, as state variables, since the former is analogous to the current in L and the latter is analogous to the voltage across C. Then writing the force on M and the velocity of the spring as functions of the state variables and the input force $f(t)$, we have

Force on mass:

$$M\frac{dv(t)}{dt} = -Bv(t) - f_k(t) + f(t) \tag{4-61}$$

Velocity of spring:

$$\frac{1}{K}\frac{df_k(t)}{dt} = v(t) \tag{4-62}$$

The state equations are obtained by dividing both sides of Eq. (4-61) by M and multiplying Eq. (4-62) by K.

This simple example illustrates that the state equations and state variables of a dynamic system are not unique.

The transfer function between $Y(s)$ and $F(s)$ is obtained by taking the Laplace transform on both sides of Eq. (4-57) with zero initial conditions:

$$\frac{Y(s)}{F(s)} = \frac{1}{Ms^2 + Bs + K} \tag{4-63}$$

The same result is obtained by applying the gain formula to Fig. 4-18(c).

EXAMPLE 4-6   As another example of writing the dynamic equations of a mechanical system with translational motion, consider the system shown in Fig. 4-19(a). Since the spring is deformed when it is subject to a force $f(t)$, two displacements, y_1 and y_2, must be assigned to the end points of the spring. The free-body diagrams of the system are shown in Fig. 4-19(b). The force equations are

$$f(t) = K[y_1(t) - y_2(t)] \tag{4-64}$$

$$K[y_1(t) - y_2(t)] = M\frac{d^2y_2(t)}{dt^2} + B\frac{dy_2(t)}{dt} \tag{4-65}$$

These equations are rearranged as

$$y_1(t) = y_2(t) + \frac{1}{K}f(t) \tag{4-66}$$

$$\frac{d^2y_2(t)}{dt^2} = -\frac{B}{M}\frac{dy_2(t)}{dt} + \frac{K}{M}[y_1(t) - y_2(t)] \tag{4-67}$$

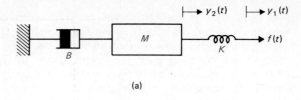

(a)

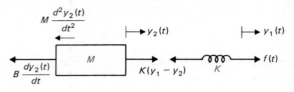

(b)

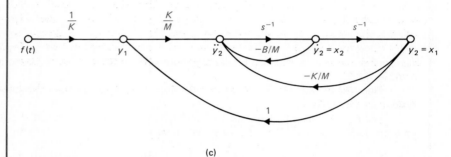

(c)

FIGURE 4-19 Mechanical system for Example 4-6. (a) Mass–spring–
friction system. (b) Free-body diagram. (c) State diagram.

By using the last two equations, the state diagram of the system is drawn in Fig. 4-19(c). The state variables are defined as $x_1(t) = y_2(t)$ and $x_2(t) = dy_2(t)/dt$. The state equations are written directly from the state diagram:

$$\frac{dx_1(t)}{dt} = x_2(t) \tag{4-68}$$

$$\frac{dx_2(t)}{dt} = -\frac{B}{M}x_2(t) + \frac{1}{M}f(t) \tag{4-69}$$

As an alternative, we can assign the velocity $v(t)$ of the mass M as one state variable and the force $f_k(t)$ on the spring as the other state variable. We have

$$\frac{dv(t)}{dt} = -\frac{B}{M}v(t) + \frac{1}{M}f_k(t) \tag{4-70}$$

and

$$f_k(t) = f(t) \tag{4-71}$$

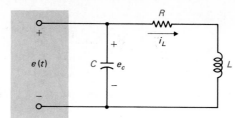

FIGURE 4-20 Electric network analogous to the mechanical system in Fig. 4-19.

energy-storage elements

One may wonder why there is only one state equation in Eq. (4-61), whereas there are two state variables in $v(t)$ and $f_k(t)$. The two state equations of Eqs. (4-68) and (4-69) clearly show that the system is of the second order. The situation is better explained by referring to the analogous electric network of the system shown in Fig. 4-20. Although the network has two energy-storage elements in L and C, and thus there should be two state variables, the voltage across the capacitance, $e_c(t)$, in this case is redundant, since it is equal to the applied voltage $e(t)$. Equations (4-70) and (4-71) can provide only the solutions to the velocity of M, $v(t)$, which is the same as $dy_2(t)/dt$, once $f(t)$ is specified. Then $y_2(t)$ is determined by integrating $v(t)$ with respect to t. The displacement $y_1(t)$ is then found using Eq. (4-64). On the other hand, Eqs. (4-68) and (4-69) give the solutions to $y_2(t)$ and $dy_2(t)/dt$ directly, and $y_1(t)$ is obtained from Eq. (4-64).

transfer function

The transfer functions of the system are obtained by applying the gain formula to the state diagram in Fig. 4-19(c).

$$\frac{Y_2(s)}{F(s)} = \frac{1}{s(Ms + B)} \qquad (4\text{-}72)$$

$$\frac{Y_1(s)}{F(s)} = \frac{Ms^2 + Bs + K}{Ks(Ms + B)} \qquad (4\text{-}73)$$

EXAMPLE 4-7

state diagram

In this example, the equations for the mechanical system in Fig. 4-21(a) are to be written. Then we are to draw the state diagram and derive the transfer functions for the system.

The free-body diagrams for the two masses are shown in Fig. 4-21(b), with the reference directions of the displacements $y_1(t)$ and $y_2(t)$ as indicated. The Newton's force equations for the system are written directly from the free-body diagrams:

$$f(t) + M_1 g = M_1\frac{d^2 y_1(t)}{dt^2} + B_1\left[\frac{dy_1(t)}{dt} - \frac{dy_2(t)}{dt}\right] + K_1[y_1(t) - y_2(t)] \qquad (4\text{-}74)$$

$$M_2 g = -B_1\left[\frac{dy_1(t)}{dt} - \frac{dy_2(t)}{dt}\right] - K_1[y_1(t) - y_2(t)]$$

$$+ M_2\frac{d^2 y_2(t)}{dt^2} + B_2\frac{dy_2(t)}{dt} + K_2 y_2(t) \qquad (4\text{-}75)$$

We may now decompose these two second-order simultaneous differential equations into four

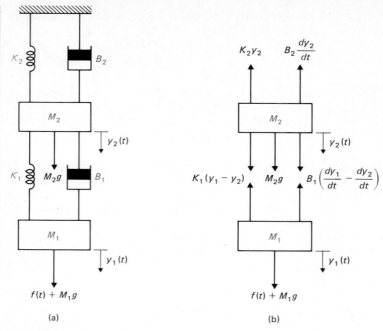

FIGURE 4-21 Mechanical system for Example 4-7.

state equations by defining the following state variables:

$$x_1(t) = y_1(t) \tag{4-76}$$

$$x_2(t) = \frac{dy_1(t)}{dt} = \frac{dx_1(t)}{dt} \tag{4-77}$$

$$x_3(t) = y_2(t) \tag{4-78}$$

$$x_4(t) = \frac{dy_2(t)}{dt} = \frac{dx_3(t)}{dt} \tag{4-79}$$

Equations (4-77) and (4-79) form two state equations naturally; the other two are obtained by substituting Eqs. (4-76) through (4-79) into Eqs. (4-74) and (4-75). Rearranging, we have

$$\frac{dx_1(t)}{dt} = x_2(t) \tag{4-80}$$

$$\frac{dx_2(t)}{dt} = -\frac{K_1}{M_1}[x_1(t) - x_3(t)] - \frac{B_1}{M_1}[x_2(t) - x_4(t)] + \frac{1}{M_1}f(t) + g \tag{4-81}$$

$$\frac{dx_3(t)}{dt} = x_4(t) \tag{4-82}$$

$$\frac{dx_4(t)}{dt} = \frac{K_1}{M_2}x_1(t) + \frac{B_1}{M_2}x_2(t) - \frac{K_1 + K_2}{M_2}x_3(t) - \frac{1}{M_2}(B_1 + B_2)x_4(t) + g \tag{4-83}$$

If we are interested in the displacements $y_1(t)$ and $y_2(t)$, the output equations are

$$y_1(t) = x_1(t) \tag{4-84}$$

$$y_2(t) = x_3(t) \tag{4-85}$$

The state diagram of the system, according to the equations written above, is as shown in Fig. 4-22. Alternately, the state diagram can be drawn by first rearranging Eqs. (4-74) and (4-75). The output transforms $Y_1(s)$ and $Y_2(s)$ are written in terms of the input $F(s)$ and g by applying Mason's gain formula to the state diagram:

$$Y_1(s) = \frac{M_2 s^2 + (B_1 + B_2)s + (K_1 + K_2)}{\Delta} F(s)$$

$$+ \frac{M_1 M_2 s^2 + (M_1 B_1 + M_1 B_2 + M_2 B_1)s + M_1 K_1 + M_1 K_2 + M_2 K_1}{\Delta} \frac{g}{s} \tag{4-86}$$

$$Y_2(s) = \frac{B_1 s + K_1}{\Delta} F(s) + \frac{M_1 M_2 s^2 + B_1(M_1 + M_2)s + K_1(M_1 + M_2)}{\Delta} \frac{g}{s} \tag{4-87}$$

where

$$\Delta = M_1 M_2 s^4 + [M_1(B_1 + B_2) + B_1 M_2]s^3 + [M_1(K_1 + K_2) + K_1 M_2 + B_1 B_2]s^2$$

$$+ (K_1 B_2 + B_1 K_2)s + K_1 K_2 \tag{4-88}$$

The transfer functions $Y_1(s)/F(s)$ and $Y_2(s)/F(s)$ can be obtained from Eqs. (4-86) and (4-87), respectively, by setting the gravitational force g to zero. This is equivalent to assuming that the system is initially at the equilibrium state and that the gravitational force is balanced by the preloading of the springs.

The state equations can also be written directly from the mechanical system by assigning the state variables as $v_1(t) = dy_1(t)/dt$ and $v_2(t) = dy_2(t)/dt$, and the forces on the two springs

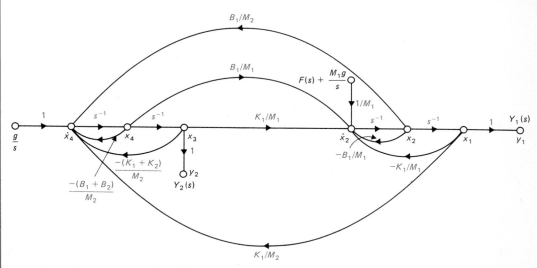

FIGURE 4-22 State diagram for the mechanical system of Fig. 4-21.

as $f_{k1}(t)$ and $f_{k2}(t)$. Then, if we write the forces acting on the masses and the velocities of the springs as functions of the four state variables and the external forces, the state equations are

$$\text{Force on } M_1: \quad M_1\frac{dv_1(t)}{dt} = -B_1v_1(t) + B_1v_2(t) - f_{k1}(t) + f(t) + M_1g \quad (4\text{-}89)$$

$$\text{Force on } M_2: \quad M_2\frac{dv_2(t)}{dt} = B_1 v_1(t) - (B_1 + B_2)v_2(t)$$
$$+ f_{k1}(t) - f_{k2}(t) + M_2g \quad (4\text{-}90)$$

$$\text{Velocity of } K_1: \quad \frac{df_{k1}(t)}{dt} = K_1[v_1(t) - v_2(t)] \quad (4\text{-}91)$$

$$\text{Velocity of } K_2: \quad \frac{df_{k2}(t)}{dt} = K_2v_2(t) \quad (4\text{-}92)$$

EXAMPLE 4-8
*rotational
system*

The rotational system shown in Fig. 4-23(a) consists of a disk mounted on a shaft that is fixed at one end. The moment of inertia of the disk about the axis of rotation is J. The edge of the disk is riding on a surface, and the viscous friction coefficient between the two surfaces is B. The inertia of the shaft is negligible, but the torsional spring constant is K.

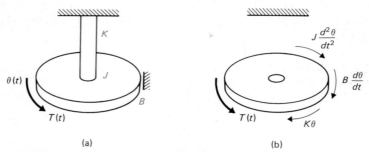

(a)   (b)

FIGURE 4-23 Rotational system for Example 4-8.

Assume that a torque is applied to the disk, as shown; then the torque or moment equation about the axis of the shaft is written from the free-body diagram of Fig. 4-23(b):

$$T(t) = J\frac{d^2\theta(t)}{dt^2} + B\frac{d\theta(t)}{dt} + K\theta(t) \quad (4\text{-}93)$$

Notice that this system is analogous to the translational system of Fig. 4-18. The state equations may be written by defining the state variables as $x_1(t) = \theta(t)$ and $x_2(t) = dx_1(t)/dt$.

EXAMPLE 4-9
motor system

Figure 4-24(a) shows the diagram of a motor coupled to an inertial load through a shaft with a spring constant K. A nonrigid coupling between two mechanical components in a control system often causes torsional resonances that can be transmitted to all parts of the system.

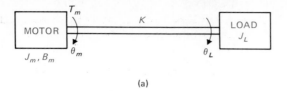

(a)

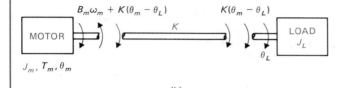

(b)

FIGURE 4-24 (a) Motor–load system. (b) Free-body diagram

The system variables and parameters are defined as follows:

$T_m(t)$ = motor torque J_m = motor inertia
B_m = motor viscous friction coefficient J_L = load inertia
K = spring constant of the shaft $\theta_m(t)$ = motor displacement
$\theta_L(t)$ = load displacement $\omega_m(t)$ = motor velocity
$\omega_L(t)$ = load velocity

The free-body diagrams of the system are shown in Fig. 4.24(b). The torque equations of the system are

$$T_m(t) = J_m \frac{d^2 \theta_m(t)}{dt^2} + B_m \frac{d\theta_m(t)}{dt} + K[\theta_m(t) - \theta_L(t)] \qquad (4\text{-}94)$$

$$K[\theta_m(t) - \theta_L(t)] = J_L \frac{d^2 \theta_L(t)}{dt^2} \qquad (4\text{-}95)$$

In this case, the system contains three energy-storage elements in J_m, J_L, and K. Thus, there should be three state variables. Care should be taken in constructing the state diagram and assigning the state variables so that a minimum number of the latter are incorporated. Equations (4-94) and (4-95) are rearranged as

$$\frac{d^2 \theta_m(t)}{dt^2} = -\frac{B_m}{J_m} \frac{d\theta_m(t)}{dt} - \frac{K}{J_m}[\theta_m(t) - \theta_L(t)] + \frac{1}{J_m} T_m(t) \qquad (4\text{-}96)$$

$$\frac{d^2 \theta_L(t)}{dt^2} = \frac{K}{J_L}[\theta_m(t) - \theta_L(t)] \qquad (4\text{-}97)$$

The state diagram with three integrators is shown in Fig. 4-25. The clue given by Eqs. (4-96) and (4-97) is that $\theta_m(t)$ and $\theta_L(t)$ appear only as the difference, $\theta_m(t) - \theta_L(t)$, in these equations. From the state diagram in Fig. 4-25, the state variables are defined as

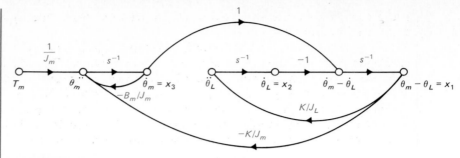

FIGURE 4-25 State diagram of the system of Fig. 4-24.

$x_1(t) = \theta_m(t) - \theta_L(t)$, $x_2(t) = d\theta_L(t)/dt$, and $x_3(t) = d\theta_m(t)/dt$. The state equations are

$$\frac{dx_1(t)}{dt} = x_3(t) - x_2(t) \tag{4-98}$$

$$\frac{dx_2(t)}{dt} = \frac{K}{J_L}x_1(t) \tag{4-99}$$

$$\frac{dx_3(t)}{dt} = -\frac{K}{J_m}x_1(t) - \frac{B_m}{J_m}x_3(t) + \frac{1}{J_m}T_m(t) \tag{4-100}$$

The transfer functions between $\Theta_m(s)$ and $T_m(s)$ and $\Theta_L(s)$ and $T_m(s)$ are written by applying the gain formula to the state diagram in Fig. 4-25:

$$\frac{\Theta_m(s)}{T_m(s)} = \frac{X_3(s)}{sT_m(s)} = \frac{J_Ls^2 + K}{s[J_mJ_Ls^3 + B_mJ_Ls^2 + K(J_m + J_L)s + B_mK]} \tag{4-101}$$

$$\frac{\Theta_L(s)}{T_m(s)} = \frac{X_2(s)}{sT_m(s)} = \frac{K}{s[J_mJ_Ls^3 + B_mJ_Ls^2 + K(J_m + J_L)s + B_mK]} \tag{4-102}$$

4-5

SENSORS AND ENCODERS IN CONTROL SYSTEMS

Sensors and encoders are important components used to monitor the performance and for feedback in control systems. In this section, the principle of operation and applications of some of the sensors and encoders that are commonly used in control systems are described.

Potentiometer

A potentiometer is an electromechanical transducer that converts mechanical energy into electrical energy. The input to the device is in the form of a mechanical displacement, either linear or rotational. When a voltage is applied across the fixed terminals of the potentiometer, the output voltage, which is measured across the variable terminal and ground, is proportional to the input displacement, either linearly or according to some nonlinear relation.

Rotary potentiometers are available commercially in single-revolution or multi-revolution form, with limited or unlimited rotational motion. The potentiometers commonly are made with wirewound or conductive plastic resistance material. Figure 4-26 shows a cutaway view of a rotary potentiometer, and Fig. 4-27 shows a linear potentiometer that also contains a built-in operational amplifier. For precision control, the conductive plastic potentiometer is more preferable, since it has infinite resolution, long rotational life, good output smoothness, and low static noise.

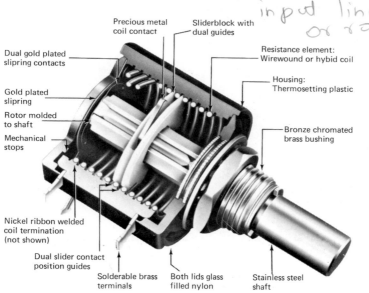

input linear or rotational

Precious metal coil contact
Sliderblock with dual guides
Dual gold plated slipring contacts
Resistance element: Wirewound or hybid coil
Gold plated slipring
Housing: Thermosetting plastic
Rotor molded to shaft
Mechanical stops
Bronze chromated brass bushing
Nickel ribbon welded coil termination (not shown)
Dual slider contact position guides
Solderable brass terminals
Both lids glass filled nylon
Stainless steel shaft

FIGURE 4-26 Ten-turn rotary potentiometer. (Courtesy of Helipot Division of Beckman Instruments, Inc.)

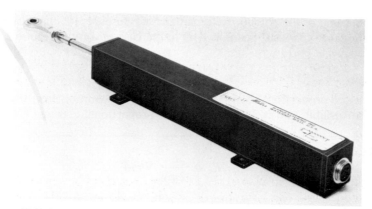

FIGURE 4-27 Linear motion potentiometer with built-in operational amplifier. (Courtesy of Waters Manufacturing, Inc.)

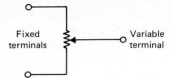

FIGURE 4-28 Circuit representation of a potentiometer.

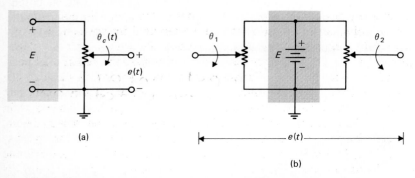

(a)

(b)

FIGURE 4-29 (a) Potentiometer used as a position indicator. (b) Two potentiometers used to sense the positions of two shafts.

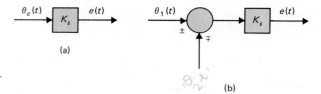

(a)

(b)

FIGURE 4-30 Block diagram representations of potentiometer arrangements in Fig. 4-29.

Figure 4-28 shows the equivalent circuit representation of a potentiometer, linear or rotary. Since the voltage across the variable terminal and reference is proportional to the shaft displacement of the potentiometer, when a voltage is applied across the fixed terminals, the device can be used to indicate the absolute position of a system or the relative position of two mechanical outputs. Figure 4-29(a) shows the arrangement when the housing of the potentiometer is fixed at reference; the output voltage $e(t)$ will be proportional to the shaft position $\theta_c(t)$ in the case of a rotary motion. Then

$$e(t) = K_s\,\theta_c(t) \tag{4-103}$$

where K_s is the proportional constant. For an N-turn potentiometer, the total displacement of the variable arm is $2\pi N$ radians. The proportional constant K_s is given by

$$K_s = \frac{E}{2\pi N}\ \text{V/rad} \tag{4-104}$$

where E is the magnitude of the reference voltage applied to the fixed terminals. A more flexible arrangement is obtained by using two potentiometers connected in parallel, as shown in Fig. 4-29(b). This arrangement allows the comparison of two remotely located shaft positions. The output voltage is taken across the variable terminals of the two potentiometers and is given by

$$e(t) = K_s[\theta_1(t) - \theta_2(t)] \tag{4-105}$$

Figure 4-30 illustrates the block diagram representations of the setups in Fig. 4-29.

In dc-motor control systems, potentiometers are often used for position feedback. Figure 4-31(a) shows the schematic diagram of a typical dc-motor position-control system. The potentiometers are used in the feedback path to compare the actual load position with the desired reference position. If there is a discrepancy between the load position and the reference input, an error signal is generated by the potentiometers that will drive the motor in such a way that this error is minimized quickly. As shown in Fig. 4-31(a), the error signal is amplified by a dc amplifier whose output drives the armature of a permanent-magnet dc motor. Typical waveforms of the signals in the system when the input $\theta_r(t)$ is a step function are shown in Fig. 4-31(b). Note that the electric signals are all unmodulated. *In control-systems terminology, a dc signal usually*

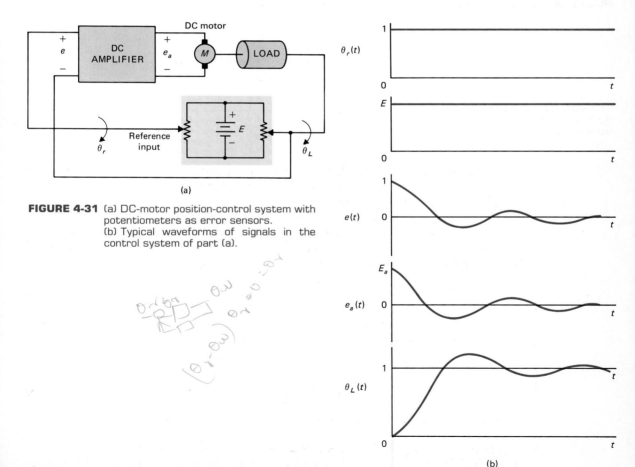

FIGURE 4-31 (a) DC-motor position-control system with potentiometers as error sensors.
(b) Typical waveforms of signals in the control system of part (a).

refers to an unmodulated signal. On the other hand, an ac signal refers to signals that are modulated by a modulation process. These definitions are different from those commonly used in electrical engineering, where dc simply refers to unidirectional signals and ac indicates alternating signals.

Figure 4-32(a) illustrates a control system that serves essentially the same purpose as that of the system in Fig. 4-31(a), except that ac signals prevail. In this case, the voltage applied to the error detector is sinusoidal. The frequency of this signal is usually much higher than that of the signal that is being transmitted through the system. Typical signals of the ac control system are shown in Fig. 4-32(b). The signal $v(t)$ is referred to as the carrier whose frequency is ω_c, or

$$v(t) = E \sin \omega_c t \tag{4-106}$$

Analytically, the output of the error signal is given by

$$e(t) = K_s \theta_e(t) v(t) \tag{4-107}$$

where $\theta_e(t)$ is the difference between the input displacement and the load displacement, or

$$\theta_e(t) = \theta_r(t) - \theta_L(t) \tag{4-108}$$

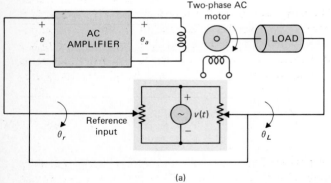

(a)

FIGURE 4-32 (a) AC control system with potentiometers as error detectors. (b) Typical waveforms of signals in the control system of part (a).

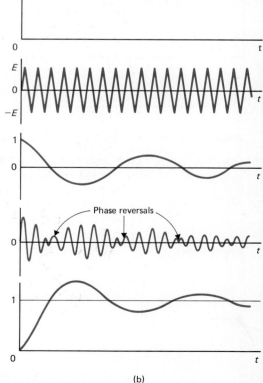

(b)

*suppressed
carrier-
modulated
signal*

For the $\theta_e(t)$ shown in Fig. 4-32(b), $e(t)$ becomes a **suppressed-carrier-modulated** signal. A reversal in phase of $e(t)$ occurs whenever the signal crosses the zero-magnitude axis. This reversal in phase causes the ac motor to reverse in direction according to the desired sense of correction of the error $\theta_e(t)$. *The name "suppressed-carrier modulation" stems from the fact that when a signal $\theta_e(t)$ is modulated by a carrier signal $v(t)$ according to Eq. (4-107), the resultant signal $e(t)$ no longer contains the original carrier frequency ω_c.* To illustrate this, let us assume that $\theta_e(t)$ is also a sinusoid given by

$$\theta_e(t) = \sin \omega_s t \qquad (4\text{-}109)$$

where, normally, $\omega_s \ll \omega_c$. By use of familiar trigonometric relations, substituting Eqs. (4-106) and (4-109) into Eq. (4-107), we get

$$e(t) = \tfrac{1}{2} K_s E[\cos(\omega_c - \omega_s)t - \cos(\omega_c + \omega_s)t] \qquad (4\text{-}110)$$

Therefore, $e(t)$ no longer contains the carrier frequency ω_c or the signal frequency ω_s, but has only the two sidebands $\omega_c + \omega_s$ and $\omega_c - \omega_s$.

demodulator

When the modulated signal is transmitted through the system, the motor acts as a demodulator, so that the displacement of the load will be of the same form as the dc signal before modulation. This is clearly seen from the waveforms of Fig. 4-32(b). It should be pointed out that a control system need not contain all-dc or all-ac components. It is quite common to couple a dc component to an ac component through a modulator or an ac device to a dc device through a demodulator. For instance, the dc amplifier of the system in Fig. 4-31(a) may be replaced by an ac amplifier that is preceded by a modulator and followed by a demodulator.

Synchros

Synchros are used widely in control systems as detectors and encoders because of their ruggedness in construction and high reliability. Basically, a synchro is a rotary device that operates on the same principle as a transformer and produces a correlation between an angular position and a voltage or set of voltages. Therefore, synchros are ac devices. There are many types of synchros and the applications vary. In this section, only the **synchro transmitter** and the **synchro control transformer** are discussed.

synchros

SYNCHRO TRANSMITTER.   A synchro transmitter has a Y-connected stator winding that resembles the stator of a three-phase induction motor. The rotor is a salient-pole dumbbell-shaped magnet with a single winding. Schematic diagrams of synchro transmitters are shown in Fig. 4-33. A single-phase ac voltage is applied to the rotor through two slip rings. Let the ac voltage applied to the rotor of a synchro transmitter be

$$e_R(t) = E_R \sin \omega_c t \qquad (4\text{-}111)$$

where E_R is the magnitude of the voltage. We can show that the magnitudes of the sta-

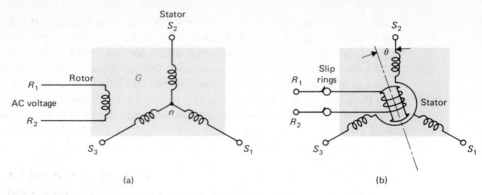

FIGURE 4-33 Schematic diagrams of synchro transmitters.

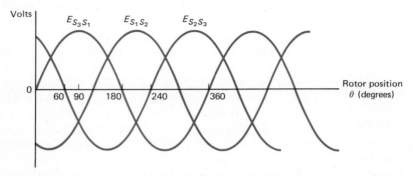

FIGURE 4-34 Variation of the terminal voltages of a synchro transmitter as a function of the rotor position. θ is measured counterclockwise from the electric zero.

tor terminal voltages with respect to a position θ relative to the electric zero reference are

$$E_{S_1 S_2} = \sqrt{3}KE_R \sin (\theta + 240°) \tag{4-112}$$

$$E_{S_2 S_3} = \sqrt{3}KE_R \sin (\theta + 120°) \tag{4-113}$$

$$E_{S_3 S_1} = \sqrt{3}KE_R \sin \theta \tag{4-114}$$

A plot of these terminal voltages as a function of the rotor-shaft position is shown in Fig. 4-34. Notice that each rotor position corresponds to one unique set of stator voltages. This leads to the use of the synchro transmitter to identify angular positions by measuring and identifying the set of voltages at the three stator terminals.

SYNCHRO CONTROL TRANSFORMER.   Since the function of an error detector is to convert the difference of two shaft positions into an electrical signal, a single synchro transmitter is apparently inadequate. A typical arrangement of a synchro error detector involves the use of two synchros: a transmitter and a control transformer, as shown in Fig. 4-35. Basically, the principle of operation of a synchro control transformer is identical to that of the synchro transmitter, except that the rotor is cylindrically shaped so that the air-gap flux is uniformly distributed around the rotor. This feature is essential for a control transformer, since its rotor terminals are usually connected to an amplifier, or similar electrical device, in order that the latter sees a constant impedance. Referring to the arrangement shown in Fig. 4-35, the voltages given in Eqs. (4-112), (4-113), and (4-114) are impressed across the corresponding stator terminals of the control transformer. When the rotor positions of the two synchros are in perfect alignment, the voltage generated across the terminals of the rotor windings is zero. When the two rotor shafts are not in alignment, the rotor voltage of the control transformer is approximately a sine function of the difference between the two shaft angles, as shown in Fig. 4-36. From this figure, it is apparent that the synchro error detector is a nonlinear device. However, for small angular deviations of up to 15 degrees in the vicinity of the two null positions, the rotor voltage of the control transformer is approximately proportional to the position difference. Therefore, for

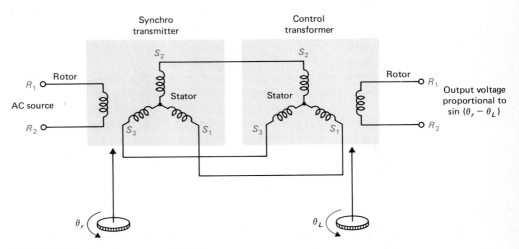

FIGURE 4-35 Synchro error detector.

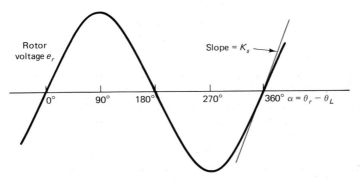

FIGURE 4-36 Rotor voltage of control transformer as a function of the difference of rotor positions.

small angular deviations, the transfer function of the synchro error detector can be approximated by a constant K_s:

$$K_s = \frac{E}{\theta_r - \theta_L} = \frac{E}{\theta_e}$$ (4-115)

where

> E = error voltage, volts
>
> θ_r = shaft position of synchro transmittor, radians
>
> θ_L = shaft position of synchro control transformer, radians
>
> θ_e = error in shaft positions, radians
>
> K_s = sensitivity of the error detector, volts/rad

The schematic diagram of a positional-control system employing a synchro error detector is shown in Fig. 4-37. The purpose of the control system is to make the controlled shaft follow the angular displacement of the reference input shaft as closely as possible. The rotor of the control transformer is connected to the controlled shaft, and the rotor of the synchro transmitter is connected to the reference input shaft. When the controlled shaft is aligned with the reference shaft, the error voltage is zero and the motor does not turn. When an angular misalignment exists, an error voltage of relative polarity appears at the amplifier input, and the output of the amplifier will drive the motor in such a direction as to reduce the error. For small deviations between the controlled and the reference shafts, the synchro error detector can be represented by the constant K_s, given by Eq. (4-115). From the characteristic of the error detector shown in Fig. 4-36, it is clear that K_s has opposite signs at the two null positions. However, in closed-loop systems, only one of the two null positions is a true null; the other one corresponds to an unstable operating point.

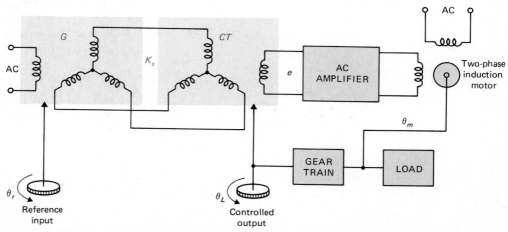

FIGURE 4-37 AC control system employing synchro error detector.

It should be pointed out that the error signal at the rotor terminals of the synchro control transformer that corresponds to the input signal in Eq. (4-111) should be expressed as

$$e(t) = K_s \theta_e(t) \sin \omega_c t \qquad (4\text{-}116)$$

which is a suppressed-carrier-modulated signal. Since ω_c is much higher than the frequency components contained in $\theta_e(t)$, $e(t)$ is directly proportional to $\theta_e(t)$ through the error constant K_s.

In the preceding sections, although we have illustrated the potentiometers as being used in a dc (unmodulated) control system, and the synchros in an ac (modulated) control system with an ac motor, in general, the applications of these sensing devices can be quite broad and versatile. For example, the output signal of a synchro control transformer can be demodulated and the dc signal used to drive a dc motor for control purposes. In modern control systems, the outputs of these sensing devices can all be conditioned by interfacing devices and fed into microcomputers for digital control.

Tachometers

tachometer

Tachometers are electromechanical devices that convert mechanical energy into electrical energy. The device works essentially as a generator, with the output voltage proportional to the magnitude of the angular velocity.

In control systems, most of the tachometers used are of the dc variety, i.e., the output voltage is a dc signal. DC tachometers are used in control systems in many ways; they can be used as velocity indicators to provide shaft-speed readout or to provide velocity feedback for speed control or stabilization. Figure 4-38 is a block diagram of a typical velocity-control system in which the tachometer output is compared with the reference voltage, which represents the desired speed. The difference between the two signals, or the error, is amplified and used to drive the motor, so that the speed will eventually reach the desired value. In this type of application, the accuracy of the tachometer is highly critical, as the accuracy of the speed control depends on it.

In a position-control system, velocity feedback is often used to improve the stability or the damping of the closed-loop system. Figure 4-39 shows the block diagram

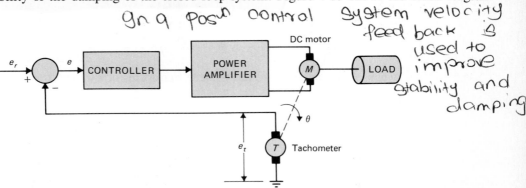

FIGURE 4-38 Velocity-control system with tachometer feedback.

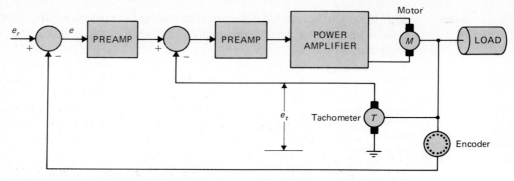

FIGURE 4-39 Position-control system with tachometer feedback.

of such an application. In this case, the tachometer feedback forms an inner loop to improve the damping characteristics of the system, and the accuracy of the tachometer is not so critical.

The third and most traditional use of dc tachometers is in providing the visual speed readout of a rotating shaft. Tachometers used in this capacity are generally connected directly to a voltmeter calibrated in revolutions per minute (rpm).

MATHEMATICAL MODELING OF TACHOMETERS.   The dynamics of the tachometer can be represented by the equation

$$e_t(t) = K_t \frac{d\theta(t)}{dt} = K_t \omega(t) \tag{4-117}$$

where $e_t(t)$ is the output voltage, $\theta(t)$ the rotor displacement in radians, $\omega(t)$ the rotor velocity in rad/s, and K_t the tachometer constant in V/rad/s. *The value of K_t is usually given as a catalog parameter in volts per 1000 rpm (V/krpm).*

The transfer function of a tachometer is obtained by taking the Laplace transform on both sides of Eq. (4-117). We have

$$\frac{E_t(s)}{\Theta(s)} = K_t s \tag{4-118}$$

where $E_t(s)$ and $\Theta(s)$ are the Laplace transforms of $e_t(t)$ and $\theta(t)$, respectively.

Incremental Encoder

absolute encoders

Incremental encoders are frequently found in modern control systems for converting linear or rotary displacement into digitally coded or pulse signals. The encoders that output a digital signal are known as **absolute encoders**. In the simplest terms, absolute encoders provide as output a distinct digital code indicative of each particular least

incremental encoders

significant increment of resolution. **Incremental encoders**, on the other hand, provide a pulse for each increment of resolution, but do not make distinctions between the increments. In practice, the choice of which type of encoder to use depends on economics and control objectives. For the most part, the need for absolute encoders has much to do with the concern for data loss during power failure or the applications involving periods of mechanical motion without the readout under power. However, the incremental encoder's simplicity in construction, low cost, ease of application, and versatility have made it by far one of the most popular encoders in control systems.

Incremental encoders are available in rotary and linear forms. Figures 4-40 and 4-41 show typical rotary and linear incremental encoders, respectively.

A typical incremental encoder has four basic parts: a light source, a rotary disk, a stationary mask, and a sensor, as shown in Fig. 4-42. The disk has alternate opaque

FIGURE 4-40 Rotary incremental encoder. (Courtesy DISC Instruments, Inc.)

FIGURE 4-41 Linear incremental encoder. (Courtesy DISC Instruments, Inc.)

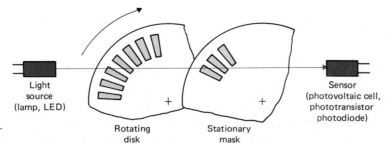

FIGURE 4-42 Typical increment optomechanics.

Light source (lamp, LED)

Rotating disk

Stationary mask

Sensor (photovoltaic cell, phototransistor photodiode)

and transparent sectors. Any pair of these sectors represents an incremental period. The mask is used to pass or block the light beam between the light source and the photosensor located behind the mask. For encoders with relatively low resolution, the mask is not necessary. For fine-resolution encoders (up to thousands of increments per revolution), a multiple-slit mask is often used to maximize reception of the shutter light.

The waveforms of the sensor outputs are generally triangular or sinusoidal, depending on the resolution required. Square-wave signals compatible with digital logic are derived by using a linear amplifier followed by a comparator. Figure 4-43(a) shows a typical rectangular output waveform of a single-channel incremental encoder. In this case pulses are produced for both directions of shaft rotation. A dual-channel encoder with two sets of output pulses is necessary for direction sensing and other control functions. When the phase of the two-output pulse train is 90 degrees apart electrically, the two signals are said to be in quadrature, as shown in Fig. 4-43(b); the signals uniquely define 0-to-1 and 1-to-0 logic transitions with respect to the direction of rotation of the encoder disk, so that a direction-sending logic circuit can be constructed to decode the signals. Figure 4-44 shows the single-channel output and the quadrature outputs with

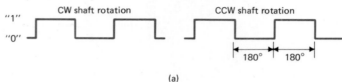

(a)

FIGURE 4-43 (a) Typical rectangular output waveform of a single-channel encoder device (bidirectional). (b) Typical dual-channel encoder signals in quadrature (bidirectional).

(b)

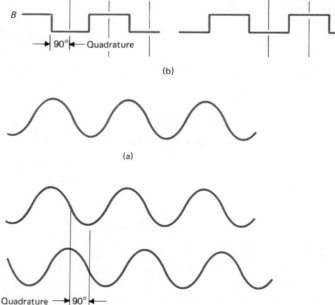

(a)

FIGURE 4-44 (a) Typical sinusoidal output waveform of a single-channel encoder device. (b) Typical dual-channel encoder signals in quadrature.

(b)

sinusoidal waveforms. Just as in the case of the synchros, the sinusoidal signals from the incremental encoder can be used for fine-position control in feedback control systems. The following example illustrates some applications of the incremental encoder in control systems.

EXAMPLE 4-10   Consider an incremental encoder that generates two sinusoidal signals in quadrature as the encoder disk rotates. The output signals of the two channels are shown in Fig. 4-45 over one cycle. Note that the two encoder signals generate four zero crossings per cycle. These zero crossings can be used for position indication, position control, or speed measurements in control systems. Let us assume that the encoder shaft is coupled directly to the rotor shaft of a motor that directly drives the printwheel of an electronic typewriter or word processor. The printwheel has 96 character positions on its periphery, and the encoder has 480 cycles. Thus, there are $480 \times 4 = 1920$ zero crossings per revolution. For the 96-character printwheel, this corresponds to $1920/96 = 20$ zero crossings per character; that is, there are 20 zero crossings between two adjacent characters.

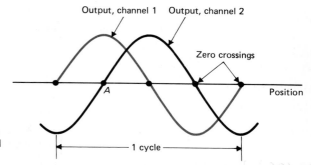

FIGURE 4-45 One cycle of the output signals of a dual-channel incremental encoder.

One way of measuring the velocity of the printwheel is to count the number of pulses generated by an electronic clock that occur between consecutive zero crossings of the encoder outputs. Let us assume that a 500-kHz clock is used, i.e., the clock generates 500,000 pulses/sec. If the counter counts, say, 500 clock pulses while the encoder rotates from one zero crossing to the next, the shaft speed is

$$\frac{500{,}000 \text{ pulses/s}}{500 \text{ pulses/zero crossing}} = 1000 \text{ zero crossings/s}$$

$$= \frac{1000 \text{ zero crossings/s}}{1920 \text{ zero crossings/s}} = 0.52083 \text{ rev/s} \qquad (4\text{-}119)$$

$$= 31.25 \text{ rpm}$$

The encoder arrangement described can be used for fine-position control of the printwheel. Let the zero crossing A of the waveforms in Fig. 4-45 correspond to a character position on the printwheel (the next character position is 20 zero crossings away), and the point corresponds to a stable equilibrium point. The coarse-position control of the system must first drive the printwheel position to within one zero crossing on either side of position A; then by using the slope of the sine wave at position A, the control system should null the error quickly.

4-6

DC MOTORS IN CONTROL SYSTEMS

dc motors

Direct-current motors are one of the most widely used prime movers in industry today. Years ago a majority of the small servomotors used for control purposes were of the ac variety. As we shall see, ac motors are more difficult to control, especially for position control, and their characteristics are quite nonlinear, which makes the analytical task more difficult. DC motors, on the other hand, are more expensive, because of the brushes and commutators, and variable-flux dc motors are suitable only for certain types of control applications. Before permanent-magnet technology was fully developed, the torque per unit volume or weight of a dc motor with a permanent-magnet (PM) field was far from desirable. Today, with the development of the rare-earth magnet, it is possible to achieve very high torque-to-volume PM dc motors at reasonable costs. Furthermore, the advances made in brush-and-commutator technology have made these wearable parts practically maintenance-free. The advancements made in power electronics have made brushless dc motors quite popular in high-performance control systems. Advanced manufacturing techniques have also produced dc motors with ironless rotors and rotors with very low inertia, thus achieving very high torque-to-inertia ratios, and low-time-constant properties have opened new applications for dc motors in computer peripheral equipment such as tape drives, printers, disk drives, and word processors, as well as in the automation and machine-tool industries.

Basic Operational Principles of DC Motors

The dc motor is basically a torque transducer that converts electric energy into mechanical energy. The torque developed on the motor shaft is directly proportional to the field flux and the armature current. As shown in Fig. 4-46, a current-carrying conductor is established in a magnetic field with flux ϕ, and the conductor is located at a distance r from the center of rotation. The relationship among the developed torque, flux ϕ, and current i_a is

$$T_m = K_m \phi i_a \qquad (4\text{-}120)$$

where T_m is the motor torque (N-m), ϕ the magnetic flux (webers), i_a the armature current (amperes), and K_m is a proportional constant.

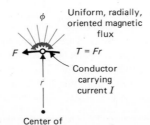

Uniform, radially, oriented magnetic flux

ϕ

F $T = Fr$

Conductor carrying current I

r

Center of rotation

FIGURE 4-46 Torque production in a dc motor.

back emf

In addition to the torque developed by the arrangement shown in Fig. 4-46, when the conductor moves in the magnetic field, a voltage is generated across its terminals. This voltage, the **back emf**, which is proportional to the shaft velocity, tends to oppose the current flow. The relationship between the back emf and the shaft velocity is

$$e_b = K_m \phi \omega_m \tag{4-121}$$

where e_b denotes the back emf (volts), and ω_m is the shaft velocity (rad/s) of the motor. Equations (4-120) and (4-121) form the basis of dc-motor operation.

Basic Classifications of DC Motors

[handwritten: based on the way the magnetic field is produced, and on the way the armature is constructed]

classifications

DC motors can be classified into several broad categories based on the way the magnetic field is produced, and on the basic design and construction of the armature. In terms of the magnetic field, dc motors can be classified as **variable-magnetic-flux motors** and **constant-magnetic-flux motors**. In variable-magnetic-flux motors, the magnetic field is produced by field windings that are connected to external sources. These motors are also divided into two subclasses: the **series-field motor**, in which the field winding is connected in series with the armature, as shown in Fig. 4-47(a), and the **separately excited-field motor**, in which the field winding is separate from the armature, as shown in Fig. 4-47(b).

For the series-field type, since the magnetic flux in the motor is proportional to the field current, which varies, the relationship between the torque and speed is generally nonlinear. Thus, series-field-type dc motors are useful only for specific applications where high torque at low speeds are called for. The motor torque usually drops off rapidly as the motor speed increases.

For the separately excited dc motor, since the magnetic flux is independent of the armature current, it can be controlled externally over a wide range.

The constant-magnetic-flux dc motor is also known as the permanent-magnet (PM) dc motor. In this case, the magnetic field is produced by a permanent magnet and is constant. This allows the torque–speed characteristics of the motor to be relatively linear.

PM dc motors can be further classified according to commutation scheme and armature design. Conventional dc motors have mechanical brushes and commutators, but there are dc motors in which the commutation is done electronically; this type of motor is called the **brushless dc motor**.

brushless dc motor

According to the armature construction, the PM dc motor can be broken down into three types of armature design: **iron core**, **surface wound**, and **moving-coil** motors.

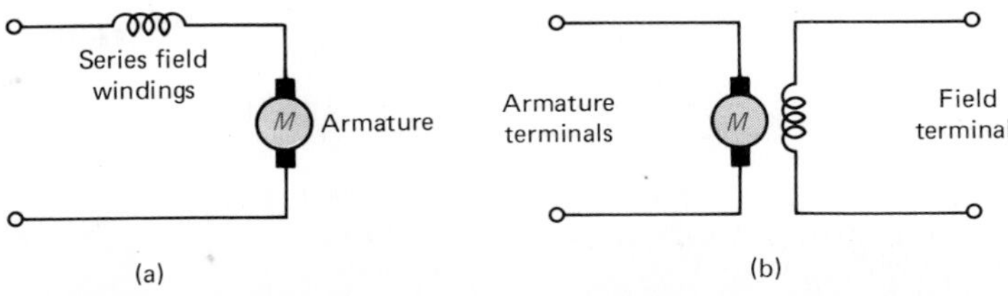

[handwritten left margin: $T_m = \phi \omega_m$; $e_b = T_m \omega_m / i_a$]

[handwritten right: iron core, surface wound, moving coil]

FIGURE 4-47 (a) Series dc-motor armature and field connection. (b) DC motor with separately variable field.

(a) Series field windings — M Armature

(b) Armature terminals — M — Field terminals

IRON-CORE PM DC MOTORS.   The rotor and stator configuration of an iron-core PM dc motor is shown in Fig. 4-48. The permanent-magnet material can be barium-ferrite, Alnico, or a rare-earth compound. The magnetic flux produced by the permanent magnet passes through a laminated rotor structure that contains slots. The armature conductors are placed in the rotor slots. This type of dc motor is characterized by relatively high rotor inertia, high inductance, low cost, and high reliability.

SURFACE-WOUND DC MOTORS.   Figure 4-49 shows the rotor construction of a surface-wound PM dc motor. The armature conductors are bonded to the surface of a cylindrical rotor structure, which is made of laminated disks fastened to the motor shaft. Since no slots are used on the rotor in this design, the armature has no "cogging" effect. Since the conductors are laid out in the air gap between the rotor disks and the permanent-magnet field, this type of motor has lower inductance than that of the iron-core structure. However, since the air gap between the magnets and the low-inductance rotor is larger than the iron-core motor, a large magnet is required in order to provide a magnetic flux equivalent to that of the iron-core motor. Therefore, surface-wound dc motors are more expensive to produce and have larger outside diameters than equivalent iron-core motors.

MOVING-COIL DC MOTORS.   Moving-coil motors are designed to have very low moments of inertia and very low armature inductance. This is achieved by placing the armature conductors in the air gap between a stationary flux return path and the permanent-magnet structure, as shown in Fig. 4-50. In this case, the conductor structure is supported by nonmagnetic material—usually epoxy resins and fiber glass—to form a hollow cylinder. One end of the cylinder forms a hub, which is attached to the motor shaft. A cross-sectional view of such a motor is shown in Fig. 4-51. Since all unnecessary elements have been removed from the armature of the moving-coil motor,

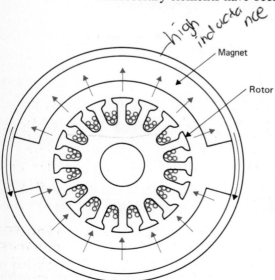

FIGURE 4-48 Cross-sectional view of a permanent-magnet iron-core dc motor.

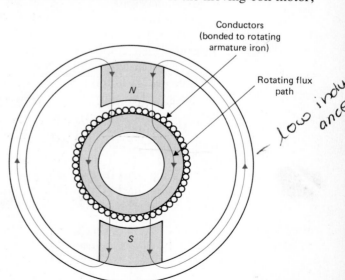

FIGURE 4-49 Cross-sectional view of a surface-wound permanent-magnet dc motor.

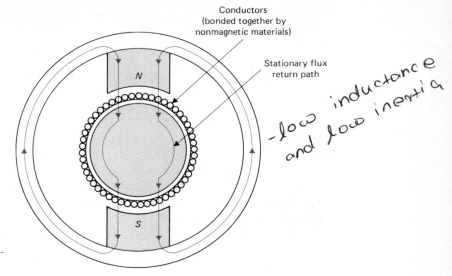

- low inductance and low inertia

FIGURE 4-50 Cross-sectional view of a moving- coil permanent-magnet dc motor.

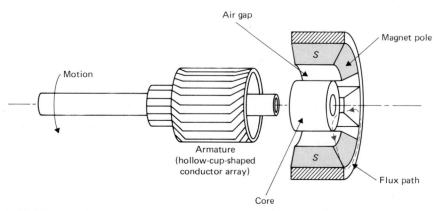

FIGURE 4-51 Cross-sectional side view of a moving-coil dc motor.

its moment of inertia is very low. However, it has a larger air gap than the two motors discussed earlier, and therefore requires an even larger magnetic structure than do the other two types of motors to produce an equivalent air-gap flux. Since the conductors in the moving-coil armature are not in direct contact with iron, the motor inductance is very low; values of less than 100 μH are common in this type of motor. The low-inertia and low-inductance properties make the moving-coil motor the best actuator choice for high-performance control systems.

brushless dc motor

BRUSHLESS DC MOTORS.   Brushless dc motors differ from the previously mentioned dc motors in that they employ electrical (rather than mechanical) commutation of the armature current. The configuration of the brushless dc motor most commonly used—especially for incremental-motion applications—is one in which the rotor consists of magnets and "back iron" support, and whose commutated windings are located external to the rotating parts, as shown in Fig. 4-52. Compared to the conventional dc motors, such as the one shown in Fig. 4-48, it is an "inside-out" configuration. De-

increment al

motion applications

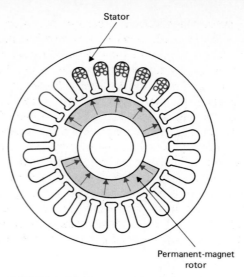

FIGURE 4-52 Cross-sectional view of a permanent-magnet brushless dc motor.

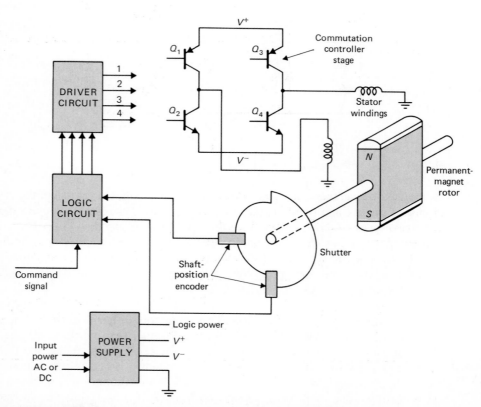

FIGURE 4-53 The essential components of a brushless dc motor.

pending on the specific application, brushless dc motors can be used when a low moment of inertia is called for. In order to achieve a low moment of inertia in this type of motor, the rotor structure must be manufactured from a magnetic material that offers a high ratio of magnetic flux and a high coercive force in relation to its mass, such as used for rare-earth magnets. The brushless dc-motor winding commutation requires special attention since economic considerations generally dictate that fewer commutations be made in brushless motors than in conventional dc motors.

Because the commutation arrangement in brushless motors exists external to the motor, we must examine the switching arrangement in conjunction with the motor in order to understand the electronic control requirements. Figure 4-53 is a block diagram with the essential parts of a brushless dc motor. The shaft encoder shown in the figure is necessary to synchronize the commutation switching with the angular position of the rotor. Figures 4-54 and 4-55 show some alternative controller configurations commonly used in brushless dc motors. Many variations of the circuits shown are possible, but it is beyond the scope of this brief overview to fully discuss all electronic commutation schemes.

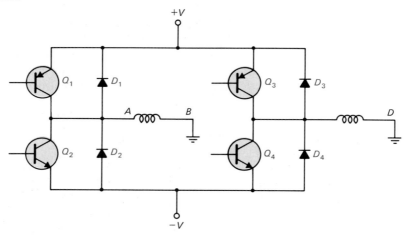

FIGURE 4-54 A two-phase brushless dc-motor controller using two power supplies.

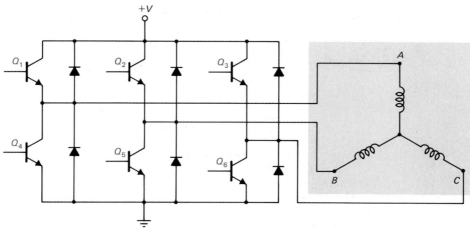

FIGURE 4-55 A three-phase brushless dc-motor controller using a single power supply.

Mathematical Modeling of DC Motors

modeling

Since dc motors are used extensively in control systems, for analytical purposes, it is necessary to establish mathematical models for dc motors for controls applications. The mathematical models of the separately excited and the PM dc motors differ only in the characterization of the magnetic fields. We use the equivalent circuit diagram in Fig. 4-56 to represent a dc motor. The armature is modeled as a circuit with a resistance R_a connected in series with an inductance L_a, and a voltage source e_b representing the back emf in the armature when the rotor rotates. The air-gap flux is designated by $\phi(t)$. The following summarizes the variables and parameters:

$i_a(t)$ = armature current	L_a = armature inductance
R_a = armature resistance	$e_a(t)$ = armature voltage
$e_b(t)$ = back emf	K_b = back-emf constant
$T_L(t)$ = load torque	$\phi(t)$ = magnetic flux in the air gap
$T_m(t)$ = motor torque	$\omega_m(t)$ = rotor angular velocity
$\theta_m(t)$ = rotor displacement	J_m = rotor inertia of motor
K_i = torque constant	B_m = viscous-friction coefficient

With reference to the circuit diagram of Fig. 4-56, the control of the dc motor is applied at the armature terminals in the form of the applied voltage $e_a(t)$. Regardless of how the magnetic field is established by the field structure, we assume that the air-gap flux $\phi(t)$ is constant, and is denoted by ϕ. For linear analysis, we assume that the torque developed by the motor is proportional to the air-gap flux and the armature current. Thus,

$$T_m(t) = K_m \phi i_a(t) = K_i i_a(t) \qquad (4\text{-}122)$$

torque constant

where K_i is the torque constant in N-m/A, lb-ft/A, or oz-in./A.

Starting with the control input voltage $e_a(t)$, the cause-and-effect equations for the system of Fig. 4-56 are

$$\frac{di_a(t)}{dt} = \frac{1}{L_a} e_a(t) - \frac{R_a}{L_a} i_a(t) - \frac{1}{L_a} e_b(t) \qquad (4\text{-}123)$$

$$T_m(t) = K_i i_a(t) \qquad (4\text{-}124)$$

$$e_b(t) = K_b \frac{d\theta_m(t)}{dt} = K_b \omega_m(t) \qquad (4\text{-}125)$$

$$\frac{d^2\theta_m(t)}{dt^2} = \frac{1}{J_m} T_m(t) - \frac{1}{J_m} T_L(t) - \frac{B_m}{J_m} \frac{d\theta_m(t)}{dt} \qquad (4\text{-}126)$$

where $T_L(t)$ represents a load frictional torque such as Coulomb friction.

Equations (4-123) through (4-126) consider that $e_a(t)$ is the cause of all causes; Eq. (4-123) considers that $di_a(t)/dt$ is the immediate effect due to $e_a(t)$, then in Eq.

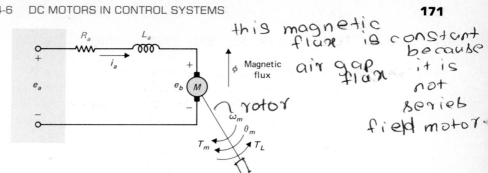

FIGURE 4-56 Model of a
separately excited dc motor.

back emf

(4-124), $i_a(t)$ causes the torque $T_m(t)$ to be generated, Eq. (4-125) defines the back emf, and, finally, in Eq. (4-126), the torque causes the angular displacement $\theta_m(t)$.

The state variables of the system can be defined as $i_a(t)$, $\omega_m(t)$, and $\theta_m(t)$. By direct substitution and eliminating all the nonstate variables from Eqs. (4-123) through (4-126), the state equations of the dc-motor system are written in vector-matrix form:

$$
\begin{bmatrix} \dfrac{di_a(t)}{dt} \\[2ex] \dfrac{d\omega_m(t)}{dt} \\[2ex] \dfrac{d\theta_m(t)}{dt} \end{bmatrix} = \begin{bmatrix} -\dfrac{R_a}{L_a} & -\dfrac{K_b}{L_a} & 0 \\[2ex] \dfrac{K_i}{J_m} & -\dfrac{B_m}{J_m} & 0 \\[2ex] 0 & 1 & 0 \end{bmatrix} \begin{bmatrix} i_a(t) \\[2ex] \omega_m(t) \\[2ex] \theta_m(t) \end{bmatrix} + \begin{bmatrix} \dfrac{1}{L_a} \\[2ex] 0 \\[2ex] 0 \end{bmatrix} e_a(t) - \begin{bmatrix} 0 \\[2ex] \dfrac{1}{J_m} \\[2ex] 0 \end{bmatrix} T_L(t)
\qquad (4\text{-}127)
$$

Notice that in this case, $T_L(t)$ is treated as a second input in the state equations.

The state diagram of the system is shown in Fig. 4-57, using Eq. (4-127). The transfer function between the motor displacement and the input voltage is obtained from the state diagram as

$$
\frac{\Theta_m(s)}{E_a(s)} = \frac{K_i}{L_a J_m s^3 + (R_a J_m + B_m L_a)s^2 + (K_b K_i + R_a B_m)s}
\qquad (4\text{-}128)
$$

where T_L has been set to zero.

Figure 4-58 shows a block-diagram representation of the dc-motor system. The advantage of using the block diagram is that it gives a clear picture of the transfer-

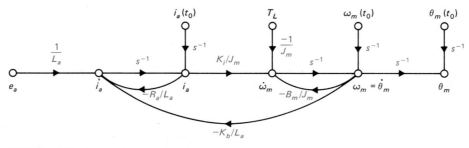

FIGURE 4-57 State diagram of a dc motor.

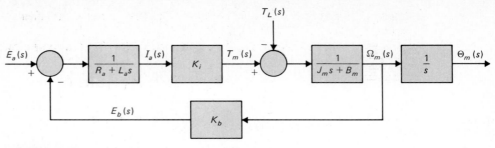

FIGURE 4-58 Block diagram of a dc-motor system.

function relation between each block of the system. Since an s can be factored out from the denominator of Eq. (4-128), *the significance of the transfer function $\Theta_m(s)/E_a(s)$ is that the dc motor is essentially an integrating device between these two variables.* This is expected, since if e_a is a constant input, the output motor displacement will behave as the output of an integrator; that is, it will increase linearly with time.

Although a dc motor by itself is basically an open-loop system, the state diagram of Fig. 4-57 and the block diagram of Fig. 4-58 show that the motor has a "built-in" feedback loop caused by the back emf. Physically, the back emf represents the feedback of a signal that is proportional to the negative of the speed of the motor. As seen from Eq. (4-128), the back-emf constant K_b represents an added term to the resistance R_a and the viscous-friction coefficient B_m. Therefore, *the back-emf effect is equivalent to an "electric friction," which tends to improve the stability of the motor and, in general, the stability of the system.*

back-emf constant

RELATION BETWEEN K_i AND K_b.   Although functionally the torque constant K_i and back-emf constant K_b are two separate paremeters, for a given motor, their values are closely related. To show the relationship, we write the mechanical power developed in the armature as

$$P = e_b(t)i_a(t) \qquad (4\text{-}129)$$

The mechanical power is also expressed as

$$P = T_m(t)\omega_m(t) \qquad (4\text{-}130)$$

where, in SI units, $T_m(t)$ is in N-m, and $\omega_m(t)$ is in rad/s. Now substituting Eqs. (4-124) and (4-125) in Eq. (4-129), we get

$$P = T_m(t)\omega_m(t) = K_b\omega_m(t)\frac{T_m(t)}{K_i} \qquad (4\text{-}131)$$

from which we have

$$K_b \text{ (V/rad/s)} = K_i \text{ (N-m/A)} \qquad (4\text{-}132)$$

[handwritten margin notes:] K_b, electric friction increases the stability of the motor and the stability of the system.

Thus, we see that in SI unit system, the values of K_b and K_i are identical if K_b is represented in V/rad/s and K_i is in N-m/A.

In the British unit system, we convert Eq. (4-129) into horsepower (hp); that is,

$$P = \frac{e_b(t)i_a(t)}{746} \text{ hp} \qquad (4\text{-}133)$$

In terms of torque and angular velocity, P is

$$P = \frac{T_m(t)\omega_m(t)}{550} \text{ hp} \qquad (4\text{-}134)$$

where $T_m(t)$ is in ft-lb, and $\omega_m(t)$ is in rad/s. Using Eqs. (4-124) and (4-125) and equating the last two equations, we get

$$\frac{K_b\omega_m(t)T_m(t)}{746K_i} = \frac{T_m(t)\omega_m(t)}{550} \qquad (4\text{-}135)$$

Thus,

$$K_b = \frac{746}{550}K_i = 1.356K_i \qquad (4\text{-}136)$$

where K_b is in V/rad/sec, and K_i is in ft-lb/A.

Torque–Speed Curves of a DC Motor

torque–speed curves

The torque–speed curves of a dc motor describe the static-torque-producing capability of the motor with respect to the applied voltage and motor speed. With reference to Fig. 4-56, in the steady state, the effect of the inductance is zero, and the torque equation of the motor is

$$T_m = K_i I_a = \frac{K_i(E_a - K_b \Omega_m)}{R_a} \qquad (4\text{-}137)$$

where T_m, I_a, E_a, and Ω_m represent the steady-state values of the motor torque, current, applied voltage, and speed, respectively.

For a given applied voltage E_a, Eq. (4-137) describes the straight-line relation of the torque–speed characteristics of the dc motor. In reality, the motor may be subject to two types of saturation or limitations. One limitation is that as the armature current increases from the increase in E_a, the magnetic circuit will saturate, so that the motor torque cannot exceed a certain maximum value. The second limitation is due to the maximum current that the motor can handle due to the heat-dissipation rating of the motor.

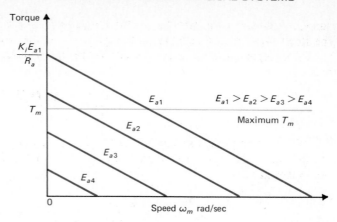

FIGURE 4-59 Typical torque–speed curves of a dc motor.

Figure 4-59 shows a typical set of torque–speed curves for various applied voltages. The slope of these curves is determined from Eq. (4-137), and is expressed as

$$k = \frac{dT_m}{d\Omega_m} = -\frac{K_i K_b}{R_a} \tag{4-138}$$

The limit on the torque due to magnetic saturation is shown by the dashed line in the figure. In practice, the torque–speed curves of a dc motor can be determined experimentally by a dynamometer.

Torque-Speed Curves of an Amplifier–DC-Motor System

In practice, since the dc motor is always driven by a power amplifier that acts as a source of energy, it is more practical to present the torque–speed curves of the amplifier–motor combination, especially if the amplifier gain is subject to saturation. Figure 4-60 shows the equivalent block diagram of an amplifier–motor system. The amplifier has a gain of K_1. The saturation level of the output voltage of the amplifier is E_L.

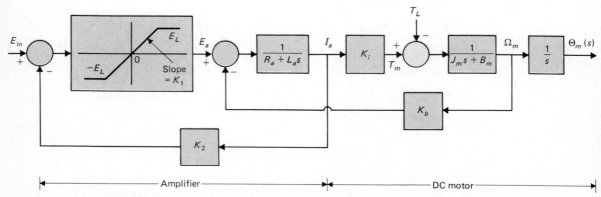

FIGURE 4-60 Equivalent block diagram of an amplifier–dc-motor system.

amplifier saturation

Current feedback with gain K_L is also introduced for the stabilization of the system. The steady-state torque equation of the system is written from Fig. 4-60 by treating E_{in} and Ω_m as inputs, and T_m as the output. When the amplifier is not saturated, the motor torque is

$$T_m = \frac{K_i(K_1 E_{in} - K_b \Omega_m)}{R_a + K_1 K_2} \qquad (4\text{-}139)$$

Without amplifier saturation, the torque–speed curves described by the last equation are again a family of straight lines, as illustrated in Fig. 4-59. The slope of the torque–speed curves is

$$k = \frac{dT_m}{d\Omega_m} = -\frac{K_i K_b}{R_a + K_1 K_2} \qquad (4\text{-}140)$$

Comparing Eq. (4-140) with Eq. (4-138), we see that the effect of the current feedback of the amplifier is that the gain K_2 reduces the slope of the torque–speed curves of the motor.

When the amplifier is subject to saturation, $|E_a| \le E_L$. When $|E_a| = E_L$, the torque equation in the steady state is

$$T_m = \frac{K_i}{R_a}(E_L - K_b \Omega_m) \qquad (4\text{-}141)$$

This gives the maximum blocked-rotor ($\Omega_m = 0$) torque to be $T_m = K_i E_L / R_a$. The slope of the torque–speed curve under amplifier saturation is $-K_i K_b / R_a$. Figure 4-61 shows the torque–speed curves of the amplifer–motor combination when the amplifier is subject to saturation.

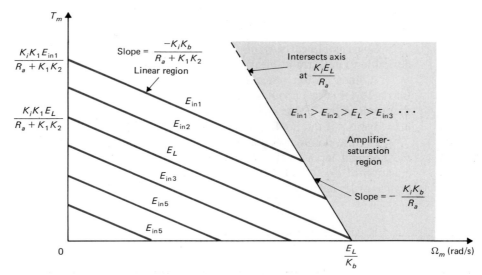

FIGURE 4-61 Torque–speed curve of amplifier–dc-motor system with amplifier saturation and current feedback.

LINEARIZATION OF NONLINEAR SYSTEMS

From the discussions given in the preceding sections, we should realize that most components and actuators found in physical systems have nonlinear characteristics. In practice, we may find that some devices have moderate nonlinear characteristics, or the nonlinear properties would occur if they are driven into certain operating regions. For these devices, the modeling by linear system models may give quite accurate analytical results over a relatively wide range of operating conditions. However, there are numer-

nonlinear
system

ous physical devices that possess strong nonlinear characteristics. For these devices, strictly, a linearized model is valid only for a limited range of operation, and often only at the operating point at which the linearization is carried out. More importantly, when a nonlinear system is linearized at an operating point, the linear model may contain time-varying elements.

Let us represent a nonlinear system by the following vector-matrix state equations:

$$\frac{d\mathbf{x}(t)}{dt} = \mathbf{f}[\mathbf{x}(t), \mathbf{r}(t)] \tag{4-142}$$

where $\mathbf{x}(t)$ represents the $n \times 1$ state vector, $\mathbf{r}(t)$ the $p \times 1$ input vector, and $\mathbf{f}[\mathbf{x}(t), \mathbf{r}(t)]$ denotes an $n \times 1$ function vector. In general, $\mathbf{f}$ is a function of the state vector and the input vector.

Being able to represent a nonlinear and/or time-varying system by state equations is a distinct advantage of the state-variable approach over the transfer-function method, since the latter is defined strictly only for linear time-invariant systems.

As a simple illustrative example, the following state equations are nonlinear:

$$\frac{dx_1(t)}{dt} = x_1(t) + x_2^2(t) \tag{4-143}$$

$$\frac{dx_2(t)}{dt} = x_1(t) + r(t) \tag{4-144}$$

Since nonlinear systems are usually difficult to analyze and design, it would be desirable to perform a linearization whenever the situation justifies.

A linearization process that depends on expanding the nonlinear state equation into a Taylor series about a nominal operating point or trajectory is now described. All the terms of the Taylor series of order higher than the first are discarded, and linear approximation of the nonlinear state equation at the nominal point results.

Let the nominal operating trajectory be denoted by $\mathbf{x}_0(t)$, which corresponds to the nominal input $\mathbf{r}_0(t)$ and some fixed initial states. Expanding the nonlinear state equation of Eq. (4-142) into a Taylor series about $\mathbf{x}(t) = \mathbf{x}_0(t)$ and neglecting all the higher-order terms yields

$$\dot{x}_i(t) = f_i(\mathbf{x}_0, \mathbf{r}_0) + \sum_{j=1}^{n} \left. \frac{\partial f_i(\mathbf{x}, \mathbf{r})}{\partial x_j} \right|_{\mathbf{x}_0, \mathbf{r}_0} (x_j - x_{0j})$$

$$+ \sum_{j=1}^{p} \left. \frac{\partial f_i(\mathbf{x}, \mathbf{r})}{\partial r_j} \right|_{\mathbf{x}_0, \mathbf{r}_0} (r_j - r_{0j}) \tag{4-145}$$

where $i = 1, 2, \ldots, n$. Let

$$\Delta x_i = x_i - x_{0i} \tag{4-146}$$

and

$$\Delta r_j = r_j - r_{0j} \tag{4-147}$$

Then

$$\Delta \dot{x}_i = \dot{x}_i - \dot{x}_{0i} \tag{4-148}$$

Since

$$\dot{x}_{0i} = f_i(\mathbf{x}_0, \mathbf{r}_0) \tag{4-149}$$

Equation (4-145) is written

$$\Delta \dot{x}_i = \sum_{j=1}^{n} \left. \frac{\partial f_i(\mathbf{x}, \mathbf{r})}{\partial x_j} \right|_{\mathbf{x}_0, \mathbf{r}_0} \Delta x_j + \sum_{j=1}^{p} \left. \frac{\partial f_i(\mathbf{x}, \mathbf{r})}{\partial r_j} \right|_{\mathbf{x}_0, \mathbf{r}_0} \Delta r_j \tag{4-150}$$

Equation (4-150) may be written in vector-matrix form:

$$\Delta \dot{\mathbf{x}} = \mathbf{A}^* \, \Delta \mathbf{x} + \mathbf{B}^* \, \Delta r \tag{4-151}$$

where

$$\mathbf{A}^* = \begin{bmatrix} \dfrac{\partial f_1}{\partial x_1} & \dfrac{\partial f_1}{\partial x_2} & \cdots & \dfrac{\partial f_1}{\partial x_n} \\[2mm] \dfrac{\partial f_2}{\partial x_1} & \dfrac{\partial f_2}{\partial x_2} & \cdots & \dfrac{\partial f_2}{\partial x_n} \\[1mm] \cdots & \cdots & \cdots & \cdots \\[1mm] \dfrac{\partial f_n}{\partial x_1} & \dfrac{\partial f_n}{\partial x_2} & \cdots & \dfrac{\partial f_n}{\partial x_n} \end{bmatrix} \tag{4-152}$$

$$\mathbf{B}^* = \begin{bmatrix} \dfrac{\partial f_1}{\partial r_1} & \dfrac{\partial f_1}{\partial r_2} & \cdots & \dfrac{\partial f_1}{\partial r_p} \\[2mm] \dfrac{\partial f_2}{\partial r_1} & \dfrac{\partial f_2}{\partial r_2} & \cdots & \dfrac{\partial f_2}{\partial r_p} \\[1mm] \cdots & \cdots & \cdots & \cdots \\[1mm] \dfrac{\partial f_n}{\partial r_1} & \dfrac{\partial f_n}{\partial r_2} & \cdots & \dfrac{\partial f_n}{\partial r_p} \end{bmatrix} \tag{4-153}$$

where it should be reiterated that $\mathbf{A}^*$ and $\mathbf{B}^*$ are evaluated at the nominal point. Thus, we have linearized the nonlinear system of Eq. (4-142) at a nominal operating point. However, in general, although Eq. (4-151) is linear, $\mathbf{A}^*$ and $\mathbf{B}^*$ may contain time-varying elements.

The following examples serve to illustrate the linearization procedure just described.

EXAMPLE 4-11

saturation

Figure 4-62 shows the block diagram of a control system with a saturation nonlinearity. The state equations of the system are

$$\dot{x}_1(t) = f_1(t) = x_2(t) \tag{4-154}$$

$$\dot{x}_2(t) = f_2(t) = u(t) \tag{4-155}$$

where the input–output relation of the saturation nonlinearity is represented by

$$u(t) = (1 - e^{-K|x_1(t)|}) \; \text{SGN} \; x_1(t) \tag{4-156}$$

where

$$\text{SGN} \; x_1(t) = \begin{cases} +1 & x_1(t) > 0 \\ -1 & x_1(t) < 0 \end{cases} \tag{4-157}$$

Substituting Eq. (4-156) into Eq. (4-155) and using Eq. (4-150), we have the linearized state equations:

$$\Delta \dot{x}_1(t) = \frac{\partial f_1(t)}{\partial x_2(t)} \Delta x_2(t) = \Delta x_2(t) \tag{4-158}$$

$$\Delta \dot{x}_2(t) = \frac{\partial f_2(t)}{\partial x_1(t)} \Delta x_1(t) = Ke^{-K|x_{01}|} \Delta x_1 \tag{4-159}$$

where x_{01} denotes a nominal value of $x_1(t)$. Notice that the last two equations are linear and are valid only for small signals. In vector-matrix form, these linearized state equations are written as

$$\begin{bmatrix} \Delta \dot{x}_1(t) \\ \Delta \dot{x}_2(t) \end{bmatrix} = \begin{bmatrix} 0 & 1 \\ a & 0 \end{bmatrix} \begin{bmatrix} \Delta x_1(t) \\ \Delta x_2(t) \end{bmatrix} \tag{4-160}$$

where

$$a = Ke^{-K|x_{01}|} = \text{constant} \tag{4-161}$$

It is of interest to check the significance of the linearization. If x_{01} is chosen to be at the origin of the nonlinearity, $x_{01} = 0$, then $a = K$; Eq. (4-159) becomes

$$\Delta \dot{x}_2(t) = K \Delta x_1(t) \tag{4-162}$$

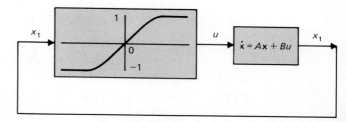

FIGURE 4-62 Nonlinear control system.

Thus, the linearized model is equivalent to having a linear amplifier with a constant gain K. On the other hand, if x_{01} is a large number, the nominal operating point will lie on the saturated portion of the nonlinearity, and $a = 0$. This means that any small variation in $x_1(t)$ [small $\Delta x_1(t)$] will give rise to practically no change in $\Delta \dot{x}_2(t)$.

EXAMPLE 4-12   In Example 4-11, the linearized system turns out to be time-invariant. In general, *linearization of a nonlinear system often leads to a linear time-varying system.* Consider the following nonlinear system:

$$\dot{x}_1(t) = \frac{-1}{x_2^2(t)} \tag{4-163}$$

$$\dot{x}_2(t) = u(t)x_1(t) \tag{4-164}$$

These equations are to be linearized about the nominal trajectory $[x_{01}(t), x_{02}(t)]$, which is the solution to the equations with the initial conditions $x_1(0) = x_2(0) = 1$ and the input $u(t) = 0$.

Integrating both sides of Eq. (4-155) with respect to t under the above condition, we have

$$x_2(t) = x_2(0) = 1 \tag{4-165}$$

Then Eq. (4-163) gives

$$x_1(t) = -t + 1 \tag{4-166}$$

Therefore, the nominal trajectory about which Eqs. (4-163) and (4-164) are to be linearized is described by

$$x_{01}(t) = -t + 1 \tag{4-167}$$

$$x_{02}(t) = 1 \tag{4-168}$$

Now evaluating the coefficients of Eq. (4-150), we get

$$\frac{\partial f_1(t)}{\partial x_1(t)} = 0 \qquad \frac{\partial f_1(t)}{\partial x_2(t)} = \frac{2}{x_2^3(t)} \qquad \frac{\partial f_2(t)}{\partial x_1(t)} = u(t) \qquad \frac{\partial f_2(t)}{\partial u(t)} = x_1(t)$$

Equation (4-150) gives

$$\Delta \dot{x}_1(t) = \frac{2}{x_{02}^3(t)} \Delta x_2(t) \tag{4-169}$$

$$\Delta \dot{x}_2(t) = u_0(t) \Delta x_1(t) + x_{01}(t) \Delta u(t) \tag{4-170}$$

By substituting Eqs. (4-167) and (4-168) into Eqs. (4-169) and (4-170), the linearized equations are

$$\begin{bmatrix} \Delta \dot{x}_1(t) \\ \Delta \dot{x}_2(t) \end{bmatrix} = \begin{bmatrix} 0 & 2 \\ 0 & 0 \end{bmatrix} \begin{bmatrix} \Delta x_1(t) \\ \Delta x_2(t) \end{bmatrix} + \begin{bmatrix} 0 \\ 1 - t \end{bmatrix} \Delta u(t) \tag{4-171}$$

which is a set of linear state equations with time-varying coefficients.

EXAMPLE 4-13

*magnetic-ball-
suspension
system*

Figure 4-63 shows the diagram of a magnetic-ball suspension system. The objective of the system is to control the position of the steel ball by adjusting the current in the electromagnet through the input voltage $e(t)$. The differential equations of the system are

$$M\frac{d^2y(t)}{dt^2} = Mg - \frac{i^2(t)}{y(t)} \tag{4-172}$$

$$e(t) = Ri(t) + L\frac{di(t)}{dt} \tag{4-173}$$

where

$e(t)$ = input voltage	$y(t)$ = ball position
$i(t)$ = winding current	R = winding resistance
L = winding inductance	M = ball mass
g = gravitational acceleration	

Let us define the state variables as $x_1(t) = y(t)$, $x_2(t) = dy(t)/dt$, and $x_3(t) = i(t)$. The state equations of the system are

$$\frac{dx_1(t)}{dt} = x_2(t) \tag{4-174}$$

$$\frac{dx_2(t)}{dt} = g - \frac{1}{M}\frac{x_3^2(t)}{x_1(t)} \tag{4-175}$$

$$\frac{dx_3(t)}{dt} = -\frac{R}{L}x_3(t) + \frac{1}{L}e(t) \tag{4-176}$$

Let us linearize the system about the equilibrium point $y_0(t) = x_{01} = $ constant. Then,

$$x_{02}(t) = \frac{dx_{01}(t)}{dt} = 0 \tag{4-177}$$

$$\frac{d^2y_0(t)}{dt^2} = 0 \tag{4-178}$$

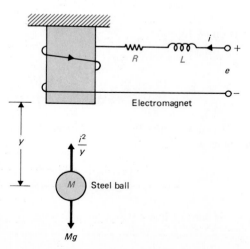

FIGURE 4-63 Magnetic-ball suspension system.

The nominal value of $i(t)$ is determined by substituting Eq. (4-178) into Eq. (4-172). Thus,

$$i_0(t) = x_{03}(t) = \sqrt{Mgx_{01}} \qquad (4\text{-}179)$$

The linearized state equation is expressed in the form of Eq. (4-151), with the coefficient matrices $\mathbf{A}^*$ and $\mathbf{B}^*$ evaluated as

$$\mathbf{A}^* = \begin{bmatrix} 0 & 1 & 0 \\ \dfrac{x_{03}^2}{Mx_{01}^2} & 0 & \dfrac{-2x_{03}}{Mx_{01}} \\ 0 & 0 & -\dfrac{R}{L} \end{bmatrix} = \begin{bmatrix} 0 & 1 & 0 \\ \dfrac{g}{x_{01}} & 0 & -2\left(\dfrac{g}{Mx_{01}}\right)^{1/2} \\ 0 & 0 & -\dfrac{R}{L} \end{bmatrix} \qquad (4\text{-}180)$$

$$\mathbf{B}^* = \begin{bmatrix} 0 \\ 0 \\ 1 \\ \dfrac{1}{L} \end{bmatrix} \qquad (4\text{-}181)$$

4-8

SYSTEMS WITH TRANSPORTATION LAGS

Thus far the systems considered all have transfer functions that are quotients of polynomials. In practice, pure time delays may be encountered in various types of systems, especially systems with hydraulic, pneumatic, or mechanical transmissions. Systems with computer control also have time delays, since it takes time for the computer to execute numerical operations. In these systems, the output will not begin to respond to an input until after a given time interval. Figure 4-64 illustrates systems in which transportation lags or pure time delays are observed. Figure 4-64(a) outlines an arrangement in which

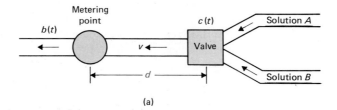

(a)

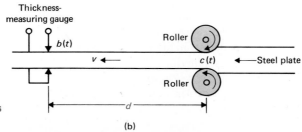

FIGURE 4-64 Physical systems with transportation lags.

(b)

two different fluids are to be mixed in appropriate proportions. To assure that a homogeneous solution is measured, the monitoring point is located some distance from the mixing point. A time delay therefore exists between the mixing point and the place where the change in concentration is detected. If the rate of flow of the mixed solution is v inches per second and d is the distance between the mixing and the metering points, the time lag is given by

$$T_d = \frac{d}{v} \text{ seconds} \tag{4-182}$$

If it is assumed that the concentration at the mixing point is $c(t)$ and that it is reproduced without change T_d seconds later at the monitoring point, the measured quantity is

$$b(t) = c(t - T_d) \tag{4-183}$$

The Laplace transform of Eq. (4-183) is

$$B(s) = e^{-T_d s}C(s) \tag{4-184}$$

where $C(s)$ is the Laplace transform of $c(t)$. The transfer function between $b(t)$ and $c(t)$ is

---

$$\frac{B(s)}{C(s)} = e^{-T_d s} \tag{4-185}$$

Figure 4-64(b) illustrates the control of thickness of rolled steel plates. The transfer function between the thickness at the rollers and the measuring point is again given by Eq. (4-185).

The operation of the sample-and-hold device of a sampled-data system, under the small sampling-period conditions, can be approximated by a pure time delay. In pure digital control systems, we can use the transfer function z^{-N} to represent the pure time delay resulting from the time required to execute the digital processing. For analytical purposes, N is a positive integer, so that the time delay is approximated as an integral multiple of the sampling period T.

Approximation of the Time-Delay Function by Rational Functions

In the analysis and design of control systems, systems that are described inherently by transcendental transfer functions are more difficult to handle. Many analytical tools such as the Routh–Hurwitz criterion (Chapter 6) are restricted to rational transfer functions. The root-locus technique (Chapter 8) is also more easily applied only to systems with rational transfer functions.

There are many ways of approximating $e^{-T_d s}$ by a rational function. One way is to approximate the exponential function by a Maclaurin series; that is,

$$e^{-T_ds} \cong 1 - T_ds + \frac{T_d^2 s^2}{2} \qquad (4\text{-}186)$$

series
approximation

or

$$e^{-T_ds} \cong \frac{1}{1 + T_ds + T_d^2 s^2/2} \qquad (4\text{-}187)$$

Padé
approximation

where only three terms of the series are used. Apparently, the approximations are not valid when the magnitude of T_ds is large.

A better approximation is to use the Padé approximation [7, 8], which is given in the following for a two-term approximation:

$$e^{-T_ds} \cong \frac{1 - T_ds/2}{1 + T_ds/2} \qquad (4\text{-}188)$$

The characteristic of the approximation in Eq. (4-188) is that the transfer function contains a zero in the right-half s-plane so that the step response of the approximating system may exhibit a small negative undershoot near $t = 0$.

4-9
SUMMARY

This chapter is devoted to the mathematical modeling of physical systems. The basic mathematical relations of linear electrical and mechanical systems are described. For linear systems, differential equations, state equations, and transfer functions are the fundamental tools of modeling. The operations and mathematical descriptions of some of the commonly used components in control systems, such as the error detectors, tachometers, and dc motors, are introduced in this chapter.

You are reminded again that due to space limitations and the intended scope of this text, only a few of the physical devices used in practice are described. The main purpose of this chapter is to illustrate the methods of system modeling.

Since nonlinear systems cannot be ignored in the real world, and this text is not devoted to the subject, Section 4-7 introduces a method of producing a linearized model of a nonlinear system at a nominal operating point. Once the linearized model is determined, the performance of the nonlinear system can be investigated under the small-signal conditions at the designated operating point.

REVIEW QUESTIONS

1. Define state variables.

2. The state variables of a dynamic system are not equal to the number of energy-storage elements under what condition?

3. Among the three types of friction described, which type is governed by a linear mathematical relation?

4. Given a two-gear system with angular displacements θ_1 and θ_2, numbers of teeth N_1 and N_2, and torques T_1 and T_2, write the mathematical relations between these variables and parameters.

5. How are potentiometers used in control systems?

6. How are synchros applied in control systems?

7. Digital encoders are used in control systems for position and speed detection. Consider that an encoder is set up to output 3600 zero crossings per revolution. What is the angular rotation of the encoder shaft in degrees if 16 zero crossings are detected?

8. The same encoder described in Review Question 7 and an electronic clock with a frequency of 1 MHz are used for speed measurement. What is the average speed of the encoder shaft in rpm if 500 clock pulses are detected between two consecutive zero crossings of the encoder?

9. Give the advantages of dc motors for control-systems applicaions.

10. What are the sources of nonlinearities in a dc motor?

11. What are the effects of inductance and inertia in a dc motor?

12. What is the back emf in a dc motor, and how does it affect the performance of a control system?

13. What are the electrical and mechanical time constants of an electric motor?

14. Under what condition is the torque constant K_i of a dc motor valid, and how is it related to the back-emf constant K_b?

15. An inertial and frictional load is driven by a dc motor with torque T_m. The dynamic equation of the system is

$$T_m(t) = J_m\dot{\omega}_m + B\omega_m$$

If the inertia is doubled, how will it affect the steady-state speed of the motor? How will the steady-state speed be affected if, instead, the frictional coefficient B_m is doubled? What is the mechanical time constant of the system?

16. The linearization technique described in Section 4-7 always results in a linear time-invariant system.   **(T)   (F)**

17. Give the transfer function of a pure time delay T_d.

REFERENCES

State-Variable Analysis of Electric Networks

1. B. C. KUO, *Linear Circuits and Systems*, McGraw-Hill Book Company, New York, 1967.
2. R. ROHRER, *Circuit Analysis: An Introduction to the State Variable Approach*, McGraw-Hill Book Company, New York, 1970.

Mechanical Systems

3. R. CANNON, *Dynamics of Physical Systems,* McGraw-Hill Book Company, New York, 1967.

DC Motors

4. B. C. KUO and J. TAL, eds., *Incremental Motion Control,* Vol. 1, *DC Motors and Control Systems,* SRL Publishing Co., Champaign, IL, 1979.

Rolling Dry Friction

5. P. B. DAHL, *A Solid Friction Model,* Report No. TOR-0158 (3107-18)-1, Aerospace Corporation, El Segundo, CA, May 1968.
6. N. A. OSBORN and D. L. RITTENHOUSE, "The Modeling of Friction and Its Effects on Fine Pointing Control," *AIAA Mechanics and Control of Flight Conference,* Paper No. 74-875, August 1974.

Padé Approximation

7. J. G. TRUXAL, *Automatic Feedback Control System Synthesis,* McGraw-Hill Book Company, New York, 1955.
8. H. S. WALL, *Continued Fractions,* Chapter 20, D. Van Nostrand Co, New York, 1948.

PROBLEMS

4-1. Write the force equations of the linear translational systems shown in Fig. 4P-1.

(a) Draw state diagrams using a minimum number of integrators. Write the state equations from the state diagrams.

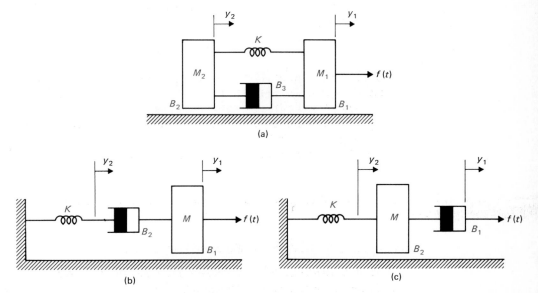

FIGURE 4P-1

(b) Define the state variables as follows:

part (i): $x_1 = y_2$, $x_2 = dy_2/dt$, $x_3 = y_1$, and $x_4 = dy_1/dt$
part (ii): $x_1 = y_2$, $x_2 = y_1$, and $x_3 = dy_1/dt$
part (iii): $x_1 = y_1$, $x_2 = y_2$, and $x_3 = dy_2/dt$

Write the state equations and draw the state diagram with these state variables. Find the transfer functions $Y_1(s)/F(s)$ and $Y_2(s)/F(s)$.

4-2. Write the force equations of the linear translational system shown in Fig. 4P-2. Draw the state diagram using a minimum number of integrators. Write the state equations from the state diagram. Find the transfer functions $Y_1(s)/F(s)$ and $Y_2(s)/F(s)$. Set $Mg = 0$ for the transfer functions.

4-3. Write the torque equations of the rotational systems shown in Fig. 4P-3. Draw state diagrams using a minimum number of integrators. Write the state equations from the state diagrams. Find the transfer function $\Theta(s)/T(s)$ for the system in (a). Find the transfer functions $\Theta_1(s)/T(s)$ and $\Theta_2(s)/T(s)$ for the systems in parts (b), (c), (d), and (e).

4-4. An open-loop motor control system is shown in Fig. 4P-4. The potentiometer has a maximum range of rotation of 10 turns (20π rad). Find the transfer functions $E_o(s)/T_m(s)$. The following parameters and variables are defined: $\theta_m(t)$ is the motor displacement, $\theta_L(t)$ the load displacement, $T_m(t)$ the motor torque, J_m the motor inertia, B_m the motor viscous-friction coefficient, B_p the potentiometer viscous-friction coefficient, $e_o(t)$ the output voltage, and K the torsional spring constant.

4-5. Write the torque equations of the gear-train system shown in Fig. 4P-5. The moments of inertia of the gears are lumped as J_1, J_2, and J_3. $T_m(t)$ is the applied torque; N_1, N_2, N_3, and N_4 are the number of gear teeth. Assume rigid shafts.
(a) Assume that J_1, J_2, and J_3 are negligible. Write the torque equations of the system. Find the total inertia the motor sees.
(b) Repeat part (a) with the moments of inertia J_1, J_2, and J_3.

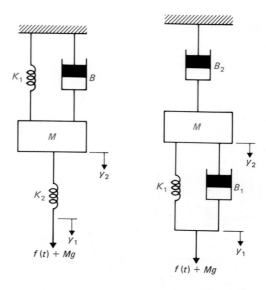

FIGURE 4P-2   (a)   (b)

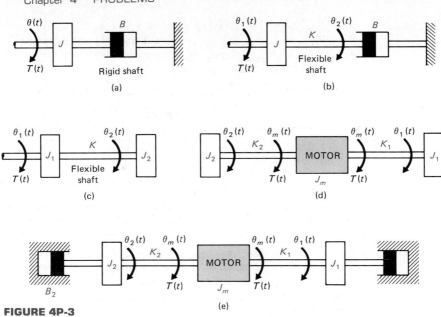

FIGURE 4P-3

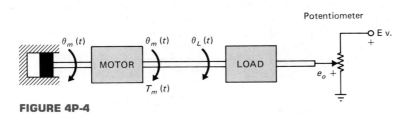

FIGURE 4P-4

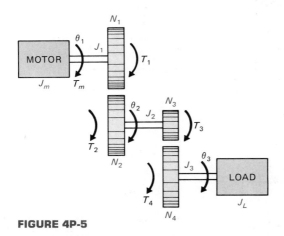

FIGURE 4P-5

4-6.   A vehicle towing a trailer through a spring-damper coupling hitch is shown in Fig. 4P-6. The following parameters and variables are defined: M is the mass of the trailer, K_h the spring constant of the hitch, B_h the viscous damping coefficient of the hitch, B_t the viscous-friction coefficient of the trailer, $y_1(t)$ the displacement of the towing vehicle, $y_2(t)$ the displacement of the trailer, and $f(t)$ the applied force of the towing vehicle.
    (a)   Write the differential equations of the system.
    (b)   Write the state equations by defining the following state variables: $x_1(t) = y_1(t) - y_2(t)$ and $x_2(t) = dy_2(t)/dt$.

4-7.   Figure 4P-7 shows a motor-load system coupled through a gear train with gear ratio $n = N_1/N_2$. The motor torque is $T_m(t)$, and $T_L(t)$ represents a load torque.
    (a)   Find the optimum gear ratio n^* such that the load acceleration $\alpha_L = d^2\theta_L/dt^2$ is maximized.
    (b)   Repeat part (a) when the load torque T_L is zero.

4-8.   Figure 4P-8 shows the simplified diagram of a printwheel system with belts and pulleys. Assume that the belts are rigid. The following parameters and variables are defined: $T_m(t)$ is the motor torque, $\theta_m(t)$ the motor displacement, J_m the motor inertia, B_m the motor viscous-friction coefficient, r the pulley radius, and M the mass of the printwheel.
    (a)   Write the differential equation of the system.
    (b)   Find the transfer function $Y(s)/T_m(s)$.

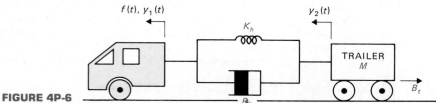

FIGURE 4P-6

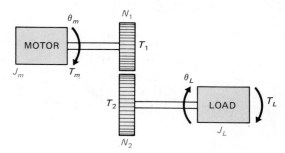

FIGURE 4P-7

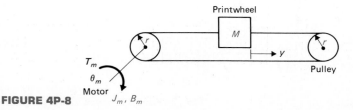

FIGURE 4P-8

4-9.   Figure 4P-9 shows the diagram of a printwheel system with belts and pulleys. The belts
       are modeled as linear springs with spring constants K_1 and K_2.
   (a)   Write the differential equations of the system using θ_m and y as the dependent
         variables.
   (b)   Write the state equations using $x_1 = r\theta_m - y$, $x_2 = dy/dt$, and $x_3 = \omega_m = d\theta_m/dt$
         as the state variables.
   (c)   Draw a state diagram for the system.
   (d)   Find the transfer function $Y(s)/T_m(s)$.
   (e)   Find the characteristic equation of the system.

4-10. The schematic diagram of a motor-load system is shown in Fig. 4P-10. The following
       parameters and variables are defined: $T_m(t)$ is the motor torque, $\omega_m(t)$ the motor veloc-
       ity, $\theta_m(t)$ the motor displacement, $\omega_L(t)$ the load velocity, $\theta_L(t)$ the load displacement,
       K the torsional spring constant, J_m the motor inertia, B_m the motor viscous-friction
       coefficient, and B_L the load viscous-friction coefficient.
   (a)   Write the torque equations of the system.
   (b)   Find the transfer functions $\Theta_L(s)/T_m(s)$ and $\Theta_m(s)/T_m(s)$.
   (c)   Find the characteristic equation.
   (d)   Let $T_m(t) = T_m$ be a constant applied torque; show that $\omega_m = \omega_L =$ constant in the
         steady state. Find the steady-state speeds ω_m and ω_L.
   (e)   Repeat part (d) when the value of J_L is doubled, but J_m stays the same.

4-11. The schematic diagram of a control system containing a motor coupled to a tachometer
       and an inertial load is shown in Fig. 4P-11. The following parameters and variables are
       defined: T_m is the motor torque, J_m the motor inertia, J_t the tachometer inertia, J_L the
       load inertia, K_1 the spring constant of the shaft, K_2 the spring constant of the shaft, θ_t
       the tachometer displacement, θ_m the motor displacement, ω_m the motor velocity, θ_L the
       load displacement, ω_t the tachometer velocity, ω_L the load velocity, and B_m the motor
       viscous-friction coefficient.
   (a)   Write the state equations of the system using θ_L, ω_L, θ_t, ω_t, θ_m, and ω_m as the
         state variables (in the listed order). The motor torque T_m is the input.
   (b)   Draw a state diagram with T_m at the left and ending with θ_L on the far right. The
         state diagram should have a total of 10 nodes. Leave out the initial states.

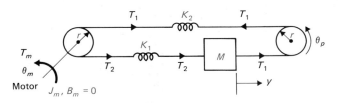

FIGURE 4P-9

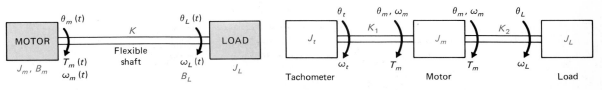

FIGURE 4P-10

FIGURE 4P-11

(c)   Find the following transfer functions:

$$\frac{\Theta_L(s)}{T_m(s)} \qquad \frac{\Theta_t(s)}{T_m(s)} \qquad \frac{\Theta_m(s)}{T_m(s)}$$

(d)   Find the characteristic equation of the system.

4-12.   The voltage equation of a dc motor is written as

$$e_a(t) = R_a i_a(t) + L_a \frac{di_a(t)}{dt} + K_b \omega_m(t)$$

where $e_a(t)$ is the applied voltage, $i_a(t)$ the armature current, R_a the armature resistance, L_a the armature inductance, K_b the back-emf constant, $\omega_m(t)$ the motor velocity, and $\omega_r(t)$ the reference input voltage.

Taking the Laplace transform on both sides of the voltage equation, with zero initial conditions, and solving for $\Omega_m(s)$, we get

$$\Omega_m(s) = \frac{E_a(s) - (R_a + L_a s)I_a(s)}{K_b}$$

which shows that the velocity information can be generated by feeding back the armature voltage and current. The block diagram in Fig. 4P-12 shows a dc-motor system, with voltage and current feedbacks, for speed control.

(a)   Let K_1 be a very high gain. Show that when $H_i(s)/H_e(s) = -(R_a + L_a s)$, the motor velocity $\omega_m(t)$ is totally independent of the load-disturbance torque T_L.

(b)   Find the transfer function between $\Omega_m(s)$ and $\Omega_r(s)$ ($T_L = 0$) when $H_i(s)$ and $H_e(s)$ are selected as in part (a).

4-13.   This problem deals with the attitude control of a guided missile. When traveling through the atmosphere, a missile encounters aerodynamic forces that tend to cause instability in the attitude of the missile. The basic concern from the flight-control standpoint is the lateral force of the air, which tends to rotate the missile about its center of gravity. If the missile centerline is not aligned with the direction in which the center of gravity C is

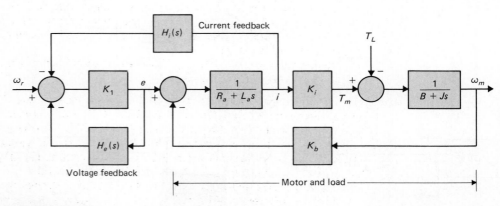

FIGURE 4P-12

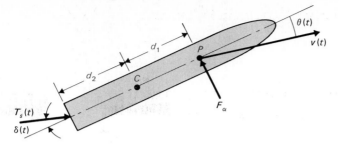

FIGURE 4P-13

traveling, as shown in Fig. 4P-13, with angle θ (θ is also called the angle of attack), a side force is produced by the resistance of the air through which the missile is traveling. The total force F_α may be considered to be applied at the center of pressure P. As shown in Fig. 4P-13, this side force has a tendency to cause the missile to tumble end over end, especially if the point P is in front of the center of gravity C. Let the angular accelera- tion of the missile about the point C, due to the side force, be denoted by α_F. Normally, α_F is directly proportional to the angle of attack θ and is given by

$$\alpha_F = \frac{K_F d_1}{J}\theta$$

where K_F is a constant that depends on such parameters as dynamic pressure, velocity of the missile, air density, and so on, and

$$J = \text{missile moment of inertia about } C$$
$$d_1 = \text{distance between points } C \text{ and } P$$

The main objective of the flight-control system is to provide the stabilizing action to counter the effect of the side force. One of the standard control means is to use gas in- jection at the tail of the missile to deflect the direction of the rocket engine thrust T_s, as shown in the figure.

(a) Write a torque differential equation to relate among T_s, δ, θ, and the system parameters given. Assume that δ is very small, so that $\sin \delta(t) = \delta(t)$.

(b) Assume that T_s is a constant torque. Find the transfer function $\Theta(s)/\Delta(s)$, where $\Theta(s)$ and $\Delta(s)$ are the Laplace transforms of $\theta(t)$ and $\delta(t)$, respectively. Assume that $\delta(t)$ is very small.

(c) Repeat parts (a) and (b) with points C and P interchanged. The d_1 in the expres- sion of α_F should be changed to d_2.

4-14. The schematic diagram of a dc-motor control system is shown in Fig. 4P-14(a). The fol- lowing parameters and variables are defined: K_s is the error-detector gain (V/rad), K_1 the torque constant (oz-in./A), K the amplifier gain (V/V), K_b the back-emf constant (V/rad/s), n is the gear train ratio = $\theta_2/\theta_m = T_m/T_2$, B_m the motor viscous-friction coefficient (oz-in.-s), J_m the motor inertia (oz-in.-s^2), K_L the torsional spring constant of the motor shaft (oz-in./rad), and J_L the load inertia (oz-in.-s^2).

(a) Write the cause-and-effect equations of the system. Rearrange these equations into the form of state equations with $x_1 = \theta_o$, $x_2 = \omega_o$, $x_3 = \theta_m$, $x_4 = \omega_m$, and $x_5 = i_a$ as state variables.

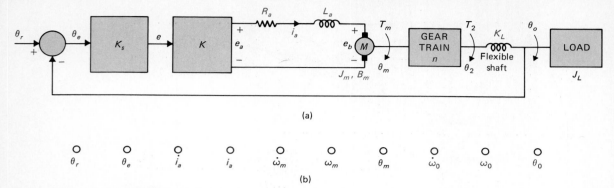

(a)

(b)

FIGURE 4P-14

(b)   Draw a state diagram using the nodes shown in Fig. 4P-14(b).

(c)   Derive the open-loop transfer function (with the outer feedback path open): $G(s) = \Theta_o(s)/\Theta_e(s)$. Find the closed-loop transfer function: $M(s) = \Theta_o(s)/\Theta_r(s)$.

(d)   Repeat part (c) when the motor shaft is rigid; i.e., $K_L = \infty$. Show that you can obtain the solutions by taking the limit of $K_L = \infty$ in the results in part (c).

4-15.   Figure 4P-15 shows the schematic diagram of an amplifier–dc-motor system with current feedback. The amplifier has a voltage-saturation characteristic shown. The torque–speed curves of the system are determined experimentally and are shown in the figure. The output voltage of the amplifier saturates at 100 volts ($E_L = 100$ V).

(a)   Determine the torque–speed curve equation (without saturation) from the block diagram, under steady state, by use of Mason's gain formula, with T_m as output, E_{in} and Ω_m as inputs. In Fig. 4P-15, Ω_m is shown to be an input to K_b, not as a feedback from the motor output. This reflects how the torque–speed curve experimentation is conducted. Determine the slope of the torque–speed curves when the amplifier output is saturated at $E_L = 100$ V. What is the effect of the current feedback gain K_2 on the torque–speed curve? Note that at steady state, all the terms that are multiplied by s in the block diagram are zero. When the amplifier is saturated, its output voltage is E_L volts, and the input voltage E_{in} no longer matters, and the current-feedback gain K_2 is also ineffective. The effective input to the system is E_L.

(b)   From the torque–speed curve data given, find the following:
(i) for the motor:

torque constant K_i (oz-in./A)
back-emf constant K_b (V/rpm)
armature resistance R_a (ohms)

(ii) for the amplifier:

forward amplifier gain K_1 (V/V)
current-feedback gain K_2 (V/A)

4-16.   The schematic diagram of a voice-coil motor (VCM), used as a linear actuator in a disk memory-storage system, is shown in Fig. 4P-16(a). The VCM consists of a cylindrical permanent magnet (PM) and a voice coil. When current is sent through the coil, the magnetic field of the PM interacts with the current-carrying conductor, causing the coil to move linearly. The voice coil of the VCM in Fig. 4P-16(a) consists of a primary coil

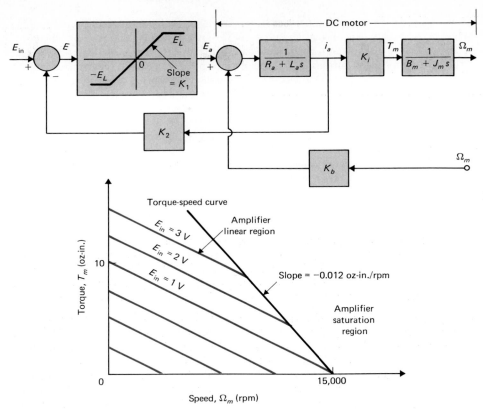

FIGURE 4P-15

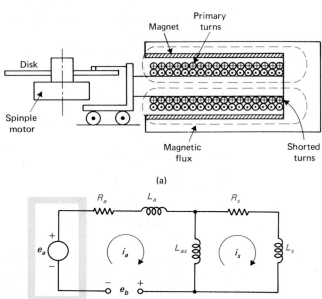

FIGURE 4P-16   (b)

and a shorted-turn coil. The latter is installed for the purpose of effectively reducing the electric constant of the device. Figure 4-16(b) shows the equivalent circuit of the coils. The following parameters and variables are defined: $e_a(t)$ is the applied coil voltage, $i_a(t)$ the primary-coil current, $i_s(t)$ the shorted-turn coil current, R_a the primary-coil resistance, L_a the primary-coil inductance, R_s the reflected shorted-turn coil resistance, L_s the reflected shorted-turn coil inductance, L_{as} the mutual inductance between the primary and shorted-turn coils, $v(t)$ the velocity of the voice coil, $y(t)$ the displacement of the voice coil, $f(t) = K_i i_a(t)$ the force of the voice coil, K_i the force constant, K_b the back-emf constant, $e_b(t) = K_b v(t)$ the back emf, M_T the total mass of the voice coil and load, and B_T the total viscous-friction coefficient of the voice coil and load.
(a)   Write the differential equations of the system.
(b)   Draw a block diagram of the system with $E_a(s)$, $I_a(s)$, $I_s(s)$, $V(s)$, and $Y(s)$ as variables.
(c)   Derive the transfer function $Y(s)/E_a(s)$.

4-17.   A dc-motor position-control system is shown in Fig. 4P-17(a). The following parameters and variables are defined: e is the error voltage, e_r the reference input, θ_L the load position, K_A the amplifier gain, e_a the motor input voltage, e_b the back emf, i_a the motor current, T_m the motor torque, J_m the motor inertia = 0.03 oz-in.-s², B_m the motor viscous-friction coefficient = 10 oz-in.-s, K_L the torsional spring constant = 50,000 oz-in./rad, J_L the load inertia = 0.05 oz-in.-s², K_i the motor torque constant = 21 oz-in./A, K_b the back-emf constant = 15.5 V/1000 rpm, K_s the error-detector gain = $E/2\pi$, E the error-detector applied voltage = 2π V, R_a the motor resistance = 1.15 ohms, $\theta_e = \theta_r - \theta_L$.
(a)   Write the state equations of the system using the following state variables: $x_1 = \theta_L$, $x_2 = d\theta_L/dt = \omega_L$, $x_3 = \theta_m$, and $x_4 = d\theta_m/dt = \omega_m$.
(b)   Draw a state diagram using the nodes shown in Fig. 4P-17(b).

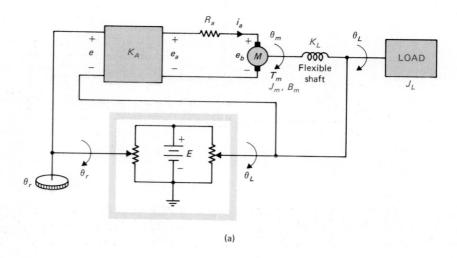

(a)

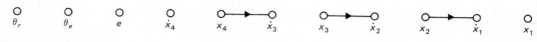

FIGURE 4P-17   (b)

     (c) Derive the open-loop transfer function $G(s) = \Theta_L(s)/\Theta_e(s)$ when the outer feedback path from θ_L is opened. Find the poles of $G(s)$.

     (d) Derive the closed-loop transfer function $M(s) = \Theta_L(s)/\Theta_r(s)$. Find the poles of $M(s)$ when $K_A = 1$, 2738, and 5476. Locate these poles in the s-plane, and comment on the significance of these values of K_A.

4-18. Figure 4P-18(a) shows the setup of the temperature control of an air-flow system. The hot-water reservoir supplies the water that flows into the heat exchanger for heating the air. The temperature sensor senses the air temperature T_{Ao} and sends it to be compared with the reference temperature T_r. The temperature error T_e is sent to the controller which has the transfer function $G_c(s)$. The output of the controller, $u(t)$, which is an electric signal, is converted to a pneumatic signal by a transducer. The output of the actuator controls the water-flow rate through the three-way valve. Figure 4P-18(b) shows the block diagram of the system.

     The following parameters and variables are defined: dM_w is the flow rate of the heating fluid $= K_M u$, $K_M = 0.054$ kg/s/V, T_w the water temperature $= K_R dM_w$, $K_R = 65$ °C/kg/s, and T_{Ao} the output air temperature.

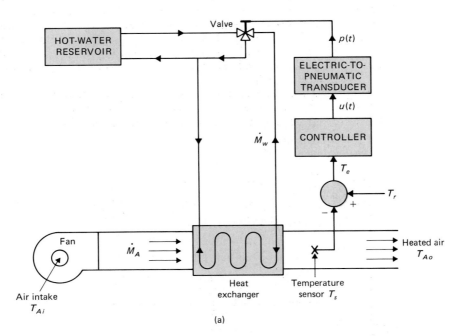

(a)

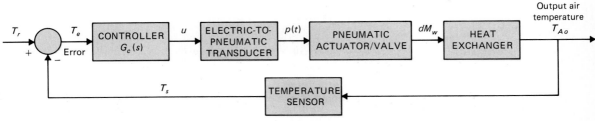

FIGURE 4P-18   (b)

Heat-transfer equation between water and air:

$$\tau_c \frac{dT_{Ao}}{dt} = T_w - T_{Ao} \qquad \tau_c = 10 \text{ seconds}$$

Temperature sensor equation:

$$\tau_s \frac{dT_s}{dt} = T_{Ao} - T_s \qquad \tau_s = 2 \text{ seconds}$$

(a)   Draw a functional block diagram that includes all the transfer functions of the system.

(b)   Derive the transfer function $T_{Ao}(s)/T_r(s)$ when $G_c(s) = 1$.

4-19.   The objective of this problem is to develop a linear analytical model of the automobile engine for idle-speed control. The input of the system is the throttle position that controls the rate of air flow into the manifold. Engine torque is developed from the buildup of manifold pressure due to air intake and the intake of the air/gas mixture into the cylinder. The engine variations are as follows:

$q_i(t)$ = amount of air flow across throttle into manifold
dq_i/dt = rate of air flow across throttle into manifold
$q_m(t)$ = average air mass in manifold
$q_o(t)$ = amount of air leaving intake manifold through intake valves
dq_o/dt = rate of air leaving intake manifold through intake valves
$T(t)$ = engine torque
T_d = disturbance torque due to application of auto-accessories
= constant
$\omega(t)$ = engine speed
$\alpha(t)$ = throttle position
τ_D = time delay in engine
J_e = inertia of engine

The following assumptions and mathematical relations between the engine variables are given:

1.   The rate of air flow into the manifold is linearly dependent on the throttle position:

$$dq_i(t)/dt = K_1\alpha(t) \qquad K_1 = \text{proportional constant}$$

2.   The rate of air flow leaving the manifold depends linearly on the air mass in the manifold and the engine speed:

$$dq_o(t)/dt = K_2 q_m(t) + K_3\omega(t) \qquad K_2, K_3 = \text{constants}$$

3.   A pure time delay of τ_D seconds exists between the change in the manifold air mass and the engine torque:

$$T(t) = K_4 q_m(t - \tau_D) \qquad K_4 = \text{constant}$$

4.   The engine drag is modeled by a viscous-friction torque $B\omega(t)$, where B is the viscous-friction coefficient.

5.   The average air mass $q_m(t)$ is determined from

$$q_m(t) = \int [\dot{q}_i(t) - \dot{q}_o(t)]\, dt$$

6.   The mechanical equation is

$$T(t) = J\dot{\omega}(t) + B\omega(t) + T_d$$

(a)   Draw a functional block diagram of the engine with $\alpha(t)$ as input, $\omega(t)$ as output, and T_d as the disturbance input. Show the transfer function of each block.

(b)   Find the transfer function $\Omega(s)/\alpha(s)$ of the system.

(c)   Find the characteristic equation and show that it is not rational with constant coefficients.

(d)   Approximate the engine time delay by

$$e^{-\tau_D s} \cong \frac{1 - \tau_D s/2}{1 + \tau_D s/2}$$

and repeat parts (b) and (c).

4-20.   Phase-locked loops are control systems used for precision motor-speed control. The basic elements of a phase-locked-loop system incorporating a dc motor is shown in Fig. 4P-20(a). An input pulse train represents the reference frequency or desired output

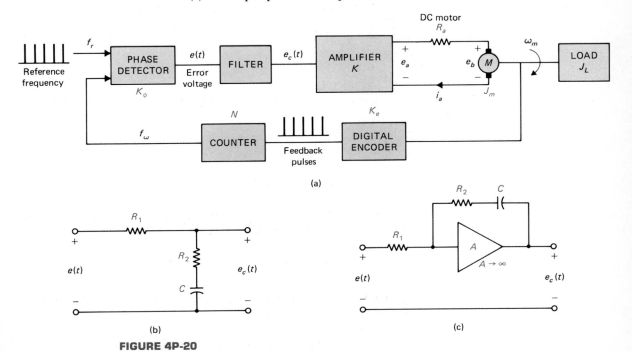

(a)

FIGURE 4P-20

speed. The digital encoder produces digital pulses that represent motor speed. The phase detector compares the motor speed and the reference frequency and sends an error voltage to the filter (controller) that governs the dynamic response of the system.

Phase detector gain = K_p, encoder gain = K_e, counter gain = $1/N$, and dc-motor torque constant = K_i. Assume zero inductance and zero friction for motor.

(a)   Derive the transfer function $E_c(s)/E(s)$ of the filter shown in Fig. 4-20(b). Assume that the filter sees infinite impedance at the output and zero impedance at the input.

(b)   Draw a functional block diagram of the system with gains or transfer functions in the blocks.

(c)   Drive the open-loop transfer function $\Omega_m(s)/E(s)$ when the feedback path is open.

(d)   Find the closed-loop transfer function $\Omega_m(s)/F_r(s)$.

(e)   Repeat parts (a), (c), and (d) for the filter shown in Fig. 4P-20(c).

(f)   The digital encoder has an output of 36 pulses per revolution. The reference frequency f_r is fixed at 120 pulses/s. Find K_e in pulses/rad. The idea of using the counter N is that with f_r fixed, various desired output speeds can be attained by changing the value of N. Find N if the desired output speed is 200 rpm. Find N if the desired output speed is 1800 rpm.

4-21.   The linearized model of a robot arm system driven by a dc motor is shown in Fig. 4P-21. The system parameters and variables are given as follows:

DC Motor:	Robot Arm:
T_m = motor torque = $K_i i_a$	J_L = inertia of arm
K_i = torque constant	T_L = disturbance torque on arm
i_a = armature current	θ_L = arm displacement
J_m = motor inertia	K = torsional spring constant
B_m = motor viscous-friction coefficient	B = viscous-friction coefficient of shaft between the motor and arm
θ_m = motor-shaft displacement	B_L = viscous-friction coefficient of the robot arm shaft

(a)   Write the differential equations of the system with $i_a(t)$ and $T_L(t)$ as input and $\theta_m(t)$ and $\theta_L(t)$ as outputs.

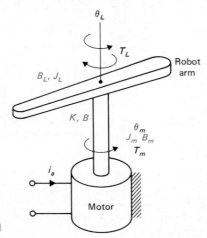

FIGURE 4P-21

(b) Draw a signal-flow graph using $I_a(s)$, $T_l(s)$, $\Theta_m(s)$, and $\Theta_L(s)$ as node variables.

(c) Express the transfer-function relations as

$$\begin{bmatrix} \Theta_m(s) \\ \Theta_L(s) \end{bmatrix} = \mathbf{G}(s) \begin{bmatrix} I_a(s) \\ -T_L(s) \end{bmatrix}$$

Find $\mathbf{G}(s)$.

4-22. The following differential equations describe the motion of an electric train in a traction system:

$$\dot{x}(t) = v(t)$$

$$\dot{v}(t) = -k(v) - g(x) + f(t)$$

where $x(t)$ = linear displacement of train
  $v(t)$ = linear velocity of train
  $k(v)$ = resistance force on train [odd function of v, with the properties: $k(0) = 0$ and $dk(v)/dv = 0$]
  $g(x)$ = gravitational force for a nonlevel track or due to curvature of track
  $f(t)$ = tractive force

The electric motor that provides the tractive force is described by the following equations:

$$e(t) = K_b\phi(t)v(t) + R_a i_a(t)$$

$$f(t) = K_i\phi(t)i_a(t)$$

where $e(t)$ is the applied voltage, $i_a(t)$ the armature current, $i_f(t)$ the field current, R_a the armature resistance, $\phi(t)$ the magnetic flux from a separately excited field = $K_f i_f(t)$, and K_i the force constant.

(a) Consider that the motor is a dc series motor with the armature and field windings connected in series, so that $i_a(t) = i_f(t)$, $g(x) = 0$, $k(v) = Bv(t)$, and $R_a = 0$. Show that the system is described by the following nonlinear state equations:

$$\dot{x}(t) = v(t)$$

$$\dot{v}(t) = -Bv(t) + \frac{K_i}{K_b^2 K_f v^2(t)}e^2(t)$$

(b) Consider that for the conditions stated in part (a), $i_a(t)$ is the input of the system [instead of $e(t)$]. Derive the state equations of the system.

(c) Consider the same conditions as in part (a), but with $\phi(t)$ as the input. Derive the state equations.

4-23. Figure 4P-23(a) shows a well-known "broom-balancing" system in control systems. The objective of the control system is to maintain the "broom" in the upright position by means of the force $u(t)$ applied to the car as shown. In practical applications, the system is analogous to the balancing of a unicycle, or a one-dimensional control problem of a missile during launching.

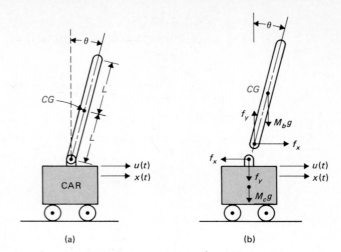

FIGURE 4P-23   (a)   (b)

The free-body diagram of the system is shown in Fig. 4P-23(b), where

f_x = force at broom base in horizontal direction
f_y = force at broom base in vertical direction
M_b = mass of broom
g = gravitational acceleration
M_c = mass of car
J_b = moment of inertia of broom about center of gravity CG
   = $M_b L^2/3$

(a)   Write the force equations in the x and the y directions at the pivot point of the broom. Write the torque equation about the center of gravity CG of the broom. Write the force equation of the car in the horizontal direction.

(b)   Express the equations obtained in part (a) as state equations by assigning the state variables as $x_1 = \theta$, $x_2 = d\theta/dt$, $x_3 = x$, and $x_4 = dx/dt$. Simplify these equations for small θ by making the approximations: $\sin \theta \cong \theta$ and $\cos \theta \cong 1$.

(c)   Obtain a small-signal linearized state-equation model for the system in the form of

$$\Delta \dot{\mathbf{x}} = \mathbf{A}^* \Delta \mathbf{x} + \mathbf{B}^* \Delta \mathbf{r}$$

at the equilibrium point $x_{01}(t) = 1$, $x_{02}(t) = 0$, $x_{03}(t) = 0$, and $x_{04}(t) = 0$.

4-24.   Figure 4P-24 shows the schematic diagram of a ball-suspension control system. The steel ball is suspended in the air by the electromagnetic force generated by the electromagnet. The objective of the control is to keep the metal ball suspended at the nominal equilibrium position by controlling the current in the magnet with the voltage $e(t)$. The resistance of the coil is R, and the inductance is $L(y) = L/y(t)$, where L is a constant. The applied voltage $e(t)$ is a constant with amplitude E.

(a)   Let E_{eq} be a nominal value of E. Find the nominal values of $y(t)$ and $dy(t)/dt$ at equilibrium.

(b)   Define the state variables at $x_1(t) = i(t)$, $x_2(t) = y(t)$, and $x_3 = dy(t)/dt$. Find the nonlinear state equations in the form of

$$\dot{\mathbf{x}} = \mathbf{f}(\mathbf{x}, e)$$

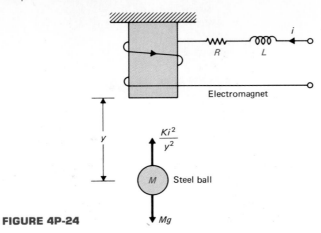

FIGURE 4P-24

(c)   Linearize the state equations about the equilibrium point and express the linearized state equations as

$$\Delta\dot{\mathbf{x}} = \mathbf{A}^* \Delta\mathbf{x} + \mathbf{B}^* \Delta e$$

The force generated by the electromagnet is $Ki^2(t)/y(t)$, where K is a proportional constant, and the gravitational force on the steel ball is Mg.

4-25.   Figure 4P-25 shows the schematic diagram of a ball-suspension control system. The steel ball is suspended in the air by the electromagnetic force generated by the electromagnet. The objective of the control is to keep the metal ball suspended at the nominal position by controlling the current in the electromagnet. When the system is at the stable equilibrium point, any small perturbation of the ball position from its floating equilibrium position will cause the control to return the ball to the equilibrium position. The

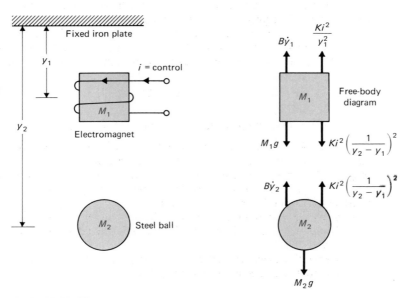

FIGURE 4P-25

free-body diagram of the system is given in Fig. 4P-25, where

M_1 = mass of electromagnet = 2.0
M_2 = mass of steel ball = 1.0
B = viscous-friction coefficient of air = 0.1
K = proportional constant of electromagnet = 1.0
g = acceleration due to gravity = 32.2

Assume all units are consistent. Let the stable equilibrium values of the variables $i(t)$, $y_1(t)$, and $y_2(t)$ be I, Y_1, and Y_2, respectively. The state variables are defined as $x_1 = y_1$, $x_2 = dy_1/dt$, $x_3 = y_2$, and $x_4 = dy_2/dt$.

(a)   Given $Y_1 = 1$, find I and Y_2.
(b)   Write the nonlinear state equations of the system in the form of $dx/dt = f(x, i)$.
(c)   Find the state equations of the linearized system about the equilibrium state I, Y_1, and Y_2 in the form

$$\Delta\dot{x} = A^* \, \Delta x + B^* \, \Delta i$$

4-26.   The schematic diagram of a steel-rolling process is shown in Fig. 4P-26. The steel plate is fed through the rollers at a constant speed of V ft/s. The distance between the rollers and the point where the thickness is measured is d ft. The rotary displacement of the motor, $\theta_m(t)$, is converted to the linear displacement $y(t)$ by the gear box and linear-actuator combination; $y(t) = n\theta_m(t)$, where n is a positive constant in ft/rad. The equivalent inertia of the load that is reflected to the motor shaft is J_L.
(a)   Draw a functional block diagram for the system.
(b)   Derive the open-loop transfer function $C(s)/E(s)$ and the closed-loop transfer function $C(s)/R(s)$.

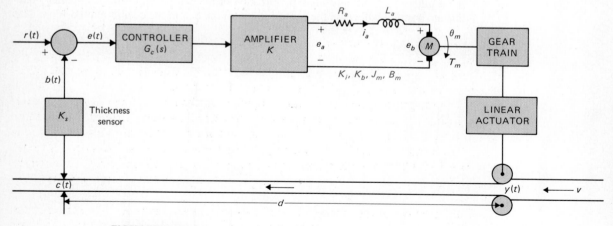

FIGURE 4P-26

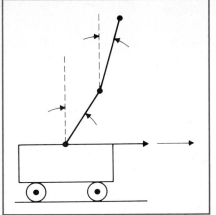

5

STATE-VARIABLE ANALYSIS OF LINEAR DYNAMIC SYSTEMS

5-1

INTRODUCTION

In Chapter 2, we defined the state equations of a dynamic system informally as a set of first-order differential equations relating the state variables among themselves and to the inputs. In Chapter 3, the signal-flow-graph method is extended to the modeling of the state equations and the result is the **state diagram**. The concept and definition of state variables and state equations are formally presented in Chapter 4, together with several illustrative examples on how the state variables are chosen and how state equations are written for linear and nonlinear dynamic systems.

In contrast to the transfer-function approach to the analysis and design of linear control systems, the state-variable method is regarded as modern, since it is the underlying force for optimal control. The basic characteristic of the state-variable formulation is that linear and nonlinear systems, time-invariant and time-invarying systems, and single-variable and multivariable systems can all be modeled in a unified manner. Transfer functions, on the other hand, are defined only for linear time-invariant systems.

The objective of this chapter is to introduce the basic methods of state variables and state equations, so that the reader can gain a working knowledge of the subject for further studies when the state approach is used for modern and optimal control design. Specifically, the closed-form solutions of linear time-invariant state equations are presented. Various transformations that may be used to facilitate the analysis and design of linear control systems in the state-variable domain are introduced. The relationship between the conventional transfer-function approach and the state-variable approach is established, so that the analyst will be able to investigate a system problem with various alternative methods. Finally, the controllability and observability of linear systems are defined and their applications are investigated.

5-2

MATRIX REPRESENTATION OF STATE EQUATIONS

Let the n state equations of an nth-order dynamic system be represented as

$$\frac{dx_i(t)}{dt} = f_i[x_1(t), x_2(t), \ldots, x_n(t), r_1(t), r_2(t), \ldots, r_p(t), \\ w_1(t), w_2(t), \ldots, w_v(t)] \tag{5-1}$$

where $i = 1, 2, \ldots, n$. The ith state variable is represented by $x_i(t)$; $r_j(t)$ denotes the jth input for $j = 1, 2, \ldots, p$; and $w_k(t)$ denotes the kth disturbance input, with $k = 1, 2, \ldots, v$.

Let the variables $c_1(t), c_2(t), \ldots, c_q(t)$ be the q output variables of the system. The output variables represent the link between the system and the outside world. In practice, not all the state variables are accessible, or, simply, measurable. However, an output variable must be measurable or accessible by some physical means. For instance, in an electric motor, such state variables as the winding current, rotor velocity and displacement can be measured physically, and these variables all qualify as output variables. On the other hand, magnetic flux can also be regarded as a state variable in an electric motor, since it represents the past, present, and future states of the motor, but it cannot be measured directly during operation, and, therefore, it does not ordinarily qualify as an output variable.

output equations

In general, the output variables are functions of the state variables and the input variables. The **output equations** of a dynamic system can be expressed as

$$c_j(t) = g_j[x_1(t), x_2(t), \ldots, x_n(t), r_1(t), r_2(t), \ldots, r_p(t), w_1(t), w_2(t), \ldots, w_v(t)] \tag{5-2}$$

where $j = 1, 2, \ldots, q$.

dynamic equations

The set of n state equations in Eq. (5-1) and q output equations in Eq. (5-2) together form the so-called **dynamic equations**.

For ease of expression and manipulation, it is convenient to represent the dynamic equations in vector-matrix form. Let us define the following column matrices, which are also called **vectors**:

$$\mathbf{x}(t) = \begin{bmatrix} x_1(t) \\ x_2(t) \\ \cdot \\ \cdot \\ \cdot \\ x_n(t) \end{bmatrix} \qquad (n \times 1) \tag{5-3}$$

$$\mathbf{r}(t) = \begin{bmatrix} r_1(t) \\ r_2(t) \\ \cdot \\ \cdot \\ \cdot \\ r_p(t) \end{bmatrix} \qquad (p \times 1) \qquad\qquad (5\text{-}4)$$

$$\mathbf{c}(t) = \begin{bmatrix} c_1(t) \\ c_2(t) \\ \cdot \\ \cdot \\ \cdot \\ c_q(t) \end{bmatrix} \qquad (q \times 1) \qquad\qquad (5\text{-}5)$$

$$\mathbf{w}(t) = \begin{bmatrix} w_1(t) \\ w_2(t) \\ \cdot \\ \cdot \\ \cdot \\ w_v(t) \end{bmatrix} \qquad (v \times 1) \qquad\qquad (5\text{-}6)$$

state vector
disturbance
vector

The $n \times 1$ column matrix $\mathbf{x}(t)$ is called the **state vector**; $\mathbf{r}(t)$ is the $p \times 1$ **input vector**, $\mathbf{c}(t)$ is the $q \times 1$ **output vector**, and $\mathbf{w}(t)$ is the $v \times 1$ **disturbance vector**. By using these vectors, the n state equations of Eq. (5-1) can be written:

$$\frac{d\mathbf{x}(t)}{dt} = \mathbf{f}[\mathbf{x}(t), \mathbf{r}(t), \mathbf{w}(t)] \qquad\qquad (5\text{-}7)$$

where $\mathbf{f}$ denotes an $n \times 1$ column matrix that contains the functions $f_1, f_2, \ldots, f_n$ as elements. Similarly, the q output equations in Eq. (5-2) become

$$\mathbf{c}(t) = \mathbf{g}[\mathbf{x}(t), \mathbf{r}(t), \mathbf{w}(t)] \qquad\qquad (5\text{-}8)$$

where $\mathbf{g}$ denotes a $q \times 1$ column matrix that contains the functions $g_1, g_2, \ldots, g_q$ as elements.

For a linear time-invariant system, the dynamic equations are written as

State equations: $\quad \dfrac{d\mathbf{x}(t)}{dt} = \mathbf{A}\mathbf{x}(t) + \mathbf{B}\mathbf{r}(t) + \mathbf{F}\mathbf{w}(t) \qquad\qquad (5\text{-}9)$

Output equations: $\quad \mathbf{c}(t) = \mathbf{D}\mathbf{x}(t) + \mathbf{E}\mathbf{r}(t) + \mathbf{H}\mathbf{w}(t) \qquad\qquad (5\text{-}10)$

where $\mathbf{A}$ is an $n \times n$ coefficient matrix with constant elements:

$$\mathbf{A} = \begin{bmatrix} a_{11} & a_{12} & \cdots & a_{1n} \\ a_{21} & a_{22} & \cdots & a_{2n} \\ \cdot & \cdot & \cdot & \cdot \\ \cdot & \cdot & \cdot & \cdot \\ \cdot & \cdot & \cdot & \cdot \\ a_{n1} & a_{n2} & \cdots & a_{nn} \end{bmatrix} \qquad\qquad (5\text{-}11)$$

B is an $n \times p$ coefficient matrix with constant elements:

$$\mathbf{B} = \begin{bmatrix} b_{11} & b_{12} & \cdots & b_{1p} \\ b_{21} & b_{22} & \cdots & b_{2p} \\ \vdots & \vdots & \ddots & \vdots \\ b_{n1} & b_{n2} & \cdots & b_{np} \end{bmatrix} \tag{5-12}$$

D is a $q \times n$ coefficient matrix with constant elements:

$$\mathbf{D} = \begin{bmatrix} d_{11} & d_{12} & \cdots & d_{1n} \\ d_{21} & d_{22} & \cdots & d_{2n} \\ \vdots & \vdots & \ddots & \vdots \\ d_{q1} & d_{q2} & \cdots & d_{qn} \end{bmatrix} \tag{5-13}$$

E is a $q \times p$ coefficient matrix with constant elements:

$$\mathbf{E} = \begin{bmatrix} e_{11} & e_{12} & \cdots & e_{1p} \\ e_{21} & e_{22} & \cdots & e_{2p} \\ \vdots & \vdots & \ddots & \vdots \\ e_{q1} & e_{q2} & \cdots & e_{qp} \end{bmatrix} \tag{5-14}$$

F is an $n \times v$ coefficient matrix with constant elements:

$$\mathbf{F} = \begin{bmatrix} f_{11} & f_{12} & \cdots & f_{1v} \\ f_{21} & f_{22} & \cdots & f_{2v} \\ \vdots & \vdots & \ddots & \vdots \\ f_{n1} & f_{n2} & \cdots & f_{nv} \end{bmatrix} \tag{5-15}$$

and **H** is a $q \times v$ coefficient matrix with constant elements:

$$\mathbf{H} = \begin{bmatrix} h_{11} & h_{12} & \cdots & h_{1v} \\ h_{21} & h_{22} & \cdots & h_{2v} \\ \vdots & \vdots & \ddots & \vdots \\ h_{q1} & h_{q2} & \cdots & h_{qv} \end{bmatrix} \tag{5-16}$$

5-3

STATE-TRANSITION MATRIX

state-transition matrix

Once the state equations of a linear time-invariant system are expressed in the form of Eq. (5-9), the next step often involves the solutions of these equations given the initial-state vector $\mathbf{x}(t_0)$, the input vector $\mathbf{r}(t)$, and the disturbance vector $\mathbf{w}(t)$, for $t \geq t_0$. The first term on the right-hand side of Eq. (5-9) is known as the homogeneous part of the state equation, and the last two terms represent the forcing functions $\mathbf{r}(t)$ and $\mathbf{w}(t)$.

The **state-transition matrix** is defined as a matrix that satisfies the linear homogeneous state equation:

$$\frac{d\mathbf{x}(t)}{dt} = \mathbf{A}\mathbf{x}(t) \tag{5-17}$$

Let $\boldsymbol{\phi}(t)$ be an $n \times n$ matrix that represents the state-transition matrix; then it must satisfy the equation

$$\frac{d\boldsymbol{\phi}(t)}{dt} = \mathbf{A}\boldsymbol{\phi}(t) \tag{5-18}$$

Furthermore, let $\mathbf{x}(0)$ denote the initial state at $t = 0$; then $\boldsymbol{\phi}(t)$ is also defined by the matrix equation

$$\mathbf{x}(t) = \boldsymbol{\phi}(t)\mathbf{x}(0) \tag{5-19}$$

which is the solution of the homogeneous state equation for $t \geq 0$.

One way of determining $\boldsymbol{\phi}(t)$ is by taking the Laplace transform on both sides of Eq. (5-17); we have

$$s\mathbf{X}(s) - \mathbf{x}(0) = \mathbf{A}\mathbf{X}(s) \tag{5-20}$$

Solving for $\mathbf{X}(s)$ from the last equation, we get

$$\mathbf{X}(s) = (s\mathbf{I} - \mathbf{A})^{-1}\mathbf{x}(0) \tag{5-21}$$

where it is assumed that the matrix $(s\mathbf{I} - \mathbf{A})$ is nonsingular. Taking the inverse Laplace transform on both sides of the last equation yields

$$\mathbf{x}(t) = \mathcal{L}^{-1}[(s\mathbf{I} - \mathbf{A})^{-1}]\mathbf{x}(0) \qquad t \geq 0 \tag{5-22}$$

By comparing Eq. (5-19) with Eq. (5-22), the state-transition matrix is identified to be

$$\boldsymbol{\phi}(t) = \mathcal{L}^{-1}[(s\mathbf{I} - \mathbf{A})^{-1}] \tag{5-23}$$

An alternative way of solving the homogeneous state equation is to assume a solution, as in the classical method of solving linear differential equations. We let the solution to Eq. (5-17) be

$$\mathbf{x}(t) = e^{\mathbf{A}t}\mathbf{x}(0) \tag{5-24}$$

for $t \geq 0$, where $e^{\mathbf{A}t}$ represents the following power series of the matrix $\mathbf{A}t$, and

$$e^{\mathbf{A}t} = \mathbf{I} + \mathbf{A}t + \frac{1}{2!}\mathbf{A}^2 t^2 + \frac{1}{3!}\mathbf{A}^3 t^3 + \cdots \tag{5-25}$$

It is easy to show that Eq. (5-24) is a solution of the homogeneous state equation, since, from Eq. (5-25),

$$\frac{de^{\mathbf{A}t}}{dt} = \mathbf{A}e^{\mathbf{A}t} \qquad (5\text{-}26)$$

Therefore, in addition to Eq. (5-23), we have obtained another expression for the state-transition matrix:

$$\boldsymbol{\phi}(t) = e^{\mathbf{A}t} = \mathbf{I} + \mathbf{A}t + \frac{1}{2!}\mathbf{A}^2 t^2 + \frac{1}{3!}\mathbf{A}^3 t^3 + \cdots \qquad (5\text{-}27)$$

Equation (5-27) can also be obtained directly from Eq. (5-23). This is left as an exercise for the reader (Problem 5-3).

EXAMPLE 5-1   As a simple illustrative example of state variables, state equations, the state-transition matrix, and the state-transition equation, let us consider the *RL* network shown in Fig. 5-1. The history of the network is completely specified by the initial current of the inductance, $i(0)$ at $t = 0$. Consider that at $t = 0$, a constant input voltage of magnitude E is applied to the network. The state equation of the network for $t \ge 0$ is

$$\frac{di(t)}{dt} = -\frac{R}{L}i(t) + \frac{1}{L}e(t) \qquad (5\text{-}28)$$

In this case, the system is of the first order, and there is only one state equation. By comparing with Eq. (5-9), $i(t)$ is the state variable $x(t)$, $e(t)$ is the input $r(t)$, $w(t) = 0$, $A = -R/L$, $B = 1/L$, and $F = 0$. Taking the Laplace transform on both sides of the last equation, we get

$$sI(s) - i(0) = -\frac{R}{L}I(s) + \frac{E}{Ls} \qquad (5\text{-}29)$$

Solving for $I(s)$ from Eq. (5-29) yields

$$I(s) = \frac{L}{R + Ls}i(0) + \frac{E}{s(R + Ls)} \qquad (5\text{-}30)$$

The current $i(t)$ for $t \ge 0$ is obtained by taking the inverse Laplace transform on both sides of

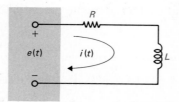

FIGURE 5-1 *RL* network.

the last equation. We have

$$i(t) = e^{-Rt/L}i(0) + \frac{E}{R}(1 - e^{-Rt/L}) \qquad t \geq 0 \qquad (5\text{-}31)$$

Equation (5-31) represents the complete solution of the state equation with the current $i(t)$ as the state variable for the specified initial state $i(0)$ and the specific input $e(t)$ for $t \geq 0$. It is apparent that $i(t)$ satisfies the basic requirements as a state variable. This is not surprising since an inductor stores electric kinetic energy, and it is the energy-storage capability that holds the information on the history of the system. Similarly, in general, the voltage across a capacitor also qualifies as a state variable, since the capacitor stores electric potential energy.

The first term on the right-hand side of Eq. (5-31) is the solution of the homogeneous state equation, that is, the state equation with $e(t) = 0$. Thus, the state-transition matrix of A, where $A = -R/L$, is

$$\phi(t) = e^{-Rt/L} \qquad (5\text{-}32)$$

Apparently, $\phi(t)$ can also be obtained directly by substituting A into Eq. (5-27).

Significance of the State-Transition Matrix

free response

Since the state-transition matrix satisfies the homogeneous state equation, it represents the **free response** of the system. In other words, it governs the response that is excited by the initial conditions only. In view of Eqs. (5-23) and (5-27), the state-transition matrix is dependent only upon the matrix **A**, and, therefore, is sometimes referred to as the **state-transition matrix of A**. As the name implies, the state-transition matrix $\phi(t)$ completely defines the transition of the states from the initial time $t = 0$ to any time t when the inputs are zero.

Properties of the State-Transition Matrix

properties

The state-transition matrix $\phi(t)$ possesses the following properties:

1.

$$\phi(0) = \mathbf{I} \qquad \text{(the identity matrix)} \qquad (5\text{-}33)$$

PROOF. Equation (5-33) follows directly from Eq. (5-27) by setting $t = 0$.

2.

$$\boldsymbol{\phi}^{-1}(t) = \boldsymbol{\phi}(-t) \qquad (5\text{-}34)$$

PROOF.   Postmultiplying both sides of Eq. (5-27) by $e^{-\mathbf{A}t}$, we get

$$\boldsymbol{\phi}(t)e^{-\mathbf{A}t} = e^{\mathbf{A}t}e^{-\mathbf{A}t} = \mathbf{I} \tag{5-35}$$

Then, premultiplying both sides of Eq. (5-35) by $\boldsymbol{\phi}^{-1}(t)$, we get

$$e^{-\mathbf{A}t} = \boldsymbol{\phi}^{-1}(t) \tag{5-36}$$

Thus,

$$\boldsymbol{\phi}(-t) = \boldsymbol{\phi}^{-1}(t) = e^{-\mathbf{A}t} \tag{5-37}$$

An interesting result from this property of $\boldsymbol{\phi}(t)$ is that Eq. (5-24) can be rearranged to read

$$\mathbf{x}(0) = \boldsymbol{\phi}(-t)\mathbf{x}(t) \tag{5-38}$$

which means that the state-transition process can be considered as bilateral in time. That is, the transition in time can take place in either direction.

3.

$$\boldsymbol{\phi}(t_2 - t_1)\boldsymbol{\phi}(t_1 - t_0) = \boldsymbol{\phi}(t_2 - t_0) \qquad \text{for any } t_0, t_1, t_2 \tag{5-39}$$

PROOF.

$$\begin{aligned}
\boldsymbol{\phi}(t_2 - t_1)\boldsymbol{\phi}(t_1 - t_0) &= e^{\mathbf{A}(t_2-t_1)}e^{\mathbf{A}(t_1-t_0)} \\
&= e^{\mathbf{A}(t_2-t_0)} = \boldsymbol{\phi}(t_2 - t_0)
\end{aligned} \tag{5-40}$$

This property of the state-transition matrix is important since it implies that a state-transition process can be divided into a number of sequential transitions. Figure 5-2 illustrates that the transition from $t = t_0$ to $t = t_2$ is equal to the transition from t_0 to t_1, and then from t_1 to t_2. In general, of course, the state-transition process can be divided into any number of parts.

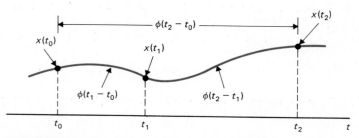

FIGURE 5-2 Property of the state-transition matrix.

Another way of proving Eq. (5-39) is to write

$$\mathbf{x}(t_2) = \boldsymbol{\phi}(t_2 - t_1)\mathbf{x}(t_1) \tag{5-41}$$

$$\mathbf{x}(t_1) = \boldsymbol{\phi}(t_1 - t_0)\mathbf{x}(t_0) \tag{5-42}$$

$$\mathbf{x}(t_2) = \boldsymbol{\phi}(t_2 - t_0)\mathbf{x}(t_0) \tag{5-43}$$

The proper result is obtained by substituting Eq. (5-42) into Eq. (5-41) and comparing the result with Eq. (5-43).

4.

$$[\boldsymbol{\phi}(t)]^k = \boldsymbol{\phi}(kt) \qquad \text{for } k = \text{positive integer} \tag{5-44}$$

PROOF.

$$[\boldsymbol{\phi}(t)]^k = e^{\mathbf{A}t}e^{\mathbf{A}t} \cdots e^{\mathbf{A}t} \qquad (k \text{ terms})$$

$$= e^{k\mathbf{A}t} \tag{5-45}$$

$$= \boldsymbol{\phi}(kt)$$

5-4
STATE-TRANSITION EQUATION

state-transition equation

The **state-transition equation** *is defined as the solution of the linear nonhomogeneous state equation.* For example, Eq. (5-28) is a state equation of the *RL* network of Fig. 5-1. Then Eq. (5-31) is the state-transition equation when the input voltage is constant of amplitude *E* for $t \geq 0$.

In general, the linear time-invariant state equation

$$\frac{d\mathbf{x}(t)}{dt} = \mathbf{A}\mathbf{x}(t) + \mathbf{B}\mathbf{r}(t) + \mathbf{F}\mathbf{w}(t) \tag{5-46}$$

can be solved by using either the classical method of solving linear differential equations or the Laplace-transform method. The Laplace-transform solution is presented in the following.

Taking the Laplace transform on both sides of Eq. (5-46), we have

$$s\mathbf{X}(s) - \mathbf{x}(0) = \mathbf{A}\mathbf{X}(s) + \mathbf{B}\mathbf{R}(s) + \mathbf{F}\mathbf{W}(s) \tag{5-47}$$

where $\mathbf{x}(0)$ denotes the initial-state vector evaluated at $t = 0$. Solving for $\mathbf{X}(s)$ in Eq. (5-47) yields

$$\mathbf{X}(s) = (s\mathbf{I} - \mathbf{A})^{-1}\mathbf{x}(0) + (s\mathbf{I} - \mathbf{A})^{-1}[\mathbf{B}\mathbf{R}(s) + \mathbf{F}\mathbf{W}(s)] \tag{5-48}$$

The state-transition equation of Eq. (5-46) is obtained by taking the inverse Laplace transform on both sides of Eq. (5-48):

$$\mathbf{x}(t) = \mathscr{L}^{-1}[(s\mathbf{I} - \mathbf{A})^{-1}]\mathbf{x}(0) + \mathscr{L}^{-1}\{(s\mathbf{I} - \mathbf{A})^{-1}[\mathbf{B}\mathbf{R}(s) + \mathbf{F}\mathbf{W}(s)]\} \qquad (5\text{-}49)$$

$$\mathbf{x}(t) = \boldsymbol{\phi}(t)\mathbf{x}(0) + \int_0^t \boldsymbol{\phi}(t - \tau)[\mathbf{B}\mathbf{r}(\tau) + \mathbf{F}\mathbf{w}(\tau)]\, d\tau \qquad t \geq 0 \qquad (5\text{-}50)$$

The state-transition equation in Eq. (5-50) is useful only when the initial time is defined to be at $t = 0$. In the study of control systems, especially discrete-data control systems, it is often desirable to break up a state-transition process into a sequence of transitions, so a more flexible initial time must be chosen. Let the initial time be represented by t_0 and the corresponding initial state by $\mathbf{x}(t_0)$, and assume that the input $\mathbf{r}(t)$ and the disturbance $\mathbf{w}(t)$ are applied at $t \geq 0$.

We start with Eq. (5-50) by setting $t = t_0$, and solving for $\mathbf{x}(0)$, we get

$$\mathbf{x}(0) = \boldsymbol{\phi}(-t_0)\mathbf{x}(t_0) - \boldsymbol{\phi}(-t_0)\int_0^{t_0} \boldsymbol{\phi}(t_0 - \tau)[\mathbf{B}\mathbf{r}(\tau) + \mathbf{F}\mathbf{w}(\tau)]\, d\tau \qquad (5\text{-}51)$$

where the property on $\boldsymbol{\phi}(t)$ of Eq. (5-34) has been applied.

Substituting Eq. (5-51) into Eq. (5-50) yields

$$\mathbf{x}(t) = \boldsymbol{\phi}(t)\boldsymbol{\phi}(-t_0)\mathbf{x}(t_0) - \boldsymbol{\phi}(t)\boldsymbol{\phi}(-t_0)\int_0^{t_0} \boldsymbol{\phi}(t_0 - \tau)[\mathbf{B}\mathbf{r}(\tau) + \mathbf{F}\mathbf{w}(\tau)]\, d\tau$$
$$+ \int_0^t \boldsymbol{\phi}(t - \tau)[\mathbf{B}\mathbf{r}(\tau) + \mathbf{F}\mathbf{w}(\tau)]\, d\tau \qquad (5\text{-}52)$$

Now by using the property of Eq. (5-39), and combining the last two integrals, Eq. (5-52) becomes

$$\mathbf{x}(t) = \boldsymbol{\phi}(t - t_0)\mathbf{x}(t_0) + \int_{t_0}^t \boldsymbol{\phi}(t - \tau)[\mathbf{B}\mathbf{r}(\tau) + \mathbf{F}\mathbf{w}(\tau)]\, d\tau \qquad t \geq t_0 \qquad (5\text{-}53)$$

It is apparent that Eq. (5-53) reverts to Eq. (5-50) when $t_0 = 0$.

Once the state-transition equation is determined, the output vector can be expressed as a function of the initial state and the input vector simply by substituting $\mathbf{x}(t)$ from Eq. (5-53) into Eq. (5-10). Thus, the output vector is

$$\mathbf{c}(t) = \mathbf{D}\boldsymbol{\phi}(t - t_0)\mathbf{x}(t_0) + \int_{t_0}^t \mathbf{D}\boldsymbol{\phi}(t - \tau)[\mathbf{B}\mathbf{r}(\tau) + \mathbf{F}\mathbf{w}(\tau)]\, d\tau$$
$$+ \mathbf{E}\mathbf{r}(t) + \mathbf{H}\mathbf{w}(t) \qquad t \geq t_0 \qquad (5\text{-}54)$$

The following example illustrates the application of the state-transition equation.

EXAMPLE 5-2   Consider the state equation

$$\begin{bmatrix} \dfrac{dx_1(t)}{dt} \\[2mm] \dfrac{dx_2(t)}{dt} \end{bmatrix} = \begin{bmatrix} 0 & 1 \\ -2 & -3 \end{bmatrix} \begin{bmatrix} x_1(t) \\ x_2(t) \end{bmatrix} + \begin{bmatrix} 0 \\ 1 \end{bmatrix} r(t) \tag{5-55}$$

The problem is to determine the state vector $\mathbf{x}(t)$ for $t \geq 0$ when the input is $r(t) = 1$ for $t \geq 0$; that is, $r(t) = u_s(t)$. The coefficient matrices are identified to be

$$\mathbf{A} = \begin{bmatrix} 0 & 1 \\ -2 & -3 \end{bmatrix} \qquad \mathbf{B} = \begin{bmatrix} 0 \\ 1 \end{bmatrix} \qquad \mathbf{F} = 0 \tag{5-56}$$

Therefore,

$$s\mathbf{I} - \mathbf{A} = \begin{bmatrix} s & 0 \\ 0 & s \end{bmatrix} - \begin{bmatrix} 0 & 1 \\ -2 & -3 \end{bmatrix} = \begin{bmatrix} s & -1 \\ 2 & s+3 \end{bmatrix} \tag{5-57}$$

The inverse matrix of $(s\mathbf{I} - \mathbf{A})$ is

$$(s\mathbf{I} - \mathbf{A})^{-1} = \frac{1}{s^2 + 3s + 2} \begin{bmatrix} s+3 & 1 \\ -2 & s \end{bmatrix} \tag{5-58}$$

The state-transition matrix of $\mathbf{A}$ is found by taking the inverse Laplace transform of the last equation. Thus,

$$\boldsymbol{\phi}(t) = \mathcal{L}^{-1}[(s\mathbf{I} - \mathbf{A})^{-1}] = \begin{bmatrix} 2e^{-t} - e^{-2t} & e^{-t} - e^{-2t} \\ -2e^{-t} + 2e^{-2t} & -e^{-t} + 2e^{-2t} \end{bmatrix} \tag{5-59}$$

The state-transition equation for $t \geq 0$ is obtained by substituting Eq. (5-59), $\mathbf{B}$, and $r(t)$ into Eq. (5-50). We have

$$\mathbf{x}(t) = \begin{bmatrix} 2e^{-t} - e^{-2t} & e^{-t} - e^{-2t} \\ -2e^{-t} + 2e^{-2t} & -e^{-t} + 2e^{-2t} \end{bmatrix} \mathbf{x}(0)$$
$$+ \int_0^t \begin{bmatrix} 2e^{-(t-\tau)} - e^{-2(t-\tau)} & e^{-(t-\tau)} - e^{-2(t-\tau)} \\ -2e^{-(t-\tau)} + e^{-2(t-\tau)} & -e^{-(t-\tau)} + 2e^{-2(t-\tau)} \end{bmatrix} \begin{bmatrix} 0 \\ 1 \end{bmatrix} d\tau \tag{5-60}$$

or

$$\mathbf{x}(t) = \begin{bmatrix} 2e^{-t} - e^{-2t} & e^{-t} - e^{-2t} \\ -2e^{-t} + 2e^{-2t} & -e^{-t} + 2e^{-2t} \end{bmatrix} \mathbf{x}(0) + \begin{bmatrix} 0.5 - e^{-t} + 0.5e^{-2t} \\ e^{-t} - e^{-2t} \end{bmatrix} \qquad t \geq 0 \tag{5-61}$$

As an alternative, the second term of the state-transition equation can be obtained by taking the inverse Laplace transform of $(s\mathbf{I} - \mathbf{A})^{-1}\mathbf{B}R(s)$. Therefore,

$$\mathcal{L}^{-1}[(s\mathbf{I} - \mathbf{A})^{-1}\mathbf{B}R(s)] = \mathcal{L}^{-1}\left(\frac{1}{s^2 + 3s + 2} \begin{bmatrix} s+3 & 1 \\ -2 & s \end{bmatrix} \begin{bmatrix} 0 \\ 1 \end{bmatrix} \frac{1}{s} \right)$$

$$= \mathcal{L}^{-1}\left(\frac{1}{s^2 + 3s + 2} \begin{bmatrix} \frac{1}{s} \\ 1 \end{bmatrix} \right) \tag{5-62}$$

$$= \begin{bmatrix} 0.5 - e^{-t} + 0.5e^{-2t} \\ e^{-t} - e^{-2t} \end{bmatrix} \quad t \geq 0$$

State-Transition Equation Determined from the State Diagram

state diagram

Equations (5-48) and (5-49) show that the Laplace-transform method of solving the state equations requires the carrying out of matrix inverse of $(s\mathbf{I} - \mathbf{A})$. We shall now show that the state diagram described in Chapter 3 and the Mason's gain formula can be used to solve for the state-transition equation by inspection. The state-transition equation in the Laplace domain is given by Eq. (5-48). Let the initial time be t_0; then Eq. (5-48) is written

$$\mathbf{X}(s) = (s\mathbf{I} - \mathbf{A})^{-1}\mathbf{x}(t_0) + (s\mathbf{I} - \mathbf{A})^{-1}[\mathbf{B}\mathbf{R}(s) + \mathbf{F}\mathbf{W}(s)] \quad t \geq t_0 \tag{5-63}$$

Therefore, the last equation can be written directly from the state diagram by using the gain formula, with $X_i(s)$, $i = 1, 2, \ldots, n$, as the output nodes, and $x_i(t_0)$, $i = 1, 2, \ldots, n$, $R_j(s)$, $j = 1, 2, \ldots, p$, as the input nodes. The following example illustrates the state-diagram method of finding the state-transition equations for the system described in Example 5-2.

EXAMPLE 5-3

The state diagram for the system described by Eq. (5-55) is shown in Fig. 5-3 with t_0 as the initial time. The outputs of the integrators are assigned as state variables. Applying the gain formula to the state diagram in Fig. 5-3, with $X_1(s)$ and $X_2(s)$ as output nodes, and $x_1(t_0)$, $x_2(t_0)$, and $R(s)$ as input nodes, we have

$$X_1(s) = \frac{s^{-1}(1 + 3s^{-1})}{\Delta}x_1(t_0) + \frac{s^{-2}}{\Delta}x_2(t_0) + \frac{s^{-2}}{\Delta}R(s) \tag{5-64}$$

$$X_2(s) = \frac{-2s^{-2}}{\Delta}x_1(t_0) + \frac{s^{-1}}{\Delta}x_2(t_0) + \frac{s^{-1}}{\Delta}R(s) \tag{5-65}$$

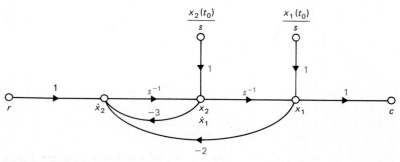

FIGURE 5-3 State diagram for Eq. (5-55).

where

$$\Delta = 1 + 3s^{-1} + 2s^{-2} \tag{5-66}$$

After simplification, Eqs. (5-64) and (5-65) are presented in matrix form:

$$\begin{bmatrix} X_1(s) \\ X_2(s) \end{bmatrix} = \frac{1}{(s+1)(s+2)} \begin{bmatrix} s+3 & 1 \\ -2 & s \end{bmatrix} \begin{bmatrix} x_1(t_0) \\ x_2(t_0) \end{bmatrix} + \frac{1}{(s+1)(s+2)} \begin{bmatrix} 1 \\ s \end{bmatrix} R(s) \tag{5-67}$$

The state-transition equation for $t \geq t_0$ is obtained by taking the inverse Laplace transform on both sides of Eq. (5-67).

Consider that the input $r(t)$ is a unit-step function applied at $t = t_0$. Then the following inverse Laplace-transform relationships are identified:

$$\mathscr{L}^{-1}\left(\frac{1}{s}\right) = u_s(t - t_0) \qquad t \geq t_0 \tag{5-68}$$

$$\mathscr{L}^{-1}\left(\frac{1}{s+a}\right) = e^{-a(t-t_0)} u_s(t - t_0) \qquad t \geq t_0 \tag{5-69}$$

Since the initial time is defined to be t_0, the Laplace-transform expressions here do not have the delay factor $e^{-t_0 s}$. The inverse Laplace transform of Eq. (5-67) is

$$\begin{bmatrix} x_1(t) \\ x_2(t) \end{bmatrix} = \begin{bmatrix} 2e^{-(t-t_0)} - e^{-2(t-t_0)} & e^{-(t-t_0)} - e^{-2(t-t_0)} \\ -2e^{-(t-t_0)} + 2e^{-2(t-t_0)} & -e^{-(t-t_0)} + 2e^{-2(t-t_0)} \end{bmatrix} \begin{bmatrix} x_1(t_0) \\ x_2(t_0) \end{bmatrix}$$
$$+ \begin{bmatrix} 0.5u_s(t - t_0) - e^{-(t-t_0)} + 0.5e^{-2(t-t_0)} \\ e^{-(t-t_0)} - e^{-2(t-t_0)} \end{bmatrix} \qquad t \geq t_0 \tag{5-70}$$

The reader should compare this result with that of Eq. (5-61), which is obtained for $t \geq 0$.

EXAMPLE 5-4   In this example, we illustrate the utilization of the state-transition method to a system with input discontinuity. Let us consider that the input voltage to the RL network of Fig. 5-1 is as shown in Fig. 5-4. The state equation of the network is

$$\frac{di(t)}{dt} = -\frac{R}{L}i(t) + \frac{1}{L}e(t) \tag{5-71}$$

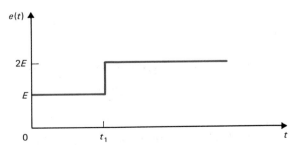

FIGURE 5-4 Input-voltage waveform for the network in Fig. 5-1.

Thus,

$$A = -\frac{R}{L} \qquad B = \frac{1}{L} \qquad F = 0 \tag{5-72}$$

The state-transition matrix is

$$\boldsymbol{\phi}(t) = e^{-Rt/L} \tag{5-73}$$

One approach to the problem of solving for $i(t)$ for $t \geq 0$ is to express the input voltage as

$$e(t) = Eu_s(t) + Eu_s(t - t_1) \tag{5-74}$$

where $u_s(t)$ is the unit-step function. The Laplace transform of $e(t)$ is

$$E(s) = \frac{E}{s}(1 + e^{-t_1 s}) \tag{5-75}$$

Then

$$(s\mathbf{I} - \mathbf{A})^{-1}\mathbf{B}R(s) = \frac{E}{Ls(s + R/L)}(1 + e^{-t_1 s})$$

$$= \frac{E}{Rs[1 + (L/R)s]}(1 + e^{-t_1 s}) \tag{5-76}$$

By substituting Eq. (5-76) into Eq. (5-49), the current for $t \geq 0$ is obtained:

$$i(t) = e^{-Rt/L}i(0)u_s(t) + \frac{E}{R}(1 - e^{-Rt/L})u_s(t) + \frac{E}{R}(1 - e^{-R(t-t_1)/L})u_s(t - t_1) \tag{5-77}$$

Using the state-transition approach, we can divide the transition period into two parts: $t = 0$ to $t = t_1$, and $t = t_1$ to $t = \infty$. First, for the time interval $0 \leq t \leq t_1$, the input is

$$e(t) = Eu_s(t) \qquad 0 \leq t < t_1 \tag{5-78}$$

Then

$$(s\mathbf{I} - \mathbf{A})^{-1}\mathbf{B}R(s) = \frac{E}{Ls(s + R/L)} = \frac{1}{Rs[1 + (L/R)s]} \tag{5-79}$$

Thus, the state-transition equation for the time interval $0 \leq t \leq t_1$ is

$$i(t) = \left(e^{-Rt/L}i(0) + \frac{E}{R}(1 - e^{-Rt/L})\right)u_s(t) \tag{5-80}$$

Substituting $t = t_1$ into the last equation, we get

$$i(t_1) = e^{-Rt_1/L}i(0) + \frac{E}{R}(1 - e^{-Rt_1/L}) \tag{5-81}$$

The value of $i(t)$ at $t = t_1$ is now used as the initial state for the next transition period of $t_1 \leq t < \infty$. The magnitude of the input for this interval is $2E$. Therefore, the state-transition equation for the second transition period is

$$i(t) = e^{-R(t-t_1)/L} i(t_1) + \frac{2E}{R}(1 - e^{-R(t-t_1)/L}) \qquad t \geq t_1 \tag{5-82}$$

where $i(t_1)$ is given by Eq. (5-81).

This example illustrates two possible ways of solving a state-transition problem. In the first approach, the transition is treated as one continuous process, whereas in the second, the transition period is divided into parts over which the input can be more easily represented. Although the first approach requires only one operation, the second method yields relatively simple results to the state-transition equation, and it often presents computational advantages. Notice that in the second method, the state at $t = t_1$ is used as the initial state for the next transition period, which begins at t_1.

5-5
RELATIONSHIP BETWEEN STATE EQUATIONS AND HIGH-ORDER DIFFERENTIAL EQUATIONS

In preceding sections, we defined the state equations and their solutions for linear time-invariant systems. In general, although it is always possible to write the state equations from the schematic diagram of a system, in practice, the system may have been described by a high-order differential equation or transfer function. Therefore, it is necessary to investigate how state equations can be written directly from the differential equation or the transfer function. The relationship between a high-order differential equation and the state equation is discussed in this section.

Let us consider that a single-variable linear time-invariant system is described by the following nth-order differential equation:

$$\frac{d^n c(t)}{dt^n} + a_n \frac{d^{n-1} c(t)}{dt^{n-1}} + a_{n-1} \frac{d^{n-2} c(t)}{dt^{n-2}} + \cdots + a_2 \frac{dc(t)}{dt} + a_1 c(t) = r(t) \tag{5-83}$$

where $c(t)$ is the output variable, and $r(t)$ is the input.

The problem is to represent Eq. (5-83) by n state equations and an output equation. This simply involves the defining of the n state variables in terms of the output $c(t)$ and its derivatives. We have shown earlier that the state variables of a given system are not unique. Therefore, in general, we seek the most convenient way of assigning the state variables as long as the definition of state variables stated in Section 5-1 is satisfied.

For the present case, it is convenient to define the state variables as

$$x_1(t) = c(t)$$

$$x_2(t) = \frac{dc(t)}{dt}$$

$$\vdots$$

$$x_n(t) = \frac{dc^{n-1}c(t)}{dt^{n-1}}$$

(5-84)

Then the state equations are

$$\frac{dx_1(t)}{dt} = x_2(t)$$

$$\frac{dx_2(t)}{dt} = x_3(t)$$

$$\vdots$$

$$\frac{dx_{n-1}(t)}{dt} = x_n(t)$$

$$\frac{dx_n(t)}{dt} = -a_1x_1(t) - a_2x_2(t) - \cdots - a_{n-1}x_{n-1}(t) - a_nx_n(t) + r(t)$$

(5-85)

where the last state equation is obtained by equating the highest-ordered derivative term to the rest of Eq. (5-83). The output equation is simply

$$c(t) = x_1(t)$$

(5-86)

In vector-matrix form, Eq. (5-85) is written

$$\frac{d\mathbf{x}(t)}{dt} = \mathbf{A}\mathbf{x}(t) + \mathbf{B}r(t)$$

(5-87)

where $\mathbf{x}(t)$ is the $n \times 1$ state vector, and $r(t)$ is the scalar input. The coefficient matrices are

$$\mathbf{A} = \begin{bmatrix} 0 & 1 & 0 & 0 & 0 & \cdots & 0 \\ 0 & 0 & 1 & 0 & 0 & \cdots & 0 \\ 0 & 0 & 0 & 1 & 0 & \cdots & 0 \\ \multicolumn{7}{c}{\cdots\cdots\cdots\cdots\cdots\cdots\cdots\cdots} \\ 0 & 0 & 0 & 0 & 0 & \cdots & 1 \\ -a_1 & -a_2 & -a_3 & -a_4 & -a_5 & \cdots & -a_n \end{bmatrix} \qquad (n \times n) \qquad \text{(5-88)}$$

$$\mathbf{B} = \begin{bmatrix} 0 \\ 0 \\ \vdots \\ 0 \\ 1 \end{bmatrix} \qquad (n \times 1) \qquad \text{(5-89)}$$

The output equation in vector-matrix form is

$$c(t) = \mathbf{D}\mathbf{x}(t) \qquad \text{(5-90)}$$

where

$$\mathbf{D} = [1 \quad 0 \quad 0 \quad \cdots \quad 0] \qquad (1 \times n) \qquad \text{(5-91)}$$

*phase-variable
canonical
form*

The state equation of Eq. (5-87) with the matrices **A** and **B** defined as in Eqs. (5-88) and (5-89), respectively, is called the **phase-variable canonical form** in the next section.

EXAMPLE 5-5   Consider the differential equation

$$\frac{d^3 c(t)}{dt^3} + 5\frac{d^2 c(t)}{dt^2} + \frac{dc(t)}{dt} + 2c(t) = r(t) \qquad \text{(5-92)}$$

Rearranging the last equation so that the highest-order derivative term is equated to the rest of the terms, we have

$$\frac{d^3 c(t)}{dt^3} = -5\frac{d^2 c(t)}{dt^2} - \frac{dc(t)}{dt} - 2c(t) + r(t) \qquad \text{(5-93)}$$

The state variables are defined as

$$x_1(t) = c(t)$$

$$x_2(t) = \frac{dc(t)}{dt} \qquad \text{(5-94)}$$

$$x_3(t) = \frac{d^2 c(t)}{dt^2}$$

Then the state equations are represented by the vector-matrix equation of Eq. (5-87) with

$$\mathbf{A} = \begin{bmatrix} 0 & 1 & 0 \\ 0 & 0 & 1 \\ -2 & -1 & -5 \end{bmatrix} \qquad (5\text{-}95)$$

and

$$\mathbf{B} = \begin{bmatrix} 0 \\ 0 \\ 1 \end{bmatrix} \qquad (5\text{-}96)$$

The output equation is

$$c(t) = x_1(t) \qquad (5\text{-}97)$$

In general, for the nth-order differential equation

$$\frac{d^n c(t)}{dt^n} + a_n \frac{d^{n-1} c(t)}{dt^{n-1}} + \cdots + a_2 \frac{dc(t)}{dt} + a_1 c(t)$$

$$(5\text{-}98)$$

$$= b_{n+1} \frac{d^n r(t)}{dt^n} + b_n \frac{d^{n-1} r(t)}{dt^{n-1}} + \cdots + b_2 \frac{dr(t)}{dt} + b_1 r(t)$$

the state variables should be defined as

$$x_1(t) = c(t) - b_{n+1} r(t)$$

$$x_2(t) = \frac{dx_1(t)}{dt} - h_1 r(t)$$

$$x_3(t) = \frac{dx_2(t)}{dt} - h_2 r(t) \qquad (5\text{-}99)$$

$$\vdots$$

$$x_n(t) = \frac{dx_{n-1}(t)}{dt} - h_{n-1} r(t)$$

where

$$h_1 = b_n - a_n b_{n+1}$$

$$h_2 = (b_{n-1} - a_{n-1} b_{n+1}) - a_n h_1$$

$$h_3 = (b_{n-2} - a_{n-2}b_{n+1}) - a_{n-1}h_1 - a_nh_{n-1} \tag{5-100}$$

$$\vdots$$

$$h_n = (b_1 - a_1b_{n+1}) - a_2h_1 - a_3h_2 - \cdots - a_{n-1}h_{n-2} - a_nh_{n-1}$$

Using Eqs. (5-99) and (5-100), we resolve the nth-order differential equation in Eq. (5-98) into the following n state equations:

$$\frac{dx_1(t)}{dt} = x_2(t) + h_1r(t)$$

$$\frac{dx_2(t)}{dt} = x_3(t) + h_2r(t)$$

$$\vdots \tag{5-101}$$

$$\frac{dx_{n-1}(t)}{dt} = x_n(t) + h_{n-1}r(t)$$

$$\frac{dx_n(t)}{dt} = -a_1x_1(t) - a_2x_2(t) - \cdots - a_{n-1}x_{n-1}(t) - a_nx_n(t) + h_nr(t)$$

The output equation is obtained by rearranging the first equation of Eq. (5-99):

$$c(t) = x_1(t) + b_{n+1}r(t) \tag{5-102}$$

In Section 5-9, we present the direct-decomposition method, which leads to a simpler procedure for getting the dynamic equations of Eq. (5-98).

EXAMPLE 5-6

Given the differential equation

$$\frac{d^3c(t)}{dt^3} + 5\frac{d^2c(t)}{dt^2} + \frac{dc(t)}{dt} + 2c(t) = \frac{dr(t)}{dt} + 2r(t) \tag{5-103}$$

we want to represent the equation by three state equations.

Comparing Eq. (5-103) with Eq. (5-98), we have

$$a_3 = 5 \qquad a_2 = 1 \qquad a_1 = 2$$

$$b_4 = 0 \qquad b_3 = 0 \qquad b_2 = 1 \qquad b_1 = 2$$

Using Eq. (5-100), we get

$$h_1 = b_3 - a_3b_4 = 0$$

$$h_2 = (b_2 - a_2b_4) - a_3h_1 = 1$$

$$h_3 = (b_1 - a_1b_4) - a_2h_1 - a_3h_2 = -3$$

Thus, according to Eq. (5-99), the state variables are defined as

$$x_1(t) = c(t)$$

$$x_2(t) = \frac{dc(t)}{dt}$$

$$x_3(t) = \frac{d^2c(t)}{dt^2} - r(t) \qquad (5\text{-}104)$$

By using Eq. (5-101), the state equations are

$$\frac{dx_1(t)}{dt} = x_2(t)$$

$$\frac{dx_2(t)}{dt} = x_3(t) + r(t) \qquad (5\text{-}105)$$

$$\frac{dx_3(t)}{dt} = -2x_1(t) - x_2(t) - 5x_3(t) - 3r(t)$$

5-6

TRANSFORMATION TO PHASE-VARIABLE CANONICAL FORM

In general, when the coefficient matrices $\mathbf{A}$ and $\mathbf{B}$ are given by Eqs. (5-88) and (5-89), respectively, the state equation of Eq. (5-87) is called the **phase-variable canonical form**. It will be shown later that *a linear time-invariant system that is representable in the phase-variable canonical form has certain unique properties with regard to controllability and pole-placement design through state feedback.*

THEOREM 5-1.   *Let the state equations of a linear time-invariant system be described by*

$$\frac{d\mathbf{x}(t)}{dt} = \mathbf{A}\mathbf{x}(t) + \mathbf{B}r(t) \qquad (5\text{-}106)$$

where $\mathbf{x}(t)$ is an $n \times 1$ state vector, $\mathbf{A}$ an $n \times n$ coefficient matrix, $\mathbf{B}$ an $n \times 1$ coefficient matrix, and $r(t)$ a scalar input. If the matrix

$$\mathbf{S} = [\mathbf{B} \quad \mathbf{AB} \quad \mathbf{A}^2\mathbf{B} \quad \cdots \quad \mathbf{A}^{n-1}\mathbf{B}] \qquad (5\text{-}107)$$

is nonsingular, then there exists a nonsingular transformation:

$$\mathbf{y}(t) = \mathbf{Q}\mathbf{x}(t) \qquad (5\text{-}108)$$

or

$$\mathbf{x}(t) = \mathbf{Q}^{-1}\mathbf{y}(t) \qquad (5\text{-}109)$$

which transforms Eq. (5-106) to the phase-variable canonical form:

$$\dot{\mathbf{y}}(t) = \mathbf{A}_1\mathbf{y}(t) + \mathbf{B}_1 r(t) \qquad (5\text{-}110)$$

where

$$\mathbf{A}_1 = \begin{bmatrix} 0 & 1 & 0 & 0 & \cdots & 0 \\ 0 & 0 & 1 & 0 & \cdots & 0 \\ 0 & 0 & 0 & 1 & \cdots & 0 \\ \cdots\cdots\cdots\cdots\cdots\cdots\cdots\cdots\cdots\cdots \\ 0 & 0 & 0 & 0 & \cdots & 1 \\ -a_1 & -a_2 & -a_3 & -a_4 & \cdots & -a_n \end{bmatrix} \qquad (5\text{-}111)$$

and

$$\mathbf{B}_1 = \begin{bmatrix} 0 \\ 0 \\ \vdots \\ 0 \\ 1 \end{bmatrix} \qquad (5\text{-}112)$$

The transforming matrix **Q** *is given by*

$$\mathbf{Q} = \begin{bmatrix} \mathbf{Q}_1 \\ \mathbf{Q}_1\mathbf{A} \\ \vdots \\ \mathbf{Q}_1\mathbf{A}^{n-1} \end{bmatrix} \qquad (5\text{-}113)$$

where

$$\mathbf{Q}_1 = \begin{bmatrix} 0 & 0 & \cdots & 1 \end{bmatrix}\begin{bmatrix} \mathbf{B} & \mathbf{AB} & \mathbf{A}^2\mathbf{B} & \cdots & \mathbf{A}^{n-1}\mathbf{B} \end{bmatrix}^{-1} \qquad (5\text{-}114)$$

PROOF.   Let

$$\mathbf{x}(t) = \begin{bmatrix} x_1(t) \\ x_2(t) \\ \vdots \\ x_n(t) \end{bmatrix} \qquad (5\text{-}115)$$

$$\mathbf{y}(t) = \begin{bmatrix} y_1(t) \\ y_2(t) \\ \vdots \\ y_n(t) \end{bmatrix} \tag{5-116}$$

and

$$\mathbf{Q} = \begin{bmatrix} q_{11} & q_{12} & \cdots & q_{1n} & Q_1 \\ q_{21} & q_{22} & \cdots & q_{2n} & Q_2 \\ \cdots & \cdots & \cdots & \cdots & \vdots \\ q_{n1} & q_{n2} & \cdots & q_{nn} & Q_n \end{bmatrix} = \begin{bmatrix} Q_1 \\ Q_2 \\ \vdots \\ Q_n \end{bmatrix} \tag{5-117}$$

where

$$\mathbf{Q}_i = [q_{i1} \quad q_{i2} \quad \cdots \quad q_{in}] \qquad i = 1, 2, \ldots, n \tag{5-118}$$

Then, from Eq. (5-108),

$$y_1(t) = q_{11}x_1(t) + q_{12}x_2(t) + \cdots + q_{1n}x_n(t)$$
$$= \mathbf{Q}_1\mathbf{x}(t) \tag{5-119}$$

By taking the time derivative on both sides of the last equation, and in view of Eqs. (5-110) and (5-111),

$$\dot{y}_1(t) = y_2(t) = \mathbf{Q}_1\dot{\mathbf{x}}(t) = \mathbf{Q}_1\mathbf{A}\mathbf{x}(t) + \mathbf{Q}_1\mathbf{B}\mathbf{r}(t) \tag{5-120}$$

Since Eq. (5-108) states that $\mathbf{y}(t)$ is a function of $\mathbf{x}(t)$ only, in Eq. (5-120), $\mathbf{Q}_1\mathbf{B} = \mathbf{0}$. Therefore,

$$\dot{y}_1(t) = y_2(t) = \mathbf{Q}_1\mathbf{A}\mathbf{x}(t) \tag{5-121}$$

Taking the time derivative of the last equation once again leads to

$$\dot{y}_2(t) = y_3(t) = \mathbf{Q}_1\mathbf{A}^2\mathbf{x}(t) \tag{5-122}$$

with $\mathbf{Q}_1\mathbf{A}\mathbf{B} = \mathbf{0}$. Repeating the procedure leads to

$$\dot{y}_{n-1}(t) = y_n(t) = \mathbf{Q}_1\mathbf{A}^{n-1}\mathbf{x}(t) \tag{5-123}$$

with $\mathbf{Q}_1\mathbf{A}^{n-2}\mathbf{B} = \mathbf{0}$. Therefore, using Eq. (5-108), we have

$$\mathbf{y}(t) = \mathbf{Q}\mathbf{x}(t) = \begin{bmatrix} \mathbf{Q}_1 \\ \mathbf{Q}_1\mathbf{A} \\ \vdots \\ \mathbf{Q}_1\mathbf{A}^{n-1} \end{bmatrix} \mathbf{x}(t) \tag{5-124}$$

or

$$\mathbf{Q} = \begin{bmatrix} \mathbf{Q}_1 \\ \mathbf{Q}_1 \mathbf{A} \\ \vdots \\ \mathbf{Q}_1 \mathbf{A}^{n-1} \end{bmatrix} \qquad (5\text{-}125)$$

and $\mathbf{Q}_1$ should satisfy the condition

$$\mathbf{Q}_1 \mathbf{B} = \mathbf{Q}_1 \mathbf{AB} = \cdots = \mathbf{Q}_1 \mathbf{A}^{n-1} \mathbf{B} = \mathbf{0} \qquad (5\text{-}126)$$

Now taking the derivative of Eq. (5-108) with respect to time, we get

$$\dot{\mathbf{y}}(t) = \mathbf{Q}\dot{\mathbf{x}}(t) = \mathbf{QAx}(t) + \mathbf{QB}r(t) \qquad (5\text{-}127)$$

Comparing Eq. (5-127) with Eq. (5-110), we obtain

$$\mathbf{A}_1 = \mathbf{QAQ}^{-1} \qquad (5\text{-}128)$$

and

$$\mathbf{B}_1 = \mathbf{QB} \qquad (5\text{-}129)$$

Then, from Eq. (5-125),

$$\mathbf{QB} = \begin{bmatrix} \mathbf{Q}_1 \mathbf{B} \\ \mathbf{Q}_1 \mathbf{AB} \\ \vdots \\ \mathbf{Q}_1 \mathbf{A}^{n-1} \mathbf{B} \end{bmatrix} = \begin{bmatrix} 0 \\ 0 \\ \vdots \\ 1 \end{bmatrix} \qquad (5\text{-}130)$$

Since $\mathbf{Q}_1$ is an $1 \times n$ row matrix, Eq. (5-130) can be written

$$\mathbf{Q}_1[\mathbf{B} \quad \mathbf{AB} \quad \mathbf{A}^2 \mathbf{B} \quad \cdots \quad \mathbf{A}^{n-1} \mathbf{B}] = [0 \quad 0 \quad \cdots \quad 1] \qquad (5\text{-}131)$$

Thus, $\mathbf{Q}_1$ is obtained as

$$\begin{aligned} \mathbf{Q}_1 &= [0 \quad 0 \quad \cdots \quad 1][\mathbf{B} \quad \mathbf{AB} \quad \mathbf{A}^2 \mathbf{B} \quad \cdots \quad \mathbf{A}^{n-1} \mathbf{B}]^{-1} \\ &= [0 \quad 0 \quad \cdots \quad 1]\mathbf{S}^{-1} \end{aligned} \qquad (5\text{-}132)$$

state control-lability

if the matrix $\mathbf{S} = [\mathbf{B} \quad \mathbf{AB} \quad \mathbf{A}^2 \mathbf{B} \quad \cdots \quad \mathbf{A}^{n-1} \mathbf{B}]$ is nonsingular. This is the condition of complete state controllability (refer to Section 5-10). Once $\mathbf{Q}_1$ is determined from Eq. (5-132), the transformation matrix $\mathbf{Q}$ is given by Eq. (5-125). ■

EXAMPLE 5-7   Let a linear time-invariant system be described by Eq. (5-87) with

$$\mathbf{A} = \begin{bmatrix} 1 & -1 \\ 0 & -1 \end{bmatrix} \qquad \mathbf{B} = \begin{bmatrix} 1 \\ 1 \end{bmatrix} \tag{5-133}$$

It is desired to transform the state equation into the phase-variable canonical form. Since the matrix

$$\mathbf{S} = [\mathbf{B} \quad \mathbf{AB}] = \begin{bmatrix} 1 & 0 \\ 1 & -1 \end{bmatrix} \tag{5-134}$$

is nonsingular, the system may be expressed in the phase-variable canonical form. Therefore, $\mathbf{Q}_1$ is obtained as a row matrix that contains the elements of the last row of $\mathbf{S}^{-1}$; that is,

$$\mathbf{Q}_1 = [1 \quad -1] \tag{5-135}$$

By using Eq. (5-125),

$$\mathbf{Q} = \begin{bmatrix} \mathbf{Q}_1 \\ \mathbf{Q}_1\mathbf{A} \end{bmatrix} = \begin{bmatrix} 1 & -1 \\ 1 & 0 \end{bmatrix} \tag{5-136}$$

Thus,

$$\mathbf{A}_1 = \mathbf{Q}\mathbf{A}\mathbf{Q}^{-1} = \begin{bmatrix} 0 & 1 \\ 1 & 0 \end{bmatrix} \tag{5-137}$$

$$\mathbf{B}_1 = \mathbf{Q}\mathbf{B} = \begin{bmatrix} 0 \\ 1 \end{bmatrix} \tag{5-138}$$

5-7
RELATIONSHIP BETWEEN STATE EQUATIONS AND TRANSFER FUNCTIONS

We have presented the methods of describing a linear time-invariant system by transfer functions and by dynamic equations. It is interesting to investigate the relationship between these two representations.

In Eq. (3-5) the transfer function of a linear single-input–output system is defined in terms of the coefficients of the system's differential equation. Similarly, Eq. (3-11) gives the transfer-function matrix relation for a multivariable system that has p inputs and q outputs. Now we investigate the transfer-function matrix relation using the dynamic-equation notation.

Consider that a linear time-invariant system is described by the dynamic equations

$$\frac{d\mathbf{x}(t)}{dt} = \mathbf{A}\mathbf{x}(t) + \mathbf{B}r(t) + \mathbf{F}\mathbf{w}(t) \tag{5-139}$$

$$\mathbf{c}(t) = \mathbf{D}\mathbf{x}(t) + \mathbf{E}r(t) + \mathbf{H}\mathbf{w}(t) \tag{5-140}$$

where $\mathbf{x}(t) = n \times 1$ state vector
  $\mathbf{r}(t) = p \times 1$ input vector
  $\mathbf{c}(t) = q \times 1$ output vector
  $\mathbf{w}(t) = v \times 1$ disturbance vector

and $\mathbf{A}, \mathbf{B}, \mathbf{D}, \mathbf{E}, \mathbf{F}$, and $\mathbf{H}$ are matrices of appropriate dimensions.

Taking the Laplace transform on both sides of Eq. (5-139) and solving for $\mathbf{X}(s)$, we have

$$\mathbf{X}(s) = (s\mathbf{I} - \mathbf{A})^{-1}\mathbf{x}(0) + (s\mathbf{I} - \mathbf{A})^{-1}[\mathbf{BR}(s) + \mathbf{FW}(s)] \qquad (5\text{-}141)$$

The Laplace transform of Eq. (5-140) is

$$\mathbf{C}(s) = \mathbf{DX}(s) + \mathbf{ER}(s) + \mathbf{HW}(s) \qquad (5\text{-}142)$$

Substituting Eq. (5-141) into Eq. (5-142), we have

$$\mathbf{C}(s) = \mathbf{D}(s\mathbf{I} - \mathbf{A})^{-1}\mathbf{x}(0) + \mathbf{D}(s\mathbf{I} - \mathbf{A})^{-1}[\mathbf{BR}(s) + \mathbf{FW}(s)] + \mathbf{ER}(s) + \mathbf{HW}(s)$$

$$(5\text{-}143)$$

Since the definition of a transfer function requires that the initial conditions be set to zero, $\mathbf{x}(0) = \mathbf{0}$; thus, Eq. (5-143) becomes

$$\mathbf{C}(s) = [\mathbf{D}(s\mathbf{I} - \mathbf{A})^{-1}\mathbf{B} + \mathbf{E}]\mathbf{R}(s) + [\mathbf{D}(s\mathbf{I} - \mathbf{A})^{-1}\mathbf{F} + \mathbf{H}]\mathbf{W}(s) \qquad (5\text{-}144)$$

Let us define

$$\mathbf{G}_r(s) = \mathbf{D}(s\mathbf{I} - \mathbf{A})^{-1}\mathbf{B} + \mathbf{E} \qquad (5\text{-}145)$$

$$\mathbf{G}_w(s) = \mathbf{D}(s\mathbf{I} - \mathbf{A})^{-1}\mathbf{F} + \mathbf{H} \qquad (5\text{-}146)$$

where $\mathbf{G}_r(s)$ is a $q \times p$ transfer-function matrix between $\mathbf{r}(t)$ and $\mathbf{c}(t)$ when $\mathbf{w}(t) = \mathbf{0}$, and $\mathbf{G}_w(s)$ is a $q \times v$ transfer-function matrix between $\mathbf{w}(t)$ and $\mathbf{c}(t)$ when $\mathbf{r}(t) = \mathbf{0}$. Then, Eq. (5-144) is written

$$\mathbf{C}(s) = \mathbf{G}_r(s)\mathbf{R}(s) + \mathbf{G}_w(s)\mathbf{W}(s) \qquad (5\text{-}147)$$

EXAMPLE 5-8

multivariable system

Consider that a multivariable system is described by the differential equations.

$$\frac{d^2 c_1(t)}{dt^2} + 4\frac{dc_1(t)}{dt} - 3c_2(t) = r_1(t) + 2w(t) \qquad (5\text{-}148)$$

$$\frac{dc_2(t)}{dt} + \frac{dc_1(t)}{dt} + c_1(t) + 2c_2(t) = r_2(t) \qquad (5\text{-}149)$$

The state variables of the system are assigned as follows:

$$x_1(t) = c_1(t)$$

$$x_2(t) = \frac{dc_1(t)}{dt}$$ (5-150)

$$x_3(t) = c_2(t)$$

These state variables have been defined by mere inspection of the two differential equations, as no particular reasons for the definitions are given other than that these are the most convenient. Now equating the first term of each of the equations of Eqs. (5-148) and (5-149) to the rest of the terms and using the state-variable relations of Eq. (5-150), we arrive at the following state equations and output equations in matrix form:

$$\begin{bmatrix} \dfrac{dx_1(t)}{dt} \\[2mm] \dfrac{dx_2(t)}{dt} \\[2mm] \dfrac{dx_3(t)}{dt} \end{bmatrix} = \begin{bmatrix} 0 & 1 & 0 \\ 0 & -4 & 3 \\ -1 & -1 & -2 \end{bmatrix} \begin{bmatrix} x_1(t) \\ x_2(t) \\ x_3(t) \end{bmatrix} + \begin{bmatrix} 0 & 0 \\ 1 & 0 \\ 0 & 1 \end{bmatrix} \begin{bmatrix} r_1(t) \\ r_2(t) \end{bmatrix} + \begin{bmatrix} 0 \\ 2 \\ 0 \end{bmatrix} w(t)$$ (5-151)

$$\begin{bmatrix} c_1(t) \\ c_2(t) \end{bmatrix} = \begin{bmatrix} 1 & 0 & 0 \\ 0 & 0 & 1 \end{bmatrix} \begin{bmatrix} x_1(t) \\ x_2(t) \\ x_3(t) \end{bmatrix} = \mathbf{D}x(t)$$ (5-152)

To determine the transfer-function matrix of the system using the state-variable formulation, we substitute the $\mathbf{A}$, $\mathbf{B}$, $\mathbf{D}$, $\mathbf{E}$, and $\mathbf{F}$ matrices into Eq. (5-144). First, we form the matrix $(s\mathbf{I} - \mathbf{A})$:

$$(s\mathbf{I} - \mathbf{A}) = \begin{bmatrix} s & -1 & 0 \\ 0 & s+4 & -3 \\ 1 & 1 & s+2 \end{bmatrix}$$ (5-153)

The determinant of $(s\mathbf{I} - \mathbf{A})$ is

$$|s\mathbf{I} - \mathbf{A}| = s^3 + 6s^2 + 11s + 3$$ (5-154)

Thus,

$$(s\mathbf{I} - \mathbf{A})^{-1} = \frac{1}{|s\mathbf{I} - \mathbf{A}|} \begin{bmatrix} s^2 + 6s + 11 & s+2 & 3 \\ -3 & s(s+2) & 3s \\ -(s+4) & -(s+1) & s(s+4) \end{bmatrix}$$ (5-155)

The transfer-function matrix between $\mathbf{r}(t)$ and $\mathbf{c}(t)$ is

$$\mathbf{G}_r(s) = \mathbf{D}(s\mathbf{I} - \mathbf{A})^{-1}\mathbf{B} = \frac{1}{s^3 + 6s^2 + 11s + 3} \begin{bmatrix} s+2 & 3 \\ -(s+1) & s(s+4) \end{bmatrix}$$ (5-156)

and that between $\mathbf{w}(t)$ and $\mathbf{c}(t)$ is

$$\mathbf{G}_w(s) = \mathbf{D}(s\mathbf{I} - \mathbf{A})^{-1}\mathbf{F} = \frac{1}{s^3 + 6s^2 + 11s + 3}\begin{bmatrix} 2(s + 2) \\ -2(s + 1) \end{bmatrix} \tag{5-157}$$

Using the conventional approach, we take the Laplace transform on both sides of Eqs. (5-148) and (5-149) and assume zero initial conditions. The resulting transformed equations are written in matrix form as

$$\begin{bmatrix} s(s + 4) & -3 \\ s + 1 & s + 2 \end{bmatrix}\begin{bmatrix} C_1(s) \\ C_2(s) \end{bmatrix} = \begin{bmatrix} R_1(s) \\ R_2(s) \end{bmatrix} + \begin{bmatrix} 2 \\ 0 \end{bmatrix}W(s) \tag{5-158}$$

Solving for $\mathbf{C}(s)$ from Eq. (5-158), we obtain

$$\mathbf{C}(s) = \mathbf{G}_r(s)R(s) + \mathbf{G}_w(s)W(s) \tag{5-159}$$

where

$$\mathbf{G}_r(s) = \begin{bmatrix} s(s + 4) & -3 \\ s + 1 & s + 2 \end{bmatrix}^{-1} \tag{5-160}$$

$$\mathbf{G}_w(s) = \begin{bmatrix} s(s + 4) & -3 \\ s + 1 & s + 2 \end{bmatrix}^{-1}\begin{bmatrix} 2 \\ 0 \end{bmatrix} \tag{5-161}$$

which are the same results as in Eqs. (5-156) and (5-157), respectively, when the matrix inverse is carried out.

5-8

CHARACTERISTIC EQUATION AND EIGENVALUES

Characteristic Equation

The characteristic equation plays an important role in the study of linear systems. It can be defined with respect to the differential equation, the transfer function, or the state equations.

Consider that a linear time-invariant system is described by the differential equation

$$\frac{d^n c(t)}{dt^n} + a_n\frac{d^{n-1}c(t)}{dt^{n-1}} + a_{n-1}\frac{d^{n-2}c(t)}{dt^{n-2}} + \cdots + a_2\frac{dc(t)}{dt} + a_1 c(t)$$

$$= b_{n+1}\frac{d^n r(t)}{dt^n} + b_n\frac{d^{n-1}r(t)}{dt^{n-1}} + \cdots + b_2\frac{dr(t)}{dt} + b_1 r(t) \tag{5-162}$$

By defining the operator s as

$$s^k = \frac{d^k}{dt^k} \qquad k = 1, 2, \ldots, n \tag{5-163}$$

Eq. (5-162) is written

$$(s^n + a_n s^{n-1} + a_{n-1} s^{n-2} + \cdots + a_2 s + a_1) c(t) \tag{5-164}$$
$$= (b_{n+1} s^n + b_n s^{n-1} + \cdots + b_2 s + b_1) r(t)$$

The characteristic equation of the system is defined as

$$s^n + a_n s^{n-1} + a_{n-1} s^{n-2} + \cdots + a_2 s + a_1 = 0 \tag{5-165}$$

which is obtained by setting the homogeneous part of Eq. (5-164) to zero. The transfer function of the system is

$$G(s) = \frac{C(s)}{R(s)} = \frac{b_{n+1} s^n + b_n s^{n-1} + \cdots + b_2 s + b_1}{s^n + a_n s^{n-1} + \cdots + a_2 s + a_1} \tag{5-166}$$

characteristic equation

Therefore, *the characteristic equation is obtained by equating the denominator of the transfer function to zero.*

From the state-variable approach, we can write Eq. (5-145) as

$$\mathbf{G}_r(s) = \mathbf{D} \frac{\text{adj } (s\mathbf{I} - A)}{|s\mathbf{I} - \mathbf{A}|} \mathbf{B} + \mathbf{E}$$
$$= \frac{\mathbf{D}[\text{adj } (s\mathbf{I} - \mathbf{A})]\mathbf{B} + |s\mathbf{I} - \mathbf{A}|\mathbf{E}}{|s\mathbf{I} - \mathbf{A}|} \tag{5-167}$$

Setting the denominator of the transfer function matrix $\mathbf{G}_r(s)$ to zero, we get the characteristic equation as

$$|s\mathbf{I} - \mathbf{A}| = 0 \tag{5-168}$$

which is an alternative form of the characteristic equation, and should lead to Eq. (5-165).

Eigenvalues

eigenvalues

The roots of the characteristic equation are often referred to as the eigenvalues of the matrix **A.** It is interesting to note that if the state equations are represented in the phase-variable canonical form with the matrix **A** given by Eq. (5-111), the coefficients of the characteristic equation are readily given by the elements in the last row of the **A** matrix, as in Eq. (5-165).

Another important property of the characteristic equation and the eigenvalues is that they are invariant under a nonsingular transformation, such as the phase-variable canonical-form transformation stated in Theorem 5-1. In other words, when the **A** matrix is transformed by a nonsingular transformation $\mathbf{x} = \mathbf{Py}$, so that

$$\overline{\mathbf{A}} = \mathbf{P}^{-1}\mathbf{AP} \tag{5-169}$$

then the characteristic equation and the eigenvalues of $\overline{\mathbf{A}}$ are identical to those of **A**. This is proved by writing

$$s\mathbf{I} - \overline{\mathbf{A}} = s\mathbf{I} - \mathbf{P}^{-1}\mathbf{AP} = s\mathbf{P}^{-1}\mathbf{P} - \mathbf{P}^{-1}\mathbf{AP} \tag{5-170}$$

The characteristic equation of $\overline{\mathbf{A}}$ is

$$|s\mathbf{I} - \overline{\mathbf{A}}| = |s\mathbf{P}^{-1}\mathbf{P} - \mathbf{P}^{-1}\mathbf{AP}| = |\mathbf{P}^{-1}(s\mathbf{I} - \mathbf{A})\mathbf{P}| \tag{5-171}$$

Since the determinant of a product of matrices is equal to the product of the determinants of the matrices, Eq. (5-171) becomes

$$|s\mathbf{I} - \overline{\mathbf{A}}| = |\mathbf{P}^{-1}||s\mathbf{I} - \mathbf{A}||\mathbf{P}| = |s\mathbf{I} - \mathbf{A}| \tag{5-172}$$

5-9

DECOMPOSITION OF TRANSFER FUNCTIONS

Up to this point, various methods of characterizing a linear system have been presented. It will be useful to summarize briefly and gather thoughts at this point before proceeding to the main topics of decomposition in this section. It has been shown that the starting point of the description of a linear system may be the system's differential equation, transfer function, or dynamic equations. It is demonstrated that all these methods are closely related. Further, the state diagram defined in Chapter 3 is shown to be a useful tool that not only can lead to the solutions of the state equations, but also serves as a vehicle of translation from one type of description to the others. The block diagram of Fig. 5-5 shows the relationships between the various ways of describing a linear system. The block diagram shows that starting, for instance, with the differential

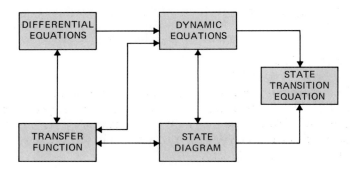

FIGURE 5-5 Block diagram showing the relationships among various methods of describing linear systems.

equation of a system, one can get to the solution by use of the transfer-function method or the state-equation method. The block diagram also shows that the majority of the relationships are bilateral, so a great deal of flexibility exists between the methods.

One subject remains to be discussed. This involves the construction of the state diagram from the transfer function. In general, it is necessary to establish a better method than using Eqs. (5-98) through (5-101) in getting from a high-order differential equation to the state equations.

The process of going from the transfer function to the state diagram or the state equations is called the **decomposition** *of the transfer function.* In general, there are three basic ways of decomposing a transfer function: **direct decomposition**, **cascade decomposition**, and **parallel decomposition**. Each of these three schemes of decomposition has its own merits and is best suited for a particular situation.

Direct Decomposition

direct decomposition

The direct decomposition scheme is applied to a transfer function that is not in factored form. Without loss of generality, the method of direct decomposition can be described by the following transfer function:

$$\frac{C(s)}{R(s)} = \frac{a_0 s^2 + a_1 s + a_2}{b_0 s^2 + b_1 s + b_2} \tag{5-173}$$

The objective is to obtain the state diagram and the state equations of the system. The following steps are outlined for the direct decomposition:

1. Express the transfer function in negative powers of s. This is accomplished by multiplying the numerator and the denominator of the transfer function by the inverse of its highest power in s. For the transfer function in Eq. (5-173), we multiply the numerator and the denominator of $C(s)/R(s)$ by s^{-2}.

2. Multiply the numerator and the denominator of the transfer function by a dummy variable $X(s)$. By implementing steps 1 and 2, Eq. (5-173) becomes

$$\frac{C(s)}{R(s)} = \frac{a_0 + a_1 s^{-1} + a_2 s^{-2}}{b_0 + b_1 s^{-1} + b_2 s^{-2}} \frac{X(s)}{X(s)} \tag{5-174}$$

3. The numerators and the denominators on both sides of the transfer function resulting from steps 1 and 2 are equated to each other, respectively. From Eq. (5-174), this step results in

$$C(s) = (a_0 + a_1 s^{-1} + a_2 s^{-2})X(s) \tag{5-175}$$

$$R(s) = (b_0 + b_1 s^{-1} + b_2 s^{-2})X(s) \tag{5-176}$$

state diagram

4. To construct a state diagram using these two equations, they must first be in the proper cause-and-effect relation. It is apparent that Eq. (5-175) already satisfies this prerequisite. However, Eq. (5-176) has the input on the left-hand side and must be rearranged. Dividing both sides of Eq. (5-176) by b_0 and writing $X(s)$ in terms of the other terms, we get

$$X(s) = \frac{1}{b_0}R(s) - \frac{b_1}{b_0}s^{-1}X(s) - \frac{b_2}{b_0}s^{-2}X(s) \qquad (5\text{-}177)$$

The state diagram is drawn in Fig. 5-6 using the expressions in Eqs. (5-175) and (5-177). For simplicity, the initial states are not drawn on the diagram. As usual, the state variables are defined as the outputs of the integrators.

By following the method described in Section 3-10, the state equations are written directly from the state diagram:

$$\begin{bmatrix} \dfrac{dx_1(t)}{dt} \\[2ex] \dfrac{dx_2(t)}{dt} \end{bmatrix} = \begin{bmatrix} 0 & 1 \\[1ex] -\dfrac{b_2}{b_0} & -\dfrac{b_1}{b_0} \end{bmatrix} \begin{bmatrix} x_1(t) \\[1ex] x_2(t) \end{bmatrix} + \begin{bmatrix} 0 \\[1ex] \dfrac{1}{b_0} \end{bmatrix} \qquad (5\text{-}178)$$

Notice that the last state equation is in the phase-variable canonical form. Thus, *direct decomposition of a transfer function leads to the state equations in the phase-variable canonical form.*

The output equation is obtained from Fig. 5-6 by applying the gain formula with $c(t)$ as the output node, and $x_1(t)$, $x_2(t)$, and $r(t)$ as the input nodes.

$$c(t) = \left(a_2 - \frac{a_0 b_2}{b_0}\right)x_1(t) + \left(a_1 - \frac{a_0 b_1}{b_0}\right)x_2(t) + \frac{a_0}{b_0}r(t) \qquad (5\text{-}179)$$

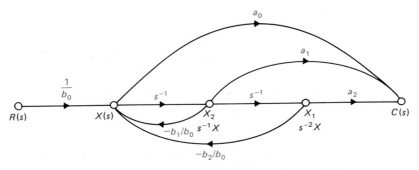

FIGURE 5-6 State diagram of the transfer function of Eq. (5-173) by direct decomposition.

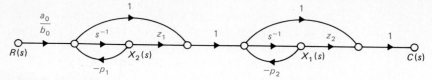

FIGURE 5-7 State diagram of the transfer function of Eq. (5-180) by cascade decomposition.

Cascade Decomposition

cascade decomposition

Cascade decomposition may be applied to a transfer function that is in factored form. The transfer function of Eq. (5-173) may be factored in the following form:

$$\frac{C(s)}{R(s)} = \frac{a_0}{b_0} \frac{s + z_1}{s + p_1} \frac{s + z_2}{s + p_2} \tag{5-180}$$

where z_1, z_2, p_1, and p_2 are real constants. Then it is possible to treat the functions as the product of two first-order transfer functions. The state diagram of each of the first-order transfer functions is realized by using the direct-decomposition method. The complete state diagram is obtained by cascading the two first-order diagrams, as shown in Fig. 5-7. As usual, the outputs of the integrators on the state diagram are assigned as the state variables. The state equations are written in matrix form:

$$\begin{bmatrix} \dfrac{dx_1(t)}{dt} \\ \dfrac{dx_2(t)}{dt} \end{bmatrix} = \begin{bmatrix} -p_2 & z_1 - p_1 \\ 0 & -p_1 \end{bmatrix} \begin{bmatrix} x_1(t) \\ x_2(t) \end{bmatrix} + \begin{bmatrix} \dfrac{a_0}{b_0} \\ \dfrac{a_0}{b_0} \end{bmatrix} r(t) \tag{5-181}$$

The output equation is

$$c(t) = (z_2 - p_2)x_1(t) + (z_1 - p_1)x_2(t) + \frac{a_0}{b_0} r(t) \tag{5-182}$$

The cascade decomposition has the advantage that the poles and zeros of the transfer function appear as isolated branch gains on the state diagram. This facilitates the study of the effects on the system when the poles and zeros are varied.

Parallel Decomposition

parallel decompositon

When the denominator of the transfer function is in factored form, it is possible to expand the transfer function by partial fractions. Consider that a second-order system is represented by the transfer function

$$\frac{C(s)}{R(s)} = \frac{P(s)}{(s + p_1)(s + p_2)} \tag{5-183}$$

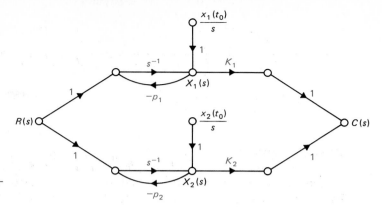

FIGURE 5-8 State diagram of the transfer function of Eq. (5-183) by parallel decomposition.

where $P(s)$ is a polynomial of order less than 2, and p_1 and p_2 are real and distinct. Although analytically p_1 and p_2 may be complex-conjugate numbers, realistically they are difficult to implement on a computer. Equation (5-183) is written

$$\frac{C(s)}{R(s)} = \frac{K_1}{s + p_1} + \frac{K_2}{s + p_2} \tag{5-184}$$

where K_1 and K_2 are real constants.

The state diagram for the system is formed by the parallel combination of the state-diagram representation of each of the first-order terms on the right-hand side of Eq. (5-184), as shown in Fig. 5-8. The state equations of the system are

$$\begin{bmatrix} \dfrac{dx_1(t)}{dt} \\ \dfrac{dx_2(t)}{dt} \end{bmatrix} = \begin{bmatrix} -p_1 & 0 \\ 0 & -p_2 \end{bmatrix} \begin{bmatrix} x_1(t) \\ x_2(t) \end{bmatrix} + \begin{bmatrix} 1 \\ 1 \end{bmatrix} r(t) \tag{5-185}$$

The output equation is

$$c(t) = \begin{bmatrix} K_1 & K_2 \end{bmatrix} \begin{bmatrix} x_1(t) \\ x_2(t) \end{bmatrix} \tag{5-186}$$

advantages   *One of the advantages of the parallel decomposition is that for transfer functions with simple poles, the resulting **A** matrix is always a diagonal matrix. Therefore, we can consider that parallel decomposition may be used for the diagonalization of the **A** matrix. When the transfer function has multiple-order poles, care must be taken that the state diagram, as obtained through the parallel decomposition, contains a minimum number of integrators.* To further clarify this point, the following example is given.

EXAMPLE 5-9   Consider the following transfer function and its partial-fraction expansion:

$$\frac{C(s)}{R(s)} = \frac{2s^2 + 6s + 5}{(s + 1)^2(s + 2)} = \frac{1}{(s + 1)^2} + \frac{1}{s + 1} + \frac{1}{s + 2} \tag{5-187}$$

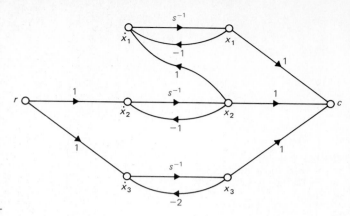

FIGURE 5-9 State diagram of the transfer function of Eq. (5-187) by parallel decomposition.

Note that the transfer function is of the third order, and although the total order of the terms on the right-hand side of Eq. (5-187) is four, only three integrators should be used in the state diagram. The state diagram for the system is shown in Fig. 5-9. The minimum number of three integrators is used, with one integrator being shared by two channels. The state equations of the system are

$$\begin{bmatrix} \dfrac{dx_1(t)}{dt} \\ \dfrac{dx_2(t)}{dt} \\ \dfrac{dx_2(t)}{dt} \end{bmatrix} = \begin{bmatrix} -1 & 1 & 0 \\ 0 & -1 & 0 \\ 0 & 0 & -2 \end{bmatrix} \begin{bmatrix} x_1(t) \\ x_2(t) \\ x_3(t) \end{bmatrix} + \begin{bmatrix} 0 \\ 1 \\ 1 \end{bmatrix} r(t) \qquad (5\text{-}188)$$

Jordan canonical form

In this case, the matrix **A** is not diagonal, but the eigenvalues are still found on the main diagonal, and the matrix is of the **Jordan canonical form**.

5-10

CONTROLLABILITY OF LINEAR SYSTEMS

The concepts of controllability and observability introduced first by Kalman [6] play an important role in both theoretical and practical aspects of modern control. The conditions on controllability and observability often govern the existence of a solution to an optimal control problem. For instance, we shall show that the condition of controllability of a system is closely related to the existence of solutions of state feedback for the purpose of placing the eigenvalues of the system arbitrarily. The concept of observability relates to the condition of observing or estimating the state variables from the output variables, which are generally measurable.

controllability

One way of illustrating the motivation of investigating controllability and observability can be made by referring to the block diagram shown in Fig. 5-10. Figure 5-10(a) shows a closed-loop system with the process dynamics described by

$$\dot{\mathbf{x}}(t) = \mathbf{A}\mathbf{x}(t) + \mathbf{B}\mathbf{u}(t) \qquad (5\text{-}189)$$

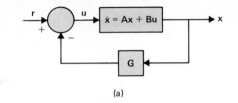

(a)

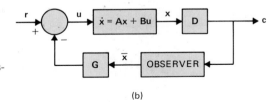

(b)

FIGURE 5-10 (a) Control system with state feedback. (b) Control system with observer and state feedback.

observability
The closed-loop system is formed by feeding back the state variables through a constant matrix **G**. Thus,

$$\mathbf{u}(t) = -\mathbf{Gx}(t) + \mathbf{r}(t) \tag{5-190}$$

where **G** is a $p \times n$ feedback matrix with constant real elements. The closed-loop system is thus described by

$$\dot{\mathbf{x}}(t) = (\mathbf{A} - \mathbf{BG})\mathbf{x}(t) + \mathbf{Br}(t) \tag{5-191}$$

The design objective in this case is to find the feedback matrix **G** such that the eigenvalues of $(\mathbf{A} - \mathbf{BG})$, or of the closed-loop system, are of certain prescribed values. This
pole-placement design
problem is also known as the **pole-placement design** through state feedback. The word "pole" refers here to the poles of the closed-loop transfer function, which are the same as the eigenvalues of $(\mathbf{A} - \mathbf{BG})$.

We shall show later that the existence of the solution to the pole-placement design through state feedback is directly based on the controllability of $[\mathbf{A}, \mathbf{B}]$. Thus, we can state that *if the system of Eq. (5-189) is controllable, then there exists a constant feedback matrix* **G** *that allows the eigenvalues of* $(\mathbf{A} - \mathbf{BG})$ *to be arbitrarily assigned.*

Once the closed-loop system is designed, one has to deal with the practical problem of feeding back the state variables. There are two practical problems with state-feedback control. One is that the number of state variables may be excessive, so that the cost of sensing each of these state variables for feedback may be prohibitive. Another problem is that not all the state variables may be accessible directly from the system. In reality, only the output variables are guaranteed to be accessible. Figure
observer
5-10(b) shows the block diagram of a closed-loop system with an **observer** that estimates the state vector from the output vector $\mathbf{c}(t)$. The observed or the estimated state vector is designated as $\bar{\mathbf{x}}(t)$, which is then used to generate the control $\mathbf{u}(t)$ through the feedback matrix **G**. *The condition that such an observer exists for the system of Eq. (5-189), together with the output equation* $\mathbf{c}(t) = \mathbf{Dx}(t)$, *is called the observability of the system.*

FIGURE 5-11 Linear time-invariant system.

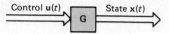

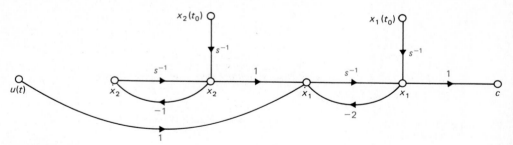

FIGURE 5-12 State diagram of the system that is not state controllable.

General Concept of Controllability

The concept of controllability can be stated with reference to the block diagram of Fig. 5-11. *The process is said to be completely controllable if* every *state variable of the process can be affected or controlled to reach a certain objective in* finite *time by some unconstrained control* **u**(t). Intuitively, it is simple to understand that if any one of the state variables is independent of the control **u**(t), there would be no way of driving this particular state variable to a desired state in finite time by means of a control effort. Therefore, this particular state is said to be uncontrollable, and as long as there is at least one uncontrollable state, the system is said to be not completely controllable, or simply uncontrollable.

As a simple example of an uncontrollable system, Fig. 5-12 illustrates the state diagram of a linear system with two state variables. Since the control $u(t)$ affects only the state $x_1(t)$, the state $x_2(t)$ is uncontrollable. In other words, it would be impossible to drive $x_2(t)$ from an initial state $x_2(t_0)$ to a desired state $x_2(t_f)$ in a finite time interval $t_f - t_0$ by the control $u(t)$. Therefore, the entire system is said to be uncontrollable.

The concept of controllability just given refers to the states and is sometimes referred to as the **state controllability**. Controllability can also be defined for the outputs of the system, so there is a difference between state controllability and output controllability.

Definition of State Controllability

Consider that a linear time-invariant system is described by the following dynamic equations:

$$\dot{\mathbf{x}}(t) = \mathbf{A}\mathbf{x}(t) + \mathbf{B}\mathbf{u}(t) \tag{5-192}$$

$$\mathbf{c}(t) = \mathbf{D}\mathbf{x}(t) + \mathbf{E}\mathbf{u}(t) \tag{5-193}$$

where

$$\mathbf{x}(t) = n \times 1 \text{ state vector}$$
$$\mathbf{u}(t) = r \times 1 \text{ input vector}$$
$$\mathbf{c}(t) = p \times 1 \text{ output vector}$$
$$\mathbf{A} = n \times n \text{ coefficient matrix}$$
$$\mathbf{B} = n \times r \text{ coefficient matrix}$$
$$\mathbf{D} = p \times n \text{ coefficient matrix}$$
$$\mathbf{E} = p \times r \text{ coefficient matrix}$$

The state $\mathbf{x}(t)$ is said to be controllable at $t = t_0$ if there exists a piecewise continuous input $\mathbf{u}(t)$ that will drive the state to any final state $\mathbf{x}(t_f)$ for a finite time $(t_f - t_0) \geq 0$. If every state $\mathbf{x}(t_0)$ of the system is controllable in a finite time interval, the system is said to be completely state controllable or simply controllable.

The following theorem shows that the condition of controllability depends on the coefficient matrices $\mathbf{A}$ and $\mathbf{B}$ of the system. The theorem also gives one method of testing for state controllability.

condition

THEOREM 5-2.   *For the system described by the state equation of Eq. (5-192) to be **completely state controllable**, it is necessary and sufficient that the following $n \times nr$ matrix has a rank of n:*

$$\mathbf{S} = [\mathbf{B} \quad \mathbf{AB} \quad \mathbf{A}^2\mathbf{B} \cdots \mathbf{A}^{n-1}\mathbf{B}] \qquad (5\text{-}194)$$

Since the matrices $\mathbf{A}$ and $\mathbf{B}$ are involved, sometimes we say that the pair $[\mathbf{A}, \mathbf{B}]$ is controllable, which implies that $\mathbf{S}$ is of rank n. ■

The proof of this theorem is given in any standard textbook on optimal control systems [7]. The idea is to start with the state-transition equation of Eq. (5-53) and then proceed to show that Eq. (5-194) must be satisfied in order that all the states are accessible by the input control.

Although the criterion of state controllability given in Theorem 5-2 is quite straightforward, manually, it is not very easy to test for multiple-input systems. Even with $r = 2$, there are $2n$ columns in $\mathbf{S}$, and there would be a large number of possible combinations of $n \times n$ matrices. A practical way may be to use one column of $\mathbf{B}$ at a time, each time giving an $n \times n$ matrix for $\mathbf{S}$. However, failure to find an $\mathbf{S}$ with a rank for n this way does not mean that the system is uncontrollable, until all the columns of $\mathbf{B}$ are used. An easier way would be to form the matrix $\mathbf{SS}'$, which is $n \times n$; then if $\mathbf{SS}'$ is nonsingular, $\mathbf{S}$ has rank n.

The program LINSYS in the software package that accompanies this text contains a subprogram that checks the state controllability of linear time-invariant systems. The user simply enters the elements of the matrices **A** and **B**, and the program will tell whether the system is state controllable or not.

EXAMPLE 5-10   Consider the system shown in Fig. 5-12, which was reasoned earlier to be uncontrollable. Let us investigate the same problem using the condition of Eq. (5-194). The state equations of the system are written, from Fig. 5-12,

$$\begin{bmatrix} \dfrac{dx_1(t)}{dt} \\ \dfrac{dx_2(t)}{dt} \end{bmatrix} = \begin{bmatrix} -2 & 1 \\ 0 & -1 \end{bmatrix} \begin{bmatrix} x_1(t) \\ x_2(t) \end{bmatrix} + \begin{bmatrix} 1 \\ 0 \end{bmatrix} u(t) \tag{5-195}$$

Therefore, from Eq. (5-194),

$$\mathbf{S} = [\mathbf{B} \quad \mathbf{AB}] = \begin{bmatrix} 1 & -2 \\ 0 & 0 \end{bmatrix} \tag{5-196}$$

which is singular, and the system is not state controllable.

EXAMPLE 5-11   Determine the state controllability of the system described by the state equation

$$\begin{bmatrix} \dfrac{dx_1(t)}{dt} \\ \dfrac{dx_2(t)}{dt} \end{bmatrix} = \begin{bmatrix} 0 & 1 \\ -1 & 0 \end{bmatrix} \begin{bmatrix} x_1(t) \\ x_2(t) \end{bmatrix} + \begin{bmatrix} 0 \\ 1 \end{bmatrix} u(t) \tag{5-197}$$

From Eq. (5-194),

$$\mathbf{S} = [\mathbf{B} \quad \mathbf{AB}] = \begin{bmatrix} 0 & 1 \\ 1 & 0 \end{bmatrix} \tag{5-198}$$

which is nonsingular. Therefore, the system is completely state controllable.

EXAMPLE 5-12   Consider a linear system whose input–output relationship is described by the differential equation

$$\frac{d^2c(t)}{dt^2} + 2\frac{dc(t)}{dt} + c(t) = \frac{du(t)}{dt} + u(t) \tag{5-199}$$

We shall show that the state controllability of the system depends upon how the state variables are defined.

Let the state variables be defined as

$$x_1(t) = c(t)$$

$$x_2(t) = \frac{dc(t)}{dt} - u(t)$$

The state equations of the system are expressed in matrix form as

$$\begin{bmatrix} \dot{x}_1(t) \\ \dot{x}_2(t) \end{bmatrix} = \begin{bmatrix} 0 & 1 \\ -1 & -2 \end{bmatrix} \begin{bmatrix} x_1(t) \\ x_2(t) \end{bmatrix} + \begin{bmatrix} 1 \\ -1 \end{bmatrix} u(t) \qquad (5\text{-}200)$$

The state controllability matrix is

$$\mathbf{S} = [\mathbf{B} \quad \mathbf{AB}] = \begin{bmatrix} 1 & -1 \\ -1 & 1 \end{bmatrix} \qquad (5\text{-}201)$$

which is singular. The system is not state controllable.

Now let us define the state variables in a different way. By the method of direct decomposition, the **A** and the **B** matrices are

$$\mathbf{A} = \begin{bmatrix} 0 & 1 \\ -1 & -2 \end{bmatrix} \qquad \mathbf{B} = \begin{bmatrix} 0 \\ 1 \end{bmatrix} \qquad (5\text{-}202)$$

Now the system is completely controllable, since **S** is nonsingular.

5-11
OBSERVABILITY OF LINEAR SYSTEMS

observability
The concept of observability is quite similar to that of controllability. Essentially, *a system is completely observable if every state variable of the system affects some of the outputs.* In other words, it is often desirable to obtain information on the state variables from measurements of the outputs and the inputs. If any one of the states cannot be observed from the measurements of the outputs, the state is said to be unobservable, and the system is not completely observable, or simply unobservable. Figure 5-13 shows the

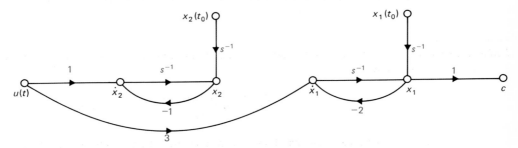

FIGURE 5-13 State diagram of a system that is not observable.

state diagram of a linear system in which the state x_2 is not connected to the output c in any way. Once we have measured c, we can observe the state x_1, since $x_1 = c$. However, the state x_2 cannot be observed from the information on c. Thus, the system is unobservable.

definition

DEFINITION OF OBSERVABILITY.   *Given a linear time-invariant system that is described by the dynamic equations of Eqs. (5-192) and (5-193), the state $\mathbf{x}(t_0)$ is said to be observable if given any input $u(t)$, there exists a finite time $t_f \geq t_0$ such that the knowledge of $\mathbf{u}(t)$ for $t_0 \leq t < t_f$; matrices $\mathbf{A}$, $\mathbf{B}$, $\mathbf{D}$, and $\mathbf{E}$; and the output $\mathbf{c}(t)$ for $t_0 \leq t < t_f$ are sufficient to determine $\mathbf{x}(t_0)$. If every state of the system is observable for a finite t_f, we say that the system is completely observable, or simply observable.*

The following theorem shows that the condition of observability depends on the coefficient matrices $\mathbf{A}$ and $\mathbf{D}$ of the system. The theorem also gives one method of testing observability.

THEOREM 5-4.   *For the system described by the dynamic equations of Eqs. (5-192) and (5-193) to be completely observable, it is necessary and sufficient that the following in $n \times np$ matrix has a rank of n:*

$$\mathbf{V} = [\mathbf{D}' \quad \mathbf{A}'\mathbf{D}' \quad (\mathbf{A}')^2\mathbf{D}' \quad \cdots \quad (\mathbf{A}')^{n-1}\mathbf{D}'] \qquad (5\text{-}203)$$

The condition is also referred to as the pair $[\mathbf{A}, \mathbf{D}]$ being observable. In particular, if the system has only one output, $\mathbf{D}$ is a $1 \times n$ matrix; $\mathbf{V}$ of Eq. (5-203) is an $n \times n$ square matrix. Then the system is completely observable if $\mathbf{V}$ is nonsingular. ∎

The proof of this theorem is not given here, but it is based on the principle that Eq. (5-203) must be satisfied so that $\mathbf{x}(t_0)$ can be uniquely determined from the output vector $\mathbf{c}(t)$.

EXAMPLE 5-13   Let us consider the system described by the differential equation of Eq. (5-199). In Example 5-12, we showed that state controllability of a system depends on how the state variables are defined. We shall now show that the observability also depends on the definition of the state variables. Let the state equations of the system be defined as in Eq. (5-200), and the output equation is $c(t) = x_1(t)$. Then, $\mathbf{D} = [1 \quad 0]$. The observability matrix is

$$\mathbf{V} = [\mathbf{D}' \quad \mathbf{A}'\mathbf{D}'] = \begin{bmatrix} 1 & 0 \\ 0 & 1 \end{bmatrix} \qquad (5\text{-}204)$$

Thus, the system is completely observable.

Let the state equations be written as in Eq. (5-202), and the output equation is $c(t) = x_1(t) + x_2(t)$. Then, $\mathbf{D} = [1 \quad 1]$. The observability matrix is

$$\mathbf{V} = [\mathbf{D}' \quad \mathbf{A}'\mathbf{D}'] = \begin{bmatrix} 1 & -1 \\ 1 & -1 \end{bmatrix} \tag{5-205}$$

which is singular. Thus the system is unobservable.

We have shown that given the differential equation of a linear system, *the controllability and observability of the system depend on how the state variables are defined.* The state model of Eq. (5-200) corresponds to a system that is observable but uncontrollable. On the other hand, the state model of Eq. (5-202) corresponds to a system that is controllable but unobservable. The reason behind this phenomenon is that the transfer function of the system has pole-zero cancellations.

pole-zero
cancellation

In general, the transfer function of a linear system gives indication on the controllability and observability of the system modelled by dynamic equations. *If the transfer function does not have pole-zero cancellation, the system can always be represented by dynamic equations as a completely controllable and observable system. If the transfer function has pole-zero cancellation, the system will be either uncontrollable or unobservable or both, depending on how the state variables are defined.*

5-12
STATE EQUATIONS OF LINEAR DISCRETE-DATA SYSTEMS

Just as for continuous-data systems, the modern way of modeling a discrete-data system is by discrete state equations. As described earlier, when dealing with discrete-data systems, we often encounter two different situations. The first one is that the system contains continuous-data components, but the signals at certain points of the system are discrete with respect to time, due to sample-and-hold operations. In this case, the components of the system are still described by differential equations, but because of the discrete-time data, the differential equations are discretized to yield a set of difference equations. The second situation involves systems that are completely discrete with respect to time, and the system dynamics should be described by difference equations from the outset.

Discrete State Equations

Let us consider the open-loop discrete-data control system with a sample-and-hold device, as shown in Fig. 5-14. Typical signals that appear at various points in the system are shown in the figure. The output signal, $c(t)$, ordinarily is a continuous-data signal. The output of the sample-and-hold, $h(t)$, is a sequence of steps. Therefore, we can write

$$h(t) = h(kT) = r(kT) \tag{5-206}$$

for $kT \le t < (k + 1)T$, $k = 0, 1, 2, \ldots$.

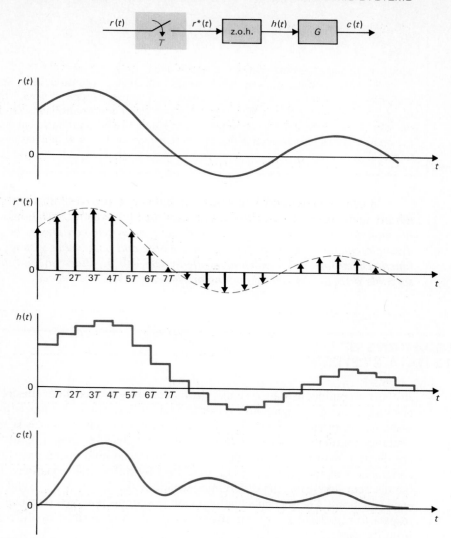

FIGURE 5-14 Discrete-data system with sample-and-hold.

Now we let the linear process G be described by the state equation and output equation:

$$\frac{d\mathbf{x}(t)}{dt} = \mathbf{A}\mathbf{x}(t) + \mathbf{B}h(t) \tag{5-207}$$

$$c(t) = \mathbf{D}\mathbf{x}(t) + \mathbf{E}h(t) \tag{5-208}$$

where $\mathbf{x}(t)$ is the $n \times n$ state vector, and $h(t)$ and $c(t)$ are the scalar input and output, respectively. The matrices $\mathbf{A}$, $\mathbf{B}$, $\mathbf{D}$, and $\mathbf{E}$ are coefficient matrices. By using Eq.

(5-53), the state-transition equation is

$$\mathbf{x}(t) = \boldsymbol{\phi}(t - t_0)\mathbf{x}(t_0) + \int_{t_0}^{t} \boldsymbol{\phi}(t - \tau)\mathbf{B}h(\tau)\, d\tau \tag{5-209}$$

for $t \geq t_0$. If we are interested only in the responses at the sampling instants, we let $t = (k + 1)T$ and $t_0 = kT$. Then Eq. (5-209) becomes

$$\mathbf{x}[(k + 1)T] = \boldsymbol{\phi}(T)\mathbf{x}(kT) + \int_{kT}^{(k+1)T} \boldsymbol{\phi}[(k + 1)T - \tau]\mathbf{B}h(\tau)\, d\tau \tag{5-210}$$

Since $h(t)$ is piecewise constant, as defined in Eq. (5-206), the input $h(\tau)$ in Eq. (5-210) can be taken outside of the integral sign. Equation (5-210) is written

$$\mathbf{x}[(k + 1)T] = \boldsymbol{\phi}(T)\mathbf{x}(kT) + \int_{kT}^{(k+1)T} \boldsymbol{\phi}[(k + 1)T - \tau]\mathbf{B}\, d\tau\, h(kT) \tag{5-211}$$

or

$$\mathbf{x}[(k + 1)T] = \boldsymbol{\phi}(T)\mathbf{x}(kT) + \boldsymbol{\theta}(T)h(kT) \tag{5-212}$$

where

$$\boldsymbol{\theta}(T) = \int_{kT}^{(k+1)T} \boldsymbol{\phi}[(k + 1)T - \tau]\mathbf{B}\, d\tau = \int_{0}^{T} \boldsymbol{\phi}(T - \tau)\mathbf{B}\, d\tau \tag{5-213}$$

Equation (5-212) is of the form of a set of first-order linear difference equations in vector-matrix form, and is referred to as the **vector-matrix discrete state equation**.

Solutions of the Discrete State Equations: Discrete State-Transition Equations

discrete state equation

The discrete state equations represented by Eq. (5-212) can be solved by means of a simple recursion procedure. By setting $k = 0, 1, 2, \ldots$ successively in Eq. (5-212), the following equations result:

$$k = 0: \qquad \mathbf{x}(T) = \boldsymbol{\phi}(T)\mathbf{x}(0) + \boldsymbol{\theta}(T)h(0) \tag{5-214}$$

$$k = 1: \qquad \mathbf{x}(2T) = \boldsymbol{\phi}(T)\mathbf{x}(T) + \boldsymbol{\theta}(T)h(T) \tag{5-215}$$

$$k = 2: \qquad \mathbf{x}(3T) = \boldsymbol{\phi}(T)\mathbf{x}(2T) + \boldsymbol{\theta}(T)h(2T) \tag{5-216}$$

$$\vdots \qquad\qquad \vdots$$

$$k = n - 1: \qquad \mathbf{x}(nT) = \boldsymbol{\phi}(T)\mathbf{x}[(n - 1)T] + \boldsymbol{\theta}(T)h[(n - 1)T] \tag{5-217}$$

Substituting Eq. (5-214) into Eq. (5-215), and then Eq. (5-215) into Eq. (5-216), and so on, we obtain the following solution for Eq. (5-212):

$$\mathbf{x}(nT) = \boldsymbol{\phi}^n(T)\mathbf{x}(0) + \sum_{i=0}^{n-1} \boldsymbol{\phi}^{n-i-1}(T)\boldsymbol{\theta}(T)h(iT) \qquad (5\text{-}218)$$

discrete state-transition equation

where, from Eq. (5-44), $\boldsymbol{\phi}^n(T) = [\boldsymbol{\phi}(T)]^n = \boldsymbol{\phi}(nT)$.

Equation (5-218) is defined as the **discrete state-transition equation** of the discrete-data system. It is interesting to note that Eq. (5-218) is analogous to its continuous-data counterpart in Eq. (5-50). The state-transition equation of Eq. (5-209) describes the state of the system of Fig. 5-14 for all values of t, whereas the discrete state-transition equation in Eq. (5-218) describes the states only at the sampling instants $t = nT$, $n = 0, 1, 2, \ldots$.

With nT considered as the initial time, where n is any positive integer, the state-transition equation is

$$\mathbf{x}[(n + N)T] = \boldsymbol{\phi}^N(T)\mathbf{x}(nT) + \sum_{i=0}^{N-1} \boldsymbol{\phi}^{N-i-1}(T)\boldsymbol{\theta}(T)h[(n + i)T] \qquad (5\text{-}219)$$

where N is a positive integer. The output of the system at the sampling instants is obtained by substituting $t = nT$ and Eq. (5-218) into Eq. (5-208), yielding

$$\begin{aligned} c(nT) &= \mathbf{D}\mathbf{x}(nT) + \mathbf{E}h(nT) \\ &= \mathbf{D}\boldsymbol{\phi}^n(T)\mathbf{x}(0) + \mathbf{D}\sum_{i=0}^{n-1}\boldsymbol{\phi}^{n-i-1}(T)\boldsymbol{\theta}(T)h(iT) + \mathbf{E}h(nT) \end{aligned} \qquad (5\text{-}220)$$

An important advantage of the state-variable method over the z-transform method is that it can be modified easily to describe the states and the output between sampling instants. In Eq. (5-209), if we let $t = (n + \Delta)T$, where $0 < \Delta \leq 1$ and $t_0 = nT$, we get

$$\begin{aligned} \mathbf{x}[(n + \Delta)T] &= \boldsymbol{\phi}(\Delta T)\mathbf{x}(nT) + \int_{nT}^{(n+\Delta)T} \boldsymbol{\phi}[(n + \Delta)T - \tau]\mathbf{B}\, d\tau\, h(nT) \\ &= \boldsymbol{\phi}(\Delta T)\mathbf{x}(nT) + \boldsymbol{\theta}(\Delta T)h(nT) \end{aligned} \qquad (5\text{-}221)$$

By varying the value of Δ between 0 and 1, the information on the state variables between the sampling instants is completely described by Eq. (5-221).

When a linear system has only discrete data through the system, its dynamics can be described by a set of discrete state equations:

$$\mathbf{x}[(k + 1)T] = \mathbf{A}\mathbf{x}(kT) + \mathbf{B}\mathbf{r}(kT) \qquad (5\text{-}222)$$

and output equations:

$$\mathbf{c}(kT) = \mathbf{D}\mathbf{x}(kT) + \mathbf{E}\mathbf{r}(kT) \qquad (5\text{-}223)$$

where $\mathbf{A}$, $\mathbf{B}$, $\mathbf{D}$, and $\mathbf{E}$ are coefficient matrices of the appropriate dimensions. Notice that Eq. (5-222) is basically of the same form as Eq. (5-212). The only difference in the two situations is the starting point of the system representation. In the case of Eq. (5-212), the starting point is the continuous-data state equations of Eq. (5-207); $\boldsymbol{\phi}(T)$ and $\boldsymbol{\theta}(T)$ are determined from the $\mathbf{A}$ and $\mathbf{B}$ matrices, and must satisfy the conditions and properties of the state-transition matrix. In the case of Eq. (5-222), the equation itself represents an outright description of the discrete-data system, and there are no restrictions on the matrices $\mathbf{A}$ and $\mathbf{B}$.

The solution of Eq. (5-222) follows directly from that of Eq. (5-212), and is

$$\mathbf{x}(nT) = \mathbf{A}^n\mathbf{x}(0) + \sum_{i=0}^{n-1} \mathbf{A}^{n-i-1}\mathbf{B}\mathbf{r}(iT) \qquad (5\text{-}224)$$

where

$$\mathbf{A}^n = \underbrace{\mathbf{A}\mathbf{A}\mathbf{A}\mathbf{A} \cdots \mathbf{A}}_{n} \qquad (5\text{-}225)$$

5-13
z-TRANSFORM SOLUTION OF DISCRETE STATE EQUATIONS

In Section 2-12, we illustrated the solution of a simple discrete state equation by the z-transform method. In this section, the discrete state equations in vector-matrix form of an nth-order system are solved by z-transformation. Consider the discrete state equations

$$\mathbf{x}[(k + 1)T] = \mathbf{A}\mathbf{x}(kT) + \mathbf{B}\mathbf{r}(kT) \qquad (5\text{-}226)$$

Taking the z-transform on both sides of the last equation, we get

$$z\mathbf{X}(z) - z\mathbf{x}(0) = \mathbf{A}\mathbf{X}(z) + \mathbf{B}\mathbf{R}(z) \qquad (5\text{-}227)$$

Solving for $\mathbf{X}(z)$ from the last equation gives

$$\mathbf{X}(z) = (z\mathbf{I} - \mathbf{A})^{-1}z\mathbf{x}(0) + (z\mathbf{I} - \mathbf{A})^{-1}\mathbf{B}\mathbf{R}(z) \qquad (5\text{-}228)$$

Taking the inverse z-transform on both sides of the last equation, we get

$$\mathbf{x}(nT) = \mathcal{z}^{-1}[(z\mathbf{I} - \mathbf{A})^{-1}z]\mathbf{x}(0) + \mathcal{z}^{-1}[(z\mathbf{I} - \mathbf{A})^{-1}\mathbf{B}\mathbf{R}(z)] \qquad (5\text{-}229)$$

In order to carry out the inverse z-transform operation of the last equation, we write the z-transform of $\mathbf{A}^n$ as

$$\mathcal{z}(\mathbf{A}^n) = \sum_{n=0}^{\infty} \mathbf{A}^n z^{-n} = \mathbf{I} + \mathbf{A}z^{-1} + \mathbf{A}z^{-2} + \cdots \qquad (5\text{-}230)$$

Premultiplying both sides of the last equation by $\mathbf{A}z^{-1}$ and subtracting the result from the last equation, we get

$$(\mathbf{I} - \mathbf{A}z^{-1})\mathcal{z}(\mathbf{A}^n) = \mathbf{I} \qquad (5\text{-}231)$$

Therefore, solving for $\mathcal{z}(\mathbf{A}^n)$ from the last equation yields

$$\mathcal{z}(\mathbf{A}^n) = (\mathbf{I} - \mathbf{A}z^{-1})^{-1} = (z\mathbf{I} - \mathbf{A})^{-1}z \qquad (5\text{-}232)$$

or

$$\mathbf{A}^n = \mathcal{z}^{-1}[(z\mathbf{I} - \mathbf{A})^{-1}z] \qquad (5\text{-}233)$$

Equation (5-233) represents a way of finding $\mathbf{A}^n$ by using the z-transform method. Similarly, we can prove that

$$\mathcal{z}^{-1}[(z\mathbf{I} - \mathbf{A})^{-1}\mathbf{B}\mathbf{R}(z)] = \sum_{i=0}^{n-1} \mathbf{A}^{n-i-1}\mathbf{B}\mathbf{r}(iT) \qquad (5\text{-}234)$$

Now we substitute Eqs. (5-233) and (5-234) into Eq. (5-229), and $\mathbf{x}(kT)$ becomes

$$\mathbf{x}(nT) = \mathbf{A}^n\mathbf{x}(0) + \sum_{i=0}^{n-1} \mathbf{A}^{n-i-1}\mathbf{B}\mathbf{r}(iT) \qquad (5\text{-}235)$$

which is identical to the expression in Eq. (5-224).

Transfer-Function Matrix and the Characteristic Equation

Once a discrete-data system is modeled by the dynamic equations of Eqs. (5-222) and (5-223), the transfer-function relation of the system can be expressed in terms of the coefficient matrices. By setting the initial state $\mathbf{x}(0)$ to zero, Eq. (5-228) becomes

$$\mathbf{X}(z) = (z\mathbf{I} - \mathbf{A})^{-1}\mathbf{B}\mathbf{R}(z) \qquad (5\text{-}236)$$

z-transform solution

Substituting the last equation into the z-transformed version of Eq. (5-223), we have

$$\mathbf{C}(z) = [\mathbf{D}(z\mathbf{I} - \mathbf{A})^{-1}\mathbf{B} + \mathbf{E}]\mathbf{R}(z) = \mathbf{G}(z)\mathbf{R}(z) \tag{5-237}$$

where the transfer-function matrix of the system is defined as

$$\mathbf{G}(z) = \mathbf{D}(z\mathbf{I} - \mathbf{A})^{-1}\mathbf{B} + \mathbf{E} \tag{5-238}$$

or

$$\mathbf{G}(z) = \frac{\mathbf{D}[\text{adj}\,(z\mathbf{I} - \mathbf{A})]\mathbf{B} + |z\mathbf{I} - \mathbf{A}|\mathbf{E}}{|z\mathbf{I} - \mathbf{A}|} \tag{5-239}$$

characteristic equation

The characteristic equation of the system is defined as

$$|z\mathbf{I} - \mathbf{A}| = 0 \tag{5-240}$$

In general, a linear time-invariant discrete-data system with one input and one output can be described by the following linear difference equation with constant coefficients:

$$\begin{aligned}
c[(k + n)T] &+ a_1 c[(k + n - 1)T] + a_2 c[(k + n - 2)T] \\
&+ \cdots + a_{n-1} c[(k + 1)T] + a_n c(kT) \\
&= b_0 r[(k + m)T] + b_1 r[(k + m - 1)T] \\
&+ \cdots + b_{m-1} r[(k + 1)T] + b_m r(kT) \qquad n \geq m
\end{aligned} \tag{5-241}$$

Taking the z-transform on both sides of the last equation and setting zero initial conditions, the transfer function of the system is

$$\frac{C(z)}{R(z)} = \frac{b_0 z^m + b_1 z^{m-1} + \cdots + b_{m-1} z + b_m}{z^n + a_1 z^{n-1} + \cdots + a_{n-1} z + a_n} \tag{5-242}$$

The characteristic equation is obtained by equating the denominator polynomial of the transfer function to zero:

$$z^n + a_1 z^{n-1} + \cdots + a_{n-1} z + a_n = 0 \tag{5-243}$$

EXAMPLE 5-14   Consider that a discrete-data system is described by the difference equation

$$c(k + 2) + 5c(k + 1) + 3c(k) = r(k + 1) + 2r(k) \qquad (5\text{-}244)$$

The transfer function of the system is simply

$$\frac{C(z)}{R(z)} = \frac{z + 2}{z^2 + 5z + 3} \qquad (5\text{-}245)$$

The characteristic equation is

$$z^2 + 5z + 3 = 0 \qquad (5\text{-}246)$$

The state variables of the system may be defined as

$$x_1(k) = c(k) \qquad (5\text{-}247)$$
$$x_2(k) = x_1(k + 1) - r(k) \qquad (5\text{-}248)$$

Substituting the last two relations into Eq. (5-244) gives the two state equations:

$$x_1(k + 1) = x_2(k) + r(k) \qquad (5\text{-}249)$$
$$x_2(k + 1) = -3x_1(k) - 5x_2(k) - 3r(k) \qquad (5\text{-}250)$$

from which we have the matrices **A** and **B** of the system:

$$\mathbf{A} = \begin{bmatrix} 0 & 1 \\ -3 & -5 \end{bmatrix} \quad \mathbf{B} = \begin{bmatrix} 1 \\ -3 \end{bmatrix} \qquad (5\text{-}251)$$

The same characteristic equation as in Eq. (5-246) is obtained by using $|z\mathbf{I} - \mathbf{A}| = 0$.

5-14
STATE DIAGRAMS OF DISCRETE-DATA SYSTEMS

state diagram   When a discrete-data system is described by difference or discrete state equations, a **discrete state diagram** may be constructed for the system. Similar to the relations between the analog-computer diagram and the state diagram for continuous-data systems, the elements of a discrete state diagram resemble the computing elements of a digital computer. Some of the operations of a digital computer are **multiplication by a constant, addition of several variables**, and **time delay or shifting**. The discrete state diagram can be used to determine the transfer-function relations as well as for digital implementation of the system. The mathematical descriptions of these basic digital computations and their corresponding z-transform expressions are as follows:

1. *Multiplication by a constant:*

$$x_2(kT) = ax_1(kT) \qquad (5\text{-}252)$$

$$X_2(z) = aX_1(z) \qquad (5\text{-}253)$$

2. *Summing:*

$$x_2(kT) = x_1(kT) + x_3(kT) \qquad (5\text{-}254)$$

$$X_2(z) = X_1(z) + X_3(z) \qquad (5\text{-}255)$$

3. *Shifting or time delay:*

$$x_2(kT) = x_1[(k + 1)T] \qquad (5\text{-}256)$$

$$X_2(z) = zX_1(z) - zx_1(0) \qquad (5\text{-}257)$$

or

$$X_1(z) = z^{-1}X_2(z) + x_1(0) \qquad (5\text{-}258)$$

The state-diagram representations of these operations are illustrated in Fig. 5-15. The initial time $t = 0$ in Eqs. (5-257) and (5-258) can be generalized to $t = t_0$. Then the equations represent the discrete-time state transition from $t \geq t_0$.

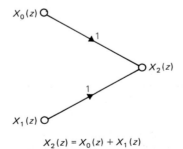

FIGURE 5-15 Basic elements of a discrete state diagram.

EXAMPLE 5-15   Consider again the difference equation in Eq. (5-244), which is

$$c(k + 2) + 5c(k + 1) + 3c(k) = r(k + 1) + 2r(k) \qquad (5\text{-}259)$$

One way of constructing the discrete state diagram for the system is to use the state equations. In this case, the state equations are already defined in Eqs. (5-249) and (5-250). By using essentially the same principle as the state diagram for continuous-data systems, the state diagram for Eqs. (5-249) and (5-250) is shown in Fig. 5-16. The time delay unit z^{-1} is used to relate $x_1(k + 1)$ to $x_1(k)$. *The state variables are defined as the outputs of the delay units on the state diagram.*

As an alternative, the discrete state diagram can also be drawn directly from the difference equation via the transfer function, using the decomposition schemes (Fig. 5-17). The decomposition of a discrete-data transfer function follows basically the same procedure as that of an analog transfer function covered in Section 5-9, and the details are not repeated.

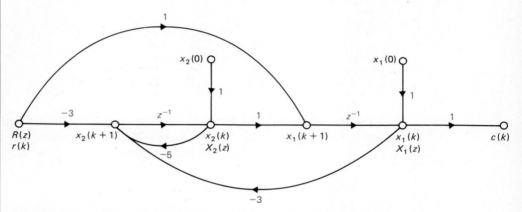

FIGURE 5-16 Discrete state diagram of the system described by the difference equation of Eq. (5-244) or by the state equations of Eqs. (5-249) and (5-250).

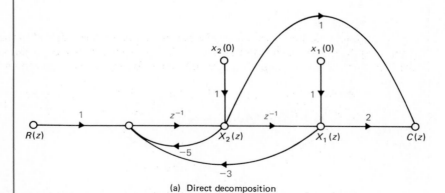

(a) Direct decomposition

FIGURE 5-17 State diagrams of the transfer function $C(z)/R(z) = (z + 2)/(z^2 + 5z + 3)$ by the three methods of decomposition.

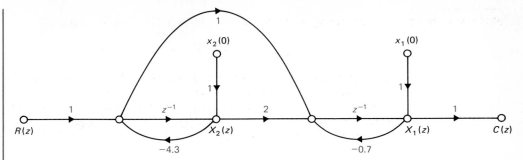

(b) Cascade decomposition

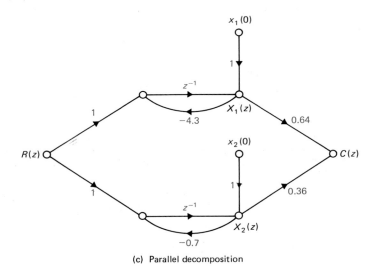

(c) Parallel decomposition

FIGURE 5-17 Continued.

The state-transition equations of the system can be obtained directly from the state diagram using the gain formula. By referring to $X_1(z)$ and $X_2(z)$ as the output nodes and $x_1(0)$, $x_2(0)$, and $R(z)$ as input nodes in Fig. 5-16, the state transition equations are written as

$$\begin{bmatrix} X_1(z) \\ X_2(z) \end{bmatrix} = \frac{1}{\Delta} \begin{bmatrix} 1 + 5z^{-1} & z^{-1} \\ -3z^{-1} & 1 \end{bmatrix} \begin{bmatrix} x_1(0) \\ x_2(0) \end{bmatrix} + \frac{1}{\Delta} \begin{bmatrix} z^{-1}(1 + 5z^{-1}) - 3z^{-2} \\ -3z^{-1} - 3z^{-2} \end{bmatrix} R(z) \qquad (5\text{-}260)$$

where

$$\Delta = 1 + 5z^{-1} + 3z^{-2} \qquad (5\text{-}261)$$

The same transfer function between $R(z)$ and $C(z)$ as in Eq. (5-245) can be obtained directly from the state diagram in Fig. 5-16 by applying the gain formula between these two nodes.

FIGURE 5-18 State-diagram representation of the zero-order hold.

$$e(kT^+) \; \circ\!\!-\!\!-\!\!-\!\!-\!\!-\!\!-\!\!-\!\!-\!\!\rightarrow\!\!-\!\!-\!\!-\!\!-\!\!-\!\!-\!\!\circ \; H(s)$$

State Diagrams for Sampled-Data Systems

sampled-data systems

When a discrete-data system has continuous-data as well as discrete-data elements, with the two types of elements separated by sample-and-hold devices, a state-diagram model for the sample-and-hold (zero-order hold) must be established.

Consider that the input of the zero-order hold is denoted by $e*(t)$, which is a train of impulses, and the output by $h(t)$. Since the zero-order hold simply holds the strength of the input impulse at the sampling instant until the next input comes along, the signal $h(t)$ is a sequence of steps. The input–output relation in the Laplace domain is

$$H(s) = \frac{1 - e^{-Ts}}{s} E*(s) \tag{5-262}$$

In the time domain, the relation is simply

$$h(t) = e(kT^+) \tag{5-263}$$

for $kT \leq t < (k + 1)T$.

In the state-diagram notation, we need the relation between $H(s)$ and $e(kT^+)$. For this purpose, we take the Laplace transform on both sides of Eq. (5-263) to give

$$H(s) = \frac{e(kT^+)}{s} \tag{5-264}$$

for $kT \leq t < (k + 1)T$. The state-diagram representation of the zero-order hold is shown in Fig. 5-18.

EXAMPLE 5-16   As an illustrative example on how the state diagram of a sampled-data system is constructed, let us consider the system in Fig. 5-19. We shall demonstrate the various available ways of modeling the input–output relations of the system. First, the Laplace transform of the output of the system is written in terms of the input to the zero-order hold.

$$C(s) = \frac{1 - e^{-Ts}}{s} \frac{1}{s + 1} E*(s) \tag{5-265}$$

FIGURE 5-19 Sampled-data system.

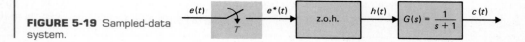

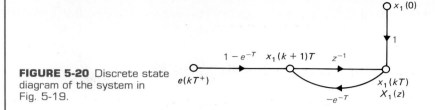

FIGURE 5-20 Discrete state diagram of the system in Fig. 5-19.

Taking the z-transform on both sides of the last equation yields

$$C(z) = \frac{1 - e^{-T}}{z - e^{-T}} E(z) \qquad (5\text{-}266)$$

Figure 5-20 shows the state diagram for Eq. (5-266). The discrete dynamic equations of the system are written directly from the state diagram:

$$x_1[(k + 1)T] = -e^{-T}x_1(kT) + (1 - e^{-T})e(kT^+) \qquad (5\text{-}267)$$

$$c(kT) = x_1(kT) \qquad (5\text{-}268)$$

5-15

A FINAL ILLUSTRATIVE EXAMPLE: MAGNETIC-BALL-SUSPENSION SYSTEM

As a final example to illustrate some of the material presented in this chapter, let us consider the magnetic-ball-suspension system shown in Fig. 5-21. The objective of the system is to regulate the current of the electromagnet so that the ball will be suspended

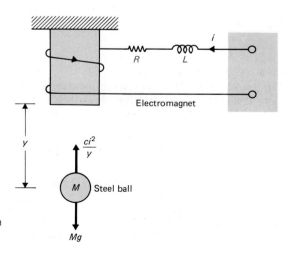

FIGURE 5-21 Ball-suspension system.

at a fixed distance from the end of the magnet. The dynamic equations of the system are

$$M\frac{d^2y(t)}{dt^2} = Mg - \frac{ci^2(t)}{y(t)} \qquad (5\text{-}269)$$

$$v(t) = Ri(t) + L\frac{di(t)}{dt} \qquad (5\text{-}270)$$

where Eq. (5-269) is nonlinear. The system variables and parameters are as follows:

$v(t)$ = input voltage (V)	$y(t)$ = ball position (m)
$i(t)$ = winding current (A)	c = proportional constant = 1.0
R = winding resistance = 1 Ω	L = winding inductance = 0.01 H
M = ball mass = 1.0 Kg	g = gravitational acceleration
	= 32.2 m/sec^2

The state variables are defined as

$$\begin{aligned}
x_1(t) &= y(t) \\
x_2(t) &= dy(t)/dt \\
x_3(t) &= i(t)
\end{aligned} \qquad (5\text{-}271)$$

The state equations are

$$\frac{dx_1(t)}{dt} = x_2(t) \qquad (5\text{-}272)$$

$$\frac{dx_2(t)}{dt} = g - \frac{c}{M}\frac{x_3^2(t)}{x_1(t)} \qquad (5\text{-}273)$$

$$\frac{dx_3(t)}{dt} = -\frac{R}{L}x_3(t) + \frac{v(t)}{L} \qquad (5\text{-}274)$$

These nonlinear state equations are linearized about the equilibrium point, $x_1(t) = y(t) = 0.5$ m, using the method described in Chapter 4. After substituting the parameter values, the linearized equations are written

$$\Delta\dot{x}(t) = \mathbf{A}^* \, \Delta\mathbf{x}(t) + \mathbf{B}^* \, \Delta v(t) \qquad (5\text{-}275)$$

where $\Delta\mathbf{x}(t)$ denotes the state vector, and $\Delta v(t)$ the input voltage of the linearized system. The coefficient matrices are

$$\mathbf{A}^* = \begin{bmatrix} 0 & 1 & 0 \\ 64.4 & 0 & -16 \\ 0 & 0 & -100 \end{bmatrix} \quad \mathbf{B}^* = \begin{bmatrix} 0 \\ 0 \\ 100 \end{bmatrix} \qquad (5\text{-}276)$$

All the analysis performed in the following can be carried out using the LINSYS program of the ACSP software. To show the analytical method, we carry out the steps of the derivations.

THE CHARACTERISTIC EQUATION

$$|s\mathbf{I} - \mathbf{A}^*| = \begin{vmatrix} s & -1 & 0 \\ -64.4 & s & 16 \\ 0 & 0 & s+100 \end{vmatrix} = s^3 + 100s^2 - 64.4s - 6440 = 0 \quad (5\text{-}277)$$

EIGENVALUES.   The eigenvalues of $\mathbf{A}^*$, or the roots of the characteristic equation, are

$$s = -100 \qquad s = -8.025 \qquad s = 8.025$$

THE STATE-TRANSITION MATRIX.   The state-transition matrix of $\mathbf{A}^*$ is

$$\boldsymbol{\phi}(t) = \mathcal{L}^{-1}[(s\mathbf{I} - \mathbf{A}^*)^{-1}] = \mathcal{L}^{-1}\left(\begin{bmatrix} s & -1 & 0 \\ -64.4 & s & 16 \\ 0 & 0 & s+100 \end{bmatrix}^{-1}\right) \quad (5\text{-}278)$$

or

$$\boldsymbol{\phi}(t) = \mathcal{L}^{-1}\left(\frac{1}{(s+100)(s+8.025)(s-8.025)}\right)$$

$$(5\text{-}279)$$

$$\times \begin{bmatrix} s(s+100) & s+100 & -16 \\ 64.4(s+100) & s(s+100) & -16s \\ 0 & 0 & s^2-64.4 \end{bmatrix}$$

By performing the partial-fraction expansion and carrying out the inverse Laplace transform, the state-transition matrix is

$$\boldsymbol{\phi}(t) = \begin{bmatrix} 0 & 0 & -0.0016 \\ 0 & 0 & 0.16 \\ 0 & 0 & 1 \end{bmatrix}e^{-100t} + \begin{bmatrix} 0.5 & -0.062 & 0.0108 \\ -4.012 & 0.5 & -0.087 \\ 0 & 0 & 0 \end{bmatrix}e^{-8.025t}$$

$$+ \begin{bmatrix} 0.5 & 0.062 & -0.0092 \\ 4.012 & 0.5 & -0.074 \\ 0 & 0 & 0 \end{bmatrix}e^{0.8025t}$$

$$(5\text{-}280)$$

TRANSFER FUNCTION.   Since the output variable is $y(t)$ and the input is $v(t)$, the input–output transfer function of the system is

$$\frac{Y(s)}{V(s)} = \mathbf{D}^*(s\mathbf{I} - \mathbf{A}^*)^{-1}\mathbf{B}^* = [1 \quad 0 \quad 0](s\mathbf{I} - \mathbf{A}^*)^{-1}\mathbf{B}^*$$

$$(5\text{-}281)$$

$$= \frac{-1600}{(s+100)(s+8.025)(s-8.025)}$$

CONTROLLABILITY. The controllability matrix is

$$S = [B^* \quad A^*B^* \quad A^{*2}B^*] = \begin{bmatrix} 0 & 0 & -1600 \\ 0 & -1600 & 160,000 \\ 100 & -10000 & 1,000,000 \end{bmatrix} \quad (5\text{-}282)$$

Since the rank of S is three, the system is completely controllable.

OBSERVABILITY. The observability of the system depends on which variable is defined as the output. For state-feedback control, the full controller requires feeding back all three state variables, x_1, x_2, and x_3. However, for reasons of economy, we may want to feedback only one of the three state variables. To make the problem more general, we may want to investigate which state if chosen as the output would render the system unobservable.

1. $c(t) = y(t)$: $\quad D^* = [1 \quad 0 \quad 0]$
   The observability matrix is

$$V = [D^{*\prime} \quad A^{*\prime}D^{*\prime} \quad (A^{*\prime})^2 D^{*\prime}] = \begin{bmatrix} 1 & 0 & 64.4 \\ 0 & 1 & 0 \\ 0 & 0 & -16 \end{bmatrix} \quad (5\text{-}283)$$

   which has a rank of three. Thus, the system is completely observable.
2. $c(t) = dy(t)/dt$: $\quad D^* = [0 \quad 1 \quad 0]$
   The observability matrix is

$$V = \begin{bmatrix} 0 & 64.4 & 0 \\ 1 & 0 & 64.4 \\ 0 & -16 & 1600 \end{bmatrix} \quad (5\text{-}284)$$

   which has a rank of three. Thus, the system is completely observable.
3. $c(t) = i(t)$: $\quad D^* = [0 \quad 0 \quad 1]$

$$V = \begin{bmatrix} 0 & 0 & 0 \\ 0 & 0 & 0 \\ 1 & -100 & -10000 \end{bmatrix} \quad (5\text{-}285)$$

Since V has a rank of one, the system is unobservable. The physical meaning of this result is that if we choose the current $i(t)$ as the measurable output, we would not be able to reconstruct the state variables from the measured information.

The interested reader should enter the data of this system to the LINSYS program and verify the results obtained.

5-16

SUMMARY

This chapter is devoted to the state-variable analysis of linear systems. The fundamentals on state variables and state equations are introduced in Chapters 2 and 4. Formal discussions on these subjects are covered in this chapter. Specifically, the state-transition matrix and the state-transition equations are introduced. The relationship between the state equations and the transfer functions is established. Given the transfer function of a linear system, the state equations of the system can be obtained by means of decomposition of the transfer function. Given the state equations and the output equations, the transfer function can be determined either analytically or directly from the state diagram.

State equations in the phase-variable canonical form have certain unique properties that simplifies the analysis and design problem in the state-variable formulation. A system that is representable in the phase-variable canonical form also possesses certain unique characteristics such as controllability and observability. State controllability and observability of linear time-invariant systems are defined and illustrated.

The state formulation of a discrete-data system is covered in the chapter. The analysis closely parallels the state-variable formulation of continuous-data systems. A final example on the magnetic-ball-suspension system summarizes the important elements of the state-variable analysis of linear systems.

REVIEW QUESTIONS

1. What are the components of the dynamic equations of a linear system?
2. Given the state equations of a linear system as

$$\dot{\mathbf{x}}(t) = \mathbf{A}\mathbf{x}(t) + \mathbf{B}\mathbf{u}(t)$$

   write the state-transition matrix $\boldsymbol{\phi}(t)$.
3. Give two expressions of the state-transition matrix in terms of the matrix $\mathbf{A}$.
4. List the properties of the state-transition matrix $\boldsymbol{\phi}(t)$.
5. Given the state equations as in Review Question 2, write the state-transition equation.
6. List the advantages of expressing a linear system in the phase-variable canonical form (PVCF). Give an example of $\mathbf{A}$ and $\mathbf{B}$ in PVCF.
7. Given the state equations in the form as in Review Question 2, give the conditions for $\mathbf{A}$ and $\mathbf{B}$ to be transformable into PVCF.
8. Express the characteristic equation in terms of the matrix $\mathbf{A}$.
9. List the three methods of decomposition of a transfer function.
10. What special form will the state equations be in if the transfer function is decomposed by direct decomposition?
11. What special form will the state equations be in if the transfer function is decomposed by parallel decomposition?
12. What is the advantage of using cascade decomposition?
13. State the relationship between the phase-variable canonical form and controllability.
14. For controllability, does the magnitude of the inputs have to be finite?
15. Give the condition of controllability in terms of the matrices $\mathbf{A}$ and $\mathbf{B}$.
16. What is the motivation behind the concept of observability?
17. Give the condition of observability in terms of the matrices $\mathbf{A}$ and $\mathbf{D}$.
18. What can be said about the controllability and observability conditions if the transfer function has a pole-zero cancellation?
19. If a system is completely controllable, then its state variables can never become infinitely large as time increases.   **(T)**   **(F)**

REFERENCES

State Variables and State Equations

1. L. A. ZADEH, "An Introduction to State Space Technique," *Proceedings of the Joint Automatic Control Conference,* 1962.
2. B. C. KUO, *Linear Networks and Systems,* McGraw-Hill Book Company, New York, 1967.
3. D. W. WIBERG, *Theory and Problems of State Space and Linear Systems,* McGraw-Hill Book Company, New York, 1971.

State-Transition Matrix

4. M. VIDYASAGAR, "A Novel Method of Evaluating e^{At} in Closed Form," *IEEE Trans. Automatic Control,* Vol. AC-15, pp. 600–601, Oct. 1970.
5. M. HEALEY, "Study of Methods of Computing Transition Matrices," *Proc. IEE,* Vol. 120, No. 8, pp. 905–912, Aug. 1973.

Controllability and Observability

6. R. E. KALMAN, "On the General Theory of Control Systems," *Proc. IFAC,* Vol. 1, pp. 481–492, Butterworths, London, 1961.
7. H. KWAKERNAAK and R. SIVAN, *Linear Optimal Control Systems,* John Wiley & Sons, New York, 1972.

PROBLEMS

5-1. Write state equations for the electric networks shown in Fig. 5P-1. Keep in mind that the number of state variables should be kept to a minimum. Assign the voltages across capacitors and currents through inductances as state variables. Write the currents through the capacitors and the voltages across the inductors as functions of the state variables and the inputs.

5-2. The following differential equations represent linear time-invariant systems. Write the dynamic equations (state equations and output equations) in vector-matrix form.

(a) $\dfrac{d^2 c(t)}{dt^2} + 4\dfrac{dc(t)}{dt} + c(t) = 5r(t)$

(b) $2\dfrac{d^3 c(t)}{dt^3} + 3\dfrac{d^2 c(t)}{dt^2} + 5\dfrac{dc(t)}{dt} + 2c(t) = r(t)$

(c) $\dfrac{d^3 c(t)}{dt^3} + 5\dfrac{d^2 c(t)}{dt^2} + 4\dfrac{dc(t)}{dt} + c(t) + \displaystyle\int_0^t c(\tau)\, d\tau = 5r(t)$

(d) $\dfrac{d^4 c(t)}{dt^4} + 1.5\dfrac{d^3 c(t)}{dt^3} + 3\dfrac{dc(t)}{dt} + 2c(t) = r(t)$

5-3. By use of Eq. (5-23), show that

$$\boldsymbol{\phi}(t) = \mathbf{I} + \mathbf{A}t + \frac{1}{2!}\mathbf{A}^2 t^2 + \frac{1}{3!}\mathbf{A}^3 t^3 + \cdots$$

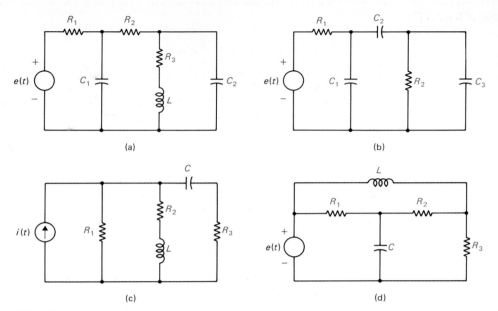

FIGURE 5P-1

5-4.   The state equations of a linear time-invariant system are represented by

$$\dot{\mathbf{x}}(t) = \mathbf{A}\mathbf{x}(t) + \mathbf{B}u(t)$$

Find the state-transition matrix $\boldsymbol{\phi}(t)$, the characteristic equation, and the eigenvalues of
$\mathbf{A}$ for the following cases:

(a) $\mathbf{A} = \begin{bmatrix} 0 & 1 \\ -2 & -1 \end{bmatrix}$ $\mathbf{B} = \begin{bmatrix} 0 & 1 \\ 1 & 0 \end{bmatrix}$

(b) $\mathbf{A} = \begin{bmatrix} 0 & 1 \\ -4 & -5 \end{bmatrix}$ $\mathbf{B} = \begin{bmatrix} 1 \\ 1 \end{bmatrix}$

(c) $\mathbf{A} = \begin{bmatrix} -3 & 0 \\ 0 & -3 \end{bmatrix}$ $\mathbf{B} = \begin{bmatrix} 0 \\ 1 \end{bmatrix}$

(d) $\mathbf{A} = \begin{bmatrix} 3 & 0 \\ 0 & -3 \end{bmatrix}$ $\mathbf{B} = \begin{bmatrix} 0 \\ 1 \end{bmatrix}$

(e) $\mathbf{A} = \begin{bmatrix} 0 & 2 \\ -2 & 0 \end{bmatrix}$ $\mathbf{B} = \begin{bmatrix} 0 \\ 1 \end{bmatrix}$

(f) $\mathbf{A} = \begin{bmatrix} -1 & 0 & 0 \\ 0 & -2 & 1 \\ 0 & 0 & -2 \end{bmatrix}$ $\mathbf{B} = \begin{bmatrix} 0 \\ 1 \\ 0 \end{bmatrix}$

(g) $\mathbf{A} = \begin{bmatrix} -5 & 1 & 0 \\ 0 & -5 & 1 \\ 0 & 0 & -5 \end{bmatrix}$ $\mathbf{B} = \begin{bmatrix} 0 \\ 0 \\ 1 \end{bmatrix}$

Find $\boldsymbol{\phi}(t)$ and the characteristic equation using a computer program; e.g., LINSYS of
the ACSP software package.

5-5. Find the state-transition equation of each of the systems described in Problem 5-4 for $t \geq 0$. Assume that $\mathbf{x}(0)$ is the initial state vector, and the components of the input vector $\mathbf{u}(t)$ are all unit-step functions.

5-6. Find out if the matrices given in the following can be state-transition matrices. [Hint: check the properties of $\boldsymbol{\phi}(t)$.]

(a) $\begin{bmatrix} -e^{-t} & 0 \\ 0 & 1 - e^{-t} \end{bmatrix}$ (b) $\begin{bmatrix} 1 - e^{-t} & 0 \\ 1 & e^{-t} \end{bmatrix}$

(c) $\begin{bmatrix} 1 & 0 \\ 1 - e^{-t} & e^{-t} \end{bmatrix}$ (d) $\begin{bmatrix} e^{-2t} & te^{-2t} & t^2 e^{-2t}/2 \\ 0 & e^{-2t} & te^{-2t} \\ 0 & 0 & e^{-2t} \end{bmatrix}$

5-7. Given the state equation

$$\dot{\mathbf{x}}(t) = \mathbf{A}\mathbf{x}(t) + \mathbf{B}\mathbf{u}(t)$$

(a) $\mathbf{A} = \begin{bmatrix} 0 & 2 & 0 \\ 1 & 2 & 0 \\ -1 & 0 & 1 \end{bmatrix}$ $\mathbf{B} = \begin{bmatrix} 0 \\ 1 \\ 1 \end{bmatrix}$

(b) $\mathbf{A} = \begin{bmatrix} 0 & 2 & 0 \\ 1 & 2 & 0 \\ -1 & 1 & 1 \end{bmatrix}$ $\mathbf{B} = \begin{bmatrix} 1 \\ 1 \\ 0 \end{bmatrix}$

(c) $\mathbf{A} = \begin{bmatrix} -2 & 1 & 0 \\ 0 & -2 & 0 \\ -1 & -2 & -3 \end{bmatrix}$ $\mathbf{B} = \begin{bmatrix} 1 \\ 1 \\ 1 \end{bmatrix}$

(d) $\mathbf{A} = \begin{bmatrix} -1 & 1 & 0 \\ 0 & -1 & 1 \\ 0 & 0 & -1 \end{bmatrix}$ $\mathbf{B} = \begin{bmatrix} 0 \\ 1 \\ 1 \end{bmatrix}$

Find the transformation $\mathbf{y} = \mathbf{Q}\mathbf{x}$ such that the system $d\mathbf{y}(t)/dt = \mathbf{A}_1 \mathbf{y}(t) + \mathbf{B}_1 u(t)$ is in the phase-variable canonical form. If the ACSP software is available, check the answers using the LINSYS program.

5-8. Explain why the following state equations cannot be transformed into the phase-variable canonical form: $d\mathbf{x}(t)/dt = \mathbf{A}\mathbf{x}(t) + \mathbf{B}\mathbf{u}(t)$.

(a) $\mathbf{A} = \begin{bmatrix} -2 & 0 \\ 0 & -1 \end{bmatrix}$ $\mathbf{B} = \begin{bmatrix} 1 \\ 0 \end{bmatrix}$

(b) $\mathbf{A} = \begin{bmatrix} -1 & 0 & 0 \\ 0 & -1 & 0 \\ 0 & 0 & -1 \end{bmatrix}$ $\mathbf{B} = \begin{bmatrix} 1 \\ 2 \\ 3 \end{bmatrix}$

(c) $\mathbf{A} = \begin{bmatrix} 1 & 2 \\ 1 & 1 \end{bmatrix}$ $\mathbf{B} = \begin{bmatrix} 2 \\ \sqrt{2} \end{bmatrix}$

(d) $\mathbf{A} = \begin{bmatrix} -2 & 1 & 0 \\ 0 & -2 & 0 \\ -1 & -2 & -3 \end{bmatrix}$ $\mathbf{B} = \begin{bmatrix} 1 \\ 0 \\ 1 \end{bmatrix}$

5-9. Given a system described by the dynamic equations:

$$\dot{\mathbf{x}}(t) = \mathbf{A}\mathbf{x}(t) + \mathbf{B}u(t) \qquad c(t) = \mathbf{D}\mathbf{x}(t)$$

(a)
$$\mathbf{A} = \begin{bmatrix} 0 & 1 & 0 \\ 0 & 0 & 1 \\ -1 & -2 & -3 \end{bmatrix} \qquad \mathbf{B} = \begin{bmatrix} 0 \\ 0 \\ 1 \end{bmatrix} \qquad \mathbf{D} = [1 \quad 0 \quad 0]$$

(b)
$$\mathbf{A} = \begin{bmatrix} -1 & 1 \\ 0 & -1 \end{bmatrix} \qquad \mathbf{B} = \begin{bmatrix} 0 \\ 1 \end{bmatrix} \qquad \mathbf{D} = [1 \quad 1]$$

(c)
$$\mathbf{A} = \begin{bmatrix} 0 & 1 & 0 \\ 0 & 0 & 1 \\ 0 & -1 & -2 \end{bmatrix} \qquad \mathbf{B} = \begin{bmatrix} 0 \\ 0 \\ 1 \end{bmatrix} \qquad \mathbf{D} = [1 \quad 1 \quad 0]$$

(1)   Find the eigenvalues of **A**. If a computer program such as LINSYS of ACSP is available, check the answers with the computer solutions. You may also get the characteristic equation and solve for the roots using a root-finding program such as POLYROOT.

(2)   Find the transfer-function relation between $\mathbf{X}(s)$ and $U(s)$.

(3)   Find the transfer function $C(s)/U(s)$.

5-10.   Given the dynamic equations of a time-invariant system:

$$\dot{\mathbf{x}}(t) = \mathbf{A}\mathbf{x}(t) + \mathbf{B}\mathbf{u}(t) \qquad c(t) = \mathbf{D}\mathbf{x}(t)$$

where

$$\mathbf{A} = \begin{bmatrix} 0 & 1 & 0 \\ 0 & 0 & 1 \\ -1 & -2 & -3 \end{bmatrix} \qquad \mathbf{B} = \begin{bmatrix} 0 \\ 0 \\ 1 \end{bmatrix} \qquad \mathbf{D} = [1 \quad 1 \quad 0]$$

Find the matrices $\mathbf{A}_1$ and $\mathbf{B}_1$ so that the state equations are written as

$$\dot{\mathbf{y}}(t) = \mathbf{A}_1\mathbf{y}(t) + \mathbf{B}_1 u(t)$$

where

$$\mathbf{y}(t) = \begin{bmatrix} x_1(t) \\ c(t) \\ \dot{c}(t) \end{bmatrix}$$

5-11.   The equations that describe the dynamics of a motor control system are

$$e_a(t) = R_a i_a(t) + L_a \frac{di_a(t)}{dt} + K_b \frac{d\theta_m(t)}{dt}$$

$$T_m(t) = K_i i_a(t)$$

$$T_m(t) = J\frac{d^2\theta_m(t)}{dt^2} + B\frac{d\theta_m(t)}{dt} + K\theta_m(t)$$

$$e_a(t) = K_a e(t)$$

$$e(t) = K_s[\theta_r(t) - \theta_m(t)]$$

(a)   Assign the state variables as $x_1(t) = \theta_m(t)$, $x_2(t) = d\theta_m(t)/dt$, and $x_3(t) = i_a(t)$.

Write the state equations in the form of

$$\dot{x}(t) = Ax(t) + B\theta_r(t)$$

Write the output equation in the form of $c(t) = Dx(t)$, where $c(t) = \theta_m(t)$.

(b) Find the transfer functions $G(s) = \Theta_m(s)/E(s)$ when the feedback path from $\Theta_m(s)$ to $E(s)$ is broken. Find $M(s) = \Theta_m(s)/\Theta_r(s)$.

5-12. Given the matrix A of a linear state equation $dx(t) = Ax(t) + Bu(t)$.

(a) $A = \begin{bmatrix} 0 & 1 \\ -1 & 0 \end{bmatrix}$ (b) $A = \begin{bmatrix} -1 & 0 \\ 0 & -2 \end{bmatrix}$ (c) $A = \begin{bmatrix} 0 & 1 \\ 1 & 0 \end{bmatrix}$

Find the state transition matrix $\phi(t)$ using the following methods:
(1) Infinite-series expansion of e^{At} and express it in a closed form.
(2) The inverse Laplace transform of $(sI - A)^{-1}$.

5-13. The schematic diagram of a feedback control system using a dc motor is shown in Fig. 5P-13. The torque developed by the motor is $T_m(t) = K_i i_a(t)$. The constants of the system are

$$K_s = 2 \qquad K = 10 \qquad R = 2 \text{ ohms} \qquad R_s = 0.1 \text{ ohm}$$
$$K_b = 5 \text{ V/rad/s} \qquad K_i = 5 \text{ N-m/A} \qquad L_a \cong 0 \text{ H}$$
$$J_m + J_L = 0.1 \text{ N-m-s}^2 \qquad B_m \cong 0$$

Assume that all the units are consistent so that no conversion is necessary.

(a) Let the state variables be assigned as $x_1 = \theta_c$ and $x_2 = d\theta_c/dt$. Let the output be $c = \theta_c$. Write the state equations in vector-matrix form. Show that the matrices A and B are in phase-variable canonical form.

(b) Let $\theta_r(t)$ be a unit-step function. Find $x(t)$ in terms of $x(0)$, the initial state. Use the Laplace-transform table.

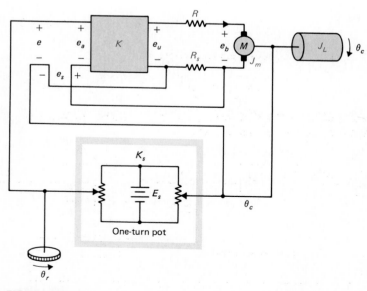

FIGURE 5P-13

(c) Find the characteristic equation of **A** and the eigenvalues of **A**.
(d) Comment on the purpose of the feedback resistor R_s.

5-14. Repeat Problem 5-13 with the following system parameters:

$K_s = 1$ $K = 9$ $R_a = 0.1$ ohm $R_s = 0.1$ ohm
$K_b = 1$ V/rad/s $K_i = 1$ N-m/A $L_a \cong 0$ H
$J_m + J_L = 0.01$ N-m-s^2 $B_m \cong 0$

5-15. The block diagram of a feedback control system is shown in Fig. 5P-15.
(a) Find the open-loop transfer function $C(s)/E(s)$ and the closed-loop transfer function $C(s)/R(s)$.
(b) Write the dynamic equations in the form of

$$\dot{\mathbf{x}}(t) = \mathbf{A}\mathbf{x}(t) + \mathbf{B}r(t) \qquad c(t) = \mathbf{D}\mathbf{x}(t) + \mathbf{E}r(t)$$

Find **A**, **B**, **D**, and **E** in terms of the system parameters.
(c) Apply the final-value theorem to find the steady-state value of the output $c(t)$ when the input $r(t)$ is a unit-step function. Assume that the closed-loop system is stable.

5-16. For the linear time-invariant system whose state equations have the coefficient matrices given by Eqs. (5-88) and (5-89) (phase-variable canonical form), show that

$$\text{adj}\,(s\mathbf{I} - \mathbf{A})\mathbf{B} = \begin{bmatrix} 1 \\ s \\ s^2 \\ \vdots \\ s^{n-1} \end{bmatrix}$$

and the characteristic equation of **A** is

$$s^n + a_n s^{n-1} + a_{n-1} s^{n-2} + \cdots + a_2 s + a_1 = 0$$

5-17. A closed-loop control system is described by

$$\dot{\mathbf{x}}(t) = \mathbf{A}\mathbf{x}(t) + \mathbf{B}u(t) \qquad u(t) = -\mathbf{G}\mathbf{x}(t)$$

where $\mathbf{x}(t) = n \times 1$ state vector, $u(t) = r \times 1$ input vector, **A** is $n \times n$, **B** is $n \times r$, and **G** is the $r \times n$ feedback matrix.

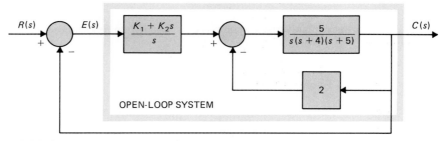

OPEN-LOOP SYSTEM

FIGURE 5P-15

(a)   Show that the roots of the characteristic equation of the closed-loop system are the eigenvalues of $\mathbf{A} - \mathbf{BG}$.

(b)   Let

$$
\mathbf{A} = \begin{bmatrix} 0 & 1 & 0 & 0 \\ 0 & 0 & 1 & 0 \\ 0 & 0 & 0 & 1 \\ 0 & -3 & -4 & -10 \end{bmatrix} \qquad \mathbf{B} = \begin{bmatrix} 0 \\ 0 \\ 0 \\ 1 \end{bmatrix}
$$

$$
\mathbf{G} = [g_1 \quad g_2 \quad g_3 \quad g_4]
$$

where the elements of $\mathbf{G}$ are real constants. Find the characteristic equation of the closed-loop system. Determine the elements of $\mathbf{G}$ so that the eigenvalues of $\mathbf{A} - \mathbf{BG}$ are at -1, -1, $-1 + j$, and $-1 - j$. Check the results by use of the pole-placement design option in LINSYS of the ACSP software if it is available. Can all the eigenvalues of $\mathbf{A} - \mathbf{BG}$ be arbitrarily assigned for this system and still be assured that the elements of $\mathbf{G}$ be uniquely determined?

5-18.   A linear time-invariant system is described by the differential equation

$$
\frac{d^3c(t)}{dt^3} + 3\frac{d^2c(t)}{dt^2} + 3\frac{dc(t)}{dt} + c(t) = r(t)
$$

(a)   Let the state variables be defined as $x_1 = c$, $x_2 = dc/dt$, and $x_3 = d^2c/dt^2$. Write the state equations of the system in vector-matrix form.

(b)   Find the state-transition matrix $\boldsymbol{\phi}(t)$ of $\mathbf{A}$.

(c)   Let $c(0) = 1$, $dc(0)/dt = 0$, $d^2c(0)/dt^2 = 0$, and $r(t) = u_s(t)$. Find the state-transition equation of the system.

(d)   Find the characteristic equation of $\mathbf{A}$ and its eigenvalues.

5-19.   A linear time-invariant system is described by the differential equation

$$
\frac{d^2c(t)}{dt^2} + 2\frac{dc(t)}{dt} + c(t) = r(t)
$$

(a)   Define the state variables as $x_1(t) = c(t)$ and $x_2(t) = dc(t)/dt$. Write the state equations in vector-matrix form. Find the state-transition matrix $\boldsymbol{\phi}(t)$ of $\mathbf{A}$.

(b)   Define the state variables as $x_1(t) = c(t)$ and $x_2(t) = c(t) + dc(t)/dt$. Write the state equations in vector-matrix form. Find the state-transition matrix $\boldsymbol{\phi}(t)$ of $\mathbf{A}$.

(c)   Show that the characteristic equations, $|s\mathbf{I} - \mathbf{A}| = 0$, for parts (a) and (b) are identical.

5-20.   A linear multivariable system is described by the following set of differential equations:

$$
\frac{d^2c_1(t)}{dt^2} + \frac{dc_1(t)}{dt} + 3c_1(t) - 5c_2(t) = r_1(t)
$$

$$
\frac{d^2c_2(t)}{dt^2} + 2c_1(t) + c_2(t) = r_2(t)
$$

(a)   Define the state variables as $x_1(t) = c_1(t)$, $x_2(t) = dc_1(t)/dt$, $x_3(t) = c_2(t)$, and $x_4(t) = dc_2(t)/dt$. Write the state equations in vector-matrix form.

(b)   Write the transfer-function relations between the inputs and the outputs in matrix form.

5-21.   Given the state equations $d\mathbf{x}(t)/dt = \mathbf{A}\mathbf{x}(t)$, where

$$\mathbf{A} = \begin{bmatrix} \sigma & -\omega \\ \omega & \sigma \end{bmatrix} \qquad \sigma \text{ and } \omega \text{ are real numbers}$$

(a)   Find the state-transition matrix of $\mathbf{A}$.

(b)   Find the eigenvalues of $\mathbf{A}$.

5-22.   (a)   Show that the input–output transfer functions of the two systems shown in Fig. 5P-22 are the same.

(b)   Write the dynamic equations of the system in Fig. 5P-22(a) as

$$\dot{\mathbf{x}}(t) = \mathbf{A}_1 \mathbf{x}(t) + \mathbf{B}_1 u_1(t)$$

$$c_1(t) = \mathbf{D}_1 \mathbf{x}(t)$$

and those of the system in Fig. 5P-22(b) as

$$\dot{\mathbf{y}}(t) = \mathbf{A}_2 \mathbf{y}(t) + \mathbf{B}_2 u_2(t)$$

$$c_2(t) = \mathbf{D}_2 \mathbf{y}(t)$$

Find $\mathbf{A}_1$, $\mathbf{B}_1$, $\mathbf{D}_1$, $\mathbf{A}_2$, $\mathbf{B}_2$, and $\mathbf{D}_2$. Show that $\mathbf{A}_2 = \mathbf{A}_1'$.

5-23.   Draw the state diagrams for the following systems.

$$\dot{\mathbf{x}}(t) = \mathbf{A}\mathbf{x}(t) + \mathbf{B}u(t)$$

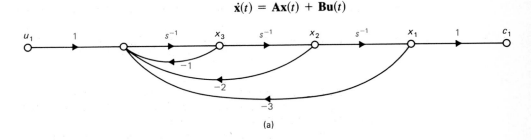

(a)

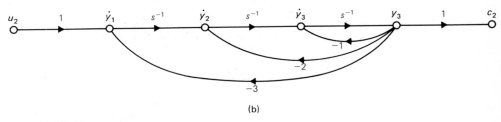

(b)

FIGURE 5P-22

(a)
$$A = \begin{bmatrix} -3 & 2 & 0 \\ -1 & 0 & 1 \\ -2 & -3 & -4 \end{bmatrix} \qquad B = \begin{bmatrix} 0 \\ 0 \\ 1 \end{bmatrix}$$

(b)   Same **A** as in part (a), but with
$$B = \begin{bmatrix} 0 & 1 \\ 1 & 0 \\ 1 & 0 \end{bmatrix}$$

5-24.   Draw state diagrams for the following transfer functions by means of direct decomposition.

(a) $G(s) = \dfrac{10}{s^3 + 8.5s^2 + 20.5s + 15}$ (b) $G(s) = \dfrac{10(s + 2)}{s^2(s + 1)(s + 3.5)}$

(c) $G(s) = \dfrac{5(s + 1)}{s(s + 2)(s + 10)}$ (d) $G(s) = \dfrac{1}{s(s + 5)(s^2 + 2s + 2)}$

Assign the state variables from right to left in ascending order. Write the state equations from the state diagram and show that the equations are in the phase-variable canonical form.

5-25.   Draw state diagrams for the systems described in Problem 5-24 by means of parallel decomposition. Make certain that the state diagrams contain a minimum number of integrators. The constant branch gains must be real. Write the state equations from the state diagram.

5-26.   Draw state diagrams for the systems described in Problem 5-24 by means of cascade decomposition. Position the transfer functions in any appropriate way. Assign the state variables in ascending order from right to left. Write the state equations from the state diagram.

5-27.   The block diagram of a feedback control system is shown in Fig. 5P-27.
(a)   Draw a state diagram for the system by first decomposing $G(s)$ by direct decomposition. Assign the state variables in ascending order from right to left. In addition to the state-variable-related nodes, the state diagram should contain nodes for $R(s)$, $E(s)$, and $C(s)$.
(b)   Write the dynamic equations of the system in vector-matrix form.
(c)   Find the state-transition equations of the system using the state equations found in part (b). The initial state vector is $x(0)$, and $r(t) = u_s(t)$.
(d)   Find the output $c(t)$ for $t \geq 0$ with the initial state $x(0)$, and $r(t) = u_s(t)$.

5-28.   (a)   Find the closed-loop transfer function $C(s)/R(s)$ of the system shown in Fig. 5P-27.
(b)   Perform a direct decomposition to $C(s)/R(s)$, and draw the state diagram.
(c)   Assign the state variables from right to left in ascending order, and write the state equations in vector-matrix form.
(d)   Find the state-transition equations of the system using the state equations found in part (c). The initial state vector is $x(0)$, and $r(t) = u_s(t)$.
(e)   Find the output $c(t)$ for $t \geq 0$ with the initial state $x(0)$, and $r(t) = u_s(t)$.

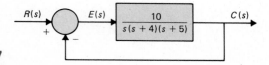

FIGURE 5P-27

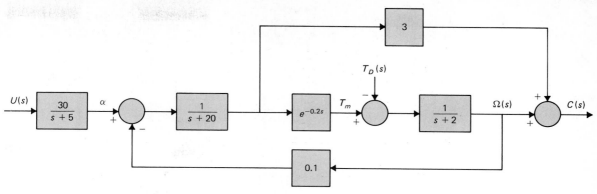

FIGURE 5P-29

5-29. The block diagram of a linearized idle-speed engine-control system is shown in Fig. 5P-29. The system is linearized about a nominal operating point, so that all the variables represent linear-perturbed quantities. The following parameters and variables are defined: m is the average air mass in the manifold, T_m the engine torque, T_D the load-disturbance torque, c the manifold pressure, ω the engine speed, u the input-voltage-to-throttle actuator, and α the throttle angle. The time delay in the engine model can be approximated by

$$e^{-0.2s} \cong \frac{1 - 0.1s}{1 + 0.1s}$$

(a) Draw a state diagram for the system by decomposing each block individually. Assign the state variables from right to left in ascending order.

(b) Write the state equations from the state diagram obtained in part (a), in the form of

$$\dot{\mathbf{x}}(t) = \mathbf{A}\mathbf{x}(t) + \mathbf{B}\begin{bmatrix} u(t) \\ T_D(t) \end{bmatrix}$$

(c) Write $C(s)$ as a function of $U(s)$ and $T_D(s)$. Write $\Omega(s)$ as a function of $U(s)$ and $T_D(s)$.

5-30. The state diagram of a linear system is shown in Fig. 5P-30.

(a) Assign state variables on the state diagram from right to left in ascending order. Create additional artificial nodes if necessary so that the state-variable nodes satisfy as "input nodes" after the integrator branches are deleted.

(b) Write the dynamic equations of the system from the state diagram in part (a).

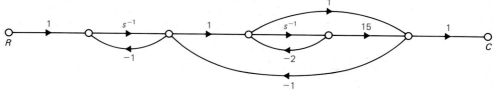

FIGURE 5P-30

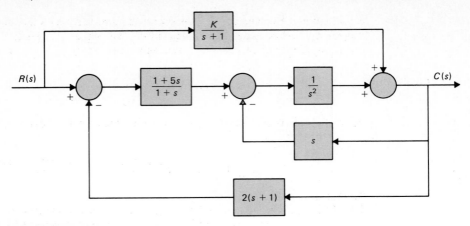

FIGURE 5P-31

5-31. The block diagram of a linear control system is shown in Fig. 5P-31.
  (a) Determine the transfer function $C(s)/R(s)$.
  (b) Find the characteristic equation of the system. Find the roots of the characteristic equation. Show that these roots are not functions of K.
  (c) When $K = 1$, draw a state diagram for the system by decomposing $C(s)/R(s)$, using a minimum number of integrators.
  (d) Repeat part (c) when $K = 4$.
  (e) Determine the values of K that must be avoided if the system is to be both state controllable and observable.

5-32. A considerable amount of effort is being spent by automobile manufacturers to meet the exhaust-emission-performance standards set by the government. Modern automobile-power-plant systems consist of an internal combustion engine that has an internal cleanup device called a catalytic converter. Such a system requires control of such variables as the engine air–fuel (A/F) ratio, ignition-spark timing, exhaust-gas recirculation, and injection air. The control-system problem considered in this problem deals with the control of the A/F ratio. In general, depending on fuel composition and other factors, a typical stoichiometric A/F is 14.7:1, that is, 14.7 grams of air to each gram of fuel. An A/F greater or less than stoichiometry will cause high hydrocarbons, carbon monoxide, and nitrous oxides in the tailpipe emission. The control system shown in Fig. 5P-32 is devised to control the air–fuel ratio so that a desired output is achieved for a given input command. The sensor senses the composition of the exhaust-gas mixture entering the catalytic converter. The electronic controller detects the difference or the

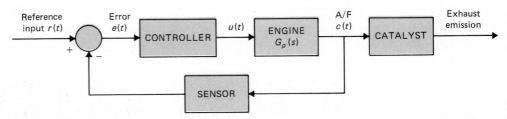

FIGURE 5P-32

error between the command and the error and computes the control signal necessary to achieve the desired exhaust-gas composition. The output $c(t)$ denotes the effective air–fuel ratio. The transfer function of the engine is given by

$$G_p(s) = \frac{C(s)}{U(s)} = \frac{e^{-T_d s}}{1 + 0.5s}$$

where T_d is the time delay (0.2 second), which can be approximated by

$$e^{-T_d s} = \frac{1}{e^{T_d s}} = \frac{1}{1 + T_d s + T_d^2 s^2/2! + \cdots} \cong \frac{1}{1 + T_d s + T_d^2 s^2/2!}$$

The gain of the sensor is 1.

(a)   Using the approximation for $e^{-T_d s}$ given, find the expression for $G_p(s)$. Decompose $G_p(s)$ by direct decomposition, and draw the state diagram with $u(t)$ as the input and $c(t)$ as the output. Assign state variables from right to left in ascending order, and write the state equations in vector-matrix form.

(b)   Assuming that the controller is a simple amplifier with a gain of 1, i.e., $u(t) = e(t)$, find the characteristic equation and its roots of the closed-loop system.

(c)   Define the state feedback control as

$$u(t) = -\mathbf{G}\mathbf{x}(t) + r(t)$$

where $r(t)$ is the reference input, and $\mathbf{G}$ is the feedback matrix,

$$\mathbf{G} = \begin{bmatrix} g_1 & g_2 & g_3 \end{bmatrix}$$

where g_1, g_2, and g_3 are real constants. Find the elements of $\mathbf{G}$ so that the eigenvalues of the closed-loop system are at -10, $-1 + j$ and $-1 - j$. With this controller, find the steady-state value of $c(t)$ when the input $r(t)$ is a unit-step function.

(d)   Check the results in part (c) by using the pole-placement design option of LINSYS of the ACSP software package. Compute and plot $c(t)$ using LINSYS or any other appropriate computer program when $\mathbf{x}(0) = \mathbf{0}$ and $r(t) = u_s(t)$.

(e)   If it is required that the steady-state value of $c(t)$ be identical to the input unit-step function (zero steady-state error), find the elements of the feedback matrix $\mathbf{G}$ so that the two complex characteristic equation roots are at $-1 + j$ and $-1 - j$. Determine the value of the third root. Compute and plot $c(t)$ using LINSYS or any other appropriate computer program with $\mathbf{x}(0) = \mathbf{0}$ and $r(t) = u_s(t)$.

5-33.   Repeat all parts of Problem 5-32 when the time delay of the automobile engine is approximated as

$$e^{-T_d s} \cong \frac{1 - T_d s/3}{1 + \frac{2}{3}T_d s + \frac{1}{6}T_d^2 s^2} \qquad T_d = 0.2 \text{ second}$$

5-34.   The schematic diagram in Fig. 5P-34(a) shows a permanent-magnet dc-motor-control system with a viscous-inertia damper. A mechanical damper such as the viscous-inertia type is sometimes used in practice as a simple and economical way of stabilizing a control system. The damping effect is achieved by a rotor suspended in a viscous fluid. The

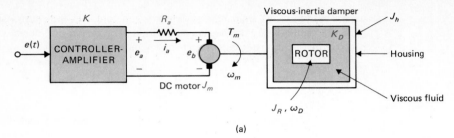

(a)

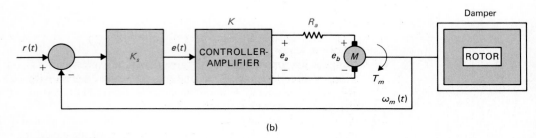

(b)

FIGURE 5P-34

differential and algebraic equations that describe the dynamics of the system are as follows:

$$e(t) = K_s[\omega_r(t) - \omega_m(t)]$$
$$e_a(t) = Ke(t) = R_a i_a(t) + e_b(t)$$
$$e_b(t) = K_b \omega_m(t)$$
$$T_m(t) = Jd\omega_m(t)/dt + K_D[\omega_m(t) - \omega_D(t)]$$
$$T_m(t) = K_i i_a(t)$$
$$K_D[\omega_m(t) - \omega_D(t)] = J_R d\omega_D(t)/dt$$
$$R_a = 1 \text{ ohm}$$

$$K_s = 1 \text{ V/rad/s}$$
$$K = 10$$
$$K_b = 0.0706 \text{ V/rad/s}$$
$$J = J_h + J_m = 0.1 \text{ oz-in.-s}^2$$
$$K_i = 10 \text{ oz-in./A}$$
$$J_R = 0.05 \text{ oz-in.-s}^2$$
$$K_D = 1 \text{ oz-in.-s}$$

(a) Let the state variables be defined as $x_1(t) = \omega_m(t)$ and $x_2(t) = \omega_D(t)$. Write the state equations for the open-loop system with $e(t)$ as the input. (The feedback path from ω_m to e is open.)

(b) Draw the state diagram for the overall system using the state equations found in part (a) and $e = K_s(\omega_r - \omega_m)$.

(c) Derive the open-loop transfer function $\Omega_m(s)/E(s)$ and the closed-loop transfer function $\Omega_m(s)/\Omega_r(s)$.

(d) Find the characteristic equation and its roots of the closed-loop system.

5-35. Determine the state controllability of the system shown in Fig. 5P-35.

(a) $a = 1, b = 2, c = 2,$ and $d = 1.$

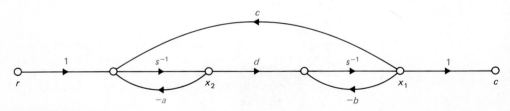

FIGURE 5P-35

(b)   Are there any nonzero value for a, b, c, and d such that the system is uncontrollable?

5-36. Determine the controllability of the following systems:

(a)
$$\mathbf{A} = \begin{bmatrix} -1 & 0 & 0 \\ 0 & -1 & 0 \\ 0 & 0 & -1 \end{bmatrix} \qquad \mathbf{B} = \begin{bmatrix} 1 \\ 1 \\ 1 \end{bmatrix}$$

(b)
$$\mathbf{A} = \begin{bmatrix} -1 & 0 & 0 \\ 0 & -2 & 0 \\ 0 & 0 & -3 \end{bmatrix} \qquad \mathbf{B} = \begin{bmatrix} 1 \\ 1 \\ 1 \end{bmatrix}$$

5-37. Determine the controllability and observability of the system shown in Fig. 5P-37 by the following methods:
(a)   Conditions on the $\mathbf{A}$, $\mathbf{B}$, $\mathbf{D}$, and $\mathbf{E}$ matrices.
(b)   Conditions on the pole-zero cancellation of the transfer functions.

5-38. The transfer function of a linear control system is

$$\frac{C(s)}{R(s)} = \frac{s + \alpha}{s^3 + 7s^2 + 14s + 8}$$

(a)   Determine the value(s) of α so that the system is either uncontrollable or unobservable.
(b)   With the value(s) of α found in part (a), define the state variables so that one of them is uncontrollable.
(c)   With the value(s) of α found in part (a), define the state variables so that one of them is unobservable.

5-39. Consider the system described by the state equation $dx/dt = \mathbf{A}x + \mathbf{B}u$, where

$$\mathbf{A} = \begin{bmatrix} 0 & 1 \\ -1 & a \end{bmatrix} \qquad \mathbf{B} = \begin{bmatrix} 1 \\ b \end{bmatrix}$$

Find the region in the a–b plane such that the system is completely controllable.

5-40. Determine the condition on b_1, b_2, d_1, and d_2 so that the following system is completely controllable and observable.

$$\dot{x}(t) = \mathbf{A}x(t) + \mathbf{B}u(t) \qquad c(t) = \mathbf{D}x(t)$$

$$\mathbf{A} = \begin{bmatrix} 1 & 1 \\ 0 & 1 \end{bmatrix} \qquad \mathbf{B} = \begin{bmatrix} b_1 \\ b_2 \end{bmatrix} \qquad \mathbf{D} = [d_1 \quad d_2]$$

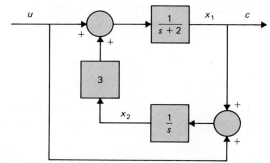

FIGURE 5P-37

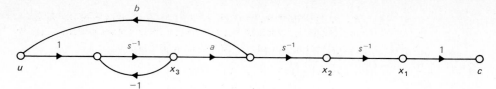

FIGURE 5P-41

5-41.   The state diagram of a linear control system is shown in Fig. 5P-41.
    (a)   Determine the relation between the parameters a and b that should be avoided in order that the system be completely controllable.
    (b)   Let $a = b = 1$ [so you know, these cannot be included in the answers to part (a)]. Apply state feedback $u = -\mathbf{Gx}$, where $\mathbf{G} = \begin{bmatrix} g_1 & g_2 & g_3 \end{bmatrix}$. Find $\mathbf{G}$ so that the eigenvalues of the closed-loop system are at -1, -1, and -1.

5-42.   The schematic diagram of Fig. 5P-42 represents a control system whose purpose is to hold the level of the liquid in the tank at a desired level. The liquid level is controlled by a float whose position $h(t)$ is monitored. The input signal of the open-loop system is $e(t)$. The system parameters and equations are as follows.

Motor resistance $R_a = 10$ ohms	Motor inductance $L_a \cong 0$ H
Torque constant $K_i = 10$ oz-in./A	Rotor inertia $J_m = 0.005$ oz-in.-s^2
Back-emf constant $K_b = 0.0706$ V/rad/s	Gear ratio $n = N_1/N_2 = 1/100$
Load inertia $J_L = 10$ oz-in.-s^2	Load and motor friction = negligible
Amplifier gain $K_a = 50$	Area of tank $A = 50$ ft^2

$$e_a(t) = R_a i_a(t) + K_b \omega_m(t) \qquad \omega_m(t) = d\theta_m(t)/dt$$
$$T_m(t) = K_i i_a(t) = (J_m + n^2 J_L)\, d\omega_m(t)/dt \qquad \theta_c(t) = n\theta_m(t)$$

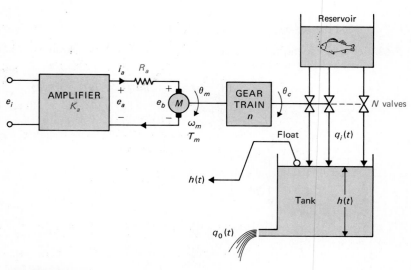

FIGURE 5P-42

The number of valves connected to the tank from the reservoir is $N = 8$. All the valves have the same characteristics, and are controlled simultaneously by θ_c. The equations that govern the volume of flow are as follows:

$$q_i(t) = K_I N \theta_c(t) \qquad K_I = 10 \text{ ft}^3/\text{s-rad}$$
$$q_o(t) = K_o h(t) \qquad K_o = 50 \text{ ft}^2/\text{s}$$
$$h(t) = \frac{\text{volume of tank}}{\text{area of tank}} = \frac{1}{A} \int [q_1(t) - q_o(t)]\, dt$$

(a) Define the state variables as $x_1(t) = h(t)$, $x_2(t) = \theta_m(t)$, and $x_3(t) = d\theta_m(t)/dt$. Write the state equations of the system in the form of $d\mathbf{x}(t)/dt = \mathbf{A}\mathbf{x}(t) + \mathbf{B}e_i(t)$. Draw a state diagram for the system.

(b) Find the characteristic equation and the eigenvalues of the $\mathbf{A}$ matrix found in part (a).

(c) Show that the open-loop system is completely controllable; that is, the pair $[\mathbf{A}, \mathbf{B}]$ is controllable.

(d) For reason of economy, only one of the three state variables is measured and fed back for control purposes. The output equation is $c = \mathbf{D}\mathbf{x}$, where $\mathbf{D}$ can be one of the following forms:

$$(1)\ \mathbf{D} = \begin{bmatrix} 1 & 0 & 0 \end{bmatrix} \qquad (2)\ \mathbf{D} = \begin{bmatrix} 0 & 1 & 0 \end{bmatrix} \qquad (3)\ \mathbf{D} = \begin{bmatrix} 0 & 0 & 1 \end{bmatrix}$$

Determine which case (or cases) corresponds to a completely observable system.

5-43. The "broom-balancing" control system described in Problem 4-23 has the following parameters: $M_b = 1$ kg, $M_c = 10$ kg, $L = 1$ m, and $g = 32.2$ ft/s^2. The small-signal linearized state equation model of the system is

$$\Delta \dot{\mathbf{x}} = \mathbf{A}^* \Delta\mathbf{x} + \mathbf{B}^* \Delta r$$

where

$$\mathbf{A}^* = \begin{bmatrix} 0 & 1 & 0 & 0 \\ 25.92 & 0 & 0 & 0 \\ 0 & 0 & 0 & 1 \\ -2.36 & 0 & 0 & 0 \end{bmatrix} \qquad \mathbf{B}^* = \begin{bmatrix} 0 \\ -0.0732 \\ 0 \\ 0.0976 \end{bmatrix}$$

(a) Find the characteristic equation of $\mathbf{A}^*$ and its roots.

(b) Determine the controllability of $[\mathbf{A}^*, \mathbf{B}^*]$.

(c) For reason of economy, only one of the state variables are to be measured for feedback. The output equation is written

$$\Delta c = \mathbf{D}^* \Delta\mathbf{x}$$

where

$$(1)\ \mathbf{D}^* = \begin{bmatrix} 1 & 0 & 0 & 0 \end{bmatrix} \qquad (2)\ \mathbf{D}^* = \begin{bmatrix} 0 & 1 & 0 & 0 \end{bmatrix}$$

$$(3)\ \mathbf{D}^* = \begin{bmatrix} 0 & 0 & 1 & 0 \end{bmatrix} \qquad (4)\ \mathbf{D}^* = \begin{bmatrix} 0 & 0 & 0 & 1 \end{bmatrix}$$

Determine which $\mathbf{D}^*$ corresponds to an observable system.

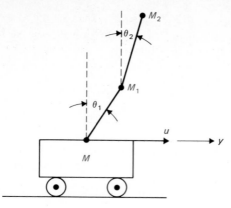

FIGURE 5P-44

5-44.   The double-inverted pendulum shown in Fig. 5P-44 is approximately modeled by the following linear state equation:

$$\dot{\mathbf{x}}(t) = \mathbf{A}\mathbf{x}(t) + \mathbf{B}u(t)$$

where

$$\mathbf{x}(t) = \begin{bmatrix} \theta_1(t) \\ \dot{\theta}_1(t) \\ \theta_2(t) \\ \dot{\theta}_2(t) \\ y(t) \\ \dot{y}(t) \end{bmatrix}$$

$$\mathbf{A} = \begin{bmatrix} 0 & 1 & 0 & 0 & 0 & 0 \\ 16 & 0 & -8 & 0 & 0 & 0 \\ 0 & 0 & 0 & 1 & 0 & 0 \\ -16 & 0 & 16 & 0 & 0 & 0 \\ 0 & 0 & 0 & 0 & 0 & 1 \\ 0 & 0 & 0 & 0 & 0 & 0 \end{bmatrix} \qquad \mathbf{B} = \begin{bmatrix} 0 \\ -1 \\ 0 \\ 0 \\ 0 \\ 1 \end{bmatrix}$$

Determine the controllability of the states.

5-45.   The block diagram of a simplified control system for the Large Space Telescope (LST) is shown in Fig. 5P-45. For simulation and control purposes, it would be desirable to model the system by state equations and by a state diagram.
    (a)   Draw a state diagram for the system and write the state equations in vector-matrix form. The state diagram should contain a minimum number of state variables, so it would be helpful if the transfer function of the system is written first.
    (b)   Find the characteristic equation of the system.

5-46.   The state diagram shown in Fig. 5P-46 represents two subsystems connected in cascade.
    (a)   Determine the controllability and observability of the system.

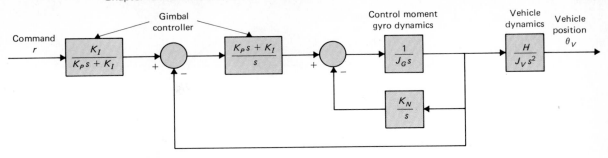

FIGURE 5P-45

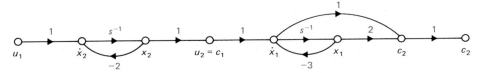

FIGURE 5P-46

(b) Consider that output feedback is applied by feeding back c_2 to u_2; that is, $u_2 = -gc_2$, where g is a real constant. Determine how the value of g affects the controllability and observability of the system.

5-47. Given the system

$$\dot{x}(t) = \mathbf{A}x(t) + \mathbf{B}u(t) \qquad c(t) = \mathbf{D}x(t)$$

where

$$\mathbf{A} = \begin{bmatrix} 0 & 1 \\ -1 & -3 \end{bmatrix} \qquad \mathbf{B} = \begin{bmatrix} 1 \\ 2 \end{bmatrix} \qquad \mathbf{D} = \begin{bmatrix} 1 & 1 \end{bmatrix}$$

(a) Determine the state controllability and observability of the system.
(b) Let $u = -\mathbf{G}x$, where $\mathbf{G} = \begin{bmatrix} g_1 & g_2 \end{bmatrix}$. Determine if and how controllability and observability of the closed-loop system are affected by the elements of $\mathbf{G}$.

5-48. Draw a state diagram for the digital control system represented by the following dynamic equations:

$$\mathbf{x}(k + 1) = \mathbf{A}\mathbf{x}(k) + \mathbf{B}u(k) \qquad c(k) = x_1(k)$$

$$\mathbf{A} = \begin{bmatrix} 0 & 1 & -1 \\ 0 & 1 & 2 \\ 5 & 3 & -1 \end{bmatrix} \qquad \mathbf{B} = \begin{bmatrix} 0 \\ 0 \\ 1 \end{bmatrix}$$

Find the characteristic equation of the system.

5-49. The state diagram of a digital control system is shown in Fig. 5P-49. Write the dynamic equations. Assign the state variables on the state diagram from right to left in ascending order.

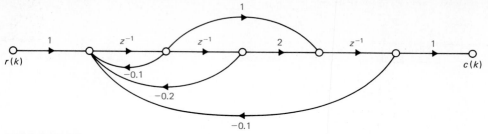

FIGURE 5P-49

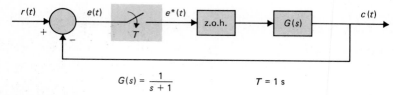

$$G(s) = \frac{1}{s+1} \qquad\qquad T = 1 \text{ s}$$

FIGURE 5P-50

5-50. The block diagram of a sampled-data control system is shown in Fig. 5P-50. Write the discrete state equations of the system. Draw a state diagram for the system.

ADDITIONAL COMPUTER PROBLEMS

5-51. The torque equation for part (a) of Problem 4-13 is

$$J\frac{d^2\theta(t)}{dt^2} = K_F d_1 \theta(t) + T_s d_2 \delta(t)$$

Define the state variables as $x_1 = \theta$ and $x_2 = d\theta/dt$. Let $K_F d_1 = 1$ and $J = 1$. Find the state-transition matrix $\boldsymbol{\phi}(t)$ using the LINSYS or any other applicable computer program.

5-52. Starting from the state equation $d\mathbf{x}(t)/dt = \mathbf{A}\mathbf{x} + \mathbf{B}\theta_r$ obtained in Problem 5-13, use the LINSYS or any other appropriate computer program to find the following:
   (a)   The state-transition matrix of $\mathbf{A}$, $\boldsymbol{\phi}(t)$.
   (b)   The characteristic equation of $\mathbf{A}$.
   (c)   The eigenvalue of $\mathbf{A}$.
   (d)   Compute and plot the unit-step response of $c = \theta_c$ for 3 seconds. Set all the initial conditions to zero.

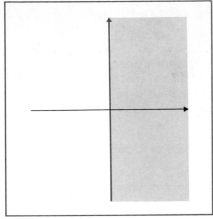

STABILITY
OF LINEAR
CONTROL SYSTEMS

6-1
INTRODUCTION

From the studies of linear differential equations with constant coefficients, we learned that the homogeneous solution that corresponds to the transient response of the system is governed by the roots of the characteristic equation. Basically, the design of linear control systems may be regarded as a problem of arranging the location of the poles and zeros of the closed-loop transfer function such that the corresponding system will perform according to the prescribed specifications.

Among the many forms of performance specifications used in the design of control systems, the most important requirement is that the system must be stable. Generally speaking, an unstable system is considered to be useless.

When all types of systems are considered—linear, nonlinear, time-invariant, and time-varying—the definition of stability can be given in many different forms. We shall deal only with the stability of linear time-invariant systems in the following discussions.

absolute stability *relative stability*

For analysis and design purposes, we can classify the stability of control systems as **absolute stability** and **relative stability.** Absolute stability refers to the condition whether the system is stable or unstable; it is a *yes* or *no* answer. Once the system is found to be stable, it is of interest to determine how stable it is, and this degree of stability is a measure of relative stability.

279

In preparation for the definitions of stability, we define the two following types of responses for linear time-invariant systems:

zero-state response

zero-input response

1. *Zero-state response.*   The zero-state response is due to the input only; all the initial conditions of the system are zero.
2. *Zero-input response.*   The zero-input response is due to the initial conditions only; all the inputs are zero.

From the principle of superposition, when a system is subject to both inputs and initial conditions, the total response is written

Total response = zero-state response + zero-input response

The above definitions apply to continuous-data as well as discrete-data systems.

6-2

BOUNDED-INPUT BOUNDED-OUTPUT (BIBO) STABILITY CONTINUOUS-DATA SYSTEMS

Let $r(t)$, $c(t)$, and $g(t)$ be the input, output, and the impulse response of a linear time-invariant system, respectively. *With zero initial conditions, the system is said to be BIBO stable, or simply stable, if its output $c(t)$ is bounded to a bounded input $r(t)$.* We

BIBO stability

shall show that BIBO stability is related to the roots of the characteristic equation.

The convolution integral relating $r(t)$, $c(t)$, and $g(t)$ is

$$c(t) = \int_0^\infty r(t - \tau)g(\tau)\,d\tau \tag{6-1}$$

Taking the absolute value on both sides of the equation, we get

$$|c(t)| = \left| \int_0^\infty r(t - \tau)g(\tau)\,d\tau \right| \tag{6-2}$$

or

$$|c(t)| \leq \int_0^\infty |r(t - \tau)|\,|g(\tau)|\,d\tau \tag{6-3}$$

If $r(t)$ is bounded,

$$|r(t)| \leq M \tag{6-4}$$

where M is a finite positive number. Then,

$$|c(t)| \leq M \int_0^\infty |g(\tau)| \, d\tau \qquad (6\text{-}5)$$

Thus, if $c(t)$ is to be bounded, or

$$|c(t)| \leq N < \infty \qquad (6\text{-}6)$$

where N is a positive finite number, the following condition must hold:

$$M \int_0^\infty |g(\tau)| \, d\tau \leq N < \infty \qquad (6\text{-}7)$$

Or, for any positive finite Q,

$$\int_0^\infty |g(\tau)| \, d\tau \leq Q < \infty \qquad (6\text{-}8)$$

The condition given in the last equation implies that the area under the $|g(\tau)|$-versus-τ curve must be finite.

To show the relation between the roots of the characteristic equation and the condition in Eq. (6-8), we write the transfer function $G(s)$ as

$$G(s) = \mathcal{L}[g(t)] = \int_0^\infty g(t)e^{-st} \, dt \qquad (6\text{-}9)$$

Taking the absolute value on both sides of the last equation, we have

$$|G(s)| = \left| \int_0^\infty g(t)e^{-st} \, dt \right| \leq \int_0^\infty |g(t)| \, |e^{-st}| \, dt \qquad (6\text{-}10)$$

Since

$$|e^{-st}| = |e^{-\sigma t}|$$

where σ is the real part of s, when s assumes a value of the poles of $G(s)$, $G(s) = \infty$, Eq. (6-10) becomes

$$\infty \leq \int_0^\infty |g(t)| \, |e^{-\sigma t}| \, dt \qquad (6\text{-}11)$$

If one or more roots of the characteristic equation are in the right-half s-plane or on the $j\omega$-axis, $\sigma \geq 0$, then

$$|e^{-\sigma t}| \leq M = 1 \qquad (6\text{-}12)$$

Equation (6-11) becomes

$$\infty \le \int_0^\infty M|g(t)|\, dt = \int_0^\infty |g(t)|\, dt \tag{6-13}$$

characteristic equation

which violates the BIBO stability requirement. Thus, *for BIBO stability, the roots of the characteristic equation, or the poles of $G(s)$, must all lie in the left-half s-plane. A system is said to be unstable if it is not BIBO stable.* When a system has roots on the $j\omega$-axis, say, at $s = j\omega_0$ and $s = -j\omega_0$, if the input is a sinusoid, $\sin \omega_0 t$, then the output will be of the form $t \sin \omega_0 t$, which is unbounded, and the system is unstable.

6-3
ZERO-INPUT STABILITY OF CONTINUOUS-DATA SYSTEMS

Zero-input stability refers to the stability condition when the input is zero, and the system is driven only by its initial conditions. We shall show that the zero-input stability of a linear time-invariant system also depends on the roots of the characteristic equation.

Let the input of an nth-order linear time-invariant system be zero, and the output due to the initial conditions be $c(t)$. Then $c(t)$ can be expressed as

$$c(t) = \sum_{k=0}^{n-1} g_k(t) c^{(k)}(t_0) \tag{6-14}$$

where

$$c^{(k)}(t_0) = \left.\frac{d^k c(t)}{dt^k}\right|_{t=0} \tag{6-15}$$

zero-input stability

and $g_k(t)$ denotes the zero-input response component due to $c^{(k)}(t_0)$. The zero-input stability is defined as follows: *If the zero-input response $c(t)$, subject to the finite initial conditions, $c^{(k)}(t_0)$, reaches zero as t approaches infinity, the system is said to be zero-input stable, or stable; otherwise, the system is unstable.*

Mathematically, the foregoing definition can be stated: *A linear time-invariant system is zero-input stable if for any set of finite $c^{(k)}(t_0)$, there exists a positive number M, which depends on $c^{(k)}(t_0)$, such that*

$$(1) \quad |c(t)| \le M < \infty \quad \text{for all } t \ge t_0 \tag{6-16}$$

and

$$(2) \quad \lim_{t\to\infty} |c(t)| = 0 \tag{6-17}$$

asymptotic stability

Because the condition in the last equation requires that the magnitude of $c(t)$ reaches zero as time approaches infinity, the zero-input stability is also known as the **asymptotic stability**.

Taking the absolute value on both sides of Eq. (6-14), we get

$$|c(t)| = \left| \sum_{k=0}^{n-1} g_k(t)c^{(k)}(t_0) \right| \leq \sum_{k=0}^{n-1} |g_k(t)| \, |c^{(k)}(t_0)| \tag{6-18}$$

Since all the initial conditions are assumed to be finite, the condition in Eq. (6-16) requires that the following be true:

$$\sum_{k=0}^{n-1} |g_k(t)| < \infty \qquad \text{for all } t \geq 0 \tag{6-19}$$

Let the n characteristic equation roots be expressed as $s_i = \sigma_i + j\omega_i$, $i = 1, 2,$ . . . , n. Then, if m of the n roots are simple, and the rest are of the multiple order, $c_k(t)$ will be of the following form:

$$c_k(t) = \sum_{i=1}^{m} K_i e^{s_i t} + \sum_{i=0}^{n-m-1} L_i t^i e^{s_i t} \tag{6-20}$$

where K_i and L_i are constant coefficients. Since the exponential terms $e^{s_i t}$ in the last equation control the response $c_k(t)$ as $t \to \infty$, to satisfy the two conditions in Eqs. (6-16) and (6-17), the real parts of s_i must be negative. In other words, the roots of the characteristic equation must all be in the left-half s-plane.

From the previous discussions, we see that *for linear time-invariant systems, BIBO and zero-input stability and asymptotic stability all have the same requirement that the roots of the characteristic equations must all be located in the left-half s-plane. Thus, if a system is BIBO stable, it must also be zero-input or asymptotically stable.* For this reason, we shall simply refer to the stability condition of a linear system as **stable** or **unstable**. The latter refers to the condition that at least one of the characteristic equation roots is not in the left-half s-plane. For practical reasons, we often refer to the situation when the characteristic equation has simple roots on the $j\omega$-axis and none in the right-half s-plane as **marginally stable** or **marginally unstable**. The following ex-

marginal stability

ample illustrates the stability conditions of systems with reference to the poles of the transfer functions that are the roots of the characteristic equation.

EXAMPLE 6-1

$$M(s) = \frac{20}{(s+1)(s+2)(s+3)}$$

BIBO and asymptotically stable, or simply stable.

$$M(s) = \frac{20(s+1)}{(s-1)(s^2+2s+2)}$$

unstable due to the pole at $s = 1$

$$M(s) = \frac{20(s+1)}{(s+2)(s^2+4)}$$

unstable or marginally stable due to the poles on the $j\omega$-axis

6-4

METHODS OF DETERMINING STABILITY OF LINEAR CONTINUOUS-DATA SYSTEMS

The discussions in the preceding sections lead to the conclusion that the stability of linear time-invariant systems can be determined by checking on the location of the roots of the characteristic equation of the system. For all practical purposes, there is no need to compute the complete system response to determine stability. The regions of stability and instability in the s-plane are illustrated in Fig. 6-1. When the system parameters are all known, the roots of the characteristic equation can be solved by means of a root-finding computer program, such as POLYROOT of the ACSP software that accompanies this text. For design purposes, there will be unknown or variable parameters imbedded in the characteristic equation, and it would not be possible to use the root-finding program. In addition, for absolute stability, we need only information on the characteristic equation roots with respect to the jω-axis of the s-plane. It is not necessary to solve for the exact roots. The methods to be outlined are well known for the determination of the stability of linear continuous-data systems, without involving root solving.

POLYROOT

Routh–Hurwitz criterion

1. *Routh–Hurwitz criterion:* This criterion is an algebraic method that provides information on the absolute stability of a linear time-invariant system that has a characteristic equation with constant coefficients. The criterion tests whether any of the roots of the characteristic equation lies in the right-half s-plane. The number of roots that lie on the jω-axis and in the right-half s-plane are also indicated.

Nyquist criterion

2. *Nyquist criterion:* This criterion is a semigraphical method that gives information on the difference between the number of poles and zeros of the closed-loop transfer function that are in the right-half s-plane by observing the behavior of the Nyquist plot of the loop transfer function. The poles of

[handwritten note in left margin:] algebraic method which allows to find whether any of the poles are on the right region.

[handwritten note in right margin:] this is a semigraphic method that gives information on the difference between poles and zeros the Right-half s

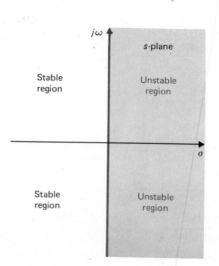

FIGURE 6.1 Stable and unstable regions in the s-plane.

the closed-loop transfer function are the roots of the characteristic equation. Data for the Nyquist plot can be obtained using the FREQRP program in the ACSP software.

root locus
diagram

3. *Root Locus diagram:*   This diagram is the loci of the characteristic equation roots when a certain system parameter varies. The root loci provide a clear picture of the stability with reference to the variable parameter. By following the rules of construction of the root loci, the latter can be roughly sketched without actually solving for the roots of the characteristic equation. The root loci can be plotted accurately with the ROOTLOCI program in the ACSP software.

Bode diagram

4. *Bode diagram:*   This diagram is a plot of the magnitude of the loop transfer function $G(j\omega)H(j\omega)$ in dB and the phase of $G(j\omega)H(j\omega)$ in degrees, all versus frequency ω. The Bode diagram can be plotted using the

FREQRP

FREQRP program in the ACSP software.

Details of the Routh–Hurwitz stability criterion are presented in the following section.

6-5
ROUTH–HURWITZ CRITERION

The Routh–Hurwitz criterion represents a method of determining the location of zeros of a polynomial with constant real coefficients with respect to the left half and the right half of the *s*-plane, without actually solving for the zeros. Since root-finding computer programs can solve for the zeros of a polynomial with ease, the value of the Routh–Hurwitz criterion is at best limited to equations with at least one unknown parameter. For the latter case, the criterion represents a convenient way of determining the range(s) of the parameter(s) for stability. The root-locus diagram covered in Chapter 8 represents an alternative way of finding the roots of the characteristic equation when one or more parameters vary, and it can be implemented by a digital computer. As mentioned before, control-system studies represent a collection of solutions all for the same common problem. The designer simply has to make the choice of the best analytical tool, depending on the given situation and the system configuration.

Consider that the characteristic equation of a linear time-invariant system is of the form

$$F(s) = a_0 s^n + a_1 s^{n-1} + a_2 s^{n-2} + \cdots + a_{n-1} s + a_n = 0 \qquad (6\text{-}21)$$

where all the coefficients are real. In order that the last equation does not have roots with positive real parts, it is *necessary but not sufficient* that the following hold:

1. All the coefficients of the polynomial have the same sign.
2. None of the coefficients vanishes.

The above requirements are based on the laws of algebra, which relate the coefficients of Eq. (6-21) as follows:

$$\frac{a_1}{a_0} = -\sum \text{all roots} \tag{6-22}$$

$$\frac{a_2}{a_0} = \sum \text{products of the roots taken two at a time} \tag{6-23}$$

$$\frac{a_3}{a_0} = -\sum \text{products of the roots taken three at a time} \tag{6-24}$$

$$\vdots$$

$$\frac{a_n}{a_0} = (-1)^n \text{ products of all roots} \tag{6-25}$$

Thus, all these ratios must be positive and nonzero unless at least one of the roots has a positive real part.

The two necessary conditions for Eq. (6-21) to have no roots in the right-half s-plane can easily be checked by inspection of the equation. However, these conditions are not sufficient, for it is quite possible that an equation with all its coefficients nonzero and of the same sign still may not have all the roots in the left half of the s-plane.

The Hurwitz Criterion

Hurwitz determinants

The necessary and sufficient condition that all roots of Eq. (6-21) lie in the left half of the s-plane is that the equation's Hurwitz determinants, D_k, $k = 1, 2, \ldots, n$, must all be positive.

The Hurwitz determinants of Eq. (6-21) are given by

These determinants should be +ve.

$$D_1 = a_1 \qquad D_2 = \begin{vmatrix} a_1 & a_3 \\ a_0 & a_2 \end{vmatrix} \qquad D_3 = \begin{vmatrix} a_1 & a_3 & a_5 \\ a_0 & a_2 & a_4 \\ 0 & a_1 & a_3 \end{vmatrix} \quad \cdots$$

$$D_n = \begin{vmatrix} a_1 & a_3 & a_5 & \cdots & a_{2n-1} \\ a_0 & a_2 & a_4 & \cdots & a_{2n-2} \\ 0 & a_1 & a_3 & \cdots & a_{2n-3} \\ 0 & a_0 & a_2 & \cdots & a_{2n-4} \\ \cdots\cdots\cdots\cdots\cdots\cdots \\ 0 & 0 & 0 & \cdots & a_n \end{vmatrix} \tag{6-26}$$

where the coefficients with indices larger than n or with negative indices are replaced with zeros. At first glance, the application of the Hurwitz determinants may seem to be formidable for high-order equations, because of the amount of work involved in evaluating the determinants in Eq. (6-26). Fortunately, Routh simplified the process by introducing a tabulation method in place of the Hurwitz determinants.

Routh's Tabulation

The first step in the simplification of the Hurwitz criterion, now called the Routh–Hurwitz criterion, is to arrange the coefficients of the equation in Eq. (6-21) into two rows. The first row consists of the first, third, fifth, . . . coefficients, and the second row consists of the second, the fourth, sixth, . . . coefficients, as shown in the following tabulation:

Routh's tabulation

	a_0	a_2	a_4	a_6	a_8	$\cdots$	(even)
	a_1	a_3	a_5	a_7	a_9	$\cdots$	(odd)

The next step is to form the following array of numbers by the indicated operations, illustrated here for a sixth-order equation:

s^6	a_0	a_2	a_4	a_6
s^5	a_1	a_3	a_5	0
s^4	$\dfrac{a_1 a_2 - a_0 a_3}{a_1} = A$	$\dfrac{a_1 a_4 - a_0 a_5}{a_1} = B$	$\dfrac{a_1 a_6 - a_0 \times 0}{a_1} = a_6$	0
s^3	$\dfrac{A a_3 - a_1 B}{A} = C$	$\dfrac{A a_5 - a_1 a_6}{A} = D$	$\dfrac{A \times 0 - a_1 \times 0}{A} = 0$	0
s^2	$\dfrac{BC - AD}{C} = E$	$\dfrac{C a_6 - A \times 0}{C} = a_6$	$\dfrac{C \times 0 - A \times 0}{C} = 0$	0
s^1	$\dfrac{ED - C a_6}{E} = F$	0	0	0
s^0	$\dfrac{F a_6 - E \times 0}{F} = a_6$	0	0	0

The array of numbers and operations just given is known as **Routh's tabulation** or **Routh's array.** The column of s's on the left side is used for identification purposes. The reference column keeps track of the calculations, and the last row of Routh's tabulation should always be the s^0 row.

Once Routh's tabulation has been completed, the last step in the application of the criterion is to investigate the *signs* of the coefficients in the *first column* of the tabulation which contains information on the roots of the equation. The following conclusions are made:

The roots of the equation are all in the left half of the s-plane if all the elements of the first column of Routh's tabulation are of the same sign. The number of changes of signs in the elements of the first column equals the number of roots with positive real parts or in the right-half s-plane.

The reason for the foregoing conclusion is simple, based on the requirements of the Hurwitz determinants. The relations between the elements in the first column of

Routh's tabulation and the Hurwitz determinants are

$$
\begin{array}{cl}
a^6 & a_0 = a_0 \\[2mm]
a^5 & a_1 = D_1 \\[2mm]
a^4 & A = \dfrac{D_2}{D_1} \\[3mm]
a^3 & C = \dfrac{D_3}{D_2} \\[3mm]
s^2 & E = \dfrac{D_4}{D_3} \\[3mm]
s^1 & F = \dfrac{D_5}{D_4} \\[3mm]
s^0 & a_6 = \dfrac{D_6}{D_5}
\end{array}
$$

Therefore, if all the Hurwitz determinants are positive, the elements in the first column of Routh's tabulation would also be of the same sign.

The following examples illustrate the applications of the Routh–Hurwitz criterion when the tabulation terminates without complications.

EXAMPLE 6-2   Consider the equation

$$(s - 2)(s + 1)(s - 3) = s^3 - 4s^2 + s + 6 = 0 \qquad (6\text{-}27)$$

which has one negative coefficient. Thus, from the necessary condition, we know without applying Routh's test that not all the roots of the equation are in the left-half s-plane. In fact, from the factored form of the equation, we know that there are two roots in the right-half s-plane, at $s = 2$ and $s = 3$. For the purpose of illustrating the Routh–Hurwitz criterion, Routh's tabulation is formed as follows:

Sign change	s^3	1	1
Sign change	s^2	-4	6
	s^1	$\dfrac{(-4)(1) - (6)(1)}{-4} = 2.5$	0
	s^0	$\dfrac{(2.5)(6) - (-4)(0)}{2.5} = 6$	0

Since there are two sign changes in the first column of the tabulation, the equation has two roots located in the right half of the s-plane. This agrees with the known result.

EXAMPLE 6-3   Consider the equation

$$2s^4 + s^3 + 3s^2 + 5s + 10 = 0 \tag{6-28}$$

Since the equation has no missing terms and the coefficients are all of the same sign, it satisfies the necessary condition for not having roots in the right half or on the imaginary axis of the s-plane. However, the sufficient condition must still be checked. Routh's tabulation is made as follows:

s^4	2	3	10
s^3	1	5	0
s^2	$\dfrac{(1)(3) - (2)(5)}{1} = -7$	10	0
s^1	$\dfrac{(-7)(5) - (1)(10)}{-7} = 6.43$	0	0
s^0	10		

Sign change — at s^2 row.
Sign change — at s^1 row.

Since there are two changes in sign in the first column of the tabulation, the equation has two roots in the right half of the s-plane. Solving for the roots of Eq. (6-28), we have the four roots at $s = -1.0055 \pm j0.93311$ and $s = 0.7555 \pm j1.4444$. Clearly, the last two roots are in the right-half s-plane, which cause the system to be unstable.

We must point out that the equations in the two preceding examples can all be solved easily to yield the exact values of the roots using a root-finding computer routine, and should be the preferred method for stability analysis. Application of the Routh–Hurwitz criterion is perhaps not justified in these cases, unless a root-finding program is not available.

Special Cases When Routh's Tabulation Terminates Prematurely

special cases   The equations considered in the two preceding examples are designed so that Routh's tabulation can be carried out without any complications. Depending on the coefficients of the equation, the following difficulties may occur that prevent Routh's tabulation from completing properly:

> 1.   The first element in any one row of Routh's tabulation is zero, but the other elements are not.
> 2.   The elements in one row of Routh's tabulation are all zero.

In the first case, if a zero appears in the first element of a row, the elements in the next row will all become infinite, and Routh's tabulation cannot continue. To remedy the situation, we replace the zero element in the first column by an arbitrary small positive number ϵ, and then proceed with Routh's tabulation. This is illustrated by the following example.

EXAMPLE 6-4   Consider the characteristic equation of a linear system:

$$s^4 + s^3 + 2s^2 + 2s + 3 = 0 \tag{6-29}$$

Since all the coefficients are nonzero and of the same sign, we need to apply the Routh–Hurwitz criterion. Routh's tabulation is carried out as follows:

s^4	1	2	3
s^3	1	2	0
s^2	0	3	

Since the first element of the s^2 row is a zero, the elements in the s^1 row would all be infinite. To overcome this difficulty, we replace the zero in the s^2 row by a small positive number ϵ, and then proceed with the tabulation. Starting with the s^2 row, the results are as follows:

Sign change	s^2	ϵ	3
Sign change	s^1	$\dfrac{2\epsilon - 3}{\epsilon} \cong -\dfrac{3}{\epsilon}$	0
	s^0	3	0

Since there are two sign changes in the first column of Routh's tabulation, the equation in Eq. (6-29) has two roots in the right-half s-plane. Solving for the roots of Eq. (6-29), we get $s = -0.09057 \pm j0.902$ and $s = 0.4057 \pm j1.2928$; the last two roots are clearly in the right-half s-plane.

It should be noted that the ϵ method described may not give correct results if the equation has pure imaginary roots [2, 3].

In the second special case, when all the elements in one row of Routh's tabulation are zeros before the tabulation is properly terminated, it indicates that one or more of the following conditions may exist:

1. The equation has at least one pair of real roots with equal magnitude but opposite signs.
2. The equation has one or more pairs of imaginary roots.
3. The equation has pairs of complex-conjugate roots forming symmetry about the origin of the s-plane; e.g., $s = -1 \pm j1, s = 1 \pm j1$.

auxiliary equation   The situation with the entire row of zeros can be remedied by using the **auxiliary equation** $A(s) = 0$, which is formed from the coefficients of the row just above the row of zeros in Routh's tabulation. The auxiliary equation is always an even polynomial, that is, only even powers of s appear. *The roots of the auxiliary equation are also the roots of the original equation.* Thus, we have by-products that are some of the roots of the original equation when we form and solve the auxiliary equation. To continue with Routh's tabulation when a row of zeros appears, we conduct the following steps:

[handwritten margin note: ϵ can be used as long as the roots are not solely in the imaginary axis.]

1. Form the auxiliary equation $A(s) = 0$ by use of the coefficients from the row just preceding the row of zeros.
2. Take the derivative of the auxiliary equation with respect to s; this gives $dA(s)/ds = 0$.
3. Replace the row of zeros with the coefficients of $dA(s)/ds$.
4. Continue with Routh's tabulation in the usual manner with the newly formed row of coefficients replacing the row of zeros.
5. Interpret the signs of the coefficients in the first column of the Routh's tabulation in the usual manner.

EXAMPLE 6-5   Consider the following equation, which may be the characteristic equation of a linear control system:

$$s^5 + 4s^4 + 8s^3 + 8s^2 + 7s + 4 = 0 \qquad (6\text{-}30)$$

Routh's tabulation is

s^5	1	8	7
s^4	4	8	4
s^3	6	6	0
s^2	4	4	
s^1	0	0	

Since a row of zeros appears prematurely, we form the auxiliary equation using the coefficients of the s^2 row. The auxiliary equation is written

$$A(s) = 4s^2 + 4 = 0 \qquad (6\text{-}31)$$

The derivative of $A(s)$ with respect to s is

$$\frac{dA(s)}{ds} = 8s = 0 \qquad (6\text{-}32)$$

from which the coefficients 8 and 0 replace the zeros in the s^1 row of the original tabulation. The remaining portion of the Routh's tabulation is

s^1	8	0	coefficients of $dA(s)/ds$
s^0	4		

Since there are no sign changes in the first column of the entire Routh's tabulation, the equation in Eq. (6-30) does not have any root in the right-half s-plane. Solving the auxiliary equation in Eq. (6-31), we get the two roots at $s = j$ and $s = -j$, which are also the roots of Eq. (6-30). Thus, the equation has two roots on the $j\omega$-axis, and the system can be regarded as marginally stable. These imaginary roots caused the initial Routh's tabulation to have the entire row of zeros in the s^1 row.

Since it was marginally stable the entire row of zeros appeared in the Routh table.

Since all zeros occurring in a row that corresponds to an odd power of s creates an auxiliary equation that has only even powers of s, the roots of the auxiliary equation may all lie on the $j\omega$-axis. For design purposes, we can use the all-zero-row condition to solve for the marginal value of a parameter for system stability. The following example illustrates this feature, which is quite typical in simple design problems.

EXAMPLE 6-6   Consider that a third-order control system has the characteristic equation

$$s^3 + 3408.3s^2 + 1,204,000s + 1.5 \times 10^7 K = 0 \qquad (6\text{-}33)$$

The Routh–Hurwitz criterion is best suited to determine the critical value of K for stability, that is, the value of K for which at least one root will lie on the $j\omega$-axis and none in the right-half s-plane. Routh's tabulation of Eq. (6-33) is conducted as follows:

s^3	1	1,204,000
s^2	3408.3	$1.5 \times 10^7 K$
s^1	$\dfrac{410.36 \times 10^7 - 1.5 \times 10^7 K}{3408.3}$	0
s^0	$1.5 \times 10^7 K$	

For the system to be stable, all the roots of Eq. (6-33) must be in the left-half s-plane, and, thus, all the coefficients in the first column of Routh's tabulation must have the same sign. This leads to the following conditions:

$$\frac{410.36 \times 10^7 - 1.5 \times 10^7 K}{3408.3} > 0 \qquad (6\text{-}34)$$

and

$$1.5 \times 10^7 K > 0 \qquad (6\text{-}35)$$

From the inequality of Eq. (6-34), we have $K < 273.57$ and the condition in Eq. (6-35) gives $K > 0$. Therefore, the condition on K for the system to be stable is

$$0 < K < 273.57 \qquad (6\text{-}36)$$

If we let $K = 273.57$, the characteristic equation in Eq. (6-33) will have two roots on the $j\omega$-axis. To find these roots, we substitute $K = 273.57$ in the auxiliary equation, which is obtained from Routh's tabulation by using the coefficients of the s^2 row. Thus,

$$A(s) = 3408.3s^2 + 4.1036 \times 10^9 = 0 \qquad (6\text{-}37)$$

which has roots at $s = j1097.27$ and $s = -j1097.27$, and the corresponding value of K at these roots is 273.57. Also, if the system is operated with $K = 273.57$, the zero-input response of the system will be an undamped sinusoid with a frequency of 1097.27 rad/s.

EXAMPLE 6-7   As another example of using the Routh–Hurwitz criterion for simple design problems, consider that the characteristic equation of a closed-loop control system is

$$s^3 + 3Ks^2 + (K + 2)s + 4 = 0 \qquad (6\text{-}38)$$

It is desired to find the range of K so that the system is stable. Routh's tabulation of Eq. (6-38) is

s^3	1	$K + 2$
s^2	$3K$	4
s^1	$\dfrac{3K(K + 2) - 4}{3K}$	0
s^0	4	

From the s^2 row, the condition of stability is $K > 0$, and from the s^1 row, the condition of stability is

$$3K^2 + 6K - 4 > 0 \quad \text{or} \quad K < -2.528 \quad \text{or} \quad K > 0.528$$

When the conditions of $K > 0$ and $K > 0.528$ are compared, it is apparent that the latter requirement is more stringent. Thus, for the closed-loop system to be stable, K must satisfy

$$K > 0.528$$

The requirement of $K < -2.528$ is disregarded since K cannot be negative.

It should be reiterated that the Routh–Hurwitz criterion is valid only if the characteristic equation is algebraic with real coefficients. If any one of the coefficients of the equation is complex, or if the equation is not algebraic, such as containing exponential functions or sinusoidal functions of s, the Routh–Hurwitz criterion simply cannot be applied.

limitation   Another limitation of the Routh–Hurwitz criterion is that it is valid only for the determination of roots of the characteristic equation with respect to the left-half or the right-half of the complex s-plane. The stability boundary is the $j\omega$-axis of the s-plane. The criterion *cannot* be applied to any other boundaries in a complex plane, such as the unit circle in the z-plane, which is the stability boundary of discrete-data systems.

6-6

STABILITY OF DISCRETE-DATA SYSTEMS

discrete-data   The definitions of BIBO and zero-input stability can be readily extended to linear time-
systems   invariant digital and discrete-data control systems.

BIBO Stability

Let $r(kT)$, $c(kT)$, and $g(kT)$ be the input, output, and impulse sequence of a linear time-invariant discrete-data system, respectively. *With zero initial conditions, the system is said to be BIBO stable, or simply stable, if its output sequence $c(kT)$ is bounded to a bounded input $r(kT)$.* As with the treatment in Section 6-1, we can show that for the system to be BIBO stable, the following condition must be met:

$$\sum_{k=0}^{\infty} |g(kT)| < \infty \qquad (6\text{-}39)$$

Zero-Input Stability

For zero-input stability, the output sequence of the system must satisfy the following conditions:

$$(1) \quad |c(kT)| \le M < \infty \qquad (6\text{-}40)$$

$$(2) \quad \lim_{k \to \infty} |c(kT)| = 0 \qquad (6\text{-}41)$$

asymptotic stability

Thus, zero-input stability can also be referred to as **asymptotic stability.** We can show that *both the BIBO stability and the zero-input stability of discrete-data systems require that the roots of the characteristic equation lie inside the unit circle $|z| = 1$ in the z-plane.* This is not surprising, since the $j\omega$-axis of the s-plane is mapped onto the unit circle in the z-plane. The regions of stability and instability for discrete-data systems in the z-plane are shown in Fig. 6-2. The following example illustrates the relationship between the closed-loop transfer-function poles, which are the characteristic equation roots, and the stability condition of the system.

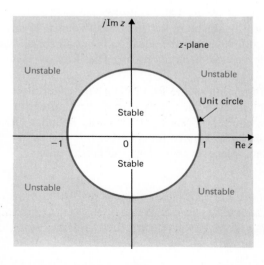

FIGURE 6.2 Stable and unstable regions for discrete-data systems in the z-plane.

EXAMPLE 6-8

$$M(z) = \frac{5z}{(z - 0.2)(z - 0.8)}$$ stable system

$$M(z) = \frac{5z}{(z + 1.2)(z - 0.8)}$$ unstable system due to the pole at $z = -1.2$

$$M(z) = \frac{5(z + 1)}{z(z - 1)(z - 0.8)}$$ one pole on the unit circle; unstable or marginally stable

6-7

STABILITY TESTS OF DISCRETE-DATA SYSTEMS

We pointed out in the last section that the stability test of a linear discrete-data control system is essentially a problem of investigating whether all the roots of the characteristic equation are inside the unit circle $|z| = 1$ in the z-plane. The Nyquist criterion, root-locus diagram, and the Bode diagram, mentioned in Section 6-4, can all be applied to the stability studies of discrete-data systems. One exception is the Routh–Hurwitz criterion, which is restricted to only the imaginary axis as the stability boundary, and thus can only be applied to continuous-data systems.

The Bilinear Transformation Method [9]

bilinear transformation   We can still apply the Routh–Hurwitz criterion to discrete-data systems if we could find a transformation that transforms the unit circle in the z-plane onto the imaginary axis of another complex plane. We cannot use the z-transform relation, $z = \exp(Ts)$, since it would transform an algebraic characteristic equation in z into a transcendental function of s, and the Routh test still cannot be applied. However, there are many bilinear transformations of the form of

$$z = \frac{aw + b}{cw + d} \qquad (6\text{-}42)$$

that will transform circles in the z-plane onto straight lines in the w-plane, where a, b, c, and d are real constants. One such transformation that transforms the interior of the unit circle of the z-plane onto the left half of the w-plane is

$$z = \frac{w + 1}{w - 1} \qquad (6\text{-}43)$$

w-transformation   which is referred to as the **w-transformation.** Once the characteristic equation in z is transformed into the w domain using Eq. (6-43), the Routh–Hurwitz criterion can again be applied to the equation in w. Let us illustrate this procedure by means of an example.

EXAMPLE 6-9   Consider that the characteristic equation of a discrete-data control system is

$$z^3 + 5.94z^2 + 7.7z - 0.368 = 0 \qquad (6\text{-}44)$$

Substituting Eq. (6-43) into the last equation and simplifying, we get

$$14.27w^3 + 2.344w^2 - 11.74w + 3.128 = 0 \qquad (6\text{-}45)$$

Routh's tabulation of the last equation is

	w^3	14.27	-11.74
Sign change	w^2	2.344	3.128
Sign change	w^1	-30.78	0
	w^0	3.128	

Since there are two sign changes in the first column of the tabulation, Eq. (6-45) has two roots in the right half of the w-plane. This corresponds to Eq. (6-44) having two roots outside the unit circle in the z-plane. This result can be checked by solving the two equations in z and w. For Eq. (6-44), the roots are at $z = -2.0$, $z = -3.984$, and $z = 0.461$. The three corresponding roots in the w-plane are at $w = 0.333$, $w = 0.5988$, and $w = -1.097$, respectively.

EXAMPLE 6-10   Let us consider a design problem using the bilinear transformation method. The characteristic equation of a linear digital control system is given as

$$F(z) = z^3 + z^2 + z + K = 0 \qquad (6\text{-}46)$$

where K is a real constant. The problem is to find the range of values of K so that the system is stable. We first transform $F(z)$ into an equation in w using the bilinear transformation defined in Eq. (6-43). The result is

$$(3 + K)w^3 + 3(1 - K)w^2 + (1 + 3K)w + 1 - K = 0 \qquad (6\text{-}47)$$

Routh's tabulation of the last equation is

w^3	$3 + K$	$1 + 3K$
w^2	$3(1 - K)$	$1 - K$
w^1	$8K/3$	
w^0	$1 - K$	

The conclusion is that for $F(z)$ to have no roots outside or on the unit circle, the value of K must satisfy the following conditions:

$$0 < K < 1 \qquad (6\text{-}48)$$

We can show that when $K = 0$, $F(z)$ becomes a second-order equation with two roots on the unit circle, and when $K = 1$, all three roots of $F(z)$ are on the unit circle.

Direct Stability Tests

Jury's test

There are stability tests that can be applied directly to the characteristic equation in z with reference to the unit circle in the z-plane. One of the first methods that gives the necessary and sufficient conditions for the characteristic equation roots to lie inside the unit circle is the Schur–Cohn criterion [5]. A simpler tabulation method was devised by Jury and Blanchard [6, 7], and is called **Jury's stability criterion** [10]. R. H. Raible [8] devised an alternate tabular form of Jury's stability test. Unfortunately, these analytical tests all become very tedious for equations higher than the second order. Weighing all the pros and cons, this author believes that the bilinear transformation method described before is still the best for determining stability of linear discrete-data systems, especially when the equation has an unknown parameter. However, it is useful to introduce the necessary conditions of stability that can be checked by inspection. Consider that the characteristic equation of a linear time-invariant discrete-data system is

$$F(z) = a_n z^n + a_{n-1} z^{n-1} + \cdots + a_2 z^2 + a_0 = 0 \qquad (6\text{-}49)$$

where all the coefficients are real. Among all the conditions provided in Jury's test, the following ones can be checked by inspection:

necessary conditions

$$F(1) > 0$$
$$F(-1) > 0 \quad \text{if } n = \text{even integer}$$
$$F(-1) < 0 \quad \text{if } n = \text{odd integer} \qquad (6\text{-}50)$$
$$|a_0| < a_n$$

Thus, these can be regarded as *necessary* conditions for $F(z)$ in Eq. (6-49) to have no roots on or outside the unit circle. If an equation of the form of Eq. (6-49) violates any one of the conditions in Eq. (6-50), then not all of the roots are inside the unit circle.

EXAMPLE 6-11

$$F(z) = z^3 + z^2 + 0.5z + 0.25 = 0 \qquad (6\text{-}51)$$

$$F(1) = 2.75 > 0 \quad \text{and} \quad F(-1) = -0.25 < 0 \quad \text{for } n = 3 = \text{odd}$$
$$|a_0| = 0.25 < a_3 = 1$$

The conditions in Eq. (6-49) are all satisfied, but nothing can be said about the roots of the equation.

EXAMPLE 6-12

$$F(z) = z^3 + z^2 + 0.5z + 1.25 = 0 \tag{6-52}$$

$$F(-1) = 0.75 > 0 \quad \text{for } n = 3 = \text{odd}$$

The equation has at least one root outside the unit circle. The condition that $|a_0|$ (= 1.25) must be less than a_3 (= 1) is also violated.

Second-Order Systems

The conditions in Eq. (6-50) are *necessary and sufficient* for a second-order system. The following examples illustrate the applications of the conditions in Eq. (6-50).

EXAMPLE 6-13

$$F(z) = z^2 + z + 0.25 = 0 \tag{6-53}$$

$$F(1) = 2.25 > 0 \quad \text{and} \quad F(-1) = 0.25 > 0 \quad \text{for } n = 2 = \text{even}$$

$$|a_0| = 0.25 < a_2 = 1$$

Thus, the conditions in Eq. (6-50) are all satisfied; the two roots of Eq. (6-53) are all inside the unit circle, and the system is stable.

SUMMARY

In this chapter, the definitions of BIBO stability, zero-input or asymptotic stability of linear time-invariant continuous-data and discrete-data systems are given. It is shown that the condition of these types of stability is related directly to the roots of the characteristic equation. For a continuous-data system to be stable, the roots of the characteristic equation must all be located in the left half of the s-plane. For a discrete-data system to be stable, the roots of the characteristic equation must all be inside the unit circle $|z| = 1$ in the z-plane.

The necessary condition for a polynomial $F(s)$ to have no zeros on the $j\omega$-axis and in the right-half s-plane is that all its coefficients must be of the same sign and none can vanish. The necessary and sufficient conditions for $F(s)$ to have zeros only in the left-half s-plane are checked with the Routh–Hurwitz criterion. The value of the Routh–Hurwitz criterion is diminished if the characteristic equation can be solved using a root-finding routine on the computer.

For discrete-data systems, the characteristic equation $F(z)$ must be checked for roots on and outside the unit circle in the z-plane. The Routh–Hurwitz criterion cannot be applied directly to this situation. Although there are direct methods similiar to the Routh–Hurwitz criterion that can be used to check the roots in the z-plane with respect to the stability boundary, these turn out to be quite cumbersome, except for systems of the second order. A reliable method is to use a bilinear transformation, which transforms the unit circle in the z-plane onto the imaginary axis of another complex-variable plane, so that the Routh–Hurwitz criterion can be applied to the transformed equation.

REVIEW QUESTIONS

1. Can the Routh–Hurwitz criterion be directly applied to the stability analysis of the following problems?
   (a) Continuous-data system with the characteristic equation:

   $$s^4 + 5s^3 + 2s^2 + s + e^{-2s} = 0$$

   (b) Continuous-data system with the characteristic equation:

   $$s^4 + 5s^3 + 3s^2 + Ks + K^2 = 0$$

   (c) Discrete-data system with the characteristic equation:

   $$z^4 + 2z^3 + z^2 + 0.5z + 0.1 = 0$$

2. If the numbers in the first column of Routh's tabulation turn out to be all negative, then the equation for which the tabulation is made has at least one root not in the left-half of the s-plane.   **(T)   (F)**

3. The first two rows of Routh's tabulation of a third-order equation are

   $$\begin{array}{ccc} s^3 & 2 & 2 \\ s^2 & 4 & 4 \end{array}$$

   Select the correct answer from the following choices:
   (a) The equation has one root in the right-half s-plane.
   (b) The equation has two roots on the $j\omega$ axis at $s = j$ and $-j$. The third root is in the left-half plane.
   (c) The equation has two roots on the $j\omega$-axis at $s = 2j$ and $s = -2j$. The third root is in the left-half plane.
   (d) The equation has two roots on the $j\omega$-axis at $s = 2j$ and $s = -2j$. The third root is in the right-half plane.

4. The roots of the auxiliary equation, $A(s) = 0$, of Routh's tabulation of a characteristic equation must also be the roots of the latter.   **(T)   (F)**

5. The following characteristic equation of a discrete-data system represents an unstable system since it contains a negative coefficient.

   $$z^3 - z^2 + 0.1z + 0.5 = 0$$ **(T)   (F)**

6. The following characteristic equation of a continuous-data system represents an unstable system since it contains a negative coefficient.

   $$s^3 - s^2 + 5s + 10 = 0$$ **(T)   (F)**

7. The following characteristic equation of a discrete-data system represents an unstable system since the coefficient of the z term is zero.

   $$z^2 - 0.5 = 0$$ **(T)   (F)**

8. The following characteristic equation of a continuous-data system represents an unstable system since there is a zero coefficient.

   $$s^3 + 5s^2 + 4 = 0$$ **(T)   (F)**

9. When a row of Routh's tabulation contains all zeros before the tabulation ends, this means that the equation has roots on the imaginary axis of the *s*-plane.   **(T)   (F)**

10. Without conducting detailed tests, it is easy to see that the following *z*-transform characteristic equation must have at least one root outside the unit circle in the *z*-plane.

$$z^4 + 5z^3 + z^2 + z + 2 = 0$$ **(T)   (F)**

REFERENCES

1. G. V. S. S. Raju, "The Routh Canonical Form," *IEEE Trans. Automatic Control*, Vol. AC-12, pp. 463–464, Aug. 1967.
2. K. J. Khatwani, "On Routh–Hurwitz Criterion," *IEEE Trans. Automatic Control*, Vol. AC-26, p. 583, April 1981.
3. S. K. Pillai, "The ϵ Method of the Routh–Hurwitz Criterion," *IEEE Trans. Automatic Control*, Vol. AC-26, 584, April 1981.
4. E. I. Jury, "The Number of Roots of a Real Polynomial Inside (or Outside) the Unit Circle Using the Determinant Method," *IEEE Trans. Automatic Control*, Vol. AC-10, pp. 371–372, July 1965.
5. M. L. Cohen, "A Set of Stability Constraints on the Denominator of a Sampled-Data Filter," *IEEE Trans. Automatic Control*, Vol. AC-11, pp. 327–328, April 1966.
6. E. I. Jury and J. Blanchard, "A Stability Test for Linear Discrete Systems in Table Form," *IRE Proc.*, Vol. 49, No. 12, pp. 1947–1948, Dec. 1961.
7. E. I. Jury and B. D. O. Anderson, "A Simplified Schur–Cohen Test," *IEEE Trans. Automatic Control*, Vol. AC-18, pp. 157–163, April 1973.
8. R. H. Raible, "A Simplification of Jury's Tabular Form," *IEEE Trans. Automatic Control*, Vol. AC-19, pp. 248–250, June 1974.
9. B. C. Kuo, *Digital Control Systems*, Holt, Rinehart and Winston, New York, 1980.
10. E. I. Jury, *Theory and Application of the z-Transform Method*, John Wiley & Sons, New York, 1964.

PROBLEMS

6-1. Without using the Routh–Hurwitz criterion, determine if the following systems are asymptotically stable, marginally stable, or unstable. In each case, the closed-loop transfer function is given.

(a) $M(s) = \dfrac{10(s + 5)}{s^3 + 4s^2 + s}$ (b) $M(s) = \dfrac{s - 1}{(s + 1)(s^2 + 4)}$

(c) $M(s) = \dfrac{K}{s^3 + 3s + 4}$ (d) $M(s) = \dfrac{10(s - 1)}{(s + 1)(s^2 + 2s + 2)}$

(e) $M(s) = \dfrac{10}{s^3 - 2s^2 + s + 1}$

6-2. Using the Routh–Hurwitz criterion, determine the stability of the closed-loop control systems that have the following characteristic equations. Determine the number of roots of each equation that are in the right-half *s*-plane or on the *j\omega*-axis.

     (a) $s^3 + 20s^2 + 10s + 400 = 0$
     (b) $s^3 + 20s^2 + 10s + 100 = 0$
     (c) $2s^4 + 10s^3 + 5s^2 + 5s + 10 = 0$
     (d) $s^4 + 2s^3 + 6s^2 + 8s + 1 = 0$
     (e) $s^6 + 2s^5 + 8s^4 + 15s^3 + 20s^2 + 16s + 16 = 0$
     (f) $s^4 + 2s^3 + 10s^2 + 20s + 5 = 0$

6-3. For each of the characteristic equations of feedback control systems given, determine the range of K so that the system is asymptotically stable. Determine the value of K so that the system is marginally stable and the frequency of sustained oscillation if applicable.

     (a) $s^4 + 20s^3 + 15s^2 + 2s + K = 0$
     (b) $s^4 + 2Ks^3 + 2s^2 + (1 + K)s + 2 = 0$
     (c) $s^3 + (K + 1)s^2 + Ks + 50 = 0$
     (d) $s^3 + Ks^2 + 5s + 10 = 0$

6-4. The loop transfer function of a single-loop feedback control system is given as

$$G(s)H(s) = \frac{K(s + 2)}{s(1 + Ts)(1 + 2s)}$$

The parameters K and T may be represented in a plane with K as the horizontal axis and T as the vertical axis. Determine the regions in the T-versus-K plane in which the closed-loop system is asymptotically stable, and unstable. Indicate the boundary on which the system is marginally stable.

6-5. Given the open-loop transfer functions of unity-feedback control systems,

  (a) $G(s) = \dfrac{K(s + 5)(s + 40)}{s^3(s + 200)(s + 1000)}$ (b) $G(s) = \dfrac{K(s + 10)(s + 20)}{s^2(s + 2)}$

apply the Routh–Hurwitz criterion to determine the stability of the closed-loop system as a function of K. Determine the values of K that will cause sustained constant-amplitude oscillations in the system. Determine the frequency of oscillation.

6-6. A controlled process is modeled by the following state equations.

$$\dot{x}_1 = x_1 - 2x_2 \qquad \dot{x}_2 = 10x_1 + u$$

The control is obtained from state feedback, such that

$$u = -g_1 x_1 - g_2 x_2$$

where g_1 and g_2 are real constants. Determine the region in the g_2-versus-g_1 plane in which the closed-loop system is asymptotically stable.

6-7. A linear time-invariant system is described by the following state equations.

$$\dot{\mathbf{x}}(t) = \mathbf{A}\mathbf{x}(t) + \mathbf{B}u(t)$$

where

$$\mathbf{A} = \begin{bmatrix} 0 & 1 & 0 \\ 0 & 0 & 1 \\ 0 & -4 & -3 \end{bmatrix} \qquad \mathbf{B} = \begin{bmatrix} 0 \\ 0 \\ 1 \end{bmatrix}$$

The closed-loop system is implemented by state feedback, so that $u(t) = -\mathbf{G}\mathbf{x}(t)$, where $\mathbf{G} = [g_1 \quad g_2 \quad g_3]$; and g_1, g_2, and g_3 are real constants. Determine the constraints on the elements of $\mathbf{G}$ so that the closed-loop system is asymptotically stable.

6-8.   Given the system $d\mathbf{x}(t)/dt = \mathbf{A}\mathbf{x}(t) + \mathbf{B}u(t)$, where

(a)
$$\mathbf{A} = \begin{bmatrix} 1 & 0 & 0 \\ 0 & -2 & 0 \\ 0 & 0 & 3 \end{bmatrix} \qquad \mathbf{B} = \begin{bmatrix} 1 \\ 0 \\ 1 \end{bmatrix}$$

(b)
$$\mathbf{A} = \begin{bmatrix} 1 & 0 & 0 \\ 0 & -2 & 0 \\ 0 & 0 & 3 \end{bmatrix} \qquad \mathbf{B} = \begin{bmatrix} 0 \\ 1 \\ 1 \end{bmatrix}$$

Can the system be stabilized by state feedback $u = -\mathbf{G}\mathbf{x}$, where $\mathbf{G} = [g_1 \quad g_2 \quad g_3]$?

6-9.   The block diagram of a motor-control system with tachometer feedback is shown in Fig. 6P-9. Find the range of K_t so that the system is asymptotically stable.

6-10.   The block diagram of a control system is shown in Fig. 6P-10. Find the region in the K-versus-α plane for the system to be asymptotically stable. (Use K as the vertical and α as the horizontal axis.)

6-11.   The attitude control of the missile shown in Fig. 4P-13 is accomplished by thrust vectoring. The transfer function between the thrust angle $\Delta(s)$ and the angle of attack $\Theta(s)$ is represented by

$$G_p(s) = \frac{\Theta(s)}{\Delta(s)} = \frac{K}{s^2 - \alpha}$$

where K and α are positive real constants. The block diagram of the control system is shown in Fig. 6P-11.

(a)   In Fig. 6P-11, consider that only the attitude-sensor loop is in operation, but $K_t = 0$. Determine the relationship between K, K_s, and α so that the missile will oscillate back and forth (marginally stable). Find the condition when the missile will tumble end over end (unstable).

(b)   Consider that both loops are in operation. Determine relationships between K, K_s, K_t, and α so that the missile will oscillate back and forth. Repeat for the missile tumbling end over end.

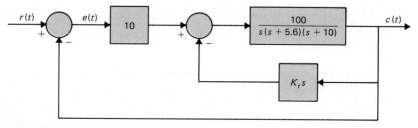

FIGURE 6P-9

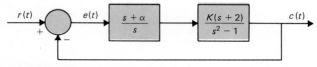

FIGURE 6P-10

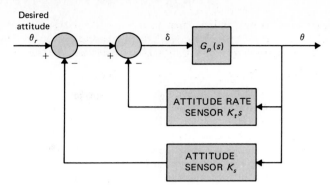

Desired attitude

FIGURE 6P-11

6-12. The conventional Routh–Hurwitz criterion gives information only on the location of the zeros of a polynomial $F(s)$ with respect to the left-half and the right-half of the s-plane. Devise a linear transformation $s = f(p, \alpha)$, where p is a complex variable, so that the Routh–Hurwitz criterion can be applied to determine whether $F(s)$ has zeros to the right of the line $s = -\alpha$, where α is a positive real number. Apply the transformation to the following characteristic equations to determine how many roots are to the right of the line $s = -1$ in the s-plane.

(a) $F(s) = s^2 + 6s + 1 = 0$ (b) $F(s) = s^3 + 3s^2 + 3s + 1 = 0$

(c) $F(s) = s^3 + 5s^2 + 3s + 10 = 0$

6-13. Figure 6P-13 shows the closed-loop version of the liquid-level control system described in Problem 5-42. All the system parameters and equations are given in Problem 5-42.

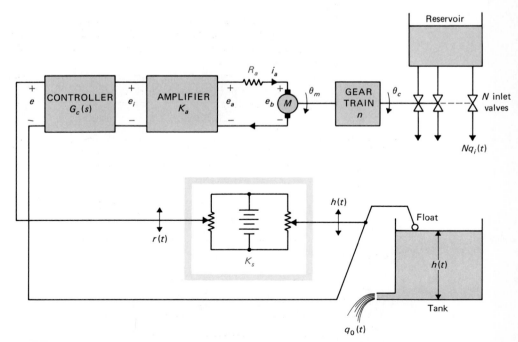

FIGURE 6P-13

The error detector is modeled as

$$K_s = 1 \text{ V/ft} \qquad e(t) = K_s[r(t) - h(t)]$$

(a) Draw a functional block diagram for the overall system, showing the functional relationship between the transfer functions.

(b) Let the transfer function of the controller, $G_c(s)$, be unity. Find the open-loop transfer function $G(s) = H(s)/E(s)$ and the closed-loop transfer function $M(s) = H(s)/R(s)$. Find the characteristic equation of the closed-loop system. Note that all the system parameters are specified except for N, the number of inlets.

(c) Apply the Routh–Hurwitz criterion to the characteristic equation and determine the maximum number of inlets (positive integer) so that the system is asymptotically stable. Set $G_c(s) = 1$.

6-14. The block diagram shown in Fig. 6P-14 represents the liquid-level system described in Problem 6-13. The following data are given: $K_a = 50$, $K_I = 50$, $K_b = 0.0706$, $J = 0.006$, $R_a = 10$, $K_i = 10$, and $n = 1/100$. The values of A, N, and K_o will be assigned in the following.

(a) Let the number of inlets, N, and the tank area, A, be the variable parameters, and $K_o = 50$. Find the ranges of N and A so that the closed-loop system is asymptotically stable. Indicate the region in the N-versus-A plane in which the system is stable. Use the fact that N is an integer. If the cross-sectional area of the tank A were infinitely large, what is the maximum value of N for stability?

(b) It is useful to investigate the relationship between the gear ratio n and K_o, which control the liquid inflow and outflow, respectively. Find the stable region in the N-versus-K_o plane for K_o up to 100. Let $A = 50$.

(c) Let $N = 10$ and $A = 50$. K_I and K_o are variable parameters. Find the stable region in the K_I-versus-K_o plane.

6-15. The payload of a space-shuttle-pointing control system is modeled as a pure mass M. The payload is suspended by magnetic bearings so that no friction is encountered in the control. The attitude of the payload in the y direction is controlled by magnetic actua-

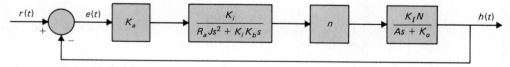

FIGURE 6P-14

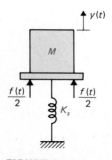

FIGURE 6P-15

tors located at the base. The total force produced by the magnetic actuators is $f(t)$. The controls of the other degrees of motion are independent, and are not considered here. Since there are experiments located on the payload, electric power must be brought to the payload through cables. The linear spring with spring constant K_s is used to model the cable attachment. The dynamic-system model for the control of the y-axis motion is shown in Fig. 6P-15. The force equation of motion in the y-direction is

$$f(t) = K_s y(t) + M d^2 y(t)/dt^2$$

where $K_s = 0.5$ N-m/m, and $M = 500$ kg.

The magnetic actuators are controlled through state feedback, so that

$$f(t) = -K_P y(t) - K_D \, dy(t)/dt$$

(a) Draw a functional block diagram for the system.
(b) Find the characteristic equation of the closed-loop system.
(c) Find the region in the K_D-versus-K_P plane in which the system is asymptotically stable.

6-16. An inventory-control system is modeled by the following differential equations:

$$\dot{x}_1(t) = -x_2(t) + u(t)$$
$$\dot{x}_2(t) = -Ku(t)$$

where $x_1(t)$ is the level of inventory, $x_2(t)$ the rate of sales of product, $u(t)$ the production rate, and K a real constant. Let the output of the system be $c(t) = x_1(t)$, and $r(t)$ be the reference set point for the desired inventory level. Let $u(t) = r(t) - c(t)$. Determine the constraint on K so that the closed-loop system is asymptotically stable.

6-17. Apply the w-transform to the following characteristic equations of discrete-data control systems, and determine the conditions of stability (asymptotically stable, and marginally stable, or unstable) by means of the Routh–Hurwitz criterion.
(a) $z^2 + 1.5z - 1 = 0$ (b) $z^3 + z^2 + 3z + 0.2 = 0$
(c) $z^3 - 1.2z^2 - 2z + 3 = 0$ (d) $z^3 - 1.2z^2 - 2z + 0.5 = 0$

Check the answers by solving for the roots of the equations using a root-finding computer program.

6-18. A digital control system is described by the state equation

$$x(k+1) = (0.368 - 0.632K)x(k) + Kr(k)$$

where $r(k)$ is the input, and $x(k)$ is the state variable. Determine the values of K for the system to be asymptotically stable.

6-19. The characteristic equation of a linear digital control system is

$$z^3 + z^2 + 1.5Kz - (K + 0.5) = 0$$

Determine the values of K for the system to be asymptotically stable.

6-20. The block diagram of a discrete-data control system is shown in Fig. 6P-20.
(a) For $T = 0.1$ second, find the values of K so that the system is asymptotically stable at the sampling instants.

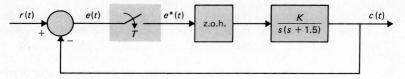

FIGURE 6P-20

    (b) Repeat part (a) when the sampling period is 0.5 second.
    (c) Repeat part (a) when the sampling period is 1.0 second.

ADDITIONAL COMPUTER PROBLEMS

  6-21. Use a root-finding program such as the POLYROOT in the ACSP package to find the roots of the following characteristic equations of linear continuous-data systems, and determine the stability condition of the systems.
    (a) $s^3 + 10s^2 + 11s + 125 = 0$
    (b) $s^4 + 12.5s^3 + s^2 + s + 10 = 0$
    (c) $s^4 + 12.5s^3 + 10s^2 + 10s + 1 = 0$
    (d) $s^4 + 12.5s^3 + s^2 + 10s + 1 = 0$
    (e) $s^6 + 6s^5 + 125s^4 + 100s^3 + 100s^2 + 20s + 10 = 0$
    (f) $s^5 + 125s^4 + 100s^3 + 100s^2 + 20s + 10 = 0$

  6-22. Use a root-finding program such as the POLYROOT in the ACSP package to find the roots of the following characteristic equations of linear discrete-data systems, and determine the stability condition of the systems.
    (a) $z^3 + 2z^2 + 1.25z + 0.5 = 0$
    (b) $z^3 + z^2 + z + 0.5 = 0$
    (c) $0.5z^3 + z^2 + 1.5z + 0.5 = 0$
    (d) $z^4 + 0.5z^3 + 0.25z^2 + 0.1z - 0.25 = 0$

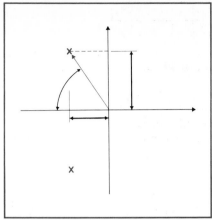

7

TIME-DOMAIN ANALYSIS OF CONTROL SYSTEMS

7-1

TIME RESPONSE OF CONTINUOUS-DATA SYSTEMS: INTRODUCTION

Since time is used as an independent variable in most control systems, it is usually of interest to evaluate the state and output responses with respect to time, simply, the time response. In the analysis problem, a reference input signal is applied to a system, and the performance of the system is evaluated by studying the system response in the time domain. For instance, if the objective of the control system is to have the output variable track the input signal, starting at some initial time and initial condition, it is necessary to compare the input and the output responses as functions of time. Therefore, for most control system problems, the final evaluation of the performance of the system is based on the time responses.

transient response The time response of a control system is usually divided into two parts: the **transient response** and the **steady-state response**. Let $c(t)$ denote the time response of a continuous-data system; then, in general, it can be written as

usually divided

$$c(t) = c_t(t) + c_{ss}(t) \tag{7-1}$$

steady-state response where $c_t(t)$ denotes the transient response, and $c_{ss}(t)$ denotes the steady-state response.

307

In control systems, transient response is defined as the part of the time response that goes to zero as time becomes very large. Thus, $c_t(t)$ has the property

$$\lim_{t \to \infty} c_t(t) = 0 \qquad (7\text{-}2)$$

The steady-state response is simply the part of the total response that remains after the transient has died out. Thus, the steady-state response can still vary in a fixed pattern, such as a sine wave, or a ramp function that increases with time.

All real control systems exhibit transient phenomena to some extent before the steady state is reached. Since inertia, mass, and inductance are unavoidable in physical systems, the responses of a typical control system cannot follow sudden changes in the input instantaneously, and transients are usually observed. Therefore, the control of the transient response is necessarily important, as it is a significant part of the dynamic behavior of the system; and the deviation between the output response and the input or the desired response, before the steady state is reached, must be closely watched.

The steady-state response of a control system is also very important, since it indicates where the system output ends up at when time becomes large. For a position-control system, the steady-state response when compared with the reference input gives an indication of the final accuracy of the system. If the steady-state response of the output does not agree with the steady-state of the input exactly, the system is said to have a **steady-state error**.

The study of a control system in the time domain essentially involves the evaluation of the transient and the steady-state responses of the system. In the design problem, specifications are usually given in terms of the transient and the steady-state performances, and controllers are designed so that the specifications are all met by the designed system.

7-2

TYPICAL TEST SIGNALS FOR THE TIME RESPONSE OF CONTROL SYSTEMS

Unlike electric circuits and communication systems, the inputs to many practical control systems are not exactly known ahead of time. In many cases, the actual inputs of a control system may vary in random fashion with respect to time. For instance, in a radar tracking system for antiaircraft missiles, the position and speed of the target to be tracked may vary in an unpredictable manner, so that they cannot be expressed deterministically by a mathematical expression. This poses a problem for the designer, since it is difficult to design a control system so that it will perform satisfactorily to all possible input signals. For the purposes of analysis and design, it is necessary to assume some basic types of test inputs so that the performance of a system can be evaluated. By selecting these basic test signals properly, not only the mathematical treatment of the problem is systematized, but the responses due to these inputs allow the prediction of the system's performance to other more complex inputs. In the design problem, performance criteria may be specified with respect to these test signals so that the system may be designed to meet the criteria.

When the response of a linear time-invariant system is analyzed in the frequency domain, a sinusoidal input with variable frequency is used. When the input frequency is swept from zero to beyond the significant range of the system characteristics, curves in terms of the amplitude ratio and phase between the input and the output are drawn as functions of frequency. It is possible to predict the time-domain behavior of the system from its frequency-domain characteristics.

To facilitate the time-domain analysis, the following deterministic test signals are used.

step input
STEP-FUNCTION INPUT. The step-function input represents an instantaneous change in the reference input. For example, if the input is the angular position of a mechanical shaft, a step input represents the sudden rotation of the shaft. The mathematical representation of a step function is

$$r(t) = R \qquad t \geq 0$$
$$= 0 \qquad t < 0 \qquad (7\text{-}3)$$

where R is a real constant. Or

$$r(t) = Ru_s(t) \qquad (7\text{-}4)$$

where $u_s(t)$ is the unit-step function. The step function as a function of time is shown in Fig. 7-1(a).

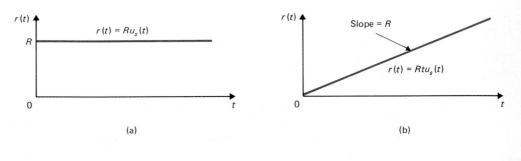

(a) (b)

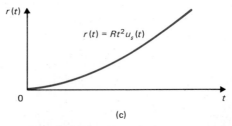

(c)

FIGURE 7-1 Basic time-domain test signals for control systems. (a) Step function. (b) Ramp function. (c) Parabolic function.

ramp input

RAMP-FUNCTION INPUT.   In the case of the ramp function, the signal has a constant change in value with respect to time. Mathematically, a ramp function is represented by

$$r(t) = Rtu_s(t) \qquad\qquad (7\text{-}5)$$

where R is a real constant. The ramp function is shown in Fig. 7-1(b). If the input variable represents the angular displacement of a shaft, the ramp input denotes the constant-speed rotation of the shaft.

parabolic input

PARABOLIC-FUNCTION INPUT.   The parabolic function defined in what follows represents still one order of magnitude faster than the ramp function.

$$r(t) = Rt^2 u_s(t) \qquad\qquad (7\text{-}6)$$

where R is a real constant. The graphical representation of the parabolic function is shown in Fig. 7-1(c).

These test signals all have the common feature that they are simple to describe mathematically. From the step function to the parabolic function, they become progressively faster with respect to time.

The step function is very useful as a test signal since its initial instantaneous jump in amplitude reveals a great deal about a system's quickness in responding to inputs with abrupt changes. Also, since the step function contains, in principle, a wide band of frequencies in its spectrum, as a result of the jump discontinuity, it is equivalent to the application of numerous sinusoidal signals with a wide range of frequencies.

sinusoidal signals

The ramp function has the ability to test how the system would respond to a signal that changes linearly with time. A parabolic function is still one degree faster than the ramp. If necessary, inputs with speeds faster than the ramp function can be defined. However, in practice, we seldom find it feasible to use a test signal faster than a parabolic function. This is because, as we shall show later, in order to track or follow a high-order input accurately, the system must have high-order integrations in the loop, which may cause serious stability problems.

7-3

TIME-DOMAIN PERFORMANCE OF CONTINUOUS-DATA CONTROL SYSTEMS: THE STEADY-STATE ERROR

Steady-state error is a measure of system accuracy when a specific type of input is applied to a control system, and the objective of the system is for the output response to follow this input accurately. In a real system, because of friction and other imperfections, and the composition of the system, the steady state of the output response seldom agrees exactly with the reference input. Therefore, steady-state errors in control systems are almost unavoidable. In a design problem, one of the objectives is to keep the error to a minimum, or below a certain tolerable value.

In practice, the type of error and the relative tolerance of errors found in control systems could vary over a wide range. For instance, in a velocity-control system, the steady-state value of the difference between the actual velocity and the desired velocity of the system is an error in velocity. Numerous control systems are devised for the purpose of controlling position. In this case, the difference between the desired position and the actual controlled position of the output is a position error.

accuracy
requirement

The accuracy requirement on control systems depends to a great extent on the control objectives of the system. For instance, if the controlled variable is the position of an elevator, then the steady-state error can be tolerable if it is kept under a fraction of an inch. On the other hand, the error requirements on certain control systems can be extremely stringent. For instance, the pointing accuracy on the control of the Large Space Telescope (LST), which is a telescope mounted on board a space shuttle, must be kept in terms of microradians.

Large Space
Telescope

Steady-State Error Caused by Nonlinear Systems Elements

In many instances, steady-state errors of control systems are attributed to some nonlinear system characteristics such as nonlinear friction or dead zone. For instance, if an amplifier used in a control system has the input–output characteristics shown in Fig. 7-2, then, when the amplitude of the amplifier input signal falls within the dead zone, the output of the amplifier would be zero, and the control would not be able to correct the error if any exists. Dead-zone nonlinearity characteristics shown in Fig. 7-2 are not limited to amplifiers. The flux-to-current relation of the magnetic field of an electric motor may exhibit a similar characteristic, as shown in Fig. 7-2. Thus, as the current of the motor falls below the dead zone D, no magnetic flux, and, thus, no torque will be produced by the motor. On the other hand, when the input-signal magnitude exceeds a certain level, the amplifier may be subject to saturation, so that the output voltage no longer increases with the input, as shown by the saturation characteristics in Fig. 7-2. Similarly, when the magnetic field of the motor is saturated, increasing the armature current will no longer produce additional torque.

The output signals of digital components used in control systems, such as a microprocessor, can take on only discrete or quantized levels. This property is illustrated

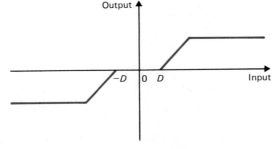

FIGURE 7-2 Typical input-output characteristics of an amplifier with dead zone and saturation.

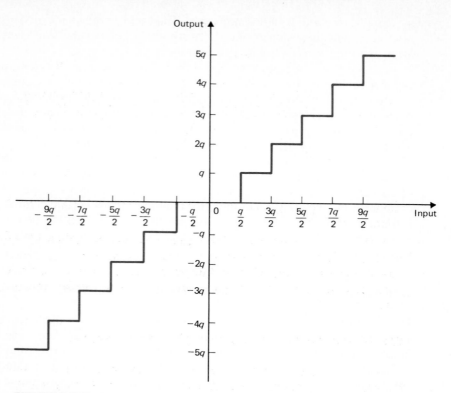

FIGURE 7-3 Typical input-output characteristics of a quantizer.

quantization error

by the quantization characteristics shown in Fig. 7-3. When the input to the quantizer is within $\pm q/2$, the output is zero, and the system may genrate an error in the output whose magnitude is related to $\pm q/2$. This type of error is also known as the **quantization error** of digital control systems.

When the control of physical objects is involved, friction is almost always unavoidable. Coulomb friction is a common cause of steady-state position errors in control systems. Figure 7-4 shows a restoring-torque-versus-position curve of a control system. The torque curve typically could be generated by a step motor or a switched-reluctance motor, or from a closed-loop system with a position encoder. The point 0 designates a stable equilibrium point on the torque curve, as well as the other periodic points along the axis where the slope on the torque curve is negative. The torque on either side of the point 0 represents a restoring torque that tends to return the output to

error due to friction

the equilibrium point when some angular-displacement disturbance takes place. When there is no friction, the position error should be zero, since there is always a restoring torque so long as the position is not at the stable equilibrium point. If the rotor of the motor sees a Coulomb friction torque T_F, then the motor torque must first overcome this frictional torque before producing any motion. Thus, as the motor torque falls below T_F as the rotor position approaches the stable equilibrium point, it may stop at any position inside the shaded band shown in Fig. 7-4, and the error band is bounded by $\pm\theta_e$.

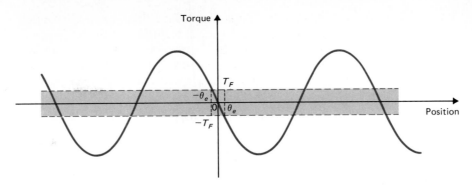

FIGURE 7-4 Torque-angle curve of a motor or closed-loop system with Coulomb friction.

Although it is relatively simple to comprehend the effects of nonlinearities on errors, and to establish maximum upper bounds on the error magnitudes, it is difficult to establish general or closed-form solutions for nonlinear systems. Usually, exact and detailed analysis of errors in nonlinear control systems can be carried out only by computer simulations.

Steady-State Error of Linear Control Systems

Linear control systems are subject to steady-state errors for somewhat different causes than nonlinear systems, although the reason is still that the system no longer "sees" the error, and no corrective effort is exerted. In general, the steady-state errors of linear control systems depend on the types of input and system.

DEFINITION OF THE STEADY-STATE ERROR WITH RESPECT TO SYSTEM CONFIGURA-TION.   In a control system, if the reference input $r(t)$ and the controlled output $c(t)$ are of the same dimension, for example, an input voltage controlling a voltage, a position input controlling a position, and these signals are at the same level, the error is simply defined as

$$e(t) = r(t) - c(t) \qquad (7\text{-}7)$$

Sometimes it may be impossible or inconvenient to provide a reference input that is at the same level or of the same dimension as the controlled variable. For instance, it may be necessary to use a low-voltage source for the control of the output of a high-voltage power source. For a velocity-control system, it is more practical to use a voltage source or position input to control the velocity of the output shaft. Under these

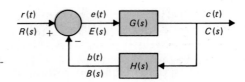

FIGURE 7-5 Nonunity-feedback control system.

conditions, the error signal cannot be defined simply as the difference between the reference input and the controlled output, as in Eq. (7-7). The input and output signals must be of the same dimension and at the same level before comparison. Therefore, a nonunity element, $H(s)$, must be incorporated in the feedback path, as shown in Fig. 7-5, for conformity of $r(t)$ and $c(t)$. The error of this nonunity-feedback control system is defined as

$$e(t) = r(t) - b(t) \tag{7-8}$$

or

$$E(s) = R(s) - B(s) = R(s) - H(s)C(s) \tag{7-9}$$

EXAMPLE 7-1   If a 10-volt reference is used to regulate a 100-volt voltage supply, $H(s)$ is a constant and is equal to 0.1. When the output voltage is exactly 100 volts, the error signal is

$$e(t) = 10 - 0.1 \times 100 = 0 \tag{7-10}$$

EXAMPLE 7-2   As another example, consider that the system in Fig. 7-5 is a velocity-control system in which $r(t)$ is used as a reference to control the output velocity of the system. Let $c(t)$ denote the output displacement. Then, we need a device such as a tachometer in the feedback path, so that $H(s) = K_t s$. Thus, the error is defined as

$$e(t) = r(t) - b(t)$$
$$= r(t) - K_t \frac{dc(t)}{dt} \tag{7-11}$$

The error becomes zero when the output velocity $dc(t)/dt$ is equal to $r(t)/K_t$.

By using the system notation defined before, the steady-state error of a feedback control system is defined as

$$\text{steady-state error} = e_{ss} = \lim_{t \to \infty} e(t) \tag{7-12}$$

With reference to Fig. 7-5, the Laplace transform of the error signal is

$$E(s) = \frac{R(s)}{1 + G(s)H(s)} \tag{7-13}$$

final-value theorem   By use of the final-value theorem of the Laplace transform, the steady-state error of the system is

$$e_{ss} = \lim_{t \to \infty} e(t) = \lim_{s \to 0} sE(s) \tag{7-14}$$

where $sE(s)$ must not have any poles on the imaginary axis and in the right-half of the s-plane. Substituting Eq. (7-13) into Eq. (7-14), we have

$$e_{ss} = \lim_{s \to 0} \frac{sR(s)}{1 + G(s)H(s)} \tag{7-15}$$

steady-state error

which shows that the steady-state error depends on the reference input $R(s)$ and the loop transfer function $G(s)H(s)$.

We should point out that although the error function is defined with reference to the system configuration shown in Fig. 7-5, in general, the signal at any point in a system may be referred to as the error. For instance, when a system is subject to a disturbance input, then the output due to this disturbance acting alone can be regarded as an error.

type of system

THE TYPE OF CONTROL SYSTEMS.   If the control system can be represented by or simplified to the block diagram in Fig. 7-5, and the steady-state error is given by Eq. (7-15), the latter depends on the type of the control system. Let us first establish the **type** of control system by referring to the form of the loop transfer function $G(s)H(s)$. In general, $G(s)H(s)$ can be expressed as

$$G(s)H(s) = \frac{K(1 + T_1 s)(1 + T_2 s) \cdots (1 + T_m s)}{s^j (1 + T_a s)(1 + T_b s) \cdots (1 + T_n s)} e^{-T_d s} \tag{7-16}$$

where K and all the T's are real constants. The **type** of the closed-loop system refers to the **order** of the pole of $G(s)H(s)$ at $s = 0$. Therefore, the closed-loop system that has the loop transfer function of Eq. (7-16) is type j, where $j = 0, 1, 2, \ldots$. The total number of terms in the numerator and the denominator and the values of the coefficients are not important to the system type. The following example illustrates the system types with reference to the form of $G(s)H(s)$.

$G(s)H(s) = $ loop transfer function.

EXAMPLE 7-3

$$G(s)H(s) = \frac{K(1 + 0.5s)}{s(1 + s)(1 + 2s)} \quad \text{type 1} \tag{7-17}$$

$$G(s)H(s) = \frac{K}{s^3} \quad \text{type 3} \tag{7-18}$$

Now let us investigate the effects of the types of inputs on the steady-state error. We shall consider only the step, ramp, and parabolic inputs.

STEADY-STATE ERROR OF SYSTEM WITH A STEP-FUNCTION INPUT.   When the reference input to the control system of Fig. 7-5 is a step function with magnitude R, the Laplace transform of the input is R/s. Equation (7-15) becomes

$$e_{ss} = \lim_{s \to 0} \frac{sR(s)}{1 + G(s)H(s)} = \lim_{s \to 0} \frac{R}{1 + G(s)H(s)} = \frac{R}{1 + \lim_{s \to 0} G(s)H(s)} \qquad (7\text{-}19)$$

For convenience, we define

$$K_p = \lim_{s \to 0} G(s)H(s) \qquad (7\text{-}20)$$

*step-error
constant*

where K_p is the **step-error constant.** Then Eq. (7-19) is written

$$e_{ss} = \frac{R}{1 + K_p} \qquad (7\text{-}21)$$

A typical e_{ss} due to a step input when K_p is finite and nonzero is shown in Fig. 7-6. We see that for e_{ss} to be zero, when the input is a step function, K_p must be infinite. If $G(s)H(s)$ is described by Eq. (7-16), we see that for K_p to be infinite, j must be at least equal to unity; that is, $G(s)H(s)$ must have at least one pure integration. Therefore, we can summarize the steady-state error due to a step function input as follows:

Type 0 system: $e_{ss} = \dfrac{R}{1 + K_p} = \text{constant}$ (7-22)

Type 1 or higher system: $e_{ss} = 0$

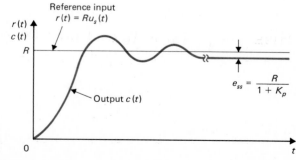

FIGURE 7-6 Typical steady-state error due to a step input.

STEADY-STATE ERROR OF SYSTEM WITH A RAMP-FUNCTION INPUT.   When the input to the control system of Fig. 7-5 is

$$r(t) = Rtu_s(t) \tag{7-23}$$

where R is a real constant, the Laplace transform of $r(t)$ is

$$R(s) = \frac{R}{s^2} \tag{7-24}$$

Substituting Eq. (7-24) into Eq. (7-15), we have

$$e_{ss} = \lim_{s \to 0} \frac{R}{s + sG(s)H(s)} = \frac{R}{\lim_{s \to 0} sG(s)H(s)} \tag{7-25}$$

ramp-error constant   We define the **ramp-error constant** as

$$K_v = \lim_{s \to 0} sG(s)H(s) \tag{7-26}$$

Then, Eq. (7-25) becomes

$$e_{ss} = \frac{R}{K_v} \tag{7-27}$$

which is the steady-state error when the input is a ramp function. A typical e_{ss} due to a ramp input when K_v is finite and nonzero is illustrated in Fig. 7-7.

Equation (7-27) shows that for e_{ss} to be zero when the input is a ramp function, K_v must be infinite. Using Eqs. (7-16) and (7-26), we obtain

$$K_v = \lim_{s \to 0} sG(s)H(s) = \lim_{s \to 0} \frac{K}{s^{j-1}} \quad j = 0, 1, 2, \ldots \tag{7-28}$$

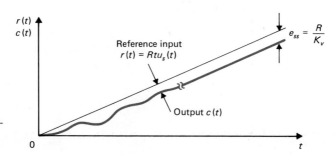

FIGURE 7-7 Typical steady-state error due to a ramp input.

Therefore, in order for K_v to be infinite, j must be at least equal to 2, or the system must be of type 2 or higher. The following conclusions may be stated with regard to the steady-state error of a system with ramp input:

Type 0 system:	$e_{ss} = \infty$
Type 1 system:	$e_{ss} = \dfrac{R}{K_v}$
Type 2 or higher system:	$e_{ss} = 0$

STEADY-STATE ERROR OF SYSTEM WITH A PARABOLIC INPUT.   When the input is described by

$$r(t) = \frac{Rt^2}{2} u_s(t) \qquad (7\text{-}29)$$

the Laplace transform of $r(t)$ is

$$R(s) = \frac{R}{s^3} \qquad (7\text{-}30)$$

The steady-state error of the system in Fig. 7-5 is

$$e_{ss} = \frac{R}{\lim_{s \to 0} s^2 G(s)H(s)} \qquad (7\text{-}31)$$

A typical e_{ss} of a system with a nonzero and finite K_a due to a parabolic input is shown in Fig. 7-8.

parabolic-error constant   Defining the **parabolic-error constant** as

$$K_a = \lim_{s \to 0} s^2 G(s)H(s) \qquad (7\text{-}32)$$

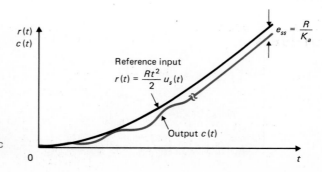

FIGURE 7-8 Typical steady-state error due to a parabolic input.

the steady-state error becomes

$$e_{ss} = \frac{R}{K_a} \tag{7-33}$$

The following conclusions are made with regard to the steady-state error of a system with parabolic input:

Type 0 system: $e_{ss} = \infty$

Type 1 system: $e_{ss} = \infty$

Type 2 system: $e_{ss} = \dfrac{R}{K_a} = \text{constant}$

Type 3 or higher system: $e_{ss} = 0$

By using the method described, the steady-state error of any linear system subject to an input with order higher than the parabolic function can also be derived if necessary. As a summary of the error analysis, Table 7-1 shows the relations among the error constants, the types of systems with reference to Eq. (7-16), and the input types.

TABLE 7-1   Summary of the Steady-State Errors Due to Step, Ramp, and Parabolic Inputs

TYPE OF SYSTEM j	K_p	K_v	K_a	STEP INPUT $e_{ss} = \dfrac{R}{1 + K_p}$	RAMP INPUT $e_{ss} = \dfrac{R}{K_v}$	PARABOLIC INPUT $e_{ss} = \dfrac{R}{K_a}$
0	K	0	0	$e_{ss} = \dfrac{R}{1 + K}$	$e_{ss} = \infty$	$e_{ss} = \infty$
1	∞	K	0	$e_{ss} = 0$	$e_{ss} = \dfrac{R}{K}$	$e_{ss} = \infty$
2	∞	∞	K	$e_{ss} = 0$	$e_{ss} = 0$	$e_{ss} = \dfrac{R}{K}$
3	∞	∞	∞	$e_{ss} = 0$	$e_{ss} = 0$	$e_{ss} = 0$

As a summary, the following points should be noted when applying the error-constant analysis.

1. The step-, ramp-, and parabolic-error constants are significant for the error analysis only when the input signal is a step function, ramp function, and parabolic function, respectively.
2. Since the error constants are defined with respect to the loop transfer function $G(s)H(s)$, strictly, the method is applicable to only the system configuration shown in Fig. 7-5. Since the error analysis relies on

Error constants e defined only for the case when there is a loop transfer function.

summary

the use of the final-value theorem of Laplace transforms, it is important to first check to see if $sE(s)$ has any poles on the $j\omega$ axis or in the
right half of the s-plane.

3. The steady-state error of a system with an input that is a linear combi-
    nation of the three basic types of inputs can be determined simply by
    superimposing the errors due to each input component.

4. When the system configuration differs from that of Fig. 7-5, we can
    simply establish the error signal and apply the final-value theorem.
    The error constants may or may not apply, depending on the individ-
    ual situation.

One of the disadvantages of the error-constant method occurs when the steady-
state error is infinite, which actually, is due to the error increases with time; the error-
constant method does not indicate how the error varies with time. The error-constant
error series method also does not apply to systems with inputs that are sinusoidal, since the final-
value theorem cannot be applied to these cases. We shall present the **error series** in the
following section, which gives a more versatile representation of the steady-state error.

The Error Series

In this section, the error-constant concept is generalized to include inputs of almost any
arbitrary function of time. We start with the transformed error function $E(s)$, which
could be of the form of Eq. (7-13), or any function that is defined as the system error.
Let us expresses $E(s)$ as

$$E(s) = W_e(s)R(s) \tag{7-34}$$

error transfer function where $R(s)$ is the Laplace transform of the input $r(t)$, and $W_e(s)$ is the **error transfer function**. By using the real convolution integral discussed in Section 3-2, the error signal $e(t)$ is

$$e(t) = \int_{-\infty}^{t} w_e(\tau)r(t - \tau)\, d\tau \tag{7-35}$$

where $w_e(\tau)$ is the inverse Laplace transform of $W_e(s)$.

Taylor series If the first n derivatives of $r(t)$ exist for all values of t, the function $r(t - \tau)$ can
be expanded into a Taylor series; that is,

$$r(t - \tau) = r(t) - \tau r^{(1)}(t) + \frac{\tau^2}{2!}r^{(2)}(t) - \frac{\tau^3}{3!}r^{(3)}(t) + \cdots \tag{7-36}$$

where $r^{(i)}(t)$ denotes the ith-order derivative of $r(t)$ with respect to time.

Since $r(t)$ is considered to be zero for negative time, the limits of the convolution
integral in Eq. (7-35) may be taken from 0 to t. Substituting Eq. (7-36) into Eq.
(7-35), we have

$$e(t) = \int_0^t w_e(\tau)\left[r(t) - \tau r^{(1)}(t) + \frac{\tau^2}{2!}r^{(2)}(t) - \frac{\tau^3}{3!}r^{(3)}(t) + \cdots \right] d\tau$$

$$= r(t)\int_0^t w_e(\tau)\, d\tau - r^{(1)}(t)\int_0^t \tau w_e(\tau)\, d\tau + r^{(2)}(t)\int_0^t \frac{\tau^2}{2!}w_e(\tau)\, d\tau - \cdots \quad (7\text{-}37)$$

Let $r_s(t)$ represent the steady-state part of the input $r(t)$, and $e_s(t)$ be the steady-state part of $e(t)$ that is due to $r_s(t)$. Then, the steady-state error function, $e_s(t)$, may be written as

$$e_s(t) = r_s(t)\int_0^\infty w_e(\tau)\, d\tau - r_s^{(1)}(t)\int_0^\infty \tau w_e(\tau)\, d\tau + r_s^{(2)}(t)\int_0^\infty \frac{\tau^2}{2!}w_e(\tau)\, d\tau - \cdots \quad (7\text{-}38)$$

Let us define

$$C_0 = \int_0^\infty w_e(\tau)\, d\tau$$

$$C_1 = -\int_0^\infty \tau w_e(\tau)\, d\tau$$

$$C_2 = \int_0^\infty \tau^2 w_e(\tau)\, d\tau \qquad\qquad (7\text{-}39)$$

$$\vdots$$

$$C_n = (-1)^n \int_0^\infty \tau^n w_e(\tau)\, d\tau$$

Equation (7-38) is written

$$e_s(t) = C_0 r_s(t) + C_1 r_s^{(1)}(t) + \frac{C_2}{2!}r_s^{(2)}(t) + \cdots + \frac{C_n}{n!}r_s^{(n)}(t) + \cdots \quad (7\text{-}40)$$

error coefficients

which is called the **error series**, and the coefficients $C_0, C_1, C_2, \ldots, C_n$ are called the **generalized error coefficients**, or simply the **error coefficients**.

The error coefficients may be readily evaluated from the error transfer function $W_e(s)$. Since $W_e(s)$ and $w_e(\tau)$ are related through the Laplace transform, we have

$$W_e(s) = \int_0^\infty w_e(\tau)e^{-\tau s}\, d\tau \qquad\qquad (7\text{-}41)$$

Taking the limit on both sides of Eq. (7-41) as s approaches zero, we have

$$\lim_{s\to 0} W_e(s) = \lim_{s\to 0}\int_0^\infty w_e(\tau)e^{-\tau s}\, d\tau \qquad\qquad (7\text{-}42)$$

Taking the derivative of $W_e(s)$ of Eq. (7-41) with respect to s, we get

$$\frac{dW_e(s)}{ds} = -\int_0^\infty \tau w_e(\tau) e^{-\tau s} \, d\tau \tag{7-43}$$

from which the error coefficient C_1 is determined as

$$C_1 = \lim_{s \to 0} \frac{dW_e(s)}{ds} \tag{7-44}$$

The rest of the error coefficients are obtained in a similar fashion by taking successive differentiations of Eq. (7-44) with respect to s. Thus,

$$C_2 = \lim_{s \to 0} \frac{d^2 W_e(s)}{ds^2} \tag{7-45}$$

$$C_3 = \lim_{s \to 0} \frac{d^3 W_e(s)}{ds^3} \tag{7-46}$$

$$\vdots$$

$$C_n = \lim_{s \to 0} \frac{d^n W_e(s)}{ds^n} \tag{7-47}$$

The following examples illustrate the application of the error series and its advantages over the error-constant method.

EXAMPLE 7-4   In this illustrative example, the steady-state error of a feedback control system will be evaluated by use of the error-constant and the error-series methods. Consider a unity-feedback control system with the open-loop transfer function given as

$$G(s) = \frac{K}{s + 1} \tag{7-48}$$

Since the system is of type 0, the error constants are $K_p = K$, $K_v = 0$, and $K_a = 0$. Thus, the steady-state errors of the system due to the three basic types of inputs are as follows:

Unit-step input $u_s(t)$: $e_{ss} = \dfrac{1}{1 + K}$

Unit-ramp input $tu_s(t)$: $e_{ss} = \infty$

Unit-parabolic input $t^2 u_s(t)$: $e_{ss} = \infty$

Notice that when the input is faster than a step function, such as either a ramp or a parabolic function, the steady-state error is infinite, since it apparently increases with time. Although the error-constant method gives the correct answers in this case, it fails to indicate the exact manner in which the steady-state error increases with time. Therefore, if more detailed information on the steady-state behavior of the system is desired, the differential equation of the system must be solved. We now show that the steady-state response of the system can actually be determined from the error series.

The error transfer function of the system described is

$$W_e(s) = \frac{E(s)}{R(s)} = \frac{1}{1 + G(s)} = \frac{s + 1}{s + K + 1} \tag{7-49}$$

The error coefficients of the system are

$$C_0 = \lim_{s \to 0} W_e(s) = \frac{1}{1 + K} \tag{7-50}$$

$$C_1 = \lim_{s \to 0} \frac{dW_e(s)}{ds} = \frac{K}{(1 + K)^2} \tag{7-51}$$

$$C_2 = \lim_{s \to 0} \frac{d^2 W_e(s)}{ds^2} = \frac{-2K}{(1 + K)^3} \tag{7-52}$$

Similarly, higher-order coefficients can be obtained by taking the derivatives of $W_e(s)$ continuously. These would be needed only if the corresponding derivatives of $r_s(t)$ are nonzero.

The error series for the system given is

$$e_s(t) = \frac{1}{1 + K} r_s(t) + \frac{K}{(1 + K)^2} r_s^{(1)}(t) + \frac{-K}{(1 + K)^3} r_s^{(2)}(t) + \cdots \tag{7-53}$$

Now let us consider the three basic types of inputs.

1. *Unit-step input.* $r(t) = r_s(t) = u_s(t)$. All the derivatives of $r_s(t)$ are zero. The error series in Eq. (7-53) becomes

$$e_s(t) = \frac{1}{1 + K} u_s(t) \tag{7-54}$$

which agrees with the result found by the error-constant method.

2. *Unit-ramp input.* $r(t) = r_s(t) = t u_s(t)$. Then, $r_s^{(1)}(t) = u_s(t)$, and all higher-order derivatives of $r_s(t)$ are zero. The error series in Eq. (7-53) becomes

$$e_s(t) = \left[\frac{1}{1 + K} t + \frac{K}{(1 + K)^2} \right] u_s(t) \tag{7-55}$$

which indicates that the steady-state error increases linearly with time. The error-constant method yields only the result that the steady-state error is infinite, but fails to give details of the time dependence of $e_s(t)$.

3. *Parabolic input.* $r(t) = r_s(t) = (t^2/2)u_s(t)$. Then, $r_s^{(1)}(t) = tu_s(t)$, $r_s^{(2)}(t) = u_s(t)$, and all higher-order derivatives of $r_s(t)$ are zero. The error series in Eq. (7-53) becomes

$$e_s(t) = \left[\frac{1}{2(1 + K)}t^2 + \frac{K}{(1 + K)^2}t - \frac{K}{(1 + K)^3}\right]u_s(t) \qquad (7\text{-}56)$$

In this case, the steady-state error increases as the second power of t.

4. Consider that the input signal is represented by a polynomial of t and an exponential term:

$$r(t) = \left(a_0 + a_1 t + \frac{a_2 t^2}{2} + e^{-a_3 t}\right)u_s(t) \qquad (7\text{-}57)$$

where a_0, a_1, a_2, and a_3 are real constants. Then, the steady-state portion of $r(t)$ is

$$r_s(t) = \left(a_0 + a_1 t + \frac{a_2 t^2}{2}\right)u_s(t) \qquad (7\text{-}58)$$

In this case, the error series is

$$e_s(t) = \frac{1}{1 + K}r_s(t) + \frac{K}{(1 + K)^2}r_s^{(1)}(t) - \frac{K}{(1 + K)^3}r_s^{(2)}(t) \qquad (7\text{-}59)$$

where $r_s^{(1)}(t)$ and $r_s^{(2)}(t)$ are easily determined from Eq. (7-58).

EXAMPLE 7-5

In this example, we shall deal with a situation in which the error-constant method is totally useless in providing a solution to the steady-state error. Consider that the input to the system described in Example 7-4 is a sinusoid:

$$r(t) = \sin \omega_0 t \qquad (7\text{-}60)$$

where $\omega_0 = 2$ rad/s. Then,

$$r_s(t) = \sin \omega_0 t$$
$$r_s^{(1)}(t) = \omega_0 \cos \omega_0 t$$
$$r_s^{(2)}(t) = -\omega_0^2 \sin \omega_0 t$$
$$r_s^{(3)}(t) = -\omega_0^3 \cos \omega_0 t \qquad (7\text{-}61)$$

The error series in Eq. (7-40) becomes

$$e_s(t) = \left(C_0 - \frac{C_2}{2!}\omega_0^2 + \frac{C_4}{4!}\omega_0^4 - \cdots\right)\sin \omega_0 t + \left(C_1\omega_0 - \frac{C_3}{3!}\omega_0^3 + \cdots\right)\cos \omega_0 t \qquad (7\text{-}62)$$

sinusoidal input

Because of the sinusoidal input, the error series is now an infinite series. For $K = 100$, the error coefficients become

$$C_0 = \frac{1}{1 + K} = 0.0099$$

$$C_1 = \frac{K}{(1 + K)^2} = 0.0098$$

$$C_2 = -\frac{2K}{(1 + K)^3} = -0.000194$$

$$C_3 = \frac{6K}{(1 + K)^4} = 5.766 \times 10^{-6}$$

Since the magnitude of C_i decreases rapidly as i increases, we can approximate the error series using only the first four error coefficients, C_0, C_1, C_2 and C_3. Equation (7-62) becomes

$$e_s(t) \cong 0.01029 \sin 2t + 0.0196 \cos 2t$$

$$= 0.02214 \sin (2t + 62.3°) \tag{7-63}$$

The steady-state error in this case is also a sinusoid with an amplitude of 0.02214 and a phase of 62.3 degrees, relative to the input sinusoid of unity magnitude and zero phase. The fact that Eq. (7-63) gives the steady-state solution of the error $e(t)$, which is $r(t) - c(t)$, can be verified by use of the conventional sinusoidal steady-state analysis. Setting $s = j\omega$ in Eq. (7-49), we have

$$W_e(j\omega) = \frac{E(j\omega)}{R(j\omega)} = \frac{j\omega + 1}{j\omega + K + 1} \tag{7-64}$$

With $\omega = \omega_0 = 2$ rad/s and $K = 100$, the last equation gives

$$\frac{E(j\omega)}{R(j\omega)} = \frac{1 + j2}{101 + j2}$$

$$= 0.02214 \angle 62.3° \tag{7-65}$$

which describes the sinusoidal steady-state properties of the error signal relative to the sinusoidal input.

7-4

TIME-DOMAIN PERFORMANCE OF CONTINUOUS-DATA CONTROL SYSTEMS: TRANSIENT RESPONSE

transient response

As defined earlier, the transient portion of the time response is that part which goes to zero as time becomes large. Nevertheless, the transient response of a control system is necessarily important, since both the amplitude and the time duration of the transient response must be kept within tolerable or prescribed limits. For example, in the automobile idle-speed control system described in Chapter 1, in addition to striving for a desirable idle speed in the steady-state, the transient droop in engine speed must not be excessive, and the recovery in speed should be made as quickly as possible.

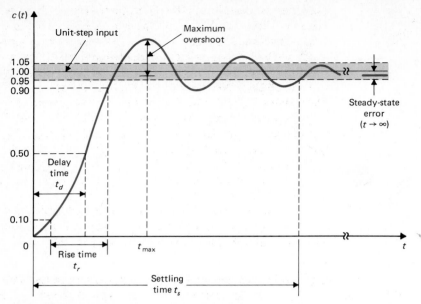

FIGURE 7-9 Typical unit-step response of a control system.

The Unit-Step Response and Time-Domain Specifications.

unit-step response

For linear control systems, the characterization of the transient response is often done by use of the unit-step function as an input. The response of a control system when the input is a unit-step function is called the **unit-step response**. Figure 7-9 illustrates a typical unit-step response of a linear control system. With reference to the unit-step response, performance criteria commonly used for the characterization of linear control systems in the time domain are defined as follows.

1. *Maximum Overshoot.*   Let $c(t)$ be the unit-step response. Let c_{max} denote the maximum value of $c(t)$, and c_{ss} be the steady-state value of $c(t)$, and $c_{max} \geq c_{ss}$. The maximum overshoot of $c(t)$ is defined as

maximum overshoot

$$\text{maximum overshoot} = c_{max} - c_{ss} \qquad (7\text{-}66)$$

The maximum overshoot is often represented as a percentage of the final value of the step response; that is,

$$\text{percent maximum overshoot} = \frac{\text{maximum overshoot}}{c_{ss}} \times 100\% \qquad (7\text{-}67)$$

The maximum overshoot is often used as a measure of the relative stability of a control system. A system with a large overshoot is usually undesirable. For design purposes, the maximum overshoot is of-

[handwritten margin notes: or some systems overshoot may occur at a later time. negative undershoot may occur ø if transfer function has odd number of poles in the ove g-plane.]

ten given as a time-domain specification. The unit-step response illustrated in Fig. 7-9 shows that the maximum overshoot occurs at the first overshoot. It should be noted that for some systems, the maximum overshoot may occur at a later peak, and if the system transfer function has an odd number of zeros in the right-half s-plane, a negative undershoot may occur [3, 4] (Problem 7-21).

[handwritten margin: delay time]

2. *Delay Time.*   The delay time t_d is defined as the time required for the step response to reach 50 percent of its final value (see Fig. 7-9).

[handwritten margin: rise time]

3. *Rise Time.*   The rise time t_r is defined as the time required for the step response to rise from 10 to 90 percent of its final value (see Fig. 7-9). An alternative measure is to represent the rise time as the reciprocal of the slope of the step response at the instant that the response is equal to 50 percent of its final value.

settling time

4. *Settling Time.*   The settling time t_s is defined as the time required for the step response to decrease and stay within a specified percentage of its final value. A frequently used figure is 5 percent.

The four quantities just defined give a direct measure of the transient characteristics of a control system in terms of the unit-step response. These time-domain specifications are relatively easy to measure when the step response is well defined, as shown in Fig. 7-9. Analytically, these quantities are difficult to establish, except for simple systems lower than the third order.

[handwritten: these quantities are difficult to establish except for systems low order]

7-5

TRANSIENT RESPONSE OF A PROTOTYPE SECOND-ORDER SYSTEM

Although true second-order control systems are rare in practice, their analysis generally helps to form a basis for the understanding of analysis and design of higher-order systems.

Consider that a second-order control system is represented by the block diagram shown in Fig. 7-10. The open-loop transfer function of the system is

$$G(s) = \frac{C(s)}{E(s)} = \frac{\omega_n^2}{s(s + 2\zeta\omega_n)} \tag{7-68}$$

where ζ and ω_n are real constants. The closed-loop transfer function of the system is

$$\frac{C(s)}{R(s)} = \frac{\omega_n^2}{s^2 + 2\zeta\omega_n s + \omega_n^2} \tag{7-69}$$

prototype second-order system

The system in Fig. 7-10 with the transfer functions given by Eqs. (7-68) and (7-69) is defined as the **prototype second-order system**.

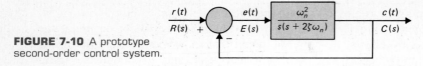

FIGURE 7-10 A prototype second-order control system.

The characteristic equation of the closed-loop system is obtained by setting the denominator of Eq. (7-69) to zero:

$$\Delta(s) = s^2 + 2\zeta\omega_n s + \omega_n^2 = 0 \qquad (7\text{-}70)$$

For a unit-step function input, $R(s) = 1/s$, the output response of the system is obtained by taking the inverse Laplace transform of

$$C(s) = \frac{\omega_n^2}{s(s^2 + 2\zeta\omega_n s + \omega_n^2)} \qquad (7\text{-}71)$$

This can be done by referring to the Laplace-transform table in Appendix B. The result is

$$c(t) = 1 - \frac{e^{-\zeta\omega_n t}}{\sqrt{1 - \zeta^2}} \sin\left(\omega_n\sqrt{1 - \zeta^2}\, t + \cos^{-1}\zeta\right) \qquad t \geq 0 \qquad (7\text{-}72)$$

The Damping Ratio and the Damping Factor

damping ratio

It is of interest to study the relationship between the roots of the characteristic equation and the behavior of the step response $c(t)$. The two roots of Eq. (7-70) are

$$s_1, s_2 = -\zeta\omega_n \pm j\omega_n\sqrt{1 - \zeta^2}$$
$$= -\alpha \pm j\omega \qquad (7\text{-}73)$$

$\omega_n \zeta$ damping factor or damping constant.

The physical significance of the constants ζ and α is now described. As seen from Eq. (7-73), $\alpha = \zeta\omega_n$, and α appears as the constant that is multiplied to t in the exponential term of Eq. (7-72). Therefore, α controls the rate of rise or decay of the unit-step response $c(t)$. In other words, α controls the "damping" of the system, and is called the **damping factor**, or the **damping constant**. The inverse of α, $1/\alpha$, is proportional to the time constant of the system.

damping factor

critical damping

When the two roots of the characteristic equation are real and equal, we call the system **critically damped**. From Eq. (7-73), we see that critical damping occurs when $\zeta = 1$. Under this condition, the damping factor is simply $\alpha = \omega_n$. Therefore, we can regard ζ as the **damping ratio**, that is,

$$\zeta = \text{damping ratio} = \frac{\alpha}{\omega_n} = \frac{\text{actual damping factor}}{\text{damping factor at critical damping}} \qquad (7\text{-}74)$$

The Natural Undamped Frequency

natural undamped frequency

The parameter ω_n is defined as the **natural undamped frequency.** As seen from Eq. (7-73), when the damping is zero, $\zeta = 0$, the roots of the characteristic equation are imaginary, and Eq. (7-72) shows that the unit-step response is purely sinusoidal. Therefore, ω_n corresponds to the frequency of the undamped sinusoidal response.

Equation (7-73) shows that when $0 < \zeta < 1$, the imaginary parts of the roots have the magnitude of

[handwritten: when there no damping here is no exponential decay term.]

[handwritten: conditional frequency or damped frequency]

$$\omega = \omega_n\sqrt{1 - \zeta^2} \qquad (7\text{-}75)$$

conditional frequency

Since when $\zeta \neq 0$, the response of $c(t)$ is not a periodic function, ω defined in Eq. (7-75) is not a frequency. For the purpose of reference, ω is sometimes defined as the **conditional frequency**, or the **damped frequency**.

Figure 7-11 illustrates the relationships between the location of the characteristic-equation roots and α, ζ, ω_n, and ω. For the complex-conjugate roots shown, ω_n is the radial distance from the roots to the origin of the s-plane. The damping factor α is the real part of the roots; the damped frequency ω is the imaginary part of the roots, and the damping ratio ζ is equal to the cosine of the angle between the radial line to the roots and the negative axis when the roots are in the left-half s-plane, or,

$$\zeta = \cos\theta \qquad (7\text{-}76)$$

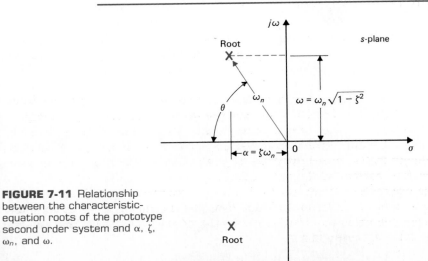

FIGURE 7-11 Relationship between the characteristic-equation roots of the prototype second order system and α, ζ, ω_n, and ω.

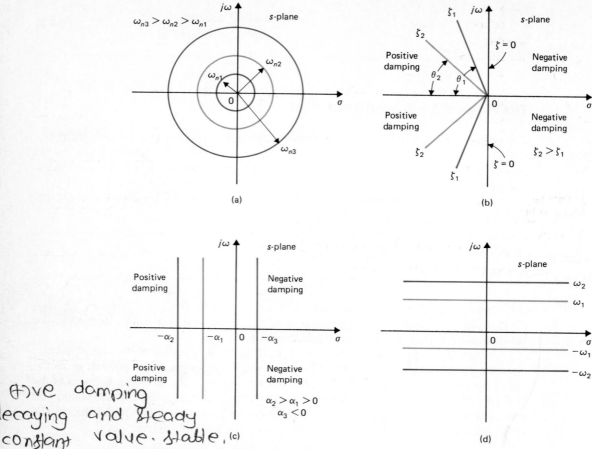

(a)

(b)

(c)

(d)

FIGURE 7-12 (a) Constant-natural-undamped-frequency loci. (b) Constant-damping-ratio loci. (c) Constant-damping-factor loci. (d) Constant-conditional-frequency loci.

(+)ve damping
decaying and steady
constant value. stable.

(-)ve damping
unboundedly increasing
and unstable.

Zero damping
sustained oscill-
ations, margin-
ally
stable.

Figure 7-12 shows (a) the constant-ω_n loci, (b) the constant-ζ loci, (c) the constant-α loci, and (d) the constant-ω loci in the s-plane. Note that the left-half s-plane corresponds to positive damping, i.e., the damping factor or damping ratio is positive, and the right-half s-plane corresponds to negative damping. The imaginary axis corresponds to zero damping ($\alpha = 0$ or $\zeta = 0$). As shown by Eq. (7-72), when the damping is positive, the unit-step response will settle to its constant final value because of the negative exponent of $\exp(-\zeta\omega_n t)$. Negative damping gives rise to a response that grows without bound with time, and the system is unstable. Zero damping results in a sustained sinusoidal oscillation, and the system is marginally stable or unstable. Therefore,

marginal stability

we have demonstrated with the help of the simple prototype second-order system that the location of the characteristic-equation roots plays an important role in the transient response of the system.

The effect of the characteristic-equation roots on the damping of the second-order system is further illustrated by Figs. 7-13 and 7-14. In Fig. 7-13, ω_n is held constant

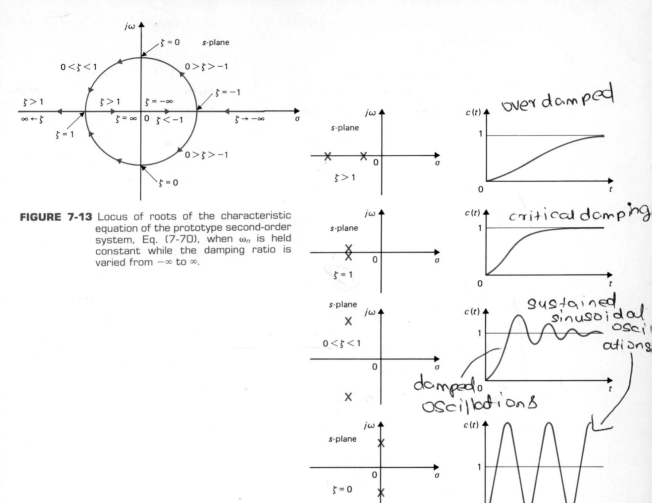

FIGURE 7-13 Locus of roots of the characteristic equation of the prototype second-order system, Eq. (7-70), when ω_n is held constant while the damping ratio is varied from $-\infty$ to ∞.

FIGURE 7-14 Step-response comparison for various characteristic-equation-root locations in the s-plane.

while the damping ratio ζ is varied from $-\infty$ to $+\infty$. The following classification of the system dynamics with respect to the value of ζ is made:

$$
\begin{aligned}
& 0 < \zeta < 1: \quad s_1, s_2 = -\zeta\omega_n \pm j\omega_n\sqrt{1 - \zeta^2} \quad (-\zeta\omega_n < 0) \quad \textit{underdamped} \\
& \zeta = 1: \quad s_1, s_2 = -\omega_n \qquad\qquad\qquad\qquad\qquad\qquad\quad \textit{critically damped} \\
& \zeta > 1: \quad s_1, s_2 = -\zeta\omega_n \pm \omega_n\sqrt{\zeta^2 - 1} \qquad\qquad\quad \textit{overdamped} \\
& \zeta = 0: \quad s_1, s_2 = \pm j\omega_n \qquad\qquad\qquad\qquad\qquad\quad\; \textit{undamped} \\
& \zeta < 0: \quad s_1, s_2 = -\zeta\omega_n \pm j\omega_n\sqrt{1 - \zeta^2} \quad (-\zeta\omega_n > 0) \quad \textit{negatively damped}
\end{aligned}
$$

Figure 7-14 illustrates typical unit-step responses that correspond to the various root locations discussed before.

In practical applications, only stable systems that correspond to $\zeta > 0$ are of interest. Figure 7-15 gives the unit-step responses of Eq. (7-72) plotted as functions of the normalized time $\omega_n t$ for various values of the damping ratio ζ. As seen, the response becomes more oscillatory as ζ decreases in value. When $\zeta \geq 1$, the step response does not exhibit any overshoot; that is, $c(t)$ never exceeds its final value during the transient.

The exact relation between the damping ratio and the amount of overshoot can be obtained by taking the derivative of Eq. (7-72) with respect to t and setting the result to zero. Thus,

$$\frac{dc(t)}{dt} = \frac{\omega_n e^{-\zeta\omega_n t}}{\sqrt{1 - \zeta^2}}[\zeta \sin(\omega t + \theta) - \sqrt{1 - \zeta^2} \cos(\omega t + \theta)] \qquad t \geq 0 \qquad (7\text{-}77)$$

where ω and θ are defined in Eqs. (7-75) and (7-76), respectively. We can show that the quantity inside the square brackets in Eq. (7-77) can be reduced to $\sin \omega t$. Thus, Eq. (7-77) is simplified to

$$\frac{dc(t)}{dt} = \frac{\omega_n}{\sqrt{1 - \zeta^2}} e^{-\zeta\omega_n t} \sin \omega_n\sqrt{1 - \zeta^2}\, t \qquad t \geq 0 \qquad (7\text{-}78)$$

Setting $dc(t)/dt$ to zero, we have the solutions: $t = \infty$, and

$$\omega_n\sqrt{1 - \zeta^2}\, t = n\pi \qquad n = 0, 1, 2, \ldots \qquad (7\text{-}79)$$

from which we get

$$t = \frac{n\pi}{\omega_n\sqrt{1 - \zeta^2}} \qquad n = 0, 1, 2, \ldots \qquad (7\text{-}80)$$

The solution at $t = \infty$ is a maximum of $c(t)$ only when $\zeta \geq 1$. For the unit-step responses shown in Fig. 7-15, the first overshoot is the maximum overshoot. This corresponds to $n = 1$ in Eq. (7-80). Thus, the time at which the maximum overshoot occurs is

$$t_{max} = \frac{\pi}{\omega_n \sqrt{1 - \zeta^2}} \tag{7-81}$$

With reference to Fig. 7-15, the overshoots occur at odd values of n, that is, $n = 1, 3, 5, \ldots$, and the undershoots occur at even values of n. Whether the extremum is an overshoot or an undershoot, the time at which it occurs is given by Eq.

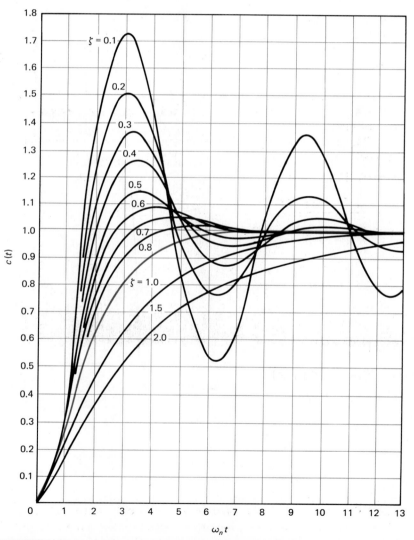

FIGURE 7-15 Unit-step responses of the prototype second-order system with various damping ratios.

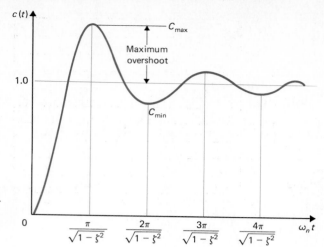

FIGURE 7-16 Unit-step responses illustrating that the maxima and minima occur at periodic intervals.

overshoots
undershoots
periodicity

(7-80). It is interesting to note that *although the unit-step response for $\zeta \neq 0$ is not periodic, the overshoots and the undershoots of the response do occur at periodic intervals, as shown in Fig. 7-16.*

The magnitudes of the overshoots and the undershoots can be determined by substituting Eq. (7-80) into Eq. (7-72). The result is

$$c(t)\big|_{\text{max or min}} = 1 - \frac{e^{-n\pi\zeta/\sqrt{1-\zeta^2}}}{\sqrt{1-\zeta^2}} \sin(n\pi + \theta) \qquad n = 1, 2, 3, \ldots \qquad (7\text{-}82)$$

or

$$c(t)\big|_{\text{max or min}} = 1 + (-1)^{n-1} e^{-n\pi\zeta/\sqrt{1-\zeta^2}} \qquad n = 1, 2, 3, \ldots \qquad (7\text{-}83)$$

The maximum overshoot is obtained by letting $n = 1$ in Eq. (7-83). Therefore,

maximum overshoot
depends only on the
damping ratio

$$\text{maximum overshoot} = c_{\text{max}} - 1 = e^{-\pi\zeta/\sqrt{1-\zeta^2}} \qquad (7\text{-}84)$$

and

$$\text{percent maximum overshoot} = 100 e^{-\pi\zeta/\sqrt{1-\zeta^2}} \qquad (7\text{-}85)$$

Equation (7-84) shows that the maximum overshoot of the step response of the prototype second-order system is a function of only the damping ratio ζ. The relationship between the percent maximum overshoot and damping ratio given in Eq. (7-85) is plotted in Fig. 7-17. The time t_{max} in Eq. (7-81) is a function of both ζ and ω_n.

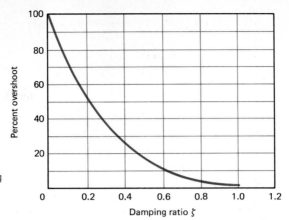

FIGURE 7-17 Percent overshoot as a function of damping ratio for the step response of the prototype second-order system.

Delay Time and Rise Time

delay time
rise time

It is more difficult to determine the exact analytical expressions of the delay time t_d, rise time t_r, and settling time t_s, even for just the simple prototype second-order system. For instance, for the delay time, we would have to set $c(t) = 0.5$ in Eq. (7-72) and solve for t. An easier way would be to plot $\omega_n t_d$ versus ζ, as shown in Fig. 7-18, and

approximation

then approximate the curve by a straight line over the range of $0 < \zeta < 1.0$. From Fig. 7-18, the delay time for the prototype second-order system is approximated as

$$t_d \cong \frac{1 + 0.7\zeta}{\omega_n} \quad 0 < \zeta < 1.0 \qquad (7\text{-}86)$$

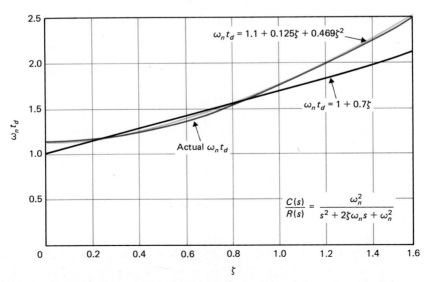

FIGURE 7-18 Normalized delay time versus ζ for the prototype second-order system.

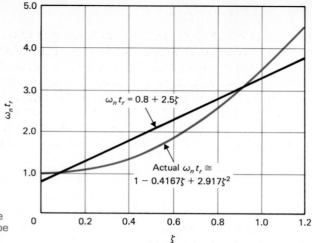

FIGURE 7-19 Normalized rise time versus ζ for the prototype second-order system.

We can obtain a better approximation by using a second-order equation for t_d:

$$t_d \cong \frac{1.1 + 0.125\zeta + 0.469\zeta^2}{\omega_n} \tag{7-87}$$

For the rise time t_r, which is the time for the step response to reach from 10 to 90 percent of its final value, the exact values can be determined directly from the responses of Fig. 7-15. The plot of $\omega_n t_r$ versus ζ is shown in Fig. 7-19. In this case, the relation can again be approximated by a straight line over a limited range of ζ:

$$t_r = \frac{0.8 + 2.5\zeta}{\omega_n} \qquad 0 < \zeta < 1 \tag{7-88}$$

Or a better approximation can be obtained by using a second-order equation:

$$t_r = \frac{1 - 0.4167\zeta + 2.917\zeta^2}{\omega_n} \qquad 0 < \zeta < 1 \tag{7-89}$$

From this discussion, it is clear that *the delay time and the rise time of the prototype second-order system are all directly proportional to ζ and inversely proportional to ω_n. Thus, increasing the damping of the system will also increase the delay and rise times. On the other hand, increasing the natural undamped frequency ω_n will reduce t_d and t_r.*

Settling Time

settling time

The settling time of the step response of the prototype second-order system is measurable only if the maximum overshoot exceeds 5 percent, which corresponds to a damping ratio that lies in the range of $0 < \zeta < 0.69$. The exact analytical description of the settling time t_s is also difficult to determine. We can obtain an approximation for the case of $0 < \zeta < 0.69$ by using the envelope of the damped sinusoid of $c(t)$, as shown in Fig. 7-20. The figure shows that the same result is obtained with the approximation whether the upper envelope or the lower envelope of $c(t)$ is used. Setting the time at which the upper envelope of $c(t)$ equals the value of 1.05 as the settling time t_s, we have

$$1 + \frac{1}{\sqrt{1 - \zeta^2}} e^{-\zeta \omega_n t_s} = 1.05 \tag{7-90}$$

Solving for $\omega_n t_s$ from the last equation, we have

$$\omega_n t_s = -\frac{1}{\zeta} \ln (0.05 \sqrt{1 - \zeta^2}) \tag{7-91}$$

approximation

The value of $\ln [0.05(1 - \zeta^2)^{1/2}]$ in Eq. (7-91) varies between -3.0 and -3.32 as ζ varies from 0 to 0.69. Thus, whenever settling time is applicable, we can approximate it for the prototype second-order system as

$$t_s \cong \frac{3.2}{\zeta \omega_n} \qquad 0 < \zeta < 0.69 \tag{7-92}$$

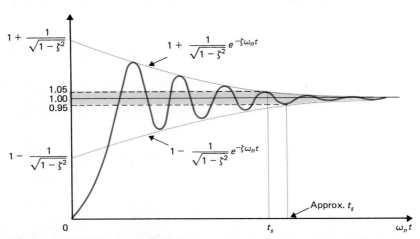

FIGURE 7-20 Approximation of settling time using the envelope of the decaying unit-step response of the prototype second-order system ($0 < \zeta < 1$).

The last equation shows that a fast settling time would require either a large ζ or a large ω_n, or both, keeping in mind that a large ζ would produce a large delay time or rise time.

7-6

TIME-DOMAIN ANALYSIS OF A POSITION-CONTROL SYSTEM

aircraft attitude control

In this section, we study the time-domain performance of a control system whose objective is to control the position of the control fins of a modern airship. Due to the requirements of improved response and reliability, the control surfaces of modern aircraft and airships are controlled by electric actuators with electronic controls. The "fly-by-wire" control system implies that the attitude control of aircrafts is no longer controlled entirely by mechanical linkages. Figure 7-21 illustrates the block diagram of one axis of such a position-control system. Figure 7-22 shows the analytical block diagram of the system using the amplifier–dc-motor model given in Fig. 4-60. The system is simplified to the extent that saturation of the amplifier gain and motor torque, gear backlash, and shaft compliances have all been neglected.

The objective of the system is to have the output of the system, $\theta_c(t)$, follow the input $\theta_i(t)$. The following system parameters are given:

Gain of encoder	$K_s = 1$ V/rad
Gain of preamplifier	$K = $ variable
Gain of power amplifier	$K_1 = 10$ V/V
Gain of current feedback	$K_2 = 0.5$ V/A
Gain of tachometer feedback	$K_t = 0$ V/rad/s
Armature resistance of motor	$R_a = 5.0\ \Omega$
Armature inductance of motor	$L_a = 0.003$ H
Torque constant of motor	$K_i = 9.0$ oz-in./A
Back-emf constant of motor	$K_b = 0.0636$ V/rad/s
Inertia of motor rotor	$J_m = 0.0001$ oz-in.-s^2
Inertia of load	$J_L = 0.01$ oz-in.-s^2
Viscous-friction coefficient of motor	$B_m = 0.005$ oz-in.-s
Viscous-friction coefficient of load	$B_L = 1.0$ oz-in.-s
Gear-train ratio	$N = \theta_c/\theta_m = 1/10$

Since the motor shaft is coupled to the load through a gear train with a gear ratio of N, $\theta_c = N\theta_m$, the total inertia and viscous-friction coefficient seen by the motor are

$$J_t = J_m + N^2 J_L = 0.0001 + 0.01/100 = 0.0002 \text{ oz-in.-s}^2 \qquad (7\text{-}93)$$

$$B_t = B_m + N^2 B_L = 0.005 + 1/100 = 0.015 \text{ oz-in.-s} \qquad (7\text{-}94)$$

respectively. The open-loop transfer function of the system is written from Fig. 7-22

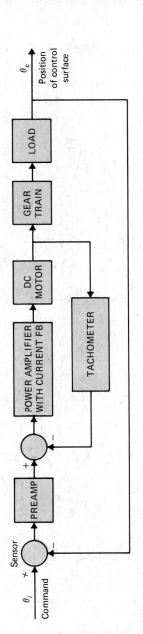

FIGURE 7-21 Block diagram of an attitude-control system of an aircraft.

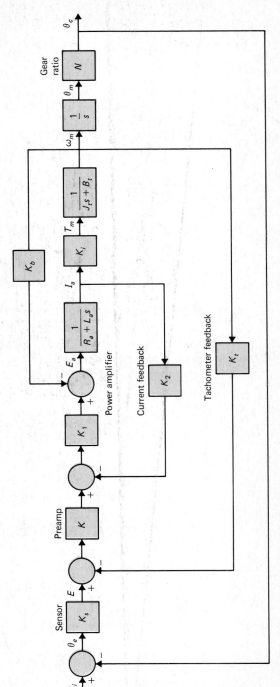

FIGURE 7-22 Transfer-function block diagram of the system shown in Fig. 7.21.

using Mason's gain formula:

*transfer
functions*

$$G(s) = \frac{\Theta_c(s)}{\Theta_e(s)}$$

$$= \frac{K_s K_1 K_i KN}{s[L_a J_t s^2 + (R_a J_t + L_a B_t + K_1 K_2 J_t)s + R_a B_t + K_1 K_2 B_t + K_i K_b + KK_1 K_t K_i]}$$

(7-95)

The system is of the third order, since the highest-order term in $G(s)$ is s^3. The electrical time constant of the amplifier–motor system is

$$\tau_a = \frac{L_a}{R_a + K_1 K_2} = \frac{0.003}{5 + 5} = 0.0003 \text{ second}$$

(7-96)

The mechanical time constant of the motor–load system is

$$\tau_t = \frac{J_t}{B_t} = \frac{0.0002}{0.015} = 0.01333 \text{ second}$$

(7-97)

Since the electrical time constant is much smaller than the mechanical time constant, we can perform a crude approximation by neglecting the armature inductance L_a. Later we will show that this is not a reliable method of approximating a high-order system by a low-order one. The result is a second-order approximation of the third-order system, with the open-loop transfer function

$$G(s) = \frac{\Theta_c(s)}{\Theta_e(s)} = \frac{K_s K_1 K_i KN}{s[(R_a J_t + K_1 K_2 J_t)s + R_a B_t + K_1 K_2 B_t + K_i K_b + KK_1 K_i K_t]}$$

$$= \frac{\dfrac{K_s K_1 K_i KN}{R_a J_t + K_1 K_2 J_t}}{s\left(s + \dfrac{R_a B_t + K_1 K_2 B_t + K_i K_b + KK_1 K_i K_t}{R_a J_t + K_1 K_2 J_t}\right)}$$

(7-98)

Substituting the system parameters in the last equation, we get

$$G(s) = \frac{4500K}{s(s + 361.2)}$$

(7-99)

Comparing Eqs. (7-98) and (7-99) with the prototype second-order transfer function of Eq. (7-68), we have

natural undamped frequency $\omega_n = \pm\sqrt{\dfrac{K_s K_1 K_i KN}{R_a J_t + K_1 K_2 J_t}} = \pm\sqrt{4500K}$ rad/s

(7-100)

damping ratio $\zeta = \dfrac{R_a B_t + K_1 K_2 B_t + K_i K_b + KK_1 K_i K_t}{2\sqrt{K_s K_1 K_i KN (R_a J_t + K_1 K_2 J_t)}} = \dfrac{2.692}{\sqrt{K}}$

(7-101)

Thus, we see that the natural undamped frequency ω_n is proportional to the square root of K, whereas the damping ratio ζ is inversely proportional to the square root of K. The characteristic equation of the unity-feedback control system is

characteristic equation

$$s^2 + 361.2s + 4500K = 0 \qquad (7\text{-}102)$$

Time Response to a Unit-Step-Input Transient Response

For time-domain analysis, it is informative to analyze the system performance by applying a unit-step input with zero initial conditions. In this way, it is possible to characterize the system performance in terms of the maximum overshoot and some of the other measures, such as rise time, delay time, settling time, if necessary.

unit-step response

Let the reference input be a unit-step function, $\theta_r(t) = u_s(t)$ rad; then $\Theta_r(s) = 1/s$. The output of the system, with zero initial conditions, is

$$\theta_c(t) = \mathcal{L}^{-1}\left[\frac{4500K}{s(s^2 + 361.2s + 4500K)}\right] \qquad (7\text{-}103)$$

The inverse Laplace transform on the right-hand side of the last equation is carried out using the Laplace-transform table in Appendix B, or Eq. (7-72),

$$\theta_c(t) = 1 - \frac{1}{\sqrt{1 - \zeta^2}}e^{-\zeta\omega_n t}\sin\left(\omega_n\sqrt{1 - \zeta^2}\,t + \cos^{-1}\zeta\right) \qquad (7\text{-}104)$$

where ω_n and ζ are given in Eqs. (7-100) and (7-101), respectively. The output response $\theta_c(t)$ is computed for three different values of K: $K = 7.248$ ($\zeta = 1.0$, critical damping), $K = 14.5$ ($\zeta = 0.707$), and $K = 181.17$ ($\zeta = 0.2$), using a computer program such as the CLRSP in the ACSP software package that accompanies this text. The responses are plotted in Fig. 7-23. Table 7-2 gives the comparison of the characteristics of the three unit-step responses for the three values of K used. When $K = 181.17$, $\zeta = 0.2$, the system is underdamped, and the overshoot is 52.7 percent, which is excessive. When the value of K is set at 7.248, $\zeta = 1.0$, and the system is critically damped. The unit-step response approaches its final values without any oscillation and overshoot. Clearly, the system will be overdamped when K is greater than 7.248, and the step response will be slower in reaching its final value. When K is set at the value of 14.5, the damping ratio is 0.707, and the overshoot is 4.3 percent.

It should be pointed out that in practice, it would be time consuming to compute the time response for each change in the system parameter for either analysis or design purposes. Indeed, one of the main objectives of studying control-systems theory using either the conventional or the modern approach is to establish methods so that the total reliance on computer simulation can be avoided or reduced. The motivation behind this discussion is to show that the performance of some control systems can be predicted by investigating the roots of the characteristic equation of the system. For the characteristic equation of Eq. (7-102), the roots are

$$s_1 = -180.6 + \sqrt{32{,}616.36 - 4500K} \qquad (7\text{-}105)$$

$$s_2 = -180.6 - \sqrt{32{,}616.36 - 4500K} \qquad (7\text{-}106)$$

*performance
comparison*

TABLE 7-2 Comparison of the Performance of the Second-Order
Position-Control System When the Gain K Varies

GAIN K	ζ	ω_n (rad/s)	MAXIMUM OVERSHOOT	t_d (s)	t_r (s)	t_s (s)	t_{max} (s)
7.248	1.000	180.62	0	0.011	0.02	—	—
14.5	0.707	255.44	0.043	0.0065	0.009	—	0.0174
181.2	0.200	903.00	0.527	0.0013	0.001	0.018	0.0036

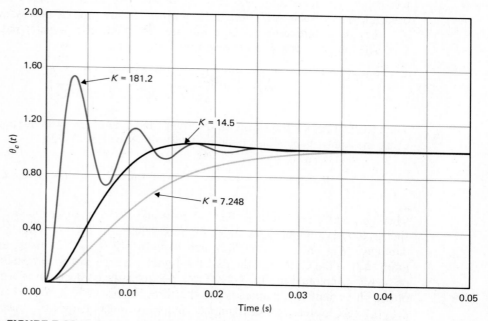

FIGURE 7-23 Unit-step responses of the attitude-control system in Fig. 7.22; $L_a = 0$.

*characteristic-
equation roots*

For $K = 7.248$, 14.5, and 181.2, the roots of the characteristic equation are tabulated
as follows:

K	s_1	s_2
7.248	-180.6	-180.6
14.5	$-180.6 + j180.6$	$-180.6 - j180.6$
181.2	$-180.6 + j884.75$	$-180.6 - j884.75$

The roots are marked in the s-plane, as shown in Fig. 7-24. The trajectories of the two
characteristic-equation roots when K varies continuously between $-\infty$ and ∞ are also

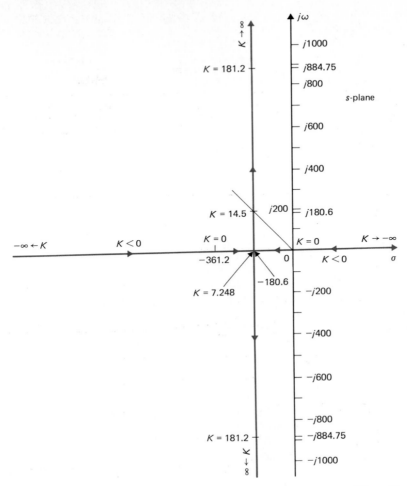

FIGURE 7-24 Root loci of the characteristic equation in Eq. (7-102) as K varies.

root loci

shown in Fig. 7-24. In general, these trajectories are called the **root loci** (see Chapter 8) of Eq. (7-102), and are used extensively for the analysis and design of linear control systems.

From Eqs. (7-105) and (7-106), we see that the two roots are real and negative for values of K between 0 and 7.248. This means that the system is overdamped, and the step response will have no overshoot for this range of K. For values of K greater than 7.248, the roots are complex-conjugate with the real parts of the roots equal to −180.6, and the system is underdamped. The damping factor is always equal to 180.6, and is independent of K. The root loci also clearly show that as K increases beyond 7.248, the natural undamped frequency will increase with the square root of K. When K is negative, one of the roots is positive, which corresponds to a time response that increases monotonically with time, and the system is unstable. The dynamic characteristics of the transient step response as determined from the root loci of Fig. 7-24 are summarized as follows:

performance comparison

AMPLIFIER GAIN	CHARACTERISTIC-EQUATION ROOTS	SYSTEM DYNAMICS
$0 < K < 7.248$	Two negative distinct real roots	Overdamped ($\zeta > 1$)
$K = 7.248$	Two negative equal real roots	Critically damped ($\zeta = 1$)
$7.248 < K < \infty$	Two complex-conjugate roots with negative real roots	Underdamped ($\zeta < 1$)
$-\infty < K < 0$	Two distinct real roots, one positive and one negative	Unstable system ($\zeta < 0$)

Steady-State Response

Since the open-loop transfer function in Eq. (7-99) has a simple pole at $s = 0$, the system is of type 1. This means that the steady-state error of the system is zero for all positive values of K when the input is a step function, and the step-error constant K_p is to be used. Substituting Eq. (7-99) into Eq. (7-20), with $H(s) = 1$, we have

$$K_p = \lim_{s \to 0} \frac{4500K}{s(s + 361.2)} = \infty \qquad (7\text{-}107)$$

position-error constant

Thus, the steady-state error of the system due to a step input, as given by Eq. (7-21), is zero. The unit-step responses in Fig. 7-23 verify this result. The zero-steady-state condition is achieved because only viscous friction is considered in the simplified system model. In the practical case, Coulomb friction is almost always present, so that the steady-state positioning accuracy of the system can never be perfect.

Time Response to a Unit-Ramp Input

The control of position may often be effected by the control of the profile of the output, rather than just applying a step input. In other words, the system may be designed to follow a reference profile that represents the desired trajectory. It may be necessary to investigate the ability of the position-control system to follow a ramp-function input.

For a unit-ramp input, $\theta_r(t) = tu_s(t)$. The output response of the system in Fig. 7-22 is

$$\theta_c(t) = \mathcal{L}^{-1}\left[\frac{4500K}{s^2(s^2 + 361.2s + 4500K)}\right] \qquad (7\text{-}108)$$

The expression of $\theta_c(t)$ in the last equation is determined from the Laplace-transform table in Appendix B.

$$\theta_c(t) = t - \frac{2\zeta}{\omega_n} + \frac{1}{\omega_n\sqrt{1 - \zeta^2}}e^{-\zeta\omega_n t}\sin(\omega_n\sqrt{1 - \zeta^2}\,t + \theta) \qquad t \geq 0 \qquad (7\text{-}109)$$

where

$$\theta = \cos^{-1}(2\zeta^2 - 1) \qquad (\zeta < 1) \tag{7-110}$$

The values of ζ and ω_n are given in Eqs. (7-101) and (7-100), respectively.

The ramp responses for $K = 7.248$, 14.5, and 181.2 are computed and plotted as shown in Fig. 7-25. Notice that in this case, the steady-state error of the system is not equal to zero. The last term in Eq. (7-109) represents the transient response. The steady-state portion of the unit-ramp response is

$$\lim_{t \to \infty} \theta_c(t) = \lim_{t \to \infty} \left(t - \frac{2\zeta}{\omega_n} \right) \tag{7-111}$$

Thus, the steady-state error of the system due to the unit-step input is

$$e_{ss} = \frac{2\zeta}{\omega_n} = \frac{0.0803}{K} \tag{7-112}$$

which is a constant.

ramp-error constant

A more direct method of determining the steady-state error due to a ramp input is to use the ramp-error constant K_v. From Eq. (7-28),

$$K_v = \lim_{s \to 0} sG(s) = \lim_{s \to 0} \frac{4500K}{s + 361.2} = 12.46K \tag{7-113}$$

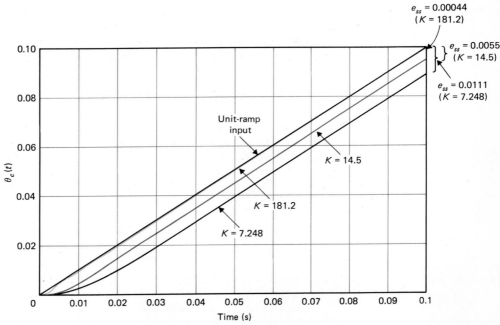

FIGURE 7-25 Unit-ramp responses of the attitude-control system in Fig. 7.22; $L_a = 0$.

Therefore,

$$e_{ss} = \frac{1}{K_v} = \frac{0.0803}{K} \tag{7-114}$$

steady-state error

which agrees with the result in Eq. (7-112). The result in Eq. (7-114) shows that the steady-state error is inversely proportional to K. For $K = 14.5$, which corresponds to a damping ratio of 0.707, the steady-state error is 0.0055 rad, or, more appropriately, 0.55 percent of the ramp-input magnitude. Apparently, if we attempt to improve the steady-state accuracy of the system by increasing the value of K, the transient step response will become more oscillatory and with higher overshoot. This phenomenon is rather typical in all control systems. For higher-order systems, if the loop gain of the system is too high, the system can become unstable.

Time Response of a Third-Order System

transfer functions

In the preceding section, we have shown that the prototype second-order system, obtained by neglecting the armature inductance, is always stable for all positive values of K. Let us now investigate the system performance with the armature inductance $L_a = 0.003$ H. The open-loop transfer function of Eq. (7-95) becomes

$$G(s) = \frac{1.5 \times 10^7 K}{s(s^2 + 3408.3s + 1,204,000)}$$

$$= \frac{1.5 \times 10^7 K}{s(s + 400.26)(s + 3008)} \tag{7-115}$$

The closed-loop transfer function is

$$\frac{\Theta_c(s)}{\Theta_r(s)} = \frac{1.5 \times 10^7 K}{s^3 + 3408.3s^2 + 1,204,000s + 1.5 \times 10^7 K} \tag{7-116}$$

The system is now of the third order, and the characteristic equation is

$$s^3 + 3408.3s^2 + 1,204,000s + 1.5 \times 10^7 K = 0 \tag{7-117}$$

characteristic equation

The roots of the characteristic equation are tabulated for the three values of K used earlier for the second-order system:

characteristic-equation roots

K	s_1	s_2	s_3
7.248	-156.21	-230.33	-3021.8
14.5	$-186.53 + j192$	$-186.53 - j192$	-3035.2
181.2	$-57.49 + j906.6$	$-57.49 - j906.6$	-3293.3

Comparing these results with those of the approximating second-order system, we see that when $K = 7.428$, the second-order system is critically damped, whereas the third-order system has three distinct real roots, and the system is slightly overdamped. The root at -3021.8 corresponds to a time constant of 0.33 milliseconds, which is over 13 times faster than the next fastest time constant due to the pole at -230.33. Thus, the transient response due to the pole at -3021.8 decays rapidly, and the pole can be neglected from the transient standpoint. The output transient response is dominated by the two roots at -156.21 and -230.33. This analysis is verified by writing the transformed output response as

$$\Theta_c(s) = \frac{10.87 \times 10^7}{s(s + 156.21)(s + 230.33)(s + 3021.8)} \tag{7-118}$$

unit-step response

Taking the inverse Laplace transform of the last equation, we get

$$\theta_c(t) = 1 - 3.28e^{-156.21t} + 2.28e^{-230.33t} - 0.0045e^{-3021.8t} \qquad t > 0 \tag{7-119}$$

Thus, the root at -3021.8 gives the last term in the last equation, which decays to zero very rapidly, and, furthermore, the magnitude of the term at $t = 0$ is very small compared to that of the other two transient terms. This simply demonstrates that, in general, the contribution of roots that lie relatively far to the left in the s-plane to the time response will be small. The roots that are closer to the imaginary axis will dominate *dominant roots* the transient response, and these are defined as the **dominant roots** of the characteristic equation.

When $K = 14.5$, the second-order system has a damping ratio of 0.707, since the real and imaginary parts of the two characteristic-equation roots are identical. For the third-order system, again, the root at -3035.2 corresponds to a fast-decaying transient response. The two roots that dominate the transient response correspond to a damping ratio of 0.697. Thus, for $K = 14.5$, the second-order approximation by setting L_a to zero is not a bad one.

When $K = 181.2$, the two complex-conjugate roots of the third-order system again dominate the transient response, and the equivalent damping ratio due to the two roots is only 6.33 percent, which is far less than the 20 percent of the second-order system.

critical K

marginal stability

By using the Routh–Hurwitz criterion, the marginal value of K for stability is found to be 273.57. With this critical value of K, the closed-loop transfer function becomes

$$\frac{\Theta_c(s)}{\Theta_r(s)} = \frac{1.0872 \times 10^8}{(s + 3408.3)(s^2 + 1.204 \times 10^6)} \tag{7-120}$$

The roots of the characteristic equation are at $s = -3408.3, -j1097.3$, and $j1097.3$. The unit-step response of the system is

$$\theta_c(t) = 1 - 0.094e^{-3408.3t} - 0.952 \sin(1097.3t + 72.16°) \qquad t \geq 0 \tag{7-121}$$

The steady-state response is an undamped sinusoid with a frequency of 1097.3 rad/s, and the system is on the verge of instability. When K becomes greater than 273.57, the two complex-conjugate roots of the characteristic equation will have positive real parts, the sinusoidal component of the time response will increase with time, and the system is unstable. Thus, we see that the third-order system is capable of being unstable, whereas the second-order system obtained with $L_a = 0$ is always stable for all finite positive values of K.

Figure 7-26 shows the unit-step responses of the third-order system for $K = 7.248$, 14.5, 181.2, and 273.57. The responses for $K = 7.248$ and $K = 14.5$ are very close to those of the second-order system with the same values of K, as shown in Fig. 7-23. This is because one of the characteristic-equation roots of the third-order system is far to the left in the s-plane, so that the transient response can be approximated by the two roots that are closer to the $j\omega$-axis.

Figure 7-27 illustrates the root loci of the third-order characteristic equation in Eq. (7-117) as K varies. The loci clearly show that when K is greater than 273.57, the two complex-conjugate roots are in the right half of the s-plane. The portion of the root loci for small values of K of the third-order system is very similar to those of the second-order system shown in Fig. 7-24.

From Eq. (7-115), we see that when the inductance is restored, the third-order system is still of type 1. The value of K_v is the same as that given in Eq. (7-113). Thus, the inductance of the motor does not affect the steady-state performance of the system. This is expected, since L_a is related only to motor-current variations.

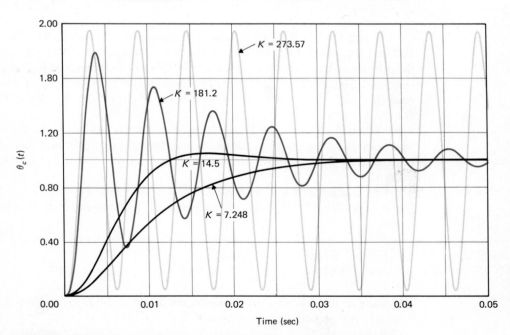

FIGURE 7-26 Unit-step responses of the third-order attitude-control system.

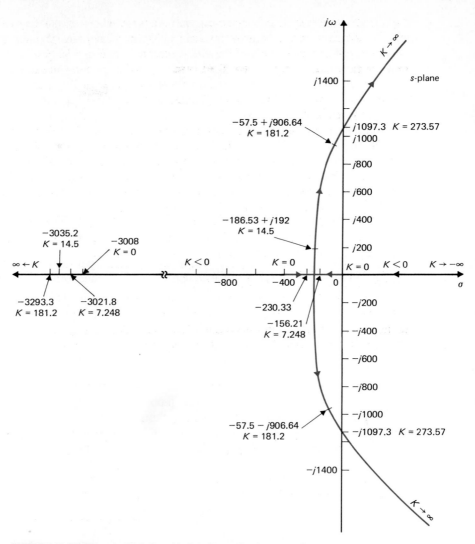

FIGURE 7-27 Root loci of the third-order attitude-control system.

7-7

EFFECTS OF ADDING POLES AND ZEROS TO TRANSFER FUNCTIONS

The position-control system discussed in the preceding section reveals important properties of the time responses of typical second- and third-order closed-loop systems. Specifically, the effects on the transient response relative to the location of the roots of the characteristic equation are demonstrated. However, in practice, successful design of a control system cannot depend only on choosing values of the system parameters so that the characteristic-equation roots are properly placed. In fact, we shall show that, *although the roots of the characteristic equation, which are the poles of the closed-loop transfer function, affect the transient response of linear time-invariant control systems, particularly the stability, the zeros of the transfer function, if there are any, are also important*. Thus, the addition of poles and zeros and/or cancellation of undesirable poles and zeros of the transfer function often are necessary in achieving satisfactory time-domain performance of control systems.

In this section, we shall show that the addition of poles and zeros to open-loop and closed-loop transfer functions has varying effects on the transient response of the closed-loop system.

Addition of a Pole to the Open-Loop Transfer Function

For the position-control system described in Section 7-6, when the motor inductance is neglected, the system is of the second order, and the open-loop transfer function is of the prototype given by Eq. (7-99). When the motor inductance is restored, the system is of the third order, and the open-loop transfer function is given in Eq. (7-115). Comparing the two transfer functions of Eqs. (7-99) and (7-115), we see that the effect of the motor inductance is equivalent to adding a pole at $s = -3008$ to the open-loop transfer function of Eq. (7-99), while shifting the pole at -361.2 to -400.26, and the proportional constant is also increased. The apparent effect of adding a pole to the open-loop transfer function is that the third-order system can now become unstable if the value of the amplifier gain K exceeds 273.57. As shown by the root-locus diagrams of Figs. 7-24 and 7-27, the new open-loop pole at $s = -3008$ essentially "pushes" and "bends" the complex-conjugate portion of the root loci of the second-order system toward the right-half s-plane. Actually, owing to the specific value of the inductance chosen, the additional pole of the third-order system is far to the left of the pole at $s = -400.26$, so that its effect is small except when the value of K is relatively large.

To study the general effect of the addition of a pole, and its relative location, to an open-loop transfer function, consider the transfer function

$$G(s) = \frac{\omega_n^2}{s(s + 2\zeta\omega_n)(1 + T_p s)} \qquad (7\text{-}122)$$

The pole at $s = -1/T_p$ is considered to be added to the prototype second-order transfer function. The closed-loop transfer function of the unity-feedback control system is

$$M(s) = \frac{C(s)}{R(s)} = \frac{G(s)}{1 + G(s)} = \frac{\omega_n^2}{T_p s^3 + (1 + 2\zeta\omega_n T_p)s^2 + 2\zeta\omega_n s + \omega_n^2} \qquad (7\text{-}123)$$

Figure 7-28 illustrates the unit-step responses of the closed-loop system when $\omega_n = 1$, $\zeta = 1$, and $T_p = 0$, 1, 2, and 5. These responses again show that the *addition of a pole to the open-loop transfer function generally has the effect of increasing the maximum overshoot of the closed-loop step response*. As the value of T_p increases, the pole at $-1/T_p$ moves closer to the origin in the s-plane, and the maximum overshoot increases. These responses also show that the *added pole increases the rise time of the*

effect on overshoot

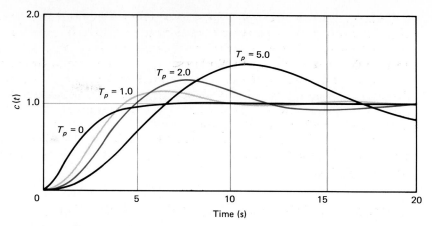

FIGURE 7-28 Unit-step responses of the system with the closed-loop transfer function in Eq. (7-123): $\zeta = 1$, $\omega_n = 1$, and $T_p = 0$, 1, 2, and 5.

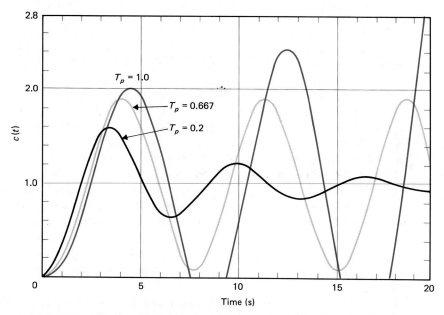

FIGURE 7-29 Unit-step responses of the system with the closed-loop transfer function in Eq. (7-123): $\zeta = 0.25$, $\omega_n = 1$, and $T_p = 0$, 0.2, 0.667, and 1.0.

effect on bandwidth

step response. This is not surprising, since the *additional pole has the effect of reducing the bandwidth (see Chapter 10) of the system,* thus cutting out the high-frequency components of the input signal.

The same conclusions can be drawn from the unit-step responses of Fig. 7-29, which are obtained with $\omega_n = 1$, $\zeta = 0.25$, and $T_p = 0$, 0.2, 0.667, and 1.0. In this case, when T_p is greater than 0.667, the amplitude of the unit-step response increases with time, and the system is unstable.

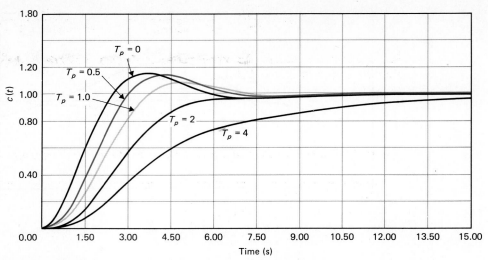

FIGURE 7-30 Unit-step responses of the system with the closed-loop transfer function in Eq. (7-124): $\zeta = 0.5$, $\omega_n = 1$, and $T_p = 0$, 0.5, 1.0, 2.0, and 4.0.

Addition of a Pole to the Closed-Loop Transfer Function

Since the poles of the closed-loop transfer function are roots of the characteristic equation, they control the transient response of the system directly. Consider the closed-loop transfer function

$$M(s) = \frac{C(s)}{R(s)} = \frac{\omega_n^2}{(s^2 + 2\zeta\omega_n s + \omega_n^2)(1 + T_p s)} \qquad (7\text{-}124)$$

effect on overshoot

where the term $(1 + T_p s)$ is added to a prototype second-order transfer function. Figure 7-30 illustrates the unit-step response of the system with $\omega_n = 1$, $\zeta = 0.5$, and $T_p = 0, 0.5, 1, 2$, and 4. As the pole at $s = -1/T_p$ is moved toward the origin in the s-plane, the rise time increases, and the maximum overshoot decreases. *Thus, as far as the overshoot is concerned, adding a pole to the closed-loop transfer function has just the opposite effect to that of adding a pole to the open-loop transfer function.*

Addition of a Zero to the Closed-Loop Transfer Function

Figure 7-31 shows the unit-step responses of the closed-loop system with the transfer function

$$M(s) = \frac{C(s)}{R(s)} = \frac{\omega_n^2(1 + T_z s)}{(s^2 + 2\zeta\omega_n s + \omega_n^2)} \qquad (7\text{-}125)$$

effect on rise time and overshoot

where $\omega_n = 1$, $\zeta = 0.5$, and $T_z = 0, 1, 3, 6$, and 10. In this case, we see that *adding a zero at $s = -1/T_z$ to the closed-loop transfer function decreases the rise time and increases the maximum overshoot of the step response.*

We can analyze the general case by writing Eq. (7-125) as

$$M(s) = \frac{C(s)}{R(s)} = \frac{\omega_n^2}{s^2 + 2\zeta\omega_n s + \omega_n^2} + \frac{T_z\omega_n^2 s}{s^2 + 2\zeta\omega_n s + \omega_n^2} \qquad (7\text{-}126)$$

For a unit-step input, let the output response that corresponds to the first term of the right-hand side of Eq. (7-126) be $c_1(t)$. Then, the total output is

$$c(t) = c_1(t) + T_z \frac{dc_1(t)}{dt} \qquad (7\text{-}127)$$

Figure 7-32 shows why the addition of the zero at $s = -1/T_z$ reduces the rise time and

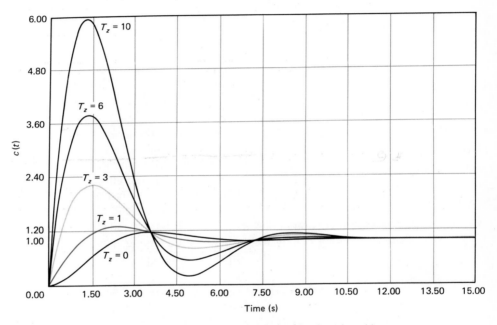

FIGURE 7-31 Unit-step responses of the system with the closed-loop transfer function in Eq. (7-125): $T_z = 0$, 1, 3, 6, and 10.

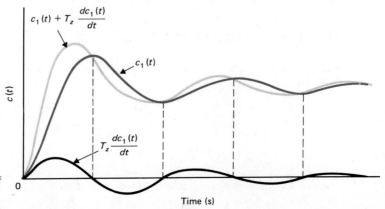

FIGURE 7-32 Unit-step responses showing the effect of adding a zero to the closed-loop transfer function.

*the system
is always
stable as long as ζ is positive*

increases the maximum overshoot of the step response, according to Eq. (7-127). In fact, as T_z approaches infinity, the maximum overshoot also approaches infinity, and yet the system is still stable as long as ζ is positive.

Addition of a Zero to the Open-Loop Transfer Function

Let us consider that a zero at $-1/T_z$ is added to an open-loop transfer function, so that the third-order transfer function is

$$G(s) = \frac{6(1 + T_z s)}{s(s + 1)(s + 2)} \qquad (7\text{-}128)$$

The closed-loop transfer function is

$$M(s) = \frac{C(s)}{R(s)} = \frac{6(1 + T_z s)}{s^3 + 3s^2 + (2 + 6T_z)s + 6} \qquad (7\text{-}129)$$

The difference between this case and that of adding a zero to the closed-loop transfer function is that in the present case, not only the term $(1 + T_z s)$ appears in the numerator of $M(s)$, but the denominator of $M(s)$ also contains T_z. The term $(1 + T_z s)$ in the numerator of $M(s)$ increases the maximum overshoot, but T_z appears in the coefficient of the s term in the denominator, which has the effect of improving the damping, or reducing the maximum overshoot. Figure 7-33 illustrates the unit-step responses when $T_z = 0$, 0.2, 0.5, 2, 5, and 10. Notice that when $T_z = 0$, the closed-loop system is on the verge of becoming unstable. When $T_z = 0.2$ and 0.5, the maximum overshoots are reduced, owing mainly to the improved damping. As T_z increases beyond 2, although the damping is still further improved, the $(1 + T_z s)$ term in the numerator becomes

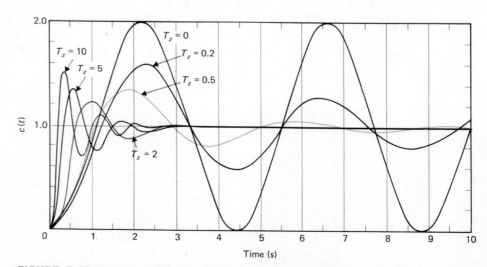

FIGURE 7-33 Unit-step repsonses of the system with the closed-loop transfer function in Eq. (7-129): $T_z = 0$, 0.2, 0.5, 2, 5, and 10.

more dominant, so that the maximum overshoot actually becomes greater as T_z is increased further.

An important finding from these discussions is that although *the characteristic-equation roots are generally used to study the relative damping and stability of linear control systems, the zeros of the transfer function should not be overlooked in their effects on the overshoot and rise time of the step response.*

7-8

DOMINANT POLES OF TRANSFER FUNCTIONS

From the discussions given in the preceding sections, it becomes apparent that the location of the poles of a transfer function in the s-plane has great effects on the transient response of the system. For analysis and design purposes, it is important to sort out the poles that have a dominant effect on the transient response and call these the **dominant poles**.

Since most control systems found in practice are of a high order, it would be useful to establish guidelines on the approximation of high-order systems by lower-order systems insofar as the transient response is concerned. In design, we can use the dominant poles to control the dynamic performance of the system, whereas the **insignificant** poles are used for the purpose of ensuring that the controller transfer function can be realized by physical components.

For all practical purposes, we can qualitatively sectionalize the s-plane into regions in which the dominant and insignificant poles can lie, as shown in Fig. 7-34. We intentionally do not assign specific values to the coordinates, since these are all relative to a given system.

The poles that are close to the imaginary axis in the left-half s-plane give rise to transient responses that will decay relatively slowly, whereas the poles that are far away from the axis (relative to the dominant poles) correspond to fast-decaying time responses. The distance D between the dominant region and the least significant region shown in Fig. 7-34 will be subject to discussion. The question is: "How large a pole is

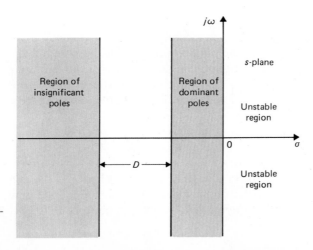

FIGURE 7-34 Regions of dominant and insignificant poles in the s-plane.

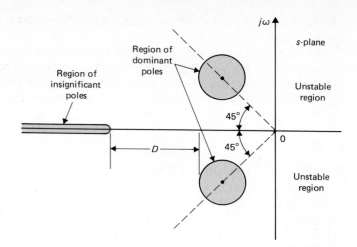

FIGURE 7-35 Regions of dominant and insignficant poles in the s-plane.

only

as far as

transient response

is concerned

pole-placement design

considered to be really large?" It has been recognized in practice and in the literature that if the magnitude of the real part of a pole is at least 5 to 10 times that of a dominant pole or a pair of complex dominant poles, then the pole may be regarded as insignificant insofar as the transient response is concerned.

We must point out that the regions shown in Fig. 7-34 are selected merely for the definitions of dominant and insignificant poles. For design purposes, such as in the pole-placement design, the dominant poles and the insignificant poles, as selected by the designer, should most likely be located in the tinted regions in Fig. 7-35. Again, we do not show any absolute coordinates, except that the desired region of the dominant poles is centered around the line that corresponds to $\zeta = 0.707$. It should also be noted that in design, we cannot place the insignificant poles arbitrarily far to the left in the s-plane or these may require unrealistic parameter values when the pencil-and-paper design is to be implemented by physical components.

The Relative Damping Ratio

relative damping

When a system is higher than the second order, strictly, we can no longer use the damping ratio ζ and the natural undamped frequency ω_n, which are defined for the prototype second-order systems. However, if the system dynamics can be accurately represented by a pair of dominant poles, then we can still use ζ and ω_n to indicate the location of these poles, and the damping ratio in this case is referred to as the **relative damping ratio** of the system. For example, consider the closed-loop transfer function

$$\frac{C(s)}{R(s)} = \frac{10}{(s + 10)(s^2 + 2s + 2)} \qquad (7\text{-}130)$$

The pole at $s = -10$ is 10 times the real part of the complex conjugate poles. Thus, the dominant poles of the system are at $s = -1 + j1$ and $s = -1 - j1$. We can refer to the relative damping ratio of the system to be 0.707.

The Proper Way of Neglecting the Insignificant Poles

Once we have found that it is justifiable to cast away a certain insignificant pole or poles of a transfer function, analytically, this must be done properly from the standpoint of both the transient and the steady-state responses. Consider the transfer function in Eq. (7-130); the pole at -10 can be neglected from the transient standpoint. To do this, we should first express Eq. (7-130) as

$$\frac{C(s)}{R(s)} = \frac{10}{10(s/10 + 1)(s^2 + 2s + 2)} \tag{7-131}$$

Then we reason that $|s/10| \ll 1$ when the absolute value of s is much smaller than 10, owing to the dominant nature of the complex poles. The term $s/10$ can be neglected when compared with 1. Equation (7-131) can be approximated by

$$\frac{C(s)}{R(s)} \cong \frac{10}{10(s^2 + 2s + 2)} \tag{7-132}$$

This way, the steady-state performance of the third-order system will not be affected by the approximation.

Although we have used a closed-loop transfer function to demonstrate the approximation procedure, the same process can be applied to an open-loop transfer function.

7-9

APPROXIMATION OF HIGH-ORDER SYSTEMS BY LOW-ORDER SYSTEMS: CONTINUOUS-DATA SYSTEMS

In the preceding sections, we have shown that a high-order control system often contains insignificant poles that have little effect on the transient response. Thus, given a high-order system, it is desirable, and frequently possible, to find a low-order approximating system so that the analysis or design effort is reduced. This means that given a high-order transfer function $M(s)$, how can we find a low-order transfer function $L(s)$ as an approximation, in the sense that the responses of the two systems are similar according to some prescribed criterion? In the last section, we set up guidelines for neglecting the poles that are far to the left in the s-plane relative to the other poles and zeros in the s-plane. However, in general, the transfer function $M(s)$ may not have the so-called **dominant poles** so that it may not be obvious that there are poles that may be neglected in order to arrive at a low-order equivalent.

In this section, we shall introduce a method [2] of approximating a high-order system by a low-order one in the sense that the frequency responses of the two systems are similar.

Let the high-order system transfer function be represented by

$$M(s) = K\frac{1 + a_1s + a_2s^2 + \cdots + a_ms^m}{1 + b_1s + b_2s^2 + \cdots + b_ns^n} \qquad (7\text{-}133)$$

where $n \geq m$. Let the transfer function of the approximating low-order system be

$$L(s) = K\frac{1 + c_1s + c_2s^2 + \cdots + c_qs^q}{1 + d_1s + d_2s^2 + \cdots + d_ps^p} \qquad (7\text{-}134)$$

where $n \geq p \geq q$. Notice that the zero-frequency ($s = 0$) gain K of the two transfer functions is the same. This will ensure that the steady-state behavior of the high-order system is preserved in the low-order system. Furthermore, we assume that the poles of $M(s)$ and $L(s)$ are all in the left-half s-plane, since we are not interested in unstable systems.

The transfer functions $M(s)$ and $L(s)$ generally refer to the closed-loop transfer function, but if necessary, they can be treated as open-loop transfer functions.

Approximation Criterion

The criterion of finding the low-order $L(s)$, given $M(s)$, is that the following relation should be satisfied as closely as possible:

$$\frac{|M(j\omega)|^2}{|L(j\omega)|^2} = 1 \qquad \text{for } 0 \leq \omega \leq \infty \qquad (7\text{-}135)$$

The last condition implies that the amplitude characteristics of the two systems in the frequency domain ($s = j\omega$) are similar. It is hoped that this will lead to similar time responses for the two systems.

Given the transfer function $M(s)$, the determination of the low-order approximating transfer function $L(s)$ involves the following two steps:

approximation steps

1. Choose the appropriate orders of the numerator polynomial, q, and the denominator polynomial, p, of $L(s)$.
2. Determine the coefficients c_i, $i = 1, 2, \ldots, q$, and d_j, $j = 1, 2, \ldots, p$, so that the condition in Eq. (7-135) is approached.

By using Eqs. (7-133) and (7-134), the ratio of $M(s)$ to $L(s)$ is

$$\frac{M(s)}{L(s)} = \frac{(1 + a_1s + a_2s^2 + \cdots + a_ms^m)(1 + d_1s + d_2s^2 + \cdots + d_ps^p)}{(1 + b_1s + b_2s^2 + \cdots + b_ns^n)(1 + c_1s + c_2s^2 + \cdots + c_qs^q)}$$

$$= \frac{1 + m_1s + m_2s^2 + \cdots + m_us^u}{1 + l_1s + l_2s^2 + \cdots + l_vs^v} \qquad (7\text{-}136)$$

where $u = m + p$, and $v = n + q$.

Equation (7-135) can be written as

$$\frac{|M(j\omega)|^2}{|L(j\omega)|^2} = \frac{M(s)M(-s)}{L(s)L(-s)}\bigg|_{s=j\omega} \tag{7-137}$$

where $M(s)M(-s)$ and $L(s)L(-s)$ are even polynomials of s; i.e., they contain only even powers of s. Thus, Eq. (7-137) can be written as

$$\frac{|M(j\omega)|^2}{|L(j\omega)|^2} = \frac{1 + e_2 s^2 + e_4 s^4 + \cdots + e_{2u} s^{2u}}{1 + f_2 s^2 + f_4 s^4 + \cdots + f_{2v} s^{2v}}\bigg|_{s=j\omega} \tag{7-138}$$

Dividing the numerator by the denominator once on the right-hand side of the last equation, we have

$$\frac{|M(j\omega)|^2}{|L(j\omega)|^2} = 1 + \frac{(e_2 - f_2)s^2 + (e_4 - f_4)s^4 + \cdots}{1 + f_2 s^2 + f_4 s^4 + \cdots + f_{2v} s^{2v}}\bigg|_{s=j\omega} \tag{7-139}$$

If $u = v$, the last term in the numerator of Eq. (7-139) will be $(e_{2u} - f_{2u})s^{2u}$. However, if $u < v$, as in most practical cases, then beyond the term $(e_{2u} - f_{2u})s^{2u}$, in addition, there will be

$$- f_{2(u+1)} s^{2(u+1)} - \cdots - f_{2v} s^{2v}$$

in the numerator of Eq. (7-139).

We see that to satisfy the condition of Eq. (7-135), one possible set of approximating solutions is obtained from Eq. (7-139):

$$e_2 = f_2$$
$$e_4 = f_4$$
$$e_6 = f_6$$
$$\vdots$$
$$e_{2u} = f_{2u}$$

$$\tag{7-140}$$

if $u = v$. If $u < v$, then the error generated by the low-order model is

error criterion

$$|\epsilon| = \frac{|M(j\omega)|^2}{|L(j\omega)|^2} - 1 = \frac{- f_{2(u+1)} s^{2(u+1)} - \cdots - f_{2v} s^{2v}}{1 + f_2 s^2 + f_4 s^4 + \cdots + f_{2v} s^{2v}}\bigg|_{s=j\omega} \tag{7-141}$$

The conditions in Eq. (7-140) are used to solve for the unknown coefficients in $L(s)$ once $M(s)$ is given. This is done by writing

$$\frac{M(s)M(-s)}{L(s)L(-s)} = \frac{[1 + m_1 s + m_2 s^2 + \cdots + m_u s^u][1 - m_1 s + m_2 s^2 - \cdots + (-1)^u m_u s^u]}{[1 + l_1 s + l_2 s^2 + \cdots + l_v s^v][1 - l_1 s + l_2 s^2 - \cdots + (-1)^v l_v s^v]}$$

$$= \frac{1 + e_2 s^2 + e_4 s^4 + \cdots + e_{2u} s^{2u}}{1 + f_2 s^2 + f_4 s^4 + \cdots + f_{2v} s^{2v}} \tag{7-142}$$

Equating both sides of Eq. (7-142), we can express e_2, e_4, . . . , e_{2u} in terms of m_1, m_2, . . . , m_u. Similar relationships can be obtained for f_2, f_4, . . . , f_{2v} in terms of l_1, l_2, . . . , l_v.

solutions

As an illustrative example, for $u = 8$, Eq. (7-142) gives

$$e_2 = 2m_2 - m_1^2$$
$$e_4 = 2m_4 - 2m_1m_3 + m_2^2$$
$$e_6 = 2m_6 - 2m_1m_5 + 2m_2m_4 - m_3^2$$
$$e_8 = 2m_8 - 2m_1m_7 + 2m_2m_6 - 2m_3m_5 + m_4^2 \qquad (7\text{-}143)$$
$$e_{10} = 2m_2m_8 - 2m_3m_7 + 2m_4m_6 - m_5^2$$
$$e_{12} = 2m_4m_8 - 2m_5m_7 + m_6^2$$
$$e_{14} = 2m_6m_8 - m_7^2$$
$$e_{16} = m_8^2$$

In general,

$$e_{2x} = \sum_{i=0}^{x-1} (-1)^i 2m_i m_{2x-i} + (-1)^x m_x^2 \qquad (7\text{-}144)$$

for $x = 1, 2, \ldots , u$, and $m_0 = 1$. Similarly,

$$f_{2y} = \sum_{i=0}^{y-1} (-1)^i 2l_i l_{2y-i} + (-1)^y l_y^2 \qquad (7\text{-}145)$$

for $y = 1, 2, \ldots , v$, and $l_0 = 1$.

illustrative examples

The following examples will illustrate the method of simplification of linear systems outlined in this section.

EXAMPLE 7-6

Consider that the open-loop transfer function of a unity-feedback control system is given as

$$G(s) = \frac{8}{s(s^2 + 6s + 12)} \qquad (7\text{-}146)$$

The closed-loop transfer function is

$$M(s) = \frac{C(s)}{R(s)} = \frac{8}{s^3 + 6s^2 + 12s + 8}$$

Divide Top/bottom by 8 gives

$$= \frac{1}{1 + 1.5s + 0.75s^2 + 0.125s^3} \qquad (7\text{-}147)$$

The poles of the closed-loop transfer function are at $s = -2$, -2, and -2. The low-order approximating system is considered for the following two cases.

Case 1

Consider that the simplified low-order system is of the second order, and the transfer function is

$$L(s) = \frac{1}{1 + d_1 s + d_2 s^2} \tag{7-148}$$

Equation (7-136) gives

$$\frac{M(s)}{L(s)} = \frac{1 + d_1 s + d_2 s^2}{1 + 1.5s + 0.75s^2 + 0.125s^3}$$

$$= \frac{1 + m_1 s + m_2 s^2}{1 + l_1 s + l_2 s^2 + l_3 s^3} \tag{7-149}$$

Thus,

$$l_1 = 1.5 \qquad l_2 = 0.75 \qquad l_3 = 0.125 \qquad d_1 = m_1 \qquad d_2 = m_2$$

Using Eq. (7-142), we have

$$\frac{M(s)M(-s)}{L(s)L(-s)} = \frac{1 + e_2 s^2 + e_4 s^4}{1 + f_2 s^2 + f_4 s^4 + f_6 s^6} \tag{7-150}$$

Now using Eqs. (7-140) and (7-143), the following nonlinear equations are obtained:

$$e_2 = f_2 = 2m_2 - m_1^2 = 2d_2 - d_1^2$$

$$e_4 = f_4 = 2m_4 - 2m_1 m_3 + m_2^2 = m_2^2 = d_2^2$$

Similarly,

$$f_2 = 2l_2 - l_1^2 = 1.5 - (1.5)^2 = -0.75$$

$$f_4 = 2l_4 - 2l_1 l_3 + l_2^2 = 0.1875$$

$$f_6 = 2l_6 - 2l_1 l_5 + 2l_2 l_4 - l_3^2 = -l_3^2 = -0.156$$

Solving the above equations, we have

$$d_1^2 = 1.616 \qquad d_2^2 = 0.1875$$

Thus,

$$d_1 = 1.271 \qquad d_2 = 0.433$$

where we have taken the positive values so that $L(s)$ will represent a stable system. Thus, the second-order simplified system transfer function is

$$L(s) = \frac{1}{1 + 1.271s + 0.433s^2} = \frac{2.31}{s^2 + 2.936s + 2.31} \tag{7-151}$$

The poles of $L(s)$ are at $s = -1.468 + j0.384$ and $-1.468 - j0.384$. Thus, the third-order system with the transfer function $M(s)$, which has real poles at -2, -2, and -2, is approxi-

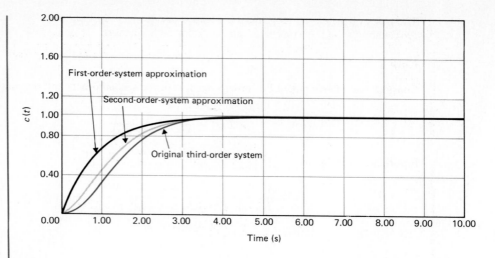

Original system: $\qquad M(s) = \dfrac{1}{1 + 1.5s + 0.75s^2 + 0.125s^3}$

Second-order system: $\qquad L(s) = \dfrac{1}{1 + 1.271s + 0.433s^2}$

First-order system: $\qquad L(s) = \dfrac{1}{1 + 0.866s}$

FIGURE 7-36 Approximation of a third-order system by a first-order system and a second-order system.

mated by the second-order system with the two complex-conjugate poles. The open-loop transfer function of the second-order system is

$$G_L(s) = \frac{2.31}{s(s + 2.936)} \tag{7-152}$$

Viewing it from the open-loop-system standpoint, the third-order system with open-loop poles at $s = 0$, $-3 + j1.732$ and $-3 - j1.732$ and a gain of 8 is approximated by a second-order system with the open-loop poles at $s = 0$ and -2.936 and a gain of 2.31.

Figure 7-36 shows the unit-step responses of the systems with the closed-loop transfer functions $M(s)$ and $L(s)$. Notice that the second-order approximating system has a faster rise time. This is due to the two complex-conjugate poles of $L(s)$.

Case 2

Now let us use a first-order system to approximate the closed-loop transfer of Eq. (7-147). We choose

$$L(s) = \frac{1}{1 + d_1 s} \tag{7-153}$$

Then

$$\frac{M(s)}{L(s)} = \frac{1 + d_1 s}{1 + 1.5s + 0.75s^2 + 0.125s^3} = \frac{1 + m_1 s}{1 + l_1 s + l_2 s^2 + l_3 s^3} \tag{7-154}$$

Thus,

$$l_1 = 1.5 \qquad l_2 = 0.75 \qquad l_3 = 0.125 \qquad d_1 = m_1$$

and

$$\frac{M(s)M(-s)}{L(s)L(-s)} = \frac{1 + e_2 s^2}{1 + f_2 s^2 + f_4 s^4 + f_6 s^6} \tag{7-155}$$

Thus,

$$e_2 = f_2 = 2d_2 - d_1^2 = -d_1^2 = -0.75$$
$$f_4 = 0.1875$$
$$f_6 = -0.156$$

This gives $d_1 = 0.866$. The transfer function of $L(s)$ is

$$L(s) = \frac{1}{1 + 0.866s} = \frac{1.1547}{s + 1.1547} \tag{7-156}$$

The open-loop transfer function of the first-order system is

$$G_L(s) = \frac{1.1547}{s} \tag{7-157}$$

The unit-step response of the first-order system is shown in Figure 7-36. As expected, the first-order system gives a more inferior approximation to the third-order system than the second-order system.

EXAMPLE 7-7

Consider the following closed-loop transfer function:

$$M(s) = \frac{1}{(1 + s + 0.5s^2)(1 + Ts)} = \frac{1}{1 + (1 + T)s + (0.5 + T)s^2 + 0.5s^3} \tag{7-158}$$

where T is a positive variable parameter.

We shall investigate the approximation of $M(s)$ by a second-order-system model when T takes on various values.

Let the second-order system be modeled by the transfer function

$$L(s) = \frac{1}{1 + d_1 s + d_2 s^2} \tag{7-159}$$

Substituting Eqs. (7-158) and (7-159) into Eq. (7-136), we get

$$\frac{M(s)}{L(s)} = \frac{1 + d_1 s + d_2 s^2}{1 + (1 + T)s + (0.5 + T)s^2 + 0.5Ts^3} = \frac{1 + m_1 s + m_2 s^2}{1 + l_1 s + l_2 s^2 + l_3 s^3} \quad (7\text{-}160)$$

Then,

$$l_1 = 1 + T \qquad l_2 = 0.5 + T \qquad l_3 = 0.5T$$

From Eq. (7-144), we get

$$e_2 = f_2 = 2d_2 - d_1^2 \qquad e_4 = f_4 = m_2^2 = d_2^2$$

From Eq. (7-145), we get

$$f_2 = 2l_2 - l_1^2 = 2(0.5 + T) - (1 + T)^2 = -T^2$$

$$f_4 = 2l_4 - 2l_1 l_3 + l_2^2 = -2(1 + T)(0.5T) + (0.5 + T)^2 = 0.25$$

$$f_6 = -l_3^2 = -(0.5T)^2 = -0.25T^2$$

From these previous equations, we solve for the values of d_1 and d_2, and the results are

$$d_1 = \sqrt{1 + T^2} \qquad d_2 = 0.5$$

Thus, the transfer function of the second-order approximating system is

$$L(s) = \frac{1}{1 + \sqrt{1 + T^2}\, s + 0.5 s^2} \quad (7\text{-}161)$$

The two poles of $L(s)$ are at

$$s = -\sqrt{1 + T^2} + j\sqrt{1 - T^2} \qquad \text{and} \qquad s = -\sqrt{1 + T^2} - j\sqrt{1 - T^2}$$

The poles of $M(s)$ are at $s = -1 + j$, $-1 - j$, and $-1/T$. The two poles of $L(s)$ are tabulated for $T = 0$ to $T = 6$:

$T = 0$	$s = -1.000 + j$	$s = -1.000 - j$
$T = 0.1$	$s = -1.005 + j0.995$	$s = -1.005 - j0.995$
$T = 0.25$	$s = -1.031 + j0.968$	$s = -1.031 - j0.968$
$T = 0.5$	$s = -1.118 + j0.866$	$s = -1.118 - j0.866$
$T = 0.75$	$s = -1.250 + j0.661$	$s = -1.250 - j0.661$
$T = 1.0$	$s = -1.414$	$s = -1.414$
$T = 2.0$	$s = -3.968$	$s = -0.504$
$T = 4.0$	$s = -7.996$	$s = -0.250$
$T = 6.0$	$s = -11.999$	$s = -0.167$

Notice that as the value of T decreases, the original system approaches a second-order system, as the pole at $-1/T$ moves toward $-\infty$. Thus, when $T = 0$, $L(s)$ is identical to $M(s)$, and the approximate solution is exact. As the value of T increases, the pole at $-1/T$ moves toward the origin, and becomes more dominant; the poles of $L(s)$ move toward the real axis, and eventually when $T \geq 1$, the poles become real. As T increases, one of the real poles of the second-order system moves toward $-\infty$, and the other approaches $-1/T$. Thus, the second-order approximation is good for small and large values of T.

Figures 7-37 through 7-40 illustrate the unit-step responses of the third-order system and the second-order approximating system when $T = 0.1, 0.5, 1.0,$ and 6, respectively. When the

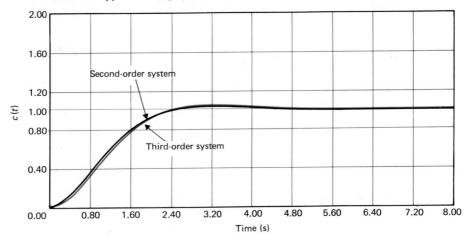

Original system: $M(s) = \dfrac{1}{(1 + s + 0.5s^2)(1 + 0.1s)}$

Second-order system: $L(s) = \dfrac{2}{s^2 + 2.01s + 2}$

FIGURE 7-37 Approximation of a third-order system by a second-order system.

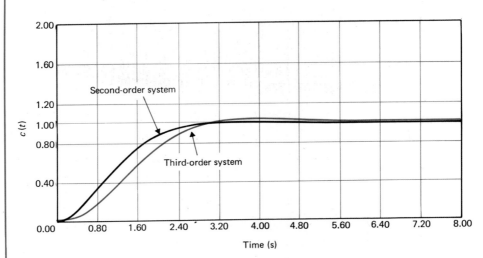

Original system: $M(s) = \dfrac{1}{(1 + s + 0.5s^2)(1 + 0.5s)}$

Second-order system: $L(s) = \dfrac{2}{s^2 + 2.2361s + 2}$

FIGURE 7-38 Approximation of a third-order system by a second-order system.

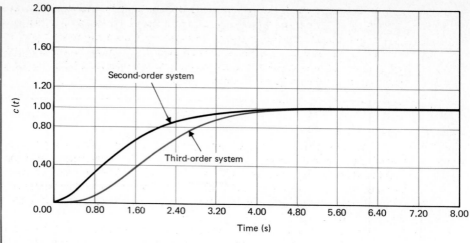

Original system: $M(s) = \dfrac{1}{(1 + s + 0.5s^2)(1 + s)}$

Second-order system: $L(s) = \dfrac{2}{s^2 + 2.828s + 2}$

FIGURE 7-39 Approximation of a third-order system by a second-order system.

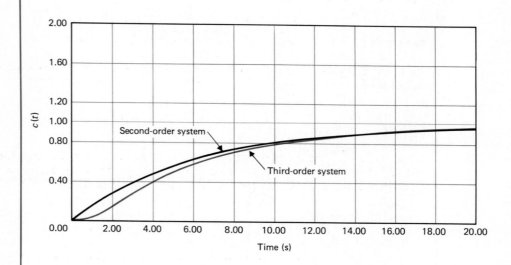

Original system: $M(s) = \dfrac{1}{(1 + s + 0.5s^2)(1 + 6s)}$

Second-order system: $L(s) = \dfrac{2}{s^2 + 12.17s + 2}$

FIGURE 7-40 Approximation of a third-order system by a second-order system.

value of T is very small, the unit-step responses of the third-order and the second-order systems are very close. The error between the two responses increases as T is increased to 0.5 and 1.0. Figure 7-40 shows that the response of the two systems again move closer to each other when $T = 6$. As the value of T increases, the real pole at $s = -1/T$ becomes dominant relative to the two complex-conjugate poles, and the second-order model for $L(s)$ in Eq. (7-159) would give a better approximation to the third-order system.

EXAMPLE 7-8
*position-control
system*

The third-order position-control system described in Section 7-6 is approximated by a second-order system, since the small value of the motor inductance can be neglected. It should be noted that the second-order model arrived at in Eq. (7-99) is intended for all values of K. We have demonstrated that the approximation is relatively good for low values of K and inferior for large values of K. Strictly speaking, we should find an approximating second-order model for all positive values of K.

Now let us apply the approximation method discussed in this section to the third-order position-control system. The closed-loop transfer function of the system is given in Eq. (7-116) and is repeated:

$$M(s) = \frac{1.5 \times 10^7 K}{s^3 + 3408.3s^2 + 1,204,000s + 1.5 \times 10^7 K} \qquad (7\text{-}162)$$

For a given value of K, this third-order transfer function is to be approximated by a second-order transfer function of the form

$$L(s) = \frac{1}{1 + d_1 s + d_2 s^2} \qquad (7\text{-}163)$$

Once $L(s)$ is determined, the open-loop transfer function of the approximating second-order system is written

$$G(s) = \frac{d_1/d_2}{s(s + d_1/d_2)} \qquad (7\text{-}164)$$

The coefficients d_1 and d_2 are determined for $K = 7.248$, 14.5, and 100, using the method described in this section. The results are tabulated as follows, where the first column of $G(s)$ denotes the approximation with $L_a = 0$, and the second column represents $G(s)$ with the values of d_1 and d_2:

*comparison of
results*

K	$G(s)$	$G(s)$
7.248	$\dfrac{32616}{s(s + 361.2)}$	$\dfrac{35829.5}{s(s + 385.5)}$
14.5	$\dfrac{65,250}{s(s + 361.2)}$	$\dfrac{71,684.6}{s(s + 378.2)}$
100.0	$\dfrac{450,000}{s(s + 361.2)}$	$\dfrac{494,300}{s(s + 188.84)}$

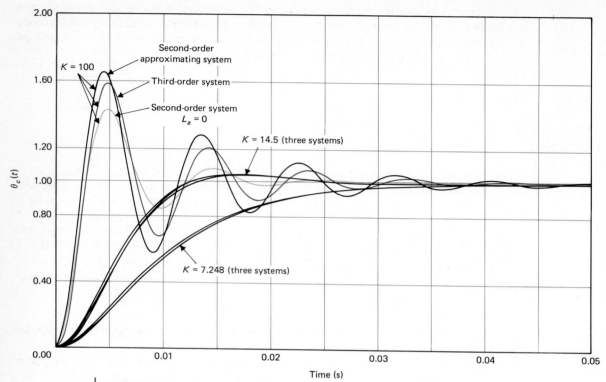

FIGURE 7-41 Comparison of the unit-step responses of the systems in Example 7-8.

Comparisons of the unit-step responses of the third-order system, and the second-order systems with $L_a = 0$ and with the approximation method are shown in Fig. 7-41 for the three values of K indicated. When the value of K is relatively small, both second-order approximations are quite good. For $K = 100$, the second-order system with $L_a = 0$ has a lower maximum overshoot and is less oscillatory than the true third-order system. The second-order system with the approximation method has a step response that is closer to the actual response, although the maximum overshoot is slightly larger.

One of the problems that surfaces from this investigation is that the low-order-approximation method described in this section provides a system that is less damped than the high-order system. This means that when the original system is lightly damped, it is possible that the dominant roots of the low-order system will end up in the right-half s-plane. For the present system, we run into this difficulty when $K = 181.2$, one of the values chosen in Section 7-6. Since d_1 and d_2 are solved from $d_1^2 = 2d_2 - f_2$, if f_2 is greater than $2d_2$, no real solution could be obtained for d_1. For the present system, we can show that when $K = 129.3$, the two poles of the second-order approximating system will be on the $j\omega$-axis. To approximate the third-order system when K is large would require an $L(s)$ that contains a zero.

In conclusion, we would like to point out that the problem of reduced-order approximation of a high-order system is a complex subject. Most of the existing methods have flaws of one type or another. The serious reader should refer to the literature [1, 2] for more in-depth treatments of the subject.

7-10
TIME RESPONSE OF DISCRETE-DATA SYSTEMS: INTRODUCTION

We have learned in Chapter 3 that the output responses of most discrete-data control systems are functions of the continuous time variable t. Thus, the time-domain specifications such as the maximum overshoot, rise time, damping ratio, etc. can still be applied to discrete-data systems. The only difference is that in order to make use of the analytical tools such as the z-transforms, the continuous data found in a discrete-data system are sampled so that the independent time variable is kT, where T is the sampling period in seconds. Also, instead of working in the s-plane, the transient performance of a discrete-data system is characterized by poles and zeros of the transfer function in the z-plane.

The objectives of these following sections are as follows:

1. To present methods of finding the discretized time responses of discrete-data control systems.
2. To describe the important characteristics of the discretized time response $c(kT)$.
3. To establish the significance of pole and zero locations in the z-plane.
4. To provide comparison between time responses of continuous-data and discrete-data control systems.

Figure 7-42 shows the block diagram of a simple discrete-data control system. The output $c(t)$ is sampled by the fictitious sampler to provide the discretized output $c^*(t)$. The closed-loop transfer function of the system is written

$$\frac{C(z)}{R(z)} = \frac{G(z)}{1 + GH(z)} \tag{7-165}$$

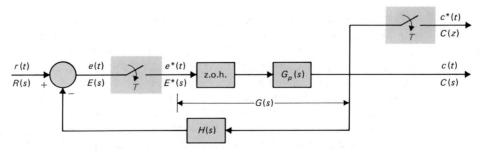

FIGURE 7-42 Block diagram of the discrete-data control system.

where $GH(z)$ denotes the z-transform of $G(s)H(s)$. Once the input $R(z)$ is given, the output sequence $c(kT)$ can be determined using one of the following two methods:

z-transform

> **1.** Take the inverse z-transform of $C(z)$ using the z-transform table.
> **2.** Expand $C(z)$ into a power series of z^{-k}.

$C(z)$ is defined as

$$C(z) = \sum_{k=0}^{\infty} c(kT)z^{-k} \tag{7-166}$$

sampling period

The discrete-time response $c(kT)$ can be determined by referring to the coefficient of z^{-k} for $k = 0, 1, 2, \ldots$. Remember that $c^*(t)$ contains only the sampled information on $c(t)$ at the sampling instants. If the sampling period is large relative to the most significant time constant of the system, $c^*(t)$ may not be an accurate representation of $c(t)$.

EXAMPLE 7-9   Consider that the position-control system described in Section 7-6 has discrete data in the forward path, so that the system is now described by the block diagram of Fig. 7-42. For $K = 14.5$, the transfer function of the controlled process is

$$G_p(s) = \frac{65{,}250}{s(s + 361.2)} \tag{7-167}$$

The open-loop transfer function of the discrete-data system is

$$G_{h0}G_p(z) = \mathscr{z}[G_{h0}(s)G_p(s)] = (1 - z^{-1})\mathscr{z}\left[\frac{G_p(s)}{s}\right] \tag{7-168}$$

SDCS

The z-transform in the last equation can be evaluated analytically using the z-transform table or with the SDCS program of the ACSP software. For a sampling period of $T = 0.001$ second, the result is

$$G_{h0}G_p(z) = \frac{0.029z + 0.0257}{z^2 - 1.697z + 0.697} \tag{7-169}$$

The closed-loop transfer function is

$$\frac{C(z)}{R(z)} = \frac{G_{h0}G_p(z)}{1 + G_{h0}G_p(z)} = \frac{0.029z + 0.0257}{z^2 - 1.668z + 0.7226} \tag{7-170}$$

where $R(z)$ and $C(z)$ represent the z-transforms of the input and the output, respectively. For a unit-step input, $R(z) = z/(z - 1)$. The output transform $C(z)$ becomes

$$C(z) = \frac{z(0.029z + 0.0257)}{(z - 1)(z^2 - 1.668z + 0.7226)} \tag{7-171}$$

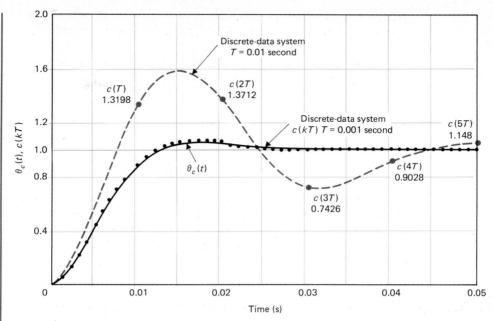

FIGURE 7-43 Comparison of unit-step responses of discrete-data and continuous-data systems.

The output sequence $c(kT)$ can be determined by dividing the numerator polynomial of $C(z)$ by its denominator polynomial to yield a power series in z^{-1}, or, more conveniently, it can be computed by the SDCS program. Figure 7-43 shows the plot of $c(kT)$, as dots, versus kT when $T = 0.001$ second. For comparison, the unit-step response of the continuous-data system in Section 7-6 with $K = 14.5$ is shown in the same figure. As seen from Fig. 7-43, when the sampling period is small, the output responses of the discrete-data and the continuous-data systems are

maximum overshoot

very similar. The maximum value of $c(kT)$ is 1.0731, or a 7.31 percent maximum overshoot, as against the 4.3 percent maximum overshoot for the continuous-data system. When the sampling period is increased to 0.01 second, the open-loop transfer function of the discrete-data system is

$$G_{h0}G_p(z) = \frac{1.3198z + 0.4379}{z^2 - 1.027z + 0.027} \tag{7-172}$$

and the closed-loop transfer function becomes

$$\frac{C(z)}{R(z)} = \frac{1.3198z + 0.4379}{z^2 + 0.2929z + 0.4649} \tag{7-173}$$

sampling-period effect

The output sequence $c(kT)$ with $T = 0.01$ s is shown in Fig. 7-43 with $k = 0, 1, 2, 3, 4,$ and 5. The true continuous-time output of the discrete-data system is shown as the dotted curve. Notice that the maximum value of $c(kT)$ is 1.3712, but the true maximum overshoot is considerably higher than that. Thus, the larger sampling period not only makes the system less stable, but the sampled output no longer gives an accurate measure of the true output.

When the sampling period is increased to 0.01658 second, the characteristic equation of the discrete-data system is

$$z^2 + 1.4938z + 0.4939 = 0 \tag{7-174}$$

which has roots at $z = -0.494$ and $z = -1.000$. The root at $z = -1.000$ causes the step response of the system to oscillate with a constant amplitude, and the system is marginally stable. Thus, for all sampling periods greater than 0.01658 second, the discrete-data system will be unstable. From Section 7-6, we learned that the second-order continuous-data system is always stable for finite positive values of K. For the discrete-data system, the sample-and-hold has the effect of making the system less stable, and if the value of T is too large, the second-order system can become unstable. Figure 7-44 shows the trajectories of the two characteristic-equation roots of the discrete-data system as the sampling period T varies. Notice that when the sampling period is very small, the two characteristic-equation roots are very close to the $z = 1$ point, and are complex. When $T = 0.01608$ second, the two roots become equal and real and are negative. Unlike the continuous-data system, the case of two identical roots on the *negative* real axis in the z-plane *does not* correspond to critical damping. For discrete-data systems, when one or more characteristic-equation roots lie on the negative real axis of the z-plane, the system response will oscillate with positive and negative peaks. Figure 7-45 shows the oscillatory re-

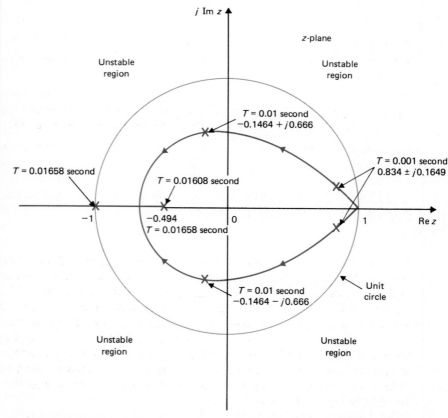

FIGURE 7-44 Trajectories of roots of a second-order discrete-data control system as the sampling period T varies.

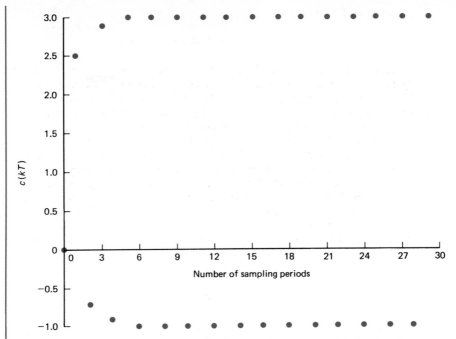

FIGURE 7-45 Oscillatory response of a discrete-data system with the sampling period $T = 0.01658$ second.

sponse for $c(kT)$ when $T = 0.01658$ second, which is the critical value for stability. Beyond this value of T, one root will move outside the unit circle, and the system becomes unstable.

7-11

MAPPING BETWEEN s-PLANE AND z-PLANE TRAJECTORIES

It is important to study the relation between the location of the characteristic-equation roots in the z-plane and the time response of the discrete-data system. In Chapter 3, the periodic property of the Laplace transform of the sampled signal, $R^*(s)$, is established by Eq. (3-118); i.e., $R^*(s + jm\omega_s) = R^*(s)$, where m is an integer. In other words, given any point s_1 in the s-plane, the function $R^*(s)$ has the same value at all periodic points $s = s_1 + jm\omega_s$. Thus, the s-plane is divided into an infinite number of periodic strips, as shown in Fig. 7-46(a). The strip between $\omega = \omega_s/2$ is called the **primary strip** and all others at higher frequencies are called the **complementary strips**. Figure 7-46(b) shows the mapping of the periodic strips from the s-plane to the z-plane, and the details are explained as follows.

periodic strips

1. The $j\omega$-axis in the s-plane is mapped onto the unit circle $|z| = 1$ in the z-plane.

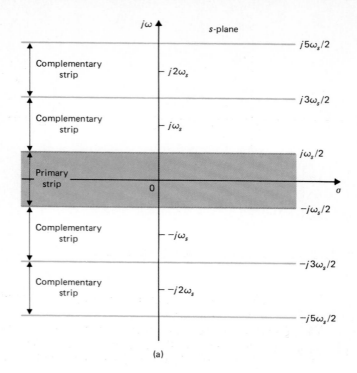

(a)

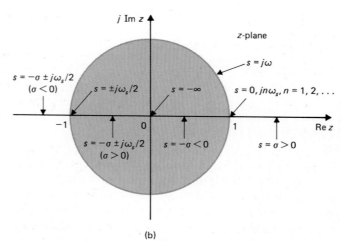

FIGURE 7-46 Periodic strips in the s-plane and the corresponding points and lines between the s-plane and the z-plane.

(b)

2. The boundaries of the periodic strips, $s = jm\omega_s/2$, $m = \pm 1$, ± 3, ± 5, . . . , are mapped onto the negative real axis of the z-plane. The portion inside the unit circle corresponds to $\sigma < 0$, and the portion outside the unit circle corresponds to $\sigma > 0$.

3. The center lines of the periodic strips, $s = jm\omega_s$, $m = 0$, ± 2, ± 4, . . . , are mapped onto the positive real axis of the z-plane. The portion inside

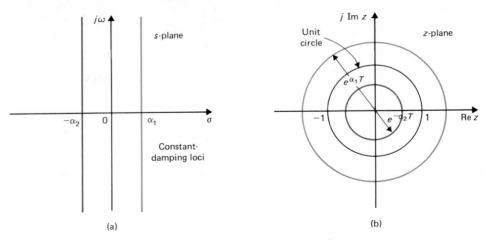

FIGURE 7-47 Constant-damping loci in the s-plane and the z-plane.

the unit circle corresponds to $\sigma < 0$, and the portion outside the unit circle corresponds to $\sigma > 0$.

4. Regions shown in the periodic strips in the left-half s-plane are mapped onto the interior of the unit circle in the z-plane.
5. The point $z = 1$ in the z-plane corresponds to the origin, $s = 0$, in the s-plane.
6. The origin, $z = 0$, in the z-plane corresponds to $s = -\infty$ in the s-plane.

In the time-domain analysis of continuous-data systems, we devised the damping factor α, the damping ratio ζ, and the natural undamped frequency ω_n to characterize the system dynamics. The same parameters can be defined for discrete-data systems with respect to the characteristic-equation roots in the z-plane. The loci of the constant-α, constant-ζ, constant-ω, and constant-ω_n in the z-plane are described in the following.

CONSTANT-DAMPING LOCI.   For a constant-damping factor $\sigma = \alpha$ in the s-plane, the corresponding trajectory in the z-plane is described by

$$z = e^{\alpha T} \tag{7-175}$$

which is a circle centered at the origin with a radius of $e^{\alpha T}$, as shown in Fig. 7-47.

CONSTANT-FREQUENCY LOCI.   The constant-frequency $\omega = \omega_1$ locus in the s-plane is a horizontal line parallel to the σ-axis. The corresponding z-plane locus is a straight line emanating from the origin at an angle of $\theta = \omega_1 T$ radian, measured from the real axis, as shown in Fig. 7-48.

CONSTANT-NATURAL-UNDAMPED-FREQUENCY LOCI.   The constant-ω_n loci in the s-plane are concentric circles with the center at the origin, and the radius is ω_n. The

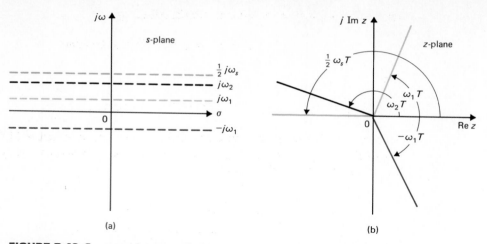

FIGURE 7-48 Constant-frequency loci in the s-plane and the z-plane.

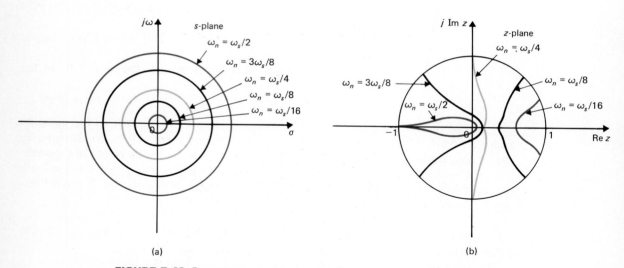

FIGURE 7-49 Constant-natural-undamped-frequency loci in the s-plane and the z-plane.

corresponding constant-ω_n loci in the z-plane are shown in Fig. 7-49 for $\omega_n = \omega_s/16$ to $\omega_n = \omega_s/2$. Only the loci inside the unit circle are shown.

CONSTANT-DAMPING RATIO LOCI.   For a constant-damping ratio ζ, the s-plane loci are described by

$$s = -\omega \tan \beta + j\omega \tag{7-176}$$

The constant-ζ loci in the z-plane are described by

$$z = e^{Ts} = e^{-2\pi\omega(\tan\beta)/\omega_s} \angle 2\pi\omega/\omega_s \qquad (7\text{-}177)$$

where

$$\beta = \sin^{-1}\zeta = \text{constant} \qquad (7\text{-}178)$$

For a given value of β, the constant-ζ locus in the z-plane, described by Eq. (7-177), is a logarithmic spiral for $0° < \beta < 90°$. Figure 7-50 shows several typical constant-ζ loci in the top half of the z-plane.

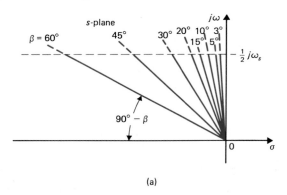

(a)

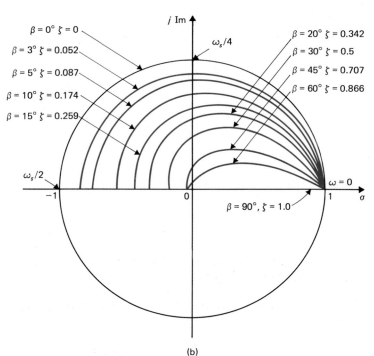

(b)

FIGURE 7-50 Constant-damping-ratio loci in the s-plane and the z-plane.

7-12

RELATION BETWEEN CHARACTERISTIC-EQUATION ROOTS AND TRANSIENT RESPONSE

Based on the discussions given in the last section, we can establish the basic relation between the characteristic-equation roots and the transient response of a discrete-data system, keeping in mind that, in general, the zeros of the closed-loop transfer function will also play an important role on the response, but not on the stability, of the system.

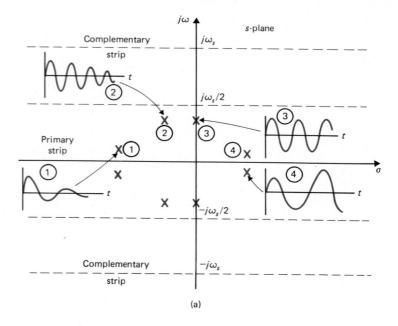

(a)

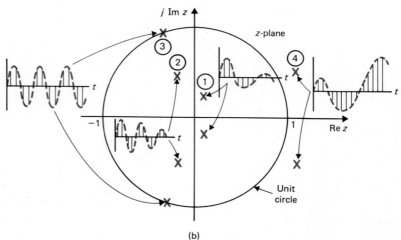

FIGURE 7-51 (a) Transient responses corresponding to various pole locations of $C^*(s)$ in the s-plane (complex-conjugate poles only). (b) Transient-response sequence corresponding to various pole locations of $C(z)$ in the z-plane.

(b)

Roots on the Positive Real Axis in the z-Plane

Roots on the positive real axis inside the unit circle of the z-plane give rise to responses that decay exponentially with an increase of kT. Typical responses relative to the root locations are shown in Fig. 7-51. The roots closer to the unit circle will decay slower. When the root is at $z = 1$, the response has a constant amplitude. Roots outside the unit circle correspond to unstable systems, and the responses will increase with kT.

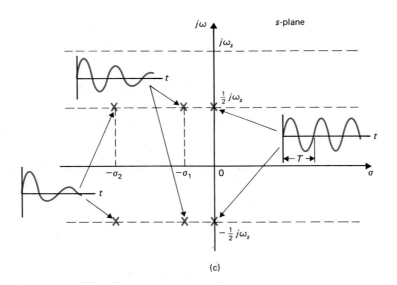

(c)

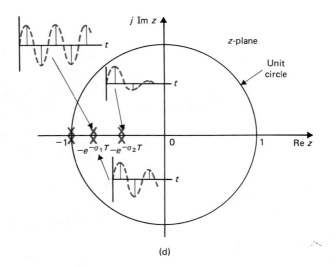

FIGURE 7-51 (cont.) (c) Transient responses corresponding to various pole locations of $C^*(s)$ in the s-plane (complex-conjugate poles on the boundaries between periodic strips). (d) Transient-response sequence corresponding to various pole locations of $C(z)$ in the z-plane.

(d)

Roots on the Negative Real Axis in the z-Plane

The negative real axis of the z-plane corresponds to the boundaries of the periodic strips in the s-plane. For example, when $s = -\sigma_1 \pm j\omega_s/2$, the complex-conjugate points are on the boundaries of the primary strip in the s-plane. The corresponding z-plane points are

$$z = e^{-\sigma_1 T} e^{\pm j\omega_s T/2} = -e^{-\sigma_1 T} \tag{7-179}$$

periodic strips

which are on the negative real axis of the z-plane. For the frequency of $\omega_s/2$, the output sequence will have exactly one sample in each one-half period of the envelope. Thus, the output sequence will occur in alternating positive and negative pulses, as shown in Fig. 7-51.

Complex-Conjugate Roots in the z-Plane

Complex-conjugate roots inside the unit circle in the z-plane correspond to oscillatory responses that decay with an increase in kT. Roots that are closer to the unit circle will decay slower. As the roots move toward the second and the third quadrants, the frequency of oscillation of the response increases. Refer to Fig. 7-51 for typical examples.

7-13

STEADY-STATE-ERROR ANALYSIS OF DISCRETE-DATA SYSTEMS

Since the input and output signals of a typical discrete-data control system are continuous-time functions, as shown in the block diagram of Fig. 7-52, the error signal should still be defined as

$$e(t) = r(t) - c(t) \tag{7-180}$$

where $r(t)$ is the input, and $c(t)$ is the output. Due to the discrete data that appear inside the system, z-transform or difference equations are often used, so that the input and output are represented in sampled form, $r(kT)$ and $c(kT)$, respectively. Thus, the

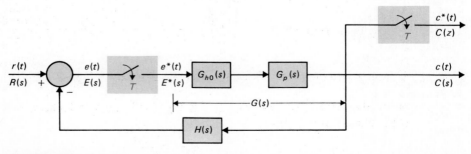

FIGURE 7-52 A discrete-data control system.

error signal is more appropriately represented by $e*(t)$ or $e(kT)$. That is,

$$e*(t) = r*(t) - c*(t) \tag{7-181}$$

or

$$e(kT) = r(kT) - c(kT) \tag{7-182}$$

The steady-state error at the sampling instants is defined as

$$e_{ss}^* = \lim_{t \to \infty} e*(t) = \lim_{k \to \infty} e(kT) \tag{7-183}$$

By using the final-value theorem of the z-transform, the steady-state error is

final-value
theorem

$$e_{ss}^* = \lim_{k \to \infty} e(kT) = \lim_{z \to \infty} (1 - z^{-1})E(z) \tag{7-184}$$

provided that the function $(1 - z^{-1})E(z)$ does not have any pole on or outside the unit circle in the z-plane. It should be pointed out that since the true error of the system is $e(t)$; e_{ss}^* predicts only the steady-state error of the system at the sampling instants.

By expressing $E(z)$ in terms of $R(z)$ and $G(z)$, Eq. (7-184) is written

$$e_{ss}^* = \lim_{k \to \infty} e(kT) = \lim_{z \to 1} (1 - z^{-1}) \frac{R(z)}{1 + G_{h0} G_p(z)} \tag{7-185}$$

This expression shows that the steady-state error depends on the reference input $R(z)$ as well as the open-loop transfer function $G(z)$. Just as in the continuous-data systems, we *error constant* shall consider three basic types of input signals and the associated error constants and relate the e_{ss}^* to these and the type of the system.

Let the transfer function of the controlled process in the system in Fig. 7-52 be of the form

$$G_p(s) = \frac{K(1 + T_a s)(1 + T_b s) \cdots (1 + T_m s)}{s^j(1 + T_1 s)(1 + T_2 s) \cdots (1 + T_n s)} \tag{7-186}$$

where $j = 0, 1, 2, \ldots$. The transfer function $G_{h0} G_p(z)$ is

$$G_{h0} G_p(z) = (1 - z^{-1}) \mathcal{Z} \left[\frac{K(1 + T_a s)(1 + T_b s) \cdots (1 + T_m s)}{s^{j+1}(1 + T_1 s)(1 + T_2 s) \cdots (1 + T_n s)} \right] \tag{7-187}$$

Steady-State Error Due to a Step-Function Input

When the input to the system, $r(t)$, in Fig. 7-52 is a step function with magnitude R, the z-transform of $r(t)$ is

$$R(z) = \frac{Rz}{z - 1} \tag{7-188}$$

Substituting $R(z)$ into Eq. (7-185), we get

$$e_{ss}^* = \lim_{z \to 1} \frac{R}{1 + G_{h0} G_p(z)} = \frac{R}{1 + \lim\limits_{z \to 1} G_{h0} G_p(z)} \qquad (7\text{-}189)$$

Let the **step-error constant** be defined as

$$K_p^* = \lim_{z \to 1} G_{h0} G_p(z) \qquad (7\text{-}190)$$

Equation (7-189) becomes

$$e_{ss}^* = \frac{R}{1 + K_p^*} \qquad (7\text{-}191)$$

Thus, we see that the steady-state error of the discrete-data control system in Fig. 7-52 is related to the step-error constant K_p^* in the same way as in the continuous-data case, except that K_p^* is given by Eq. (7-190).

We can relate K_p^* to the type of the system as follows.

For a type-0 system, $j = 0$ in Eq. (7-187), and the equation becomes

$$G_{h0} G_p(z) = (1 - z^{-1})\mathcal{z}\left[\frac{K(1 + T_a s)(1 + T_b s) \cdots (1 + T_m s)}{s(1 + T_1 s)(1 + T_2 s) \cdots (1 + T_n s)}\right] \qquad (7\text{-}192)$$

Performing the partial-fraction expansion to the function inside the square brackets of the last equation, we get

$$G_{h0} G_p(z) = (1 - z^{-1})\mathcal{z}\left(\frac{K}{s} + \text{terms due to the nonzero poles}\right)$$

$$= (1 - z^{-1})\left(\frac{Kz}{z - 1} + \text{terms due to the nonzero poles}\right) \qquad (7\text{-}193)$$

Since the terms due to the nonzero poles do not contain the term $(z - 1)$ in the denominator, the step-error constant is written

$$K_p^* = \lim_{z \to 1} G_{h0} G_p(z) = \lim_{z \to 1} (1 - z^{-1})\frac{Kz}{z - 1} = K \qquad (7\text{-}194)$$

Similarly, for a type-1 system, $G_{h0} G_p(z)$ will have an s^2 term in its denominator that corresponds to a term $(z - 1)^2$. This causes the step-error constant K_p^* to be infinite. The same is true for any system type greater than 1. The summary of the error constants and the steady-state error due to a step input is as follows:

TYPE OF SYSTEM	K_p^*	e_{ss}^*
0	K	$R/(1 + K)$
1	∞	0
2	∞	0

Steady-State Error Due to a Ramp-Function Input

When the reference input to the system in Fig. 7-52 is a ramp function of magnitude R, $r(t) = Rtu_s(t)$. The steady-state error in Eq. (7-185) becomes

$$e_{ss}^* = \lim_{z \to 1} \frac{RT}{(z - 1)[1 + G_{h0}G_p(z)]}$$

$$= \frac{R}{\lim\limits_{z \to 1} \dfrac{z - 1}{T} G_{h0}G_p(z)} \tag{7-195}$$

Let the **ramp-error constant** be defined as

$$K_v^* = \frac{1}{T} \lim_{z \to 1} [(z - 1)G_{h0}G_p(z)] \tag{7-196}$$

Then, Eq. (7-195) becomes

$$e_{ss}^* = \frac{R}{K_v^*} \tag{7-197}$$

The ramp-error constant K_v^* is meaningful only when the input $r(t)$ is a ramp function, and if the function $(z - 1)G_{h0}G_p(z)$ in Eq. (7-196) does not have any poles on or outside the unit circle $|z| = 1$. The relations between the steady-state error e_{ss}^*, K_v^*, and the type of system when the input is a ramp function with magnitude R are summarized as follows.

TYPE OF SYSTEM	K_v^*	e_{ss}^*
0	0	∞
1	K	R/K
2	∞	0

Steady-State Error Due to a Parabolic-Function Input

When the input is a parabolic function, $r(t) = Rtu_s(t)/2$; the z-transform of $r(t)$ is

$$R(z) = \frac{RT^2 z(z + 1)}{2(z - 1)^3} \tag{7-198}$$

From Eq. (7-185), the steady-state error at the sampling instants is

$$e_{ss} = \frac{T^2}{2} \lim_{z \to 1} \frac{R(z + 1)}{(z - 1)^2[1 + G_{h0}G_p(z)]}$$

$$= \frac{R}{\dfrac{1}{T^2} \lim_{z \to 1} (z - 1)^2 G_{h0} G_p(z)} \tag{7-199}$$

Now by defining the **parabolic-error constant** as

$$K_a^* = \frac{1}{T^2} \lim_{z \to 1} [(z - 1)^2 G_{h0} G_p(z)] \tag{7-200}$$

the steady-state error due a parabolic-function input is

$$e_{ss}^* = \frac{R}{K_a^*} \tag{7-201}$$

7-14
SUMMARY

This chapter is devoted to the time-domain analysis of linear continuous-data and discrete-data control systems. The time response of control systems is divided into the transient and the steady-state responses. The steady-state error is a measure of the accuracy of the system when time approaches infinity. For the step, ramp, and parabolic inputs, the steady-state error is characterized by the error constants K_p, K_v and K_a, respectively, as well as the **type** of the system. For inputs that are described by polynomials or sinusoids, the error series should be used. When applying the steady-state-error analysis, the final-value theorem of the Laplace transform is the basis; it should be ascertained that the closed-loop system is stable or the analysis is invalid.

The transient response is characterized by such criteria as the **maximum overshoot, rise time, delay time, and settling time**, and such parameters as **damping ratio, natural undamped frequency**, and **time constant**. The analytical expressions of these parameters can all be related to the system parameters simply if the transfer function is of the second-order prototype. For second-order systems that are not of the prototype, and higher-order systems, the analytical relationships between the transient parameters and the system constants are difficult to determine. Computer simulations are recommended for these systems.

Time-domain analysis of a position-control system is conducted. The transient and steady-state analyses are carried out first by approximating the system as a second-order system. The effect of varying the amplifer gain K on the transient and steady-state performance is demonstrated. The root-locus technique is introduced. The system is then analyzed as a third-order system, and it is shown that the second-order approximation is accurate only for low values of K.

The effects of adding poles and zeros to the open-loop and closed-loop transfer functions are demonstrated. The dominant poles of transfer functions are also discussed. This established the significance of the location of the poles of a transfer function in the s-plane, and under what condition the insignificant poles (and zeros) can be neglected as far as the transient response is concerned.

In Section 7-9, a method of approximating a high-order system by an equivalent low-order system is introduced. The low-order system tracks the high-order one in the sense that the frequency responses of the two systems are similar.

Finally, the last portion of this chapter is devoted to the time-domain analysis of discrete-data control systems. It is shown that practically all the steady-state and transient analyses of the continuous-data systems can all be extended to discrete-data systems. Remember that when the z-transform method is used, the time response addressed is the sampled signal $c(kT)$. The interpretation of the pole-zero configuration in the z-plane should be made with respect to the unit circle $|z| = 1$.

REVIEW QUESTIONS

1. Give the definitions of the error constants, K_p, K_v, and K_a.

2. Specify the type of input to which the error constant K_p is dedicated.

3. Specify the type of input to which the error constant K_v is dedicated.

4. Specify the type of input to which the error constant K_a is dedicated.

5. Can you define an error constant if the input to a control system is described by $r(t) = t^3 u_s(t)/6$?

6. Can you give the definition of the *type* of a linear time-invariant system?

7. If a control system is of type 2, then it is certain that the steady-state error of the system to a step input or a ramp input will be zero.   **(T)**   **(F)**

8. Linear and nonlinear frictions will generally degrade the steady-state error of a control system.   **(T)**   **(F)**

9. The maximum overshoot of the unit-step response of the second-order prototype system will never exceed 100 percent when the damping ratio ζ and the natural frequency ω_n are all positive.   **(T)**   **(F)**

10. For the second-order prototype system, when the undamped natural frequency ω_n increases, the maximum overshoot of the output stays the same.   **(T)**   **(F)**

11. The maximum overshoot of the following system will never exceed 100 percent when ζ, ω_n, and T are all positive.

$$\frac{C(s)}{R(s)} = \frac{\omega_n^2(1 + Ts)}{s^2 + 2\zeta\omega_n s + \omega_n^2}$$ **(T)**   **(F)**

12. Increasing the undamped natural frequency will generally reduce the rise time of the step response.   **(T)**   **(F)**

13. Increasing the undamped natural frequency will generally reduce the settling time of the step response.   **(T)**   **(F)**

14. Adding a zero to the open-loop transfer function will generally improve the system damping, and thus will always reduce the maximum overshoot of the system.   **(T)**   **(F)**

15. Given the following characteristic equation of a linear control system, increasing the value of K will increase the frequency of oscillation of the system

$$s^3 + 3s^2 + 5s + K = 0$$

(T)   **(F)**

16. For the characteristic equation given in Review Question 15, increasing the coefficient of the s^2 term will generally improve the damping of the system.   **(T)**   **(F)**

17. The location of the roots of the characteristic equation in the s-plane will give a definite indication on such a characteristic as the maximum overshoot of the transient response of a system.   **(T)**   **(F)**

18. The following transfer function $G(s)$ can be approximated by $G_L(s)$ since the pole at -20 is much larger than the dominant pole at $s = -1$.

$$G(s) = \frac{10}{s(s+1)(s+20)} \qquad G_L(s) = \frac{10}{s(s+1)}$$

(T)   **(F)**

19. Give definitions of the error constants K_p^*, K_v^*, and K_a^* of discrete-data control systems.

20. Point out the difference between the time responses of a system with a pole at $z = 1$ and a system with a pole at $z = -1$.

21. When the characteristic-equation roots of a discrete-data system are found in the second and the third quadrants of the z-plane, the frequency of oscillation of the system will generally be higher than if the roots are in the first and the fourth quadrants.   **(T)**   **(F)**

REFERENCES

Simplification of Linear Systems

1. E. J. DAVISON, "A Method for Simplifying Linear Dynamic Systems," *IEEE Trans. Automatic Control,* Vol. AC-11, pp. 93–101, Jan. 1966.
2. T. C. HSIA, "On the Simplification of Linear Systems," *IEEE Trans. Automatic Control,* Vol. AC-17, pp. 372–374, June 1972.

Undershoot in Step Response

3. M. VIDYASAGAR, "On Undershoot and Nonminimum Phase Zeros," *IEEE Trans. Automatic Control,* Vol. AC-31, p. 440, May 1986.
4. T. NORIMATSU and M. ITO, "On the Zero Non-regular Control System," *J. Inst. Elec. Eng. Japan,* Vol. 81, pp. 566–575, 1961.

Discrete-Data Control Systems

See the references at the end of Chapter 3.

PROBLEMS

7-1. A pair of complex-conjugate poles in the s-plane is required to meet the various specifications that follow. For each specification, sketch the region in the s-plane in which the poles should be located.

(a) $\zeta \geq 0.707$ $\omega_n \geq 2$ rad/s   (positive damping)
(b) $0 \leq \zeta \leq 0.707$ $\omega_n \leq 2$ rad/s   (positive damping)
(c) $\zeta \leq 0.5$ $1 \leq \omega_n \leq 5$ rad/s   (positive damping)
(d) $0.5 \leq \zeta \leq 0.707$ $\omega_n \leq 5$ rad/s   (positive and negative damping)

7-2. Determine the type of the following systems for which the loop transfer functions are given.

(a) $G(s)H(s) = \dfrac{K}{(1 + s)(1 + 10s)(1 + 20s)}$

(b) $G(s)H(s) = \dfrac{10}{(1 + s)(1 + 10s)(1 + 20s)} e^{-0.2s}$

(c) $G(s)H(s) = \dfrac{10(s + 1)}{s^2(s + 5)(s + 6)}$

(d) $G(s)H(s) = \dfrac{10(s + 1)}{s^3(s + 5)(s + 6)}$

7-3. Determine the step, ramp, and parabolic error constants of the following unity-feedback control systems. The open-loop transfer functions are given.

(a) $G(s) = \dfrac{1000}{(1 + 0.1s)(1 + 10s)}$ (b) $G(s) = \dfrac{100}{s(s^2 + 10s + 100)}$

(c) $G(s) = \dfrac{K}{s(1 + 0.1s)(1 + 0.5s)}$ (d) $G(s) = \dfrac{100}{s^2(s^2 + 10s + 100)}$

(e) $G(s) = \dfrac{1000}{s(s + 10)(s + 100)}$ (f) $G(s) = \dfrac{K(1 + 2s)(1 + 4s)}{s^2(s^2 + s + 1)}$

7-4. For the control systems described in Problem 7-2, determine the steady-state error for a unit-step input, a unit-ramp input, and a parabolic input, $(t^2/2)u_s(t)$, Check the stability of the system before applying the final-value theorem.

7-5. The open-loop transfer function of a unity-feedback control system is

$$G(s) = \frac{1000}{s(1 + 0.1s)}$$

Evaluate the error series for the system. Determine the steady-state error of the system when the following inputs are applied.

(a) $r(t) = (2 + t + 5t^2)u_s(t)$
(b) $r(t) = (1 + \sin 5t)u_s(t)$
(c) $r(t) = \cos 10t u_s(t)$

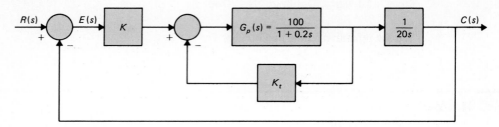

7-6.   The block diagram of a control system is shown in Fig. 7P-6. Find the step-, ramp-, and parabolic-error constants. The error signal is defined to be $e(t)$. Find the steady-state errors of the system in terms of K and K_t when the following inputs are applied.
   (a) $r(t) = u_s(t)$ (b) $r(t) = tu_s(t)$ (c) $r(t) = (t^2/2)u_s(t)$

7-7.   Repeat Problem 7-6 when the transfer function of the process is instead

$$G_p(s) = \frac{100}{(1 + 0.1s)(1 + 0.5s)}$$

What constraints must be made, if any, on the values of K and K_t so that the answers are valid? Determine the minimum steady-state error that can be achieved with a unit-ramp input by varying the values of K and K_t.

7-8.   For the position-control system shown in Fig. 3P-18, determine the following.
   (a)   Find the steady-state value of the error signal $\theta_e(t)$ in terms of the system parameters when the input is a unit-step function.
   (b)   Repeat part (a) when the input is a unit-ramp function. Assume that the system is stable.

7-9.   The block diagram of a feedback control system is shown in Fig. 7P-9. The error signal is defined to be $e(t)$.
   (a)   Find the steady-state error of the system in terms of K and K_t when the input is a unit-ramp function. Give the constraints on the values of K and K_t so that the answer is valid. Let $n(t) = 0$ for this part.
   (b)   Find the steady-state value of $c(t)$ when $n(t)$ is a unit-step function. Let $r(t) = 0$. Assume that the system is stable.

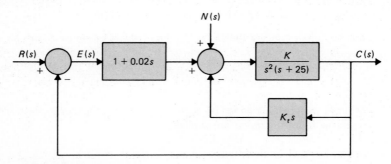

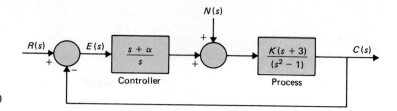

FIGURE 7P-10

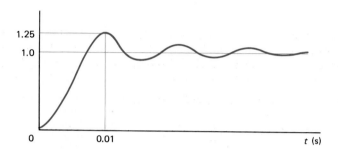

FIGURE 7P-11

7-10. The block diagram of a linear control system is shown in Fig. 7P-10, where $r(t)$ is the reference input, and $n(t)$ is the disturbance.
(a) Find the steady-state value of $e(t)$ when $n(t) = 0$ and $r(t) = tu_s(t)$. Find the conditions on the values of α and K so that the solution is valid.
(b) Find the steady-state value of $c(t)$ when $r(t) = 0$ and $n(t) = u_s(t)$.

7-11. The unit-step response of a linear control system is shown in Fig. 7P-11. Find the transfer function of a second-order prototype system to model the system.

7-12. For the control system shown in Fig. 7P-6, find the values of K and K_t so that the maximum overshoot of the output is approximately 4.3 percent and the rise time t_r is approximately 0.2 s. Use Eq. (7-89) for the rise-time relationship. Simulate the system with the CLRSP program or any other simulation program to check on the accuracy of your solutions.

7-13. Repeat Problem 7-12 with a maximum overshoot = 10 percent and a rise time t_r = 0.1 s.

7-14. Repeat Problem 7-12 with a maximum overshoot = 20 percent and a rise time t_r = 0.05 s.

7-15. For the control system shown in Fig. 7P-6, find the values of K and K_t so that the maximum overshoot of the output is approximately 4.3 percent and the delay time t_d is approximately 0.1 s. Use Eq. (7-87) for the delay-time relationship. Simulate the system with the CLRSP program or any other simulation program to check on the accuracy of your solutions.

7-16. Repeat Problem 7-15 with a maximum overshoot = 10 percent and a delay time t_d = 0.05 s.

7-17. Repeat Problem 7-15 with a maximum overshoot = 20 percent and a delay time t_d = 0.01 s.

7-18. For the control system shown in Fig. 7P-6, find the values of K and K_t so that the damping ratio of the system is 0.6 and the settling time of the unit-step response is 0.1 s. Use Eq. (7-92) for the settling-time relationship. Simulate the system with the CLRSP program or any other simulation program to check on the accuracy of your results.

7-19. **(a)** Repeat Problem 7-18 with a maximum overshoot = 10 percent and a settling time $t_s = 0.05$ s.

  (b) Repeat Problem 7-18 with a maximum overshoot = 20 percent and a settling time $t_s = 0.01$ s.

7-20. The open-loop transfer function of a control system with unity feedback is

$$G(s) = \frac{K}{s(s + a)(s + 30)}$$

where a and K are real constants.

  (a) Find the values of a and K so that the relative damping ratio of the complex roots of the characteristic equation is 0.5 and the rise time of the unit-step response is approximately 1 s. Use Eq. (7-89) as an approximation to the rise time. With the values of a and K found, determine the actual rise time.

  (b) With the values of a and K found in part (a), find the steady-state errors of the system when the reference input is (i) a unit-step function, and (ii) a unit-ramp function.

7-21. The block diagram of a linear control system is shown in Fig. 7P-21.

  (a) By means of trial and error, find the value of K so that the characteristic equation has two equal real roots and the system is stable. You may use the program POLYROOT or any root-finding computer program for this.

  (b) Find the unit-step response of the system when K has the value found in part (a). Use the program CLRSP or any simulation program for this. Set all the initial conditions to zero.

  (c) Repeat part (b) when $K = -1$. What is peculiar about the step responses for small t, and what may have caused it?

7-22. A controlled process is represented by the following dynamic equations:

$$\dot{x}_1(t) = -x_1(t) + 5x_2(t)$$
$$\dot{x}_2(t) = -6x_1(t) + u(t)$$
$$c(t) = x_1(t)$$

The control is obtained through state feedback with

$$u(t) = -g_1 x_1(t) + g_2 x_2(t) + r(t)$$

where g_1 and g_2 are real constants, and r is the reference input.

  (a) Find the locus in the g_1-versus-g_2 plane (g_1 = vertical axis) on which the overall system has a natural undamped frequency of 10 rad/s.

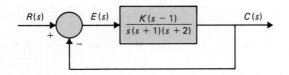

FIGURE 7P-21

(b)   Find the locus in the g_1-versus-g_2 plane on which the overall system has a damping ratio of 0.707.

(c)   Find the values of g_1 and g_2 such that $\zeta = 0.707$ and $\omega_n = 10$ rad/s.

(d)   Let the error signal be defined as $e(t) = r(t) - c(t)$. Find the steady-state error when $r(t) = u_s(t)$ and g_1 and g_2 are at the values found in part (c).

(e)   Find the locus in the g_1-versus-g_2 plane on which the steady-state error due to a unit-step input is zero.

7-23.   The block diagram of a linear control system is shown in Fig. 7P-23. Construct a parameter plane of K_p versus K_d (K_p is the vertical axis) and show the following trajectories or regions in the plane.

(a)   Unstable and stable regions.

(b)   Trajectories on which the damping is critical ($\zeta = 1$).

(c)   Region in which the system is overdamped ($\zeta > 1$).

(d)   Region in which the system is underdamped ($\zeta < 1$).

(e)   Trajectory on which the parabolic-error constant K_a is 1000 s^{-2}.

(f)   Trajectory on which the natural undamped frequency ω_n is 50 rad/s.

(g)   Trajectory on which the system is either uncontrollable or unobservable. (Hint: look for pole-zero cancellation.)

7-24.   The block diagram of a linear control system is shown in Fig. 7P-24. The fixed parameters of the system are given as $T = 0.1$, $J = 0.01$, and $K_i = 10$.

(a)   When $r(t) = tu_s(t)$ and $T_d(t) = 0$, determine how the values of K and K_t affect the steady-state value of $e(t)$. Find the restrictions on K and K_t so that the system is stable.

(b)   Let $r(t) = 0$. Determine how the values of K and K_t affect the steady-state value of $c(t)$ when the disturbance input $T_d(t) = u_s(t)$.

(c)   Let $K_t = 0.01$ and $r(t) = 0$. Find the minimum steady-state value of $c(t)$ that can be obtained by varying K, when $T_d(t)$ is a unit-step function. Find the value of this K. From the transient standpoint, would you operate the system at this value of K? Explain.

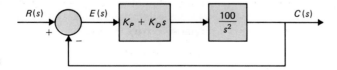

FIGURE 7P-23

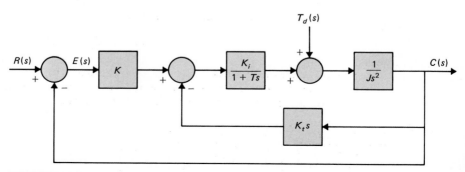

FIGURE 7P-24

(d)   Assume that it is desired to operate the system with the value of K as selected in part (c). Find the value of K_t so that the complex roots of the characteristic equation will have a real part of -2.5. Find all the three roots of the characteristic equation.

7-25.   The dc-motor-control system described in Problem 4-14 has the open-loop transfer function

$$G(s) = \frac{\Theta_o(s)}{\Theta_e(s)} = \frac{nK_sK_iK_LK}{\Delta(s)}$$

where $\Delta(s) = s[L_aJ_mJ_Ls^4 + J_L(R_aJ_m + B_mL_a)s^3 + (n^2K_LL_aJ_L + K_LL_aJ_m$

$$+ K_iK_bJ_L + R_aB_mJ_L)s^2 + (n^2R_aK_LJ_L + R_aK_LJ_m + B_mK_LL_a)s$$

$$+ R_aB_mK_L + K_iK_bK_L]$$

where $K_i = 9$ oz-in./A, $K_b = 0.0636$ V/rad/s, $R_a = 5\ \Omega$, $L_a = 1$ mH, $K_s = 1$ V/rad, $n = 1/10$, $J_m = J_L = 0.001$ oz-in.-s^2, and $B_m \cong 0$. The characteristic equation of the closed-loop system is

$$L_aJ_mJ_Ls^5 + J_L(R_aJ_m + B_mL_a)s^4 + (n^2K_LL_aJ_L + K_LL_aJ_m + K_iK_bJ_L + R_aB_mJ_L)s^3$$

$$+ (n^2R_aK_LJ_L + R_aK_LJ_m + B_mK_LL_a)s^2 + (R_aB_mK_L + K_iK_bK_L)s + nK_sK_iK_LK = 0$$

(a) . Let $K_L = 10{,}000$ oz-in./rad. Write the open-loop transfer function $G(s)$ and find the poles of $G(s)$. Find the critical value of K for the closed-loop system to be stable. Find the roots of the characteristic equation of the closed-loop system when K is at marginal stability.

(b)   Repeat part (a) when $K_L = 1000$ oz-in./rad.

(c)   Repeat part (a) when $K_L = \infty$; i.e., the motor shaft is rigid.

(d)   Compare the results of parts (a), (b) and (c), and comment on the effects of the values of K_L on the poles of $G(s)$ and the roots of the characteristic equation.

7-26.   The block diagram of the guided-missile attitude-control system described in Problem 4-13 is shown in Fig. 7P-26. $r(t)$ is the command input and $d(t)$ represents disturbance input. The objective of this problem is to study the effects of the controller $G_c(s)$.

(a)   Let $G_c(s) = 1$. Find the steady-state error of the system when $r(t)$ is a unit-step function. Set $d(t) = 0$.

(b)   Let $G_c(s) = (s + \alpha)/s$. Find the steady-state error when $r(t)$ is a unit-step function.

(c)   Obtain the unit-step response of the system for $0 \le t \le 0.5$ s with $G_c(s)$ as given

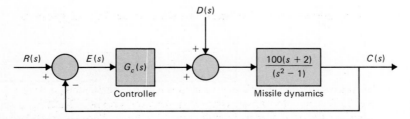

FIGURE 7P-26

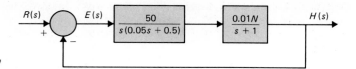

FIGURE 7P-27

in part (b), and $\alpha = 5, 50$, and 500. Assume zero initial conditions. Record the maximum overshoot of $c(t)$ for each case. Use the program CSRLP of the ACSP software or any applicable computer program. Comment on the effect of varying the value of α of the controller on the transient response.

(d)   Set $r(t) = 0$, and $G_c(s) = 1$. Find the steady-state value of $c(t)$ when $d(t) = u_s(t)$.

(e)   Let $G_c(s) = (s + \alpha)/s$. Find the steady-state value of $c(t)$ when $d(t) = u_s(t)$.

(f)   Obtain the output response for $0 \le t \le 0.5$ s, with $G_c(s)$ as given in part (e) when $r(t) = 0$ and $d(t) = u_s(t)$; $\alpha = 5, 50$, and 500. Use zero initial conditions.

(g)   Comment on the effect of varying the value of α of the controller on the transient response of $c(t)$ and $d(t)$.

7-27.   The block diagram shown in Fig. 7P-27 represents the liquid-level control system described in Problem 6-13. The liquid level is represented by $h(t)$, and N denotes the number of inlets.

(a)   Since one of the poles of the open-loop transfer function is relatively far to the left on the real axis of the s-plane at $s = -10$, it is suggested that this pole can be neglected. Approximate the system by a second-order system by neglecting the pole of $G(s)$ at $s = -10$. The approximation should be valid for both the transient and the steady-state responses. Apply the formulas for the maximum overshoot and the peak time t_{max} to the second-order model for $N = 1$ and $N = 10$.

(b)   Obtain the unit-step response (with zero initial conditions) of the original third-order system with $N = 1$ and then with $N = 10$. Compare the responses of the original system with those of the second-order approximating system. Comment on the accuracy of the approximation as a function of N.

7-28.   The open-loop transfer function of a unity-feedback control system is

$$G(s) = \frac{1 + T_z s}{s(s + 1)^2}$$

Compute and plot the unit-step responses of the closed-loop system for $T_z = 0, 0.5, 1, 10$, and 50. Assume zero initial conditions. Use the program CLRSP in the ACSP software, if available, or any other simulation program. Comment on the effects of the various values of T_z on the step response.

7-29.   The open-loop transfer function of a unity-feedback control system is

$$G(s) = \frac{1}{s(s + 1)^2(1 + T_p s)}$$

Compute and plot the unit-step responses of the closed-loop system for $T_p = 0, 0.5$, and 0.707. Assume zero initial conditions. Use the program CLRSP, if available, or any other simulation program. Find the critical value of T_p so that the closed-loop system is marginally stable. Comment on the effects of the pole at $s = -1/T_p$ in $G(s)$.

7-30.   The liquid-level control system described in Problems 6-13 and 7-27 has the open-loop transfer function

$$G(s) = \frac{10N}{s(s + 1)(s + 10)}$$

This problem is devoted to the approximation of the third-order system by a second-order one, using the method described in Section 7-9.

(a)   For $N = 1$, find the transfer function $G_L(s)$ of a second-order system that approximates the third-order system. Compute the unit-step responses of the systems, and observe the closeness of the approximation. Compare the roots of the characteristic equations of the two closed-loop systems.

(b)   Repeat part (a) with $N = 2$.

(c)   Repeat part (a) with $N = 3$.

(d)   Repeat part (a) with $N = 4$.

(e)   Repeat part (a) with $N = 5$.

7-31.   The dc-motor-control system described in Problem 4-14 has the following open-loop transfer function when $K_L = \infty$.

$$G(s) = \frac{891,100K}{s(s^2 + 5000s + 566,700)}$$

The system is to be approximated by a second-order system using the method described in Section 7-9.

(a)   For $K = 1$, find the second-order transfer function $G_L(s)$ that approximates $G(s)$. Compute and plot the unit-step responses of the systems, and evaluate the closeness of the approximation. Compare the roots of the characteristic equation of the two closed-loop systems.

(b)   Repeat part (a) with $K = 100$.

(c)   Repeat part (a) with $K = 1000$.

7-32.   Approximate the control system shown in Fig. 7P-21 when $K = -1$ by a second-order system with the closed-loop transfer function

$$M(s) = \frac{C(s)}{R(s)} = \frac{1 + c_1 s}{1 + d_1 s + d_2 s^2}$$

where c_1, d_1, and d_2 are real constants. Use the method outlined in Section 7-9. Compute and plot the unit-step responses of the original third-order system ($K = -1$) and the second-order approximating system. Compare the roots of the characteristic equations of the two systems.

7-33.   A unity-feedback control system has the open-loop transfer function

$$G(s) = \frac{K}{s(s + 1)(s + 2)(s + 20)}$$

(a)   For $K = 10$, the fourth-order system is to be approximated by a second-order system with the closed-loop transfer function

$$L(s) = \frac{G_L(s)}{1 + G_L(s)} = \frac{1}{1 + d_1 s + d_2 s^2}$$

Find d_1 and d_2 using the method outlined in Section 7-9. Compute and plot the unit-step responses of the fourth-order and the second-order systems and compare. Find the roots of the characterstic equations of the two systems.

(b)   For $K = 10$, approximate the system by a third-order model with

$$L(s) = \frac{1}{1 + d_1 s + d_2 s^2 + d_3 s^3}$$

Carry out the tasks specified in part (a).

(c)   Repeat parts (a) and (b) when $K = 40$.

7-34.   The block diagram of a sampled-data control system is shown in Fig. 7P-34.

(a)   Derive the open-loop and closed-loop transfer functions of the system in z-transforms. The sampling period is 0.1 s.

(b)   Compute and plot the unit-step response $c(kT)$ for $k = 0$ to 100.

(c)   Repeat parts (a) and (b) for $T = 0.05$ s.

7-35.   The block diagram of a sampled-data control system is shown in Fig. 7P-35.

(a)   Find the error constants K_p^*, K_v^*, and K_a^*.

(b)   Derive the open-loop and closed-loop transfer functions in z-transforms.

(c)   For $T = 0.1$ s, find the critical value of K_t for system stability.

(d)   Compute the unit-step response $c(kT)$ for $k = 0$ to 50 for $T = 0.1$ s and $K_t = 5$.

(e)   Repeat part (d) for $T = 0.1$ s and $K_t = 1$.

7-36.   The open-loop dc-motor-control system described in Problem 5-34 is now incorporated in a digital control system, as shown in Fig. 7P-36(a). The microprocessor takes the information from the encoder and computes the velocity information. This generates the sequence of numbers, $\omega(kT)$, $k = 0, 1, 2, \ldots$. The microprocessor then generates the error signal $e(kT) = r(kT) - \omega(kT)$. The digital control is modeled by the block diagram shown in Fig. 7P-36(b). Use the parameter values given in Problem 5-34.

(a)   Find the open-loop transfer function $\Omega(s)/E(z)$ with the sampling period $T = 0.1$ second.

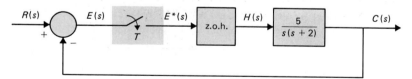

FIGURE 7P-34

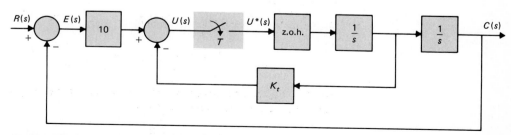

FIGURE 7P-35

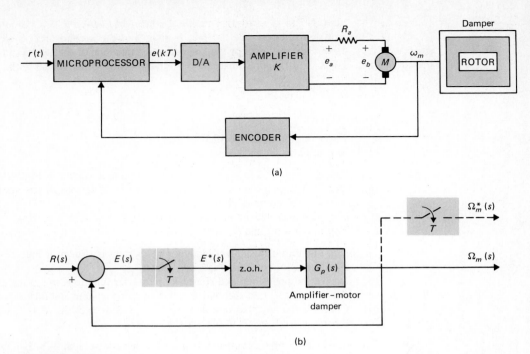

FIGURE 7P-36

(b)   Find the closed-loop transfer function $\Omega(z)/R(z)$. Find the characteristic equation and its roots. Locate these roots in the z-plane. Show that the closed-loop system is unstable when $T = 0.1$ second.

(c)   Repeat parts (a) and (b) for $T = 0.01$ and 0.001 s. The computer program SDCS of the ACSP software package or any other suitable computer program for the solution of digital control systems may be used.

(d)   Find the error constants K_p^*, K_v^*, and K_a^*. Find the steady-state error $e(kT)$ as $k \to \infty$ when the input $r(t)$ is a unit-step function $u_s(t)$, a unit-ramp function $tu_s(t)$, and a parabolic function $t^2 u_s(t)/2$.

ADDITIONAL COMPUTER PROBLEMS

Compute and plot the unit-step responses of the closed-loop systems with the open-loop transfer functions given. Assume zero initial conditions. Use the program CLRSP in the ACSP software package, if available, or any other applicable simulation program.

7-37.   (a)   For $T_z = 0, 1, 5, 20$,
$$G(s) = \frac{1 + T_z s}{s(s + 0.55)(s + 1.5)}$$

(b)   For $T_z = 0, 1, 5, 20$,
$$G(s) = \frac{1 + T_z s}{(s^2 + 2s + 2)}$$

7-38. **(a)** For $T_p = 0, 0.5, 1.0$,

$$G(s) = \frac{2}{(s^2 + 2s + 2)(1 + T_p s)}$$

(b) For $T_p = 0, 0.5, 1.0$,

$$G(s) = \frac{10}{s(s + 5)(1 + T_p s)}$$

7-39.

$$G(s) = \frac{K}{s(s + 1.25)(s^2 + 2.5s + 10)}$$

(a) For $K = 5$.   **(b)** For $K = 10$.   **(c)** For $K = 30$.

7-40.

$$G(s) = \frac{K(s + 2.5)}{s(s + 1.25)(s^2 + 2.5s + 10)}$$

(a) For $K = 5$.   **(b)** For $K = 10$.   **(c)** For $K = 30$.

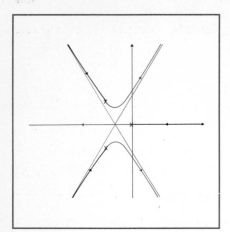

8

ROOT-LOCUS TECHNIQUE

8-1

INTRODUCTION

In preceding chapters, we have demonstrated the importance of the poles and the zeros of the closed-loop transfer function of a linear control system on the dynamic performance of the system. The poles of the closed-loop transfer function are the roots of the characteristic equation, which determine the stability of the system. Studying the behavior of the roots of the characteristic equation will reveal the relative and absolute stability of the system, keeping in mind that the transient behavior of the system is also governed by the zeros of the closed-loop transfer function.

An important study in linear control systems is the investigation of the trajectories of the roots of the characteristic equation—or, simply, the **root loci**—when a certain system parameter varies. In fact, several examples in Chapter 7 already illustrated the usefulness of the root loci in the study of linear control systems.

The basic properties and the systematic construction of the root loci are first due to W. R. Evans [1–3]. In this chapter, we show how to construct these loci by following some simple rules. For complex systems, we can always rely on a root-locus program and a digital computer to solve and plot the root loci automatically. The program ROOTLOCI in the ACSP package that accompanies this text is designed for this purpose.

ROOTLOCI

The root-locus technique is not confined to the study of linear control systems. In general, the technique can be applied to study the behavior of roots of any algebraic equation with constant coefficients.

The general root-locus problem can be formulated by referring to the following equation of the complex variable s:

$$F(s) = P(s) + KQ(s) = 0 \qquad (8\text{-}1)$$

where $P(s)$ is an nth-order polynomial of s,

$$P(s) = s^n + a_1 s^{n-1} + \cdots + a_{n-1}s + a_n \qquad (8\text{-}2)$$

and $Q(s)$ is an mth-order polynomial of s; n and m are positive integers.

$$Q(s) = s^m + b_1 s^{m-1} + \cdots + b_{m-1}s + b_m \qquad (8\text{-}3)$$

K is a real constant that can vary from $-\infty$ to $+\infty$.

The coefficients $a_1, a_2, \ldots, a_n, b_1, b_2, \ldots, b_m$ are considered to be real and fixed.

root contours

Root loci of multiple-variable parameters can be treated by varying one parameter at a time. The resultant loci are called the **root contours**, and the subject is treated in Section 8-5.

By replacing s with z in Eqs. (8-1) through (8-3), we can construct the root loci of the characteristic equation of a linear discrete-data control system in a similar fashion.

Although the loci of roots of Eq. (8-1) when K varies from $-\infty$ to ∞ are generally referred to as the **root loci** in control-systems literature, for the purpose of reference, we define the following categories based on the sign of K:

1. *Root Loci (RL):* the portion of the root loci when K varies from 0 to ∞; i.e., K is positive.
2. *Complementary Root Loci (CRL):* the portion of the root loci when K varies from $-\infty$ to 0; i.e., K is negative.
3. *Root Contours (RC):* loci of roots when more than one parameter varies.
4. *Complete Root Loci:* the combination of the root loci and the complementary root loci; i.e., $-\infty < K < \infty$.

8-2

BASIC PROPERTIES OF THE ROOT LOCI

Since our main interest is in the study of control systems, let us refer to the closed-loop transfer function of a single-loop control system:

*basic
properties*

$$\frac{C(s)}{R(s)} = \frac{G(s)}{1 + G(s)H(s)} \qquad (8\text{-}4)$$

keeping in mind that the transfer functions of multiple-loop systems can also be expressed in a similar form. The characteristic equation of the closed-loop system is obtained by setting the denominator polynomial of $C(s)/R(s)$ to zero, which is the same as setting the numerator of $1 + G(s)H(s)$ to zero. Thus, the roots of the characteristic equation must satisfy

*characteristic
equation*

$$1 + G(s)H(s) = 0 \qquad (8\text{-}5)$$

Suppose that $G(s)H(s)$ contains a variable parameter K as a factor, such that the rational function can be written as

$$G(s)H(s) = \frac{KQ(s)}{P(s)} \qquad (8\text{-}6)$$

where $P(s)$ and $Q(s)$ are polynomials as defined in Eqs. (8-2) and (8-3), respectively. Equation (8-5) is written

$$1 + \frac{KQ(s)}{P(s)} = \frac{P(s) + KQ(s)}{P(s)} = 0 \qquad (8\text{-}7)$$

The numerator polynomial of Eq. (8-7) is identical to Eq. (8-1). Thus, by considering that the loop transfer function $G(s)H(s)$ of a closed-loop system can be written in the form of Eq. (8-6), we have matched the root loci of a control system with the form of the general root-locus problem. In general, we can regard Eq. (8-7) to be the denominator of the closed-loop transfer function of any linear control system, and use Eq. (8-1) as the basis of the root-locus analysis.

When the variable parameter K does not appear as a multiplying factor of $G(s)H(s)$, we can always condition the functions in the form of Eq. (8-1). As an illustrative example, consider that the denominator of the closed-loop transfer function of a control system is of the form

$$1 + \frac{s^2 + (3 + 2K)s + 5}{s(s + 1)(s + 2)} \qquad (8\text{-}8)$$

The characteristic equation of the closed-loop system is

$$s(s + 1)(s + 2) + s^2 + (3 + 2K)s + 5 = 0 \qquad (8\text{-}9)$$

To express the last equation in the form of Eq. (8-7), we *divide both sides of the equation by the terms that do not contain K*, and we get

$$1 + \frac{2Ks}{s(s + 1)(s + 2) + s^2 + 3s + 5} = 0 \qquad (8\text{-}10)$$

Comparing the last equation with Eq. (8-7), we get

$$Q(s) = 2s \qquad (8\text{-}11)$$

and

$$P(s) = s^3 + 4s^2 + 5s + 5 \qquad (8\text{-}12)$$

The conclusion is that given any loop transfer function $G(s)H(s)$ with one variable parameter K imbedded in it, we can always find the characteristic equation by equating the numerator polynomial of $1 + G(s)H(s)$ to zero, as in the step from Eq. (8-8) to Eq. (8-9). To isolate the variable parameter K as a multiplying factor, we first factor the terms in the characteristic equation with K, and then without K. This forms the $P(s)$ and the $Q(s)$ polynomials, as in Eq. (8-1). Finally, we divide both sides of the characteristic equation by $P(s)$, which is the polynomial without K. This last step is essential in conditioning the characteristic equation so that the properties of the root loci of Eq. (8-1) can be obtained. As it turns out, all the properties of the root loci of Eq. (8-1) are derived from the properties of the function $Q(s)/P(s)$.

Since $G(s)H(s) = KQ(s)/P(s)$, this is another example in which the characteristics of the closed-loop system, in this case represented by the roots of the characteristic equation, are determined from the knowledge of the loop transfer function $G(s)H(s)$.

Now we are ready to investigate the conditions under which Eq. (8-5) or Eq. (8-7) is satisfied.

Let us express $G(s)H(s)$ as

$$G(s)H(s) = KG_1(s)H_1(s) \qquad (8\text{-}13)$$

where $G_1(s)H_1(s)$ is equal to $Q(s)/P(s)$, and it does not contain the variable parameter K. Then, Eq. (8-5) is written

$$G_1(s)H_1(s) = -\frac{1}{K} \qquad (8\text{-}14)$$

root-loci conditions

To satisfy the last equation, the following conditions must be met simultaneously:

CONDITION ON MAGNITUDE

$$\left| G_1(s)H_1(s) \right| = \frac{1}{|K|} \qquad -\infty < K < \infty \qquad (8\text{-}15)$$

CONDITION ON ANGLES

$$\angle\, G_1(s)H_1(s) = (2k + 1)\pi \quad K \ge 0$$

$$= \text{odd multiples of } \pi \text{ radians or } 180° \qquad (8\text{-}16)$$

$$\angle\, G_1(s)H_1(s) = 2k\pi \quad K \le 0$$

$$= \text{even multiples of } \pi \text{ radians or } 180° \qquad (8\text{-}17)$$

where $k = 0, \pm 1, \pm 2, \ldots$ (any integer).

In practice, the conditions stated in Eqs. (8-15) through (8-17) play different roles in the construction of the complete root loci. *The conditions on angles in Eq. (8-16) or Eq. (8-17) are used to determine the trajectories of the root loci in the s-plane. Once the root loci are drawn, the values of K on the loci are determined by using the condition on magnitude in Eq. (8-15).*

The construction of the root loci is basically a graphical problem, although some of the rules of construction are arrived at analytically. The start of the graphical construction of the root loci is based on the knowledge of the poles and zeros of the function $G(s)H(s)$. In other words, $G(s)H(s)$ must first be written as

$$G(s)H(s) = KG_1(s)H_1(s) = \frac{K(s + z_1)(s + z_2) \cdots (s + z_m)}{(s + p_1)(s + p_2) \cdots (s + p_n)} \qquad (8\text{-}18)$$

where the zeros and poles of $G(s)H(s)$ are real or in complex-conjugate pairs.

Applying the conditions in Eqs. (8-15), (8-16), and (8-17) to Eq. (8-18), we have

$$|G_1(s)H_1(s)| = \frac{\prod\limits_{i=1}^{m} |s + z_i|}{\prod\limits_{j=1}^{n} |s + p_j|} = \frac{1}{|K|} \quad -\infty < K < \infty \qquad (8\text{-}19)$$

For $0 \le K < \infty$ (RL),

$$\angle G_1(s)H_1(s) = \sum_{i=1}^{m} \angle(s + z_i) - \sum_{j=1}^{n} \angle(s + p_j) = (2k + 1)\pi \qquad (8\text{-}20)$$

For $-\infty < K \le 0$ (CRL),

$$\angle G_1(s)H_1(s) = \sum_{i=1}^{m} \angle(s + z_i) - \sum_{j=1}^{n} \angle(s + p_j) = 2k\pi \qquad (8\text{-}21)$$

where $k = 0, \pm 1, \pm 2, \ldots$.

The graphical interpretation of Eq. (8-20) is that any point s_1 on the RL that corresponds to a positive value of K must satisfy the condition:

> *The difference between the sums of the angles of the vectors drawn from the zeros and those from the poles of $G(s)H(s)$ to s_1 is an odd multiple of 180°.* Similarly, Eq. (8-21) may be used to construct the CRL; that is, any point on the CRL $(-\infty < K \leq 0)$ must satisfy: *the difference between the sums of the angles of the vectors drawn from the zeros and those from the poles of $G(s)H(s)$ to s_1 is an even multiple of 180°, including zero degrees.*

Once the complete root loci are constructed, the values of K along the loci can be determined by writing Eq. (8-19) as

$$|K| = \frac{\prod\limits_{i=1}^{n} |s + p_1|}{\prod\limits_{i=1}^{m} |s + z_1|} \tag{8-22}$$

The value of K at any point s_1 on the RL is obtained from Eq. (8-22) by substituting the value of s_1 into the equation. Graphically, the numerator of Eq. (8-22) represents the product of the lengths of the vectors drawn from the poles of $G(s)H(s)$ to s_1, and the denominator represents the product of the lengths of the vectors drawn from the zeros of $G(s)H(s)$ to s_1. If the point s_1 is on the RL, the value of K is positive, and if s_1 is on the CRL, then the value of K is negative.

To illustrate the use of Eqs. (8-20) to (8-22) for the construction of the root loci, let us consider the function

$$G_1(s)H_1(s) = \frac{K(s + z_1)}{s(s + p_2)(s + p_3)} \tag{8-23}$$

The locations of the poles and zeros of $G(s)H(s)$ are arbitrarily assigned as shown in Fig. 8-1. Let us select an arbitrary trial point s_1 in the s-plane and draw vectors directing from the poles and zeros of $G(s)H(s)$ to the point. If s_1 is indeed a point on the RL $(0 \leq K < \infty)$, it must satisfy Eq. (8-20); that is, the angles of the vectors shown in

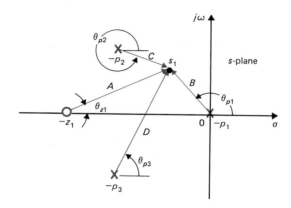

FIGURE 8-1 Pole-zero configuration of $G(s)H(s) = K(s + z_1)/[s(s + p_2)(s + p_3)]$.

Fig. 8-1 must satisfy

$$\angle(s_1 + z_1) - \angle s_1 - \angle(s_1 + p_2) - \angle(s_1 + p_3)$$

$$= \theta_{z1} - \theta_{p1} - \theta_{p2} - \theta_{p3} = (2k + 1)\pi \qquad (8\text{-}24)$$

where $k = 0, \pm 1, \pm 2, \ldots$. As shown in Fig. 8-1, the angles of the vectors are measured with the positive real axis as reference. Similarly, if s_1 is a point on the CRL ($-\infty < K \le 0$), it must satisfy Eq. (8-21); that is,

$$\angle(s_1 + z_1) - \angle s_1 - \angle(s_1 + p_2) - \angle(s_1 + p_3)$$

$$= \theta_{z1} - \theta_{p1} - \theta_{p2} - \theta_{p3} = 2k\pi \qquad (8\text{-}25)$$

where $k = 0, \pm 1, \pm 2, \ldots$.

If s_1 is found to satisfy either Eq. (8-24) or Eq. (8-25), Eq. (8-22) is used to determine the value of K at the point. As shown in Fig. 8-1, the lengths of the vectors are represented by A, B, C, and D. The magnitude of K is

$$|K| = \frac{|s_1||s_1 + p_2||s_1 + p_3|}{|s_1 + z_1|} = \frac{BCD}{A} \qquad (8\text{-}26)$$

The sign of K, of course, depends on whether s_1 is on the RL or the CRL. Consequently, given the function $G(s)H(s)$ with K as a multiplying factor and the poles and zeros known, the construction of the complete root-locus diagram involves the following two steps:

construction steps

1. A search for all the s_1 points in the s-plane that satisfy Eqs. (8-20) and (8-21).
2. The determination of the values of K at points on the complete root loci by use of Eq. (8-22).

We have established the basic conditions on the construction of the root-locus diagram. However, if we were to use the trial-and-error method just described, the search for all the root-locus points in the s-plane that satisfy Eqs. (8-21) and (8-22) would be a very tedious task. Years ago, when Evans first invented the root-locus technique, digital computer technology was still at its infancy; he had to devise a special tool, called the **Spirule,** which can be used to assist in adding and subtracting angles of vectors efficiently, according to Eq. (8-21) or Eq. (8-22). Even with the Spirule, the user still has to know the general proximity of the roots in the s-plane in order for the device to be effective.

With the availability of digital computers and efficient root-finding subroutines, the Spirule and the trial-and-error method have long become obsolete. Nevertheless, even with a high-speed computer, the analyst should still have an in-depth understanding of the properties of the root loci in order to be able to manually sketch the loci of simple and moderately complex systems and interpret the computer results correctly, and apply the root-locus techniques for the analysis **and** design of linear control systems.

8-3

PROPERTIES AND CONSTRUCTION
OF THE COMPLETE ROOT LOCI

root-loci
construction

The following properties of the root loci are introduced for the purpose of constructing the root loci manually. The properties are developed based on the relation between the poles and zeros of $G(s)H(s)$ and the zeros of $1 + G(s)H(s)$, which are the roots of the characteristic equation.

$K = 0$ Points

THEOREM 8-1.   *The $K = 0$ points on the complete root locus are at the poles of $G(s)H(s)$.*

PROOF.   From Eq. (8-19),

$$|G_1(s)H_1(s)| = \frac{1}{|K|} \qquad (8\text{-}27)$$

As K approaches zero, $|G_1(s)H_1(s)|$ approaches infinity, so s must approach the poles of $G_1(s)H_1(s)$ or of $G(s)H(s)$. This property applies to the RL and the CRL, since the sign of K is of no concern in Eq. (8-27).   ■

EXAMPLE 8-1

Consider the following equation:

$$s(s + 2)(s + 3) + K(s + 1) = 0 \qquad (8\text{-}28)$$

When $K = 0$, the three roots of the equation are at $s = 0$, $s = -2$, and $s = -3$. These three points are also the poles of the function $G(s)H(s)$ if we divide both sides of Eq. (8-28) by the terms that do not contain K, resulting in

$$1 + G(s)H(s) = 1 + \frac{K(s + 1)}{s(s + 2)(s + 3)} = 0 \qquad (8\text{-}29)$$

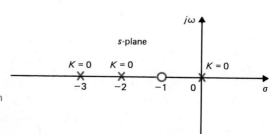

FIGURE 8-2 Points at which $K = 0$ on the complete root loci of $s(s + 2)(s + 3) + K(s + 1) = 0$.

Thus,

$$G(s)H(s) = \frac{K(s+1)}{s(s+2)(s+3)} \qquad (8\text{-}30)$$

The three points at which $K = 0$ are shown in Fig. 8-2.

$K = \pm\infty$ Points

THEOREM 8-2.   The $K = \pm\infty$ points on the complete root loci are at the zeros of $G(s)H(s)$.

PROOF.   Referring again to Eq. (8-27), as K approaches $\pm\infty$, the equation approaches zero. This corresponds to s approaching the zeros of $G(s)H(s)$. ∎

EXAMPLE 8-2   Consider again the equation

$$s(s+2)(s+3) + K(s+1) = 0 \qquad (8\text{-}31)$$

It is apparent that when K is very large in magnitude, the equation can be approximated by

$$K(s+1) = 0 \qquad (8\text{-}32)$$

which has the root $s = -1$. Notice that this is also the zero of $G(s)H(s)$ in Eq. (8-30). Therefore, Fig. 8-3 shows $s = -1$ as the point at which $K = \pm\infty$. It should be pointed out that in this case, $G(s)H(s)$ also has two other zeros at $s = \infty$, because for a rational function, the total number of poles and zeros must be equal if the poles and zeros at infinity are included. Thus, for the equation in Eq. (8-31), the $K = \pm\infty$ points are at $s = -1$, ∞, and ∞. It is useful to consider that infinity in the s-plane is a point concept. We can visualize that the finite s-plane is only a small part of a sphere with an infinite radius. Then, infinity in the s-plane is a point on the opposite side of the sphere that we face.

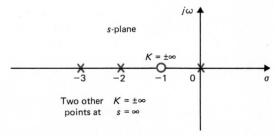

FIGURE 8-3 Points at which $K = \pm\infty$ on the complete root loci of $s(s+2)(s+3) + K(s+1) = 0$.

Number of Branches on the Complete Root Loci

number of branches

A branch of the complete root loci is the locus of one root when K takes on values between $-\infty$ and ∞. Since the number of branches of the complete root loci must equal the number of roots of the equation, the following theorem results:

> **THEOREM 8-3.** *The number of branches of the root loci of Eq. (8-1) is equal to the order of the polynomial.* ■

Keeping track of the individual branches and the total number of branches of the root-locus diagram is often helpful to ensure that the plot is done correctly.

EXAMPLE 8-3

The number of branches of the complete root loci of

$$s(s + 2)(s + 3) + K(s + 1) = 0 \qquad (8\text{-}33)$$

is three, since the equation is of the third order. The equation has three roots, and thus three root loci.

Symmetry of the Complete Root Loci

symmetry

> **THEOREM 8-4.** *The complete root loci are symmetrical with respect to the real axis of the s-plane. In general, the loci are symmetrical with respect to the axes of symmetry of the pole-zero configuration of $G(s)H(s)$.*

> **PROOF.** The proof of the first statement is self-evident, since, for real coefficients in Eq. (8-1), the roots must be real or in complex-conjugate pairs. The reasoning behind the second statement on symmetry is also simple, since if the poles and zeros of $G(s)H(s)$ are symmetrical to an axis in addition to the real axis in the s-plane, we can regard this axis of symmetry as if it were the real axis of a new complex plane obtained through a linear transformation. ■

EXAMPLE 8-4

Consider the equation

$$s(s + 1)(s + 2) + K = 0 \qquad (8\text{-}34)$$

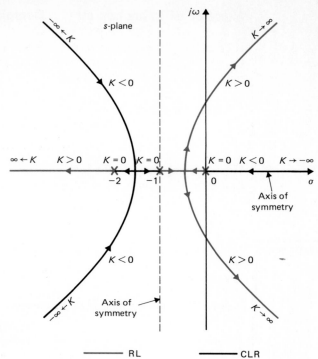

FIGURE 8-4 Complete root loci of $s(s + 1)(s + 2) + K = 0$, showing the properties of symmetry.

Dividing both sides of the equation by the terms that do not contain K leads to

$$G(s)H(s) = \frac{K}{s(s + 1)(s + 2)} \tag{8-35}$$

The complete root loci of Eq. (8-34) are shown in Fig. 8-4. Notice that since the pole-zero configuration of $G(s)H(s)$ is symmetrical with respect to the real axis as well as the $s = -1$ axis, the complete root loci are symmetrical to these axes.

As a review of Theorems 8-1 to 8-3, the points at which $K = 0$ are at the poles of $G(s)H(s)$, $s = 0$, $s = -1$, and $s = -2$. The function $G(s)H(s)$ has three zeros at $s = \infty$ at which $K = \pm\infty$. The reader should try to trace out the three separate branches of the complete root loci by starting from one of the $K = -\infty$ points, through a CRL, then an RL, and ending at $K = \infty$ at $s = \infty$.

EXAMPLE 8-5   When the pole-zero configuration of $G(s)H(s)$ is symmetrical with respect to a point in the s-plane, the complete root loci will also be symmetrical to that point. This is illustrated by the root-locus plot of

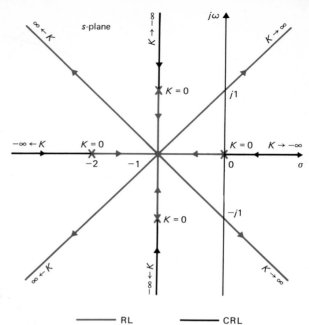

FIGURE 8-5 Complete root loci of $s(s + 2)(s^2 + 2s + 2) + K = 0$, showing the properties of symmetry.

$$s(s + 1)(s + 1 + j)(s + 1 - j) + K = 0 \qquad (8\text{-}36)$$

shown in Fig. 8-5.

Asymptotes of the Complete Root Loci

(Behavior of Root Loci at $|s| = \infty$)

As shown by the root loci in Figs. 8-4 and 8-5, when n, the order of $P(s)$ is not equal to m, the order of $Q(s)$, $2|n - m|$ of the loci will approach infinity in the s-plane. The properties of the complete root loci near infinity in the s-plane are described by the *asymptotes* of the loci when $|s| \to \infty$. The angles of the asymptotes and their intersect with the real axis of the s-plane are described by the following two theorems. The proofs of the theorems are too lengthy to be given here.

THEOREM 8-5. *For large values of s, the RL $(K \geq 0)$ are asymptotic to asymptotes with angles given by*

$$\theta_k = \frac{(2k + 1)\pi}{|n - m|} \qquad n \neq m \qquad (8\text{-}37)$$

where $k = 0, 1, 2, \ldots, |n - m| - 1$; n and m are the number of finite poles and zeros of $G(s)H(s)$, respectively.

For the CRL $(K \leq 0)$, the angles of the asymptotes are

$$\theta_k = \frac{2k\pi}{|n - m|} \qquad n \neq m \qquad\qquad (8\text{-}38)$$

where $k = 0, 1, 2, \ldots, |n - m| - 1$. ∎

Intersect of the Asymptotes (Centroid)

THEOREM 8-6.   (a) The intersect of the $2|n - m|$ asymptotes of the complete root loci lies on the real axis of the s-plane.

(b) The intersect of the asymptotes is given by

$$\sigma_1 = \frac{\Sigma \text{ finite poles of } G(s)H(s) - \Sigma \text{ finite zeros of } G(s)H(s)}{n - m}$$

$$(8\text{-}39)$$

where n is the number of finite poles of $G(s)H(s)$, and m is the number of finite zeros of $G(s)H(s)$. ∎

Since the poles and zeros of $G(s)H(s)$ are either real or in complex-conjugate pairs, the imaginary parts in the numerator of Eq. (8-39) always cancel each other. Thus, in Eq. (8-39), the terms in the summations may be replaced by the real parts of the poles and zeros of $G(s)H(s)$, respectively. Further, the intersect of the asymptotes represents the physical center of gravity of the complete root loci, and σ_1 is always a real number.

EXAMPLE 8-6   Consider the transfer function

$$G(s)H(s) = \frac{K(s + 1)}{s(s + 4)(s^2 + 2s + 2)} \qquad\qquad (8\text{-}40)$$

which corresponds to the **characteristic equation**

$$s(s + 4)(s^2 + 2s + 2) + K(s + 1) = 0 \qquad\qquad (8\text{-}41)$$

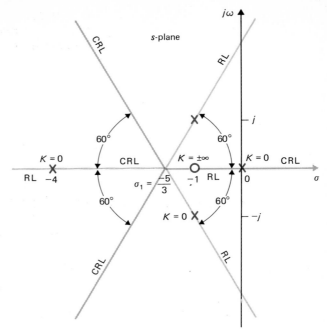

FIGURE 8-6 Asymptotes of the complete root loci of $s(s + 4)(s^2 + 2s + 2) + K(s + 1) = 0$.

The pole-zero configuration of $G(s)H(s)$ is shown in Fig. 8-6. From the six theorems on the construction of the root-locus diagram described so far, the following information concerning the RL and the CRL of Eq. (8-41) is obtained:

1. $K = 0$: The points at which $K = 0$ on the complete root loci are at the poles of $G(s)H(s)$: $s = 0$, $s = -4$, $s = -1 + j$, and $s = -1 - j$.
2. $K = \pm\infty$: The points at which $K = \pm\infty$ on the complete root loci are at the zeros of $G(s)H(s)$: $s = -1$, $s = \infty$, $s = \infty$, and $s = \infty$.
3. There are four complete root-loci branches, since Eq. (8-41) is of the fourth order.
4. The complete root loci are symmetrical to the real axis.
5. Since the number of finite poles of $G(s)H(s)$ exceeds the number of finite zeros of $G(s)H(s)$ by three ($n - m = 4 - 1 = 3$), when $K = \pm\infty$, three complete root loci approach $s = \infty$.

    The angles of the asymptotes of the RL ($K \geq 0$) are given by Eq. (8-37):

$$k = 0 \qquad \theta_0 = \frac{180°}{3} = 60°$$

$$k = 1 \qquad \theta_1 = \frac{540°}{3} = 180°$$

$$k = 2 \qquad \theta_2 = \frac{900°}{3} = 300°$$

The angles of the asymptotes of the CRL ($K \leq 0$) are given by Eq. (8-38), and are calculated to be 0°, 120°, and 240°.

6. The asymptotes of the complete root loci are calculated using Eq. (8-39):

$$\sigma_1 = \frac{\Sigma \text{ finite poles of } G(s)H(s) - \Sigma \text{ finite zeros of } G(s)H(s)}{n - m}$$

$$= \frac{(0 - 4 - 1 + j - 1 - j) - (-1)}{4 - 1} = -\frac{5}{3} \tag{8-42}$$

The asymptotes of the complete root loci are shown in Fig. 8-6.

EXAMPLE 8-7

As further illustrations, the asymptotes of the complete root loci of several equations are shown in Fig. 8-7.

Root Loci on the Real Axis

THEOREM 8-7.   *The entire real axis of the s-plane is occupied by the complete root loci; i.e., either the RL or the CRL.*

A.   RL: On a given section of the real axis, RL are found in the section only if the total number of poles and zeros of $G(s)H(s)$ to the right of the section is *odd*.
B.   CRL: On a given section of the real axis, CRL are found in the section only if the total number of real poles and zeros of $G(s)H(s)$ to the right of the section is *even*.

Complex poles and zeros of $G(s)H(s)$ do not affect the distribution of the root loci on the real axis.

PROOF.   The proof of the theorem is based on the following observations:

1.   At any point s_1 on the real axis, the angles of the vectors drawn from the complex-conjugate poles and zeros of $G(s)H(s)$ add up to zero. Therefore, the only contribution to the angular relations in Eqs. (8-20) and (8-21) is from the real poles and zeros of $G(s)H(s)$.
2.   Only the real poles and zeros of $G(s)H(s)$ that lie to the right of the point s_1 may contribute to Eqs. (8-20) and (8-21), since real poles and zeros that lie to the left of the point contribute zero degrees.
3.   Each real pole of $G(s)H(s)$ to the right of s_1 contributes -180 degrees, and each real zero of $G(s)H(s)$ to the right of s_1 contributes 180 degrees to Eqs. (8-20) and (8-21).

The last observation shows that for s_1 to be a point on the RL, there must be an odd number of poles and zeros of

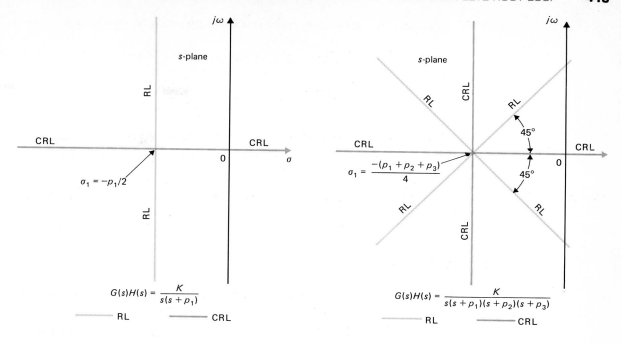

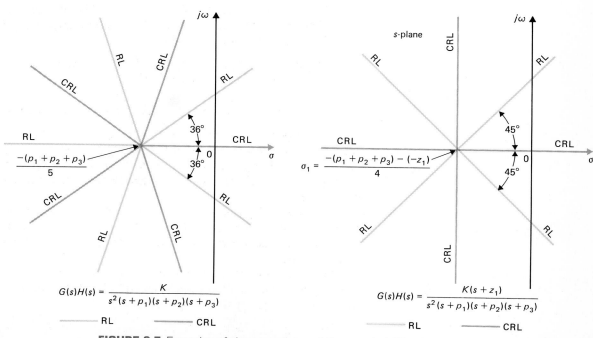

FIGURE 8-7 Examples of the asymptotes of the root loci (RL and CRL).

$G(s)H(s)$ to the right of the point, and for s_1 to be a point of the CRL, the total number of poles and zeros of $G(s)H(s)$ to the right of the point must be even. The following example illustrates the determination of the properties of the root loci on the real axis of the s-plane. ∎

EXAMPLE 8-8   The complete root loci for two pole-zero configurations of $G(s)H(s)$ are shown in Fig. 8-8. Notice that the entire real axis is occupied by either the RL or the CRL

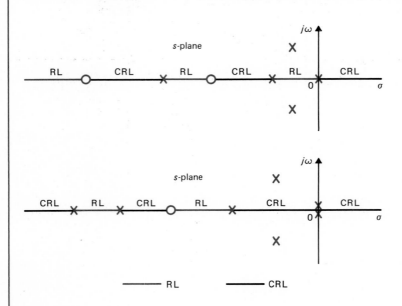

FIGURE 8-8 Properties of root loci on the real axis.

Angles of Departure and Angles of Arrival of the Complete Root Loci

angles of departure or arrival

The angle of departure or arrival of a root locus at a pole or zero of $G(s)H(s)$ denotes the angle of the tangent to the locus near the point.

EXAMPLE 8-9   For the root-locus diagram shown in Fig. 8-9, the root locus near the pole at $s = -1 + j$ may be more accurately sketched by knowing the angle at which the root locus leaves the pole.

The angle of arrival or departure of RL may be determined by use of Eq. (8-20), and that of CRL is determined by use of Eq. (8-21). As shown in Fig. 8-9, the angle of departure of the root locus at $s = -1 + j$ is represented by θ_2, measured with respect to the real axis. Let us assign s_1 to be a point on the root locus leaving the pole at $-1 + j$ and is very close to the pole.

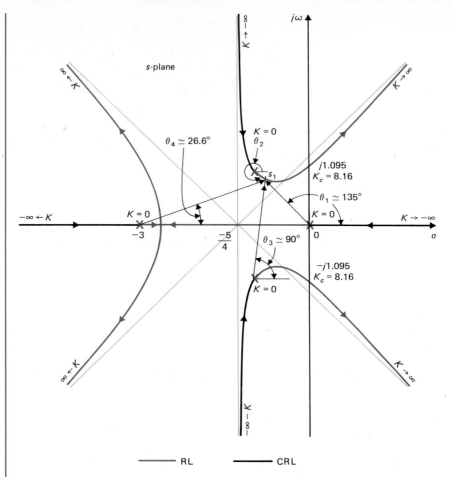

— RL   — CRL

FIGURE 8-9 Complete root loci of $s(s + 3)(s^2 + 2s + 2) + K = 0$ to illustrate the angles of departure or arrival.

Then s_1 must satisfy Eq. (8-20). Thus,

$$\angle G(s_1)H(s_1) = -(\theta_1 + \theta_2 + \theta_3 + \theta_4) = (2k + 1)180° \qquad (8\text{-}43)$$

where k is any integer. Since s_1 is assumed to be very close to the pole at $-1 + j$, the angles of the vectors drawn from the other three poles are approximated by considering that s_1 is at $+1 + j$. From Fig. 8-9, Eq. (8-43) is written

$$-(135° + \theta_2 + 90° + 26.6°) = (2k + 1)180° \qquad (8\text{-}44)$$

where θ_2 is the only unknown angle. In this case, we can set k to be -1, and the result for θ_2 is $-71.6°$.

When the angle of departure or arrival of the RL at a pole or zero $G(s)H(s)$ is determined, the angle of arrival or departure of the CRL at the same point differs from this angle by

180°, since Eq. (8-21) is now used. Figure 8-9 shows that the angle of arrival of the CRL at $-1 + j$ is 108.4°, which is $180° - 71.6°$. Similarly, for the root-locus diagram in Fig. 8-9, we can show that CRL arrives at the pole $s = -3$ with an angle of 180°, and the RL leaves the same pole at 0°. For the pole at $s = 0$, the angle of arrival of the CRL is 180°, whereas the angle of departure of the RL is 180°. These angles are also determined from the knowledge of the type of root loci on sections of the real axis separated by the poles and zeros of $G(s)H(s)$. Since the total angles of the vectors drawn from complex poles and zeros to any point on the real axis add up to be zero, the angles of arrival or departure of root loci on the real axis are not affected by complex poles and zeros of $G(s)H(s)$.

EXAMPLE 8-10   Consider a $G(s)H(s)$ that has a multiple-order pole on the real axis, as shown in Fig. 8-10. Only the real poles and zeros of $G(s)H(s)$ are shown, since the complex-conjugate ones do not affect the type or the angles of arrival and departure of the root loci on the real axis. For the third-order pole at $s = -2$, there are three RL leaving and three CRL arriving at the point. To find the angles of departure of the RL, we assign a point s_1 on one of the RL near $s = -2$, and apply Eq. (8-20). The result is

$$-\theta_1 - 3\theta_2 + \theta_3 = (2k + 1)180° \qquad (8\text{-}45)$$

where θ_1 and θ_3 denote the angles of the vectors drawn from the pole at 0 and the zero at -3, respectively, to s_1. The angle θ_2 is multiplied by 3, since there are three poles at $s = -2$, so that there are three vectors drawn from -2 to s_1. Setting $k = 0$ in Eq. (8-45), and since $\theta_1 = 180°$, $\theta_3 = 0°$, we have $\theta_2 = 0°$, which is the angle of departure of the RL that lies between $s = 0$ and $s = -2$. For the angles of departure of the other two RL's, we set $k = 1$ and $k = 2$ successively in Eq. (8-45), and we have $\theta_2 = 120°$ and $-120°$. Similarly, for the three CRL that arrive at $s = -2$, Eq. (8-21) is used, and the angles of arrivals are found to be 60°, 180°, and $-60°$.

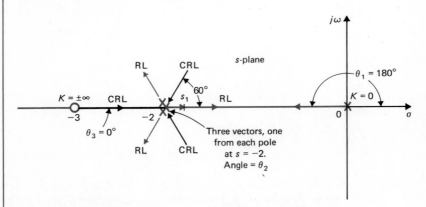

FIGURE 8-10 Angles of departure and arrival at a third-order pole.

Intersection of the Root Loci with the Imaginary Axis

The points where the complete root loci intersect the imaginary axis of the s-plane, and
Routh–Hurwitz   the corresponding values of K, may be determined by means of the Routh–Hurwitz
criterion. For complex situations when the root loci have multiple number of intersec-

tions on the imaginary axis, the intersects and the critical values of K can be determined by properly selecting the resolution of the root-locus computer solution. The Bode diagram described in Chapter 10, associated with the frequency response, can also be used for this purpose.

EXAMPLE 8-11

The root-locus diagram shown in Fig. 8-9 is for the equation

$$s(s + 3)(s^2 + 2s + 2) + K = 0 \qquad (8\text{-}46)$$

auxiliary equation

The RL intersect the $j\omega$ axis at two points. Applying the Routh–Hurwitz criterion to Eq. (8-46), and by solving the auxiliary equation, we have the critical value of K for stability as $K_c = 8.16$ and the corresponding frequency is $\omega_c = \pm 1.095$ rad/s.

Breakaway Points (Saddle Points) on the Complete Root Loci

Breakaway points, or saddle points, on the root loci of an equation correspond to multiple-order roots of the equation. Figure 8-11(a) illustrates a case in which two branches of the root loci meet at the breakaway point on the real axis and then depart from the axis in opposite directions. In this case, the breakaway point represents a double root of

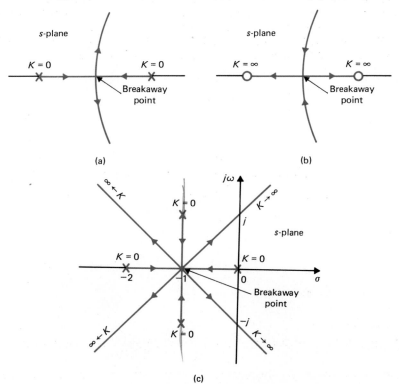

FIGURE 8-11 Examples of breakaway points on the real axis in the s-plane.

breakaway points

the equation when the value of K is assigned the value corresponding to the point. Figure 8-11 (b) shows another common situation when two complex-conjugate root loci approach the real axis, meet at the breakaway point , and then depart in opposite directions along the real axis. In general, a breakaway point may involve more than two root loci. Figure 8-11 (c) illustrates a situation when the breakaway point represents a fourth-order root.

A root-locus diagram can have, of course, more than one breakaway point. Moreover, the breakaway points need not always be on the real axis, as shown by the previous examples. Because of the conjugate symmetry of the root loci, the breakaway points not on the real axis must be in complex-conjugate pairs. Refer to Fig. 8-14 for an example with complex breakaway points. The properties of the breakaway points are given in the following theorem with the proof omitted.

THEOREM 8-8. *The breakaway points on the complete root loci of $1 + KG_1(s)H_1(s) = 0$ must satisfy*

$$\frac{dG_1(s)H_1(s)}{ds} = 0 \qquad (8\text{-}47)$$

necessary condition

It is important to point out that the condition for the breakaway point given in the last equation is *necessary* but *not sufficient*. In other words, all breakaway points on the RL and the CRL must satisfy Eq. (8-47), but not all solutions of Eq. (8-47) are breakaway points. To be a breakaway point, the solution of Eq. (8-47) must also satisfy the equation $1 + KG_1(s)H_1(s) = 0$, that is, a point of the root loci for some real K. In general, the following conclusions can be made with regard to the solutions of Eq. (8-47):

1. All *real* solutions of Eq. (8-47) are breakaway points on the root loci $(-\infty < K < \infty)$, since the entire real axis of the s-plane is occupied by the complete root loci.
2. The complex-conjugate solutions of Eq. (8-47) are breakaway points only if they satisfy the characteristic equation, or be points on the complete root loci.
3. Since the condition of the root loci is

$$K = -\frac{1}{G_1(s)H_1(s)} \qquad (8\text{-}48)$$

taking the derivative on both sides of the equation with respect to s, we have

$$\frac{dK}{ds} = \frac{dG_1(s)H_1(s)/ds}{[G_1(s)H_1(s)]^2} \qquad (8\text{-}49)$$

Thus, the breakaway-points condition can also be written as

$$\frac{dK}{ds} = 0 \qquad (8\text{-}50)$$

THE ANGLE OF ARRIVAL AND DEPARTURE AT THE BREAKAWAY POINTS. The angles at which the root loci arrive or depart from a breakaway point depends on the number of loci that are involved at the point. For example, the root loci shown in Figs. 8-11(a) and 8-11(b) all arrive and break away at 180° apart, whereas in Fig. 8-11(c), the four root loci arrive and depart with angles 90° apart. *In general, n root loci (RL or CRL) arrive or leave a breakaway point at 180/n degrees apart.*

ROOTLOCI

The ROOTLOCI program in the ACSP package contains features that will obtain Eq. (8-47) from the expression $G_1(s)H_1(s)$, and then solve for the roots of the equation for the possible breakaway points. The following examples are devised to illustrate the application of Eq. (8-47) for the determination of breakaway points on the root loci.

illustrative examples

EXAMPLE 8-12 Consider the second-order equation

$$s(s + 2) + K(s + 4) = 0 \qquad (8\text{-}51)$$

Based on some of the properties of the root loci described thus far, the root loci of Eq. (8-51) are sketched as shown in Fig. 8-12 for $-\infty < K < \infty$. It can be proven that the complex portion of the root loci is described by a circle. The two breakaway points are all on the real axis, one between 0 and -2 and the other between -4 and $-\infty$. From Eq. (8-51), we have

$$G_1(s)H_1(s) = \frac{s + 4}{s(s + 2)} \qquad (8\text{-}52)$$

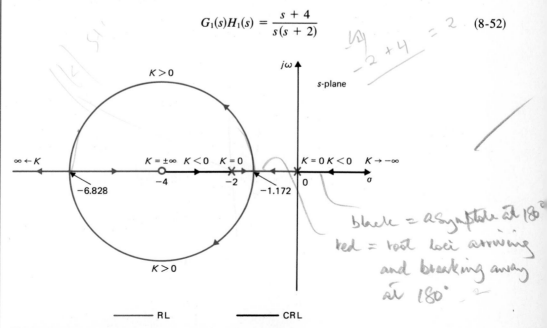

FIGURE 8-12 Complete root loci of $s(s + 2) + K(s + 4) = 0$.

Applying Eq. (8-47), the breakaway points of the root loci must satisfy

$$\frac{dG_1(s)H_1(s)}{ds} = \frac{s(s + 2) - 2(s + 1)(s + 4)}{s^2(s + 2)^2} = 0 \tag{8-53}$$

or

$$s^2 + 8s + 8 = 0 \tag{8-54}$$

Solving Eq. (8-54), we find that the two breakaway points of the root loci are at $s = -1.172$ and $s = -6.828$. Note also that the breakaway points happen to occur all on the RL ($K > 0$).

EXAMPLE 8-13   Consider the equation

$$s^2 + 2s + 2 + K(s + 2) = 0 \tag{8-55}$$

The equivalent $G(s)H(s)$ is obtained by dividing both sides of Eq. (8-55) by $s^2 + 2s + 2$:

$$G(s)H(s) = \frac{K(s + 2)}{s^2 + 2s + 2} \tag{8-56}$$

Based on the poles and zeros of $G(s)H(s)$, the complete root loci of Eq. (8-55) are plotted as shown in Fig. 8-13. The diagram shows that the RL and the CRL each has a breakaway point. These breakaway points are determined from

$$\frac{dG_1(s)H_1(s)}{ds} = \frac{d}{ds}\left(\frac{s + 2}{s^2 + 2s + 2}\right) = \frac{s^2 + 2s + 2 - 2(s + 1)(s + 2)}{(s^2 + 2s + 2)^2} = 0 \tag{8-57}$$

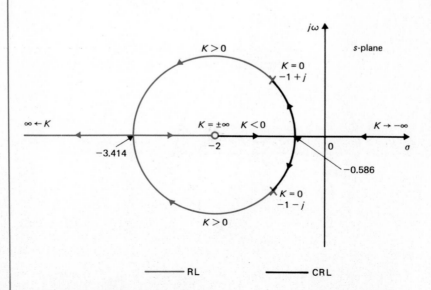

FIGURE 8-13 Complete root loci of $s^2 + 2s + 2 + K(s + 2) = 0$.

or

$$s^2 + 4s + 2 = 0 \qquad (8\text{-}58)$$

The solution of the equation gives the breakaway point at $s = -0.586$, which is on the CRL, and $s = -3.414$, which is a breakaway point on the RL.

EXAMPLE 8-14   Figure 8-14 shows the complete root loci of the equation

$$s(s + 4)(s^2 + 4s + 20) + K = 0 \qquad (8\text{-}59)$$

Dividing both sides of the last equation by the terms that do not contain K, we have

$$1 + KG_1(s)H_1(s) = 1 + \frac{K}{s(s + 4)(s^2 + 4s + 20)} = 0 \qquad (8\text{-}60)$$

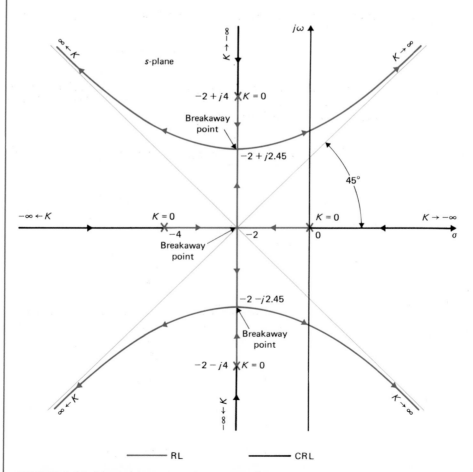

FIGURE 8-14 Complete root loci of $s(s + 4)(s^2 + 4s + 20) + K = 0$.

Since the poles of $G_1(s)H_1(s)$ are symmetrical about the axes $\sigma = -2$ and $\omega = 0$ in the s-plane, the complete root loci of the equation are also symmetrical with respect to these two axes. Taking the derivative of $G_1(s)H_1(s)$ with respect to s, we get

$$\frac{dG_1(s)H_1(s)}{ds} = -\frac{4s^3 + 24s^2 + 72s + 80}{[s(s+4)(s^2 + 4s + 20)]^2} = 0 \tag{8-61}$$

or

$$s^3 + 6s^2 + 18s + 20 = 0 \tag{8-62}$$

The solutions of the last equation are $s = -2$, $-2 + j2.45$, and $-2 - j2.45$. In this case, Fig. 8-14 shows that all the solutions of Eq. (8-62) are breakaway points on the root loci, and two of these points are complex.

EXAMPLE 8-15   In this example, we shall show that not all the solutions of Eq. (8-47) necessarily represent breakaway points on the root loci. The complete root loci of the equation

$$s(s^2 + 2s + 2) + K = 0 \tag{8-63}$$

are shown in Fig. 8-15; neither the RL nor the CRL have any breakaway point in this case.

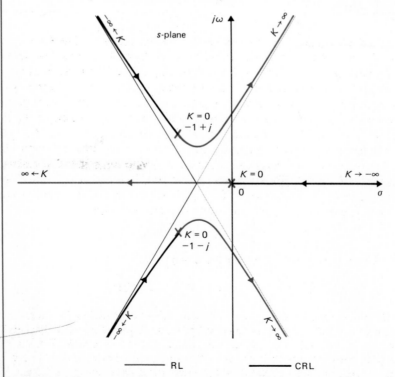

FIGURE 8-15 Complete root loci of $s(s^2 + 2s + 2) + K = 0$.

However, writing Eq. (8-63) as

$$1 + KG_1(s)H_1(s) = 1 + \frac{K}{s(s^2 + 2s + 2)} = 0 \tag{8-64}$$

and applying Eq. (8-47), we have the equation for the breakaway points:

$$3s^2 + 4s + 2 = 0 \tag{8-65}$$

The roots of Eq. (8-65) are $s = -0.667 + j0.471$ and $s = -0.667 - j0.471$. These two roots do not represent breakaway points on the root loci, since they do not satisfy Eq. (8-63) for any real values of K.

root sensitivity **THE ROOT SENSITIVITY [20, 21, 22].** The condition on the breakaway points on the root loci in Eq. (8-50) gives a by-product on the properties of the root loci. The sensitivity of the roots of the characteristic equation when K varies is defined as the **root sensitivity**, and is given by the following expression:

$$S_k = \frac{ds/s}{dK/K} = \frac{K}{s}\frac{ds}{dK} \tag{8-66}$$

the K value of the break point should be avoided, therwise the root sensitivity is very high and system

robust system can become unstable.

Thus, the last equation shows that *the root sensitivity at the breakaway points is infinite*. From the root-sensitivity standpoint, we should avoid selecting the value of K to operate at the breakaway points, which correspond to multiple-order roots for the characteristic equation. In the design of control systems, not only is it important to arrive at a system that has the desired performance characteristics, but, ideally, the system must be insensitive to parameter variations. For instance, a system may perform satisfactorily at a certain forward gain K, but if it is very sensitive to the variation of K, it may get into the undesirable performance region or become unstable if K varies by only a small amount. In formal control-system terminology, a system that is insensitive to parameter variations is called a **robust system**. Therefore, the root-locus study of control systems must involve not only the shape of the loci with respect to the variable parameter K, but also how the roots along the loci vary with the variation of K.

EXAMPLE 8-16   Figure 8-16 illustrates the root locus diagram of

$$s(s + 1) + K = 0 \tag{8-67}$$

with K incremented uniformly over 100 values from -20 to 20. The root loci were computed with the ROOTLOCI program and plotted with the Hewlett-Packard 7470A digital plotter. Each dot on the root-locus plot represents one root for a given value of K. Thus, we see that the root sensitivity is low when the magnitude of K is large. As the magnitude of K decreases, the movements of the roots become larger for the same incremental change in K, and at the breakaway point, $s = -0.5$, the root sensitivity is infinite. Figure 8-17 shows the root-locus diagram of the

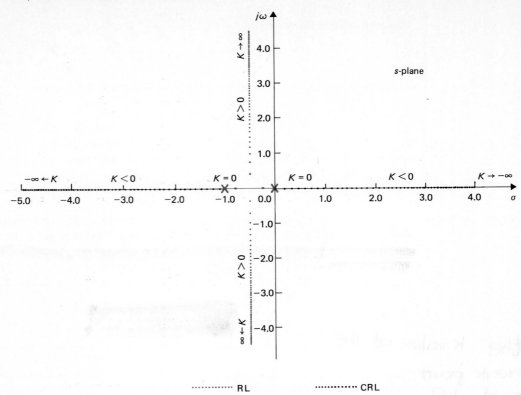

FIGURE 8-16 Complete root loci of $s(s + 1) + K = 0$, showing the root sensitivity with respect to K.

equation

$$s^2(s + 1)^2 + K(s + 2) = 0 \qquad (8\text{-}68)$$

with K incremented uniformly over 200 values from -40 to 50. Again, the loci show that the root sensitivity increases as the roots approach the breakaway points at $s = 0$, -0.543, -1.0, and -2.457.

We can investigate the root sensitivity further by using the analytical expression in Eq. (8-50). For the second-order equation in Eq. (8-67),

$$\frac{dK}{ds} = -2s - 1 \qquad (8\text{-}69)$$

From Eq. (8-67), $K = -s(s + 1)$, the root sensitivity becomes

$$S_k = \frac{ds}{dK}\frac{K}{s} = \frac{s + 1}{2s + 1} \qquad (8\text{-}70)$$

where $s = \sigma + j\omega$, and s must take on the values of the roots of Eq. (8-67). For the roots on

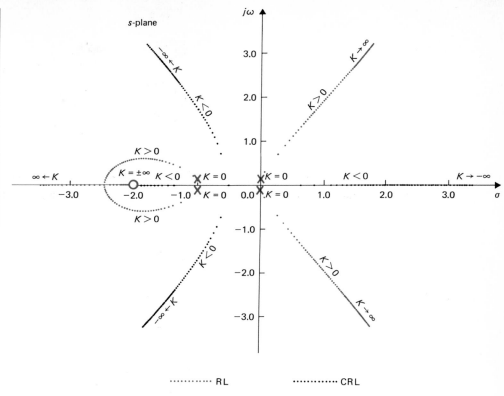

FIGURE 8-17 Complete root loci of $s^2(s + 1) + K(s + 2) = 0$, showing the root sensitivity with respect to K.

the real axis, $\omega = 0$; thus, Eq. (8-70) leads to

$$|S_k|_{\omega=0} = \left| \frac{\sigma + 1}{2\sigma + 1} \right| \tag{8-71}$$

When the two roots are complex, $\sigma = -0.5$ for all values of ω; Eq. (8-70) gives

$$|S_k|_{\sigma=-0.5} = \left(\frac{0.25 + \omega^2}{4\omega^2} \right)^{1/2} \tag{8-72}$$

From the last equation, it is apparent that the sensitivities of the pairs of complex-conjugate roots are the same, since ω appears only as ω^2 in the equation. Equation (8-71) indicates that the sensitivities of the two real roots are different for a given value of K. Table 8-1 gives the magnitudes of the sensitivities of the two roots of Eq. (8-67) for several values of K, where $|S_{k1}|$ denotes the root sensitivity of the first root and $|S_{k2}|$ denotes that of the second root. These values indicate that although the two real roots reach $\sigma = -0.5$ for the same value of $K = 0.25$, and each root travels the same distance from $s = 0$ and $s = -1$, the sensitivities of the two real roots are not the same.

TABLE 8-1

K	ROOT NO. 1	$|S_{k1}|$	ROOT NO. 2	$|S_{k2}|$
0	0	1.000	-1.000	0
0.04	-0.042	1.045	-0.958	0.454
0.16	-0.200	1.333	-0.800	0.333
0.24	-0.400	3.000	-0.600	2.000
0.25	-0.500	∞	-0.500	∞
0.28	$-0.5 + j0.173$	1.527	$-0.5 - j0.173$	1.527
0.36	$-0.5 + j0.332$	0.905	$-0.5 - j0.332$	0.905
0.40	$-0.5 + j0.387$	0.817	$-0.5 - j0.387$	0.817
0.80	$-0.5 + j0.742$	0.603	$-0.5 - j0.742$	0.603
1.20	$-0.5 + j0.975$	0.562	$-0.5 - j0.975$	0.562
2.00	$-0.5 + j1.323$	0.535	$-0.5 - j1.323$	0.535
4.00	$-0.5 + j1.937$	0.516	$-0.5 - j1.937$	0.516
∞	$-0.5 + j\infty$	0.500	$-0.5 - j\infty$	0.500

Calculation of K on the Root Loci

calculation of K

Once the complete root loci have been constructed, the values of K at any point s_1 on the loci can be determined by use of the defining equation of Eq. (8-22). Graphically, the magnitude of K can be written as

$$|K| = \frac{\text{product of lengths of vectors drawn from the poles of } G_1(s)H_1(s) \text{ to } s_1}{\text{product of lengths of vectors drawn from the zeros of } G_1(s)H_1(s) \text{ to } s_1} \quad (8\text{-}73)$$

EXAMPLE 8-17   As an illustration, the root loci of the equation

$$s^2 + 2s + 2 + K(s + 2) = 0 \quad (8\text{-}74)$$

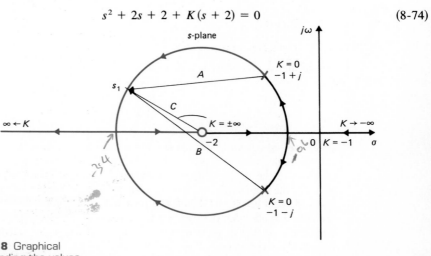

FIGURE 8-18 Graphical method of finding the values of K on the root loci.

———— RL   ———— CRL

are shown in Fig. 8-18. The value of K at the point s_1 is given by

$$K = \frac{A \cdot B}{C} \tag{8-75}$$

where A and B are the lengths of the vectors drawn from the poles of $G(s)H(s) = K(s + 2)/(s^2 + 2s + 2)$ to the point s_1, and C is the length of the vector drawn from the zero of $G(s)H(s)$ to s_1. In this case, s_1 is on the RL, so K is positive. In general, the value of K at the point where the root loci intersect the imaginary axis can also be found by the method just described. Figure 8-18 shows that the value of K at $s = 0$ is -1. The Routh–Hurwitz criterion represents a more convenient method of finding this critical value of K.

In summary, except for extremely complex cases, the properties described should be adequate for making a reasonably accurate sketch of the root-locus diagram short of plotting it point by point. The computer program can be used to solve for the exact root locations, the breakaway points, and some of the other more specific details of the root loci. One cannot rely on the computer program completely, since the user still has to decide on the range and resolution of K so that the root-locus plot has a reasonable appearance. For quick reference, the important properties described are summarized in Table 8-2.

The following example illustrates the construction of a root-locus diagram manually, step by step, using the root-locus properties given in Table 8-2.

TABLE 8-2   Properties of the Complete Root Loci

1.	$K = 0$ points	The $K = 0$ points on the complete root loci are at the poles of $G(s)H(s)$. (The poles include those at $s = \infty$.)
2.	$K = \pm\infty$ points	The $K = \pm\infty$ points on the complete root loci are at the zeros of $G(s)H(s)$. (The zeros include those at $s = \infty$.)
3.	Number of separate root loci	The total number of complete root loci is equal to the order of the equation $F(s) = 0$.
4.	Symmetry of root loci	The complete root loci are symmetrical with respect to the axes of symmetry of the pole-zero configuration of $G(s)H(s)$.
5.	Asymptotes of root loci as $s \to \infty$	For large values of s, the RL $(K > 0)$ are asymptotic to straight lines with angles given by

$$\theta_k = \frac{(2k + 1)\pi}{|n - m|}$$

For CRL $(K < 0)$,

$$\theta_k = \frac{2k\pi}{|n - m|}$$

where $k = 0, 1, 2, \ldots, |n - m| - 1$,
n = number of finite poles of $G(s)H(s)$, and
m = number of finite zeros of $G(s)H(s)$.

6.	Intersection of the asymptotes	(a) The intersection of the asymptotes lies only on the real axis in the s-plane.
		(b) The point of intersection of the asymptotes on the real axis is given by (for all values of K)

$$\sigma_1 = \frac{\Sigma \text{ poles of } G(s)H(s) - \Sigma \text{ zeros of } G(s)H(s)}{n - m}$$

TABLE 8-2   Continued

7.	Root loci on the real axis	RL $(K > 0)$ are found in a section of the real axis only if the total number of real poles and zeros of $G(s)H(s)$ to the **right** of the section is **odd**. If the total number of real poles and zeros to the right of a given section is **even**, CRL $(K < 0)$ are found.
8.	Angles of departure	The angle of departure or arrival of the complete root locus from a pole or a zero of $G(s)H(s)$ can be determined by assuming a point s_1 that is on the root locus associated with the pole, or zero, and that is very close to the pole, or zero, and applying the equation,

$$\angle G(s_1)H(s_1) = \sum_{i=1}^{m} \angle(s_1 + z_1) - \sum_{j=1}^{n} \angle(s_1 + p_j)$$

$$= 2(k + 1)\pi \qquad \text{RL} \quad (K \geq 0)$$

$$= 2k\pi \qquad\qquad \text{CRL} \ (K \leq 0)$$

		where $k = 0, \pm 1, \pm 2, \ldots$.
9.	Intersection of the root loci with the imaginary axis	The values of ω and K at the crossing points of the root loci $(-\infty < K < \infty)$ on the imaginary axis of the s-plane may be determined by use of the Routh–Hurwitz criterion. The Bode diagram of $G(s)H(s)$ discussed in Chapter 10 may also be used.
10.	Breakaway points	The breakaway points on the complete root loci are determined by finding the roots of $dK/ds = 0$, or $dG(s)H(s)/ds = 0$. These are necessary conditions only.
11.	Calculation of the value of K on the complete root loci	The absolute value of K at any point s_1 on the complete root loci is determined from the equation

$$|K| = \frac{1}{|G(s_1)H(s_1)|}$$

EXAMPLE 8-18   Consider the equation

$$s(s + 5)(s + 6)(s^2 + 2s + 2) + K(s + 3) = 0 \tag{8-76}$$

The following properties of the complete root loci are determined:

K = 0 points

1. The $K = 0$ points are at $s = 0, -5, -6, -1 + j$, and $-1 - j$. These are the poles of $G(s)H(s)$, where

$$G(s)H(s) = \frac{K(s + 3)}{s(s + 5)(s + 6)(s^2 + 2s + 2)} \tag{8-77}$$

K = ±∞ points

2. The $K = \pm\infty$ points are at $s = -3, \infty, \infty, \infty, \infty$, which are the zeros of $G(s)H(s)$.

number of branches

3. There are five separate branches on the complete root loci.

symmetry

4. The complete root loci are symmetrical with respect to the real axis of the s-plane.

asymptotes

5. Since $G(s)H(s)$ has five finite poles and one finite zero, four RL and four CRL should approach infinity in the s-plane. The angles of the asymptotes of the RL at $|s| = \infty$ are given by [Eq. (8-37)]

$$\theta_k = \frac{(2k + 1)\pi}{|n - m|} = \frac{(2k + 1)\pi}{|5 - 1|} \qquad 0 \leq K < \infty \tag{8-78}$$

for $k = 0, 1, 2, 3$. Thus, the four root loci that approach infinity as K approaches $+\infty$ should approach asymptotes with angles of $45°$, $-45°$, $135°$, and $-135°$, respectively. The angles of the asymptotes of the CRL at infinity are given by [Eq. (8-38)]

$$\theta_k = \frac{2k\pi}{|n - m|} = \frac{2k\pi}{|5 - 1|} \qquad -\infty < K \le 0 \qquad (8\text{-}79)$$

Therefore, as K approaches $-\infty$, four CRL should approach infinity along asymptotes with angles of $0°$, $90°$, $180°$, and $270°$.

intersect of asymptotes

6. The intersection of the asymptotes is given by [Eq. (8-39)]

$$\sigma_1 = \frac{\Sigma \text{ poles of } G(s)H(s) - \Sigma \text{ zeros of } G(s)H(s)}{n - m}$$

$$= \frac{(0 - 5 - 6 - 1 + j - 1 - j) - (-3)}{4} = -2.5 \qquad (8\text{-}80)$$

The results from these six step are illustrated in Fig. 8-19. It should be pointed out that in general the properties of the asymptotes do **not** indicate on which side of the

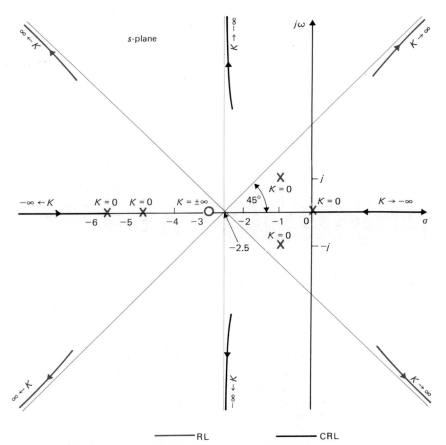

FIGURE 8-19 Preliminary calculations of the complete root loci of $s(s + 5)(s + 6)(s^2 + 2s + 2) + K(s + 3) = 0$.

asymptote the root locus will lie. The asymptotes indicate nothing more than the behavior of the root loci as $s \to \infty$. In fact, the root locus can even cross an asymptote in the finite s domain. The segments of the RL and the CRL shown in Fig. 8-19 can be accurately made only if additional information is obtained.

root loci on real axis

7. Complete root loci on the real axis: There are RL on the real axis between $s = 0$ and $s = -3$, $s = -5$ and $s = -6$. There are CRL on the remaining portions of the real axis, that is, between $s = -3$ and $s = -5$, and $s = -6$ and $s = -\infty$, as shown in Fig. 8-20.

angles of departure

8. Angles of departure: The angle of departure θ of the RL leaving the pole at $-1 + j$ is determined using Eq. (8-20). If s_1 is a point on the RL leaving the pole at $-1 + j$, and s_1 is very close to the pole, as shown in Fig. 8-21, Eq. (8-20) gives

$$\angle(s_1 + 3) - \angle s_1 - \angle(s_1 + 1 + j) - \angle(s_1 + 5) - \angle(s_1 + 6) - \angle(s_1 + 1 - j)$$

$$= (2k + 1)180° \qquad (8\text{-}81)$$

or

$$26.6° - 135° - 90° - 14° - 11.4° - \theta \cong (2k + 1)180° \qquad (8\text{-}82)$$

for $k = 0, \pm1, \pm2, \ldots$. Therefore, for $k = -2$,

$$\theta \cong -43.8° \qquad (8\text{-}83)$$

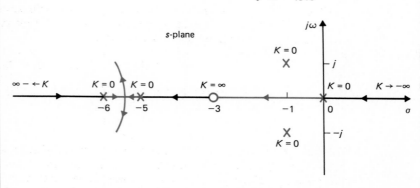

FIGURE 8-20 Complete root loci of $s(s + 5)(s + 6)(s^2 + 2s + 2) + K(s + 3) = 0$ on the real axis.

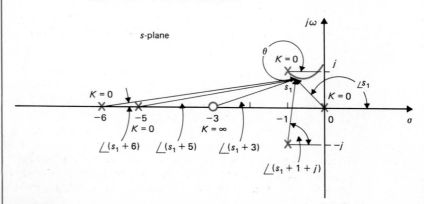

FIGURE 8-21 Computation of angle of departure of the root loci of $s(s + 5)(s + 6)(s^2 + 2s + 2) + K(s + 3) = 0$.

angle of arrival

Similarly, Eq. (8-21) is used to determine the angle of arrival θ' of the CRL arriving at the pole $-1 + j$. It is easy to see that θ' differs from θ by 180°; thus,

$$\theta' \cong 180° - 43.8° = 136.2° \tag{8-84}$$

intersection on jω-axis

9. The intersection of the root loci on the imaginary axis is determined using Routh's tabulation. Equation (8-76) is rewritten

$$s^5 + 13s^4 + 54s^3 + 82s^2 + (60 + K)s + 3K = 0 \tag{8-85}$$

Routh's tabulation is

s^5	1	54	$60 + K$
s^4	13	82	$3K$
s^3	47.7	$0.769K$	0
s^2	$65.6 - 0.212K$	$3K$	0
s^1	$\dfrac{3940 - 105K - 0.163K^2}{65.6 - 0.212K}$	0	0
s^0	$3K$	0	0

For Eq. (8-85) to have no roots on the $j\omega$-axis or in the right-half s-plane, the elements in the first column of Routh's tabulation must all be of the same sign. Therefore, the following inequalities must be satisfied:

$$65.6 - 0.212K > 0 \quad \text{or} \quad K < 309 \tag{8-86}$$

$$3940 - 105K - 0.163K^2 > 0 \quad \text{or} \quad K < 35 \tag{8-87}$$

$$K > 0 \tag{8-88}$$

Thus, all the roots of Eq. (8-85) will stay in the left half of the s-plane if K lies between 0 and 35, which means that the root loci of Eq. (8-85) cross the imaginary axis when $K = 35$ and $K = 0$. The coordinates at the crossover points on the imaginary axis that correspond to $K = 35$ is determined from the auxiliary equation:

auxiliary equation

$$A(s) = (65.6 - 0.212K)s^2 + 3K = 0 \tag{8-89}$$

which is obtained by using the coefficients from the row just above the row of zeros in the s^1 row that would have happened when K is set to 35. Substituting $K = 35$ in Eq. (8-89), we get

$$58.2s^2 + 105 = 0 \tag{8-90}$$

The roots of the last equation are $s = j1.34$ and $s = -j1.34$, which are the points at which the root loci cross the $j\omega$-axis.

breakaway points

10. Breakaway points: Based on the information gathered from the preceding nine steps, a trial sketch of the root loci indicates that there can be only one breakaway point on the entire root loci, and the point should lie between the two poles of $G(s)H(s)$ at $s = -5$ and -6. To find the breakaway point, we take the derivative on both sides of Eq. (8-77) with respect to s and set it to zero; the resulting equation is

$$s^5 + 13.5s^4 + 66s^3 + 142s^2 + 123s + 45 = 0 \tag{8-91}$$

Since there is only one breakaway point expected, only one root of the last equation is the correct solution of the breakaway point. The five roots of Eq. (8-91) are

$$s = 3.33 + j1.204 \qquad s = 3.33 - j1.204$$

$$s = -0.656 + j0.468 \qquad s = -0.656 - j0.468$$

$$s = -5.53$$

Clearly, the breakaway point is at -5.53. The other four solutions do not satisfy Eq. (8-85) and are not breakaway points. Based on the information obtained in the last 10 steps, the complete root loci of Eq. (8-85) are sketched as shown in Fig. 8-22.

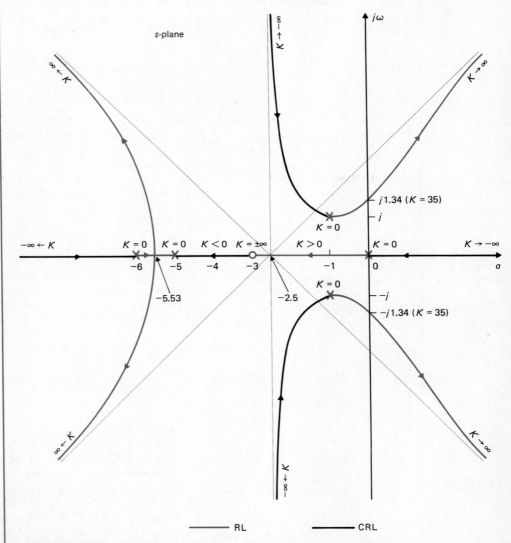

FIGURE 8-22 Complete root loci of $s(s + 5)(s + 6)(s^2 + 2s + 2) + K(s + 3) = 0$.

8-4

SOME IMPORTANT ASPECTS OF THE CONSTRUCTION OF THE ROOT LOCI

One of the important aspects of the root-locus technique is that for most control systems with moderate complexity, the analyst or designer can obtain vital information on the performance of the system by making a quick sketch of the root loci using some or all of the properties of the root loci. It is of importance to understand all the properties of the root loci even when the diagram is to be plotted with the help of a digital computer. From the design standpoint, it is useful to learn the effects on the root loci when poles and zeros of $G(s)H(s)$ are added or moved around in the s-plane. Some of these properties are helpful in the construction of the root-locus diagram.

Effects of Adding Poles and Zeros to *G(s)H(s)*

In Chapter 7, the effects of the derivative and integral control were simply illustrated by means of the root-locus diagram. The general problem of controller design may be treated as an investigation of the effects to the root loci when poles and zeros are added to the loop transfer function $G(s)H(s)$.

   ADDITION OF POLES.   In general, we may state that *adding a pole to the function $G(s)H(s)$ in the left half of the s-plane has the effect of pushing the root loci toward the right-half plane.* We can illustrate the effect qualitatively with several examples.

EXAMPLE 8-19   Let us consider the function

$$G(s)H(s) = \frac{K}{s(s + a)} \qquad a > 0 \qquad (8\text{-}92)$$

effects of adding poles

The zeros of $1 + G(s)H(s)$ are presented by the root-locus diagram of Fig. 8-23(a). These root loci are constructed based on the poles of $G(s)H(s)$, which are at $s = 0$ and $s = -a$. Now let us introduce a pole at $s = -b$, with $b > a$. The function $G(s)H(s)$ now becomes

$$G(s)H(s) = \frac{K}{s(s + a)(s + b)} \qquad (8\text{-}93)$$

Figure 8-23(b) shows that the pole at $s = -b$ causes the complex part of the root loci to bend toward the right-half s-plane. The angles of the asymptotes for the complex roots are changed from $\pm 90°$ to $\pm 60°$. The intersect of the asymptotes is also moved from $-a/2$ to $-(a + b)/2$ on the real axis. If $G(s)H(s)$ represents the loop transfer function of a feedback control system, the system with the root loci in Fig. 8-23(b) may become unstable if the value of K exceeds the critical value for stability, whereas the system represented by the root loci in Fig. 8-23(a) is always stable for $K > 0$. Figure 8-23(c) shows the root loci when another pole is added to $G(s)H(s)$ at $s = -c, c > b$. The system is now of the fourth order, and the two complex root loci are bent further to the right. The angles of the asymptotes of these two loci are now $\pm 45°$.

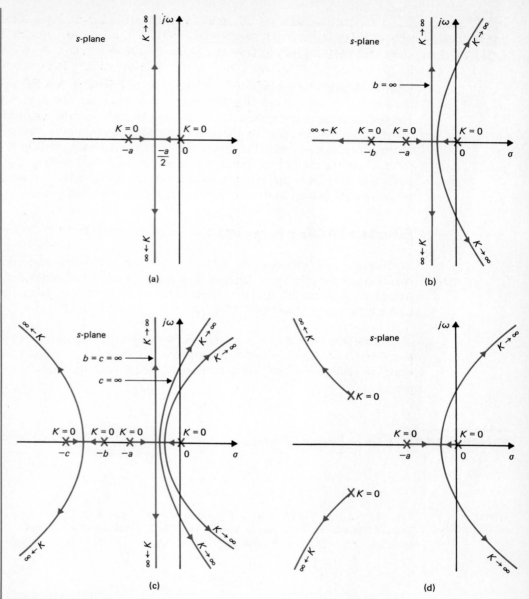

FIGURE 8-23 Root-locus diagrams that show the effects of adding poles to
$G(s)H(s)$.

For a feedback control system, the stability condition of the system becomes even more acute
than that of the third-order system. Figure 8-23(d) illustrates that the addition of a pair of
complex-conjugate poles to the transfer function of Eq. (8-92) will result in a similar effect.
Therefore, we may draw a general conclusion that the addition of poles to the function $G(s)H(s)$
has the effect of moving the root loci toward the right-half s-plane.

effects of adding zeros

ADDITION OF ZEROS.   *Adding left-half plane zeros to the function $G(s)H(s)$ generally has the effect of moving and bending the root loci toward the left-half s-plane.* For instance, Fig. 8-24(a) shows the root-locus diagram when a zero at $s = -b$ ($b > a$) is added to $G(s)H(s)$ in Eq. (8-92). The complex-conjugate root loci of the original system are bent toward the left, and in this case, the loci form a circle. There-

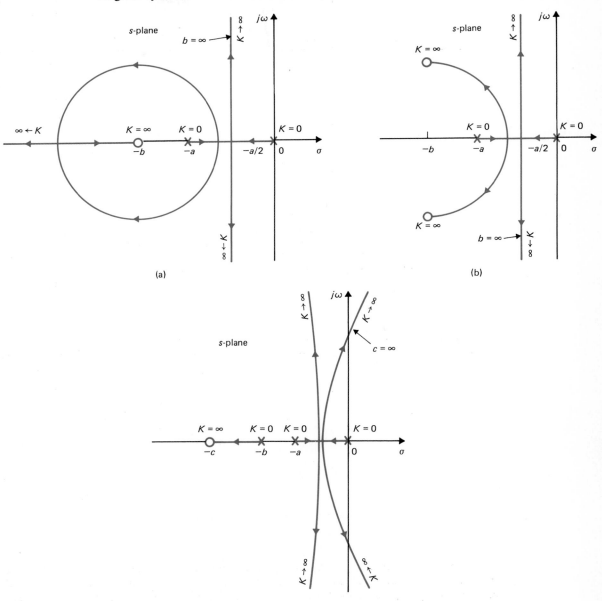

FIGURE 8-24 Root-locus diagrams that show the effects of adding a zero to $G(s)H(s)$.

fore, if $G(s)H(s)$ represents the loop transfer function of a feedback control system, the relative stability of the system is improved by the addition of the zero. Figure 8-24(b) illustrates that a similar effect will result if a pair of complex-conjugate zeros is added to the function of Eq. (8-92). Figure 8-24(c) shows the root-locus diagram when a zero at $s = -c$ is added to the transfer function of Eq. (8-93).

Effects of Movements of Poles and Zeros

In many situations, the construction of the root loci is aided by the understanding of the effects of the movement of the poles and zeros of $G(s)H(s)$. The subject is best illustrated by the following examples.

EXAMPLE 8-20   Consider the equation

$$s^2(s + a) + K(s + b) = 0 \tag{8-94}$$

which is easily converted to the form of $1 + G(s)H(s) = 0$, with

$$G(s)H(s) = \frac{K(s + b)}{s^2(s + a)} \tag{8-95}$$

Let us set $b = 1$ and investigate the root loci of Eq. (8-94) for several values of a. It can be

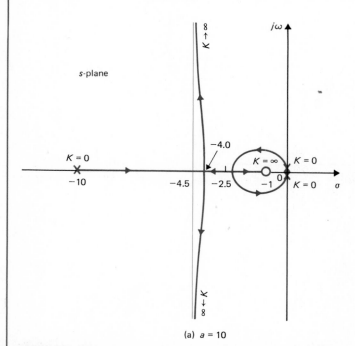

(a) $a = 10$

FIGURE 8-25 Root-locus diagrams that show the effects of moving a pole of $G(s)H(s)$. $G(s)H(s) = K(s + 1)/[s^2(s + a)]$.

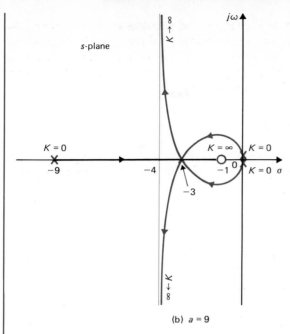

(b) $a = 9$

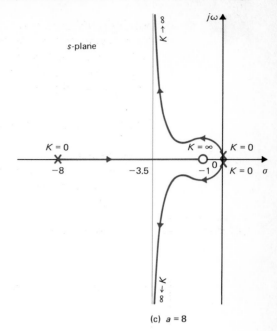

(c) $a = 8$

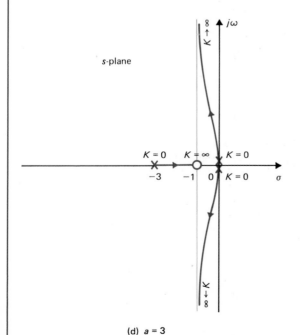

(d) $a = 3$

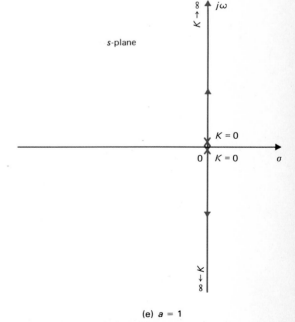

(e) $a = 1$

FIGURE 8-25 (Continued)

shown that the nonzero breakaway points depend on the value of a, and are

$$s = -\frac{a+3}{4} \pm \frac{1}{4}\sqrt{a^2 - 10a + 9} \qquad (8\text{-}96)$$

When $a = 10$, the two breakaway points are at $s = -2.5$ and -4.0. Figure 8-25(a) illustrates the root loci of Eq. (8-94) with $a = 10$. When $a = 9$, the two breakaway points given by Eq. (8-96) converge to one point at $s = -3$, and the root loci become that of Fig. 8-25(b). It is interesting to note the change in the root loci when the pole at $-a$ is moved from -10 to -9. For values of a less than 9, the values of s as given by Eq. (8-96) no longer satisfy Eq. (8-94), which means that there are no finite, nonzero, breakaway points. Figure 8-25(c) illustrates this case with $a = 8$. As the pole at $s = -a$ is moved further to the right along the real axis, the complex portion of the root loci is pushed further toward the right-half plane. Figure 8-25(d) illustrates the case when $a = -3$. When $a = b$, the pole at $s = -a$ and the zero at $-b$ cancel each other, and the root loci degenerate into a second-order case and lie entirely on the $j\omega$-axis, as shown in Fig. 8-25(e).

EXAMPLE 8-21   Consider the equation

$$s(s^2 + 2s + a) + K(s + 2) = 0 \qquad (8\text{-}97)$$

which gives the equivalent $G(s)H(s)$ as

$$G(s)H(s) = \frac{K(s+2)}{s(s^2 + 2s + a)} \qquad (8\text{-}98)$$

breakaway points   The objective is to study the root loci ($K \geq 0$) for various values of $a(> 0)$. The breakaway-point equation of the root loci is determined as

$$s^3 + 4s^2 + 4s + a = 0 \qquad (8\text{-}99)$$

As a start, let $a = 1$ so that the poles of $G(s)H(s)$ are at $s = 0$, -1, and -1. The root loci for this case are shown in Fig. 8-26(a). The breakaway points are determined from Eq. (8-99), and are at $s = -0.38$, -1.0, and -2.618, with the last point being on the CRL. As the value of a is increased from unity, the two double poles of $G(s)H(s)$ at $s = -1$ will move vertically up and down with the real parts equal to -1. The breakaway points at $s = -0.38$ and $s = -2.618$ will move to the left, whereas the breakaway point at $s = -1$ will move to the right. Figure 8-26(b) shows the root loci when $a = 1.12$. Since the real parts of the poles and zeros of $G(s)H(s)$ are not affected by the value of a, the intersect of the asymptotes is always at the origin of the s-plane. When $a = -1.12$, Eq. (8-99) gives the breakaway points at $s = -0.493$, -0.857, and -2.65. When $a = 1.185$, Eq. (8-99) gives two real equal roots, which means that the two breakaway points of the RL that lie between $s = 0$ and $s = -1$ converge to a point. The root loci for this situation are shown in Fig. 8-26(c). When a is greater than 1.185, Eq. (8-99) yields one real root and two complex-conjugate roots. We can easily show that in this case, the two complex roots do not correspond to breakaway points on the root loci. Thus, the root loci have only one breakaway point when $a > 1.185$. Figure 8-26(d) illustrates the root loci when $a = 3$. The reader may investigate the difference between the root loci in Figs. 8-26(c) and 8-26(d), and fill in the evolution of the loci when the value of a is gradually changed from 1.185 to 3 and beyond.

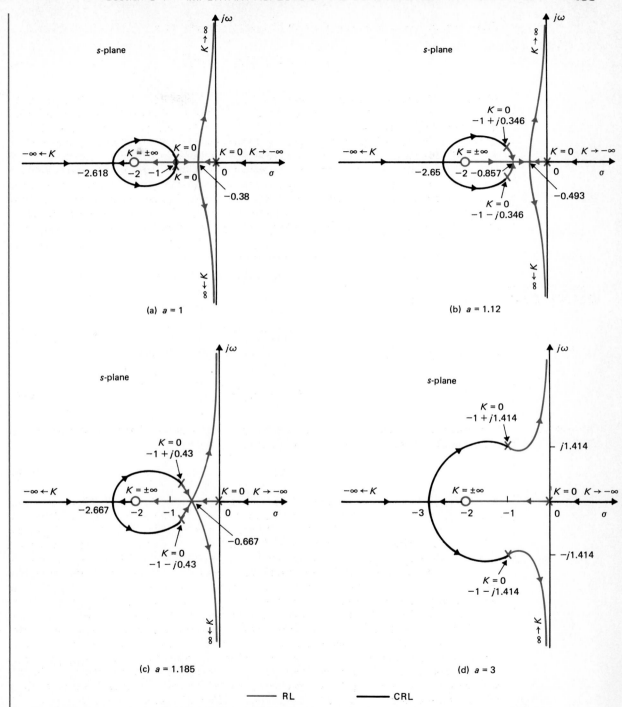

FIGURE 8-26 Root-locus diagrams that show the effects of moving a pole of $G(s)H(s) = K(s + 2)/[s(s^2 + 2s + 2)]$.

8-5

ROOT CONTOUR: MULTIPLE-PARAMETER VARIATION

The root-locus technique discussed thus far is restricted to only one variable parameter in K. In many control-systems problems, the effects of varying several parameters must be investigated. For example, when designing a controller that is represented by a transfer function with poles and zeros, it would be useful to investigate the effects on the characteristic-equation roots when these poles and zeros take on various values. In Section 8-4, the root loci of equations with two variable parameters are studied by fixing one parameter and assigning different values to the other. In this section, the multiparameter problem is investigated through a more systematic method of embedding. When more than one parameter varies continuously from $-\infty$ to ∞, the root loci

root contours   are referred to as the **root contours**. It will be shown that the root contours silll possess the same properties as the single-parameter root loci, so that the methods of construction discussed thus far are all applicable.

The principle of root contours can be described by considering the equation

$$P(s) + K_1 Q_1(s) + K_2 Q_2(s) = 0 \qquad (8\text{-}100)$$

where K_1 and K_2 are the variable parameters, and $P(s)$, $Q_1(s)$, and $Q_2(s)$ are polynomials of s. The first step involves setting the value of one of the parameters to zero. Let us set K_2 to zero. Then, Eq. (8-100) becomes

$$P(s) + K_1 Q_1(s) = 0 \qquad (8\text{-}101)$$

which now has only one variable parameter in K_1. The root loci of Eq. (8-101) may be determined by dividing both sides of the equation by $P(s)$. Thus,

$$1 + \frac{K_1 Q_1(s)}{P(s)} = 0 \qquad (8\text{-}102)$$

Note that the last equation is of the form of $1 + K_1 G_1(s)H_1(s) = 0$, so that we can construct the root loci of Eq. (8-101) based on the pole-zero configuration of $G_1(s)H_1(s)$. Next, we restore the value of K_2, while considering that the value of K_1 is fixed, and divide both sides of Eq. (8-100) by the terms that do not contain K_2. We have

$$1 + \frac{K_2 Q_2(s)}{P(s) + K_1 Q_1(s)} = 0 \qquad (8\text{-}103)$$

which is of the form of $1 + K_2 G_2(s)H_2(s) = 0$. The root contours of Eq. (8-100) when K_2 varies (while K_1 is fixed) are constructed based on the pole-zero configuration of

$$G_2(s)H_2(s) = \frac{Q_2(s)}{P(s) + K_1 Q_1(s)} \qquad (8\text{-}104)$$

It is important to note that the poles of $G_2(s)H_2(s)$ are identical to the roots of Eq. (8-101). Thus, the root contours of Eq. (8-100) when K_2 varies must all start $(K_2 = 0)$ at the points that lie on the root loci of Eq. (8l-101). This is the reason why one root-contour problem is considered to be embedded in another. The same procedure may be extended to more than two variable parameters. The following examples illustrate the construction of root contours when multiparameter-variation situations exist.

EXAMPLE 8-22   Consider the equation

$$s^3 + K_2 s^2 + K_1 s + K_1 = 0 \tag{8-105}$$

where K_1 and K_2 are the variable parameters, which can vary from 0 to ∞.
As the first step, we let $K_2 = 0$, and Eq. (8-105) becomes

$$s^3 + K_1 s + K_1 = 0 \tag{8-106}$$

Dividing both sides of the last equation by s^3, which is the term that does not contain K_1, we

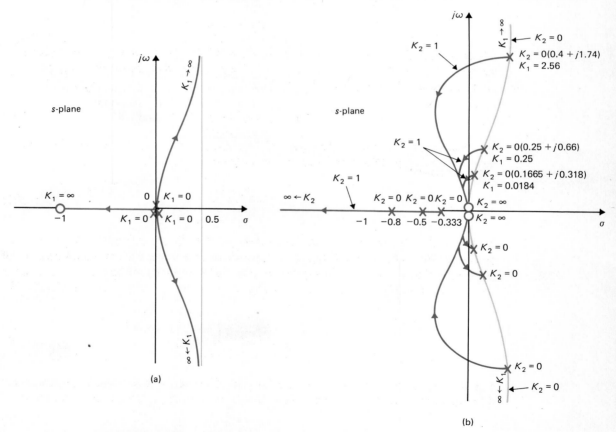

FIGURE 8-27 Root contours of $s^3 + K_2 s^2 + K_1 s + K_1 = 0$. (a) $K_2 = 0$. (b) K_2 varies and K_1 is constant.

have

$$1 + \frac{K_1(s + 1)}{s^3} = 0 \tag{8-107}$$

The root loci of Eq. (8-106) are drawn based on the pole-zero configuration of

$$G_1(s)H_1(s) = \frac{s + 1}{s^3} \tag{8-108}$$

as shown in Fig. 8-27(a).

Next, we let K_2 vary between 0 and ∞ while holding K_1 at a constant nonzero value. Dividing both sides of Eq. (8-105) by the terms that do not contain K_2, we have

$$1 + \frac{K_2 s^2}{s^3 + K_1 s + K_1} = 0 \tag{8-109}$$

Thus, the root contours of Eq. (8-105) when K_2 varies may be drawn from the pole-zero configuration of

$$G_2(s)H_2(s) = \frac{s^2}{s^3 + K_1 s + K_1} \tag{8-110}$$

The zeros of $G_2(s)H_2(s)$ are at $s = 0, 0$; but the poles are the zeros of $1 + K_1 G_1(s)H_1(s)$, which are found on the root loci of Fig. 8-27(a). Thus, for fixed K_1, the root contours when K_2 varies must all emanate from the root loci of Fig. 8-27(a). Figure 8-27(b) shows the root contours of Eq. (8-105) when K_2 varies from 0 to ∞, for $K_1 = 0.0184$, 0.25, and 2.56.

EXAMPLE 8-23   Consider the loop transfer function

$$G(s)H(s) = \frac{K}{s(1 + Ts)(s^2 + 2s + 2)} \tag{8-111}$$

of a closed-loop control system. It is desired to construct the root contours of the characteristic equation with K and T as variable parameters. The characteristic equation of the system is

$$s(1 + Ts)(s^2 + 2s + 2) + K = 0 \tag{8-112}$$

First, we set the value of T to zero. The characteristic equation becomes

$$s(s^2 + 2s + 2) + K = 0 \tag{8-113}$$

The root loci of this equation when K varies are drawn based on the pole-zero configuration of

$$G_1(s)H_1(s) = \frac{1}{s(s^2 + 2s + 2)} \tag{8-114}$$

as shown in Fig. 8-28(a). Next we let K be fixed, and consider that T is the variable parameter.

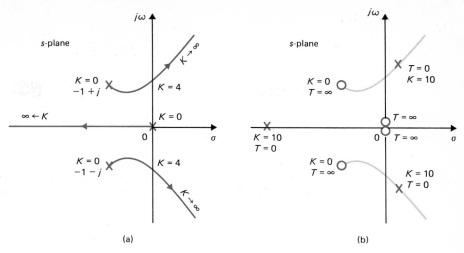

FIGURE 8-28 (a) Root loci for $s(s^2 + 2s + 2) + K = 0$. (b) Pole-zero configuration of $G_2(s)H_2(s) = Ts^2(s^2 + 2s + 2)/[s(s^2 + 2s + 2) + K]$.

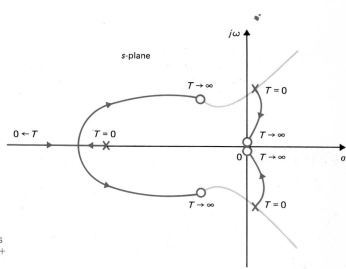

FIGURE 8-29 Root contours for $s(1 + Ts)(s^2 + 2s + 2) + K = 0$. $K > 4$.

Dividing both sides of Eq. (8-112) by the terms that do not contain T, we get

$$1 + TG_2(s)H_2(s) = 1 + \frac{Ts^2(s^2 + 2s + 2)}{s(s^2 + 2s + 2) + K} = 0 \qquad (8\text{-}115)$$

The root contours when T varies are constructed based on the pole-zero configuration of $G_2(s)H_2(s)$. When $T = 0$, the points on the root contours are at the poles of $G_2(s)H_2(s)$, which are on the root loci of Eq. (8-113); when $T = \infty$, the roots of Eq. (8-112) are at the zeros of $G_2(s)H_2(s)$, which are at $s = 0$, 0, $-1 + j$, and $-1 - j$. Figure 8-28(b) shows the pole-zero configuration of $G_2(s)H_2(s)$ for $K = 10$. Notice that $G_2(s)H_2(s)$ has three finite poles and four finite zeros. The root contours for Eq. (8-112) when T varies are shown in Figs. 8-29, 8-30, and

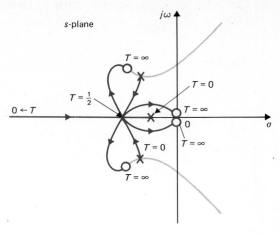

FIGURE 8-30 Root contours for $s(1 + Ts)(s^2 + 2s + 2) + K = 0$. $K = 0.5$.

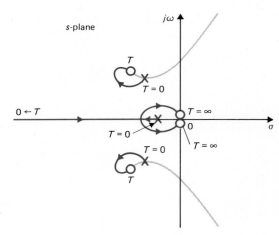

FIGURE 8-31 Root contours for $s(1 + Ts)(s^2 + 2s + 2) + K = 0$. $K < 0.5$.

8-31 for three different values of K. The root contours in Fig. 8-30 show that when $K = 0.5$ and $T = 0.5$, the characteristic equation in Eq. (8-112) has a quadruple root at $s = -1$.

EXAMPLE 8-24   As an example illustrating the effect of the variation of a zero of $G(s)H(s)$, consider

$$G(s)H(s) = \frac{K(1 + Ts)}{s(s + 1)(s + 2)} \qquad (8\text{-}116)$$

The characteristic equation of the system is

$$s(s + 1)(s + 2) + K(1 + Ts) = 0 \qquad (8\text{-}117)$$

Let us first set T to zero and consider the effect of varying K; Eq. (8-117) becomes

$$s(s + 1)(s + 2) + K = 0 \qquad (8\text{-}118)$$

This leads to

$$G_1(s)H_1(s) = \frac{1}{s(s + 1)(s + 2)} \qquad (8\text{-}119)$$

The root loci of Eq. (8-118) are drawn based on the pole-zero configuration of Eq. (8-119), and are shown in Fig. 8-32.

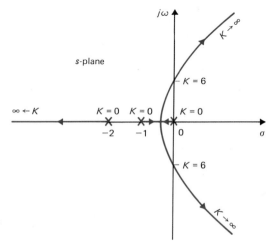

FIGURE 8-32 Root loci for $s(s + 1)(s + 2) + K = 0$.

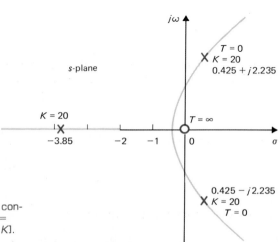

FIGURE 8-33 Pole-zero configuration of $G_2(s)H_2(s) = TKs/[s(s + 1)(s + 2) + K]$. $K = 20$.

When the value of K is fixed and is nonzero, we divide both sides of Eq. (8-117) by the terms that do not contain T, and we get

$$1 + TG_2(s)H_2(s) = 1 + \frac{TKs}{s(s+1)(s+2) + K} = 0 \tag{8-120}$$

The points that correspond to $T = 0$ on the root contours are at the poles of $G_2(s)H_2(s)$ or the zeros of $s(s+1)(s+2) + K$, whose loci are sketched as shown in Fig. 8-32 when K varies. If we choose $K = 20$ just as an illustration, the pole-zero configuration of $G_2(s)H_2(s)$ is shown in Fig. 8-33. The root contours of Eq. (8-117) for $0 \le T < \infty$ are shown in Fig. 8-34 for three different values of K. Since $G_2(s)H_2(s)$ has three poles and one zero, the angles of the asymptotes of the root contours when T varies are at $90°$ and $-90°$. We can show that the intersection

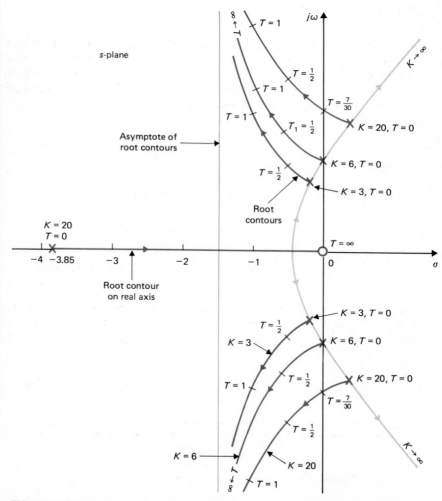

FIGURE 8-34 Root contours of $s(s+1)(s+2) + K + KTs = 0$.

of the asymptotes is always at $s = 1.5$. This is because the sum of the poles of $G_2(s)H_2(s)$, which is given by the negative of the coefficient of the s^2 term of the denominator polynomial of Eq. (8-120), is 3, the sum of the zeros of $G_2(s)H_2(s)$ is 0, and $n - m$ in Eq. (8-39) is 2.

The root contours in Fig. 8-34 show that adding a zero to the loop transfer function generally improves the relative stability of the closed-loop system by moving the characteristic-equation roots toward the left in the s-plane. As shown in Fig. 8-34, for $K = 20$, the system is stabilized for all values of T greater than 0.2333. However, the largest damping ratio that the system can have by increasing T is only approximately 30 percent.

8-6

ROOT LOCI OF DISCRETE-DATA CONTROL SYSTEMS [23, 24]

discrete-data systems

z-plane

The root-locus technique for continuous-data systems can be extended to discrete-data systems without any modification. We shall show that the root loci of discrete-data systems are more easily constructed in the z-plane rather than the s-plane. Let us consider the discrete-data control system shown in Fig. 8-35. The characteristic equation roots of the system satisfy the following equation:

characteristic equation

$$1 + GH^*(s) = 0 \qquad (8\text{-}121)$$

in the s-plane, or

$$1 + GH(z) = 0 \qquad (8\text{-}122)$$

in the z-plane. From Chapter 3, $GH^*(s)$ is written

$$GH^*(s) = \frac{1}{T} \sum_{n=-\infty}^{\infty} G(s + jn\omega_s)H(s + jn\omega_s) \qquad (8\text{-}123)$$

which is an infinite series. Thus, the poles and zeros of $GH^*(s)$ in the s-plane will be infinite in number. This evidently makes the construction of the root loci of Eq. (8-121) in the s-plane quite difficult. As an illustration, consider that for the system of

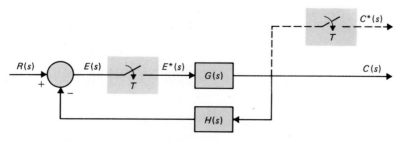

FIGURE 8-35 Discrete-data control system.

Fig. 8-35,

$$G(s)H(s) = \frac{K}{s(s+1)} \qquad (8\text{-}124)$$

Substituting Eq. (8-124) into Eq. (8-123), we get

$$GH^*(s) = \frac{1}{T} \sum_{n=-\infty}^{\infty} \frac{K}{(s + jn\omega_s)(s + jn\omega_s + 1)} \qquad (8\text{-}125)$$

which has poles at $s = -jn\omega_s$ and $s = -1 - jn\omega_s$, where n takes on all integers between $-\infty$ and ∞. The pole configuration of $GH^*(s)$ is shown in Fig. 8-36(a). By using the properties of the root loci in the s-plane, the RL of $1 + GH^*(s) = 0$ are drawn as shown in Fig. 8-36(b) for the sampling period $T = 1$ s. The RL contain an infinite number of branches, and these clearly indicate that the closed-loop system is unstable for all values of K greater than 4.32. In contrast, it is well known that the same system without sampling is stable for all positive values of K.

The root-locus problem for discrete-data systems is simplified if the root loci are

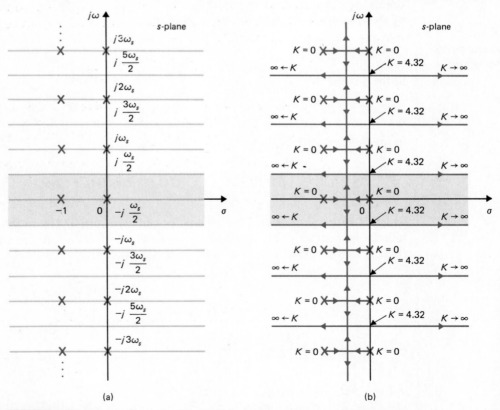

(a)   (b)

FIGURE 8-36 Pole configuration of $GH^*(s)$ and the root-locus diagram in the s-plane for the discrete-data system in Fig. 8-35 with $G(s)H(s) = K/[s(s + 1)]$. $T = 1$ second.

constructed in the z-plane using Eq. (8-122). Since Eq. (8-122) is, in general, a polynomial in z with constant coefficients, the number of root loci is finite in the z-plane, and the same procedures of construction for continuous-data systems are directly applicable.

EXAMPLE 8-25   As an illustrative example of the construction of root loci for discrete-data systems in the z-plane, let us consider the system of Fig. 8-35 with $G(s)H(s)$ given in Eq. (8-124), and $T = 1$ second. Taking the z-transform of Eq. (8-124), we have

$$GH(z) = \frac{0.632Kz}{(z - 1)(z - 0.368)} \qquad (8\text{-}126)$$

SDCS

A convenient way of obtaining $GH(z)$ is to use the program SDCS in the ACSP software package. The root loci of the closed-loop characteristic equation are constructed based on the pole-zero configuration of Eq. (8-126), and are shown in Fig. 8-37. Notice that when the value of K exceeds 4.32, one of the roots of the characteristic equation moves outside the unit circle in the z-plane, and the system becomes unstable. The constant-damping-ratio locus may be superimposed on the root loci to determine the required value of K for a specified damping ratio. In Fig. 8-37, the constant-damping-ratio locus for $\zeta = 0.5$ is drawn, and the intersection with the root loci gives the desired value of $K = 1$. For the same system, if the sampling period T is increased to 2 seconds, the z-transform of $G(s)H(s)$ becomes

*constant
damping ratio*

$$GH(z) = \frac{0.865Kz}{(z - 1)(z - 0.135)} \qquad (8\text{-}127)$$

The root loci for this case are shown in Fig. 8-38. It should be noted that although the complex part of the root loci for $T = 2$ seconds takes the form of a smaller circle than that when $T = 1$ second, the system is actually less stable, since the marginal value of K for stability is 2.624 as compared with the marginal K of 4.32 for $T = 1$ second.

Next, let us consider that a zero-order hold is inserted between the sampler and the controlled process $G(s)$ in the system of Fig. 8-35. For the loop transfer function in Eq. (8-123), the

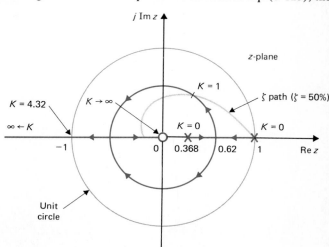

FIGURE 8-37 Root-locus diagram of a discrete-data control system without zero-order hold. $G(s)H(s) = K/[s(s + 1)]$. $T = 1$ second.

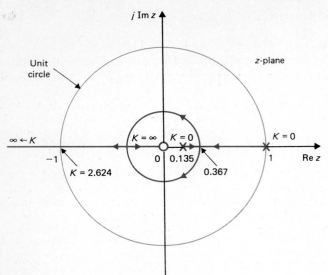

FIGURE 8-38 Root-locus diagram of a discrete-data control system without zero-order hold. $G(s)H(s) = K/[s(s + 1)]$. $T = 2$ seconds.

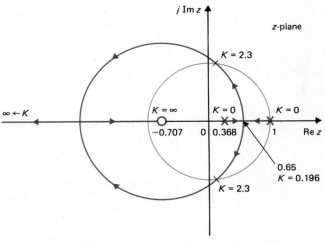

(a) Root loci for $T = 1$ second

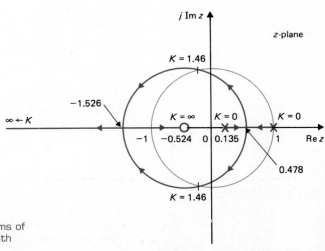

(b) Root loci for $T = 2$ seconds

FIGURE 8-39 Root-locus diagrams of discrete-data control systems with sample-and-hold. $G(s)H(s) = K/[s(s + 1)]$.

z-transform with zero-order hold is

$$G_{h0}GH(z) = \frac{K[(T - 1 + e^{-T})z - Te^{-T} + 1 - e^{-T}}{(z - 1)(z - e^{-T})} \tag{8-128}$$

The root loci of the system with sample-and-hold for $T = 1$ and $T = 2$ seconds are shown in Figs. 8-39(a) and 8-39(b), respectively. In this case, the marginal value of stability for K is 2.3 for $T = 1$ second and 1.46 for $T = 2$ seconds. Comparing the root loci of the system with and without the zero-order hold, we see that the zero-order hold reduces the stability margin of the discrete-data system.

In conclusion, the root-locus diagrams of discrete-data systems are constructed in the z-plane using essentially the same properties as those of the continuous-data systems in the s-plane. However, the stability condition of the discrete-data system must be investigated with respect to the unit circle in the z-plane. The interpretation of the root location in the z-plane with respect to the time-domain performance of the discrete-data system is also different than that of the s-plane.

8-7

SUMMARY

In this chapter, we have introduced the root-locus technique of analysis and design of linear control systems. The technique presents a graphical method of investigating the roots of the characteristic equation of a linear time-invariant system when one or more parameters vary. The subjects on time-domain designs covered in the next chapter rely heavily on the root-locus method. However, remember that the characteristic-equation roots give exact indication on the absolute stability of the system, but give only qualitative information on the relative stability, since the zeros of the closed-loop transfer function, if any, also govern the dynamic performance of the system.

The properties and guide to the manual sketching of the root loci are given. For more detailed information, the root-loci data can be generated by the ROOTLOCI program found in the ACSP software that accompanies this text. The loci can also be plotted on the computer monitor, line printer, or the HP digital plotter.

The root-locus technique can also be applied to digital control systems with the characteristic equation expressed in the z-transform. We see that the properties and construction of the root loci in the z-plane are identical to those of the continuous-data systems in the s-plane, except that the interpretation of the root locations to stability must be made with respect to the unit circle $|z| = 1$.

The root-locus technique can also be applied to linear systems with a time delay in the system loop [25, 26.] The subject is not treated here, since systems with a pure time delay are more easily studied in the frequency domain (Chapter 10).

REVIEW QUESTIONS

The following questions and true or false problems all refer to the equation $P(s) + KQ(s) = 0$, where $P(s)$ and $Q(s)$ are polynomials of s with constant coefficients.

1. Give the condition from which the root loci are constructed.

3.   Determine the points on the complete root loci at which $K = 0$, with reference to the poles and zeros of $Q(s)/P(s)$.

4.   Determine the points on the complete root loci at which $K = \pm \infty$, with reference to the poles and zeros of $Q(s)/P(s)$.

5.   Give the significance of the breakaway points with respect to the roots of $P(s) + KQ(s) = 0$.

6.   Give the equation of the intersect of the asymptotes.

7.   The asymptotes of the complete root loci refer to the angles of the root loci when $K = \pm \infty$.   **(T)**   **(F)**

8.   There is only one intersect of the asymptotes of the complete root loci.   **(T)**   **(F)**

9.   The intersect of the asymptotes must always be on the real axis.   **(T)**   **(F)**

10. The breakaway points of the complete root loci must always be on the real axis.   **(T)**   **(F)**

11. Given the equation: $1 + KG_1(s)H_1(s) = 0$, where $G_1(s)H_1(s)$ is a rational function of s, and does not contain K, the roots of $dG_1(s)H_1(s)/ds$ are all breakaway points on the root loci $(-\infty < K < \infty)$.   **(T)**   **(F)**

12. At the breakaway points on the root loci, the root sensitivity is infinite.   **(T)**   **(F)**

13. Without modification, all the rules and properties for the construction of the root loci in the s-plane can be applied to the construction of the root loci of discrete-data systems in the z-plane.   **(T)**   **(F)**

14. The determination of the intersections of the root loci in the s-plane with the $j\omega$-axis can be made by solving the auxiliary equation of Routh's tabulation of the equation.   **(T)**   **(F)**

15. Adding a pole to $Q(s)/P(s)$ has the general effect of pushing the root loci to the right, whereas adding a zero pushes the loci to the left.   **(T)**   **(F)**

General Subjects

1. W. R. EVANS, "Graphical Analysis of Control Systems," *Trans. AIEE,* Vol. 67, pp. 547–551, 1948.
2. W. R. EVANS, "Control System Synthesis by Root Locus Method," *Trans. AIEE,* Vol. 69, pp. 66–69, 1950.
3. W. R. EVANS, *Control System Dynamics,* McGraw-Hill Book Company, New York, 1954.

Construction and Properties of Root Loci

4. C. C. MacDuff, *Theory of Equations,* pp. 29–104, John Wiley & Sons, New York, 1954.
5. C. S. LORENS and R. C. TITSWORTH, "Properties of Root Locus Asymptotes," *IRE Trans. Automatic Control,* AC-5, pp. 71–72, Jan. 1960.
6. C. A. STAPLETON, "On Root Locus Breakaway Points," *IRE Trans. Automatic Control,* Vol. AC-7, pp. 88–89, April 1962.
7. M. J. REMEC, "Saddle-Points of a Complete Root Locus and an Algorithm for Their Easy Location in the Complex Frequency Plane," *Proc. Natl. Electronics Conf.,* Vol. 21, pp. 605–608, 1965.

8. C. F. CHEN, "A New Rule for Finding Breaking Points of Root Loci Involving Complex Roots," *IEEE Trans. Automatic Control,* AC-10, pp. 373–374, July 1965.

9. V. KRISHNAN, "Semi-analytic Approach to Root Locus," *IEEE Trans. Automatic Control,* Vol. AC-11, pp. 102–108, Jan. 1966.

10. R. H. LABOUNTY and C. H. HOUPIS, "Root Locus Analysis of a High-Grain Linear System with Variable Coefficients; Application of Horowitz's Method," *IEEE Trans. Automatic Control,* Vol. AC-11, pp. 255–263, April 1966.

11. A FREGOSI and J. FEINSTEIN, "Some Exclusive Properties of the Negative Root Locus," *IEEE Trans. Automatic Control,* Vol. AC-14, pp. 304–305, June 1969.

Analytical Representation of Root Loci

12. G. A. BENDRIKOV and K. F. TEODORCHIK, "The Analytic Theory of Constructing Root Loci," *Automation and Remote Control,* pp. 340–344, March 1959.

13. K. STEIGLITZ, "Analytical Approach to Root Loci," *IRE Trans. Automatic Control,* Vol. AC-6, pp. 326–332, Sept. 1961.

14. C. WOJCIK, "Analytical Representation of Root Locus," Trans. ASME, J. Basic Engineering, Ser. D. Vol. 86, March 1964.

15. C. S. CHANG, "An Analytical Method for Obtaining the Root Locus with Positive and Negative Gain," *IEEE Trans. Automatic Control,* Vol. AC-10, pp. 92–94, Jan. 1965.

16. B. P. BHATTACHARYYA, "Root Locus Equations of the Fourth Degree," *Internat. J. Control,* Vol. 1, No. 6, pp. 533–556, 1965.

Computer-Aided Plotting of Root Loci

17. D. J. DODA, "The Digital Computer Makes Root Locus Easy," *Control Eng.,* May 1958.

18. Z. KLAGSBRUNN and Y. WALLACH, "On Computer Implementation of Analytic Root-Locus Plotting," *IEEE Trans. Automatic Control,* Vol. AC-13, pp. 744–745, Dec. 1968.

19. R. H. ASH and G. R. ASH, "Numerical Computation of Root Loci Using the Newton–Raphson Technique," *IEEE Trans. Automatic Control,* Vol. AC-13, pp. 576–582, Oct. 1968.

Root Sensitivity

20. J. G. TRUXAL and M. HOROWITZ, "Sensitivity Considerations in Active Network Synthesis," *Proceedings of the Second Midwest Symposium on Circuit Theory,* East Lansing, MI, 1956.

21. R. Y. HUANG, "The Sensitivity of the Poles of Linear Closed-Loop Systems," *Trans. AIEE, Appl. Ind.,* Vol. 77, Part 2, pp. 182–187, Sept. 1958.

22. H. UR, "Root Locus Properties and Sensitivity Relations in Control Systems," *IRE Trans. Automatic Control,* Vol. AC-5, pp. 57–65, Jan. 1960.

Root Locus for Discrete-Data Systems

23. M. MORI, "Root Locus Method of Pulse Transfer Function for Sampled-Data Control Systems," *IRE Trans. Automatic Control,* Vol. AC-3, pp. 13–20, Nov. 1963.

24. B. C. KUO, *Digital Control Systems,* Holt, Rinehart and Winston, New York 1980.

Root Locus for Systems with Time Delays

25. Y. CHU, "Feedback Control System with Dead-Time Lag or Distributed Lag by Root-Locus Method," *Trans. AIEE,* Vol. 70, Part 2, p. 291, 1951.

26. B. C. KUO, *Automatic Control Systems,* 5th ed. Prentice Hall, Englewood Cliffs, NJ, 1987.

PROBLEMS

The program ROOTLOCI in the ACSP software package or any other root loci program can be used to solve the problems in this chapter. Do the problems without relying on the computer first, and then carry out the computer solutions if the root-locus program is available. Determine all the vital information and mark on the root loci, including the points at which the variable parameter, e.g., K, equals zero and $\pm\infty$, asymptotic properties, intersections on the $j\omega$-axis and the corresponding parameter value, and the breakaway points. Indicate the direction of increase of the variable parameter on the loci with arrows.

8-1. Find the angles of the asymptotes and the intersect of the asymptotes of the root loci of the following equations when K varies from $-\infty$ to ∞.

(a) $s^4 + 4s^3 + 5s^2 + (K + 10)s + K = 0$
(b) $s^3 + 5s^2 + (K + 2)s + K = 0$
(c) $s^2 + K(s^3 + 3s^2 + 2s + 10) = 0$
(d) $s^3 + 2s^2 + 3s + K(s^2 - 1)(s + 3) = 0$

8-2. For the loop transfer functions that follow, find the angle of departure or arrival of the root loci at the designated pole or zero.

(a) $G(s)H(s) = \dfrac{Ks}{(s + 1)(s^2 + 1)}$
Angle of arrival ($K < 0$) and angle of departure ($K > 0$) as $s = j$.

(b) $G(s)H(s) = \dfrac{Ks}{(s - 1)(s^2 + 1)}$
Angle of arrival ($K < 0$) and angle of departure ($K > 0$) at $s = j$.

(c) $G(s)H(s) = \dfrac{K}{s(s + 2)(s^2 + 2s + 2)}$
Angle of departure ($K > 0$) at $s = -1 + j$.

(d) $G(s)H(s) = \dfrac{K}{s^2(s^2 + 2s + 2)}$
Angle of departure ($K > 0$) at $s = -1 + j$.

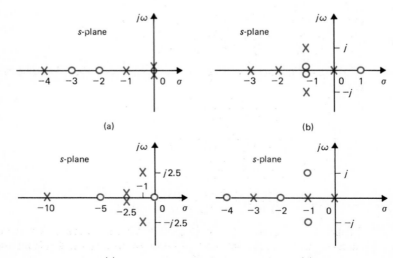

FIGURE 8P-3

8-3. Mark the $K = 0$ and $K = \pm\infty$ points and the root loci and complementary root loci on the real axis for the pole-zero configurzations shown in Fig. 8P-3. Add arrows on the root loci on the real axis in the direction of increasing K.

8-4. Find all the breakaway points of the root loci of the systems described by the pole-zero configurations shown in Fig. 8P-3.

8-5. Construct the root-locus diagram for each of the following control systems for which the poles and zeros of $G(s)H(s)$ are given. The characteristic equation is obtained by equating the numerator of $q + G(s)H(s)$ to zero.
  (a) Poles at 0, -5, -6; zero at -10.
  (b) Poles at 0, -1, -3, -5; no finite zeros.
  (c) Poles at 0, 0, -2, -2; zero at -4.
  (d) Poles at 0, $-1 + j$, $-1 - j$; zero at -1.
  (e) Poles at 0, $-1 + j$, $-1 - j$; zero at -4.
  (f) Poles at 0, $-1 + j$, $-1 - j$; no finite zeros.
  (g) Poles at 0, 0, -8, -8; zeros at -4, -4.
  (h) Poles at 0, 0, -8, -8; zeros at $-4 + j2$, $-4 - j2$.
  (i) Poles at 0, j, $-j$; zeros at -1, -1.
  (j) Poles at -2, 2; zeros at 0, 0.
  (k) Poles at j, $-j$, $j2$, $-j2$; zeros at -1, 1.
  (l) Poles at 0, 0, 0, 1; zeros at -1, -2, -3.
  (m) Poles at 0, 0, 0, -100, -200; zeros at -5, -40.
  (n) Poles at 0, -1, -2; zero at 1.

8-6. The characteristic equations of linear control systems are given as follows. Construct the root loci for $-\infty < K < \infty$.
  (a) $s^3 + 3s^2 + (K + 2)s + 5K = 0$
  (b) $s^3 + s^2 + (K + 2)s + 3K = 0$
  (c) $s^2 + 5Ks + 10 = 0$
  (d) $s^4 + (K + 3)s^3 + (K + 1)s^2 + (2K + 5)s + 10 = 0$
  (e) $s^3 + 2s^2 + 2s + K(s^2 - 1)(s + 2) = 0$
  (f) $s^3 - 2s + K(s + 1)(s + 4) = 0$
  (g) $s^4 + 8s^3 + 16s^2 + K(s^2 + 4s + 5) = 0$
  (h) $s^3 + 2s^2 + 2s + K(s^2 - 2)(s + 4) = 0$
  (i) $s(s^2 - 1) + K(s + 2)(s + 0.5) = 0$
  (j) $s^4 + 2s^3 + s^2 + 2Ks + 5K = 0$

8-7. The open-loop transfer function of a unity-feedback control system is given in the following.
  (a) $G(s) = \dfrac{K(s + 4)}{s(s^2 + 4s + 4)(s + 5)(s + 6)}$
  (b) $G(s) = \dfrac{K}{s(s + 2)(s + 5)(s + 10)}$
  (c) $G(s) = \dfrac{K(s^2 + 2s + 10)}{s(s + 5)(s + 10)}$
  (d) $G(s) = \dfrac{K(s^2 + 4)}{(s + 2)^2(s + 5)(s + 6)}$

Construct the root loci for $-\infty < K < \infty$. Find the value of K that makes the relative damping ratio of the closed-loop system (measured by the dominant complex characteristic-equation roots) equal to 0.707.

8-8. A unity-feedback control system has the open-loop transfer function given in the following.

(a) $G(s) = \dfrac{K}{s(1 + 0.02s)(1 + 0.05s)}$ (b) $G(s) = \dfrac{K}{s(s + 1)(s + 2)(s + 5)}$

Construct the root-locus diagram for $-\infty < K < \infty$. Find the values of K at all breakaway points.

8-9. The open-loop transfer function of a unity-feedback control system is

$$G(s) = \frac{K}{(s + 5)^n}$$

Construct the root loci of the characteristic equation of the closed-loop system for $-\infty < K < \infty$, with (a) $n = 1$, (b) $n = 2$, (c) $n = 3$, (d) $n = 4$, and (e) $n = 5$.

8-10. The characteristic equation of the control system shown in Fig. 7P-9 when $K = 100$ is

$$s^3 + 25s^2 + (100K_t + 2)s + 100 = 0$$

Construct the root loci of the equation for $0 < K_t < \infty$.

8-11. The block diagram of a control system with tachometer feedback is shown in Fig. 8P-11.

(a) Construct the root loci of the characteristic equation for $-\infty < K < \infty$ when $K_t = 0$.

(b) Set $K = 10$. Construct the root loci of the characteristic equation for $0 < K_t < \infty$.

8-12. The characteristic equation of the dc-motor-control system described in Problem 7-25 can be approximated as

$$2.05J_L s^3 + (1 + 10.25J_L)s^2 + 116.84s + 1843 = 0$$

when $K_L = \infty$, and the load inertia J_L is considered as a variable parameter. Construct the root loci of the characteristic equation for $0 \le J_L < \infty$.

8-13. The open-loop transfer function of the control system shown in Fig. 7P-10 is

$$G(s) = \frac{K(s + \alpha)(s + 3)}{s(s^2 - 1)}$$

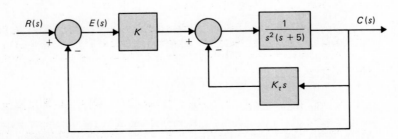

FIGURE 8P-11

(a)   Construct the root loci for $-\infty < K < \infty$, with $\alpha = 5$.
(b)   Construct the root loci for $-\infty < \alpha < \infty$, with $K = 10$.

8-14.   The characteristic equation of the liquid-level control system described in Problem 6-13 is written

$$0.06s(s + 12.5)(As + K_o) + 250N = 0$$

(a)   For $A = K_o = 50$, construct the root-locus diagram of the characteristic equation as N varies from 0 to ∞.
(b)   For $N = 10$ and $K_o = 50$, construct the root-locus diagram of the characteristic equation for $0 \leq A < \infty$.
(c)   For $A = 50$ and $N = 20$, construct the root loci for $-\infty < K_o < \infty$.

8-15.   Repeat Problem 8-14 for the following cases.
(a) $A = K_o = 100$.
(b) $N = 20$ and $K_o = 50$.
(c) $A = 100$ and $N = 20$.

8-16.   The transfer functions of a single-loop control system are given as

$$G(s) = \frac{K}{s^2(s + 1)(s + 3)} \qquad H(s) = 1$$

(a)   Construct the loci of the zeros of $1 + G(s)H(s)$ for $0 \leq K < \infty$.
(b)   Repeat part (a) when $H(s) = 1 + 5s$.

8-17.   The transfer functions of a single-loop control system are given as

$$G(s) = \frac{10}{s^2(s + 1)(s + 3)} \qquad H(s) = 1 + T_d s$$

Construct the loci of the zeros of $1 + G(s)H(s)$ for $0 \leq T_d < \infty$.

8-18.   For the characteristic equation of the dc-motor-control system given in Problem 7-25, it is of interest to study the effects of the motor-shaft compliance K_L on the system performance.
(a)   Let $K = 1$, with the other system parameters as given in Problem 7-25. Find an equivalent $G(s)H(s)$ with K_L as the gain factor. Construct the root loci of the characteristic equation for $0 \leq K_L < \infty$. The system can be aprpoximated as a fourth-order system by cancelling the large negative pole and zero of $G(s)H(s)$ that are very close to each other.
(b)   Repeat part (a) with $K = 1000$.

8-19.   The characteristic equation of the dc-motor-control system given in Problem 7-25 is written in the following when the motor shaft is considered to be rigid ($K_L = \infty$). Let $K = 1$, $J_m = 0.001$, $L_a = 0.001$, $n = 0.1$, $R_a = 5$, $K_i = 9$, $K_b = 0.0636$, $B_m \cong 0$, and $K_s = 1$.

$$L_a(J_m + n^2 J_L)s^3 + (R_a J_m + n^2 R_a J_L + B_m L_a)s^2 + (R_a B_m + K_i K_b)s + nK_s K_i K = 0$$

Construct the root-locus diagram for $0 \leq J_L < \infty$ to show the effects of variation of the load inertia on the system performance.

8-20.   Given the equation

$$s^3 + \alpha s^2 + Ks + K = 0$$

it is desired to investigate the root loci of this equation for $-\infty < K < \infty$ and for the various values of α.

(a)   Construct the root loci for $-\infty < K < \infty$ when $\alpha = 12$.
(b)   Repeat part (a) when $\alpha = 2$.
(c)   Determine the value of α so that there is only one nonzero breakaway point on the entire root loci for $-\infty < K < \infty$. Construct the root loci.

8-21.   The open-loop transfer function of a unity-feedback control system is

$$G(s) = \frac{K(s + \alpha)}{s^2(s + 3)}$$

Determine the values of α so that the root loci $(-\infty < K < \infty)$ will have zero, one, and two breakaway points, respectively, not including the one at $s = 0$. Construct the root loci for $-\infty < K < \infty$ for all three cases.

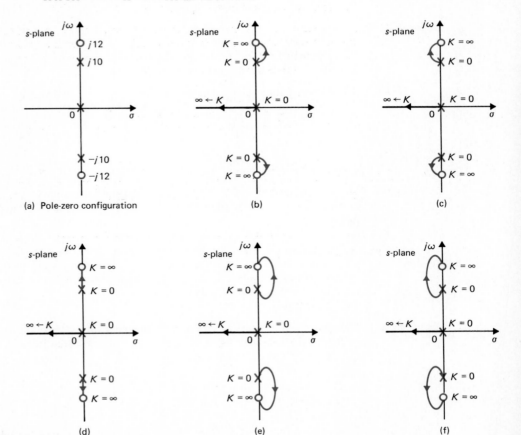

(a) Pole-zero configuration   (b)   (c)

(d)   (e)   (f)

FIGURE 8P-22

8-22. The pole-zero configuration of $G(s)H(s)$ of a single-loop control system is shown in Fig. 8P-22(a). Without actually plotting, apply the angle-of-departure property of the root loci to determine which root-locus diagram shown is the correct one.

8-23. The block diagram of a sampled-data control system is shown in Fig. 8P-23.
  (a) Construct the root loci in the z-plane for the system for $0 \leq K < \infty$, without the zero-order hold, when $T = 0.5$ second and then with $T = 0.1$ second. Find the marginal values of K for stability.

$$G(s) = \frac{K}{s(s + 5)}$$

  (b) Repeat part (a) when the system has a zero-order hold, as shown in Fig. 8P-23.

8-24. The system shown in Fig. 8P-23 has the following transfer function.

$$G(s) = \frac{Ke^{-0.1s}}{s(s + 1)(s + 2)}$$

Construct the root loci in the z-plane for $0 \leq K < \infty$, with $T = 0.1$ second.

8-25. The characteristic equations of linear discrete-data control systems are given in the following. Construct the root loci for $-\infty < K < \infty$. Determine the marginal value of K for stability.
  (a) $z^3 + Kz^2 + 1.5Kz - (K + 1) = 0$
  (b) $z^2 + (0.15K - 1.5)z + 1 = 0$
  (c) $z^2 + (0.1K - 1)z + 0.5 = 0$
  (d) $z^2 + (0.4 + 0.14K)z + (0.5 + 0.5K) = 0$
  (e) $(z - 1)(z^2 - z + 0.4) + 4 \times 10^{-5}K(z + 1)(z + 0.7) = 0$

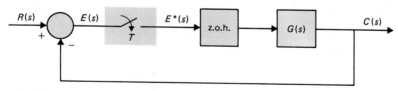

FIGURE 8P-23

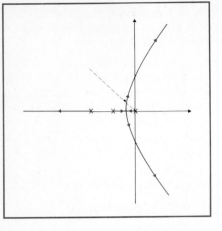

TIME-DOMAIN DESIGN OF CONTROL SYSTEMS

9-1
INTRODUCTION

The time-domain design of control systems refers to the utilization of the time-domain properties and specifications of the system to be designed. The material presented in the proceding chapters shows that there are close relationships between the time-domain properties and the s-plane characteristics of the system transfer function. Thus, the time-domain design of linear control systems is carried out by working with the poles and zeros of the transfer function in the s-plane.

As discussed in Chapter 7, the time-domain characteristics of a linear control system are represented by the transient and the steady-state responses of the system when certain test signals are applied. Depending on the objectives of the design, these test signals are usually in the form of a step function or a ramp function or other time-domain functions. For a step-function input, the percent maximum overshoot, rise time, and settling time are often used to measure the performance of the system. Qual-itatively, the damping ratio and the natural undamped frequency can be used to indicate the relative stability of the system. These quantities are defined strictly only for the prototype second-order system. For higher-order systems, these parameters are mean-ingful only if the corresponding pair of poles of the closed-loop transfer function domi-nates the dynamic response of the system. Therefore, for design in the time domain, the design criteria often include the maximum overshoot as a design parameter.

performance measure

design criteria

In general, the dynamics of a linear controlled process can be represented by the block diagram shown in Fig. 9-1. The design objective is to have the controlled vari-

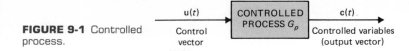

FIGURE 9-1 Controlled process.

ables, represented by the output vector $\mathbf{c}(t)$, behave in certain desirable ways. The problem essentially involves the determination of the control signal $\mathbf{u}(t)$ over the prescribed time interval so that the design objectives are all satisfied.

Most of the conventional design methods in control systems rely on the so-called **fixed-configuration design** in that the designer at the outset decides the basic configuration of the overall designed system, and the place where the controller is to be positioned relative to the controlled process. The problem then involves the design of the elements of the controller. Because most control efforts involve the modification or compensation of the system-performance characteristics, the general design using fixed configuration is also called **compensation**.

Figure 9-2 illustrates several commonly used system configurations with controller compensation. The most commonly used system configuration is shown in Fig. 9-2(a). In this case, the controller is placed in series with the controlled process, and the configuration is referred to as **series** or **cascade compensation**. In Fig. 9-2(b), the controller is placed in the minor feedback path, and the scheme is called **feedback compensation**. Figure 9-2(c) shows a system that generates the control signal by feeding back the state variables through constant real gains, and the scheme is known as **state feedback**. The problem with state-feedback control is that for high-order systems, the large number of state variables involved would require a large number of transducers to sense the state variables for feedback. Thus, the actual implementation of the state-feedback control scheme may be costly. Even for low-order systems, often not all the state variables are directly accessible, and an **observer** or **estimator** may be necessary to create the estimated state variables from measurements of the output variables.

The compensation schemes shown in Figs. 9-2(a), (b), and (c) all have one degree of freedom in that there is only one controller in each system, even though the controller may have more than one parameter that can be varied. The disadvantage with a one-degree-of-freedom controller is that the performance criteria that can be realized are limited. For example, if a system is to be designed to achieve a certain amount of relative stability, it may have poor sensitivity to parameter variations. Or if the roots of the characteristic equation are selected to provide a certain amount of relative damping, the maximum overshoot of the step response may still be excessive, owing to the zeros in the closed-loop transfer function.

Figures 9-2(d), (e), (f), and (g) show compensation schemes that have two degrees of freedom. The configuration shown in Fig. 9-2(d) is called the **series-feedback compensation**. Figures 9-2(e) and (f) show the so-called **feedforward compensation**. In Fig. 9-2(e), the controller $G_{cf}(s)$ is placed in series with the closed-loop system, which has a controller $G_c(s)$ in the forward path. In Fig. 9-2(f), the feedforward controller $G_{cf}(s)$ is in parallel with the forward path. The key to the feedforward compensation is that the controller $G_{cf}(s)$ is not in the loop of the system, so that it does not affect the roots of the characteristic equation of the original system. The poles and zeros of $G_{cf}(s)$ may be selected to add to or cancel the poles and zeros of the closed-loop transfer function.

compensation schemes

observer

relative stability

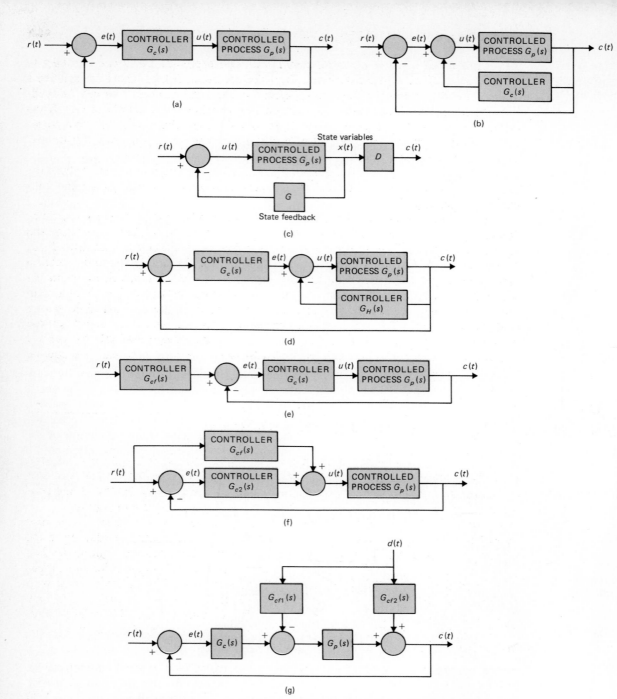

FIGURE 9-2 Various controller configurations in control-system compensation. (a) Series or cascade compensation. (b) Feedback compensation. (c) State-feedback control. (d) Series-feedback compensation (two degrees of freedom). (e) Forward compensation with series compensation (two degrees of freedom). (f) Feedback compensation (two degrees of freedom). (g) Forward load disturbance with series compensation (two degrees of freedom).

two-degree-of-freedom controllers

Figure 9-2(g) shows a control system with two-degree-of-freedom controllers to minimize the effects of the disturbance $d(t)$. The controller $G_c(s)$ is chosen to achieve a desired closed-loop transfer function $C(s)/R(s)$ in the usual manner, whereas the feed-forward controllers $G_{cf1}(s)$ and $G_{cf2}(s)$ are devised to minimize or eliminate the effects of the disturbance $d(t)$. The controller $G_{cf2}(s)$ is assigned so that $G_{cf1}(s)$ will be a physically realizable transfer function. Details of these compensation schemes will be discussed in the ensuing sections.

Although the systems illustrated in Fig. 9-2 all have continuous-data control, the same configurations can be applied to discrete-data control, in which case the controllers are all digital.

Because of the lack of straightforward or unique relationships between the time-domain specifications and the transfer functions of systems with order higher than the second, a general design procedure in the time domain is difficult to establish. J. G. Truxal [1] in his 1955 book *Automatic Feedback Control Systems Synthesis* did detail a time-domain synthesis method for linear control systems. Modern practicing engineers still rely to a great extent on the known relations between the time-domain specifications and performance characteristics, and generally use established controller configurations to solve day-to-day design problems. Many designers still depend on the trial-and-error approach. With the wealth of computer software available for control systems analysis and design, such as the ACSP accompanying this text, the routine trial-and-error method may not be tedious at all.

ACSP

9-2

TIME-DOMAIN DESIGN WITH THE PID CONTROLLER

proportional control

In all the control-systems examples we have discussed thus far, the controller has been typically a simple amplifier with a constant gain K. This type of control action is formally known as **proportional control**, since the control signal at the output of the controller is simply related to the input of the controller by a proportional constant.

From a mathematical standpoint, a linear continuous-data controller should also be able to take a time derivative or time integral of the input signal, in addition to the proportional operation. Therefore, we can consider a more general continuous-data controller that is a device that contains such components as adders (addition or subtraction), amplifiers, attenuators, differentiators, and integrators. The designer's task is to determine which of these components should be used, in what proportion, and how they are to be connected. For example, one of the best-known controllers used in practice is the PID controller, where the letters stand for **proportional**, **integral**, and **derivative**. The transfer function of the basic PID controller is

PID control

$$G_c(s) = K_P + K_D s + \frac{K_I}{s} \qquad (9\text{-}1)$$

where K_P, K_D, and K_I are real constants. The design problem involves the determination of the values of these three constants so that the performance of the system meets the design requirements. The investigation of the PID controller at this stage also enhances the understanding of the role each of the proportional, derivative, and integral components play on the time-domain performance of a control system. We shall investigate the effects of derivative control and integral control separately.

Effects of Derivative Control on the Time Response of Feedback Control Systems

Figure 9-3 shows the block diagram of a feedback control system that has a second-order prototype process with transfer function

$$G_P(s) = \frac{\omega_n^2}{s(s + 2\zeta\omega_n)} \tag{9-2}$$

PD control   The series controller is a proportional-derivative (PD) type with the transfer function

$$G_c(s) = K_P + K_D s \tag{9-3}$$

The control signal is

$$u(t) = K_P e(t) + K_D \frac{de(t)}{dt} \tag{9-4}$$

The open-loop transfer function of the overall system is

$$G(s) = G_c(s)G_P(s) = \frac{C(s)}{E(s)} = \frac{\omega_n^2(K_P + K_D s)}{s(s + 2\zeta\omega_n)} \tag{9-5}$$

which shows that the PD control is equivalent to adding a simple zero at $s = -K_P/K_D$ to the open-loop transfer function.

The effect of the derivative control on the transient response of a feedback control system can be investigated by referring to the time responses shown in Fig. 9-4. Let us assume that the unit-step response of a system with only proportional control is as shown in Fig. 9-4(a), which has a relatively high maximum overshoot and is rather os-

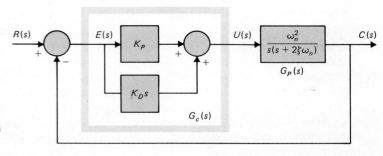

FIGURE 9-3 Control system with PD control.

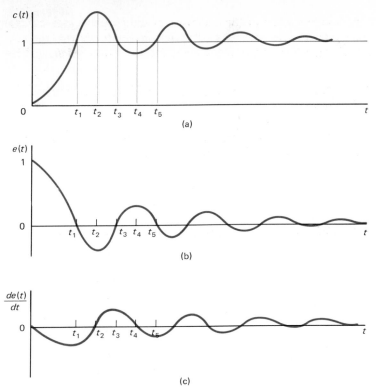

FIGURE 9-4 Waveforms of $c(t)$, $e(t)$, and $de(t)/dt$, showing the effect of derivative control. (a) Unit-step response. (b) Error signal. (c) Time rate of change of the error signal.

cillatory. The corresponding error signal $e(t)$, which is the difference between the unit-step input and $c(t)$, and its time derivative $de(t)/dt$ are shown in Figs. 9-4(b) and (c), respectively. The overshoot and oscillation characteristics are also reflected in $e(t)$ and $de(t)/dt$. Assuming that the system contains a motor of some kind with its torque proportional to $e(t)$, the large overshoot and subsequent oscillations in the output $c(t)$ are due to the excessive amount of torque developed by the motor and the lack of damping in the time interval $0 < t < t_1$, during which the error signal $e(t)$ is positive. For the time interval $t_1 < t < t_3$, $e(t)$ is negative, and the corresponding motor torque is negative. This negative torque tends to slow down the output acceleration and eventually causes the direction of the output $c(t)$ to reverse and undershoot during $t_3 < t < t_5$. During the time interval $t_3 < t < t_5$, the motor torque is again positive, thus tending to reduce the undershoot in the reponse caused by the negative torque in the previous interval. Since the system is assumed to be stable, the error amplitude is reduced with each oscillation, and the output eventually settles to its final value.

effects of PD control

Considering this explanation, we can say that the contributing factors to high overshoot are as follows: (1) the positive correcting torque in the interval $0 < t < t_1$ is too large, and (2) the retarding torque in the time interval $t_1 < t < t_2$ is inadequate. Therefore, in order to reduce the overshoot in the step response, without significantly increasing the rise time, a logical approach would be to decrease the amount of positive correcting torque during $0 < t < t_1$ and to increase the retarding torque during $t_1 < t < t_2$. Similarly, during the time interval $t_2 < t < t_4$, the negative corrective torque in $t_2 < t < t_3$ should be reduced, and the retarding torque during $t_3 < t < t_4$, which is now in the positive direction, should be increased in order to improve the undershoot of $c(t)$.

The PD control shown in Fig. 9-3 gives precisely the compensation effect described. Since the control signal of the PD control is given by Eq. (9-4), Fig. 9-4(c) shows that for $0 < t < t_1$, $de(t)/dt$ is negative; this will reduce the original torque developed due to $e(t)$ alone. For $t_1 < t < t_2$, both $e(t)$ and $de(t)/dt$ are negative, which means that the negative retarding torque developed will be greater than that with only proportional control. It is easy to see that $e(t)$ and $de(t)/dt$ have opposite signs in the time interval $t_2 < t < t_3$; therefore, the negative torque that originally contributes to the undershoot is reduced also. Therefore, all these effects will result in smaller overshoots and undershoots in $c(t)$.

Another way of looking at the derivative control is that since $de(t)/dt$ represents the slope of $e(t)$, the PD control is essentially an anticipatory kind of control. Normally, in a linear system, if the slope of $e(t)$ or $c(t)$ due to a step input is large, a high overshoot will subsequently occur. The derivative control measures the instantaneous slope of $e(t)$, predicts the large overshoot ahead of time, and makes a proper correcting effort before the overshoot actually occurs.

It is apparent that the derivative control will affect the steady-state error of a system only if the steady-state error varies with time. If the steady-state error of a system is constant with respect to time, the time derivative of this error is zero, and the derivative control would have no effect on the steady-state error. But if the steady-state error increases with time, a torque is again developed in proportion to $de(t)/dt$, which will reduce the magnitude of the error.

As with most control efforts, although the basic ingredients of the derivative control are favorable for increased damping and reduction of overshoots, the designer still has to design the PD controller properly, otherwise, a negative effect could very well result.

EXAMPLE 9-1

PD control example

transfer functions

ramp-error constant

As an illustrative example, let us consider that the gain of the preamplifier, K, of the attitude control system shown in Fig. 7-21 is chosen to be 181.17 so that the steady-state error due to a unit-ramp input is 0.000443. However, with this value for K, the damping ratio of the system is 0.2, and the maximum overshoot is 52.7 percent, as shown in Fig. 7-23. Let us consider that a PD controller is to be inserted in the forward path of the system, so that the damping and the maximum overshoot of the system are improved while maintaining the steady-state error at 0.000443. With the PD controller, the open-loop transfer function becomes

$$G(s) = \frac{\Theta_c(s)}{\Theta_e(s)} = \frac{815,265(K_P + K_D s)}{s(s + 361.2)} \qquad (9\text{-}6)$$

The closed-loop transfer function is

$$\frac{\Theta_c(s)}{\Theta_i(s)} = \frac{815,265(K_P + K_D s)}{s^2 + (361.2 + 815,265 K_D)s + 815,265 K_P} \qquad (9\text{-}7)$$

The ramp-error constant is

$$K_v = \lim_{s \to \infty} sG(s) = \frac{815{,}265K_P}{361.2} = 2257.1K_P \qquad (9\text{-}8)$$

The steady-state error due to a unit-ramp input is $e_{ss} = 1/K_v = 0.000443/K_P$.

Equation (9-7) shows that the effects of the PD controller are to add a zero at $s = -K_P/K_D$ to the closed-loop transfer function and increase the "damping term," which is the coefficient of the s term in the denominator, from 361.2 to $361.2 + 815{,}265K_D$. The characteristic equation of the system is

$$s^2 + (361.2 + 815{,}265K_D)s + 815{,}265K_P = 0 \qquad (9\text{-}9)$$

When $K_P = 1$, the damping ratio is

$$\zeta = \frac{361.2 + 815{,}265K_D}{1805.84} = 0.2 + 451.46K_D \qquad (9\text{-}10)$$

which clearly shows the positive effect of K_D on damping. If we wish to have critical damping,

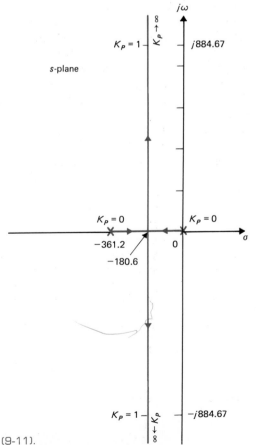

FIGURE 9-5 Root loci of Eq. (9-11).

$\zeta = 1$, Eq. (9-10) gives $K_D = 0.001772$. We should quickly point out that since Eq. (9-6) no longer represents a prototype second-order system, the transient response is also affected by the zero of the transfer function at $s = -K_P/K_D$. In general, if the value of K_D is very large, the zero will be very close to the origin in the s-plane, the overshoot could be increased substantially, and the damping ratio ζ no longer gives an accurate estimate on the maximum overshoot of the output. In the present case, the overshoot will not increase with the increase in K_D, since one of the closed-loop poles will be very close to the zero at $s = -1/K_D$ when K_D is large, and the two are effectively cancelled.

We can apply the root-contour method to the characteristic equation in Eq. (9-9) to demonstrate the effect of K_P and K_D. First, by setting K_P to zero, Eq. (9-9) becomes

$$s^2 + 361.2s + 815,265K_P = 0 \tag{9-11}$$

The root loci of the last equation as K_P varies are shown in Fig. 9-5. When $K_D \neq 0$, the characteristic equation of Eq. (9-9) is conditioned as

$$1 + G_{eq}(s) = 1 + \frac{815,265K_D s}{s^2 + 361.2s + 815,265K_P} = 0 \tag{9-12}$$

root-contour design

The root contours of Eq. (9-9) with $K_P = $ constant and as K_D varies are constructed based on the pole-zero configuration of $G_{eq}(s)$, and are shown in Fig. 9-6 for $K_P = 0.25$ and $K_P = 1$. We

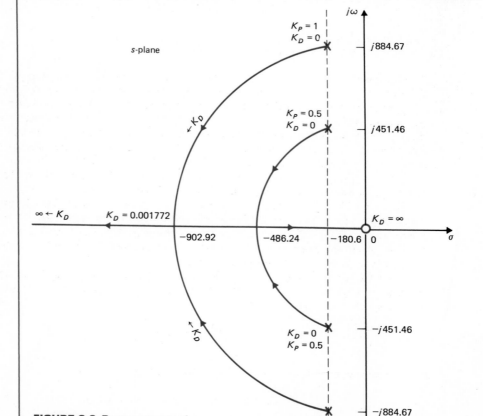

FIGURE 9-6 Root contours of Eq. (9-9) when $K_P = 0.5$ and 1.0; K_D varies.

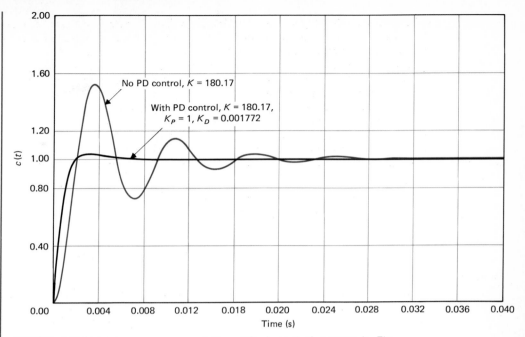

FIGURE 9-7 Unit-step responses of the attitude-control system in Fig. 7-21 with and without PD control.

see that when $K_P = 1$ and $K_D = 0$, the characteristic equation roots are at $-180.6 + j884.67$ and $-180.6 - j884.67$, and the damping ratio of the closed-loop system is 0.2. When the value of K_D is increased, the two characteristic-equation roots move toward the real axis along a circular arc. When $K_D = 0.001772$, the roots are real and equal, at -902.92, and the damping is critical. When K_D is increased beyond 0.001772, the two roots become real and unequal, and the system is overdamped. When K_P is 0.25 and $K_D = 0$, the two characteristic-equation roots are at $-180.6 + j451.46$ and $-180.6 - j451.46$. As K_D increases in value, the root contours again show the improved damping due to the PD controller. **Figure 9-7** shows the unit-step responses of the closed-loop system with and without PD control. **When $K = 180.17$** and without the PD controller, the response has a maximum overshoot of **52.7 percent**. With the PD controller, $K_P = 1$, and $K_D = 0.001772$, the maximum overshoot is **slightly over 4 percent**. In the present case, although K_D is chosen for critical damping, **the overshoot** is due to the zero at $s = -K_P/K_D$ of the closed-loop transfer function. Notice also **that the PD controller causes the** rise time to be faster, due to the derivative effect.

Another analytic way of studying the effects of the parameters K_P and K_D is to evaluate the performance characteristics in the parameter plane of K_P and K_D. From the characteristic equation of Eq. (9-9), we have

damping ratio

$$\zeta = \frac{0.2 + 451.46 K_D}{\sqrt{K_P}} \qquad (9\text{-}13)$$

Applying the stability criterion to Eq. (9-9), we find that for stability,

$$K_P > 0 \qquad \text{and} \qquad K_D > -0.000443$$

parameter plane

The boundaries of stability in the K_P-versus-K_D parameter plane is shown in Fig. 9-8. The constant-damping-ratio trajectory is given by Eq. (9-13), and is a parabola for a constant ζ. Figure

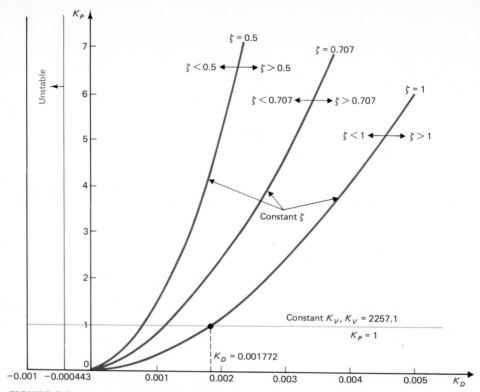

FIGURE 9-8 K_P-versus-K_D parameter plane for the attitude-control system with a PD controller.

9-8 illustrates the constant-ζ trajectories for $\zeta = 0.5$, 0.707, and 1.0. The ramp-error constant K_v is given by Eq. (9-8), which describes a horizontal line in the K_P-versus-K_D parameter plane, as shown in Fig. 9-8. The figure gives a clear picture as to how the values of K_P and K_D effect the various performance criteria of the system. For example, if K_v is set at 2257.1, which corresponds to $K_P = 1$, the constant-ζ loci clearly show that the damping is increased monotonically with an increase in K_D. The intersection between the constant-K_v and the constant-ζ locus gives the value of K_D for the desired K_v and ζ.

From the practical implementation standpoint, the PD controller described in Eq. (9-3) cannot be physically realized by passive circuit elements alone, since the transfer function $G_c(s)$ has more zeros than poles. However, for one who is familiar with analog-circuit design, it would be a simple task to construct a PD controller using operational amplifiers, resistors, and capacitors. The practical disadvantage with the analog-circuit implementation of the PD controller is that the differentiator portion is a high-pass filter, which usually accentuates high-frequency noise that enters at the input.

Effects of Integral Control on the Time Response of Feedback Control Systems

The integral part of the PID controller produces a signal that is proportional to the time integral of the input of the controller. Figure 9-9 illustrates the block diagram of a con-

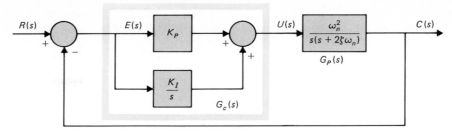

FIGURE 9-9 Control system with PI control.

PI control

trol system that has a prototype second-order process with transfer function $G_p(s)$, and a controller with proportional-integral (PI) control. The transfer function of the PI controller is

$$G_c(s) = K_P + \frac{K_I}{s} \qquad (9\text{-}14)$$

The open-loop transfer function of the overall system is

$$G(s) = G_c(s)G_p(s) = \frac{\omega_n^2(K_P s + K_I)}{s^2(s + 2\zeta\omega_n)} \qquad (9\text{-}15)$$

Clearly, the PI controller adds a zero at $s = -K_I/K_P$ and a pole at $s = 0$ to the open-loop transfer function. One obvious effect of the integral control is that it increases the order of the system by one. More important is that the *type* of the system is increased by one. Therefore, the steady-state error of the original system without integral control is improved by one order; that is, if the steady-state error to a given input is constant, the integral control reduces it to zero (provided that the final system is stable, of course). In the case of Eq. (9-14), the closed-loop system in Fig. 9-9 will now have a *zero* steady-state error when the reference input is a ramp function. However, because the system is now of the third order, it *may be less stable* than the original second-order system or even become *unstable* if the parameters K_P and K_I are not properly chosen.

steady-state error

stability

In the case of a system with PD control, the value of K_P is important because for a type-1 system, the ramp-error constant K_v is directly proportional to K_P, and thus the magnitude of the steady-state error is inversely proportional to K_P when the input is a ramp. On the other hand, if K_P is too large, the system may become unstable. Similarly, for a type-0 system, the steady-state error due to a step input will be inversely proportional to the magnitude of K_P.

When a type-1 system is converted to type-2 by a PI controller, the proportional constant K_P no longer affects the steady-state error, and the latter is always zero for a ramp-function input. The problem is then to choose the proper combination of K_P and K_I so that the transient response is satisfactory.

EXAMPLE 9-2

PI controller

For the attitude-control system, let the controller be described by the transfer function of Eq. (9-14). The open-loop transfer function of the overall system is

$$G(s) = G_c(s)G_p(s) = \frac{815{,}265K_P(s + K_I/K_P)}{s^2(s + 361.2)} \tag{9-16}$$

The characteristic equation of the closed-loop system is

design guideline

$$s^3 + 361.2s^2 + 815{,}265K_Ps + 815{,}265K_I = 0 \tag{9-17}$$

Applying Routh's test to the last equation yields the result that the system is stable for $0 < K_I < 361.2K_P$. This means that the zero of $G(s)$ at $s = -K_I/K_P$ cannot be placed too far to the left in the left-half s-plane, or the system would be unstable. *A viable method of designing the PI controller is to select the zero at $s = -K_I/K_P$ so that it is relatively close to the origin, and away from the most significant pole of $G(s)$.* For the present case, the most significant pole of $G(s)$ is at $s = -361.2$. Thus, the values of K_I and K_P should be chosen so that the following condition is satisfied:

$$\frac{K_I}{K_P} \ll 361.2 \tag{9-18}$$

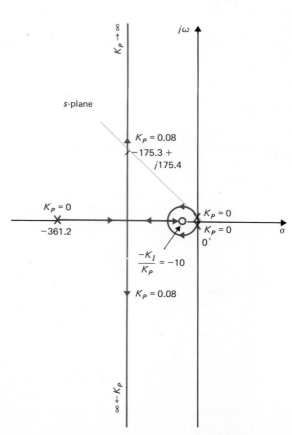

FIGURE 9-10 Root loci of Eq. (9-17) with $K_I/K_P = 10$; K_P varies.

The root loci of Eq. (9-17) with $K_I/K_P = 10$ are shown in Fig. 9-10. Notice that other than the small loop around the zero at $s = -10$, these root loci for the most part are very similar to those shown in Fig. 9-5, which are for Eq. (9-11). With the condition in Eq. (9-18) satisfied, Eq. (9-16) can be approximated by

$$G(s) \cong \frac{815,265K_P}{s(s + 361.2)} \qquad (9\text{-}19)$$

where the term K_I/K_P in the numerator is neglected when compared with the magnitude of s, which takes on values near the point where the damping ratio of 0.707 is realized. From Eq. (9-19), we get $K_P = 0.08$ when the damping ratio of the approximating second-order system is 0.707. This should also be true for the third-order system with the PI controller if the value of K_I/K_P satisfies Eq. (9-18). Thus, with $K_P = 0.08$ and $K_I = 0.8$, the root loci in Fig. 9-10 show that the relative damping ratio of the two complex roots is approximately 0.707. In fact, the three characteristic equation roots are at $s = -10.605$, $-175.3 + j175.4$, and $-175.3 - j175.4$. The reason for this is that when we "stand" at the root at $-175.3 + j175.4$ and "look" toward the neighborhood near the origin, we see that the zero at $s = -10$ is relatively close to the origin, and, thus, practically cancels one of the poles at $s = 0$. In fact, we can show that as long as $K_P = 0.08$, and the value of K_I is chosen such that Eq. (9-18) is satisfied, the relative damping ratio of the complex roots will be very close to 0.707. For example, if $K_I/K_P = 5$, the three characteristic equation roots are at $s = -5.145$, $-178.03 + j178.03$, and $-178,03 - j178.03$, and the relative damping ratio of the two complex roots is still 0.707. Although the real pole of the closed-loop transfer function is moved in each of these cases, the real pole is very close to the zero at $s = -K_I/K_P$ so that the transient due to the real pole is negligible. For

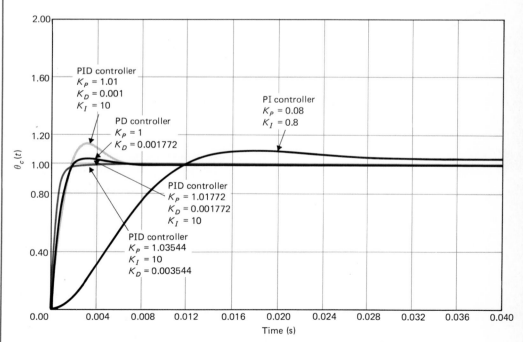

FIGURE 9-11 Unit-step responses of the attitude-control system with PD, PI, and PID controls.

example, when $K_I = 0.4$ and $K_P = 0.08$, the closed-loop transfer function is

$$\frac{\Theta_c(s)}{\Theta_i(s)} = \frac{65,221.2(s + 5)}{(s + 5.145)(s + 178.03 + j178.03)(s + 178.03 - j178.03)} \qquad (9\text{-}20)$$

rise time

Since the pole at $s = -5.145$ is very close to the zero at $s = -5$, the transient response due to this pole is negligible, and the system dynamics are essentially dominated by the two complex poles. The unit-step response of the PI-compensated system with $K_P = 0.08$ and $K_I = 0.8$ is shown in Fig. 9-11. Notice that although the response is well damped, the integral control

settling time

causes the step response to have long rise and settling times.

Design of the PID Controller

PID controller

From the previous discussions, we see that the PD controller would add damping to a system, but the steady-state response is not affected; the PI controller could add damping and improve the steady-state error at the same time, but the rise time and settling time are penalized. This leads to the motivation of using a PID controller so that the best properties of each of the PI and the PD controllers are utilized. We can outline the following approach as a possible means of designing the PID control for a control system.

1. Consider that the PID controller consists of a PI portion connected in cascade with a PD portion. The PID-controller transfer function is written as

$$G_c(s) = K_P + K_D s + K_I/s = (1 + K_{D1}s)(K_{P2} + K_{I2}/s) \qquad (9\text{-}21)$$

The proportional constant of the PD portion is set to unity, since we need only three parameters in the PID controller. Equating both sides of Eq. (9-21), we have

design procedure

$$K_P = K_{P2} + K_{D1}K_{I2} \qquad (9\text{-}22)$$

$$K_D = K_{D1}K_{P2} \qquad (9\text{-}23)$$

$$K_I = K_{I2} \qquad (9\text{-}24)$$

2. Consider that the PI portion is in effect only and select the values of K_{I2} and K_{P2} so that the requirement on the rise time of the system is satisfied. The steady-state error of the system is improved by one order. Do not be concerned about the maximum overshoot at this stage, as it may be large.
3. Use the PD portion to reduce the maximum overshoot. Select the value of K_{D1} to meet the damping requirements.
4. The values of K_P, K_D, and K_I are found using Eqs. (9-22) to (9-24).

As an alternative, we can first design the PD controller by selecting an appropriate value of K_{D1}. It is possible that the PD controller acting alone may not be adequate to achieve the desired relative stability. As a final step, the PI portion of the PID controller is designed to satisfy the desired performance specifications.

EXAMPLE 9-3

PID-control example

As an illustrative example on the design of the PID controller, let us consider again the attitude-control system of Fig. 7-21. The open-loop transfer function of the system with only the PI controller is written

$$G(s) = \frac{815{,}265 K_{P2}(s + K_{I2}/K_{P2})}{s^2(s + 361.2)} \tag{9-25}$$

Using the condition given in Eq. (9-18), the ratio, K_{I2}/K_{P2} is chosen to be 10. Instead of choosing the value of K_{P2} for good damping, we set $K_{P2} = 1$, which gives a relative damping ratio of 0.2 for the complex roots of the characteristic equation. Then, $K_{I2} = 10$. However, the rise time of the system is very short. Next, we bring in the PD portion, and select the value of K_{D1} to improve the damping. Figure 9-11 illustrates the unit-step responses with the PID controller when $K_{P2} = 1$, $K_{I2} = 10$, and $K_{D1} = 0.001772, 0.003544$, and 0.001. The true parameters of the PID controller are determined from Eqs. (9-22), (9-23), and (9-24).

EXAMPLE 9-4

This example illustrates the design of the PD, PI, and PID controllers for a third-order control system. Let the process of a control system with unity feedback be described by

$$G_p(s) = \frac{K}{s(1 + 0.1s)(1 + 0.2s)} \tag{9-26}$$

where $K = 100$, so that the ramp-error constant of the system is 100. We can show that the closed-loop system without any compensation is unstable, as the marginal value of K for stability is 15. Figure 9-12 shows the root loci of the system when K varies. When $K = 100$, the two complex roots of the characteristic equation are at $3.8 + j14.4$ and $3.8 - j14.4$.

The PD-Controller Design

PD control

Let us first use a PD controller with transfer function given in Eq. (9-3). The open-loop transfer function of the compensated system is

$$G(s) = G_c(s)G_p(s) = \frac{5000 K_D(s + K_P/K_D)}{s(s + 5)(s + 10)} \tag{9-27}$$

The objective of the design is to maintain the ramp-error constant at 100, while achieving good relative stability. The ramp-error-constant requirement is achieved by setting $K_P = 1$. We then investigate the effect of the PD controller by use of the root-contour method. The characteristic equation of the system is

$$s^3 + 15s^2 + (50 + 5000 K_D)s + 5000 = 0 \tag{9-28}$$

Dividing both sides of the last equation by the terms that do not contain K_D, we get

$$1 + \frac{5000 K_D s}{s^3 + 15s^2 + 50s + 5000} = 0 \tag{9-29}$$

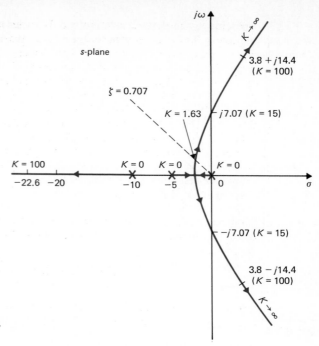

FIGURE 9-12 Root loci of the system described by Eq. (9-26).

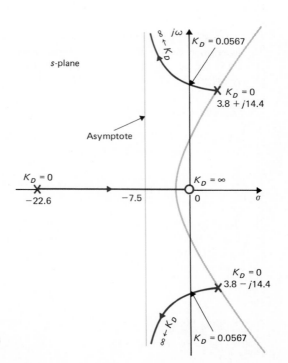

FIGURE 9-13 Root contours of Eq. (9-28).

controller performance

The root contours of Eq. (9-28) when K_D varies are shown in Fig. 9-13. The limitations of the PD controller for this system becomes clear immediately. Although the two complex roots of the characteristic equation are brought into the left-half plane as K_D increases, the effective amount of damping that can be gained by increasing K_D is limited. As K_D increases, the two roots approach 90° and −90°, respectively, toward the asymptotes that intersect at −7.5. Furthermore, as K_D becomes very large, the real root will be very close to the origin, which may contribute significantly to the overshoot of the step response. The conclusion is that although the PD control will stabilize the system, the amount of damping that can be realized is limited. We can show that by varying the value of K_D, the best maximum overshoot that can be achieved is approximately 60 percent. Also, with the higher value of K_D, the natural frequency of the system will be higher, which corresponds to a higher bandwidth.

The PI-Controller Design

PI control

Now consider that the process described in Eq. (9-26) is compensated by a PI controller described by Eq. (9-14). The open-loop transfer function of the compensated system is

$$G(s) = G_c(s)G_p(s) = \frac{5000K_P(s + K_I/K_P)}{s^2(s + 5)(s + 10)} \tag{9-30}$$

The PI controller is designed by selecting the value of K_I/K_P to be significantly smaller than the dominant pole of $G_p(s)$, which in this case is at $s = -5$. Let us set $K_I/K_P = 0.1$. The value of

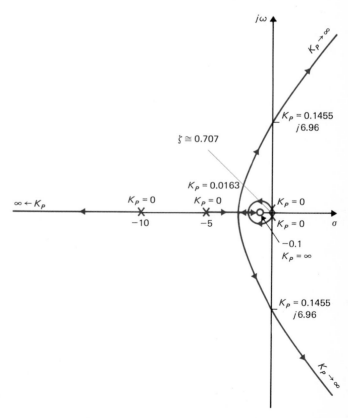

FIGURE 9-14 Root loci of the system described by Eq. (9-30).

K_P is then chosen so that the required damping of the system is satisfied. Since the zero of $G(s)$ at -0.1 practically cancels one of the poles at $s = 0$, the transfer function of Eq. (9-30) is approximately

$$G(s) \cong \frac{5000K_P}{s(s + 5)(s + 10)} \tag{9-31}$$

from the transient standpoint. Notice that this transfer function is similar to that in Eq. (9-26), except for the factor K_P. Thus, we can select the value of K_P so that the damping requirement is met. As shown in Fig. 9-12, when $K = 1.63$, the relative damping of the complex roots is 0.707. Thus, $K_P = 1.63/100 = 0.0163$, where $K = 100$ gives the desired ramp-error constant. In general, we can express K_P as

$$K_P = \frac{\text{value of } K \text{ to achieve damping}}{\text{value of } K \text{ for steady state}} \tag{9-32}$$

system performance

Figure 9-14 shows the root loci of the system described by Eq. (9-30), with $K_I/K_P = 0.1$, and K_P varying. Notice that since the zero at -0.1 is very close to the origin, relative to the nearest pole at -5, the root loci of Fig. 9-14 near the region where $K_P = 0.0163$ are very similar to those in Fig. 9-12 near the region where $K = 1.63$. The unit-step response of the system with the PI controller with $K_P = 0.0163$ and $K_I = 0.00163$ is shown in Fig. 9-15. As with the typical PI control, the step response rises and settles slowly.

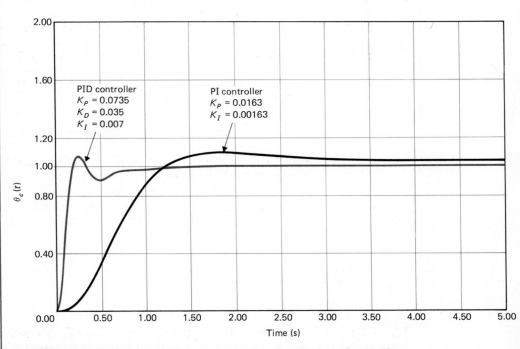

FIGURE 9-15 Unit-step responses of the system described by Eq. (9-26) with PI and PID controls.

The PID-Controller Design

PID control

To reduce the rise time and the settling time, we apply the PID control described by Eq. (9-21) to the third-order system of Eq. (9-26). The transfer function of the open-loop system becomes

$$G(s) = G_c(s)G_p(s) = \frac{5000(1 + K_{D1}s)(K_{P2}s + K_{I2})}{s^2(s + 5)(s + 10)} \qquad (9\text{-}33)$$

First, let us design the PI portion of the controller. Setting $K_{I2}/K_{P2} = 0.1$, we have ($K_{D1} = 0$)

$$G(s) = \frac{5000K_{P2}(s + 0.1)}{s^2(s + 5)(s + 10)} \qquad (9\text{-}34)$$

We realize that when $K_{P2} = 1$, the system will have a transient response that is very similar to that of the uncompensated system, since the zero of $G(s)$ at $s = -0.1$ practically cancels one of the poles at $s = 0$. When only the PI controller is used, K_{P2} should be selected as 0.0163 to give a relative damping ratio of 0.707. To realize a faster rise time, let us choose K_{P2} to be 0.07. (The system becomes unstable for $K_{P2} \geq 0.15$.) Now the value of K_{D1} can be fine tuned to reduce the maximum overshoot and maintain fast rise and settling times. Figure 9-15 shows the unit-step response of the PID-compensated system with $K_{D1} = 0.5$, $K_{P2} = 0.07$, and $K_{I2} = 0.007$. These parameters are converted to the real PID-controller parameters using Eqs. (9-22) through (9-24):

PID controller

$$K_P = 0.0735 \qquad K_D = 0.035 \qquad K_I = 0.007$$

As an alternative, we can first design the PD portion of the PID controller by somewhat arbitrarily selecting K_{D1} to be 0.5. From the root contours in Fig. 9-13, we realize that the PD controller acting alone will not be able to meet the relative stability requirement.

The PI portion of the PID controller is designed by considering the system with the PD controller as a new system. The value of K_{P2} is found by applying Eq. (9-32) to the new system. Since the new system has a zero at $s = -2$, a relative damping ratio of 0.707 may not have the same meaning as in the case before when the PI controller is designed first. In fact, when $K_{P2} = 0.07$ and $K_{I2} = 0.007$, and including the PD portion, the dominant roots of the characteristic equation are at $s = -6.59 + j12.55$ and $s = -6.59 - j12.55$, which correspond to a damping ratio of 0.46.

Since the numerator of the transfer function of the PID controller is of the second order, in general, the zeros of the controller may be placed to cancel undesirable poles of the controlled process, and at the same time, a pole is added at $s = 0$ to increase the type of the system. Complex zeros of the PID controller can also be used to realize a

robust control

robust control system, which will be discussed in Section 9-7.

9-3

TIME-DOMAIN DESIGN OF THE PHASE-LEAD CONTROLLER

The PID controller represents the simplest form of controllers that utilize the derivative and integration operations in the compensation of control systems. In general, we can regard the design of controllers for control systems as a filter design problem; then we

phase-lead controller

can come up with a great variety of possible schemes. Since root loci are used often for design, it would be advantageous to describe the controller by the poles and zeros of its transfer function. For the PD controller in Eq. (9-3), the transfer function has a zero at $s = -K_P/K_D$. For the PI controller, the transfer function in Eq. (9-14) has a pole at $s = 0$ and a zero at $s = -K_I/K_P$. The PID controller in Eq. (9-21) has a pole at $s = 0$ and two zeros from the function $K_D s^2 + K_P s + K_I$.

From the filtering standpoint, the PD controller is a high-pass filter, whereas the PI controller is a low-pass filter. The PID controller is a band-pass or band-attenuate filter, depending on the values of the controller parameters. The high-pass filter is often referred to as a phase-lead controller since positive phase is introduced to the system over some appropriate frequency range. The low-pass filter is also known as a phase-lag controller, since the corresponding phase introduced is negative. These ideas related to filtering and phase shifts are explored further in Chapter 11, where the subject of frequency-domain design is treated.

There are obvious advantages of using only passive network elements in a controller. The transfer function of a simple first-order controller that can be realized by passive resistor–capacitor (RC) network elements is

$$G_c(s) = \frac{s + z_1}{s + p_1} \tag{9-35}$$

In the last equation, the controller is high-pass or phase-lead if $p_1 > z_1$, and low-pass or phase-lag if $p_1 < z_1$.

The s-domain design of the high-pass or phase-lead controller described by Eq. (9-35) is investigated in this section.

A network realization of the phase-lead controller of Eq. (9-35) ($p_1 > z_1$) is shown in Fig. 9-16. Although a low-pass filter can be realized by a simpler network by eliminating the resistor R_1, the resulting network would not be able to pass dc signals, and transmission would be blocked in the steady state, which is not acceptable for control systems. The transfer function of the network in Fig. 9-16 is derived by assuming that the source impedance that the network sees is zero and the output load impedance is infinite. This assumption is necessary in the derivation of the transfer function of any four-terminal network. From Fig. 9-16,

$$\frac{E_2(s)}{E_1(s)} = \frac{R_2}{R_1 + R_2} \frac{1 + R_1 C s}{1 + \frac{R_1 R_2}{R_1 + R_2} C s} \tag{9-36}$$

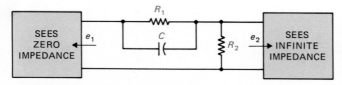

FIGURE 9-16 A phase-lead network with passive elements.

Let

$$a = \frac{R_1 + R_2}{R_2} \qquad a > 1 \qquad (9\text{-}37)$$

and

$$T = \frac{R_1 R_2}{R_1 + R_2} C \qquad (9\text{-}38)$$

Then Eq. (9-36) becomes

transfer
function
$$\frac{E_2(s)}{E_1(s)} = \frac{s + 1/aT}{s + 1/T} = \frac{1}{a} \frac{1 + aTs}{1 + Ts} \qquad a > 1 \qquad (9\text{-}39)$$

As seen from Eq. (9-39), the transfer function of the phase-lead network has a real zero at $s = -1/aT$ and a real pole at $s = -1/T$. These are represented in the s-plane, as shown in Fig. 9-17. By varying the values of a and T, the pole and zero may be located at any point on the negative real axis in the s-plane. Since $a > 1$, the zero is always located to the right of the pole, and the distance between them is determined by the constant a. It is also important to note that the phase-lead controller in Eq. (9-39) is accompanied by a zero-frequency attenuation $1/a$, since a is greater than 1. This attenuation is detrimental to the steady-state performance since it will reduce the error constant. Thus, when applying the phase-lead controller of Eq. (9-39), *it is important to first increase the loop gain by the amount a, so that the zero-frequency gain of the phase-lead controller is unity.*

The zero to the right of the pole is the reason that the phase-lead controller can improve the stability of a closed-loop control system. We shall use the following example to illustrate the design of the phase-lead controller for control systems.

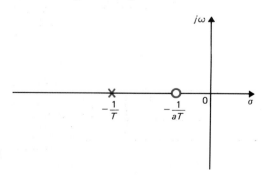

FIGURE 9-17 Pole-zero configuration of the phase-lead controller.

EXAMPLE 9-5

*phase-lead
control
example*

*sun-seeker
system*

The block diagram in Fig. 9-18 describes the components of a sun-seeker control system. The system may be mounted on a space vehicle so that it will track the sun with high accuracy. The variable θ_r represents the reference angle of the solar ray, and θ_o denotes the vehicle axis. The objective for the sun-seeker control system is to maintain the error between θ_r and θ_0, α, near zero. The parameters of the system are as follows:

$R_F = 10{,}000$	$K_b = 0.0125$ V/rad/s
$K_i = 0.0125$ N-m/A	$R_a = 6.25\Omega$
$J = 10^{-6}$ kg-m^2	$K_s = 0.1$ A/rad
$K = $ to be determined	$B = 0$
$n = 800$	

*uncompensated
system*

The open-loop transfer function of the uncompensated system is

$$\frac{\Theta_o(s)}{\alpha(s)} = \frac{K_s R_F K K_i / n}{R_a J s^2 + K_i K_b s} \tag{9-40}$$

Substituting the numerical values of the system parameters, Eq. (9-40) gives

$$\frac{\Theta_o(s)}{\alpha(s)} = \frac{2500K}{s(s+25)} \tag{9-41}$$

The specifications of the system are as follows:

*design
specifications*

1. The steady-state value of $\alpha(t)$ due to a unit-ramp function input for $\theta_r(t)$ should be ≤ 0.01 rad per rad/s of the final steady-state output velocity. In other words, the steady-state error due to a ramp input should be ≤ 1 percent.
2. The maximum overshoot of the step response should be less than 10 percent.

The loop gain of the system is determined from the steady-state error requirement. Applying the final-value theorem to $\alpha(t)$, we have

*steady-state
performance*

$$\lim_{t \to \infty} \alpha(t) = \lim_{s \to 0} s\alpha(s) = \lim_{s \to 0} \frac{s\Theta_r(s)}{1 + \Theta_o(s)/\alpha(s)} \tag{9-42}$$

where $\alpha(s)$ is the Laplace transform of $\alpha(t)$.

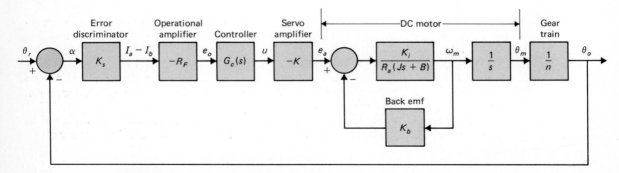

FIGURE 9-18 Block diagram of the sun-seeker control system.

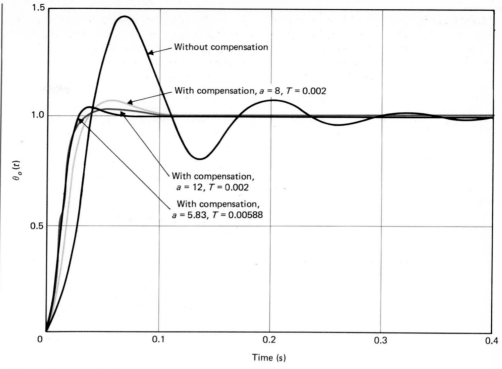

FIGURE 9-19 Unit-step responses of the sun-seeker control system.

For the unit-ramp input, $\Theta_r(s) = 1/s^2$. By using Eq. (9-41), Eq. (9-42) gives

$$\lim_{t \to \infty} \alpha(t) = \frac{0.001}{K} \tag{9-43}$$

Thus, for the steady-state error to be ≤ 0.01, K must be ≥ 1. For $K = 1$, the worst case from the steady-state-error standpoint, the characteristic equation of the uncompensated system is

$$s^2 + 25s + 2500 = 0 \tag{9-44}$$

We can show that the damping ratio of the uncompensated system is only 25 percent, which corresponds to a maximum overshoot of over 44.4 percent. Figure 9-19 shows the unit-step response of the system with $K = 1$.

A space has been reserved in the block diagram of Fig. 9-18 for a series controller with transfer function $G_c(s)$. Let us consider using the phase-lead controller of Eq. (9-39), although in the present case, a PD controller or many other types would also be effective in satisfying the performance specifications given above.

root-contour design   Let us use the root-contour method to aid the selection of the controller parameters a and T. The open-loop transfer function of the system with the phase-lead controller is written

$$G(s) = \frac{2500(1 + aTs)}{s(s + 25)(1 + Ts)} \tag{9-45}$$

We have increased the loop gain of the system to compensate for the attenuation $1/a$ of the phase-lead controller, so that the ramp-error constant K_v is maintained at 100.

To begin with the root-contour design, we first set a to zero in Eq. (9-45). Then, the characteristic equation of the compensated system becomes

$$s(s + 25)(1 + Ts) + 2500 = 0 \qquad (9\text{-}46)$$

Since T is the variable parameter, we divide both sides of Eq. (9-46) by the terms that do not contain T. We have

$$1 + \frac{Ts^2(s + 25)}{s^2 + 25s + 2500} = 0 \qquad (9\text{-}47)$$

This equation is of the form of $1 + TG_1(s) = 0$, where $G_1(s)$ is an equivalent transfer function that can be used to construct the root contours of Eq. (9-46), which are drawn in Fig. 9-20 using the poles and zeros of $G_1(s)$. Notice that the poles of $G_1(s)$ are the roots of the characteristic equation of the system when $a = 0$ and $T = 0$. As seen from the root contours, adding the factor $1 + Ts$ in the denominator of Eq. (9-45) alone would not improve the system performance at all. In fact, the characteristic-equation roots are moved toward the right-half s-plane, and the system becomes unstable when T is greater than 0.0133. To achieve the full effect of the phase-

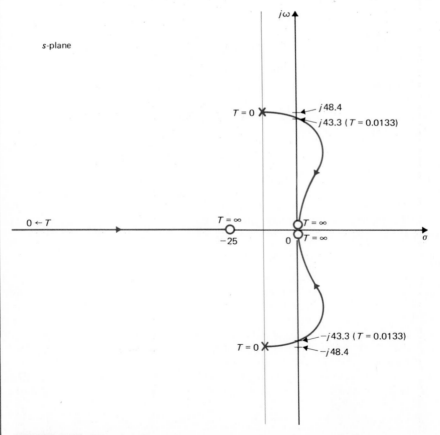

FIGURE 9-20 Root contours of the sun-seeker system with $a = 0$, and T varies from 0 to ∞.

lead compensation, we must restore the value of a in Eq. (9-45). The characteristic equation of the fully compensated system now becomes

$$s(s + 25)(1 + Ts) + 2500(1 + aTs) = 0 \qquad (9\text{-}48)$$

Now we consider that a is the variable parameter while keeping T constant. To prepare for the root contours with a as a variable parameter, we divide both sides of Eq. (9-48) by the terms that do not contain a, and arrive at an equation in the form of $1 + aG_2(s) = 0$, where

$$G_2(s) = \frac{2500Ts}{s(s + 25)(1 + Ts) + 2500} \qquad (9\text{-}49)$$

For a given T, the root contours of Eq. (9-48) when a varies are obtained based on the poles and zeros of $G_2(s)$. Notice that the poles of $G_2(s)$ are the same as the roots of Eq. (9-46). Thus, for a given T, the root contours of Eq. (9-48) when a varies must start ($a = 0$) at points on the root contours of Fig. 9-20. These root contours end ($a = \infty$) at $s = 0$, ∞, ∞, which are the zeros of $G_2(s)$. The complete root contours of Eq. (9-48) are now shown in Fig. 9-21 for several values of T, and a varies from 0 to ∞.

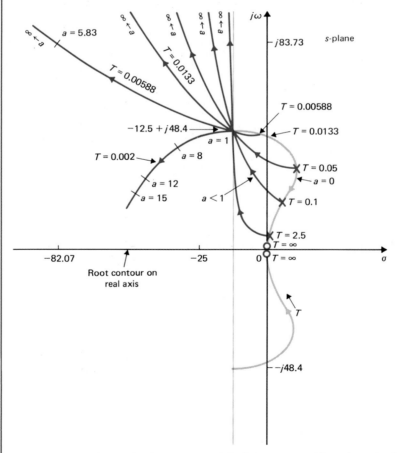

FIGURE 9-21 Root contours of the sun-seeker system with a phase-lead controller.

From the root contours of Fig. 9-21, we see that for effective phase-lead compensation, the value of T should be small. For large values of T, the natural frequency of the system increases rapidly as a increases, and very little improvement is made on the damping of the system. As seen from the root contours of Fig. 9-21, there are many sets of values of a and T that will satisfy the transient-response requirement, but in any case, the value of T must be very small. By referring to Eq. (9-45), for very small T, the term Ts in the denominator can be neglected. As a very rough approximation, the characteristic equation in Eq. (9-48) becomes

$$s^2 + 25(1 + 100aT)s + 2500 = 0 \qquad (9\text{-}50)$$

Let us select the damping ratio for the approximating second-order system to be 0.707. Then, Eq. (9-50) gives

$$\zeta = 0.707 = 25aT + 0.25 \qquad (9\text{-}51)$$

from which we have

$$aT = 0.0183$$

This gives us an idea on what the value of aT should be. For instance, if T is chosen to be 0.00588, then $a = 3.1$; if $T = 0.002$, the corresponding value of a is 9.14. Keep in mind that the actual pole at $-1/T$ that was neglected will have an adverse effect on the transient response, so that it would be safer to select a larger value for a than those previously calculated.

In the following, we present three sets of results for different combinations of a and T. The first two cases yield maximum overshoots in the compensated system that are less than the specified 10 percent. The corresponding unit-step responses are shown in Fig. 9-19.

result comparisons

T	a	MAXIMUM OVERSHOOT (%)	CHARACTERISTIC-EQUATION ROOTS	
0.00588	5.83	7.3	-30.93,	$-82.07 \pm j83.73$
0.00200	12.00	6.6	-433.60,	$-45.68 \pm j28.21$
0.00200	8.00	13.8	-460.30,	$-32.35 \pm j40.85$

If we select the case with $a = 5.83$ and $T = 0.00588$, the transfer function of the phase-lead controller is

phase-lead controller

$$G_c(s) = \frac{s + 1/aT}{s + 1/T} = \frac{s + 29.17}{s + 170} \qquad (9\text{-}52)$$

EXAMPLE 9-6   Consider the same system described in Example 9-4. The transfer function of the controlled process is

$$G_p(s) = \frac{K}{s(1 + 0.1s)(1 + 0.2s)} \qquad (9\text{-}53)$$

where $K = 100$. As pointed out in Example 9-4, the closed-loop system without compensation is unstable, and the marginal value of K for stability is 15. The root loci of the characteristic equation of the uncompensated system are shown in Fig. 9-12.

Let us attempt to improve the stability of the system by using the phase-lead controller:

$$G_c(s) = \frac{1 + aTs}{1 + Ts} \qquad a > 1 \tag{9-54}$$

Notice that the attenuation of the phase-lead network, $1/a$, has been offset by additional loop gain. The open-loop transfer function of the system is

$$G(s) = G_c(s)G_p(s) = \frac{5000(1 + aTs)}{s(s + 5)(s + 10)(1 + Ts)} \tag{9-55}$$

First, we set $a = 0$ while T is varied from zero to infinity. The characteristic equation is

$$s(s + 5)(s + 10)(1 + Ts) + 5000 = 0 \tag{9-56}$$

The root contours of the last equation are constructed by first conditioning the equation into the form of $1 + TG_1(s) = 0$, where

$$G_1(s) = \frac{s^2(s + 5)(s + 10)}{s(s + 5)(s + 10) + 5000} \tag{9-57}$$

In Fig. 9-22, the poles of $G_1(s)$ are labeled as $T = 0$ and are found on the root loci of Fig. 9-12. The zeros of $G_1(s)$ are the points where $T = \infty$, and are at $s = 0, 0, -5$, and -10.

Next, we restore the value of a in Eq. (9-55), and the characteristic equation of the overall system is

$$s(s + 5)(s + 10)(1 + Ts) + 5000(1 + aTs) = 0 \tag{9-58}$$

When a is the variable parameter, we condition the last equation into the form of $1 + aG_2(s) = 0$, where

$$G_2(s) = \frac{5000Ts}{s(s + 5)(s + 10)(1 + Ts) + 5000} \tag{9-59}$$

limitations of phase-lead controller

The root contours with a as a variable parameter start ($a = 0$) at the poles of $G_2(s)$, which for a given T are on the root contours of Fig. 9-22, and end ($a = \infty$) at the zeros of $G_2(s)$, which are at $s = 0, \infty, \infty$, and ∞. The dominant part of the root contours of Eq. (9-58) is shown in Fig. 9-22. Notice that since for the phase-lead controller the value of a is greater than 1, the corresponding root contours are mostly in the right-half plane. The reason for this is that the value of K is too large so that the original system started out to be very unstable. The purpose of this example is to bring out the limitations of the single-stage phase-lead controller, which are very similar to those of the PD controller. It should be pointed out that the process in Eq. (9-53) can be effectively compensated with two or more stages of the phase-lead controller of Eq. (9-54).

two-stage controller

In fact, by setting $a = 22.3$ and $T = 0.00444$, two identical stages of the phase-lead controller yield characteristic-equation roots of the closed-loop system at $-10, -11.6, -313.6, -65.12 + j52.16$, and $-65.12 - j52.16$. The unit-step response of the compensated system has a maximum overshoot of 10 percent at 0.065 seconds.

In general, the multiple-stage phase-lead controller may be designed one stage at a time, in which case the values of a and T for each stage would most likely be different.

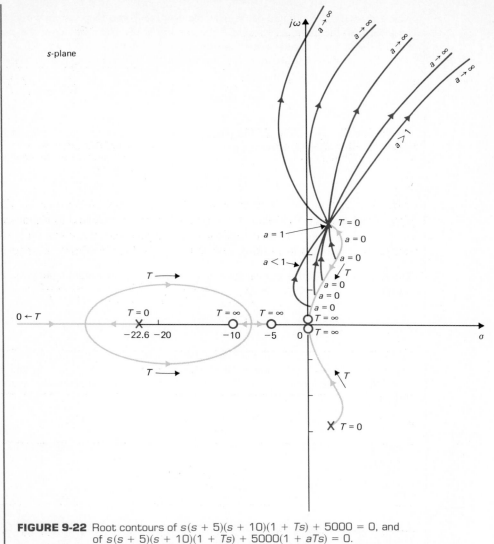

FIGURE 9-22 Root contours of $s(s + 5)(s + 10)(1 + Ts) + 5000 = 0$, and of $s(s + 5)(s + 10)(1 + Ts) + 5000(1 + aTs) = 0$.

9-4

TIME-DOMAIN DESIGN OF THE PHASE-LAG CONTROLLER

A phase-lag controller or a low-pass filter can be modeled by the transfer function

phase-lag controller

$$G_c(s) = \frac{1 + aTs}{1 + Ts} \qquad a < 1 \qquad (9\text{-}60)$$

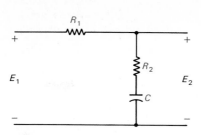

FIGURE 9-23 A phase-lag network with *RC* elements.

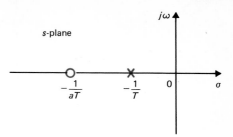

FIGURE 9-24 Pole-zero configuration of the phase-lag controller

Notice that except for the attenuation $1/a$, Eq. (9-60) is identical to Eq. (9-39), but with $a < 1$. Figure 9-23 shows a simple *RC* network that realizes the transfer function of Eq. (9-60). If we assume that the input impedance and output impedance the network sees are zero and infinite, respectively, the transfer function of the network is

$$G_c(s) = \frac{E_2(s)}{E_1(s)} = \frac{1 + R_2 Cs}{1 + (R_1 + R_2)Cs} = \frac{1 + aTs}{1 + Ts} \qquad (9\text{-}61)$$

where

$$aT = R_2 C \qquad (9\text{-}62)$$

and

$$a = \frac{R_2}{R_1 + R_2} \qquad a < 1 \qquad (9\text{-}63)$$

The transfer function of the phase-lag controller in Eq. (9-61) has a real zero at $s = -1/aT$ and a real pole at $s = -1/T$. As shown in Fig. 9-24, since a is less than unity, the pole is always located to the right of the zero, and the distance between them is determined by a.

EXAMPLE 9-7

phase-lag control example

The design principle of the phase-lag controller is similar to that of the PI controller discussed in Section 9-2, and is best illustrated by the root-locus diagram of the sun-seeker system considered in Example 9-5. Figure 9-25(a) shows the root loci of the uncompensated system, based on the open-loop transfer function of Eq. (9-41). As shown by the root loci, to achieve a damping ratio of 0.707, K has to be set at 0.125. However, to satisfy the steady-state error requirement, K must be ≥ 1; the latter value of K yields a system with low damping. The strategy of the phase-lag control is that for effective control, *the pole and zero of the controller transfer function should be placed very close together, and then for the type-1 system, the combination should be located relatively close to the origin of the s-plane.* For the present case, the pole-zero combination of the phase-lag controller should be arranged so that when viewed from the point at which $K = 0.125$ on the root loci of Fig. 9-25(a), the pole and zero are so close together that they practically cancel each other. We can use the pole of $G_p(s)$ at -25 as a guideline, and choose the value of $1/aT$ to be much smaller than 25. Figure 9-25(b) shows the root loci of the compensated system with $1/T = 0.4$ and $1/aT = 2.0$, or $a = 0.2$. As K increases from 0, the two roots that leave from $s = 0$ and $s = -0.4$, move toward each other on the real axis. After reaching the breakaway point, they separate, forming a small loop around the zero at $s = -2$, and then meet again at the breakaway point to the left of the -2 point. In the meantime, the root that leaves the open-loop pole at $s = -25$ travels to the right as K increases; it meets one of the two roots from the other two loci at the third breakaway point at $s = -11.25$, and then the two roots become complex as K approaches infinity. It is important to note that since the pole and zero of the phase-lag controller are placed close together and are very near the origin, they have very little effect on the shape of the original root loci, especially at points far away from the origin. Thus, in Fig. 9-25(b), other than the small loop near $s = 0$, the rest of the root loci are very similar to that of Fig. 9-25(a). However, the values of K that correspond to similar points on the two sets of root loci are different. It is simple to show that for the values of a and T chosen, the values of K on the root loci of the compensated system in Fig. 9-25(b) at points relatively far away from the origin are $1/a$ times greater than those values of K at similar points on the uncompensated root loci in Fig. 9-25(a). For instance, at the root $s = -11.6 + j18$ on the compensated root loci in Fig. 9-25(b), the value of K is 1, which is five times the value of $K (= 0.2)$ at the comparable point, $s = -12.5 + j18.4$, on the uncompensated root loci in Fig.

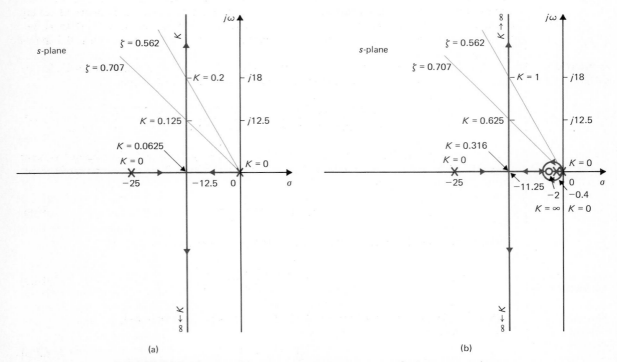

(a)   (b)

FIGURE 9-25 (a) Root loci of $G(s) = 2500K/[s(s + 25)]$ of the sun-seeker system. (b) Root loci of $G(s) = 500(s + 2)/[s(s + 0.4)]$ of the sun-seeker system.

9-25(a). Similarly, Fig. 9-25(a) shows that when $K = 0.125$ on the uncompensated root loci, the damping ratio is 0.707; the comparable point on the compensated root loci corresponds to $K = 0.625$, which is $1/a$ times 0.125.

A formal way of demonstrating the relationship between the values of K on the compensated root loci and the uncompensated root loci and the value of a is to refer to the open-loop transfer functions of the original system and the compensated system. The value of K at any point s_1 on the root loci of the uncompensated system is determined by applying the root-locus condition of Eq. (8-22) to Eq. (9-41),

$$|K| = \frac{|s_1||s_1 + 25|}{2500} \tag{9-64}$$

The value of K at s_1, a comparable point, on the root loci of the compensated system is given by

$$|K| = \frac{|s_1||s_1 + 0.4||s_1 + 25|}{500|s_1 + 2|} \tag{9-65}$$

If the point s_1 on the compensated root loci is far from the pole-zero combination of the phase-lag controller, the magnitude of K at s_1 can be approximated by

$$|K| \cong \frac{|s_1||s_1 + 25|}{500} \tag{9-66}$$

since the distance from s_1 to -0.4 will be approximately the same as that from s_1 to -2. This argument also raises the point that the exact locations of the pole and the zero of the phase-lag controller are not significant as long as they are close to the origin, and that the distance between the pole and the zero is a fixed desired quantity governed by a, which is set at 0.2 in this example.

root-locus design

Based on the development given on the principle of design of the phase-lag controller via the root loci, we set the value of a to the ratio of the two values of K as follows:

$$a = \frac{K \text{ to realize the desired damping}}{K \text{ to realize the steady-state performance}} = \frac{0.125}{1} = \frac{1}{8} \tag{9-67}$$

We can also arrive at this relationship for a with the following derivations. The desired relative damping is achieved by setting K to 0.125 in the original system. The open-loop transfer function of Eq. (9-41) becomes

$$\frac{\Theta_o(s)}{\alpha(s)} = \frac{312.5}{s(s + 25)} \tag{9-68}$$

The open-loop transfer function of the compensated system becomes

$$\frac{\Theta_o(s)}{\alpha(s)} = \frac{2500aK(s + 1/aT)}{s(s + 25)(s + 1/T)} \tag{9-69}$$

If the values of aT and T are chosen to be large, Eq. (9-69) is approximately

$$\frac{\Theta_o(s)}{\alpha(s)} = \frac{2500aK}{s(s + 25)} \tag{9-70}$$

from the transient standpoint. Note that we cannot apply this approximation from the steady-state-response standpoint, since the ramp-error constant from Eq. (9-70) does not agree with that of Eq. (9-41). Since K is necessarily equal to unity, to have the right side of Eqs. (9-68) and (9-70) equal to each other, $a = 1/8$, as is already concluded in Eq. (9-67). Theoretically, the value of T can be arbitrarily large. However, if the value of T is too large, we may encounter difficulties in realizing the phase-lag controller by physical components. The following table summarizes the results when the value of a is set at $1/8$, and $T = 5$, 10, and 20.

design summary

T	a	MAXIMUM OVERSHOOT (%)	CHARACTERISTIC-EQUATION ROOTS	
5	0.125	15.0	-1.818,	$-11.69 \pm j11.76$
10	0.125	10.0	-0.849,	$-12.13 \pm j12.14$
20	0.125	7.2	-0.412,	$-12.32 \pm j12.32$

When $T = 5$, the complex roots of the characteristic equation correspond to a relative damping ratio of approximately 0.707, but the real root at -1.818 still has some influence on the transient response, and the maximum overshoot is 15 percent. When the value of T is increased to 10, the relative damping is closer to 0.707, and the real root is at -0.849, and is closer to the zero of the closed-loop transfer function at $s = -0.8$ so that the contribution on the overshoot from the real root is smaller; the maximum overshoot is reduced to 10 percent. When $T = 20$, the real root is at -0.412, which is even closer to the zero of the closed-loop transfer function now at -0.4, and the maximum overshoot is reduced to 7.2 percent. It is interesting to realize

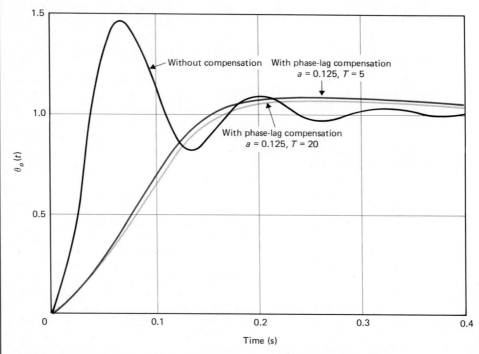

FIGURE 9-26 Unit-step responses of the sun-seeker control system with phase-lag control.

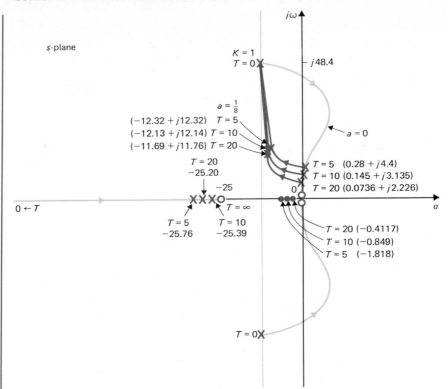

FIGURE 9-27 Root contours of $s(s + 25)(1 + Ts) + 2500K(1 + aTs) = 0$.
$K = 1$.

that when T becomes very large, the two complex roots of the characteristic equation should eventually become $-12.5 \pm j12.5$, which are the same as those of the original system when $K = 0.125$, and the maximum overshoot should approach 4.32 percent, which corresponds to that of a second-order prototype system with $\zeta = 0.707$.

Figure 9-26 illustrates the unit-step responses of the uncompensated system and the compensated system with $a = 0.125$, and $T = 5$ and 20.

root-contour design

As an alternative, the design of the phase-lag control can also be carried out by means of the root-contour method. The root-contour design conducted earlier by Eqs. (9-45) through (9-49) for the phase-lead controller and Figs. 9-20 and 9-21 is still valid for the phase-lag control, except that in the present case, $a < 1$. Thus, in Fig. 9-21, only the portions of the root contours that correspond to $a < 1$ are applicable for the phase-lag compensation. These root contours clearly show that for effective phase-lag control, the value of T should be relatively large. In Fig. 9-27, we illustrate further that the complex poles of the closed-loop transfer function are rather insensitive to the value of T when the latter is relatively large.

effects on rise time, settling time

The time responses of Fig. 9-26 point out a major disadvantage of the phase-lag control. Since the phase-lag controller is essentially a low-pass filter, the rise time and the settling time of the compensated system are usually increased. However, we shall show by the following example that the phase-lag controller is more versatile and has a wider range of effectiveness in its application of improving the stability of a control system than the single-stage phase-lead controller.

*design
procedure*

Based on the previous discussions, we can outline a root-locus design procedure for the phase-lag design of control systems as follows. Since the design will be carried out in the *s*-plane, the specifications on the transient response or the relative stability should be given in terms of the damping ratio of the dominant roots, and other quantities, such as rise time, bandwidth, and the maximum overshoot, which can be correlated with the location of the characteristic-equation roots.

1. Sketch the root loci of the characteristic equation of the uncompensated system.
2. Determine on these root loci where the desired roots should be located to achieve the desired relative stability or damping of the system. Find the value of *K* that corresponds to these roots.
3. Compare the value of *K* required for steady-state performance and the *K* found in the last step. The ratio of these two values of *K* is $a(a < 1)$, which is the desired ratio between the pole and zero of the phase-lag controller.
4. The exact value of *T* is not critical as long as it is relatively large. We may choose the value of $1/aT$ to be many orders of magnitudes smaller than the smallest pole of the transfer function of the controlled process.

EXAMPLE 9-8

Consider the system given in Example 9-6 for which the phase-lead control was proven to be ineffective. The open-loop transfer function of the original system and the performance specifications are

$$G_p(s) = \frac{K}{s(1 + 0.1s)(1 + 0.2s)}$$

specifications

$$K_v = 100 \text{ s}^{-1} \tag{9-71}$$

$$\text{relative damping ratio} = 0.707$$

The root loci of the uncompensated system are shown in Fig. 9-28. For $K_v = 100$, *K* has to equal 100, which corresponds to an unstable system. When $K = 1.635$, the roots of the characteristic equation of the uncompensated system are at -11.12, $-1.91 + j1.91$, and $-1.91 + j1.91$, which correspond to a relative damping ratio of 0.707. Thus, using the design method outlined earlier, we set

$$a = \frac{1.635}{100} = 0.01635 \tag{9-72}$$

Let us set *T* to be arbitrarily large at 100. Then, the phase-lag controller is described by the transfer function

controller

$$G_c(s) = \frac{1 + 1.635s}{1 + 100s} \tag{9-73}$$

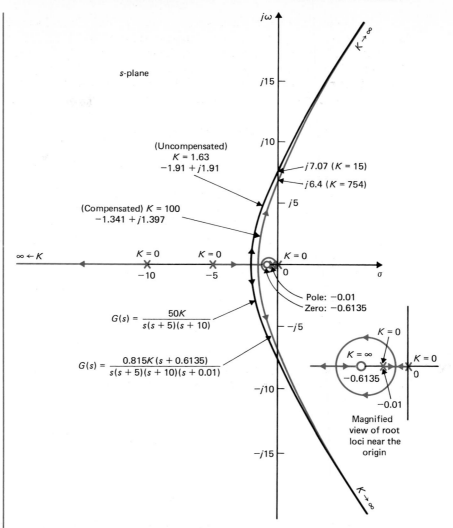

FIGURE 9-28 Root loci of compensated and uncompensated systems in Example 9-4.

and the open-loop transfer function of the compensated system is

$$G(s) = G_c(s)G_p(s) = \frac{0.815K(s + 0.6135)}{s(s + 5)(s + 10)(s + 0.01)} \qquad (9\text{-}74)$$

where $K = 100$. Figure 9-28 illustrates the root loci of the compensated system in color. The pole of the controlled process at $s = 0$ and the pole and zero of the controller at $s = -0.01$ and -0.6135, respectively, give rise to a small loop near the origin. The loci then break away at a point slightly to the right of the breakaway point of the original root loci and follow the latter very closely as K approaches infinity. When $K = 100$, the roots of the characteristic equation of the compensated system are at -11.13, -1.198, $-1.341 + j1.397$, and $-1.341 - j1.397$.

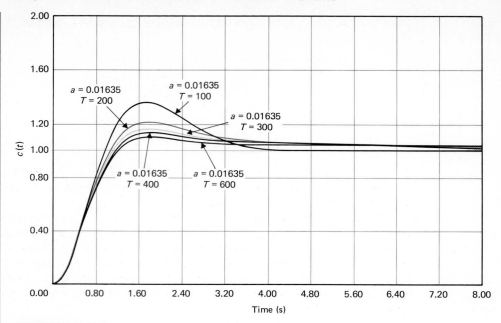

FIGURE 9-29 Unit-step responses of a control system with phase-lag compensation in Example 9-4.

The relative damping ratio of the complex roots is slightly less than 0.7. This can be improved by selecting a smaller value for a or a larger T. However, once the value of T is sufficiently large, its variation should have little effect on the transient performance. Figure 9-29 shows the unit-step responses of the phase-lag compensated system with $a = 0.01635$, and $T = 100, 200, 300, 400,$ and 600. When $T = 100$, the maximum overshoot is approximately 36 percent. When $T = 400$, the maximum overshoot is slightly over 5 percent, which means that as T increases, the relative damping ratio is approaching 0.707. Note that when the value of T exceeds 300, the further improvement on the damping is minimal. The penalty we pay with a larger value of T is a longer rise time and settling time.

9-5

LEAD-LAG CONTROLLER

We have learned from preceding sections that the phase-lead controller generally improves the rise time and the damping but increases the natural frequency of the closed-loop system. On the other hand, phase-lag control when applied properly improves damping, but usually results in a longer rise time and settling time. Therefore, each of these control schemes has its advantages, disadvantages, and limitations, and there are many systems that cannot be satisfactorily compensated by either scheme. It is natural, therefore, whenever necessary, to consider using a combination of the lead and lag controllers, so that the advantages of both schemes are utilized.

lead-lag controller

The transfer function of a simple lead-lag (or lag-lead) controller can be written as

$$G_c(s) = \left(\frac{1 + aT_1s}{1 + T_1s}\right)\left(\frac{1 + bT_2s}{1 + T_2s}\right) \tag{9-75}$$

$$|\leftarrow \text{lead} \rightarrow||\leftarrow \text{lag} \rightarrow|$$

where $a > 1$ and $b < 1$, and the attenuation factor $1/a$ of the phase-lead controller is not included in the equation if we assume that adequate loop gain is available in the system to compensate for this attenuation.

Since the lead-lag controller transfer function in Eq. (9-75) now has four unknown parameters, its design is not as straightforward as the single-stage phase-lead or phase-lag controller. We shall outline a design procedure by first establishing the phase-lead portion of the controller, and then determine the parameters for the phase-lag portion. We can *use the phase-lead portion of the controller mainly for the purpose of achieving a short rise time, and then the phase-lag portion is brought in to improve the damping of the system.*

improved rise-time damping

EXAMPLE 9-9

As an illustrative example, we can use the controlled process of Example 9-8. With $K = 100$, the transfer function of the controlled process in Eq. (9-71) is written

$$G_p(s) = \frac{5000}{s(s + 5)(s + 10)} \tag{9-76}$$

phase-lead controller

The uncompensated closed-loop system is unstable. Let us introduce the lead-lag controller of Eq. (9-75) with the objective of attaining a shorter rise time than the phase-lag-compensated system in Example 9-8. First, we set the phase-lead controller transfer function as

$$G_{c1}(s) = \frac{1 + aT_1s}{1 + T_1s} = \frac{1 + 0.2s}{1 + 0.05s} = \frac{4(s + 5)}{s + 20} \tag{9-77}$$

Since the purpose of the phase-lead controller is to reduce the rise time, this can be achieved by increasing the natural frequency of the system. Thus, we place the zero of $G_{c1}(s)$ at -5, to cancel the same pole of $G_p(s)$, and add a pole arbitrarily at -20, at a distance to the left of the pole of $G_p(s)$ at -10. As will be shown in the next section, inexact cancellation of the pole and zero will have little effect on the final response of the system, as long as the pole to be cancelled is not in the right-half s-plane. The open-loop transfer function with just the phase-lead controller is

$$G_{c1}(s)G_p(s) = \frac{20,000}{s(s + 10)(s + 20)} \tag{9-78}$$

The root loci of the characteristic equation are shown in Fig. 9-30, with the numerator of Eq. (9-78) replaced by the variable parameter K_1. When $K_1 = 20,000$, the roots are at -38.37, $4.186 + j22.44$, and $4.186 - j22.44$. The roots of the characteristic equation of the uncompensated system that corresponds to Eq. (9-76) are at -22.587, $3.794 + j14.387$, and $3.794 -$

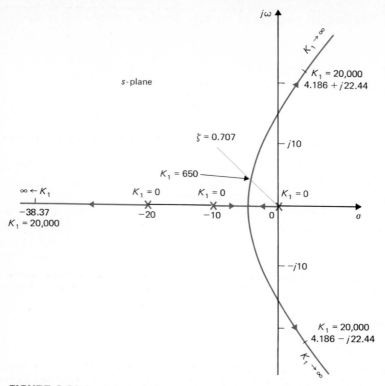

FIGURE 9-30 Root loci of the control system described by Eq. (9-78) with the gain replaced by K_1.

$j14.387$. Clearly, the natural frequency of the system represented by Eq. (9-78) is higher than that of the original system, as seen from the imaginary parts of the complex roots, although both systems are unstable. In fact, the natural frequency of the system is raised and the rise time and settling time are further reduced if the value of T_1 is reduced further, which corresponds to moving the pole of $G_{c1}(s)$ at -20 more to the left.

To design the phase-lag section, we found on the root loci of Fig. 9-30 that the damping ratio of the complex roots is approximately 0.707 when $K_1 = 650$. Thus, by using Eq. (9-67),

$$b = \frac{650}{20,000} = 0.325 \qquad (9\text{-}79)$$

We can now select T_2 to be relatively large at 300. Thus, the overall lead-lag controller has the transfer function

phase-lag controller

$$G_c(s) = \left(\frac{1 + 0.2s}{1 + 0.05s}\right)\left(\frac{1 + 9.75s}{1 + 300s}\right) \qquad (9\text{-}80)$$

The open-loop transfer function of the compensated system is

$$G(s) = \frac{650(s + 0.1026)}{s(s + 10)(s + 20)(s + 0.0033)} \qquad (9\text{-}81)$$

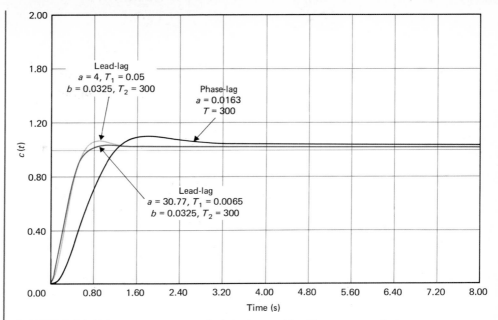

FIGURE 9-31 Unit-step responses of phase-lag and lead-lag compensated systems.

Figure 9-31 shows the step responses of the system with the phase-lag controller with $a = 0.01635$, $T = 300$, and the lead-lag controller with the above design parameters. Clearly, the system with the lead-lag controller has a smaller overshoot and shorter rise and settling times.

It is not necessary to cascade the individual phase-lead and the phase-lag controllers to realize the transfer function in Eq. (9-75) if a and b need not be specified independently. A network that has phase-lead phase-lag characteristics, but with fewer elements, is shown in Fig. 9-32. The transfer function of the network is

lead-lag
network

$$G_c(s) = \frac{E_2(s)}{E_1(s)} = \frac{(1 + R_1 C_1 s)(1 + R_2 C_2 s)}{1 + (R_1 C_1 + R_1 C_2 + R_2 C_2)s + R_1 R_2 C_1 C_2 s^2} \qquad (9\text{-}82)$$

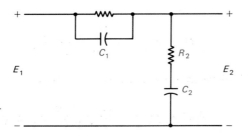

FIGURE 9-32 Lead-lag network.

Comparing Eq. (9-75) with Eq. (9-82), we have

$$aT_1 = R_1 C_1 \tag{9-83}$$

$$bT_2 = R_2 C_2 \tag{9-84}$$

$$T_1 T_2 = R_1 R_2 C_1 C_2 \tag{9-85}$$

From Eqs. (9-83) and (9-84), we have

$$abT_1 T_2 = R_1 R_2 C_1 C_2 \tag{9-86}$$

Thus,

$$ab = 1 \tag{9-87}$$

which means that a and b cannot be specified independently.

For the lead-lag controller designed above, $a = 4$ and $b = 0.0325$, and $ab \neq 1$, so that the simplified lag-lead network cannot be used. In order to utilize the simplified network, we have to modify the design parameters. This can be done by maintaining the values of b and T_2 at 0.0325 and 300, respectively, and set $a = 1/b = 30.77$. In order to cancel the pole at $s = -5$ of the original process, the value of aT_1 is set at 0.2. Thus, $T_1 = 0.0065$. The open-loop transfer function of the compensated system is now

$$G(s) = \frac{5000(s + 0.1026)}{s(s + 10)(s + 153.8)(s + 0.0033)} \tag{9-88}$$

The pole of the phase-lead controller is now at -153.8, which would have even less effect on the overall system response. The unit-step response of the system is shown in Fig. 9-31, which is even better than the system with the transfer function in Eq. (9-81). This modified design of the lead-lag controller further shows that the phase-lead portion of the controller is used to improve the rise and settling times, and the selection of its parameters is not critical, and the phase-lag portion is used to improve damping.

9-6

POLE-ZERO-CANCELLATION COMPENSATION

The transfer functions of many controlled processes contain one or more pairs of complex-conjugate poles that are very close to the imaginary axis of the s-plane. These complex poles usually cause the closed-loop system to be lightly damped or unstable. One immediate thought is to use a controller that has a transfer function with zeros so selected as to cancel the undesired complex poles of the controlled process, and the poles of the controller are placed at more desirable locations in the s-plane to achieve the desired dynamic performance. For example, if the transfer function of a process is

$$G_p(s) = \frac{K}{s(s^2 + s + 10)} \tag{9-89}$$

in which the complex-conjugate poles may cause stability problems in the closed-loop system, especially if the value of K is large, the suggested series controller may be of the form

$$G_c(s) = \frac{s^2 + s + 10}{s^2 + 2\zeta\omega_n s + \omega_n^2} \tag{9-90}$$

The constants ζ and ω_n may be selected according to the performance specifications of the closed-loop system.

There are practical difficulties with the pole-zero-cancellation design scheme that should be mentioned to prevent the method from being used indiscriminantly. The problem is that in practice *exact* cancellation of poles and zeros of transfer functions is rarely possible. One reality is that the transfer function of the process, $G_p(s)$, is usually determined through testing and physical modeling; linearization of a nonliner process and approximations of a complex process are unavoidable. Thus, the "true" poles and zeros of the transfer function of the process may not be accurately modeled. In fact, the true order of the system may even be higher than that represented by the transfer function used for modeling purposes. Another difficulty is that the dynamic properties of the process may vary, even very slowly, due to aging of the system components or changes in the operating environment, so that the poles and zeros of the transfer function may move during the operation of the system. For these and other reasons, even if we could precisely design the poles and zeros of the transfer function of the controller, exact pole-zero cancellation is almost never possible in practice. We show in the following that in most cases, exact cancellation is *not* really necessary to effectively improve the performance of a control system using the pole-zero-cancellation compensation scheme.

Let us assume that a process is represented by

inexact cancellation

$$G_p(s) = \frac{K}{s(s + p_1)(s + p_2)} \tag{9-91}$$

where p_1 and p_2 are the two complex-conjugate poles that are to be cancelled. Let the transfer function of the series controller be

$$G_c(s) = \frac{(s + p_1 + \epsilon_1)(s + p_2 + \epsilon_2)}{s^2 + 2\zeta\omega_n s + \omega_n^2} \tag{9-92}$$

where ϵ_1 is a complex number whose magnitude is very small, and ϵ_2 is its complex conjugate. The open-loop transfer function of the compensated system is

$$G(s) = G_c(s)G_p(s) = \frac{K(s + p_1 + \epsilon_1)(s + p_2 + \epsilon_2)}{s(s + p_1)(s + p_2)(s^2 + 2\zeta\omega_n s + \omega_n^2)} \tag{9-93}$$

Because of the inexact cancellation, we cannot discard the terms $(s + p_1)(s + p_2)$ in the denominator of Eq. (9-93). The closed-loop transfer function is

$$\frac{C(s)}{R(s)} = \frac{K(s + p_1 + \epsilon_1)(s + p_2 + \epsilon_2)}{s(s + p_1)(s + p_2)(s^2 + 2\zeta\omega_n s + \omega_n^2) + K(s + p_1 + \epsilon_1)(s + p_2 + \epsilon_2)}$$

(9-94)

The root-locus diagram in Fig. 9-33 explains the effect of inexact pole-zero cancellation. Notice that the two closed-loop poles as a result of inexact cancellation lie between the pairs of poles and zeros at $s = -p_1, -p_2$, and $-p_1 - \epsilon_1, -p_2 - \epsilon_2$, respectively. Thus, these closed-loop poles are very close to the open-loop poles and zeros that are meant to be cancelled. Equation (9-94) can be approximated as

$$\frac{C(s)}{R(s)} = \frac{K(s + p_1 + \epsilon_1)(s + p_2 + \epsilon_2)}{(s + p_1 + \delta_1)(s + p_2 + \delta_2)(s^3 + 2\zeta\omega_n s^2 + \omega_n^2 s + K)}$$

(9-95)

where δ_1 and δ_2 are a pair of very small complex-conjugate numbers that depend on ϵ_1, ϵ_2, and all the other parameters. The partial-fraction expansion of Eq. (9-95) is

$$\frac{C(s)}{R(s)} = \frac{K_1}{s + p_1 + \delta_1} + \frac{K_2}{s + p_2 + \delta_2} + \text{terms due to the remaining poles}$$

(9-96)

We can show that K_1 is proportional to $\epsilon_1 - \delta_1$, which is a very small number. Similarly, K_2 is also very small. This exercise simply shows that although the poles at $-p_1$ and $-p_2$ cannot be cancelled precisely, the resulting transient-response terms due to the inexact cancellation will have insignificant amplitudes, so that unless the controller zeros earmarked for cancellation are too far off target, the effect can be neglected for all

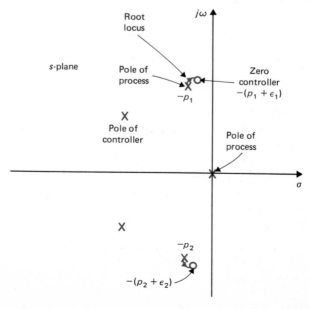

FIGURE 9-33 Pole-zero configuration and root loci of inexact pole-zero cancellation.

practical purposes. Another way of viewing this problem is that the zeros of $G(s)$ are retained as the zeros of the closed-loop transfer function $C(s)/R(s)$, so that from Eq. (9-95), we see that the two pairs of poles and zeros are close enough to be cancelled from the transient-response standpoint.

unstable poles

We must always keep in mind that we should never attempt to cancel poles in the *right-half s-plane, since any inexact cancellation will result in an unstable closed-loop system.* Inexact cancellation of poles could cause difficulties if the unwanted poles of the open-loop transfer function are very close to or right on the imaginary axis of the *s*-plane. In this case, inexact cancellation may also result in an unstable system. Figure 9-34(a) illustrates a situation in which the relative positions of the poles and zeros intended for cancellation result in a stable system, whereas in Fig. 9-34(b), the inexact cancellation is unacceptable. Although the relative distance between the poles and zeros intended for cancellation is small, which results in terms in the time response that have very small amplitudes, these responses will grow without bound as time increases.

bridged-T networks

Transfer functions with complex zeros can be realized by various types of electric networks. The bridged-T networks shown in Fig. 9-35 have the advantage of containing only RC elements. In the following discussions, the network shown in Fig. 9-35(a) is referred to as the type-1 bridged-T, and that of Fig. 9-35(b) is referred to as type-2 bridged-T.

With the assumption of zero-input source impedance and infinite-output impedance, the transfer function of the type-1 bridged-T network is

transfer functions

$$\frac{E_2(s)}{E_1(s)} = \frac{1 + 2R_1 Cs + R^2 C_1 C_2 s^2}{1 + R(C_1 + 2C_2)s + R^2 C_1 C_2 s^2} \qquad (9\text{-}97)$$

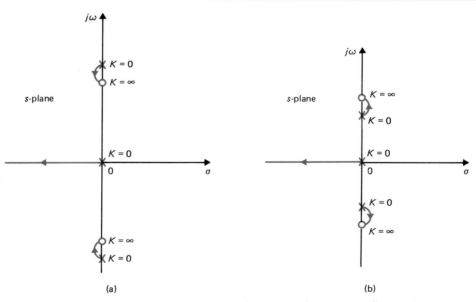

(a)   (b)

FIGURE 9-34 Root loci showing the effects of inexact pole-zero cancellations.

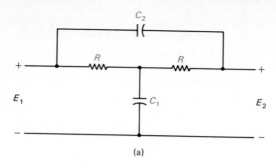

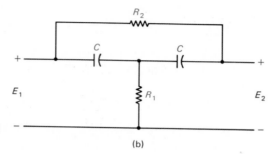

FIGURE 9-35 Two basic types of bridged-T network. (a) Type-1 network. (b) Type-2 network.

and that of the type-2 bridged-T network is

$$\frac{E_2(s)}{E_1(s)} = \frac{1 + 2R_1 Cs + C^2 R_1 R_2 s^2}{1 + C(R_2 + 2R_1)s + C^2 R_1 R_2 s^2} \qquad (9\text{-}98)$$

When these two equations are compared, it is apparent that the two networks have similar transfer-function characteristics. In fact, if R, C_1, and C_2 in Eq. (9-97) are replaced by C, R_2, and R_1, respectively, Eq. (9-97) becomes the transfer function of the type-2 network given in Eq. (9-98).

It is useful to study the properties of the poles and zeros of the transfer functions of the bridge-T networks of Eqs. (9-97) and (9-98).

Equation (9-97) is written as

$$\frac{E_2(s)}{E_1(s)} = \frac{s^2 + \dfrac{2}{RC_1}s + \dfrac{1}{R^2 C_1 C_2}}{s^2 + \dfrac{C_1 + 2C_2}{RC_1 C_2}s + \dfrac{1}{R^2 C_1 C_2}} \qquad (9\text{-}99)$$

If both the numerator and the denominator polynomials of Eq. (9-99) are written in the standard second-order prototype form,

$$s^2 + 2\zeta\omega_n s + \omega_n^2 = 0 \tag{9-100}$$

we have, for the numerator,

$$\omega_{nz} = \pm\frac{1}{R\sqrt{C_1 C_2}} \tag{9-101}$$

$$\zeta_z = \sqrt{C_2/C_1} \tag{9-102}$$

and for the denominator,

$$\omega_{np} = \pm\frac{1}{R\sqrt{C_1 C_2}} = \omega_{nz} \tag{9-103}$$

$$\zeta_p = \frac{C_1 + 2C_2}{2\sqrt{C_1 C_2}} = \frac{1 + 2\zeta_z^2}{2\zeta_z} \tag{9-104}$$

Since ζ_z can be less than or greater than one, the two zeros of Eq. (9-99) are either real or complex. We can show that ζ_p is always greater than one, so that the two poles of Eq. (9-99) are always real. Thus, as a controller, the transfer function of Eq. (9-97) can be used to cancel undesirable real or complex poles of the process transfer function, and then real poles at appropriate locations are added.

The ζ and ω_n parameters of the denominator and the numerator of the transfer function of the type-2 bridged-T network in Eq. (9-98) are obtained by replacing R, C_1, and C_2 in Eqs. (9-101) through (9-104) by C, R_2, and R_1, respectively. Again, the two zeros of Eq. (9-98) can be real or complex, but the poles are always real for all possible combinations of the network elements.

EXAMPLE 9-10

speed-control system

Complex-conjugate poles in system transfer functions are often due to compliances in the coupling between mechanical elements. For instance, if the shaft between the motor and load is nonrigid, the shaft is modeled as a torsional spring, which could lead to complex-conjugate poles in the system transfer function. Figure 9-36 shows a speed-control system in which the coupling between the motor and the load is modeled as a torsional spring. The system equations are

$$T_m(t) = J_m\frac{d\omega_m(t)}{dt} + B_m\omega_m(t) + J_L\frac{d\omega_L(t)}{dt} \tag{9-105}$$

$$K_L[\theta_m(t) - \theta_L(t)] + B_L[\omega_m(t) - \omega_L(t)] = J_L\frac{d\omega_L(t)}{dt} \tag{9-106}$$

$$T_m(t) = K\omega_e(t) \tag{9-107}$$

$$\omega_e(t) = \omega_r(t) - \omega_L(t) \tag{9-108}$$

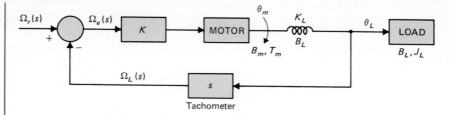

FIGURE 9-36 Block diagram of speed-control system in Example 9-10.

$T_m(t)$ = motor torque
$\omega_m(t)$ = motor angular velocity
$\omega_L(t)$ = load angular velocity
$\theta_L(t)$ = load angular displacement
$\theta_m(t)$ = motor angular displacement
J_m = motor inertia = 0.0001 oz-in.-s^2
J_L = load inertia = 0.0005 oz-in.-s^2
B_m = viscous-friction coefficient of motor = 0.01 oz-in.-s
B_L = viscous-friction coefficient of shaft = 0.001 oz-in.-s
K_L = spring constant of shaft = 100 oz-in./rad
K = amplifier gain = 1

The loop transfer function of the system is

$$G_p(s) = \frac{\Omega_L(s)}{\Omega_e(s)}$$

$$= \frac{B_L s + K_L}{J_m J_L s^3 + (B_m J_L + B_L J_m + B_L J_L)s^2 + (K_L J_L + B_m B_L + K_L J_m)s + B_m K_L} \tag{9-109}$$

By substituting the system parameters in the last equation, $G_p(s)$ becomes

process transfer function

$$G_p(s) = \frac{20,000(s + 100,000)}{s^3 + 112s^2 + 1,200,200s + 20,000,000} \tag{9-110}$$

$$= \frac{20,000(s + 100,000)}{(s + 16.69)(s + 47.66 + j1094)(s + 47.66 - j1094)} \tag{9-111}$$

Thus, the shaft compliance between the motor and the load creates two complex-conjugate poles in $G_p(s)$ that are lightly damped. The resonant frequency is approximately 1095 rad/s, and the closed-loop system is unstable. The complex-conjugate poles of $G_p(s)$ will cause the speed response to oscillate even if the system were stable.

pole-zero-cancellation design

To compensate the system, we need to get rid of the complex poles of $G_p(s)$ at $s = -47.66 + j1094$ and $-47.66 - j1094$. Let us select the type-1 bridged-T network as a series controller to improve the system. The complex-conjugate zeros of the controller should be so placed that they will cancel the undesirable poles of the controlled process. Therefore, the transfer function of the bridged-T controller should be

$$G_{c1}(s) = \frac{s^2 + 95.3s + 1,198,606.6}{s^2 + 2\zeta_p \omega_{np} s + \omega_{np}^2} \tag{9-112}$$

From the last equation, we get $\zeta_z = 0.0435$ and $\omega_{nz} = 1094.8$. Using Eqs. (9-103) and (9-104), we get

$$\zeta_p = 11.54 \quad \text{and} \quad \omega_{np} = \omega_{nz} = 1094.8 \text{ rad/s}$$

Thus,

bridged-T controller

$$G_{c1}(s) = \frac{s^2 + 95.3s + 1,198,606.6}{(s + 47.53)(s + 25,220)} \tag{9-113}$$

The loop transfer function of the system with the type-1 bridged-T controller is

$$G_{c1}(s)G_p(s) = \frac{20,000(s + 100,000)}{(s + 16.69)(s + 47.53)(s + 25,220)} \tag{9-114}$$

We can show that the closed-loop system is now stable, but the damping is still very poor. We cannot expect the bridged-T controller to cancel the unwanted complex poles of $G_p(s)$ and simultaneously place the two real poles at appropriate locations. Now we can regard the loop transfer function in Eq. (9-114) as a new design problem. There are a number of possible solutions to the problem of improving the damping of the system represented by Eq. (9-114). We can introduce a phase-lag controller or a PI controller, among other possibilities. Although the system is type-0, without a pole at the origin, we can still use the phase-lag controller design method described earlier to arrive at the controller parameters. One design yields the following transfer function:

phase-lag controller

$$G_{c2}(s) = \frac{1 + 0.0675s}{1 + 10s} \tag{9-115}$$

Thus, the loop transfer function of the compensated system is

$$G(s) = G_{c1}(s)G_{c2}(s)G_p(s) = \frac{135(s + 14.8)(s + 100,000)}{(s + 16.69)(s + 47.53)(s + 25,220)(s + 0.1)} \tag{9-116}$$

Since the system is of type-0, and the step-error constant is 100, the steady-state speed of the system will have an error of 1 percent when a step speed command is applied. We can apply a PI controller to improve the system to type-1. A PI controller is chosen as

PI controller

$$G_{c2}(s) = 0.045 + \frac{0.36}{s} = \frac{0.045(s + 8)}{s} \tag{9-117}$$

The gain and the zero of the PI controller was chosen by reasoning and trial and error. The zero at $s = -8$ was chosen to lie to the right of all the poles of the process transfer function, and the gain was adjusted to get a satisfactory maximum overshoot in the step-speed response. The zero should not be placed too close to the origin, or the settling time will be too long.

In summary, the combined type-1 bridged-T and phase-lag controller transfer function is

bridged-T phase-lag controller

$$G_c(s) = \frac{0.00675(s + 14.8)(s^2 + 95.3s + 1,198,606.6)}{(s + 0.1)(s + 47.53)(s + 25,220)} \tag{9-118}$$

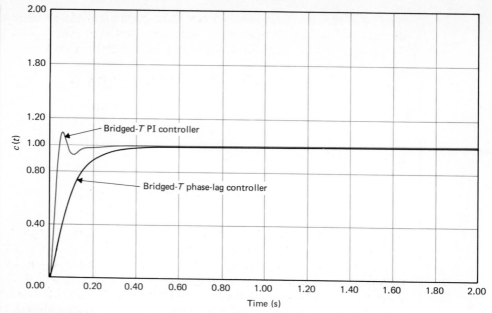

FIGURE 9-37 Unit-step responses of speed-control system in Example 9-10, with a bridged-T phase-lag controller and with a bridged-T PI controller.

and the combined type-1 bridged-T and PI controller transfer function is

bridged-T
PI controller

$$G_c(s) = \frac{0.045(s + 8)(s^2 + 95.3s + 1,198,606.6)}{s(s + 47.53)(s + 25,220)}$$

(9-119)

The unit-step responses of the output speed of the system with the two controllers are shown in Fig. 9-37. As expected, the response of the system with the bridged-T phase-lag controller has a longer rise time, and the steady-state error is approximately 1 percent. The response of the system with the bridged-T PI controller has a much faster rise time, but the maximum overshoot is approximately 8 percent. Since the system is type-1, the steady-state error to a step input is zero. If necessary, the maximum overshoot can be reduced by fine tuning the gain and the location of the zero of the PI controller.

9-7

FORWARD AND FEEDFORWARD COMPENSATION

The compensation schemes discussed in the preceding sections all have one degree of freedom in that there is essentially one controller in the system, although the controller can contain several types of compensators connected in series or in parallel. The limitations of a one-degree-of-freedom controller were discussed in Section 9-1. The two-degrees-of-freedom compensation schemes shown in Figs. 9-2(a) and (f) offer flexibility when a multiple number of design criteria have to be satisfied simultaneously.

From Fig. 9-2(e), the closed-loop transfer function is

$$\frac{C(s)}{R(s)} = \frac{G_{cf}(s)G_c(s)G_p(s)}{1 + G_c(s)G_p(s)} \qquad (9\text{-}120)$$

and the error transfer function is

$$\frac{E(s)}{R(s)} = \frac{1}{1 + G_c(s)G_p(s)} \qquad (9\text{-}121)$$

Thus, the controller $G_c(s)$ can be designed so that the error transfer function will have certain desirable characteristics, and the controller $G_{cf}(s)$ can be selected to satisfy performance requirements with reference to the input–output relationship. Another way of describing the flexibility of a two-degrees-of-freedom design is that the controller $G_c(s)$ is usually designed to provide a certain degree of system stability and performance, but since the zeros of $G_c(s)$ always become the zeros of the closed-loop transfer function, unless some of the zeros are cancelled by the poles of the process transfer function $G_p(s)$, these zeros may cause a large overshoot in the system output even when the relative damping as determined by the characteristic equation is satisfactory. In this case and for other reasons, the transfer function $G_{cf}(s)$ may be used for the control or cancellation of the undesirable zeros of the closed-loop transfer function, while keeping the characteristic equation intact. Of course, we can also introduce zeros in $G_{cf}(s)$ to cancel some of the undesirable poles of the closed-loop transfer function that are the result of the compensation by $G_c(s)$. The feedforward compensation scheme shown in Fig. 9-2(f) serves the same purpose as the forward compensation, and the difference between the two configurations depends on system and hardware implementation considerations.

two-degrees-of-freedom design

It should be kept in mind that while the forward and feedforward compensation may sound powerful, in that they can be used directly for the addition or deletion of poles or zeros of the closed-loop transfer function, there is a fundamental question involving the basic characteristics of feedback. If the forward or feedforward controller is so powerful, then why do we need feedback at all? Since $G_{cf}(s)$ in the system of Figs. 9-2(e) and 9-2(f) are outside the feedback loop, the system is susceptible to parameter variations in $G_{cf}(s)$. Therefore, in reality, these types of compensation cannot be satisfactorily applied to all situations.

external disturbance

External disturbance and parameter variations are often unavoidable in control systems. The design problems in control systems usually involve the design of controllers that will make the system sensitive to input signals while insensitive to external disturbance and parameter variations. This often calls for the use of a two-dimensional controller configuration, such as that shown in Figs. 9-38(a) and 9-38(b). In Fig. 9-38(a), the feedforward controller is applied to the disturbance input $d(t)$. In this case, the controller $G_c(s)$ is designed so that the system achieves a certain relative stability, since $G_c(s)$ appears only in the denominator of the input–output transfer function, and

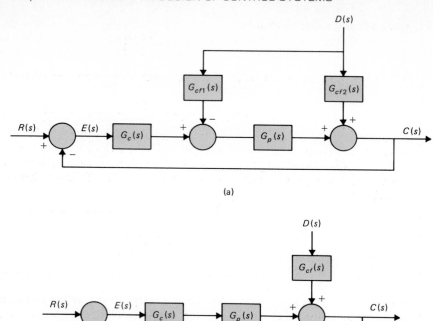

(a)

(b)

FIGURE 9-38 (a) Control system with disturbance input and feedforward controller. (b) Control system with disturbance input and forward controller.

the feedforward controllers $G_{cf1}(s)$ and $G_{cf2}(s)$ are so designed that ideally the output will not be affected by the disturbance $d(t)$ at all. Figure 9-38(b) shows a forward controller $G_{cf}(s)$ that is applied directly to the disturbance signal. In this case, it would be impossible to eliminate the effects of the disturbance completely, and the best that can be hoped for is to design the controller $G_{cf}(s)$ so that the disturbance is surpressed in the important frequency range.

disturbance elimination

EXAMPLE 9-11   As an illustration of the design of the forward and feedforward compensation, consider the system studied in Example 9-8. The unit-step responses of the system with the phase-lag controller of Eq. (9-61) are shown in Fig. 9-29. The compensated system has a relatively high overshoot when $a = 0.01635$ and $T = 100$. Increasing the value of T will decrease the maximum overshoot but at the expense of longer delay and settling times, with the latter being a more serious problem.

The closed-loop transfer function of the phase-lag compensated system with $a = 0.01635$ and $T = 100$ is

$$\frac{C(s)}{R(s)} = \frac{81.5(s + 0.6135)}{s^4 + 15.01s^3 + 50.15s^2 + 82s + 50} \tag{9-122}$$

We can show that the zero of the transfer function at $s = -0.6135$ contributes to the overshoot of the step response. We can cancel this zero by using the forward-compensation scheme of Fig. 9-2(e) or the feedforward-compensation scheme of Fig. 9-2(f). For the forward-compensation scheme shown in Fig. 9-2(e), let

$$G_{cf}(s) = \frac{0.6135}{s + 0.6135} \tag{9-123}$$

The closed-loop transfer function of the system with forward compensation is

$$\frac{C(s)}{R(s)} = \frac{50}{s^4 + 15.01s^3 + 50.15s^2 + 82s + 50} \tag{9-124}$$

Figure 9-39 shows the unit-step responses of the system with the transfer function in Eq. (9-124). Clearly, the overshoot is totally eliminated, but the rise time is increased substantially. We can further improve the step response by moving the zero of the transfer function in Eq. (9-122). Let

$$G_{cf}(s) = \frac{0.409(s + 1.5)}{s + 0.6135} \tag{9-125}$$

It is important that the gain of $G_{cf}(s)$ at $s = 0$ is unity, so that the ramp-error constant K_v of the system is maintained. Thus, the closed-loop transfer function is

$$\frac{C(s)}{R(s)} = \frac{33.333(s + 1.5)}{s^4 + 15.01s^3 + 50.15s^2 + 82s + 50} \tag{9-126}$$

Figure 9-39 shows that the unit-step response is substantially improved.

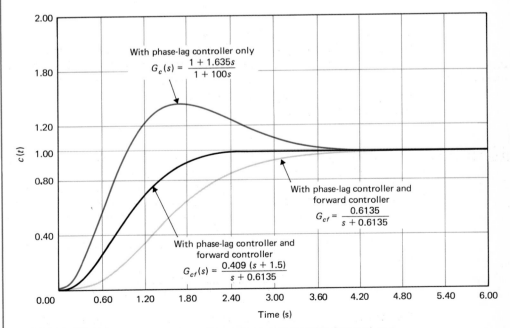

FIGURE 9-39 Unit-step responses of a control system with forward controllers.

If instead, the feedforward configuration of Fig. 9-2(f) is chosen, the transfer function of $G_{c1}(s)$ is directly related to $G_{cf}(s)$; that is, equating the closed-loop transfer functions of the two systems, we have

$$\frac{[G_{c1}(s) + G_c(s)]G_p(s)}{1 + G_c(s)G_p(s)} = G_{cf}(s)\frac{G_c(s)G_p(s)}{1 + G_c(s)G_p(s)} \tag{9-127}$$

Solving for $G_{c1}(s)$ from the last equation yields

$$G_{c1}(s) = [G_{cf}(s) - 1]G_c(s) \tag{9-128}$$

Thus, with $G_{cf}(s)$ given in Eq. (9-125) and $G_c(s)$ given in Eq. (9-73), we have

$$G_{c1}(s) = -\frac{0.00966s}{s + 0.01} \tag{9-129}$$

System with Noise and Disturbance

feedforward control

When a control system has external noise or disturbance, the output of the system should be insensitive to these undesirable signals. Figures 9-38(a) and (b) show how the feedforward control scheme can be used to minimize the noise effects. For the system configuration in Fig. 9-38(a), the transfer function between the noise signal $d(t)$ and the output $c(t)$ is

noise rejection

$$\frac{C(s)}{D(s)} = \frac{G_{cf2}(s) - G_{cf1}(s)G_p(s)}{1 + G_c(s)G_p(s)} \tag{9-130}$$

Thus, if the feedforward controller $G_{cf1}(s)$ has the transfer function

$$G_{cf1}(s) = \frac{G_{cf2}(s)}{G_p(s)} \tag{9-131}$$

the output will be completely unaffected by $d(t)$. In practice, the process transfer function $G_p(s)$ usually has more poles than zeros, so that $G_{cf}(s)$ according to Eq. (9-131) would not be physically realizable if $G_{cf2}(s) = 1$.

EXAMPLE 9-12

Consider the speed-control system described in Example 9-10. The process transfer function is

$$G_p(s) = \frac{\Omega_L(s)}{\Omega_e(s)} = \frac{20,000(s + 100,000)}{(s + 16.69)(s^2 + 95.3s + 1,198,606.6)} \tag{9-132}$$

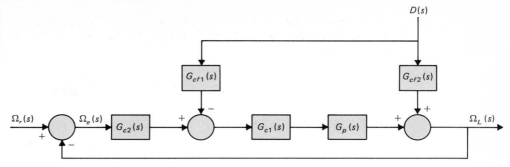

FIGURE 9-40 Speed-control system with disturbance and feedforward controller.

Let us assume that a disturbance signal $d(t)$ appears at the output so that it is additive to the output-speed response $\omega_L(t)$. We have already designed the two-stage bridged-T phase-lag controller or the bridge-T PI controller so that the output-speed response to a step input is satisfactory. Now we wish to minimize the effect of $d(t)$ on $\omega_L(t)$. Let us select the feedforward control scheme shown in Fig. 9-38(a). The block diagram of the system is shown in Fig. 9-40. The transfer functions of the controllers of the bridged-T phase-lag compensation, $G_{c1}(s)$ and $G_{c2}(s)$, are given in Eqs. (9-113) and (9-115), respectively. Without the feedforward control, the transfer function between $d(t)$ and $\omega_L(t)$ is

$$\frac{\Omega_L(s)}{D(s)} = \frac{1}{1 + G_{c1}(s)G_{c2}(s)G_p(s)}$$

$$= \frac{(s + 16.69)(s + 47.53)(s + 25,220)(s + 0.1)}{s^4 + 25,284.3s^3 + 1,623,085s^2 + 33,670,453.3s + 201,800,641}$$

$$(9\text{-}133)$$

Since when $s = 0$, the gain of the last transfer function is 0.01, the output-speed response to a unit-step disturbance $d(t)$ will have a steady-state value of 0.01, and the entire response is shown in Fig. 9-41.

When $G_{c2}(s)$ is the PI controller described in Eq. (9-117), the transfer function between $d(t)$ and $\omega_L(t)$ is

$$\frac{\Omega_L(s)}{D(s)} = \frac{s(s + 16.69)(s + 47.53)(s + 25,220)}{s^4 + 25,284.2s^3 + 1,621,321.7s^2 + 110,013,613s + 720,000,000} \qquad (9\text{-}134)$$

Notice that the last transfer function has a zero at $s = 0$, which will cause the step response of $\omega_L(t)$ due to $d(t)$ go to zero as time approaches infinity. This is due to the pure integration in the controller $G_{c2}(s)$ that makes the system type 2. Figure 9-41 shows the response of $\omega_L(t)$ when $d(t)$ is a unit-step function. Although the response initially decays rapidly, and the final steady-state value is zero, the overshoot is quite large.

feedforward
control

When the feedforward control is applied to $d(t)$, theoretically, the output $\omega_L(t)$ is independent of $d(t)$ when

$$G_{cf}(s) = \frac{G_{cf2}(s)}{G_{c1}(s)G_p(s)} \qquad (9\text{-}135)$$

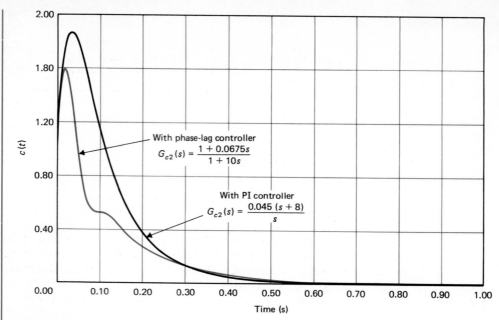

FIGURE 9-41 Output responses of a speed-control system with unit-step disturbance input.

where $G_{c1}(s)G_p(s)$ is as given in Eq. (9-114). We can simply select $G_{cf2}(s)$ to be equal to $G_{c1}(s)G_p(s)$ and $G_{cf2}(s) = 1$. Then, the output due to $d(t)$ will be zero at all times.

9-8
DESIGN OF ROBUST CONTROL SYSTEMS

robust control

In many control-system applications, the system designed must not only satisfy the damping and accuracy specifications, but the control must also yield performance that is **robust** (insensitive) to external disturbance and parameter variations. We have shown that feedback in conventional control systems has the inherent ability of reducing the effects of external disturbance and parameter variations. Unfortunately, robustness with the conventional feedback configuration is achieved only with a high loop gain, which normally is detrimental to stability. Let us consider the control system shown in Fig. 9-42. The external disturbance is denoted by the signal $d(t)$, and we assume that the amplifier gain K is subject to variation during operation. The input–output transfer function of the system is $[d(t) = 0]$

$$M(s) = \frac{C(s)}{R(s)} = \frac{KG_{cf}(s)G_c(s)G_p(s)}{1 + KG_c(s)G_p(s)} \qquad (9\text{-}136)$$

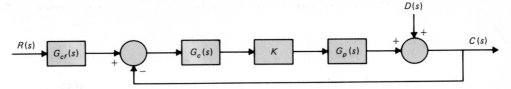

FIGURE 9-42 Control system with disturbance.

and the disturbance-output transfer function is $[r(t) = 0]$

$$T(s) = \frac{C(s)}{D(s)} = \frac{1}{1 + KG_c(s)G_p(s)} \tag{9-137}$$

In general, the design strategy is to select the controller $G_c(s)$ so that the output $c(t)$ is insensitive to the disturbance over the frequency range in which the latter is dominant, and the feedforward controller $G_{cf}(s)$ is designed to achieve the desired transfer relation between the input $r(t)$ and the output $c(t)$.

Let us define the sensitivity of $M(s)$ due to the variation of K as

$$S_K^M = \frac{\text{percent change in } M(s)}{\text{percent change in } K} = \frac{dM(s)/M(s)}{dK/K} \tag{9-138}$$

Then, for the system in Fig. 9-42,

$$S_K^M = \frac{1}{1 + KG_c(s)G_p(s)} \tag{9-139}$$

sensitivity function

which is identical to Eq. (9-137). Thus, the sensitivity function and the disturbance-output transfer function are identical, which means that disturbance suppression and robustness with respect to variations of K can be designed with the same control schemes.

feedforward control

We have shown in the last section how the feedforward control can be applied to reduce the effect of external disturbance. In this section, we shall show how the two-degrees-of-freedom control system of Fig. 9-42 can be used to achieve a high-gain system that will satisfy both the performance and robustness requirements, as well as for noise rejection.

robust system

EXAMPLE 9-13   Let us consider the third-order system in Example 9-8, which is compensated by the phase-lag controller. The transfer function of the controlled process is given in Eq. (9-71). Earlier, the value of $K = 100$ was chosen according to the steady-state-error requirement. Apparently, the phase-lag controller of Eq. (9-73) can only undermine the robustness of the system, since it adds an attenuation of $a(< 1)$ to the open-loop system starting at $\omega = 1/T$. With refer-

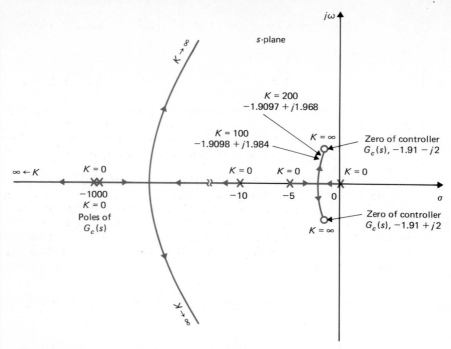

FIGURE 9-43 Root loci of the robust system described by Eq. (9-143).

ence to the root loci of the uncompensated system, Fig. 9-12, we see that in order to realize a relative damping ratio of 0.707, accompanied by a very high value of K, we can place two open-loop zeros in the neighborhood of the desired closed-loop poles, as shown in Fig. 9-43. In order that the controller $G_c(s)$ be physically realizable, two controller poles must be added far enough to the left so that their effects on the system performance are negligible. Thus, the transfer function of the series controller is selected as

robust controller

$$
\begin{aligned}
G_c(s) &= \frac{130{,}753(s + 1.91 + j2)(s + 1.91 - j2)}{(s + 1000)^2} \\
&= \frac{130{,}753(s^2 + 3.82s + 7.648)}{(s + 1000)^2}
\end{aligned}
\tag{9-140}
$$

The compensated open-loop transfer function is

$$
G(s) = G_c(s)G_p(s) = \frac{6{,}537{,}650K(s^2 + 3.82s + 7.648)}{s(s + 5)(s + 10)(s + 1000)^2}
\tag{9-141}
$$

By placing the open-loop zeros very near the desired location of the closed-loop poles, the value of K can be increased to improve robustness. As shown in Fig. 9-43, when $K = 100$, the closed-loop characteristic equation roots are at

$$-1634.6 \qquad -1.9097 \pm j1.968 \qquad -188.3 \pm j609.4$$

characteristic-equation roots

When $K = 200$, the roots are at

$$-1844.5 \qquad -1.9098 \pm j1.984 \qquad -83.3 \pm j841.4$$

Thus, the dominant roots hardly move when the value of K is increased from 100 to 200. We must point out that although the two complex roots created by the two far-away poles of the controller have negligible effect on the transient response, they do move into the right-half plane if the value of K is too large. For the present case, we can show that the compensated system becomes unstable when K is greater than 315.

Since the zeros of the open-loop transfer function are identical to the zeros of the closed-loop transfer function, we cannot consider the design complete by using only the series controller $G_c(s)$, since the closed-loop zeros will essentially cancel the dominant closed-loop poles, and the system response will be governed by the two large complex roots. This means that we must add the forward controller, as shown in Fig. 9-44, where $G_{cf}(s)$ should contain poles to cancel the zeros of $s^2 + 3.82s + 7.648$ of the closed-loop transfer function. The closed-loop transfer function becomes

$$\frac{C(s)}{R(s)} = \frac{5 \times 10^7 K}{s^5 + 2015s^4 + 1{,}030{,}050s^3 + 151 \times 10^5 s^2 + 5 \times 10^7 s + (6{,}537{,}650 s^2 + 25 \times 10^6 s + 5 \times 10^7)K}$$

(9-142)

Figure 9-45 shows that the unit-step responses of the compensated system with $K = 100$ and 200 are almost identical. However, the rise time of the responses is very slow, especially in comparison with the phase-lag and lead-lag compensated responses in Fig. 9-31. To improve on the rise time, we can add a zero to the transfer function of $G_{cf}(s)$, so that

forward controller

$$G_{cf}(s) = \frac{7.648(s + 1)}{s^2 + 3.82s + 7.648}$$

(9-143)

The unit-step responses of the compensated system with $K = 100$ and 200 are now shown in Fig. 9-45 with much improved rise time. Better responses can be obtained by fine tuning the value of the zero of $G_{cf}(s)$.

Since the system shown in Fig. 9-44 is now more robust, it is expected that the disturbance effect would be reduced. This requires more discussion. If we compute the output response when the disturbance $d(t)$ is a unit-step function, we would observe that the maximum overshoot

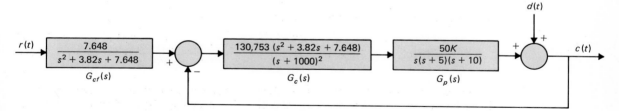

FIGURE 9-44 Robust control system with forward compensation.

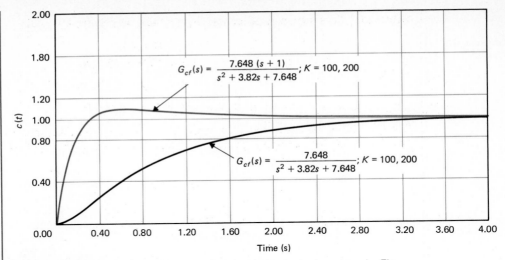

FIGURE 9-45 Unit-step responses of the robust control system in Fig. 9-44.

is approximately 400 percent. The fact is that controller $G_c(s)$ pushes the resonant frequency of the system to $d(t)$ to a very high frequency, so that when a step function is applied as $d(t)$, the response will exhibit a high overshoot, but quickly damps out. The improvement on the disturbance suppression with the system configuration of Fig. 9-42 is made on the low-frequency noise, which is difficult to show in the time domain with a step input. Since the sensitivity function in Eq. (9-139) is a function of s, in the sinusoidal steady state, it is a function of frequency ω. In Chapter 11, we shall show that the compensated system will have greatly improved responses to disturbance inputs in the relatively low-frequency range over which the latter signals occur. However, if the disturbance signal consists of step functions, then the compensated system in this example would not be effective from the noise-suppression standpoint.

EXAMPLE 9-14

dc-motor control

variable-load inertia

In this example, we consider the design of a dc-motor-control system that has a variable load inertia. This type of situation is quite common in control systems. For example, the load inertia seen by the motor in an electronic character printer will change when different printwheels are used. The process transfer function of the unity-feedback control system is

$$G_p(s) = \frac{KK_i}{s[(Js + B)(Ls + R) + K_iK_b]} \qquad (9\text{-}144)$$

Let the system parameters be defined as

K_i = motor-torque constant = 1 N-m/A
K_b = motor back-emf constant = 1 V/rad/s
R = motor resistance = 1 Ω
L = motor inductance = 0.01 H
B = motor and load viscous-friction coefficient $\cong$ 0
J = motor and load inertia, varies between 0.01 and 0.02 N-m/rad/s^2
K = amplifier gain

Substituting the system parameters into Eq. (9-144), we get, for $J = 0.01$,

$$G_p(s) = \frac{10{,}000K}{s(s^2 + 100s + 10{,}000)} \tag{9-145}$$

and for $J = 0.02$,

$$G_p(s) = \frac{5000K}{s(s^2 + 100s + 5000)} \tag{9-146}$$

The root loci of the characteristic equation for $J = 0.01$ and 0.02 are shown in Fig. 9-46. Figure 9-47 shows the unit-step responses of the closed-loop system with the two extreme values of

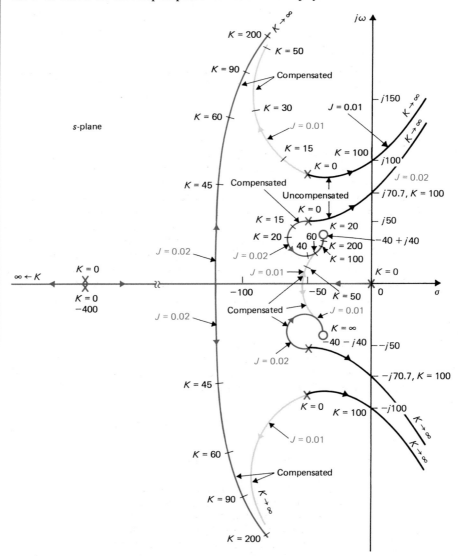

FIGURE 9-46 Root loci of compensated and uncompensated systems in Example 9-14.

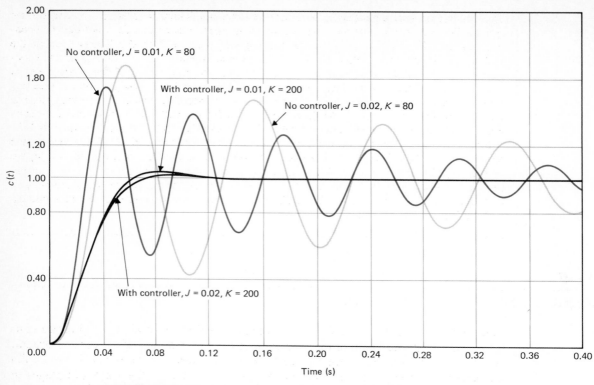

FIGURE 9-47 Unit-step responses of compensated and uncompensated systems in Exampale 9-14.

J and with $K = 80$. When $J = 0.01$, the natural frequency of the system is higher, since in general the natural frequency is proportional to the square root of J, but the maximum overshoot is smaller. When $J = 0.02$, the response oscillates with a lower frequency, but the maximum overshoot is larger. Clearly, the responses are quite different when J takes on the two extreme values. We can show that independent of the values of J, the uncompensated system is unstable for $K \geq 100$.

robust controller

    To achieve robust control, we can introduce a series controller with the transfer function

$$G_c(s) = \frac{50(s^2 + 80s + 3200)}{(s + 400)^2} \tag{9-147}$$

where the idea is to add two zeros at $-40 + j40$ and $-40 - j40$ so that the dominant poles of the closed-loop transfer function can be brought to the neighborhood of these zeros when the value of K is large. The two poles of $G_c(s)$ at -400 are for the purpose of physical realization.

    Figure 9-46 shows the root loci of the compensated system with $J = 0.01$ and $J = 0.02$. When $K = 200$, the characteristic equation roots are at:

characteristic-equation roots

$J = 0.01$:	-764.6	$-25.1 \pm j377.2$	$-42.6 \pm j34.53$
$J = 0.02$:	-676.0	$-72.0 \pm j271.8$	$-40.0 \pm j37.37$

The last two complex-conjugate roots in each case are the dominant roots of the compensated system. When $J = 0.01$, the root locus of the compensated system that departs from the open-loop pole at $-50 + j86.6$ bends upward, reaching the $j\omega$-axis at $K = 280$. The two loci on the real axis between 0 and -400 eventually break away at around -53, and then reach the open-loop zeros at $-40 \pm j40$. When $J = 0.02$, the root loci that leave the open-loop poles at $-50 \pm j50$ go directly to the open-loop zeros at $-40 \pm j40$ when K is large. Thus, we see that when K is very large, the two dominant roots of the compensated system will be very close to the open-loop zeros at $-40 \pm j40$. In fact, we can arrive at a still more robust system by using a larger value of K than 200. However, when K is greater than 280, the system with $J = 0.01$ will become unstable, so the value of K must be kept lower than this critical value. To cancel the zeros of $s^2 + 80s + 3200$ of the closed-loop transfer function of the compensated system, we must again add the forward controller $G_{cf}(s)$, as in Fig. 9-42, with a transfer function that has the same poles. Figure 9-47 illustrates the unit-step responses of the compensated system with $K = 200$ and $J = 0.01$ and 0.02. The difference between these two responses with the two extreme values of J is quite small.

9-9

MINOR-LOOP FEEDBACK CONTROL

The control schemes discussed in the preceding sections have all utilized series controllers in the forward path of the main loop or feedforward path of the control system. Although series controllers are the most common because of their simplicity in implementation, depending on the nature of the system, sometimes there are advantages in placing the controller in a minor feedback loop, as shown in Fig. 9-2(b). For example,

tachometer

a tachometer may be coupled directly to a dc motor not only for the purpose of speed indication, but more often for improving the stability of the closed-loop system by feeding back the output signal of the tachometer. The motor speed can also be gener-

back emf

ated by processing the back emf of the motor electronically, as illustrated in Chapter 4. In principle, the PID controller or the phase-lead and phase-lag controllers discussed earlier can all, with varying degree of effectiveness, be applied as minor-loop feedback controllers. We shall show later that under certain conditions, minor-loop compensation can yield systems that are more robust, that is, less sensitive to external disturbance effects and internal parameter variations.

Rate-Feedback or Tachometer-Feedback Control

The principle of using the derivative of the actuating signal to improve the damping of a closed-loop system can be applied to the output signal to achieve a similar effect. In other words, the derivative of the output signal is fed back and added algebraically to the actuating signal of the system. In practice, if the output variable is mechanical displacement, a tachometer may be used that converts the mechanical displacement into an electrical signal that is proportional to the derivative of the displacement. Figure 9-48 shows the block diagram of a control system with a secondary path that feeds back the derivative of the output. The transfer function of the tachometer is denoted by $K_t s$, where K_t is the tachometer constant, usually expressed in volts/radian per second for analytical purposes. Commercially, K_t is given in the data sheet of the tachometer typi-

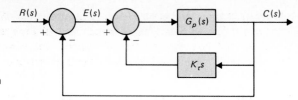

FIGURE 9-48 Control system with tachometer feedback.

cally in volts/1000 rpm. The effects of rate or tachometer feedback can be illustrated simply by means of an example. Consider that the controlled process of the system shown in Fig. 9-48 has the transfer function

$$G_p(s) = \frac{\omega_n^2}{s(s + 2\zeta\omega_n)} \tag{9-148}$$

The closed-loop transfer function of the system is

$$\frac{C(s)}{R(s)} = \frac{\omega_n^2}{s^2 + (2\zeta\omega_n + K_t\omega_n^2)s + \omega_n^2} \tag{9-149}$$

and the characteristic equation is

$$s^2 + (2\zeta\omega_n + K_t\omega_n^2)s + \omega_n^2 = 0 \tag{9-150}$$

effect of tachometer feedback

From the characteristic equation, it is apparent that the effect of the tachometer feedback is the increase of the damping of the closed-loop system, since K_t appears in the same term as the damping ratio ζ. In this respect, the rate-feedback control has exactly the same effect as the PD control. However, the closed-loop transfer function of the closed-loop system with PD control in Fig. 9-3 is

$$\frac{C(s)}{R(s)} = \frac{\omega_n^2(K_P + K_D s)}{s^2 + (2\zeta\omega_n + K_D\omega_n^2)s + \omega_n^2 K_P} \tag{9-151}$$

Comparing the two transfer functions in Eqs. (9-149) and (9-151), we see that the two characteristic equations are identical if $K_P = 1$ and $K_D = K_t$. However, Eq. (9-151) has a zero at $s = -K_P/K_D$, whereas Eq. (9-149) does not. Thus, the response of the system with tachometer feedback is uniquely defined by the characteristic equation, whereas the response of the system with the PD control also depends on the zero at $s = -K_P/K_D$, which could have a significant effect on the overshoot of the step response.

     With reference to the steady-state analysis, the open-loop transfer function of the system with tachometer feedback is

$$\frac{C(s)}{E(s)} = G(s) = \frac{\omega_n^2}{s(s + 2\zeta\omega_n + K_t\omega_n^2)} \tag{9-152}$$

Since the system is still of type 1, the basic characteristics of the steady-state error are not altered by the tachometer feedback; that is, when the input is a step function, the steady-state error is zero. However, for a unit-ramp function input, the steady-state error of the system is $(2\zeta + K_t\omega_n)/\omega_n$, whereas that of the system with the PD control in Fig. 9-3 is $2\zeta/\omega_n$. Thus, *for the type-1 system, tachometer feedback decreases the* *error constants*   *ramp-error constant K_v while not affecting the step-error constant K_p.*

Minor-Loop Feedback Control with Passive Networks

Instead of using a tachometer to reduce cost, a network with passive *RC* elements can be used in the minor feedback loop for compensation. We illustrate this approach with the following example.

EXAMPLE 9-15   Consider that for the sun-seeker system in Example 9-5, instead of placing the controller in the forward path, we adopt the minor-loop feedback control, as shown in Fig. 9-49, with

minor-loop *feedback*

$$G_p(s) = \frac{2500}{s(s + 25)} \tag{9-153}$$

and

$$H(s) = \frac{K_t s}{1 + Ts} \tag{9-154}$$

RC network   To maintain the open-loop transfer function as type 1, it is necessary that $H(s)$ contains a zero at $s = 0$. Equation (9-154) can be realized by the simple *RC* network shown in Fig. 9-50. This network cannot be applied as a series controller in the forward path, since it acts as an open circuit in the steady-state when the frequency is zero. As a minor-loop controller, the zero-transmission property to dc signals does not pose any problems. The closed-loop transfer function of the system in Fig. 9-49 is

$$\frac{\Theta_o(s)}{\Theta_r(s)} = \frac{G_p(s)}{1 + G_p(s) + G_p(s)H(s)} \tag{9-155}$$

Substituting Eqs. (9-153) and (9-154) into the last equation, we get

$$\frac{\Theta_o(s)}{\Theta_r(s)} = \frac{2500(1 + Ts)}{s(s + 25)(1 + Ts) + 2500K_t s} \tag{9-156}$$

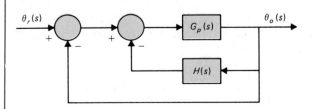

FIGURE 9-49 Sun-seeker control system with minor-loop control.

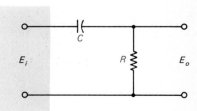

FIGURE 9-50 Simple *RC* network with phase-lead characteristics.

The characteristic equation of the system is

$$Ts^3 + (25T + 1)s^2 + (25 + 2500T + 2500K_t)s + 2500 = 0 \qquad (9\text{-}157)$$

root-contour design

To show the effects of the parameters K_t and T, we shall construct the root contours of Eq. (9-157). Consider that K_t is fixed, and with T as the variable parameter, we divide both sides of Eq. (9-157) by the terms that do not contain T. The result is

$$1 + \frac{Ts(s^2 + 25s + 2500)}{s^2 + (25 + 2500K_t)s + 5000} = 0 \qquad (9\text{-}158)$$

When the value of K_t is relatively large, the two poles of the last equation are real with one very close to the origin. It is more effective to choose K_t so that the poles of Eq. (9-158) are complex.

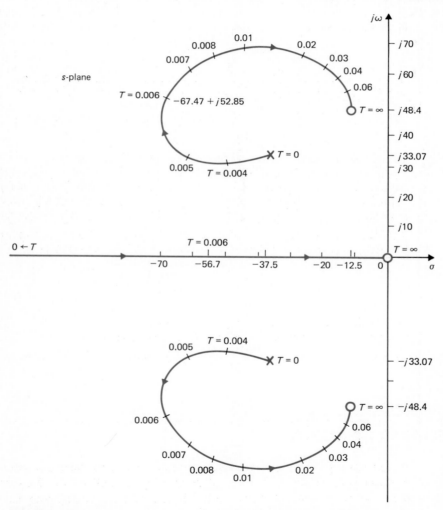

FIGURE 9-51 Root contours of $Ts^3 + (25T + 1)s^2 + (25 + 1500K_t + 2500T)s + 2500 = 0$. $K_t = 0.02$.

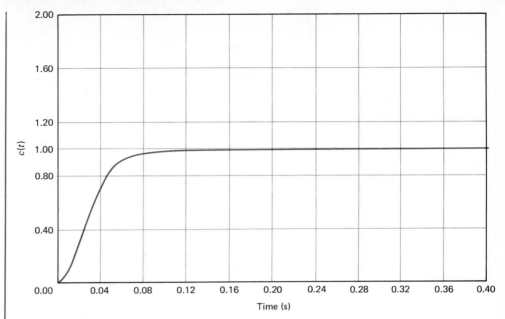

FIGURE 9-52 Unit-step responses of the sun-seeker system with the minor-loop controller.

Figure 9-51 shows the root contours of Eq. (9-158) with $K_t = 0.02$, for $T = 0$ to $T = \infty$. When $T = 0.006$, the characteristic equation roots are at -56.72, $-67.47 + j52.85$, and $-67.47 - j52.85$. Figure 9-52 shows the unit-step response of the designed system.

Just as with the tachometer feedback, the minor-loop feedback controller of Eq. (9-154) reduces the ramp-error constant K_v, although the system is still type 1. Although the forward-path gain K can be increased to improve the value of K_v, which is given as

$$K_v = \frac{100K}{1 + 100KK_t} \tag{9-159}$$

the combination of the value of K and K_t necessary to satisfy a certain K_v requirement may be too much for the controller in Eq. (9-154) to compensate.

9-10

STATE-FEEDBACK CONTROL

A majority of the design techniques in modern control theory is based on the state-feedback configuration. That is, instead of using controllers with fixed configurations in the forward or feedback path, control is achieved by feeding back the state variables through constant gains. The block diagram of a system with state-feedback control is shown in Fig. 9-2(c).

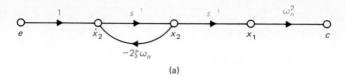

(a)

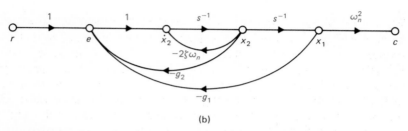

(b)

FIGURE 9-53 Control of a second-order system by state feedback.

tachometer feedback

We can show that the PID control and the tachometer-feedback control discussed earlier are all special cases of the state-feedback control scheme. In the case of tachometer-feedback control, let us consider the second-order process described in Eq. (9-148). The process is decomposed by direct decomposition to be represented by the state diagram of Fig. 9-53(a). If the states $x_1(t)$ and $x_2(t)$ are physically accessible, these variables may be fed back through constant gains $-g_1$ and $-g_2$, respectively, to form the control $u(t)$, as shown in Fig. 9-53(b). The closed-loop transfer function of the system is

$$\frac{C(s)}{R(s)} = \frac{\omega_n^2}{s^2 + (2\zeta\omega_n + g_2)s + g_1} \tag{9-160}$$

Comparing this transfer function with that of the system with tachometer feedback, Eq. (9-149), we notice that the two transfer functions are identical if $g_1 = \omega_n^2$ and $g_2 = K_t\omega_n^2$. In fact, if the system is to have zero steady-state error due to a step input, g_1 should equal ω_n^2. The value of g_2 is selected to satisfy the damping requirements.

For the system with PD control shown in Fig. 9-3, the closed-loop transfer function is given in Eq. (9-151). The characteristic equations of the system with state-feedback and with PD control are identical if $g_1 = \omega_n^2 K_P$ and $g_2 = \omega_n^2 K_D$. The numerators of the two transfer functions are different however.

regulators

If the reference input $r(t)$ is zero, the class of systems is commonly described as **regulators**. Under this condition, the control objective is to drive any arbitrary initial conditions of the system to zero in some prescribed manner, such as as quickly as possible. Then the regulator system with the PD controller is the same as the state-feedback control.

state feedback

Since the PI control increases the order of the system by one, it cannot be made equivalent to state feedback through constant gains. We show in Section 9-11 that if we combine state feedback with integral control, we can again realize PI control in the sense of state-feedback control.

9-11
POLE-PLACEMENT DESIGN THROUGH STATE FEEDBACK

pole placement

When root loci are utilized for the design of control systems, the general approach may be described as that of **pole placement**; the pole here refers to that of the closed-loop transfer function, which is also the same as the root of the characteristic equation. Knowing the relation between the closed-loop poles and the system performance, we can effectively design the system by specifying the location of these poles. The design methods discussed in the preceding sections are all characterized by the property that the poles are selected based on what can be achieved with the fixed-controller configuration and the physical range of the controller parameters. A natural question would be: *Under what condition can the poles be placed arbitrarily?* This is an entirely new design philosophy and freedom that apparently can be achieved only under certain conditions.

When we have a controlled process of the third order or higher, the PD, PI, the first-order phase-lead, or the first-order phase-lag controller would not be able to control independently all of the three or more poles of the system, since there are only two free parameters in each of these controllers.

To investigate the condition required for arbitrary pole placement in an nth-order system, let us consider that the process is described by the following state equation:

$$\dot{\mathbf{x}}(t) = \mathbf{A}\mathbf{x}(t) + \mathbf{B}u(t) \qquad (9\text{-}161)$$

where $\mathbf{x}(t)$ is the $n \times 1$ state vector, and $u(t)$ is the scalar control input. The state-feedback control is

$$u(t) = -\mathbf{G}\mathbf{x}(t) + r(t) \qquad (9\text{-}162)$$

where $\mathbf{G}$ is the $1 \times n$ feedback matrix with constant-gain elements. By substituting Eq. (9-162) into Eq. (9-161), the closed-loop system is represented by the state equation

$$\dot{\mathbf{x}}(t) = (\mathbf{A} - \mathbf{B}\mathbf{G})\mathbf{x}(t) + \mathbf{B}r(t) \qquad (9\text{-}163)$$

condition on pole placement

It will be shown in the following that if the pair $[\mathbf{A}, \mathbf{B}]$ is completely controllable, then a matrix $\mathbf{G}$ exists that can give an arbitrary set of eigenvalues of $(\mathbf{A} - \mathbf{B}\mathbf{G})$; that is, the n roots of the characteristic equation

$$|\lambda \mathbf{I} - \mathbf{A} + \mathbf{B}\mathbf{G}| = 0 \qquad (9\text{-}164)$$

can be arbitrarily placed.

It has been shown in Chapter 5 that if a system is state controllable, it can always be represented in the phase-variable canonical form; that is, in Eq. (9-161),

phase-variable canonical form

$$\mathbf{A} = \begin{bmatrix} 0 & 1 & 0 & \cdots & 0 \\ 0 & 0 & 1 & \cdots & 0 \\ \vdots & \vdots & \vdots & \ddots & \vdots \\ 0 & 0 & 0 & \cdots & 1 \\ -a_1 & -a_2 & -a_3 & \cdots & -a_n \end{bmatrix} \qquad \mathbf{B} = \begin{bmatrix} 0 \\ 0 \\ \vdots \\ 0 \\ 1 \end{bmatrix}$$

Conversely, if the system is represented in the phase-variable canonical form, it is always controllable.

The feedback gain matrix **G** is

feedback-gain matrix

$$\mathbf{G} = [g_1 \quad g_2 \quad \cdots \quad g_n] \tag{9-165}$$

where $g_1, g_2, \ldots, g_n$ are real constants. Then,

$$\mathbf{A} - \mathbf{BG} = \begin{bmatrix} 0 & 1 & 0 & \cdots & 0 \\ 0 & 0 & 1 & \cdots & 0 \\ 0 & 0 & 0 & \cdots & 0 \\ \vdots & \vdots & \vdots & \ddots & \vdots \\ 0 & 0 & 0 & \cdots & 1 \\ -a_1 - g_1 & -a_2 - g_2 & -a_3 - g_3 & \cdots & -a_n - g_n \end{bmatrix} \tag{9-166}$$

The eigenvalues of $\mathbf{A} - \mathbf{BG}$ are then found from the characteristic equation

characteristic equation

$$|\lambda \mathbf{I} - (\mathbf{A} - \mathbf{BG})| = \lambda^n + (a_n + g_n)\lambda^{n-1} + (a_{n-1} + g_{n-1})\lambda^{n-2} + \cdots + (a_1 + g_1)$$
$$= 0 \tag{9-167}$$

Clearly, the eigenvalues can be arbitrarily assigned, since the feedback gains $g_1, g_2, \ldots, g_n$ are isolated in each coefficient of the characteristic equation.

EXAMPLE 9-16   Consider that the transfer function of a linear process is

$$G_p(s) = \frac{C(s)}{E(s)} = \frac{20}{s^2(s + 1)} \tag{9-168}$$

Figure 9-54(a) shows the state diagram of $G_p(s)$, and Fig. 9-54(b) shows the state diagram of the closed-loop system with feedback from all three states. The closed-loop transfer function of the system is

$$\frac{C(s)}{R(s)} = \frac{20}{s^3 + (g_3 + 1)s^2 + g_2 s + g_1} \tag{9-169}$$

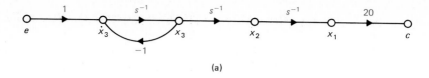

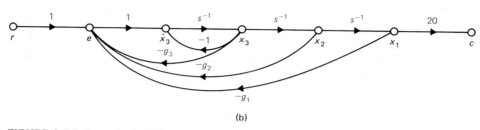

FIGURE 9-54 Control of a third-order system by state feedback.

Let us assume that we desire to have zero steady-state error with the input being a unit-step function, and, in addition, two of the closed-loop poles must be at $s = -1 + j$ and $s = -1 - j$. The steady-state requirement fixes the value of g_1 at 20, and only g_2 and g_3 need to be determined from the eigenvalue location. Since the transfer function does not have common poles and zeros, the process is completely state controllable.

The characteristic equation of the system is

$$s^3 + (g_3 + 1)s^2 + g_2 s + 20 = (s + 1 - j)(s + 1 + j)(s + a) = 0$$

$$= s^3 + (2 + a)s^2 + 2(a + 1)s + 2a = 0 \qquad (9\text{-}170)$$

Equating the coefficients of the corresponding terms in the last equation, we get

state-feedback gains

$$a = 10 \qquad g_1 = 20 \qquad g_2 = 22 \qquad \text{and} \qquad g_3 = 11$$

Thus, the third pole is at $s = -10$. Since the complex closed-loop poles have a damping ratio of 0.707, and the third pole is quite far to the left of these poles, the system acts like a second-order system, and the maximum overshoot of the step response should not exceed 5 percent.

LINSYS

The program LINSYS of the ACSP software package that accompanies this text has a subprogram for pole-placement design through state feedback.

In general, it is not necessary to transform a system first into the phase-variable canonical form for the purpose of solving for the state-feedback gains. General methods for the determination of the feedback matrix **G** are available as long as the process is completely controllable. The following development describes a method that determines **G** based on the matrices **A** and **B**, and the characteristic equations of the open-loop and the closed-loop systems with a single input. Let us define the following equations:

$$\Delta_o(s) = |s\mathbf{I} - \mathbf{A}| = \text{characteristic equation of } \mathbf{A} \text{ (open-loop system)} \qquad (9\text{-}171)$$

$$\Delta_c(s) = |s\mathbf{I} - \mathbf{A} - \mathbf{BG}| = \text{characteristic equation of } \mathbf{A} - \mathbf{BG} \text{ (closed-loop system)}$$

(9-172)

$$\Delta(s) = 1 + \mathbf{G}(s\mathbf{I} - \mathbf{A})^{-1}\mathbf{B}$$

(9-173)

where $\mathbf{G}$ is $1 \times n$, and $\mathbf{B}$ is $n \times 1$.

First, we show that

$$\Delta(s) = \frac{\Delta_c(s)}{\Delta_o(s)}$$

(9-174)

This relation is proved by writing

$$s\mathbf{I} - \mathbf{A} + \mathbf{BG} = (s\mathbf{I} - \mathbf{A})[\mathbf{I} + (s\mathbf{I} - \mathbf{A})^{-1}\mathbf{BG}]$$

(9-175)

Taking the determinant on both sides of the last equation, we get

$$\Delta_c(s) = |s\mathbf{I} - \mathbf{A} + \mathbf{BG}| = \Delta_o(s)|\mathbf{I} + (s\mathbf{I} - \mathbf{A})^{-1}\mathbf{BG}|$$

(9-176)

Since

$$|\mathbf{I} + (s\mathbf{I} - \mathbf{A})^{-1}\mathbf{BG}| = |\mathbf{I} + \mathbf{BG}(s\mathbf{I} - \mathbf{A})^{-1}| = |\mathbf{I} + \mathbf{G}(s\mathbf{I} - \mathbf{A})^{-1}B|$$

$$= 1 + \mathbf{G}(s\mathbf{I} - \mathbf{A})^{-1}\mathbf{B} = \Delta(s)$$

(9-177)

where the identity matrices are of different dimensions, Eq. (9-176) becomes

$$\Delta_c(s) = \Delta_o(s)\Delta(s)$$

(9-178)

Now we write Eq. (9-173) as

$$\Delta(s) = 1 + \mathbf{G}\frac{\text{Adj } (s\mathbf{I} - \mathbf{A})\mathbf{B}}{\Delta_o(s)}$$

(9-179)

where Adj $(s\mathbf{I} - \mathbf{A})$ denotes the adjoint matrix of $s\mathbf{I} - \mathbf{A}$. Let

$$\mathbf{k}(s) = \text{Adj } (s\mathbf{I} - \mathbf{A})\mathbf{B} \qquad n \times 1$$

(9-180)

Then, Eq. (9-179) becomes

$$\Delta(s) = \frac{\Delta_o(s) + \mathbf{Gk}(s)}{\Delta_o(s)}$$

(9-181)

and using Eq. (9-178), we get

$$\mathbf{Gk}(s) = \Delta_c(s) - \Delta_o(s) \qquad (9\text{-}182)$$

Thus, by knowing $\mathbf{k}(s)$, $\Delta_c(s)$, and $\Delta_o(s)$, the feedback-gain matrix $\mathbf{G}$ can be solved from Eq. (9-182) if the process is completely controllable.

EXAMPLE 9-17   Consider that a linear process is described by the state equation

$$\dot{\mathbf{x}}(\mathbf{t}) = \mathbf{A}\mathbf{x}(t) + \mathbf{B}u(t) \qquad (9\text{-}183)$$

where

$$\mathbf{A} = \begin{bmatrix} 1 & 0 & 0 \\ -1 & 0 & 2 \\ 0 & -1 & 1 \end{bmatrix} \qquad \mathbf{B} = \begin{bmatrix} 1 \\ 0 \\ 0 \end{bmatrix}.$$

From Eq. (9-180), we have

$$\mathbf{k}(s) = \begin{bmatrix} s^2 - s + 2 \\ -(s - 1) \\ 1 \end{bmatrix} \qquad (9\text{-}184)$$

The open-loop characteristic equation is

$$\Delta_o(s) = \left| s\mathbf{I} - \mathbf{A} \right| = s^3 - 2s^2 + 3s - 2 = 0 \qquad (9\text{-}185)$$

Let the desired closed-loop eigenvalues be $s = -2$, -1, and -1. Then the closed-loop characteristic equation is

$$\Delta_c(s) = s^3 + 4s^2 + 5s + 2 = 0 \qquad (9\text{-}186)$$

By substituting Eqs. (9-184), (9-185), and (9-186) into Eq. (9-182), the elements of $\mathbf{G}$ are found to be $g_1 = 6$, $g_2 = -8$, and $g_3 = 0$.

9-12

STATE FEEDBACK WITH INTEGRAL CONTROL [4]

The state-feedback control structured in the preceding two sections has a serious deficiency in that it does not improve the type of the system. As a result, the state-feedback control with constant-gain feedback is generally useful only for regulator systems for which the system does not track inputs. In general, there is a large class of control systems that has inputs that must be tracked, and often there are undesirable noise or

PI controller

disturbances that the system must suppress. One remedy for this problem is to introduce integral control, just as with the PI controller, together with the constant-gain state feedback.

state feedback

Let us consider the following design problem.

Given a linear system that is represented by the following dynamic equations in matrix form:

$$\dot{\mathbf{x}}(t) = \mathbf{A}\mathbf{x}(t) + \mathbf{B}\mathbf{u}(t) + \mathbf{F}\mathbf{w} \qquad (9\text{-}187)$$

where

$$\mathbf{x}(t) = n \times 1 \text{ state vector}$$

$$\mathbf{u}(t) = p \times 1 \text{ control vector}$$

$$\mathbf{w} = m \times 1 \text{ input and disturbance vector}$$

and $\mathbf{A} = n \times n$, $\mathbf{B} = n \times p$, $\mathbf{F} = n \times m$, and $\mathbf{w}$ is defined as a constant vector whose elements are composed of input signals and disturbances. In practice, the magnitudes of the input signals are given, but the magnitudes of some or all of the constant disturbances may be unknown.

design objective

The objective of the design is to find feedback controls from the state variables such that the state $\mathbf{x}(t)$ *is driven to any desired state (set point) as time approaches infinity.* For instance, we may want the state variable $x_1(t)$ to be driven to a set point $r(t) = R$ as t approaches infinity, while the overall system is asymptotically stable.

For the problem described, it is convenient to define the output equation as

$$\mathbf{c}(t) = \mathbf{D}\mathbf{x}(t) + \mathbf{H}\mathbf{w} \qquad (9\text{-}188)$$

where $\mathbf{c}(t) = q \times 1$ output vector, $\mathbf{D} = q \times n$, and $\mathbf{H} = q \times m$ are coefficient matrices. Then, to drive the state $x_1(t)$ to the set point $r(t)$ is equivalent to driving

$$\mathbf{c}(t) = w_1 - x_1(t) \qquad (9\text{-}189)$$

to zero as t approaches infinity, where

$$w_1 = r(t) = R = \text{constant} \qquad (9\text{-}190)$$

output regulation

Thus, the design problem may be regarded as **output regulation**.

As indicated earlier in the state-feedback design, the present output-regulation design problem cannot be achieved by simply using constant-gain state feedback, since the latter does not improve the **type** of a system. Since a system with constant-gain state-feedback control corresponds to a type-0 system, we must introduce integral control in order to have the output of the system track any input. Let the control $\mathbf{u}(t)$ be given by

$$\mathbf{u}(t) = -\mathbf{G}_1\mathbf{x}(t) - \mathbf{G}_2 \int \mathbf{c}(t)\, dt \qquad (9\text{-}191)$$

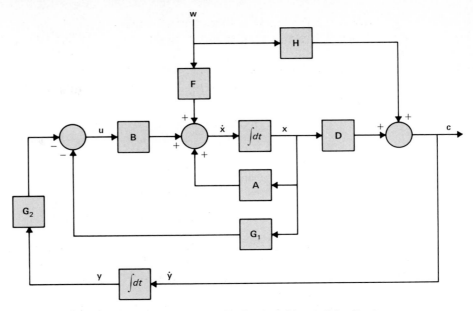

FIGURE 9-55 Control system with state feedback and integral feedback.

where $\mathbf{G}_1$ is a $p \times n$ feedback-gain matrix, and $\mathbf{G}_2$ is a $p \times q$ feedback-gain matrix, both with constant elements.

Equation (9-191) shows that the control is realized by feeding back the states $\mathbf{x}(t)$ through a constant-gain feedback matrix $\mathbf{G}_1$, and, in addition, the integrals of the "outputs" $\mathbf{c}(t)$ are fed back through the constant-gain matrix $\mathbf{G}_2$. The block diagram in Fig. 9-55 illustrates the elements of the overall feedback control system with integral feedback. Naturally, when $\mathbf{G}_2 = \mathbf{0}$, the system becomes a state regulator with state feedback.

Since the feedback system now has q additional integrators, the overall system is of the $(n + q)$th order. Let $\mathbf{y}(t)$ denote the $q \times 1$ state vector of the integral control, as shown in Fig. 9-55; the $(n + q)$ state equations are written in vector-matrix form directly from Fig. 9-55.

$$\begin{bmatrix} \dot{\mathbf{x}}(t) \\ \dot{\mathbf{y}}(t) \end{bmatrix} = \begin{bmatrix} \mathbf{A} & \mathbf{0} \\ \mathbf{D} & \mathbf{0} \end{bmatrix} \begin{bmatrix} \mathbf{x}(t) \\ \mathbf{y}(t) \end{bmatrix} - \begin{bmatrix} \mathbf{B} \\ \mathbf{0} \end{bmatrix} [\mathbf{G}_1 \quad \mathbf{G}_2] \begin{bmatrix} \mathbf{x}(t) \\ \mathbf{y}(t) \end{bmatrix} + \begin{bmatrix} \mathbf{F} \\ \mathbf{H} \end{bmatrix} \mathbf{w} \qquad (9\text{-}192)$$

Or

$$\dot{\hat{\mathbf{x}}}(t) = \hat{\mathbf{A}}\hat{\mathbf{x}}(t) - \hat{\mathbf{B}}\mathbf{G}\hat{\mathbf{x}}(t) + \hat{\mathbf{F}}\mathbf{w} \qquad (9\text{-}193)$$

where

$$\hat{\mathbf{x}}(t) = \begin{bmatrix} \mathbf{x}(t) \\ \mathbf{y}(t) \end{bmatrix} \qquad (n + q) \times 1 \qquad (9\text{-}194)$$

$$\hat{\mathbf{A}} = \begin{bmatrix} \mathbf{A} & \mathbf{0} \\ \mathbf{D} & \mathbf{0} \end{bmatrix} \qquad (n+q) \times (n+q) \qquad (9\text{-}195)$$

$$\hat{\mathbf{B}} = \begin{bmatrix} \mathbf{B} \\ \mathbf{0} \end{bmatrix} \qquad (n+q) \times p \qquad (9\text{-}196)$$

$$\mathbf{G} = [\mathbf{G}_1 \quad \mathbf{G}_2] \qquad p \times (n+q) \qquad (9\text{-}197)$$

$$\hat{\mathbf{F}} = \begin{bmatrix} \mathbf{F} \\ \mathbf{H} \end{bmatrix} \qquad (n+q) \times m \qquad (9\text{-}198)$$

Equation (9-193) is also written as

$$\dot{\hat{\mathbf{x}}}(t) = (\hat{\mathbf{A}} - \hat{\mathbf{B}}\mathbf{G})\hat{\mathbf{x}}(t) + \hat{\mathbf{F}}\mathbf{w} \qquad (9\text{-}199)$$

which is of the form of state feedback from the state $\hat{\mathbf{x}}(t)$.

The following example illustrates the application of the design with state-feedback and output integral control.

EXAMPLE 9-18   Consider a dc-motor-control system that is described by the following state equations:

$$\frac{d\omega(t)}{dt} = \frac{-B}{J}\omega(t) + \frac{K_i}{J}i_a(t) - \frac{1}{J}T_L \qquad (9\text{-}200)$$

$$\frac{di_a(t)}{dt} = \frac{-K_b}{L}\omega(t) - \frac{R}{L}i_a(t) + \frac{1}{L}e_a(t) \qquad (9\text{-}201)$$

where

$i_a(t)$ = armature current, A

$e_a(t)$ = armature applied voltage, V

$\omega(t)$ = motor velocity, rad/s

B = viscous-friction coefficient of motor and load = 0

J = moment of inertia of motor and load = 0.02 N-m/rad/s^2

K_i = motor-torque constant = 1 N-m/A

K_b = motor back-emf constant = 1 V/rad/s

T_L = constant-load torque (magnitude not known), N-m

L = armature inductance = 0.005 H

R = armature resistance = 1 Ω

The design problem is to find the control $u(t) = e_a(t)$ such that

and

(1) $\lim_{t \to \infty} i_a(t) = 0$ and $\lim_{t \to \infty} \dot{\omega}(t) = 0$ \qquad (9\text{-}202)

(2) $\lim_{t \to \infty} \omega(t)$ = constant set point r \qquad (9\text{-}203)

Let the state variables be defined as $x_1(t) = \omega(t)$ and $x_2(t) = i_a(t)$. The vector $\mathbf{w}$ is defined as

$$\mathbf{w} = \begin{bmatrix} T_L \\ r \end{bmatrix} = \begin{bmatrix} w_1 \\ w_2 \end{bmatrix} \tag{9-204}$$

Let the output variable be defined as

$$c(t) = r - \omega(t) = w_2 - x_2(t) \tag{9-205}$$

Then the condition in Eq. (9-203) is equivalent to

$$\lim_{t \to \infty} c(t) = 0$$

*output
regulation*

which is the condition of output regulation.

Since the original system has two state variables in x_1 and x_2, and one output in $c(t)$, the control is of the form

$$u(t) = -g_1 x_1(t) - g_2 x_2(t) - g_3 \int c(t)\, dt$$

$$= -\mathbf{G}_1 \mathbf{x}(t) - \mathbf{G}_2 \int c(t)\, dt \tag{9-206}$$

where

$$\mathbf{G}_1 = [g_1 \quad g_2] \qquad \mathbf{G}_2 = g_3 \tag{9-207}$$

and g_1, g_2, and g_3 are real constants.

*integral
control*

However, with the integral control, the closed-loop system is of the third order, and there are three poles to be specified. Let the closed-loop poles be placed at -300, $-10 + j10$, and $-10 - j10$. We must now find the values of g_1, g_2, and g_3 so that the conditions in Eqs. (9-202) and (9-203) and the pole-placement requirements are met simultaneously. Expressing the state equations in Eqs. (9-200) and (9-201) in the form of Eq. (9-187), we have

$$\begin{bmatrix} \dot{\omega}(t) \\ \dot{i}_a(t) \end{bmatrix} = \begin{bmatrix} -\dfrac{B}{J} & \dfrac{K_i}{J} \\[2ex] -\dfrac{K_b}{L} & -\dfrac{R}{L} \end{bmatrix} \mathbf{x}(t) + \begin{bmatrix} 0 \\[1ex] \dfrac{1}{L} \end{bmatrix} u(t) + \begin{bmatrix} -\dfrac{1}{J} & 0 \\[1ex] 0 & 0 \end{bmatrix} \begin{bmatrix} w_1 \\ w_2 \end{bmatrix} \tag{9-208}$$

The output equation is

$$c(t) = [-1 \quad 0]\mathbf{x}(t) + [0 \quad 1]\mathbf{w} \tag{9-209}$$

Substituting the values of the system parameters into Eq. (9-208), we have

$$\mathbf{A} = \begin{bmatrix} -\dfrac{B}{J} & \dfrac{K_i}{J} \\[2ex] -\dfrac{K_b}{L} & -\dfrac{R}{L} \end{bmatrix} = \begin{bmatrix} 0 & 50 \\ -200 & -200 \end{bmatrix} \tag{9-210}$$

$$\mathbf{B} = \begin{bmatrix} 0 \\[1ex] \dfrac{1}{L} \end{bmatrix} = \begin{bmatrix} 0 \\ 200 \end{bmatrix} \tag{9-211}$$

$$\mathbf{F} = \begin{bmatrix} -\dfrac{1}{J} & 0 \\ 0 & 0 \end{bmatrix} = \begin{bmatrix} -50 & 0 \\ 0 & 0 \end{bmatrix} \tag{9-212}$$

From Eq. (9-209),

$$\mathbf{D} = \begin{bmatrix} -1 & 0 \end{bmatrix} \qquad \mathbf{H} = \begin{bmatrix} 0 & 1 \end{bmatrix}$$

Define the transformation according to Eq. (9-192); then

$$\hat{\mathbf{A}} = \begin{bmatrix} \mathbf{A} & \mathbf{0} \\ \mathbf{D} & 0 \end{bmatrix} = \begin{bmatrix} 0 & 50 & 0 \\ -200 & -200 & 0 \\ -1 & 0 & 0 \end{bmatrix} \tag{9-213}$$

$$\hat{\mathbf{B}} = \begin{bmatrix} \mathbf{B} \\ 0 \end{bmatrix} = \begin{bmatrix} 0 \\ 200 \\ 0 \end{bmatrix} \tag{9-214}$$

We can show that both $[\mathbf{A}, \mathbf{B}]$ and $[\hat{\mathbf{A}}, \hat{\mathbf{B}}]$ are controllable.
Let the control of the transformed system be

$$u(t) = -\mathbf{G}_1 \mathbf{x}(t) - \mathbf{G}_2 \int c(t)\, dt$$

$$= -\mathbf{G} \begin{bmatrix} \mathbf{x}(t) \\ y(t) \end{bmatrix} \tag{9-215}$$

where

$$\mathbf{G} = \begin{bmatrix} g_1 & g_2 & g_3 \end{bmatrix} \tag{9-216}$$

and

$$\mathbf{G}_1 = \begin{bmatrix} g_1 & g_2 \end{bmatrix} \tag{9-217}$$

$$\mathbf{G}_2 = g_3 \tag{9-218}$$

The coefficient matrix of the closed-loop system becomes

$$\hat{\mathbf{A}} - \hat{\mathbf{B}}\mathbf{G} = \begin{bmatrix} 0 & 50 & 0 \\ -200 - 200g_1 & -200 - 200g_2 & -200g_3 \\ -1 & 0 & 0 \end{bmatrix} \tag{9-219}$$

The characteristic equation is

$$\Delta_c(s) = [s\mathbf{I} - \hat{\mathbf{A}} + \hat{\mathbf{B}}\mathbf{G}]$$

$$= s^3 + 200(1 + g_2)s^2 + 10{,}000(1 + g_1)s - 10{,}000g_3 = 0 \tag{9-220}$$

For the three assigned poles, Eq. (9-220) must equal

$$\Delta_c(s) = s^3 + 320s^2 + 6200s + 60{,}000 = 0 \tag{9-221}$$

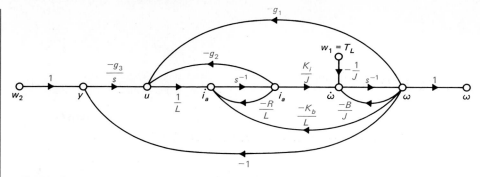

FIGURE 9-56 State diagram of the system in Example 9-18.

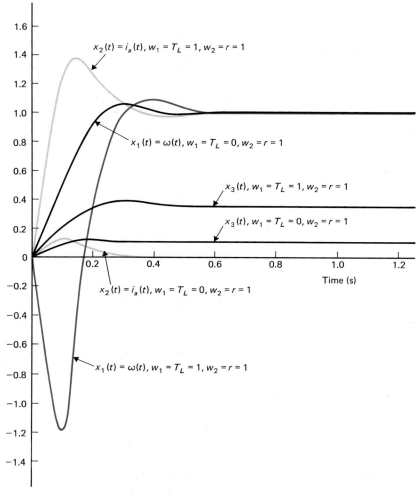

FIGURE 9-57 Time responses of the system in Example 9-18.

feedback gains

Thus, the three feedback gains are

$$g_1 = -0.38 \qquad g_2 = 0.6 \qquad g_3 = -6$$

Figure 9-56 shows the state diagram of the overall designed system. Applying Mason's gain formula between the inputs w_1 and w_2 and the states ω and i_a on the state diagram, we have

$$
\begin{bmatrix} \Omega(s) \\ I_a(s) \end{bmatrix} = \frac{1}{\Delta_c(s)}
\begin{bmatrix}
-\frac{1}{J}\left(s^2 + \frac{R}{L}s + \frac{g_2}{L}s\right) & -\frac{g_3 K_i}{JL} \\
-\frac{1}{J}\left(-\frac{K_b}{L}s - \frac{g_1}{L}s + \frac{g_3}{L}\right) & -\frac{g_3}{L}\left(s + \frac{B}{J}\right)
\end{bmatrix}
\begin{bmatrix} \dfrac{w_1}{s} \\ \dfrac{w_2}{s} \end{bmatrix}
\tag{9-222}
$$

steady-state performance

Applying the final-value theorem to the last equation, the steady-state values of the state variables are found to be

$$
\lim_{t \to \infty} \begin{bmatrix} \omega(t) \\ i_a(t) \end{bmatrix} = \lim_{s \to \infty} s \begin{bmatrix} \Omega(s) \\ I_a(s) \end{bmatrix} = \begin{bmatrix} 0 & K_i \\ 1 & B \end{bmatrix} \begin{bmatrix} w_1 \\ w_2 \end{bmatrix}
\tag{9-223}
$$

Therefore, the motor velocity $\omega(t)$ will approach the constant reference set point $r = w_2$ as t approaches infinity, independent of the disturbance torque, $w_1 = T_L$. Figure 9-57 illustrates the responses of all the three state variables. The large overshoot in the response $\omega(t)$ is caused by the disturbance torque of 1 N-m. The figure also shows the responses when $T_L = 0$.

9-13
DIGITAL IMPLEMENTATION OF CONTROLLERS

Since digital control systems have many advantages over continuous-data control systems, quite often, controllers that are designed in the analog domain are implemented digitally. Ideally, if the designer has intended to use digital control, the system should be designed so that the dynamics of the controller be described by a z-transfer function or difference equations. However, there are situations under which the controller of an existing system is analog, and the system already operates in a satisfactory fashion, but the availability and advantages of digital control suggest that the controller be implemented by digital elements. Thus, the problems discussed in this section are twofold; first, we investigate how continuous-data controllers such as the PID, the phase-lead and phase-lag controllers, and others can be approximated by digital controllers; second, the problem of implementing digital controllers by digital processors is investigated.

Digital Implementation of the PID Controller

The PID Controller in the continuous-data domain is described by

$$G_c(s) = K_P + K_D s + \frac{K_I}{s} \tag{9-224}$$

The proportional component K_P is implemented digitally by a constant gain K_P. Since a digital computer or processor has finite word length, the constant K_P cannot be realized with infinite resolution.

The time derivative of a function $f(t)$ at $t = kT$ can be approximated by the **backward-difference rule**, using the values of $f(t)$ measured at $t = kT$ and $(k - 1)T$, that is,

$$\left.\frac{df(t)}{dt}\right|_{t=kT} = \frac{1}{T}\Big(f(kT) - f[(k - 1)T]\Big) \tag{9-225}$$

To find the z-transfer function of the derivative operation described before, we take the z-transform on both sides of Eq. (9-225). We have

$$\mathfrak{z}\left(\left.\frac{df(t)}{dt}\right|_{t=kT}\right) = \frac{1}{T}(1 - z^{-1})F(z) = \frac{z - 1}{Tz}F(z) \tag{9-226}$$

Thus, the z-transfer function of the digital differentiator is

*digital
differentiator*

$$G_D(z) = K_D\frac{z - 1}{Tz} \tag{9-227}$$

where K_D is the proportional constant of the derivative controller. Replacing z by e^{Ts} in Eq. (9-227), we can show that as the sampling period T approaches zero, $G_D(z)$ approaches $K_D s$, which is the transfer function of the analog derivative controller. In general, the choice of the sampling period is extremely important. The value of T should be sufficiently small, so that the digital approximation is adequately accurate.

There are a number of numerical integration rules that can be used to digitally approximate the integral controller K_I/s. The three basic methods of approximating the area of a function numerically are **trapezoidal integration, forward-rectangular integration, and backward-rectangular integration**. These are described as follows.

TRAPEZOIDAL INTEGRATION.   The trapezoidal-integration rule approximates the area under the function $f(t)$ by a series of trapezoids, as shown in Fig. 9-58. Let the integral of $f(t)$ evaluated at $t = kT$ be designated as $u(kT)$. Then,

$$u(kT) = u[(k - 1)T] + \frac{T}{2}\{f(kT) - f[(k - 1)T]\} \tag{9-228}$$

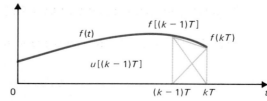

FIGURE 9-58 The trapezoidal-integration rule.

where the area under $f(t)$ for $(k - 1)T \leq t < kT$ is approximated by the area of the trapezoid in the interval. Taking the z-transform on both sides of Eq. (9-228), we have the transfer function of the digital integrator as

$$G_I(z) = K_I \frac{U(z)}{F(z)} = \frac{K_I T(z + 1)}{2(z - 1)} \qquad (9\text{-}229)$$

FORWARD-RECTANGULAR INTEGRATION.   For the forward-rectangular integration, we approximate the area under $f(t)$ by rectangles, as shown in Fig. 9-59. The integral of $f(t)$ at $t = kT$ is approximated by

$$u(kT) = u[(k - 1)T] + Tf(kT) \qquad (9\text{-}230)$$

digital integrator   By taking the z-transform on both sides of the last equation, the transfer function of the digital integrator using the forward-rectangular rule is

$$G_I(z) = K_I \frac{U(z)}{F(z)} = \frac{K_I T z}{z - 1} \qquad (9\text{-}231)$$

BACKWARD-RECTANGULAR INTEGRATION.   For the backward-rectangular integration, the digital approximation rule is illustrated in Fig. 9-60. The integral of $f(t)$ at $t = kT$ is approximated by

$$u(kT) = u[(k - 1)T] + Tf[(k - 1)T] \qquad (9\text{-}232)$$

digital integrator   The z-transfer function of the digital integrator using the backward-rectangular integration rule is

$$G_I(z) = K_I \frac{U(z)}{F(z)} = \frac{K_I T}{z - 1} \qquad (9\text{-}233)$$

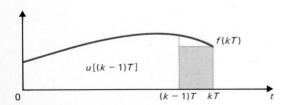

FIGURE 9-59 The forward-rectangular integration rule.

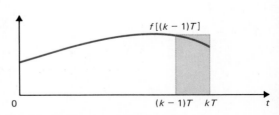

FIGURE 9-60 The backward-rectangular integration rule.

PID controller

By combining the proportional, derivative, and integration operations described before, the digital PID controller is modeled by the following transfer functions.

Trapezoidal Integration

$$G_c(z) = \frac{\left(K_P + \dfrac{TK_I}{2} + \dfrac{K_D}{T}\right)z^2 + \left(\dfrac{TK_I}{2} - K_P - \dfrac{2K_D}{T}\right)z + \dfrac{K_D}{T}}{z(z-1)} \qquad (9\text{-}234)$$

Forward-Rectangular Integration

$$G_c(z) = \frac{\left(K_P + \dfrac{K_D}{T} + TK_I\right)z^2 - \left(K_P + \dfrac{2K_D}{T}\right)z + \dfrac{K_D}{T}}{z(z-1)} \qquad (9\text{-}235)$$

Backward-Rectangular Integration

$$G_c(z) = \frac{\left(K_P + \dfrac{K_D}{T}\right)z^2 + \left(TK_I - K_P - \dfrac{2K_D}{T}\right)z + \dfrac{K_D}{T}}{z(z-1)} \qquad (9\text{-}236)$$

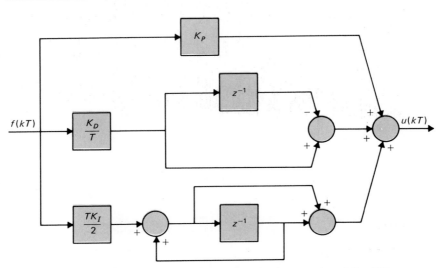

FIGURE 9-61 Block diagram of a digital-program implementation of the PID controller.

PD controller   When $K_I = 0$, the transfer function of the PD controller is

$$G_c(z) = \frac{\left(K_P + \dfrac{K_D}{T}\right)z - \dfrac{K_D}{T}}{z} \tag{9-237}$$

Once the transfer function of a digital controller is determined, the controller can be implemented by a digital processor or computer. The operator z^{-1} is interpreted as a time delay of T seconds. In practice, the time delay is implemented by storing a variable in some storage location in the computer and then taking it out after T seconds have elapsed. Figure 9-61 illustrates a block diagram representation of the digital program of the PID controller using the trapezoidal-integration rule.

Digital Implementation of Lead and Lag Controllers

In principle, any continuous-data controller can be made into a digital controller simply by adding sample-and-hold units at the input and the output terminals of the controller. Figure 9-62 illustrates the basic scheme with $G_c(s)$ being the transfer function of the continuous-data controller and $G_c(z)$ the equivalent digital controller. The sampling period T should be sufficiently small so that the dynamic characteristics of the continuous-data controller are not lost through the digitization. The system configuration shown in Fig. 9-62 actually suggests that given the continuous-data controller $G_c(s)$, the equivalent digital controller $G_c(z)$ can be obtained by the arrangement shown. On the other hand, given the digital controller $G_c(z)$, we can realize it by using an analog controller $G_c(s)$ and sample-and-hold units, as shown in Fig. 9-62.

EXAMPLE 9-19   As an illustrative example, consider that the continuous-data controller in Fig. 9-62 is represented by the transfer function

$$G_c(s) = \frac{s + 1}{s + 1.61} \tag{9-238}$$

From Fig. 9-62, the transfer function $G_c(z)$ is written

$$G_c(z) = (1 - z^{-1})\mathcal{Z}\left[\frac{s + 1}{s(s + 1.61)}\right] = \frac{z - 0.5}{z - 0.2} \tag{9-239}$$

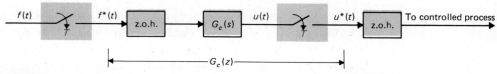

FIGURE 9-62 Realization of a digital controller by an analog controller with sample-and-hold units.

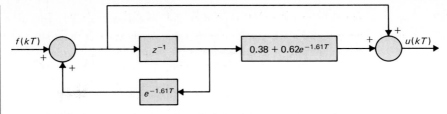

FIGURE 9-63 Digital-program realization of Eq. (9-239).

Figure 9-63 shows the digital-program implementation of Eq. (9-239).

9-14

SUMMARY

This chapter is devoted to the time-domain or the s-domain design of linear control systems. The chapter begins by first illustrating some of the fixed configurations of compensation schemes used in practice, such as the series, forward and feedforward, feedback and minor loop, and state feedback. The types of controllers considered are the PD, PI, PID, phase-lead, phase-lag, lead-lag, and pole-zero cancellation. The advantages and disadvantages of each of these types of controllers are pointed out, and the design procedures are outlined. It should be noted that, in general, the controllers are not limited by the few types illustrated in this chapter.

The time-domain design is characterized by specifications such as the relative damping ratio and the undamped natural frequency, or simply the location of the characteristic-equation roots, keeping in mind that the zeros of the closed-loop transfer function also affects the transient response. The performance is generally measured by the step response and the steady-state error.

Noise and disturbance effects are reduced or eliminated by the use of feedforward control. A section is devoted to the design of robust control systems. State feedback through constant-feedback gains and with dynamic feedback for the reduction of steady-state error are discussed. The state-feedback design is more versatile than the conventional fixed-configuration controller design, since the characteristic-equation roots are directly controlled. An unstable system that is controllable can always be stabilized by state-feedback control. The disadvantage with state feedback is that all the states must be sensed and fed back for control, which may not be practical.

REVIEW QUESTIONS

1. What is a PD controller? Write its input–output transfer function.
2. A PD controller has the constants K_D and K_P. Give the effects of these constants on the steady-state error of the system. Does the PD control change the type of a system?
3. Give the effects of the PD control on the rise time and settling time of a control system.
4. What is a PI controller? Write its input–output transfer function.
5. A PI controller has the constants K_P and K_I. Give the effects of the PI controller on the steady-state error of the system. Does the PI control change the type of a system?

6. Give the effects of the PI control on the rise time and settling time of a control system.

7. What is a PID controller? Write its input–output transfer function.

8. If a PD controller is so designed that the characteristic-equation roots have better damping than the original system, then the maximum overshoot of the system is always reduced.   **(T)**   **(F)**

9. Give the limitations of the phase-lead controller.

10. Give the general effects of the phase-lead controller on the rise time and settling time of a control system.

11. For the phase-lead controller, $G_c(s) = (1 + aTs)/(1 + Ts)$, $a > 1$. What is the effect of the controller on the steady-state performance of the system?

12. The phase-lead controller is generally less effective if the uncompensated system is very unstable to begin with.   **(T)**   **(F)**

13. For the phase-lag controller, $G_c(s) = (1 + aTs)/(1 + Ts)$, $a < 1$. What is the effect of the controller on the steady-state performance of the system?

14. Give the general effects of the phase-lag controller on the rise time and settling time of a control system.

15. For a phase-lag controller, if the value of T is large and the value of a is small, it is equivalent to adding a pure attenuation of a to the original uncompensated system.   **(T)**   **(F)**

16. Give the limitations of the pole-zero-cancellation control scheme.

17. For the bridged-T network shown in Fig. 9-35, if the zeros are complex-conjugate, then the poles will always be real.   **(T)**   **(F)**

18. What does it mean when a control system is described as being robust?

REFERENCES

1. J. G. TRUXAL, *Control Systems Synthesis,* McGraw-Hill Book Company, New York, 1955.

2. J. C. WILLEMS and S. K. MITTER, "Controllability, Observability, Pole Allocation, and State Reconstruction," *IEEE Trans. Automatic Control,* Vol. AC-16, pp. 582–595, Dec. 1971.

3. H. W. SMITH and E. J. DAVISON, "Design of Industrial Regulators," *Proc. IEE (London),* Vol. 119, pp. 1210–1216, Aug. 1972.

4. F. N. BAILEY and S. MESHKAT, "Root Locus Design of A Robust Speed Control," *Proc. Incremental Motion Control Symposium,* pp. 49–54, June 1983.

PROBLEMS

9-1. The block diagram of a control system with a series controller is shown in Fig. 9P-1. Find the transfer function of the controller $G_c(s)$ so that the following specifications are satisfied:

The ramp-error constant K_v is 9

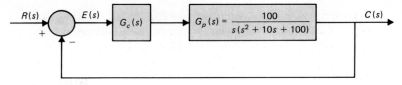

FIGURE 9P-1

The closed-loop transfer function is of the form:

$$M(s) = \frac{C(s)}{R(s)} = \frac{K}{(s^2 + 20s + 200)(s + a)}$$

where a and K are real constants.

9-2. A control system with a PD controller is shown in Fig. 9P-2.
  (a) Find the values of K_P and K_D so that the ramp-error constant K_v is 1000, and the damping ratio is 0.5.
  (b) Find the values of K_P and K_D so that the ramp-error constant K_v is 1000, and the damping ratio is 0.707.
  (c) Find the values of K_P and K_D so that the ramp-error constant K_v is 1000, and the damping ratio is 1.0.

9-3. A control system with a type-0 process $G_p(s)$ and a PI controller is shown in Fig. 9P-3.
  (a) Find the value of K_I so that the ramp-error constant K_v is 10.
  (b) Find the value of K_P so that the magnitude of the imaginary parts of the complex roots of the characteristic equation of the closed-loop system is 15 rad/s. Find the roots of the characteristic equation.
  (c) Sketch the root contours of the characteristic equation with the value of K_I as determined in part (a) and for $0 \le K_P < \infty$.

9-4. For the control system shown in Fig. 9P-3, perform the following:
  (a) Find the value of K_I so that the ramp-error constant K_v is 100.
  (b) With the value of K_I found in part (a), find the critical value of K_P so that the system is stable. Sketch the root contours of the characteristic equation for $0 \le K_P < \infty$.

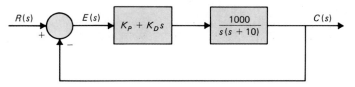

FIGURE 9P-2

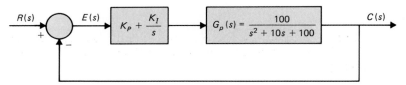

FIGURE 9P-3

(c)   If a computer program such as CLRSP of the ACSP software package is available, show that the maximum overshoot of the unit-step response is high for both large and small values of K_P. Use the value of K_I found in part (a). Find the value of K_P when the maximum overshoot is a minimum. What is the value of this maximum overshoot?

9-5.   Repeat Problem 9-4 for $K_v = 10$.

9-6.   A control system with a type-0 process $G_p(s)$ and PID controller is shown in Fig. 9P-6. Design the parameters of the PID controller so that the following specifications are satisfied:

Ramp-error constant $K_v = 100$
Rise time < 0.01 second
Maximum overshoot < 2 percent

Plot the unit-step response of the designed system.

9-7.   A considerable amount of effort is being spent by automobile manufacturers to meet the exhaust-emission-performance standards set by the government. Modern automotive-power-plant systems consist of an internal combustion engine that has an internal cleanup device called the catalytic converter. Such a system requires control of the engine air–fuel ratio (A/F), the ignition-spark timing, exhaust-gas recirculation, and injection air. The control-system problem considered in this problem deals with the control of the air-fuel ratio A/F. In general, depending on fuel composition and other factors, a typical stoichiometric A/F is 14.7:1, that is, 14.7 grams of air to each gram of fuel. An A/F greater or less than stoichiometry will cause high hydrocarbons, carbon monoxide, and nitrous oxides in the tailpipe emission. The control system whose block diagram is shown in Fig. 9P-7 is devised to control the air-fuel ratio so that a desired output variable is maintained for a given command signal. Figure 9P-7 shows that the sensor senses the composition of the exhaust-gas mixture entering the catalytic converter. The electronic controller detects the difference or the error between the command and the sensor signals and computes the control signal necessary to achieve the desired exhaust-gas composition. The output variable $c(t)$ denotes the effective air–fuel ratio A/F. The

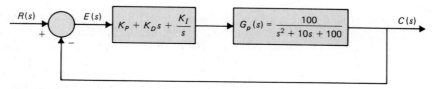

FIGURE 9P-6

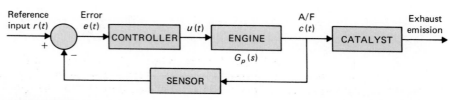

FIGURE 9P-7

transfer function of the engine is given by

$$\frac{C(s)}{U(s)} = \frac{e^{-T_d s}}{1 + \tau s} = G_p(s)$$

where T_d is the time delay and is 0.2 second. The time constant τ is 0.2 second. Approximate the time delay by a power series:

$$e^{-T_d s} \cong \frac{1}{1 + T_d s + T_d^2 s^2 / 2}$$

(a)   Let the controller be a PI controller so that

$$G_c(s) = \frac{U(s)}{E(s)} = K_P + \frac{K_I}{s}$$

Find the value of K_I so that the ramp-error constant K_v is 2. Determine the value of K_P so that the maximum overshoot of the unit-step response is a minimum. What is this maximum overshoot? Plot the unit-step response of $c(t)$.

(b)   The system performance can be further improved by using a PID controller:

$$G_c(s) = \frac{U(s)}{E(s)} = K_P + K_D s + \frac{K_I}{s}$$

Let the performance specifications be

$K_v = 2$
Maximum overshoot < 5 percent

Determine the values of K_P, K_D, and K_I so that these specifications are satisfied. Compute the plot the unit-step response of $c(t)$. How confident are you with the power-series approximation of the time delay when $T_d = 0.2$ second? (Refer to Problem 11-4 for the frequency-domain treatment of the problem.)

9-8.   The telescope for tracking stars and asteroids on the space shuttle may be modeled as a pure mass M. It is suspended by magnetic bearings so that there is no friction, and its attitude is controlled by magnetic actuators located at the base of the payload. The dynamic model for the control of the z-axis motion is shown in Fig. 9P-8(a). Since there are electrical components on the telescope, electric power must be brought to the telescope through a cable. The spring shown is used to model the wire-cable attachment, which exerts a spring force on the mass. The force produced by the magnetic actuators is denoted by $f(t)$. The force equation of motion in the z direction is

$$f(t) - K_s z(t) = M \frac{d^2 z(t)}{dt^2}$$

where $K_s = 1$ lb/ft, and $M = 150$ lb (mass); $f(t)$ is in lb and $z(t)$ in ft.

(a)   Show that the natural response of the system output $z(t)$ is oscillatory without damping. Find the natural undamped frequency of the open-loop space-shuttle system.

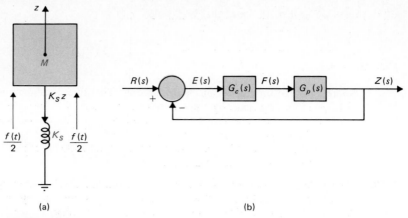

(a)

(b)

FIGURE 9P-8

(b) Design the PID controller:

$$G_c(s) = K_p + K_D s + \frac{K_I}{s}$$

shown in Fig. 9P-8(b) so that the following performance specifications are satisfied:

Ramp-error constant $K_v = 100$.
The complex roots of the characteristic equation corresponds to a relative damping ratio of 0.707 and a natural undamped frequency of 1 rad/s.

Compute and plot the unit-step response of the system. What is the maximum overshoot? Comment on the results.

(c) Design the PID controller so that the following specifications are satisfied:

Ramp-error constant $K_v = 100$
Maximum overshoot < 5 percent

Compute and plot the unit-step response of the system. Find the roots of the characteristic equation of the compensated system.

9-9. An inventory-control system is modeled by the following state equations:

$$\frac{dx_1(t)}{dt} = -2x_2(t)$$

$$\frac{dx_2(t)}{dt} = -2u(t)$$

where $x_1(t) =$ level of inventory, $x_2(t) =$ rate of sales of product, and $u(t) =$ production rate. The output equation is $c(t) = x_1(t)$. One unit of time is one day. Figure 9P-9 shows the block diagram of the closed-loop inventory-control system. The controller is a PD controller, $G_c(s) = K_P + K_D s$.

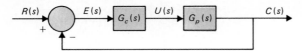

FIGURE 9P-9

      (a) Find the parameters of the PD controller, K_P and K_D, so that the roots of the characteristic equation are at $s = -1$ and $s = -2$. Plot the unit-step response of $c(t)$ and find the maximum overshoot.

      (b) Find the values of K_P and K_D so that the maximum overshoot of the unit-step response of $c(t)$ is less than 2 percent.

9-10. Figure 9P-10 shows the block diagram of the liquid-level control system described in Problems 6-13 and 7-27. The number of inlets is denoted by N. It is desired that $N = 20$. Design the PD controller so that (1) the overshoot is zero, and (2) the tank is filled to the reference level is less than 2.5 seconds.

9-11. The block diagram of a type-2 control system with a cascade controller $G_c(s)$ is shown in Fig. 9P-11. The objective is to design a PD controller so that the following performance specifications are satisfied:

Maximum overshoot ≤ 10 percent
Rise time ≤ 0.5 s.

      (a) Obtain the characteristic equation of the closed-loop system, and determine the ranges of the values of K_P and K_D for system stability.

      (b) Construct the root loci of the characteristic equation with $K_D = 0$ and $0 \leq K_P < \infty$. Then construct the root contours for $0 \leq K_D < \infty$ and several fixed values of K_P ranging from 0.001 to 0.01.

      (c) Design the PD controller to satisfy the performance specifications given above. Use the information on the root contours to help your design. Plot the unit-step response of $c(t)$.

9-12. Consider the "broom-balancing" control system described in Problems 4-23 and 5-43. The $\mathbf{A}^*$ and $\mathbf{B}^*$ matrices are given in Problem 5-43 for the small-signal linearized model.

$$\Delta \dot{\mathbf{x}} = \mathbf{A}^* \Delta \mathbf{x} + \mathbf{B}^* \Delta r$$

$$\Delta c = \mathbf{D}^* \Delta \mathbf{x} \qquad\qquad \mathbf{D}^* = \begin{bmatrix} 0 & 0 & 1 & 0 \end{bmatrix}$$

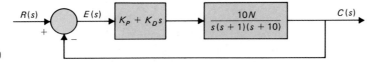

FIGURE 9P-10

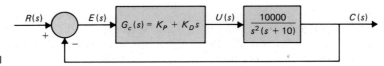

FIGURE 9P-11

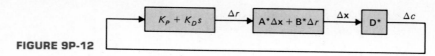

FIGURE 9P-12

Figure 9P-12 shows the closed-loop control of the system with a PD controller. Determine if the PD controller can stabilize the system; if so, find the values of K_P and K_D.

9-13. For the inventory-control system shown in Fig. 9P-9, let the controller be of the phase-lead type:

$$G_c(s) = \frac{1 + aTs}{1 + Ts} \qquad a > 1$$

Determine the values of a and T so that the maximum overshoot is ≤ 5 percent. Draw the root contours with T and a as variable parameters. Plot the unit-step response of the designed system.

9-14. For the control system shown in Fig. 9P-2, let the controller be a single-stage phase-lead controller:

$$G_c(s) = \frac{1 + aTs}{1 + Ts} \qquad a > 1$$

Determine the values of a and T so that the zero of $G_c(s)$ cancels the pole of $G_p(s)$ at $s = -10$. The damping ratio of the designed system should be unity.

9-15. Consider that the controller in the liquid-level control system shown in Fig. 9P-10 is a phase-lead controller:

$$G_c(s) = \frac{1 + aTs}{1 + Ts} \qquad a > 1$$

For $N = 20$, select the values of a and T so that the zero of $G_c(s)$ cancels the poles of $G_p(s)$ at $s = -1$, and the maximum overshoot of the system output is at a minimum. The value of a must not exceed 100. Is the phase-lead controller effective in compensating the system? Explain.

9-16. Consider that the controller in the liquid-level control system shown in Fig. 9P-10 is a phase-lag controller:

$$G_c(s) = \frac{1 + aTs}{1 + Ts} \qquad a < 1$$

For $N = 20$, select the values of a and T so that the two complex roots of the characteristic equation correspond to a relative damping ratio of approximately 0.707. Plot the unit-step response of the output $c(t)$.

9-17. The controlled process of a unity-feedback control system is

$$G_p(s) = \frac{K}{s(s + 5)^2}$$

The series controller has the transfer function

$$G_c(s) = \frac{1 + aTs}{1 + Ts}$$

The ramp-error constant K_v must equal to 10.

(a) Design a phase-lead controller $(a > 1)$ so that the maximum overshoot is less than 30 percent, but try to minimize it. Plot the unit-step response of the designed system.

(b) Design a phase-lag controller $(a < 1)$ so that the complex roots of the characteristic equation correspond to a damping ratio of 0.707 approximately. Plot the unit-step response of the designed system and compare the maximum overshoot with that of part (a).

(c) In order to obtain a better transient performance, a lead-lag controller is applied:

$$G_c(s) = \left(\frac{1 + aT_1s}{1 + T_1s}\right)\left(\frac{1 + bT_2s}{1 + T_2s}\right) \qquad a > 1, b < 1$$

Use the values of a and T found in part (a) for a and T_1, respectively. Determine the values of b and T_2 so that the dominant complex roots of the characteristic equation of the compensated system correspond to a damping ratio of 0.707. Determine the characteristic-equation roots and plot the unit-step response.

9-18. A phase-lock-loop dc-motor-speed-control system is described in Problem 4-20. The block diagram of the system is shown in Fig. 9P-18. The system parameters and transfer functions are given as follows:

Reference speed command, $f_r = 120$ pulses/s
Phase-detector gain, $K_p = 0.06$ V/pulse/s
Amplifier gain, $K_a = 20$
Encoder gain, $K_e = 5.73$ pulses/rad
Counter gain, $N = 1$
Motor transfer function,

$$\frac{\Omega_m(s)}{E_a(s)} = \frac{10}{s(1 + 0.05s)}$$

(a) Let the filter (controller) transfer function be

$$G_c(s) = \frac{E_o(s)}{E_i(s)} = \frac{1 + R_2Cs}{R_1Cs}$$

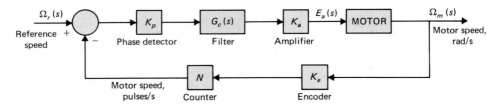

FIGURE 9P-18

where $R_1 = 2 \times 10^6 \ \Omega$, and $C = 1 \ \mu F$. Determine the value of R_2 so that the complex roots of the closed-loop characteristic equation have a maximum relative damping ratio. Sketch the root loci of the characteristic equation for $0 \le R_2 < \infty$. Compute and plot the unit-step responses of the motor speed $f_\omega(t)$ (pulses/s) with the values of R_2 found, when the input is 120 pulses/s. Convert the speed in pulses/s to rpm.

(b)   Let the filter transfer function be

$$G_c(s) = \frac{1 + aTs}{1 + Ts} \qquad a > 1$$

where $T = 0.01$. Find a so that the complex roots of the closed-loop characteristic equation have a maximum relative damping ratio. Compute and plot the unit-step response of the motor speed $f_\omega(t)$ (pulses/s) when the input is 120 pulses/s.

9-19.   The block diagram of the dc-motor-control system is shown in Fig. 9P-19. The controlled process is described by the transfer function

$$G_p(s) = \frac{608.7 \times 10^6}{s(s^3 + 423.42s^2 + 2.6667 \times 10^6 \ s + 4.2342 \times 10^8)}$$

(a)   The transfer function of the controller is

$$G_c(s) = \frac{1 + aTs}{1 + Ts} \qquad T = 0.01$$

Find the value of a so that the complex roots of the characteristic equation that are closest to the $j\omega$-axis in the s-plane correspond to a damping ratio of approximately 0.707. With the designed controller, compute and plot the unit-step response of $c(t)$.

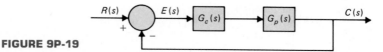

FIGURE 9P-19

(b)   The controlled process has complex-conjugate poles, due to the compliance of the load shaft. Let the transfer function of the controller be of the form:

$$G_c(s) = \frac{s^2 + 2\zeta_z \omega_{nz} s + \omega_{nz}^2}{s^2 + 2\zeta_p \omega_{np} s + \omega_{np}^2}$$

where the relationships between ζ_z, ζ_p, ω_{nz}, and ω_{np} are given in Eqs. (9-103) and (9-104). Select the values of ζ_z and ω_{nz} so that the zeros of $G_{c1}(s)$ cancel the complex poles of $G_p(s)$. Determine the values of ζ_p and ω_{np} using Eqs. (9-103) and (9-104). Compute and plot the unit-step response of the compensated system.

9-20.   Figure 9P-20 shows the block diagram of the control system in Problem 9-1 with output disturbance and feedforward control. Select the transfer function $G_{cf1}(s)$ and $G_{cf2}(s)$ so that the output $C(s)$ is not affected by the disturbance $D(s)$.

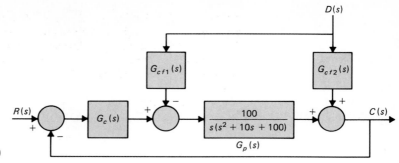

FIGURE 9P-20

9-21. Design the controllers $G_{cf}(s)$ and $G_c(s)$ for the system shown in Fig. 9P-21 so that the following specifications are satisfied:

Ramp-error constant $K_v = 50$
Dominant roots of the characteristic equation at $-5 \pm j5$ approximately
Rise time < 0.1 second
System must be robust when K varies ± 20 percent from the nominal value, with the rise time and overshoot within specifications

Compute and plot the unit-step responses to check the design.

9-22. Figure 9P-22 shows the block diagram of a motor-control system. The transfer function of the controlled process is

$$G_p(s) = \frac{1000K}{s(s + a)}$$

where K is the amplifier gain, and a is the inverse of the motor time constant. Design the controllers $G_{cf}(s)$ and $G_c(s)$ so that the following performance specifications are satisfied. The system must be robust when a varies between 8 and 12.

$K_v = 100$ when $a = 10$
Rise time < 0.3 second
Maximum overshoot < 8 percent
Dominant characteristic-equation roots $\cong -5 \pm j5$

Compute and plot the unit-step responses to verify the design.

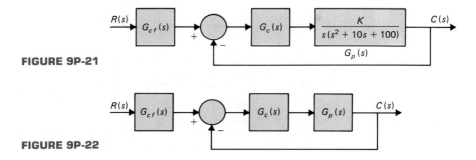

FIGURE 9P-21

FIGURE 9P-22

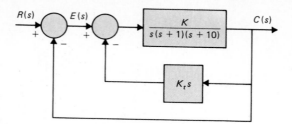

FIGURE 9P-23

9-23. Figure 9P-23 shows the block diagram of a dc-motor-control system with tachometer feedback. Find the values of K and K_t so that the following specifications are satisfied.

$K_v = 1$
Dominant characteristic-equation roots correspond to a damping ratio of approximately 0.707. If there are two solutions, select the larger value of K.

9-24. Carry out the design with the specifications given in Problem 9-23 for the system shown in Fig. 9P-24.

9-25. The block diagram of a control system with a type-2 process is shown in Fig. 9P-25. The system is to be compensated by tachometer feedback and a series controller. Find the values of a, T, K_t, and K so that the following performance specifications are satisfied:

$K_v = 100$
Dominant characteristic-equation roots correspond to a damping ratio of 0.707

For a given set of values of a and T, there are two values for K that will satisfy these requirements. Compute and plot the unit-step responses of the compensated system with the two values of K found.

9-26. A computer-tape-drive system utilizing a permanent-magnet dc motor is shown in Fig. 9P-26(a). The closed-loop system is modeled by the block diagram in Fig. 9P-26(b).

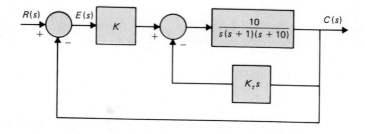

FIGURE 9P-24

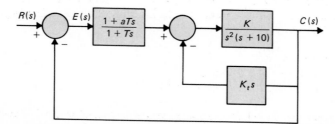

FIGURE 9P-25

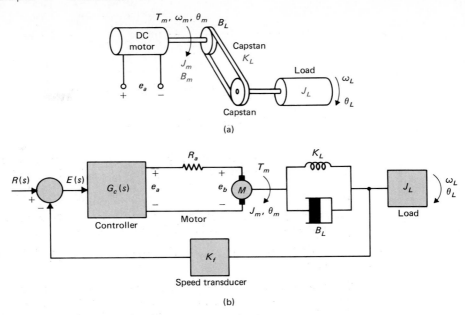

FIGURE 9P-26

(b)

The constant K_L represents the spring constant of the elastic tape, and B_L denotes the viscous-friction coefficient between the tape and the capstans. The system parameters are as follows:

K_i = torque constant = 10 oz-in./A
K_b = back-emf constant = 0.0706 V/rad/s
B_m = motor-friction coefficient = 3 oz-in./rad/s
R_a = 0.25 Ω
K_L = 3000 oz-in./rad
J_L = 6 oz-in./rad/s^2
J_m = 0.05 oz.in./rad/s^2

$L_a \cong 0$ H
B_L = 10 oz-in./rad/s
K_f = 1 V/rad/s

(a)   Write the state equations of the system between e_a and θ_L using θ_L, ω_L, θ_m, and ω_m as the state variables and e_a as the input. Draw a state diagram using the state equations. Derive the transfer functions:

$$\frac{\Omega_m(s)}{E_a(s)} \quad \text{and} \quad \frac{\Omega_L(s)}{E_a(s)}$$

(b)   The objective of the system is to control the speed of the tape, ω_L, accurately. Consider that the PI controller with the transfer function $G_c(s) = K_P + K_I/s$ is to be used. Find the values of K_P and K_I so that the following specifications are satisfied:

K_v = 100
Rise time < 0.02 second
Maximum overshoot < 3 percent

Compute and plot the unit-step response of $\omega_L(t)$ to verify the design.

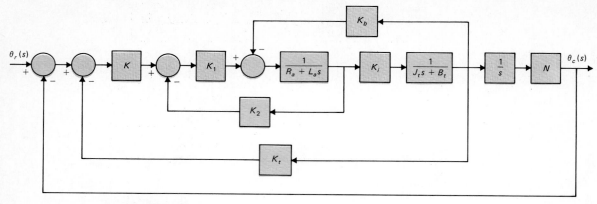

FIGURE 9P-27

9-27. The aircraft-attitude-control system described in Section 7-6 is modeled by the block diagram shown in Fig. 9P-27. Find the values of K and K_t so that the following specifications are satisfied:

$K_v = 100$

Relative damping ratio of the complex roots of the characteristic equation $\cong 0.707$

9-28. The block diagram of a control system with state feedback is shown in Fig. 9P-28. Find the real feedback gains g_1, g_2, and g_3 so that:

    1. the steady-state error e_{ss} [$e(t)$ is the error] due to a step input is zero, and

    2. the complex roots of the characteristic equation are at $-1 + j$ and $-1 - j$. Find the third root. Can all three roots be arbitrarily assigned while still meeting the steady-state requirement in 1?

9-29. The block diagram of a control system with state feedback is shown in Fig. 9P-29(a). The feedback gains g_1, g_2, and g_3 are real constants.

    (a) Find the values of g_1, g_2 and g_3 so that:

        1. the steady-state error e_{ss} [$e(t)$ is the error] is zero when $r(t)$ is a unit-step function, and

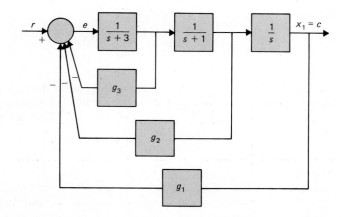

FIGURE 9P-28

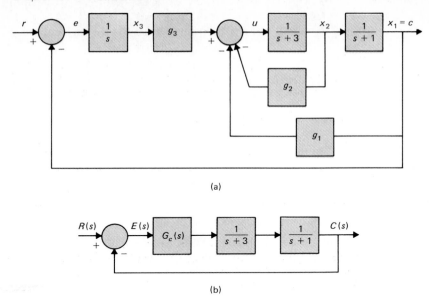

(a)

(b)

2.   the roots of the characteristic equation of the closed-loop system are at
      $-1 + j$, $-1 - j$, and -10.

(b)   Instead of using state feedback, a cascade controller is implemented, as shown in
           Fig. 9P-29(b). Find the transfer function $G_c(s)$ in terms of the g_1, g_2, and g_3 found
           in part (a) and the other system parameters.

9-30.   Figure 9P-30 shows the block diagram of the linearized idle-speed engine-control sys-
            tem described in Problem 5-29. Assume that it is possible to get access to the load-dis-
            turbance torque T_D, and the controllers $G_{cf1}(s)$ and $G_{cf2}(s)$ are for the purpose of elimi-
            nating the effects of T_D on the engine speed ω. Find the transfer functions $G_{cf1}(s)$ and

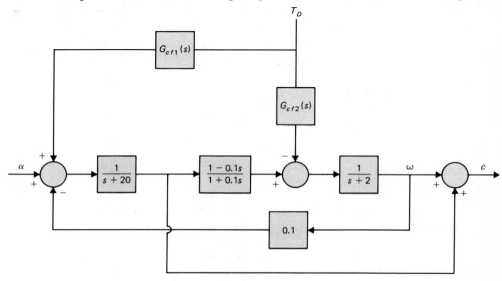

FIGURE 9P-30

$G_{c2}(s)$ in terms of the systems parameters so that the engine speed ω is not affected by T_D.

9-31. Problem 9-12 has revealed that it is impossible to stabilize the "boom-balancing" control system described in Problems 4-23 and 5-43 with a series PD controller. Consider that the system is now controlled by state feedback with $\Delta r = -\mathbf{Gx}$, where

$$\mathbf{G} = [g_1 \quad g_2 \quad g_3 \quad g_4]$$

(a)   Find the feedback gains g_1, g_2, g_3, and g_4 so that the eigenvalues of $\mathbf{A^*} - \mathbf{B^*G}$ are at $-1 + j$, $-1 - j$, -10, and -10. Compute and plot the responses of $\Delta x_1(t)$, $\Delta x_2(t)$, $\Delta x_3(t)$, and $\Delta x_4(t)$ for the initial condition, $\Delta x_1(0) = \Delta\theta(0) = 0.1$, and all other initial conditions are zero.

(b)   Repeat part (a) for the eigenvalues at $-2 + j2$, $-2 - j2$, -20, and -20. Comment on the difference between the two systems.

9-32. The linearized state equations of the ball-suspension control system described in Problem 4-25 are expressed as

$$\Delta\dot{\mathbf{x}} = \mathbf{A^*}\,\Delta\mathbf{x} + \mathbf{B^*}\,\Delta I$$

where

$$\mathbf{A^*} = \begin{bmatrix} 0 & 1 & 0 & 0 \\ 115.2 & -0.05 & -18.6 & 0 \\ 0 & 0 & 0 & 1 \\ -37.2 & 0 & 37.2 & -0.1 \end{bmatrix} \qquad \mathbf{B^*} = \begin{bmatrix} 0 \\ -6.55 \\ 0 \\ -6.55 \end{bmatrix}$$

Let the control current ΔI be derived from state feedback $\Delta I = -\mathbf{G}\,\Delta\mathbf{x}$, where

$$\mathbf{G} = [g_1 \quad g_2 \quad g_3 \quad g_4]$$

(a)   Find the elements of $\mathbf{G}$ so that the eigenvalues of $\mathbf{A^*} - \mathbf{B^*G^*}$ are at $-1 + j$, $-1 - j$, -10, and -10.

(b)   Plot the responses of $\Delta x_1(t) = \Delta y_1(t)$ (magnet displacement) and $\Delta x_3(t) = \Delta y_2(t)$ (ball displacement) with the initial condition:

$$\Delta\mathbf{x}(0) = \begin{bmatrix} 0.1 \\ 0 \\ 0 \\ 0 \end{bmatrix}$$

(c)   Repeat part (b) with the initial condition:

$$\Delta\mathbf{x}(0) = \begin{bmatrix} 0 \\ 0 \\ 0.1 \\ 0 \end{bmatrix}$$

Comment on the responses of the closed-loop system with the two sets of initial conditions used in (b) and (c).

Furnace

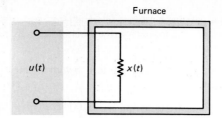

FIGURE 9P-33

9-33. The temperature $x(t)$ in the electric furnace shown in Fig. 9P-33 is described by the differential equation

$$\frac{dx(t)}{dt} = -2x(t) + u(t) + w_2(t)$$

where $u(t)$ is the control signal, and $w_2(t)$ the constant disturbance of unknown magnitude due to heat loss. It is desired that the temperature $x(t)$ follows a reference input $w_1(t)$ that is constant.

(a) Design a control system with state and dynamic feedback so that the following specifications are satisfied:

The roots of the characteristic equation of the closed-loop system are at $s = -10$ and $s = -10$, and

$$\lim_{t \to \infty} x(t) = w_1 = \text{constant}$$

Plot the responses of $x(t)$ for $t \geq 0$ with $w_1 = 1$, $w_2 = -1$, and then with $w_1 = 1$, $w_2 = 0$, all with zero initial conditions.

(b) Design a PI controller so that

$$U(s) = \left(K_P + \frac{K_I}{s} \right) E(s) \qquad E(s) = W_1(s) - X(s)$$

Find K_P and K_I so that the closed-loop characteristic-equation roots are at -10 and -10. Plot the responses of $x(t)$ for $t \geq 0$ with $w_1 = 1$, $w_2 = -1$, and $w_1 = 1$, $w_2 = 0$, all with zero initial conditions.

9-34. (a) For the controlled process shown in Fig. 9P-34, design state and dynamic feedback control so that the state variable x_1 will follow a reference input $w_1 = \text{constant}$ as t approaches infinity. The constant-noise signals w_2 and w_3 are of unknown magnitudes. The roots of the characteristic equation of the closed-loop system should be at $-1 + j$, $-1 - j$, and -10. Compute and plot the time responses of $x_1(t)$ when $w_1(t) = w_2(t) = w_3(t) = u_s(t)$, and when $w_1(t) = u_s(t)$, $w_2(t) = w_3(t) = 0$.

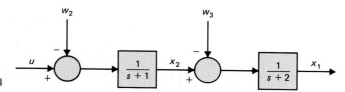

FIGURE 9P-34

(b) Repeat part (a) when the characteristic-equation roots are at -50, $-5 + j5$, and $-5 - j5$. Comment on the noise effects as related to placement of the closed-loop characteristic-equation roots.

9-35. The PI controller for the system in Problem 9-33(b) is determined to be

$$G_c(s) = 2 + \frac{200}{s}$$

Find the digital equivalent using the following: **(a)** The backward-rectangular integration rule. **(b)** The forward-rectangular integration rule. **(c)** The trapezoidal-integration rule.

9-36. The series phase-lead controller for the system in Problem 9-17 is

$$G_c(s) = \frac{1 + 0.4s}{1 + 0.01s}$$

Find the digital equivalent $G_c(z)$ of $G_c(s)$ using a sample-and-hold, as shown in Fig. 9-62.

9-37. Figure 9P-37(a) shows the block diagram of the inventory-control system is Problem 9-9, with a PD controller $G_c(s)$.
  (a) Replace the PD controller by a digital equivalent using Eq. (9-238) for $G_c(z)$ and the digital model in Fig. 9P-37(b). Select the sampling period T so that the maximum overshoot of $c(kT)$, $k = 0, 1, 2, \ldots$, is less than 1 percent.
  (b) Consider that the digital PD controller has the transfer function

$$G_c(z) = K_P + \frac{K_P(z - 1)}{Tz}$$

Find the values of K_P and K_D in terms of the sampling period T so that the two roots of the closed-loop characteristic equation are at $z = 0.5$, and 0.5. Find the

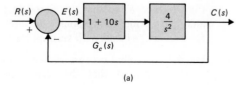

(a)

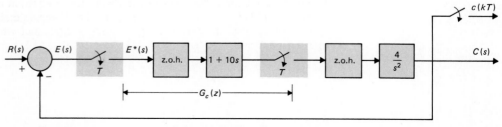

(b)

FIGURE 9P-37

other characteristic-equation root. Find $c(kT)$ for $k = 0, 1, 2, \ldots$ for $r(t) = u_s(t)$ and $T = 0.01$ second.

9-38.   Consider the missile-attitude-control system described in Problem 4-13, part (b). The transfer function between the thrust angle $\delta(t)$ and the angle of attack $\theta(t)$ is

$$G_p(s) = \frac{\Theta(s)}{\Delta(s)} = \frac{T_s d_2}{Js^2 - K_F d_1}$$

Let $T_s d_2 / J = 10$ and $K_F d_1 / J = 1$. Since the process has a pole in the right-half s-plane, it is difficult to stabilize it with an analog controller. Let us choose a digital controller with the block diagram of the digital control system as shown in Fig. 9P-38. The sampling period is chosen to be 0.1 s. The transfer function of the digital controller is

$$G_c(z) = \frac{z + z_1}{z + z_2}$$

where z_1 and z_2 are real constants.

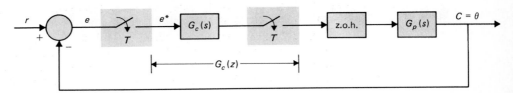

FIGURE 9P-38

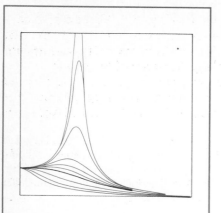

10

FREQUENCY-DOMAIN ANALYSIS OF CONTROL SYSTEMS

INTRODUCTION

In practice the performance of a control system is more realistically and directly measured by its time-domain response characteristics. The reason is that the performance of most control systems is judged based on the time responses due to certain test signals. This is in contrast to the analysis and design of communication systems for which the frequency response is of more importance, since in this case, most of the signals to be processed are either sinusoidal or composed of sinusoidal components. We learned in Chapter 7 that analytically the time response of a control system is usually more difficult to determine, especially in the case of high-order systems. In design problems, the difficulties lie in the fact that there are no unified methods of arriving at a designed system given the time-domain performance specifications, such as maximum overshoot, rise time, delay time, settling time, and so on. On the other hand, there is a wealth of graphical methods available in the frequency domain, all suitable for the analysis and design of linear control systems. It is important to realize that there are correlating relations between the frequency-domain and time-domain performances in a linear system, so that the time-domain properties of the system can be predicted based on the frequency-domain characteristics. With this in mind, we may consider that the primary motivation of conducting control-systems analysis and design in the frequency domain is because of convenience and the availability of the existing analytical tools. Another reason is that it presents an alternative dimension to the control-systems problems, so that together with the time-domain studies, one can gain additional perspective to the complex analysis and design problems of control systems. Therefore, to conduct a frequency-domain analysis of a linear control system does not imply that the system will ever be subject to a sine-wave input. Rather, from the frequency-response studies, we will be able to project the time-domain performance of the system.

The starting point of the frequency-domain analysis of a linear system is the transfer function of the system. It is well known from linear system theory that when the input to a linear time-invariant system is sinusoidal with amplitude R and frequency ω_o,

$$r(t) = R \sin \omega_o t \tag{10-1}$$

the steady-state output of the system $c(t)$ will still be a sinusoid with the same frequency ω_o, but possibly with different amplitude and phase; i.e.,

$$c(t) = C \sin (\omega_o t + \phi) \tag{10-2}$$

where C is the amplitude of the output sine wave, and ϕ is the phase shift in degrees or radians.

Let the transfer function of a single-input single-output linear system be $G(s)$; then the Laplace transforms of the input and the output are related through

$$C(s) = G(s)R(s) \tag{10-3}$$

sinusoidal
steady state

For sinusoidal steady-state analysis, we replace s by $j\omega$, and the last equation becomes

$$C(j\omega) = G(j\omega)R(j\omega) \tag{10-4}$$

By writing the function $C(j\omega)$ as

$$C(j\omega) = |C(j\omega)| \angle C(j\omega) \tag{10-5}$$

with similar definitions for $G(j\omega)$ and $R(j\omega)$, Eq. (10-4) leads to the magnitude relation between the input and the output:

$$|C(j\omega)| = |G(j\omega)||R(j\omega)| \tag{10-6}$$

and the phase relation:

$$\angle C(j\omega) = \angle G(j\omega) + \angle R(j\omega) \tag{10-7}$$

Thus, for the input and output signals described by Eqs. (10-1) and (10-2), respectively, the amplitude of the output sinusoid is

$$C = R|G(j\omega_o)| \tag{10-8}$$

and the phase of the output is

$$\phi = \angle G(j\omega_o) \tag{10-9}$$

magnitude
phase
Thus, by knowing the transfer function $G(s)$ of a linear system, the magnitude characteristics, $|G(j\omega)|$, and the phase characteristics, $\angle G(j\omega)$, completely describe the steady-state performance when the input is a sinusoid. The crux of the frequency-domain analysis of linear control systems is that the amplitude and phase characteristics of a closed-loop system can be used to predict the time-domain transient performance characteristics such as overshoot, rise time, and general damping properties, as well as the steady-state properties.

Frequency Response of Closed-loop Systems

For the single-loop control-system configuration studied in the preceding chapters, the closed-loop transfer function is

$$M(s) = \frac{C(s)}{R(s)} = \frac{G(s)}{1 + G(s)H(s)} \tag{10-10}$$

Under the sinusoidal steady state, $s = j\omega$; Eq. (10-10) becomes

$$M(j\omega) = \frac{C(j\omega)}{R(j\omega)} = \frac{G(j\omega)}{1 + G(j\omega)H(j\omega)} \tag{10-11}$$

The sinusoidal steady-state transfer function $M(j\omega)$ may be expressed in terms of its magnitude and phase; that is,

$$M(j\omega) = |M(j\omega)| \angle M(j\omega) \tag{10-12}$$

Or $M(j\omega)$ can be expressed in terms of its real and imaginary parts:

$$M(j\omega) = \text{Re}\,[M(j\omega)] + j\,\text{Im}\,[M(j\omega)] \tag{10-13}$$

The magnitude of $M(j\omega)$ is

$$|M(j\omega)| = \left| \frac{G(j\omega)}{1 + G(j\omega)H(j\omega)} \right| = \frac{|G(j\omega)|}{|1 + G(j\omega)H(j\omega)|} \tag{10-14}$$

and the phase of $M(j\omega)$ is

$$\angle M(j\omega) = \phi_M(j\omega) = \angle G(j\omega) - \angle[1 + G(j\omega)H(j\omega)] \tag{10-15}$$

For communication, the magnitude and phase of the frequency response carry great significance on the system performance. For instance, let $M(s)$ represent the input–output transfer function of an electric filter; then the properties of $|M(j\omega)|$ and $\angle M(j\omega)$ govern the filtering characteristics on the input signal. Figure 10-1 shows the gain and phase characteristics of an ideal low-pass filter that has a sharp cutoff frequency at ω_c. It is well known that such a low-pass filter characteristic is physically un-

low-pass filter realizable, since the corresponding system is noncausal. The ideal low-pass filter characteristics shown in Fig. 10-1 would be highly desirable, since it passes all signals with a flat gain below the frequency of ω_c, and cuts off completely signals with frequencies above ω_c. In many ways, the design of control systems is quite similar to filter design, and the control system is regarded as a signal processor. In fact, if the gain and phase characteristics shown in Fig. 10-1 were physically realizable, they would be ideal as the frequency response of a control system. The output of such a system would follow all inputs with frequencies up to ω_c without any error.

If ω_c is increased indefinitely, the output $C(j\omega)$ would be identical to the input $R(j\omega)$ at all frequencies. Such a system would be able to follow a step-function input in the time domain exactly. From Eq. (10-14), we see that for $|M(j\omega)|$ to be unity at all frequencies, the magnitude of the open-loop transfer function $G(j\omega)$ must be infinite. An infinite magnitude for $G(j\omega)$ is, of course, impossible to achieve in practice, nor would it be desirable, since most control systems become unstable when their loop gains become very high. Furthermore, all control systems are subject to noise during operation. Thus, in addition to responding to the input signal, the system should be able to reject and suppress noise and unwanted signals. This means that the frequency response of a control system should have a finite cutoff frequency.

The phase characteristics of the frequency response of a control system are also of importance, as they play an important role on the stability of closed-loop systems.

Figure 10-2 illustrates typical gain and phase characteristics of a feedback control system. As shown by Eqs. (10-14) and (10-15), the gain and phase characteristics of the closed-loop system can be determined from that of the open-loop transfer function $G(s)$ and the loop transfer function $G(s)H(s)$. In practice, the frequency responses of $G(s)$ and $H(s)$ can be determined by applying sine-wave inputs to the system and sweeping the frequency from 0 to a value beyond the frequency range of the system.

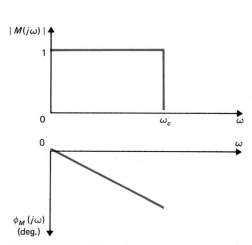

FIGURE 10-1 Gain—phase characteristics of an ideal low-pass filter.

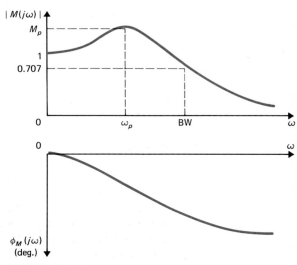

FIGURE 10-2 Typical gain—phase characteristics of a feedback control system.

Frequency-Domain Specifications

In the design of linear control systems using the frequency-domain methods, it is necessary to define a set of specifications so that the quality of the system can be properly described using the frequency-domain characteristics. The following frequency-domain specifications are often used in practice.

maximum overshoot

PEAK RESONANCE M_p.   *The peak resonance M_p is defined as the maximum value of $|M(j\omega)|$ in Eq. (10-14).* In general, the magnitude of M_p gives an indication of the relative stability of a stable feedback control system. Normally, a large M_p corresponds to a large maximum overshoot of the step response in the time domain. For most control systems, it is generally accepted in practice that the desirable value of M_p should be between 1.1 and 1.5.

RESONANT FREQUENCY ω_p.   *The resonant frequency ω_p is defined as the frequency at which the peak resonance M_p occurs.*

rise time

BANDWIDTH.   *The bandwidth, BW, is defined as the frequency at which the magnitude of $|M(j\omega)|$ drops to 70.7 percent of its zero-frequency value, or 3 dB down from the zero-frequency value.* In general, the bandwidth of a control system gives measure of the transient-response properties, in that a large bandwidth corresponds to a faster rise time, since higher-frequency signals are more easily passed on to the outputs. Conversely, if the bandwidth is small, only signals of relatively low frequencies are passed, and the time response will generally be slow and sluggish. Bandwidth also indicates the noise-filtering characteristics and the robustness of the system.

CUTOFF RATE.   Often, bandwidth alone is inadequate in the indication of the characteristics of the system in distinguishing signals from noise. Sometimes it may be necessary to specify the cutoff rate of the frequency response, which is the slope of $|M(j\omega)|$ at the high frequencies. Apparently, two systems can have the same bandwidth, but the cutoff rates of the frequency responses may be different.

The performance criteria defined for the frequency-domain analysis are illustrated in Fig. 10-2. There are other criteria that are just as important in specifying the relative stability and performance of a control system in the frequency domain. These are defined in the later sections of this chapter.

10-2

M_p, ω_p, AND BANDWIDTH OF THE PROTOTYPE SECOND-ORDER SYSTEM

Peak Resonance and Resonant Frequency

damping ratio

For the prototype second-order feedback control system without any open-loop zeros, the peak resonance M_p, the resonant frequency ω_p, and the bandwidth BW are all uniquely related to the damping ratio ζ and the natural undamped frequency ω_n of the system. Consider the closed-loop transfer function

prototype
second-order
system

$$\frac{C(s)}{R(s)} = \frac{\omega_n^2}{s^2 + 2\zeta\omega_n s + \omega_n^2} \tag{10-16}$$

The transfer function at sinusoidal steady state is obtained by substituting $s = j\omega$ in the equation. We have

$$M(j\omega) = \frac{C(j\omega)}{R(j\omega)} = \frac{\omega_n^2}{(j\omega)^2 + 2\zeta\omega_n(j\omega) + \omega_n^2}$$

$$= \frac{1}{1 + j2(\omega/\omega_n)\zeta - (\omega/\omega_n)^2} \tag{10-17}$$

We can simplify the last expression by letting $u = \omega/\omega_n$. Then, Eq. (10-17) becomes

$$M(ju) = \frac{1}{1 + j2u\zeta - u^2} \tag{10-18}$$

The magnitude and phase of $M(ju)$ are

$$|M(ju)| = \frac{1}{[(1 - u^2)^2 + (2\zeta u)^2]^{1/2}} \tag{10-19}$$

and

$$\angle M(ju) = \phi_m(u) = -\tan^{-1}\frac{2\zeta u}{1 - u^2} \tag{10-20}$$

resonant
frequency

respectively. The resonant frequency is determined by setting the derivative of $|M(u)|$ with respect to u to zero. Thus,

$$\frac{d|M(u)|}{du} = -\frac{1}{2}[(1 - u^2)^2 + (2\zeta u)^2]^{-3/2}(4u^3 - 4u + 8u\zeta^2) = 0 \tag{10-21}$$

from which

$$4u^3 - 4u + 8u\zeta^2 = 0 \tag{10-22}$$

The roots of Eq. (10-22) are $u_p = 0$ and

$$u_p = \sqrt{1 - 2\zeta^2} \tag{10-23}$$

The solution of $u_p = 0$ merely indicates that the slope of the $|M(\omega)|$-versus-ω curve is zero at $\omega = 0$; it is not a true maximum if ζ is less than 0.707. Equation (10-23) gives the resonant frequency:

$$\omega_p = \omega_n\sqrt{1 - 2\zeta^2} \tag{10-24}$$

Since frequency is a real quantity, Eq. (10-24) is valid only for $1 \geq 2\zeta^2$, or $\zeta \leq 0.707$. This means simply that for all values of ζ greater than 0.707, the resonant frequency is $\omega_p = 0$, and $M_p = 1$.

Substituting Eq. (10-23) in Eq. (10-19) for u and simplifying, we get

resonance
peak

$$M_p = \frac{1}{2\zeta\sqrt{1 - \zeta^2}} \qquad \zeta \leq 0.707 \qquad (10\text{-}25)$$

It is important to note that for the prototype second-order system described by Eq. (10-16), M_p is a function of the damping ratio ζ only, and ω_p is a function of ζ and ω_n. Furthermore, although taking the derivative of $|M(ju)|$ with respect to u is a valid

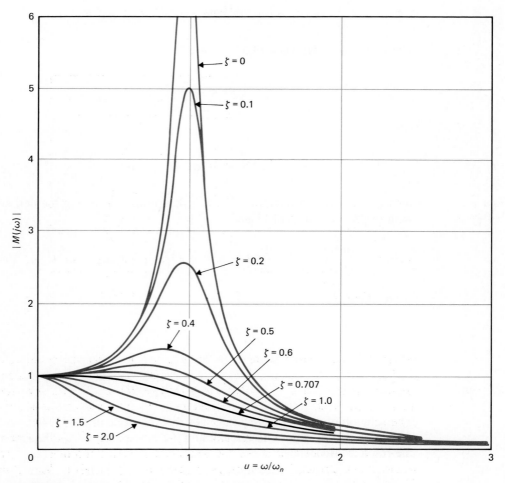

FIGURE 10-3 Magnification versus normalized frequency of a second-order closed-loop control system.

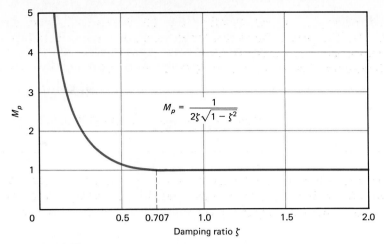

FIGURE 10-4 M_p versus damping ratio for a second-order system.

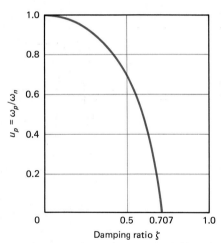

FIGURE 10-5 Normalized resonant frequency versus the damping ratio for a second-order system. $u_p = \sqrt{1 - 2\zeta^2}$.

FREQRP

method of determining M_p and ω_p, in general, for high-order systems, this analytical method is quite tedious and is not recommended. The graphical methods that will be discussed in this chapter and the computer program FREQRP that accompanies this text are much more efficient for the frequency-domain analysis and design of linear control systems.

Figure 10-3 illustrates the plots of $|M(ju)|$ of Eq. (10-19) versus u for various values of ζ. Notice that if the frequency scale were unnormalized, the value of ω_p would increase when ζ decreases, as indicated by Eq. (10-24). When $\zeta = 0$, $\omega_p = \omega_n$. Figures 10-4 and 10-5 illustrate the relationship between M_p and ζ, and $u_p(= \omega_p/\omega_n)$ and ζ, respectively.

Bandwidth

The bandwidth BW of a closed-loop system is the frequency at which $|M(j\omega)|$ drops to 70.7 percent of, or 3 dB down, from the zero-frequency value. Equating Eq. (10-19) to 0.707, we have

$$|M(ju)| = \frac{1}{[(1 - u^2)^2 + (2\zeta u)^2]^{1/2}} = 0.707 \tag{10-26}$$

Thus,

$$[(1 - u^2)^2 + (2\zeta u)^2]^{1/2} = 1.414 \tag{10-27}$$

This equation leads to

$$u^2 = (1 - 2\zeta^2) \pm \sqrt{4\zeta^4 - 4\zeta^2 + 2} \tag{10-28}$$

In the last equation, the plus sign should be chosen, since u must be a positive real quantity for any ζ. Therefore, from Eq. (10-28), the bandwidth of the prototype second-order system is determined as

bandwidth

$$BW = \omega_n[(1 - 2\zeta^2) + \sqrt{4\zeta^4 - 4\zeta^2 + 2}]^{1/2} \tag{10-29}$$

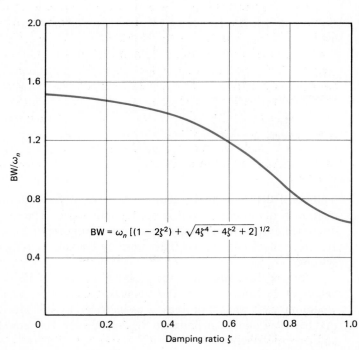

FIGURE 10-6 Bandwidth/ω_n versus the damping ratio for a second-order system.

Figure 10-6 shows the plot of BW/ω_n as a function of ζ. It is of interest to note that for a fixed ω_n, as the damping ratio ζ decreases from unity, the bandwidth increases and the resonance peak M_p also increases.

We have established some simple relationships between the time-domain response and the frequency-response characteristics of the prototype second-order system under consideration. The summary of these relationships are as follows.

summary of relations

1. The maximum overshoot of the unit-step response in the time domain depends upon ζ only [Eq. (7-84)].
2. The resonance peak of the closed-loop frequency response M_p depends on ζ only [Eq. (10-25)].
3. The rise time of the step response increases with ζ, Eq. (7-89) and Fig. 7-19. The bandwidth of the system decreases with the increase in ζ, for a fixed ω_n, Eq. (10-29) and Fig. 10-6. Therefore, bandwidth and rise time are inversely proportional to each other.
4. Bandwidth is directly proportional to ω_n.
5. Higher bandwidth corresponds to larger M_p.

10-3

EFFECTS OF ADDING A ZERO TO THE OPEN-LOOP TRANSFER FUNCTION

The relationships between the time-domain and the frequency-domain responses arrived at in the preceding section apply only to the prototype second-order system described by Eq. (10-16). When other second-order or higher-order systems are involved, the relationships are different and may be more complex. It is of interest to consider the effects on the frequency-domain characteristics of a feedback control system when poles and zeros are added to the prototype second-order transfer function. It would be a simpler procedure to study the effects of adding poles and zeros to the closed-loop transfer function. However, it is more realistic for design considerations to modify the open-loop transfer function.

The closed-loop system transfer function of Eq. (10-16) may be considered as that of a unit-feedback control system with the prototype second-order open-loop transfer function

$$G(s) = \frac{\omega_n^2}{s(s + 2\zeta\omega_n)} \qquad (10\text{-}30)$$

Let us add a zero at $s = -1/T$ to the last transfer function so that Eq. (10-30) becomes

$$G(s) = \frac{\omega_n^2(1 + Ts)}{s(s + 2\zeta\omega_n)} \qquad (10\text{-}31)$$

This corresponds to the second-order system with derivative control studied in Section

9-2. The closed-loop transfer function of the system is

$$M(s) = \frac{C(s)}{R(s)} = \frac{\omega_n^2(1 + Ts)}{s^2 + (2\zeta\omega_n + T\omega_n^2)s + \omega_n^2}$$
(10-32)

In principle, M_p, ω_p, and BW of the system can all be derived using the same steps as illustrated in the previous section. However, since there are now three parameters in ζ, ω_n, and T, the exact expression for M_p, ω_p, and BW are difficult to obtain even though the system is still of the second order. The bandwidth of the system is derived to be

$$\text{BW} = (-b + \tfrac{1}{2}\sqrt{b^2 + 4\omega_n^4})^{1/2}$$
(10-33)

bandwidth   where

$$b = 4\zeta^2\omega_n^2 + 4\zeta\omega_n^3 T - 2\omega_n^2 - \omega_n^4 T^2$$
(10-34)

It is difficult to see how each of the parameters in Eq. (10-33) affects the bandwidth. Figure 10-7 shows the relationship between BW and T for $\zeta = 0.707$ and
adding a zero $\omega_n = 1$. Notice that *the general effect of adding a zero to the open-loop transfer function is to increase the* bandwidth *of the closed-loop system.* However, over a certain range of small values of T, the bandwidth is actually decreased. Figures 10-8 and 10-9 give the plots of $|M(j\omega)|$ of the closed-loop system that has the $G(s)$ of Eq. (10-31) as its open-loop transfer function; $\omega_n = 1$, T takes on various values, and $\zeta = 0.707$ and $\zeta_2 = 0.2$, respectively. *These curves show that for large values of T, the bandwidth of*

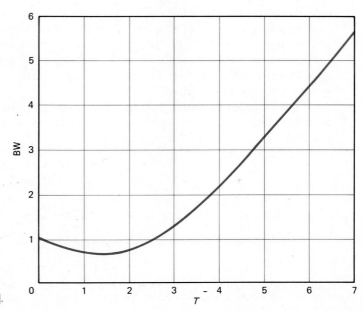

FIGURE 10-7 Bandwidth of a second-order system with open-loop transfer function $G(s) = (1 + Ts)/[s(s + 1.414)]$.

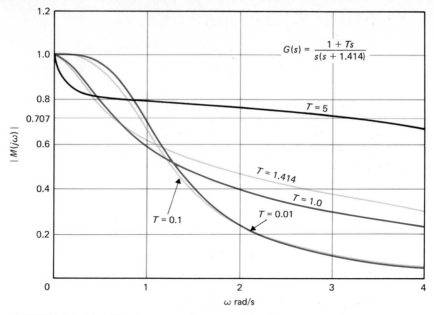

FIGURE 10-8 Magnification curves for a second-order system with an open-loop transfer function $G(s)$.

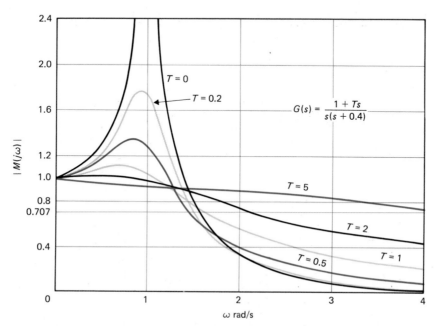

FIGURE 10-9 Magnification curves for a second-order system with an open-loop transfer function $G(s)$.

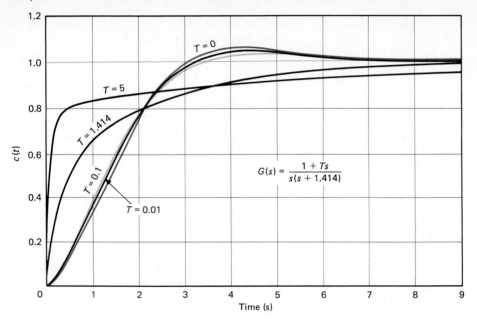

FIGURE 10-10 Unit-step responses of a second-order system with an open-loop transfer function $G(s)$.

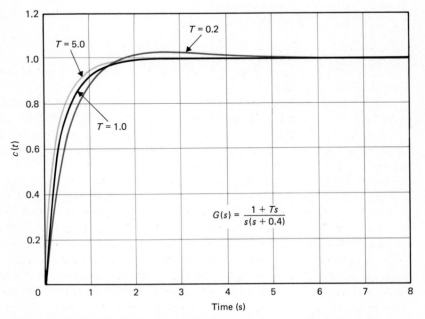

FIGURE 10-11 Unit-step responses of a second-order system with an open-loop transfer function $G(s)$.

the closed-loop system is increased, whereas there exists a range of smaller values of T for which the BW is decreased by the addition of the zero to G(s). Figure 10-10 and 10-11 show the corresponding unit-step responses of the closed-loop system. The time-domain responses indicate that a high bandwidth corresponds to a faster rise time. However, as T becomes very large, the zero of the closed-loop transfer function, which is at $s = -1/T$, moves very close to the origin, causing the system to have a large time constant. Thus, Fig. 10-10 illustrates the situation that the rise time is fast, but the large time constant of the zero near the origin of the s-plane causes the time response to drag out in reaching the final steady state.

10-4

EFFECTS OF ADDING A POLE TO THE OPEN-LOOP TRANSFER FUNCTION

The addition of a pole to the open-loop transfer function generally has the effect of decreasing the bandwidth of the closed-loop system. The following transfer function is arrived at by adding $(1 + Ts)$ to the denominator of Eq. (10-30).

$$G(s) = \frac{\omega_n^2}{s(s + 2\zeta\omega_n)(1 + Ts)} \qquad (10\text{-}35)$$

The derivation of the bandwidth of the closed-loop system that has the open-loop transfer function in Eq. (10-35) is quite tedious. The result is that the BW is the real solution of the following equation:

$$T^2\omega^6 + (1 + 4\zeta^2\omega_n^2 T^2)\omega^4 + (4\zeta^2\omega^2 - 2\omega_n^2 - 4\zeta\omega_n^3 T)\omega^2 - \omega_n^4 = 0 \quad (10\text{-}36)$$

We can obtain a qualitative indication on the bandwidth properties by referring to Fig. 10-12, which shows the plots of $|M(j\omega)|$ for $\omega_n = 1$, $\zeta = 0.707$, and various values of T. Since the system is now of the third order, it can be unstable for a certain set of system parameters. It can be easily shown by use of the Routh–Hurwitz criterion that for $\omega_n = 1$ and $\zeta = 0.707$, the system is stable for all positive values of T. The $|M(j\omega)|$-versus-ω curves of Fig. 10-12 show that for small values of T, the bandwidth of the system is slightly increased, but M_p is also increased. When T becomes large, the pole added to $G(s)$ has the effect of decreasing the bandwidth but increasing M_p.

effect on BW   Therefore, we can conclude that, *in general, the effect of adding a pole to the open-loop transfer function is to make the closed-loop system less stable, while decreasing the bandwidth.* The unit-step responses of Fig. 10-13 clearly show that for the larger values of T, $T = 1$ and $T = 5$, the rise time increases with the decrease of the bandwidth, and the larger values of M_p also correspond to greater peak overshoots in the step responses. It is important to note that the correlation between M_p and the peak overshoot of the step response is meaningful only when the closed-loop system is stable. When $G(j\omega) = -1$, $|M(j\omega)|$ is infinite, and the closed-loop system is unstable. On the other hand, the closed-loop system may be unstable, but the value of $|M(j\omega)|$ may be finite.

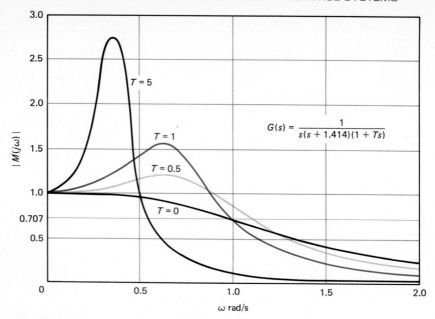

FIGURE 10-12 Magnification curves for a third-order system with an open-loop transfer function $G(s)$.

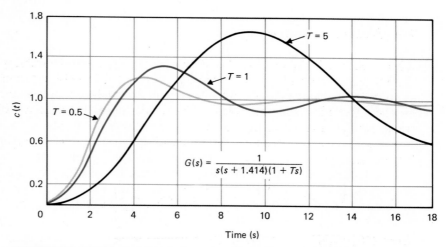

FIGURE 10-13 Unit-step responses of a third-order system with an open-loop transfer function $G(s)$.

The objective of these last two sections is to demonstrate the simple relationships between BW and M_p, and the time-domain response. Typical effects on BW of adding a pole and a zero to the open-loop transfer function are discussed. No attempt is made to include all general cases.

10-5

NYQUIST STABILITY CRITERION

In the earlier chapters, we discussed two methods of determining the stability of linear systems. The Routh–Hurwitz criterion and the root-locus method both rely on the determination of the location of the roots of the characteristic equation.

The Nyquist criterion is a semigraphical method that determines the stability of a closed-loop system by investigating the properties of the frequency-domain plot of the loop transfer function $G(s)H(s)$. The Nyquist method has attractive features that make it desirable for the analysis as well as the design of control systems. The distinctive features of the Nyquist criterion are as follows:

distinctive features

1. The Nyquist criterion provides the same information on the absolute stability of a control system as does the Routh–Hurwitz criterion.
2. In addition to absolute system stability, the Nyquist criterion indicates the degree of stability (relative stability) of a stable system, and the degree of instability of an unstable system, and gives an indication of how the system stability may be improved, if needed.
3. The frequency-domain plots of $G(j\omega)H(j\omega)$ give information on the frequency-domain characteristics of the closed-loop system.
4. The stability of a closed-loop system with a pure time delay can be studied using the Nyquist criterion.

The Stability Problem

It has been established that for a linear time-invariant system, the stability of the system depends on the roots of the characteristic equation. The Nyquist criterion is used to determine the location of these roots with respect to the left-half s-plane and the right-half s-plane. Unlike the root-locus method, the Nyquist criterion does not give the exact location of the characteristic-equation roots.

Let us consider that the closed-loop transfer function of a single-input single-output system is

$$M(s) = \frac{C(s)}{R(s)} = \frac{G(s)}{1 + G(s)H(s)} \tag{10-37}$$

where $G(s)$ and $H(s)$ can assume the following form:

$$G(s) \text{ or } H(s) = \frac{K(1 + T_1s)(1 + T_2s) \cdots (1 + T_ms)}{s^p(1 + T_as)(1 + T_bs) \cdots (1 + T_ns)} e^{-T_ds} \tag{10-38}$$

characteristic equation

Since the characteristic equation is obtained by setting the denominator polynomial of the closed-loop system transfer function to zero, it is recognized that the *roots of the characteristic equation are the zeros of* $1 + G(s)H(s)$. Or the characteristic roots must satisfy

$$\Delta(s) = 1 + G(s)H(s) = 0 \qquad (10\text{-}39)$$

In general, for a closed-loop control system with multiple number of loops, it is more appropriate to represent the denominator of the closed-loop transfer function as

$$\Delta(s) = 1 + F(s) = 0 \qquad (10\text{-}40)$$

where $F(s)$ is of the form of Eq. (10-38).

Before embarking on the details of the Nyquist criterion, it is helpful to summarize the pole-zero relationships of the various system transfer functions.

pole-zero relations

1. Identification of Poles and Zeros

loop transfer function zeros: zeros of $F(s)$

loop transfer function poles: poles of $F(s)$

closed-loop transfer function poles: zeros of $1 + F(s)$

= roots of the characteristic equation

The poles of $1 + F(s)$ = poles of $F(s)$, the loop transfer function.

2. Stability Conditions

The stability referred to here is again the BIBO and asymptotic stability defined in Chapter 6.

We can define two types of stability with respect to the closed-loop system with the transfer function given in Eq. (10-37).

open-loop stability

OPEN-LOOP STABILITY.   The system is said to be **open-loop stable** if the *poles* of the loop transfer function $G(s)H(s)$ are all in the left-half s-plane.

closed-loop stability

CLOSED-LOOP STABILITY.   The system is said to be **closed-loop stable**, or simply stable, if the poles of the closed-loop transfer function or the *zeros* of $1 + F(s)$ are all in the left-half s-plane.

An exception to the open-loop stability is a loop transfer function with a pole at the origin that is defined as stable.

Definition of "Encircled" and "Enclosed"

Since the Nyquist stability criterion is a graphical method, we need to establish the concepts of **encircled** and **enclosed**, which are used for the interpretation of the Nyquist plots for stability.

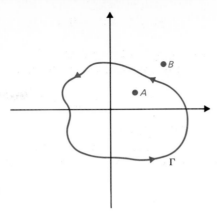

FIGURE 10-14 Definition of encirclement.

encircled

ENCIRCLED.   *A point in a complex plane is said to be encircled by a closed path if it is found inside the path.* For example, point A in Fig. 10-14 is encircled by the closed path Γ, since A is *inside* the closed path. Point B is not encircled by the closed path Γ, since it is *outside* the path. Furthermore, when the closed path Γ has a direction assigned to it, the encirclement, if made, can be in the clockwise or the counterclockwise direction. As shown in Fig. 10-14, point A is encircled by Γ in the counterclockwise direction. Thus, we can say that the region *inside* the path is encircled in the prescribed direction, and the region *outside* is not encircled.

enclosed

ENCLOSED.   *A point or region is said to be enclosed by a closed path if it is found to be encircled in the counterclockwise direction, or, the point or region lies to the left of the path when the path is traversed in any prescribed direction.* The latter definition is particularly useful if only a portion of the closed path is drawn. For instance, the shaded regions shown in Figs. 10-15(a) and (b) are considered to be *enclosed* by the closed path Γ. In other words, point A in Fig. 10-15(a) is *enclosed* by Γ, but point A in Fig. 10-15(b) is not. However, in Fig. 10-15(b), point B and all the points in the shaded region outside Γ are enclosed. Similarly, point A in Fig. 10-14 is enclosed by Γ, but point B is not.

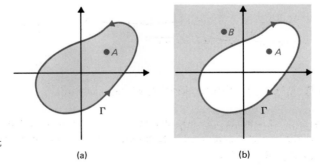

FIGURE 10-15 Definition of enclosed points and regions. (a) Point A is enclosed by Γ. (b) Point A is not enclosed, but B is enclosed by the locus Γ.

(a)   (b)

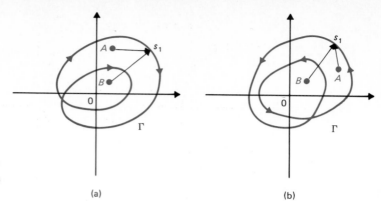

FIGURE 10-16 Definition of the number of encirclements and enclosures.

(a)   (b)

Number of Encirclements and Enclosures

When a point is encircled by a closed path Γ, a number N can be assigned to the number of times it is encircled. The magnitude of N can be determined by drawing a phasor from the point to any arbitrary point s_1 on the closed path Γ and then let s_1 follow the path in the prescribed direction until it returns to the starting point. The total *net* number of revolutions traversed by this phasor is N. Alternatively, the net angle traversed by the phasor is $2\pi N$ radians. For example, point A in Fig. 10-16(a) is *encircled once* by Γ, and point B is *encircled twice*, all in the clockwise direction. Thus, the magnitude of the angle traversed by the phasor drawn from A to the trajectory Γ as the representative point s_1 travels around the trajectory once is 2π radians, and that of B is 4π radians. In Fig. 10-16(b), point A is *enclosed once*, and point B is *enclosed twice* by Γ. Later we shall define a sign for N according to the direction of encirclement.

Principle of the Argument

The Nyquist criterion was originated as an engineering application of the well-known principle of the argument in complex-variable theory. The principle is stated in the following in a heuristic manner.

Let $\Delta(s)$ be a single-valued function of the form of the right-hand side of Eq. (10-38), and the function has a finite number of poles in the s-plane. Single valued means that for each point in the s-plane, there is one and only one corresponding point, including infinity, in the complex $\Delta(s)$-plane. As defined in Chapter 8, infinity in the complex plane is defined to be a point.

Suppose that a continuous closed path Γ_s is arbitrarily chosen in the s-plane, as shown in Fig. 10-17(a). If Γ_s does not go through any poles of $\Delta(s)$, then the trajectory Γ_Δ mapped by the function $\Delta(s)$ into the $\Delta(s)$-plane is also a closed one, as shown by the example illustrated in Fig. 10-17(b). Starting from a point s_1, the Γ_s locus is traversed in the arbitrarily chosen direction (clockwise in the illustrated case), through the points s_2 and s_3, and then returning to s_1 after going through all the points on the Γ_s locus, as shown in Fig. 10-17(a), the corresponding Γ_Δ locus will start from the point $\Delta(s_1)$ and go through points $\Delta(s_2)$ and $\Delta(s_3)$, corresponding to s_1, s_2, and s_3, respec-

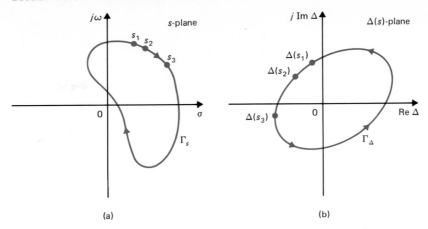

FIGURE 10-17 (a) Arbitrarily chosen closed path in the s-plane. (b) Corresponding locus Γ_Δ in the $\Delta(s)$-plane.

tively, and finally return to the starting point, $\Delta(s_1)$. The direction of traverse of Γ_Δ can be either clockwise or counterclockwise; that is, in the same direction or the opposite direction as that of Γ_s, depending on the particular function $\Delta(s)$. In Fig. 10-17(b), the direction of Γ_Δ is arbitrarily assigned, for illustration purposes, to be counterclockwise.

single-valued mapping

It should be pointed out that, although the mapping from the s-plane to the $\Delta(s)$-plane is single-valued for the function $\Delta(s)$, the reverse process is usually not a single-valued mapping. For example, consider the function

$$\Delta(s) = \frac{K}{s(s+1)(s+2)} \qquad (10\text{-}41)$$

which has poles $s = 0$, -1, and -2 in the s-plane. For each value of s in the s-plane, there is only one corresponding point found in the $\Delta(s)$-plane. However, for each point in the $\Delta(s)$-plane, the function maps into three corresponding points in the s-plane. The simplest way to illustrate this is to write Eq. (10-41) as

$$s(s+1)(s+2) - \frac{K}{\Delta(s)} = 0 \qquad (10\text{-}42)$$

For example, if $\Delta(s)$ is a constant, which represents any point on the real axis in the $\Delta(s)$-plane, the third-order equation in Eq. (10-42) gives three roots in the s-plane. The reader should recognize the parallel of this situation to the root-locus diagram that essentially represents the mapping of $\Delta(s) = -1 + j0$ onto the roots of the characteristic equation in the s-plane, for a given value of K. Thus, the root loci of Eq. (10-41) have three individual branches in the s-plane.

principle of the argument

The principle of the argument can be stated: *Let $\Delta(s)$ be a single-valued function that has a finite number of poles in the s-plane. Suppose that an arbitrary closed path Γ_s is chosen in the s-plane so that the path does not go through any one of the poles or* *zeros of $\Delta(s)$; the corresponding Γ_Δ locus mapped in the $\Delta(s)$-plane will encircle the origin as many times as the difference between the number of the zeros and the number of poles of $\Delta(s)$ that are encircled by the s-plane locus Γ_s.*

In equation form, the principle of the argument is

$$N = Z - P \qquad (10\text{-}43)$$

where

N = number of encirclements of the origin made by the $\Delta(s)$-plane locus Γ_Δ.
Z = number of zeros of $\Delta(s)$ encircled by the s-plane locus Γ_s in the s-plane.
P = number of poles of $\Delta(s)$ encircled by the s-plane locus Γ_s in the s-plane.

In general, N can be positive $(Z > P)$, zero $(Z = P)$, or negative $(Z < P)$. These three situations are described in more detail as follows.

1. $N > 0$ $(Z > P)$. If the s-plane locus encircles more zeros than poles of $\Delta(s)$ in a certain prescribed direction (clockwise or counterclockwise), N is a positive integer. In this case, the $\Delta(s)$-plane locus Γ_Δ will encircle the origin of the $\Delta(s)$-plane N times in the same direction as that of Γ_s.
2. $N = 0$ $(Z = P)$. If the s-plane locus encircles as many poles as zeros, or no poles and zeros, of $\Delta(s)$, the $\Delta(s)$-plane locus Γ_Δ will not encircle the origin of the $\Delta(s)$-plane.
3. $N < 0$ $(Z < P)$. If the s-plane locus encircles more poles than zeros of $\Delta(s)$ in a certain direction, N is a negative integer. In this case, the $\Delta(s)$-plane locus Γ_Δ will encircle the origin N times in the *opposite* direction as that of Γ_s.

A convenient way of determining N with respect to the origin (or any other point) of the $\Delta(s)$-plane is to draw a line from the point in any direction to a point as far as necessary; the number of *net* intersections of this line with the $\Delta(s)$ locus gives the magnitude of N. Figure 10-18 gives several examples of this method of determining N. In these illustrated cases, it is assumed that the Γ_s locus has a counterclockwise sense.

For convenience, we shall designate the origin of the Δ-plane as the **critical point** from which the value of N is determined. Later we shall designate other points in the complex-function plane as critical points, dependent on the way the Nyquist criterion is applied.

A rigorous proof of the principle of the argument is not given here. The following illustration may be considered as a heuristic explanation of the principle.

heuristic explanation

Let us consider the function $\Delta(s)$:

$$\Delta(s) = \frac{K(s + z_1)}{(s + p_1)(s + p_2)} \qquad (10\text{-}44)$$

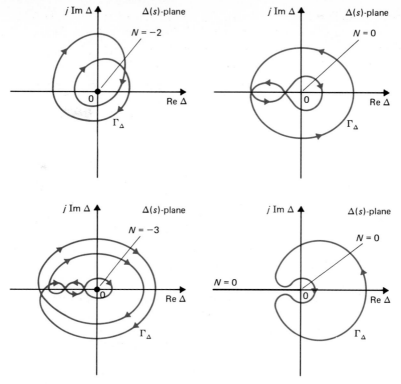

FIGURE 10-18 Examples of the determination of N in the $\Delta(s)$-plane.

where K is a positive number. The poles and the zero of $\Delta(s)$ are assumed to be as shown in Fig. 10-19(a). The function $\Delta(s)$ can be written as

$$\Delta(s) = |\Delta(s)| \angle \Delta(s)$$

$$= \frac{K|s + z_1|}{|s + p_1||s + p_2|}[\angle(s + z_1) - \angle(s + p_1) - \angle(s + p_2)] \quad (10\text{-}45)$$

Figure 10-19(a) shows an arbitrarily chosen trajectory Γ_s in the s-plane, with the arbitrary point s_1 on the path, and Γ_s does not pass through any of the poles and the zero of $\Delta(s)$. The function $\Delta(s)$ evaluated at $s = s_1$ is

$$\Delta(s_1) = \frac{K(s_1 + z_1)}{(s_1 + p_1)(s_1 + p_2)} \quad (10\text{-}46)$$

The term $(s_1 + z_1)$ can be represented graphically by the phasor drawn from $-z_1$ to s_1. Similar phasors can be drawn for $(s_1 + p_1)$ and $(s_1 + p_2)$. Thus, $\Delta(s_1)$ is represented by the phasors drawn from the finite poles and zero of $\Delta(s)$ to the point s_1, as shown in Fig. 10-19(a). Now, if the point s_1 is moved along the locus Γ_s in the prescribed counterclockwise direction until it returns to the starting point, the angles gen-

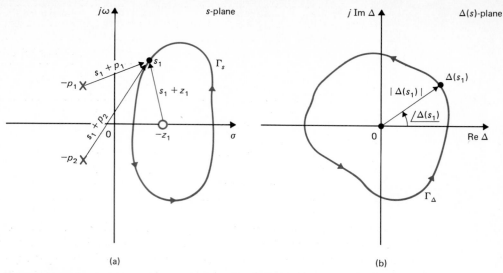

(a)   (b)

FIGURE 10-19 (a) Pole-zero configuration of $\Delta(s)$ in Eq. (10-46) and the
s-plane trajectory Γ_s. (b) $\Delta(s)$-plane locus Γ_Δ, which corresponds to the Γ_s locus of (a) through the mapping of Eq.
(10-46).

erated by the phasors drawn from the two poles that are not encircled by Γ_s when s_1
completes one round trip are zero; whereas the phasor $(s_1 + z_1)$ drawn from the zero at
$-z_1$, which is encircled by Γ_s, generates a positive angle (counterclockwise) of 2π radians. Then, in Eq. (10-45), the net angle or argument of $\Delta(s)$ as the point s_1 travels once
around Γ_s is equal to 2π, which means that the corresponding $\Delta(s)$ plot must go around
the origin 2π radians, or one revolution, in the counterclockwise direction, as shown in
Fig. 10-19(b). This is why only the poles and zeros of $\Delta(s)$ that are inside the Γ_s trajectory in the s-plane will contribute to the value of N of Eq. (10-43). Since the poles of
$\Delta(s)$ contribute to negative phase, and zeros contribute to positive phase, the value of N
depends only on the difference between Z and P. For the case illustrated in Fig.
10-19(a),

$$Z = 1 \quad \text{and} \quad P = 0$$

Thus,

$$N = Z - P = 1$$

which means that the $\Delta(s)$-plane locus Γ_Δ should encircle the origin once in the same
direction as that of the s-plane locus Γ_s. It should be kept in mind that Z and P refer
only to the zeros and poles, respectively, of $\Delta(s)$ that are encircled by Γ_s, and not the
total number of zeros and poles of $\Delta(s)$.

In general, if there are N more zeros than poles of $\Delta(s)$, which are encircled by
the s-plane locus Γ_s in a prescribed direction, the net angle traversed by the $\Delta(s)$-plane
locus, as the s-plane locus is traversed once, is equal to

$$2\pi (Z - P) = 2\pi N \qquad (10\text{-}47)$$

TABLE 10-1   Summary of All Possible Outcomes of the Principle of the Argument

		$F(s)$-PLANE-LOCUS	
$N = Z - P$	Sense of the s-plane Locus	Number of Encirclements of the Origin	Direction of Encirclement
$N > 0$	Clockwise	N	Clockwise
	Counterclockwise	N	Counterclockwise
$N < 0$	Clockwise	N	Counterclockwise
	Counterclockwise	N	Clockwise
$N = 0$	Clockwise	0	No encirclement
	Counterclockwise	0	No encirclement

This equation implies that the $\Delta(s)$-plane locus will encircle the origin N times in the *same direction* as that of Γ_s. Conversely, if N more poles than zeros are encircled by Γ_s in a given prescribed direction, N in Eq. (10-47) will be negative, and the $\Delta(s)$-plane locus must encircle the origin N times in the *opposite direction* to that of Γ_s.

A summary of all the possible outcomes of the principle of the argument is given in Table 10-1.

The Nyquist Path

At this point the reader may place himself or herself in the position of Nyquist many years ago, confronted with the problem of the stability of the closed-loop system that has the transfer function of Eq. (10-37), which is equivalent to determining whether or not the function $\Delta(s) = 1 + F(s)$ has zeros in the right-half of the s-plane. Apparently, Nyquist discovered that the principle of the argument of the complex-variable theory could be applied to solve the stability problem if the s-plane locus Γ_s is taken to

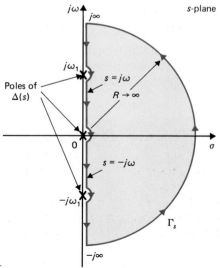

FIGURE 10-20 Nyquist path.

Nyquist path

be one that encircles the entire right half of the s-plane. Of course, as an alternative, Γ_s can be chosen to encircle the entire left-half s-plane, as the solution is a relative one. Figure 10-20 illustrates a Γ_s locus, with a counterclockwise sense, that encircles the entire right half of the s-plane. This path is chosen to be the s-plane trajectory Γ_s for the Nyquist criterion, since in mathematics, counterclockwise is traditionally defined to be the positive sense. The path Γ_s shown in Fig. 10-20 is known to be the **Nyquist path**.

Since the Nyquist path must not pass through any poles or zeros of $\Delta(s)$, the small semicircles shown along the $j\omega$-axis in Fig. 10-20 are used to indicate that the path should go around these poles and zeros if they fall on the $j\omega$-axis. It is apparent that if any pole or zero of $\Delta(s)$ lies inside the right half of the s-plane, it will be encircled by the Nyquist path.

Nyquist Criterion and the *F(s)* or the *G(s)H(s)* Plot

principle of the argument

The Nyquist criterion is a direct application of the principle of the argument when the s-plane locus is the Nyquist path. In principle, once the Nyquist path is specified, the stability of a closed-loop system can be determined by plotting the $\Delta(s) = 1 + F(s)$ locus when s takes on values along the Nyquist path, and investigating the behavior of the $\Delta(s)$ plot with respect to the critical point, which in this case is the origin of the

Nyquist plot

$\Delta(s)$-plane. The $\Delta(s)$-plot that corresponds to the Nyquist path is called the **Nyquist plot** of $\Delta(s)$. However, since the function $F(s)$, is generally known, it would be simpler to construct the Nyquist plot of $F(s)$, and the same conclusion on the stability of the closed-loop system can be obtained from the $F(s)$ plot with respect to the $(-1, j0)$ point in the $F(s)$-plane. This is because the origin of the $\Delta(s) = [1 + F(s)]$-plane corresponds to the $(-1, j0)$ point of the $F(s)$-plane. Thus, the $(-1, j0)$ point in the $F(s)$- or $G(s)H(s)$-plane becomes the critical point for the determination of closed-loop stability.

For single-loop systems, $F(s) = G(s)H(s)$, the previous development leads to the determination of the closed-loop system stability by investigating the behavior of the Nyquist plot of $G(s)H(s)$ with respect to the $(-1, j0)$ point of the $G(s)H(s)$-plane. Thus, the Nyquist stability criterion is another example of the common feature found in control-systems theory that closed-loop systems behavior is often determined from working with "open-loop" transfer functions.

With this added dimension to the stability problem, it is necessary to define two sets of N, Z, and P as follows:

N_0 = number of encirclements of the origin made by $F(s)$
Z_0 = number of zeros of $F(s)$ that are encircled by the Nyquist path, or in the right half of the s-plane
P_0 = number of poles of $F(s)$ that are encircled by the Nyquist path, or in the right half of the s-plane
N_{-1} = number of encirclements of the $(-1, j0)$ point made by $F(s)$
Z_{-1} = number of zeros of $1 + F(s)$ that are encircled by the Nyquist path, or in the right half of the s-plane
P_{-1} = number of poles of $1 + F(s)$ [same as those of $F(s)$] that are encircled by the Nyquist path, or in the right half of the s-plane

Again, when the closed-loop system has only one loop with the loop transfer function equal to $G(s)H(s)$, $F(s) = G(s)H(s)$.

Several facts become clear and should be emphasized at this point:

$$P_0 = P_{-1} \qquad (10\text{-}48)$$

stability requirements

since $F(s)$ and $1 + F(s)$ always have the same poles.

Closed-loop stability requires that

$$Z_{-1} = 0 \qquad (10\text{-}49)$$

Open-loop stability requires that

$$P_0 = 0 \qquad (10\text{-}50)$$

For an understanding of the Nyquist stability criterion, the following steps are outlined.

1. Given a feedback control system that has the closed-loop transfer function such as that given in Eq. (10-37), the denominator of the closed-loop transfer function is $\Delta(s)$, as given by Eq. (10-40). The Nyquist path is defined in the s-plane according to the pole-zero properties of $F(s)$ on the $j\omega$-axis.
2. The Nyquist plot of $F(s)$ is constructed in the $F(s)$-plane.
3. The values of N_0 and N_{-1} are determined by observing the behavior of the Nyquist plot of $F(s)$ with respect to the origin and the $(-1, j0)$ point, respectively.
4. Once N_0 and N_{-1} are determined, the value of P_0 (if it is not already known) is determined from

$$N_0 = Z_0 - P_0 \qquad (10\text{-}51)$$

if Z_0 is given or known. Once P_0 is determined, $P_{-1} = P_0$ [Eq. (10-48)], and Z_{-1} is determined from

$$N_{-1} = Z_{-1} - P_{-1} \qquad (10\text{-}52)$$

Since it has been established that for a stable closed-loop system, Z_{-1} must be zero, Eq. (10-52) gives

$$N_{-1} = -P_{-1} \qquad (10\text{-}53)$$

Therefore, the Nyquist criterion may be formally stated:

Nyquist criterion

For a closed-loop system to be stable, the Nyquist plot of $F(s)$ must encircle the $(-1, j0)$ point as many times as the number of poles of $F(s)$ that are in the right-

half of the s-plane, and the encirclement, if any, must be made in the clockwise direction (if the Nyquist path is defined in the counterclockwise sense).

A Simplified Nyquist Criterion

simplified criterion

The original Nyquist criterion discussed in the preceding sections requires the construction of the Nyquist plot that corresponds to the entire Nyquist path in the s-plane. Evidently, when the function $F(s)$ has poles or zeros on the $j\omega$-axis, small indentations around these points, such as those shown in Fig. 10-20, are necessary. These small indentations inevitably will add complexity to the construction of the Nyquist plot of

FREQRP

$F(s)$. Computer programs such as the FREQRP found in the ACSP software package accompanying this text can be used to construct only the Nyquist plot that corresponds to the positive $j\omega$-axis of the s-plane. For an $F(s)$ that is of the form of Eq. (10-38), the Nyquist plot for the negative $j\omega$-axis is symmetrical to that of the $+j\omega$-axis, but the rest of the Nyquist plot still has to be done manually. With modern computer facilities and software, the analyst should not be burdened with the chores of plotting the Nyquist plot manually. The proper learning process should place emphasis on the interpretation of the computer data and stability conditions. Yeung and Lai [6, 7] introduced a simplified version of the Nyquist criterion for closed-loop systems that requires the Nyquist plot that corresponds to only the positive $j\omega$-axis of the s-plane, and the idea is adopted in the following development.

Let us consider the two Nyquist paths shown in Figs. 10-21(a) and 10-21(b). Apparently, the Nyquist path, Γ_{s1}, in Fig. 10-21(a) is the original path defined in Fig. 10-20, whereas the path Γ_{s2}, in Fig. 10-21(b) encircles not only the entire right-half

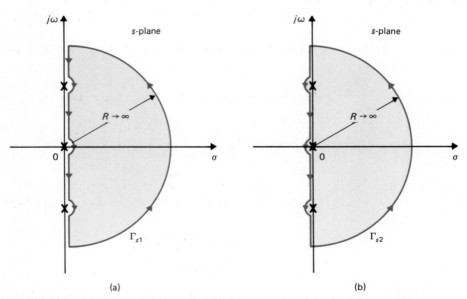

(a)   (b)

FIGURE 10-21 (a) Nyquist path. (b) An alternate Nyquist path. *Source:* K. S. Yeung, "A Reformulation of Nyquist's Criterion", *IEEE Trans. Educ.*, Vol. E-28 (© 1985, IEEE).

s-plane, but also all the poles and zeros of $F(s)$ on the $j\omega$-axis, if there are any. Let us define the following quantities.

Z_{-1} = number of zeros of $1 + F(s)$ that are in the right-half s-plane
P_{-1} = number of poles of $1 + F(s)$ that are in the right-half s-plane = P_0
[number of poles of $F(s)$ that are in the right-half s-plane]
P_ω = number of poles of $F(s)$ [or of $1 + F(s)$] that are on the $j\omega$-axis, including the origin
$N_{-1,1}$ = number of times the $(-1, j0)$ point of the $F(s)$-plane is encircled by the Nyquist plot of $F(s)$ corresponding to Γ_{s1}
$N_{-1,2}$ = number of times the $(-1, j0)$ point of the $F(s)$-plane is encircled by the Nyquist plot of $F(s)$ corresponding to Γ_{s2}

Then, with reference to the two Nyquist paths in Fig. 10-21, and according to the Nyquist criterion,

$$N_{-1,1} = Z_{-1} - P_{-1} \qquad (10\text{-}54)$$

and

$$N_{-1,2} = Z_{-1} - P_\omega - P_{-1} \qquad (10\text{-}55)$$

Let Φ_1 and Φ_2 represent the net angles traversed by the Nyquist plots of $F(s)$ with respect to the $(-1, j0)$ point, corresponding to Γ_{s1} and Γ_{s2}, respectively. Then,

$$\Phi_1 = N_{-1,1} \times 360° = (Z_{-1} - P_{-1})360° \qquad (10\text{-}56)$$

$$\Phi_2 = N_{-1,2} \times 360° = (Z_{-1} - P_\omega - P_{-1})360° \qquad (10\text{-}57)$$

Let us consider that each of the Nyquist paths Γ_{s1} and Γ_{s2} is composed of three portions:

1.   The portion from $s = -j\infty$ to $+j\infty$ along the semicircle with infinite radius.
2.   The portion along the $j\omega$-axis, excluding all the small indentations.
3.   All the small indentations on the $j\omega$-axis.

Since the Nyquist path is symmetrical about the real axis in the s-plane, the angles traversed by the Nyquist plots are identical for positive and negative values of ω. Thus, Φ_1 and Φ_2 are

$$\Phi_1 = 2\Phi_{11} + \Phi_{12} + \Phi_{13} \qquad (10\text{-}58)$$

$$\Phi_2 = 2\Phi_{11} - \Phi_{12} + \Phi_{13} \qquad (10\text{-}59)$$

where Φ_{11} = angle traversed by the Nyquist plot of $F(s)$ with respect to the $(-1, j0)$ point, corresponding to the positive $j\omega$-axis or the $-j\omega$-axis of the s-plane, excluding the small indentations.

Φ_{12} = angle traversed by the Nyquist plot of $F(s)$ with respect to the $(-1, j0)$ point, corresponding to the small indentations on the $j\omega$-axis of Γ_{s1}. Since on Γ_{s2} the directions of the small indentations are opposite to that of Γ_{s1}, the sign of Φ_{12} in Eq. (10-59) is negative.

Φ_{13} = angle traversed by the Nyquist plot of $F(s)$ with respect to the $(-1, j0)$ point, corresponding to the semicircle with infinite radius on the Nyquist paths.

For physically realizable transfer functions, $F(s)$ cannot have more zeros than poles. This means that the Nyquist plot of $F(s)$ that corresponds to the infinite semicircle must either be a point on the real axis or a trajectory around the origin of the $F(s)$-plane. Thus, the angle Φ_{13} traversed by the phasor drawn from the $(-1, j0)$ point to the Nyquist plot along the semicircle with infinite radius is always zero.

Now adding Eq. (10-58) to Eq. (10-59) and using Eqs. (10-56) and (10-57), we get

$$\Phi_1 + \Phi_2 = 4\Phi_{11}$$
$$= (2Z_{-1} - P_\omega - 2P_{-1})360° \tag{10-60}$$

Solving for Φ_{11}, we get

$$\Phi_{11} = (Z_{-1} - 0.5P_\omega - P_{-1})180° \tag{10-61}$$

The last equation states:

The total angle traversed by the phasor drawn from the $(-1, j0)$ point to the $F(s)$ Nyquist plot that corresponds to the portion on the positive $j\omega$-axis of the s-plane, excluding the small indentations, if any, equals

[the number of zeros of $1 + F(s)$ in the right-half s-plane

− the number of poles of $F(s)$ on the $j\omega$-axis/2

− the number of poles of $F(s)$ in the right-half s-plane] × 180°

Thus, *the Nyquist stability criterion can be carried out by constructing only the Nyquist plot that corresponds to the $s = j\infty$ to $s = 0$ portion on the Nyquist path.* Furthermore, if the closed-loop system is unstable, by knowing the values of Φ_{11}, P_ω, and P_{-1}, Eq. (10-61) gives the number of roots of the characteristic equation that are in the right-half s-plane.

For the closed-loop system to be stable, Z_{-1} must equal zero. Thus, the Nyquist criterion for stability of the closed-loop system is

$$\Phi_{11} = -(0.5P_\omega + P_{-1})180° \tag{10-62}$$

Since P_ω and P_{-1} cannot be negative, the last equation indicates:

For closed-loop stability, the phase traversed by the Nyquist plot of F(s) as s varies from s = j∞ to s = 0, with respect to the (−1, j0) point, cannot be positive.

It must be emphasized that the angle Φ_{11} is an angle variation, so that $\Phi_{11} = -270°$ is *not the same* as $\Phi_{11} = 90°$.

Application of the Simplified Nyquist Criterion to Noncausal Transfer Functions

It was mentioned earlier that the angle Φ_{13} is zero because a physically realizable transfer function $F(s)$ does not have more zeros than poles. However, in the broad applications of the Nyquist criterion to control systems, just as in the case of the root locus, it is often necessary to create an equivalent loop transfer function $G_{eq}(s)H_{eq}(s)$ so that the variable parameter K will appear as a multiplying factor in $G_{eq}(s)H_{eq}(s)$. Since the equivalent loop transfer function does not represent any physical entity, it can have more zeros than poles. If this situation does occur, to apply the simplified Nyquist criterion, we can simply plot the Nyquist plot of $1/G_{eq}(s)H_{eq}(s)$ and still use the $(-1, j0)$ point as the critical point for $K > 0$. The reason is that the condition of $G_{eq}(s)H_{eq}(s) = -1$ is still satisfied by $1/G_{eq}(s)H_{eq}(s) = -1$. The angular condition in Eqs. (10-61) and (10-62) are still applicable to the $1/G_{eq}(s)H_{eq}(s)$, except that the parameters P_ω and P_{-1} must all refer to the function $1/G_{eq}(s)H_{eq}(s)$. The gain factor K will appear now in the denominator of $1/G_{eq}(s)H_{eq}(s)$, and the multiplying factor becomes $1/K$.

Nyquist Criterion According to the Enclosure of the Critical Point

Thus far we have utilized only the concept of encirclement, which is a rudiment of the Nyquist stability application. Under certain restricted conditions, the stability of the closed-loop system can be investigated by observing whether the Nyquist plot of $F(s)$ that corresponds to only the $+j\omega$-axis in the s-plane *encloses* the $(-1, j0)$ point or not. This turns out to be a simpler observation, when valid, than the requirement in Eq. (10-62).

minimum-phase transfer functions

In Appendix A, the properties of the **minimum-phase transfer functions** are described. These properties are again summarized as follows:

1. A minimum-phase transfer function does not have poles or zeros in the right-half s-plane or on the $j\omega$-axis, excluding the origin.
2. For a minimum-phase transfer function $F(s)$ with m zeros and n poles, excluding the poles at $s = 0$, when $s = j\omega$, and as ω varies from ∞ to 0, the total phase variation of $F(j\omega)$ is $(n - m)\pi/2$.
3. The value of a minimum-phase transfer function cannot become zero or infinity at any finite nonzero frequency.
4. A nonminimum-phase transfer function will always have a more positive phase shift as ω is varied from ∞ to 0.

Now consider that the loop transfer function $F(s)$ is of the minimum-phase type; it does not have any poles or zeros in the right-half s-plane; $Z_0 = 0$ and $P_0 = P_{-1} = 0$.

The Nyquist criterion of Eq. (10-52) becomes

$$N_{-1} = Z_{-1} \tag{10-63}$$

Nyquist criterion

Since Z_{-1} is either zero (stable system) or a positive integer, N_{-1} must be either zero or positive. This means that if the closed-loop system is unstable, Z_{-1} is greater than zero; the $(-1, j0)$ point must be encircled by the Nyquist $F(s)$ plot in the counterclockwise direction, or, simply, **enclosed**. Thus, the simplified Nyquist criterion is stated as:

If the loop transfer function $F(s)$ is of the minimum-phase type, and the gain factor K is positive, so that the critical point is at $(-1, j0)$, the closed-loop system is stable if the $(-1, j0)$ point is not enclosed by the $F(j\omega)$ plot.

Since minimum-phase-type transfer functions are quite common in practice, the simplified Nyquist criterion using the enclosure of the critical point as a test is quite convenient. The only drawback is that the $F(j\omega)$ plot that corresponds to the $+j\omega$-axis tells only whether the $(-1, j0)$ points is enclosed or not, but not how many times. Thus, if the system is found to be unstable, the enclosure property does not give information on how many roots of the characteristic equation are in the right-half s-plane. In any case, this information is given by the angle Φ_{11} in Eq. (10-61). If $F(s)$ is of the minimum-phase type, $P_{-1} = 0$; Eq. (10-61) becomes

Nyquist criterion

$$\Phi_{11} = (Z_{-1} - 0.5P_\omega)180° \tag{10-64}$$

For closed-loop stability, the angle traversed by the phasor drawn from the $(-1, j0)$ point to the Nyquist plot of $F(s)$ that corresponds to the portion on the positive $j\omega$-axis of the s-plane must equal $-P_\omega \times 90°$.

10-6

RELATION BETWEEN THE ROOT LOCI AND THE NYQUIST PLOT

Since both the root-locus analysis and the Nyquist criterion deal with the location of the roots of the characteristic equation of a linear system, the two analyses are closely related. Exploring the relationship betwen the root loci and the Nyquist plot will enhance the understanding of both methods. Given the characteristic equation

$$1 + G(s)H(s) = 1 + KG_1(s)H_1(s) = 0 \tag{10-65}$$

the Nyquist plot of $G(s)H(s)$ is the mapping of the Nyquist path in the s-plane onto the Nyquist plot in the $G(s)H(s)$-plane.

Since the root loci of Eq. (10-65) must satisfy the conditions

$$KG_1(s)H_1(s) = (2k + 1)\pi \qquad K \geq 0 \tag{10-66}$$

$$KG_1(s)H_1(s) = 2k\pi \qquad K \leq 0 \tag{10-67}$$

for $k = 0, \pm 1, \pm 2, \ldots$, the root loci simply represent a mapping of the real axis of the $G(s)H(s)$-plane onto the s-plane. In fact, for the root loci (RL), $K \geq 0$, the mapping points are on the negative real axis of the $G(s)H(s)$-plane; and for the complementary root loci (CRL), $K \leq 0$, the mapping points are on the positive real axis of the $G(s)H(s)$-plane. It was pointed out earlier that the mapping from the s-plane to the $G(s)H(s)$-plane for a rational function is single valued, but the reverse process is multivalued. As a simple illustration, the Nyquist (polar) plot of a type-1 third-order transfer function that corresponds to points on the $j\omega$-axis of the s-plane is shown in Fig. 10-22. The complete root loci for the same system are shown in Fig. 10-23 as a mapping

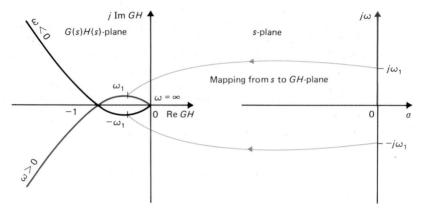

FIGURE 10-22 Polar plot of $G(s)H(s) = K/[s(s + a)(s + b)]$ interpreted as a mapping of the $j\omega$-axis of the s-plane onto the $G(s)H(s)$-plane.

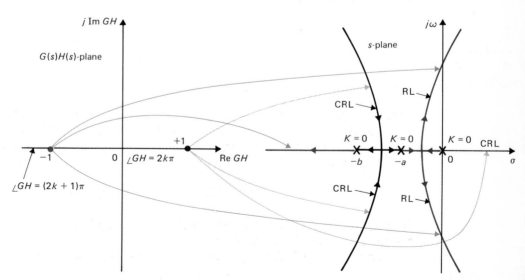

FIGURE 10-23 Root-locus diagram of $G(s)H(s) = K/[s(s + a)(s + b)]$ interpreted as a mapping of the real axis of the $G(s)H(s)$-plane onto the s-plane.

of the real axis of the $G(s)H(s)$-plane onto the s-plane. Note that in this case, each point of the $G(s)H(s)$-plane corresponds to three points in the s-plane. The $(-1, j0)$ point of the $G(s)H(s)$-plane corresponds to the two points where the root loci intersect the $j\omega$-axis and a point on the real axis.

The Nyquist plot and the root loci each represents the mapping of only a very limited portion of one domain to the other. In general, it would be helpful to consider the mapping of points other than those on the $j\omega$-axis of the s-plane and on the real axis of the $G(s)H(s)$-plane. For instance, we may use the mapping of the constant-damping-ratio lines in the s-plane onto the $G(s)H(s)$-plane for the purpose of determining relative stability of the closed-loop system. Figure 10-24 illustrates the $G(s)H(s)$ plots that correspond to different constant-damping ratio lines in the s-plane. As shown by curve (3) in Fig. 10-24, when the $G(s)H(s)$ curve passes through the $(-1, j0)$ point, it means that Eq. (10-65) is satisfied, and the corresponding trajectory in the s-plane passes through the root of the characteristic equation. Similarly, we can construct root loci that correspond to the straight lines rotated at various angles from the real axis in the $G(s)H(s)$-plane, as shown in Fig. 10-25. Notice that these root loci now satisfy the condition of

$$\angle KG_1(s)H_1(s) = (2k + 1)\pi - \theta \qquad K \geq 0 \qquad (10\text{-}68)$$

Or the root loci of Fig. 10-25 satisfy the equation

$$1 + G(s)H(s)e^{j\theta} = 0 \qquad (10\text{-}69)$$

for the various values of θ indicated.

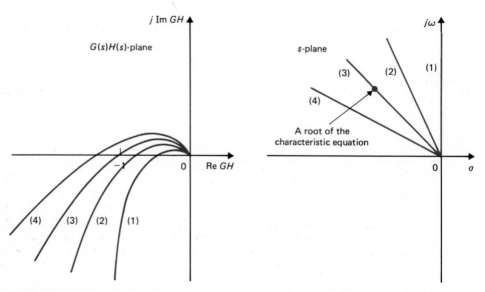

FIGURE 10--24 $G(s)H(s)$ plots that correspond to constant-damping-ratio lines in the s-plane.

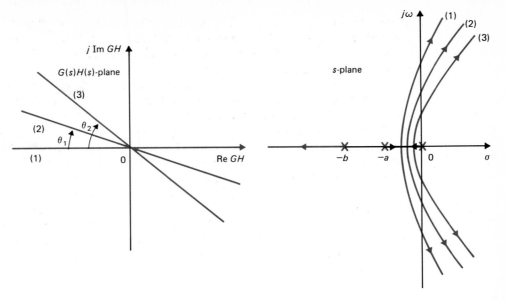

FIGURE 10-25 Root loci that correspond to different phase-angle loci in the $G(s)H(s)$-plane. (The CRL are not shown.)

10-7

ILLUSTRATIVE EXAMPLES OF THE APPLICATIONS OF THE NYQUIST CRITERION

examples
The following examples serve to illustrate the application of the Nyquist criterion to the stability study of control systems.

EXAMPLE 10-1
Consider a single-loop feedback control system with the loop transfer function given by

$$F(s) = G(s)H(s) = \frac{K}{s(s + a)} \tag{10-70}$$

where a and K are positive real constants. It is apparent that $G(s)H(s)$ does not have any poles in the right-half s-plane, but there is a pole on the $j\omega$-axis at $s = 0$. Thus, $P_0 = P_{-1} = 0$, and $P_\omega = 1$. The function is of the minimum-phase type. According to the original Nyquist criterion, the Nyquist path, Γ_s should be defined as shown in Fig. 10-26, which encircles the entire right-half s-plane, but includes a small semicircle at $s = 0$, so that the pole of $G(s)H(s)$ at that point is excluded by the Nyquist path. However, with the simplified Nyquist criterion, it is sufficient to sketch only the Nyquist plot of $G(s)H(s)$ that corresponds to the positive portion of the $j\omega$-axis of the s-plane. This is done by substituting $s = j\omega$ into $G(s)H(s)$. From Eq. (10-70),

$$G(j\omega)H(j\omega) = \frac{K}{j\omega(j\omega + a)} \tag{10-71}$$

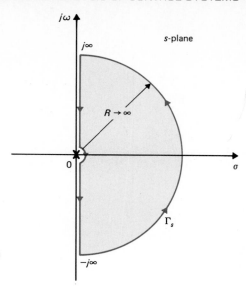

FIGURE 10-26 Nyquist path for the system in Example 10-1.

The last equation is rationalized by multiplying the numerator and the denominator by the complex conjugate of the denominator. Thus,

$$G(j\omega)H(j\omega) = \frac{K(-\omega^2 - ja\omega)}{\omega^4 + a^2\omega^2} \tag{10-72}$$

For stability studies, it is not necessary to construct the $G(j\omega)H(j\omega)$ plot accurately, since the only information needed is the angle made by the phasor drawn from the $(-1, j0)$ point to the $G(j\omega)H(j\omega)$ plot as ω varies from ∞ to 0, or, alternately, whether or not the $(-1, j0)$ point is enclosed by the $G(j\omega)H(j\omega)$ trajectory. In general, it is necessary to calculate the intersect(s), if any, of the real axis made by the $G(j\omega)H(j\omega)$ plot. This is determined by equating the imaginary part of $G(j\omega)H(j\omega)$ to zero; i.e.,

$$\text{Im}\,[G(j\omega)H(j\omega)] = \frac{-Ka\omega}{\omega^4 + a^2\omega^2} = \frac{-Ka}{\omega(\omega^2 + a^2)} = 0 \tag{10-73}$$

The solution of the last equation is $\omega = \infty$. Substituting $\omega = \infty$ into Eq. (10-71), we get

$$G(j\infty)H(j\infty) = 0\angle 180° \tag{10-74}$$

Thus, the only intersect on the real axis in the $G(s)H(s)$-plane when $\omega = \infty$ is at the origin.
    At the other extreme, $\omega = 0$; Eq. (10-71) gives

$$G(j0)H(j0) = \infty\angle -90° \tag{10-75}$$

Figure 10-27(a) shows the sketch of the Nyquist plot of $G(j\omega)H(j\omega)$ for $s = j\omega$, and $\omega = \infty$ to $\omega = 0$. The angle Φ_{11} in this case is $-90°$; the negative sign is due to the clockwise sense of rotation.
    From Eq. (10-62), the closed-loop system is stable if

stability condition

$$\Phi_{11} = (-0.5P_\omega - P_{-1})180°$$
$$= -0.5 \times 180° = -90° \tag{10-76}$$

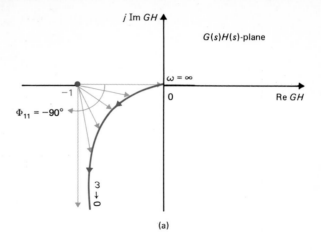

(a)

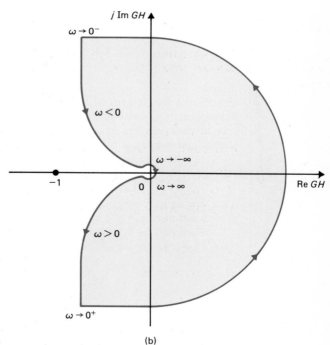

FIGURE 10-27 (a) Nyquist plot of $G(s)H(s) = K/[s(s + a)]$ that corresponds to $\omega = \infty$ to $\omega = 0$. (b) Complete Nyquist plot of $G(s)H(s) = K/[s(s + a)]$.

(b)

that is, the net angle traversed by the phasor drawn from the $(-1, j0)$ point to the Nyquist plot, from $s = j\infty$ to $s = 0$, must equal $-90°$. Thus, the closed-loop system is stable for all positive and finite values of K. Alternatively, we see from Fig. 10-27(a) that the $(-1, j0)$ point is not enclosed by the $G(j\omega)H(j\omega)$ plot.

For the sake of illustration, Fig. 10-27(b) shows the Nyquist plot that corresponds to the entire Nyquist path in Fig. 10-26. The details on how the Nyquist plot in Fig.

10-27(b) is constructed is not elaborated here, since, with the simplified Nyquist criterion, the entire plot is unnecessary.

Since $P_{-1} = 0$ [the function $1 + G(s)H(s)$ does not have any poles in the right-half s-plane], and from Fig. 10-27(b), $N_{-1} = 0$ for all finite positive K, the Nyquist criterion in Eq. (10-52) gives

$$Z_{-1} = 0 \qquad (10\text{-}77)$$

which means that $1 + G(s)H(s)$ does not have zeros in the right-half s-plane.

The stability condition arrived for the second-order system should have been anticipated, since the characteristic equation of the closed-loop system is

$$s^2 + as + K = 0 \qquad (10\text{-}78)$$

The roots of the equation will always lie in the left-half s-plane for positive a and K.

EXAMPLE 10-2

✓

minimum-phase transfer function

stability condition

Consider that a control system with a single feedback loop has the loop transfer function

$$G(s)H(s) = \frac{K(s-1)}{s(s+1)} \qquad (10\text{-}79)$$

We observe from $G(s)H(s)$ that $P_\omega = 1$, and $P_0 = P_{-1} = 0$. The function is of the nonminimum-phase type, since it has a zero at $s = 1$. Thus, the simplified stability criterion on enclosure of the critical point cannot be used in this case. From Eq. (10-62), the requirement for closed-loop stability is

$$\Phi_{11} = (-0.5P_\omega - P_{-1})180° = -90° \qquad (10\text{-}80)$$

Thus, the stability criterion is that the phasor drawn from the $(-1, j0)$ point to the $G(j\omega)H(j\omega)$ plot should traverse $-90°$ as ω varies from ∞ to 0.

To sketch the Nyquist plot of $G(s)H(s)$ that corresponds to the positive portion of the $j\omega$-axis of the s-plane, we set $s = j\omega$ in Eq. (10-79). We get

$$G(j\omega)H(j\omega) = \frac{K(j\omega - 1)}{j\omega(j\omega + 1)} = \frac{K(j\omega - 1)}{-\omega^2 + j\omega} \qquad (10\text{-}81)$$

When $\omega = \infty$,

$$G(j\infty)H(j\infty) = \frac{K}{j\omega}\bigg|_{\omega=\infty} = 0\angle -90° \qquad (10\text{-}82)$$

When $\omega = 0$,

$$G(j0)H(j0) = \frac{-K}{j\omega}\bigg|_{\omega=0} = \infty\angle 90° \qquad (10\text{-}83)$$

To find the intersect of the $G(j\omega)H(j\omega)$ plot on the real axis, we rationalize $G(j\omega)H(j\omega)$ by multiplying the numerator and the denominator of Eq. (10-81) by $-\omega^2 - j\omega$. We have

$$G(j\omega)H(j\omega) = \frac{K(j\omega - 1)(-\omega^2 - j\omega)}{\omega^4 + \omega^2} = \frac{K[2\omega + j(1 - \omega^2)]}{\omega(\omega^2 + 1)}$$ (10-84)

Setting the imaginary part of $G(j\omega)H(j\omega)$ to zero, we have

$$1 - \omega^2 = 0$$ (10-85)

or

$$\omega = \pm 1 \text{ rad/s}$$ (10-86)

For $\omega = 1$, $\rightarrow$ $\frac{K(j-1)}{j(j+1)} = \frac{K(j-1)}{(-1+j)} = \frac{K(j-1)}{(j-1)} = K \frac{(-1-j)}{}$ ⟶ see eqtn 10-81

$$G(j1)H(j1) = K$$ (10-87)

Based on the information acquired, the Nyquist plot of $G(s)H(s)$ that corresponds to the positive portion of the $j\omega$-axis is shown in Fig. 10-28 for $K > 0$. Figure 10-28 shows that as ω varies from ∞ to 0 along the Nyquist plot, the net angle Φ_{11} traversed by the phasor drawn from the $(-1, j0)$ point to the Nyquist plot is $+90°$. Thus, the closed-loop system is unstable, since Φ_{11} is positive. From Eq. (10-61),

stability condition

$$\Phi_{11} = (Z_{-1} - 0.5)180° = 90°$$ (10-88)

Thus, $Z_{-1} = 1$, which means that the characteristic equation of the closed-loop system has one root in the right-half s-plane.

The characteristic equation of the system is

$$s^2 + (1 + K)s - K = 0$$ (10-89)

We can easily verify that stability requires

stability condition

$$0 > K > -1$$ (10-90)

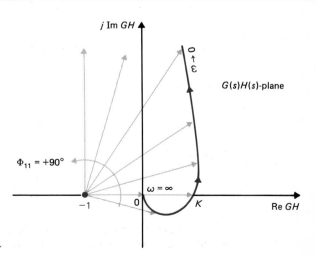

FIGURE 10-28 Nyquist plot of the system in Example 10-2. $G(s)H(s) = K(s - 1)/[s(s + 1)]$.

Figure 10-29(a) shows the Nyquist plot of $G(j\omega)H(j\omega)$ when K is negative, and $K > -1$. Notice that the plot is obtained by *rotating* the $G(j\omega)H(j\omega)$ plot of Fig. 10-28 by 180° about the origin. As ω is varied from ∞ to 0, the angle Φ_{11} in Fig. 10-29(a) has a net rotation of $-90°$, which agrees with the stability requirement in Eq. (10-80), and the closed-loop system is stable.

Figure 10-29(b) shows the Nyquist plot when $K < -1$. In this case, $\Phi_{11} = 270°$, and the closed-loop system is unstable. From Eq. (10-61),

$$\Phi_{11} = (Z_{-1} - 0.5)180° = 270° \qquad (10\text{-}91)$$

Thus,

$$Z_{-1} = 2 \qquad (10\text{-}92)$$

or the characteristic equation has two roots in the right-half s-plane.

Although it would seem that the criterion on enclosure of the critical point still works for the case in Figure 10-29, since $G(s)H(s)$ is of nonminimum phase, and K is negative, in general, because of the unpredictable behavior of the Nyquist plot that corresponds to the small indentation at $s = 0$, not enclosing the $(-1, j0)$ point does not necessarily mean that the system is stable.

When K changes sign, it is not necessary to redraw the Nyquist plot of $G(j\omega)H(j\omega)$, as shown in Fig. 10-29. Equation (10-39) can be written as

$$1 + G(s)H(s) = 1 + KG_1(s)H_1(s) = 0 \qquad (10\text{-}93)$$

where K is positive. For negative K, the last equation can be written as

$$1 - KG_1(s)H_1(s) = 0 \qquad (10\text{-}94)$$

or

$$KG_1(s)H_1(s) = 1 \qquad (10\text{-}95)$$

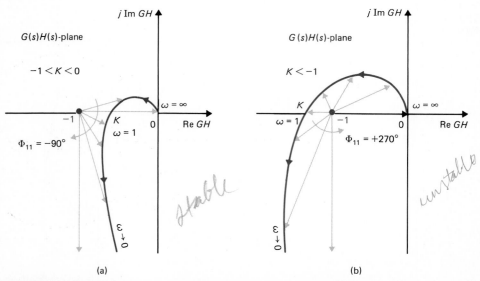

(a)   (b)

FIGURE 10-29 Nyquist plots of the system in Example 10-2. $G(s)H(s) = K(s - 1)/[s(s + 1)]$. (a) $-1 < K < 0$. (b) $K < -1$.

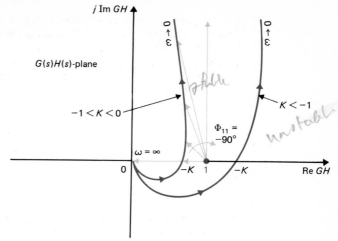

FIGURE 10-30 Nyquist plot of $G(s)H(s) = K(s - 1)/[s(s + 1)]$ with $K < 0$. The $(+1, j0)$ point is the critical point.

where K is now positive. Thus, when K is negative, we can still make use of the $G(j\omega)H(j\omega)$ plot with positive K, but designate the $(+1, j0)$ point as the critical point for stability analysis.

As shown in Fig. 10-30, for $-1 < K < 0$, the angle Φ_{11} is $-90°$, which is the required value, and the system is stable. The reader should observe that when $K < -1$, Φ_{11} is $270°$, and the system is unstable.

It is of interest to compare the Nyquist stability analysis with the root-locus analysis. Figure 10-31 shows the complete root loci of the characteristic equation of the system with the loop transfer function given in Eq. (10-79). The stability condition of the system as a function of K is clearly indicated by the root-locus diagram. The RL between 0 and $+1$ on the real axis indicates that the system is unstable for $0 < K < \infty$. The CRL indicate that for negative values of K, the

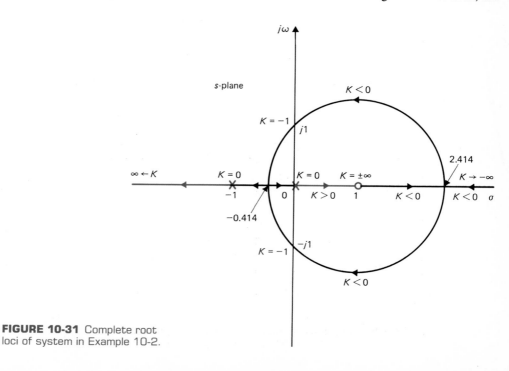

FIGURE 10-31 Complete root loci of system in Example 10-2.

system is unstable for $-\infty < K < -1$, and the system is stable only for the range of $-1 < K < 0$. The CRL cross the $j\omega$-axis at $\omega = \pm 1$ rad/s, which are the values of ω at which the Nyquist plot of $G(s)H(s)$ crosses the negative real axis.

EXAMPLE 10-3   Consider the control system shown in Fig. 10-32. It is desired to determine the range of K for which the system is stable.

The open-loop transfer function of the system is

$$\frac{C(s)}{E(s)} = G(s) = \frac{10K(s+2)}{s^3 + 3s^2 + 10} \tag{10-96}$$

The poles of $G(s)$ are found to be at $s = -3.72$, $s = 0.361 + j1.6$, and $s = 0.361 - j1.6$. Or we can use the Routh–Hurwitz tabulation to verify that $G(s)$ has two poles in the right-half s-plane. Thus, $P_0 = P_{-1} = 2$, and $P_\omega = 0$. The transfer function $G(s)$ is of the nonminimum-phase type. From Eq. (10-62), the requirement for the closed-loop system to be stable is

stability condition

$$\Phi_{11} = (-0.5P_\omega - P_{-1})180° = -360° \tag{10-97}$$

Setting $s = j\omega$, Eq. (10-96) becomes

$$G(j\omega) = \frac{10K(j\omega + 2)}{(10 - 3\omega^2) - j\omega^3} \tag{10-98}$$

At $\omega = \infty$,

$$G(j\infty) = 0\angle 180° \tag{10-99}$$

At $\omega = 0$,

$$G(j0) = 2K \tag{10-100}$$

To find the intersect on the real axis of the $G(j\omega)$-plane, we rationalize $G(j\omega)$ as

$$G(j\omega) = \frac{10K\{2(10 - 3\omega^2) - \omega^4 + j[\omega(10 - 3\omega^2) + 2\omega^3]\}}{(10 - 3\omega^2)^2 + \omega^6} \tag{10-101}$$

Setting the imaginary part of $G(j\omega)$ to zero, we have

$$\omega(10 - 3\omega^2) + 2\omega^3 = 0 \tag{10-102}$$

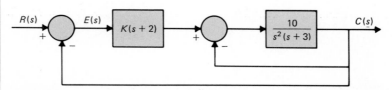

FIGURE 10-32 Block diagram of the control system in Example 10-3.

Thus,

$$\omega = 0$$

and

$$\omega = \pm\sqrt{10} = 3.16 \text{ rad/s}$$

which are the frequencies at which the $G(j\omega)$ plot intersects the real axis of the $G(j\omega)$-plane. When $\omega = 0$, we already have $G(j0) = 2K$ in Eq. (10-100). When $\omega = 3.16$ rad/s,

$$G(j3.16) = -K \qquad\qquad (10\text{-}103)$$

Figure 10-33(a) shows the Nyquist plot of $G(j\omega)$ for $1 > K > 0$. As ω varies from ∞ to 0, the net angle traversed by Φ_{11} is $0°$, not $-360°$, as required in Eq. (10-97). Thus, the closed-loop system is unstable. Figure 10-33(b) shows the Nyquist plot of $G(j\omega)$ when $K > 1$. In this case, the angle Φ_{11} rotates a total of $-360°$, and thus, the system is stable.

stability conditions   When K is negative, we can use either Fig. 10-33(a) or Fig. 10-33(b), and regard the $(1, j0)$ point as the critical point. The following stability conditions are observed:

$2K < -1$:   the $(1, j0)$ point lies between 0 and $2K$, $\Phi_{11} = -180°$; the system is unstable. Since for stability, Φ_{11} must equal $-360°$.

$0 > 2K > -1$:   the $(1, j0)$ point is to the right of $2K$, $\Phi_{11} = 0°$; the system is unstable.

The conclusion is that the system in Fig. 10-32 is stable for $K > 1$.

The complete root loci of the system are shown in Fig. 10-34. Clearly, when K is negative, one branch of the CRL will always stay in the right-half plane, and the system is unstable.

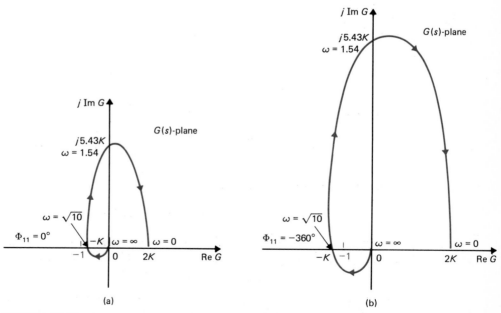

(a)   (b)

FIGURE 10-33 Nyquist plots of the system in Example 10-3.
(a) $1 > K > 0$. (b) $K > 1$.

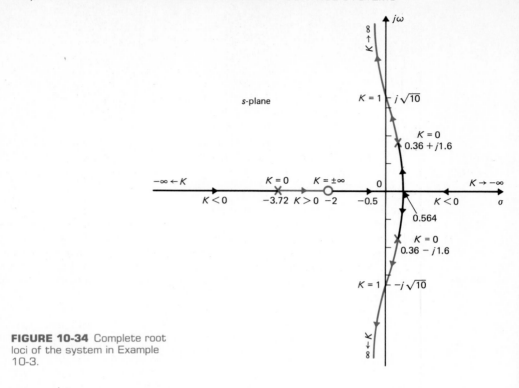

FIGURE 10-34 Complete root loci of the system in Example 10-3.

The system is stable only for $K > 1$, and the RL cross the $j\omega$-axis at $\omega = \pm 3.16$ rad/s, which corresponds to the frequency at which the $G(s)$ plot crosses the negative real axis. The value of K at the crossing points on the $j\omega$-axis is 1.

EXAMPLE 10-4   Consider a feedback control system with the loop transfer function

$$G(s)H(s) = \frac{K}{(s + 2)(s^2 + 4)} \tag{10-104}$$

which has a pair of imaginary poles at $s = j2$ and $s = -j2$. To apply the Nyquist criterion in the original form, we would have to define the Nyquist path with small indentations around these poles.

Instead of constructing the entire Nyquist plot, the portion that corresponds to $s = j\infty$ to $s = j0$ is sketched, as shown in Fig. 10-35. The data for this Nyquist plot are easily obtained from a frequency-response program, such as FREQRP, which accompanies this text.

From Eq. (10-61), the value of Φ_{11} required for stability is

stability condition

$$\Phi_{11} = (-0.5P_\omega)180° = -180° \tag{10-105}$$

As seen from Fig. 10-35, the magnitude of $G(j\omega)H(j\omega)$ goes to infinity when $\omega = 2$ rads/s. When K is positive, the critical point for closed-loop stability is at $(-1, j0)$. When ω varies from ∞ to 2, the angle Φ_{11} is $+135°$, and for the portion of $\omega = 2$ to 0, Φ_{11} is $+45°$. Thus, the total Φ_{11} is $+180°$, and the system is unstable for all positive values of K.

When K is negative, the critical point can be regarded as at $(+1, j0)$. Figure 10-35 shows that the closed-loop system is stable if K lies between 0 and -8. The summary of the Nyquist-criterion application to this system is as follows.

stability summary

Range of K	Φ_{11} for $\omega = \infty$ to 2	Φ_{11} for $\omega = 2$ to 0	Total Φ_{11}	Stability
$K > 0$	$+135°$	$+45°$	$+180°$	Unstable
$K < -8$	$-45°$	$+45°$	$0°$	Unstable
$-8 < K < 0$	$-45°$	$-135°$	$-180°$	Stable

The stability analysis just conducted can be verified by applying Routh–Hurwitz's criterion to the characteristic equation:

$$s^3 + 2s^2 + 4s + 8 + K = 0 \qquad (10\text{-}106)$$

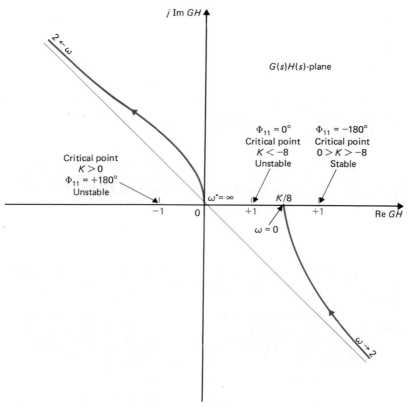

FIGURE 10-35 Nyquist plot of the control system in Example 10-4.

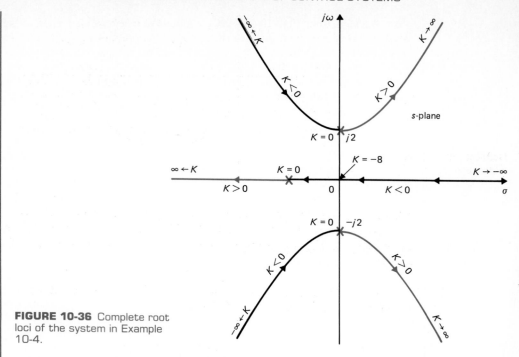

FIGURE 10-36 Complete root loci of the system in Example 10-4.

It should be reiterated that although the Routh–Hurwitz criterion is sometimes easier to apply in stability problems such as those illustrated by the previous examples, in general, the Nyquist criterion leads to a more informative solution, which also includes the relative stability of the system. More important, in practical situations, given a control system, it may be more convenient to measure the frequency response and display the Nyquist plot with hardware than to determine the accurate expression of the characteristic equation.

The root loci of Eq. (10-106) are constructed as shown in Fig. 10-36 using the pole-zero configuration of Eq. (10-104). The stability condition of $-8 < K < 0$ is easily viewed from the root-locus diagram.

10-8

EFFECTS OF ADDITION OF POLES AND ZEROS TO *G(s)H(s)* ON THE SHAPE OF THE NYQUIST LOCUS

Since the performance of a feedback control system is often affected by adding and moving poles and zeros of the open-loop or the loop transfer function, it is important to investigate how the Nyquist plot is affected when poles and zeros are added to a typical loop transfer function $G(s)H(s)$. The following investigation will also be helpful to gain further insight on the quick sketch of the Nyquist plot of a given transfer function.

Let us begin with a first-order transfer function

$$G(s)H(s) = \frac{K}{1 + T_1 s} \tag{10-107}$$

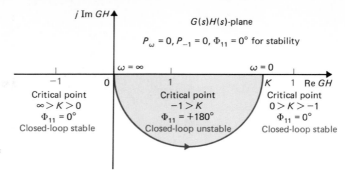

FIGURE 10-37 Nyquist plot of $G(s)H(s) = K/(1 + T_1 s)$.

where T_1 is a positive real constant. The Nyquist locus of $G(j\omega)H(j\omega)$ for $0 \leq \omega < \infty$ is a semicircle, as shown in Fig. 10-37. The figure also shows the interpretation of the closed-loop stability with respect to the critical point for all values of K between $-\infty$ and ∞.

ADDITION OF POLES AT $s = 0$.   Consider that a pole at $s = 0$ is added to the transfer from function of Eq. (10-107); then

$$G(s)H(s) = \frac{K}{s(1 + T_1 s)} \tag{10-108}$$

Since adding a pole at $s = 0$ is equivalent to dividing $G(j\omega)H(j\omega)$ by $j\omega$, the phase of $G(j\omega)H(j\omega)$ is reduced by $90°$ at both zero and infinite frequencies. In addition, the magnitude of $G(j\omega)H(j\omega)$ at $\omega = 0$ becomes infinite. Figure 10-38 illustrates the Nyquist plot of $G(s)H(s)$ in Eq. (10-108) and the closed-loop stability interpretations with respect to the critical points for $-\infty < K < \infty$. In general, adding a pole of multiplicity p at $s = 0$ to the transfer function of Eq. (10-107) will give the following

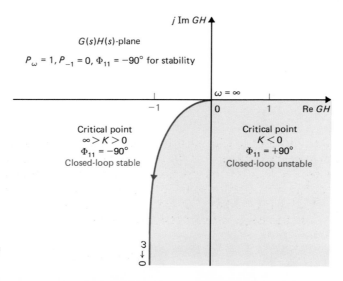

FIGURE 10-38 Nyquist plot of $G(s)H(s) = K/[s(1 + T_1 s)]$.

properties to the Nyquist plot of $G(s)H(s)$:

$$\lim_{\omega \to \infty} \angle G(j\omega)H(j\omega) = -(p+1)\pi/2 \qquad (10\text{-}109)$$

$$\lim_{\omega \to 0} \angle G(j\omega)H(j\omega) = -p\pi/2 \qquad (10\text{-}110)$$

$$\lim_{\omega \to \infty} |G(j\omega)H(j\omega)| = 0 \qquad (10\text{-}111)$$

$$\lim_{\omega \to 0} |G(j\omega)H(j\omega)| = \infty \qquad (10\text{-}112)$$

EXAMPLE 10-5   Figure 10-39 illustrates the Nyquist plot of

$$G(s)H(s) = \frac{K}{s^2(1 + T_1 s)} \qquad (10\text{-}113)$$

and the critical points, with the stability interpretations.

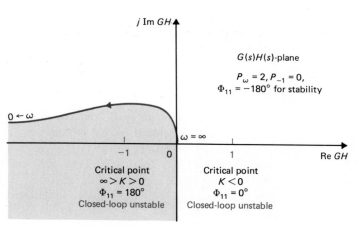

FIGURE 10-39 Nyquist plot of $G(s)H(s) = K/[s^2(1 + T_1 s)]$.

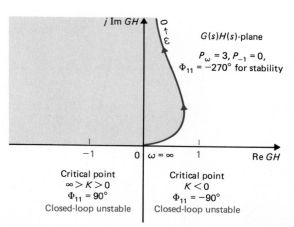

FIGURE 10-40 Nyquist plot of $G(s)H(s) = K/[s^3(1 + T_1 s)]$.

Figure 10-40 illustrates the same for

$$G(s)H(s) = \frac{K}{s^3(1 + T_1 s)} \tag{10-114}$$

effects of adding poles

The conclusion from these illustrations is that the addition of poles at $s = 0$ to a loop transfer function will affect the stability of the closed-loop system adversely; a system that has a loop transfer function with more than one pole at $s = 0$ (type 2 or higher) is likely to be unstable or nearly so.

ADDITION OF FINITE NONEZERO POLES.   When a pole at $s = -1/T_2$ is added to the function $G(s)H(s)$ of Eq. (10-107), we have

$$G(s)H(s) = \frac{K}{(1 + T_1 s)(1 + T_2 s)} \tag{10-115}$$

The Nyquist plot of $G(j\omega)H(j\omega)$ at $\omega = 0$ is not affected by the addition of the pole, since

$$\lim_{\omega \to 0} G(j\omega)H(j\omega) = K \tag{10-116}$$

The value of $G(j\omega)H(j\omega)$ at $\omega = \infty$ is

$$\lim_{\omega \to \infty} G(j\omega)H(j\omega) = \lim_{\omega \to \infty} \frac{-K}{T_1 T_2 \omega^2} = 0\angle -180° \tag{10-117}$$

Thus, the effect of adding a pole at $s = -1/T_2$ to the transfer function of Eq. (10-107) is to shift the phase of the Nyquist plot by $-90°$ at $\omega = \infty$, as shown in Fig. 10-41. The figure also shows the Nyquist plot of

$$G(s)H(s) = \frac{K}{(1 + T_1 s)(1 + T_2 s)(1 + T_3 s)} \tag{10-118}$$

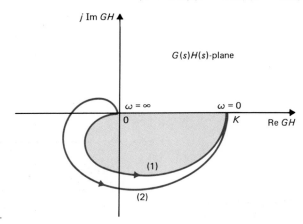

FIGURE 10-41 Nyquist plots.
Curve (1): $G(s)H(s) = K/(1 + T_1 s)(1 + T_2 s)$.
Curve (2): $G(s)H(s) = K/(1 + T_1 s)(1 + T_2 s)(1 + T_3 s)$.

where two nonzero poles have been added to the transfer function of Eq. (10-107). In this case, the Nyquist plot at $\omega = \infty$ is rotated clockwise by another 90° from that of Eq. (10-115). *These examples show the adverse effects on closed-loop stability when poles are added to the loop transfer function.* The closed-loop systems with the loop transfer functions of Eqs. (10-107) and (10-115) are all stable as long as K is positive. The system represented by Eq. (10-118) is unstable if the intersect of the Nyquist plot on the negative real axis is to the left of the $(-1, j0)$ point when K is positive.

ADDITION OF ZEROS.   It was pointed out in Chapter 9 that derivative control causes a closed-loop control system to be more stable. In terms of the Nyquist plot, this stabilization effect is easily demonstrated, since the multiplication of the derivative control term $(1 + T_d s)$ to the loop transfer function increases the phase of $G(s)H(s)$ by 90° at $\omega = \infty$.

EXAMPLE 10-6   As an illustrative example, consider that the loop transfer function of a closed-loop control system is

$$G(s)H(s) = \frac{K}{s(1 + T_1 s)(1 + T_2 s)} \tag{10-119}$$

It can be shown that the closed-loop system is stable for

$$0 \le K < \frac{T_1 + T_2}{T_1 T_2} \tag{10-120}$$

Suppose that a zero at $s = -1/T_d$ is added to the transfer function of Eq. (10-119), such as with a derivative control, or the PD components of the PID control. Then,

effects of adding zeros

$$G(s)H(s) = \frac{K(1 + T_d s)}{s(1 + T_1 s)(1 + T_2 s)} \tag{10-121}$$

The Nyquist plots of the two transfer functions of Eqs. (10-119) and (10-121) are shown in Fig. 10-42. The effect of the zero in Eq. (10-121) is to add 90° to the phase of the $G(j\omega)H(j\omega)$ in

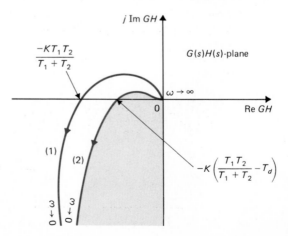

FIGURE 10-42 Nyquist plots.
Curve (1): $G(s)H(s) = K/[s(1 + T_1 s)(1 + T_2 s)]$.
Curve (2): $G(s)H(s) = K(1 + T_d s)/[s(1 + T_1 s)(1 + T_2 s)]$; $T_d < T_1, T_2$.

Eq. (10-119) at $\omega = \infty$ while not affecting its value at $\omega = 0$. The intersect on the negative real axis of the $G(s)H(s)$-plane is moved from $-KT_1T_2/(T_1 + T_2)$ to $-K(T_1T_2 - T_dT_1 - T_dT_2)/(T_1 + T_2)$. Thus, the system with the derivative control is stable for

$$0 \le K < \frac{T_1 + T_2}{T_1T_2 - T_d(T_1 + T_2)} \tag{10-122}$$

which for positive T_d and K has a higher upper bound than that of Eq. (10-120).

10-9

STABILITY ANALYSIS OF MULTILOOP SYSTEMS

The Nyquist stability analyses conducted in the preceding sections are all centered toward systems with a single feedback loop, since the loop transfer function $G(s)H(s)$ or $F(s)$ is used. With the Routh–Hurwitz criterion and the root-locus method, the loop configuration of the system is not an issue, since the starting point for stability analysis is the characteristic equation. With the Nyquist criterion, we can approach the stability of multiloop systems with one of two methods. The first one simply involves the determination of the system characteristic equation, and then the formation of a fictitious loop transfer function $G(s)H(s)$. The stability of the multiloop system is studied by applying the Nyquist criterion to this fictitious $G(s)H(s)$ plot. This is the same approach used in the root-locus analysis when only the characteristic equation is known.

EXAMPLE 10-7

As an illustrative example, consider the two-loop system shown in Fig. 10-43.

One way of applying the Nyquist stability criterion to the system in Fig. 10-43 is to first derive the characteristic equation, which is

$$(s + 10)[s(s + 1)(s + 2) + 5] + K(s + 2) \tag{10-123}$$

We can then create a fictitious loop transfer function $F(s)$ by dividing both sides of the last equation by the terms that do not contain K. The result is

$$1 + F(s) = 1 + \frac{K(s + 2)}{(s + 10)[s(s + 1)(s + 2) + 5]} \tag{10-124}$$

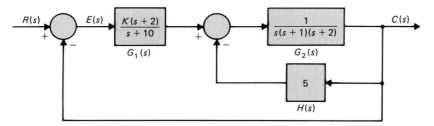

FIGURE 10-43 Multiloop feedback control system.

We can construct the Nyquist plot of $F(s)$, and knowing the number of poles of $F(s)$ in the right-half s-plane, we can determine the number of zeros of $1 + F(s)$ of the characteristic equation that are in the right-half s-plane. Thus, the Nyquist stability analysis of a multiloop system becomes a routine matter once the characteristic equation is known.

An alternate approach to the problem is to conduct a systematic stability analysis of the multiloop system, so that in the process, more information on the stability of the inner loops can be obtained.

For the system in Fig. 10-43, we can first derive the loop transfer function of the inner loop, which is

$$G_2(s)H(s) = \frac{5}{s(s + 1)(s + 2)} \tag{10-125}$$

This can also be regarded as the loop transfer function of the overall system if the outer feedback path is opened. Since $P_\omega = 1$ and $P_{-1} = 0$ for $G_2(s)H(s)$, the inner loop is stable if Φ_{11} for the Nyquist plot of $G_2(s)H(s)$ is $-90°$, or, simply, if the $(-1, j0)$ point is not enclosed by the Nyquist plot of $G_2(j\omega)H(j\omega)$. Figure 10-44 shows the Nyquist plot of $G_2(j\omega)H(j\omega)$, and the angle Φ_{11} is indeed $-90°$, or the $(-1, j0)$ point is not enclosed. Thus, the inner loop is stable, or the equation

$$s(s + 1)(s + 2) + 5 = 0 \tag{10-126}$$

does not have any roots in the right-half s-plane.

Next, we consider that the outer feedback path is closed, so that the loop transfer function of the overall system is

$$F(s) = \frac{G_1(s)G_2(s)}{1 + G_2(s)H(s)} = \frac{K(s + 2)}{(s + 10)[s(s + 1)(s + 2) + 5]} \tag{10-127}$$

Observing the last equation, and from the stability analysis of $G_2(s)H(s)$, we know that all the poles of $F(s)$ are in the left-half s-plane. Thus, for the overall system to be stable, Φ_{11} of the Nyquist plot of $F(s)$ should be $0°$, or the $(-1, j0)$ point should not be enclosed. Figure 10-45 shows the Nyquist plot of $F(j\omega)$. The intersect of the $F(j\omega)$ plot on the real axis is at $-0.02K$. Thus, the system is stable if the $(-1, j0)$ point is to the left of $-0.02K$, and the range of K for stability is $0 \leq K < 50$.

Although the second method involves the construction of two Nyquist plots, the method is more systematic, and it gives the stability of not just the overall system, but also that of the inner loop. In general, when more than two loops are involved, the proper way is to start with the sta-

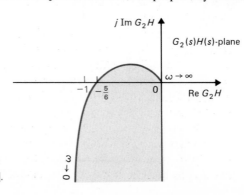

FIGURE 10-44 Nyquist plot of
$G_2(s)H(s) = 5/[s(s + 1)(s + 2)]$.

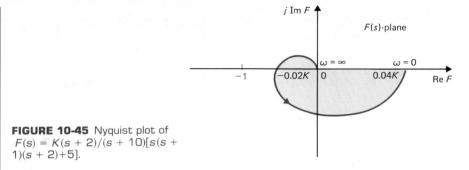

FIGURE 10-45 Nyquist plot of $F(s) = K(s + 2)/(s + 10)[s(s + 1)(s + 2)+5]$.

bility analysis of the innermost loop by opening all the outer loops, and then add one loop at a time, until the outermost loop is closed.

10-10

STABILITY OF LINEAR CONTROL SYSTEMS WITH TIME DELAYS

Systems with time delays and their modeling were discussed in Section 4-9. Closed-loop systems with time delays in the loops generally will be subject to more stability problems than systems without delays. Since a pure time delay T_d is modeled by the transfer-function relationship $e^{-T_d s}$, the characteristic equation of the system will no longer have constant coefficients. Therefore, the Routh–Hurwitz criterion is *not* applicable. In Chapter 8, we pointed out that the root-locus method can be applied to systems with pure time delays, but the analysis is usually quite complex. In this section, we shall show that the Nyquist criterion can be readily applied to a system with a pure time delay.

Let us consider that the loop transfer function of a feedback control system with pure time delay is represented by a form similar to Eq. (10-38), or

$$G(s)H(s) = G_1(s)H_1(s)e^{-T_d s} \tag{10-128}$$

where $G_1(s)H_1(s)$ is a rational function of s with constant coefficients, and T_d is the pure time delay in seconds.

In principle, the stability of the closed-loop system can be investigated by constructing the Nyquist plot of $G(s)H(s)$ and then observing its behavior with respect to the $(-1, j0)$ point. The effect of the exponential term in Eq. (10-128) is that it rotates the phasor $G_1(s)H_1(s)$ at each ω by an angle of ωT_d radians in the clockwise direction. The amplitude of $G_1(j\omega)H_1(j\omega)$ is not affected by the time delay, since the magnitude of $e^{-T_d s}$ is unity at all frequencies.

In control systems, the magnitude of $G_1(j\omega)H_1(j\omega)$ usually approaches zero as ω approaches infinity. Thus, the Nyquist plot of the transfer function of Eq. (10-128) will usually spiral toward the origin in the clockwise direction as ω approaches infinity, and there is an infinite number of intersects on the real axis of the $G(s)H(s)$-plane. Once the Nyquist plot of $G(j\omega)H(j\omega)$ is constructed, the stability of the system is determined in the usual manner by investigating the angle Φ_{11}.

EXAMPLE 10-8   As an illustrative example, Fig. 10-46 shows the Nyquist plot of

$$G(s)H(s) = G_1(s)H_1(s)e^{-T_d s} = \frac{e^{-T_d s}}{s(s+1)(s+2)} \qquad (10\text{-}129)$$

for several values of T_d. These Nyquist plots show that when T_d is zero, the system is stable. The stability condition deteriorates as T_d increases, since the intersect on the real axis by the $G(j\omega)H(j\omega)$ plot moves closer to the $(-1, j0)$ point. When $T_d = 2.09$, the $G(j\omega)H(j\omega)$ plot passes through the $(-1, j0)$ point, and the system is on the verge of instability. The system is unstable when T_d is greater than 2.09, since the angle Φ_{11} would be positive.

Unlike the rational-function case, the analytical solution of the intersects on the real axis of the $G(s)H(s)$-plane is not trivial, since the equations that govern the intersects are no longer algebraic. The loop transfer function of Eq. (10-129) can be rationalized in the usual manner by

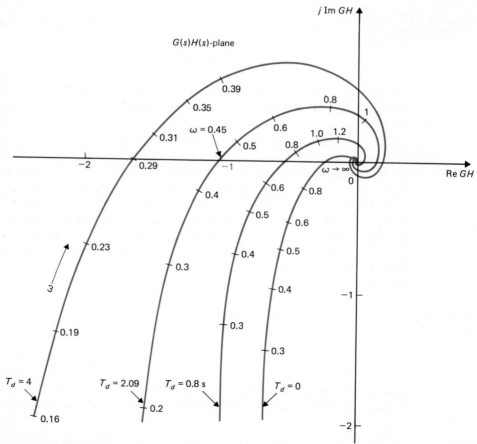

FIGURE 10-46 Nyquist plot of $G(s)H(s) = e^{-T_d s}/[s(s+1)(s+2)]$.

multiplying its numerator and denominator by the complex conjugate of the denominator. The result is

$$G(j\omega)H(j\omega) = \frac{(\cos \omega T_d - j \sin \omega T_d)[-3\omega^2 - j\omega(2 - \omega^2)]}{9\omega^4 + \omega^2(2 - \omega^2)^2} \qquad (10\text{-}130)$$

The condition of $G(j\omega)H(j\omega)$ intersecting the real axis is

$$3\omega^2 \sin \omega T_d - \omega(2 - \omega^2) \cos \omega T_d = 0 \qquad (10\text{-}131)$$

which is not easily solved for ω, given T_d.

The Critical Trajectory

Thus far we have used either the $(-1, j0)$ point or the $(+1, j0)$ point of the $F(s)$-plane as the critical point for stability analysis when the gain K is either positive or negative, respectively. We can relax the critical-point idea by extending it into a trajectory if necessary.

Referring to Eq. (10-128), we recognize that the roots of the characteristic equation satisfy

$$G_1(s)H_1(s)e^{-T_d s} = -1 \qquad (10\text{-}132)$$

The right-hand side of the last equation points to the fact that the $(-1, j0)$ is the critical point for stability analysis of the closed-loop system. Equation (10-132) can be written as

$$G_1(s)H_1(s) = -e^{T_d s} \qquad (10\text{-}133)$$

critical trajectory

When $s = j\omega$, the left-hand side of the last equation gives the Nyquist plot of the system when there is no time delay. The exponential term in Eq. (10-133) has a magnitude of one for all values of ω, and its phase is $-\omega T_d$ radians. Thus, the right-hand side of Eq. (10-133) describes a **critical trajectory**, which is a circle with unit radius, and centered at the origin of the $G_1(s)H_1(s)$-plane. When $\omega = 0$, the trajectory starts at the point $(-1, j0)$, and as ω increases, the critical point traces out the unit circle in the counterclockwise direction.

EXAMPLE 10-9   Figure 10-47 shows the Nyquist plot of

$$G_1(s)H_1(s) = \frac{1}{s(s + 1)(s + 2)} \qquad (10\text{-}134)$$

together with the critical trajectory of $-e^{j\omega T_d}$. The frequency at which the $G_1(j\omega)H_1(j\omega)$ plot intersects the critical trajectory is found by setting the magnitude of $G_1(j\omega)H_1(j\omega)$ to unity; i.e.,

$$|G_1(j\omega)H_1(j\omega)| = \frac{1}{|-3\omega^2 + j\omega(2 - \omega^2)|} = 1 \qquad (10\text{-}135)$$

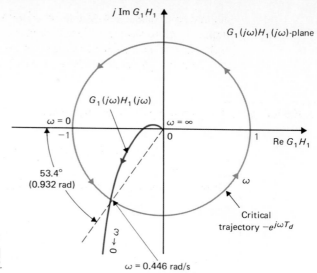

FIGURE 10-47 Nyquist plot of $G_1(s)H_1(s) = 1/[s(s + 1)(s + 2)]$ and the critical trajectory $-e^{j\omega T_d}$.

Solving the last equation yields the positive solution for ω to be 0.446 rad/s. Note that this is the frequency at which the Nyquist plot of Fig. 10-46 intersects the $(-1, j0)$ point on the negative real axis. Since the angle in radians measured from the $(-1, j0)$ point to the point of intersect between the $G_1(j\omega)H_1(j\omega)$ plot and the unit circle is ωT_d, where $\omega = 0.446$, we can find the critical value of T_d for stability by equating

$$\angle G_1(j0.446)H_1(j0.446) = 0.446T_d + \pi \qquad (10\text{:}136)$$

Or

$$-\tan^{-1}\left(\frac{2 - \omega^2}{-3\omega}\right)_{\omega = 0.446} = 0.446T_d + \pi \qquad (10\text{-}137)$$

which leads to

$$0.932 \text{ rad} = 0.446T_d \qquad (10\text{-}138)$$

Thus, the critical value of T_d is 2.09 seconds, which agrees with the result shown in Fig. 10-46.

Although the method just described still involves the calculation of the properties of the intersection between the $G_1(j\omega)H_1(j\omega)$ plot and the critical trajectory for the time delay, it does present another perspective on the application of the Nyquist stability criterion to systems with pure time delays. The valuable lesson learned is that the critical point for stability analysis not only can be the $(-1, j0)$ point, but also can be generalized to a trajectory whenever the situation requires.

Approximation of $e^{-T_d s}$

The advantage of the frequency-domain analysis is that systems with pure time delays in the loop can be conducted by plotting the frequency response of the open-loop transfer function directly with the aid of a computer. However, for analysis and design techniques that require a rational function, the series of Padé approximation of the time delay described in Chapter 4 can be applied. We can use the frequency-domain plot to evaluate the accuracy of these approximations. An illustrative example follows.

EXAMPLE 10-10   By using the series approximation of $e^{-T_d s}$ given in Eq. (4-187), Eq. (10-129) becomes

$$G(s)H(s) \cong \frac{1}{s(s + 1)(s + 2)(1 + T_d s + T_d^2 s^2/2)} \qquad (10\text{-}139)$$

Or by using the Padé approximation in Eq. (4-188), Eq. (10-129) becomes

$$G(s)H(s) = \frac{1 - T_d s/2}{s(s + 1)(s + 2)(1 + T_d s/2)} \qquad (10\text{-}140)$$

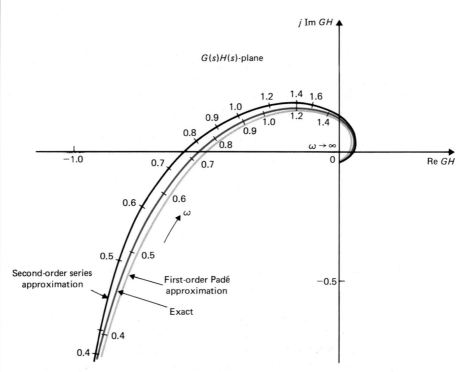

FIGURE 10-48 Approximation of Nyquist plot of Figure 10-46 by a two-term power-series approximaion; $T_d = 0.8$ s.

Figure 10-48 shows the polar plots of Eqs. (10-129), (10-139), and (10-140) for $T_d = 0.8$. The intersects of the polar plots of Eqs. (10-129), (10-139), and (10-140) are at -0.5164, -0.525, and -0.5076, respectively.

10-11

RELATIVE STABILITY: GAIN MARGIN
AND PHASE MARGIN

We have demonstrated in Sections 10-2 through 10-4 the general relationship between the resonance peak M_p of the frequency response and the maximum overshoot of the step response. Comparisons and correlations between frequency-domain and time-domain parameters such as these are useful in the prediction of performance of control systems. In general, we are interested not only in systems that are stable, but also in *relative*
stability systems that have a certain degree of stability. The latter is often called **relative stability**. In the time domain, relative stability of a control system is measured by parameters such as the maximum overshoot and the damping ratio. In the frequency domain, the resonance peak M_p can be used to indicate relative stability. Another way of measuring relative stability is by means of the Nyquist plot of the loop transfer function $G(s)H(s)$. The closeness of the $G(j\omega)H(j\omega)$ plot in the polar coordinates to the $(-1, j0)$ point gives an indication of how stable or unstable the closed-loop system is.

To demonstrate the concept of relative stability, the Nyquist plots and the corresponding step responses and frequency responses of a typical third-order system are shown in Fig. 10-49 for four different values of loop gain K. It is assumed that the $G(s)H(s)$ is a minimum-phase transfer function, so that the portion of the Nyquist plot that corresponds to $s = j\infty$ to $s = j0$ is sufficient for stability analysis. Let us consider the case shown in Fig. 10-49(a), in which the loop gain K is low, so the Nyquist plot of $G(s)H(s)$ intersects the negative real axis at a point that is quite far to the right of the $(-1, j0)$ point. The corresponding step response is shown to be quite well damped, and the value of M_p of the frequency response is low. As K is increased, Fig. 10-49(b) shows that the intersect of the $G(j\omega)H(j\omega)$ plot of the negative real axis is moved closer to the $(-1, j0)$ point; the system is still stable, since the critical point is not enclosed, but the step response has a larger maximum overshoot, and M_p is also larger. The phase curve $\phi(j\omega)$ of the closed-loop frequency response does not give as good an indication of relative stability as M_p, except that one should note that the slope of the phase curve becomes steeper as the relative stability decreases. When K is increased still further, the Nyquist plot of $G(j\omega)H(j\omega)$ now passes through the $(-1, j0)$ point, and the system is not asymptotically stable. The step response becomes oscillatory with constant amplitude, and M_p becomes infinite. For a still larger value of K, the Nyquist plot will enclose the $(-1, j0)$ point, and the system is unstable. The step response is now unbounded, as shown in Fig. 10-49(d). In this case, the magnitude curve of $|M(j\omega)|$-versus-ω ceases to have any significance, and the only indication of instability from the closed-loop frequency response is that the phase curve now has a positive slope beyond the resonant frequency.

Gain Margin

Gain margin is used to indicate the closeness of the intersect of the negative real axis made by the Nyquist $G(j\omega)H(j\omega)$ plot to the $(-1, j0)$ point. Let us first give the following definitions:

PHASE-CROSSOVER POINT.   A phase-crossover point in the $G(j\omega)$ plane is a point at which the Nyquist $G(j\omega)H(j\omega)$-plot intersects the negative real axis.

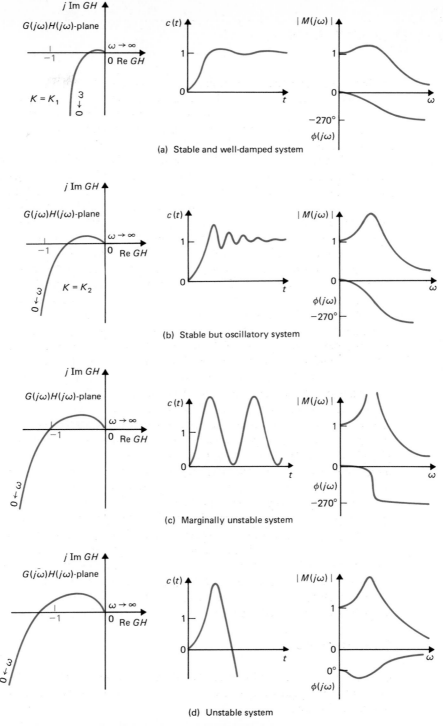

(a) Stable and well-damped system

(b) Stable but oscillatory system

(c) Marginally unstable system

(d) Unstable system

FIGURE 10-49 Correlation among Nyquist plots, step responses, and frequency responses.

PHASE-CROSSOVER FREQUENCY.   The phase-crossover frequency ω_c is the frequency at the phase-crossover point, or where

$$\angle G(j\omega_c)H(j\omega_c) = 180° \qquad (10\text{-}141)$$

minimum phase

The Nyquist plot of a loop transfer function $G(j\omega)H(j\omega)$ that is of the minimum-phase type shown is shown in Fig. 10-50. The phase-crossover frequency is denoted as ω_c, and the magnitude of $G(j\omega)H(j\omega)$ at $\omega = \omega_c$ is designated as $|G(j\omega_c)H(j\omega_c)|$. Then, the gain margin of the closed-loop system that has $G(s)H(s)$ as its loop transfer function is defined as

gain margin

$$\text{gain margin} = \text{G.M.} = 20\log_{10}\frac{1}{|G(j\omega_c)H(j\omega_c)|} \text{ dB} \qquad (10\text{-}142)$$

On the basis of this definition, we can draw the following conclusions on the gain margin of the system illustrated in Fig. 10-50, depending on the shape of the Nyquist plot.

1. The $G(j\omega)H(j\omega)$ plot does not intersect the negative real axis.

    $$|G(j\omega_c)H(j\omega_c)| = 0 \qquad \text{G.M.} = \infty \text{ dB}$$

2. The $G(j\omega)H(j\omega)$ plot intersects the negative real axis between 0 and $(-1, j0)$.

    $$0 < |G(j\omega_c)H(j\omega_c)| < 1 \qquad \text{G.M.} > 0 \text{ dB}$$

3. The $G(j\omega)H(j\omega)$ plot passes through the $(-1, j0)$ point.

    $$|G(j\omega_c)H(j\omega_c)| = 1 \qquad \text{G.M.} = 0 \text{ dB}$$

4. The $G(j\omega)H(j\omega)$ plot encloses the $(-1, j0)$ point.

    $$|G(j\omega_c)H(j\omega_c) > 1 \qquad \text{G.M.} < 0 \text{ dB}$$

Based on the foregoing analysis, the physical significance of gain margin can be stated as:

gain margin

Gain margin is the amount of gain in decibels (dB) that is allowed to be increased in the loop before the closed-loop system reaches instability.

When the $G(j\omega)H(j\omega)$ plot goes through the $(-1, j0)$ point, the gain margin is 0 dB, which implies that the loop gain can no longer be increased, as the system is at the margin of instability. When the minimum-phase $G(j\omega)H(j\omega)$ plot does not intersect the negative real axis at any finite nonzero frequency, and the Nyquist stability criterion indicates that the critical point must not be enclosed for stability, the gain margin is infinite in dB; this means that, theoretically, the value of the loop gain can be increased to infinity before instability occurs.

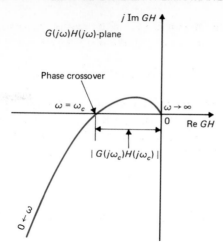

FIGURE 10-50 Definition of the gain margin in the polar coordinates.

When the phase-crossover point is to the left of the $(-1, j0)$ point, the phase margin is negative in dB, and the closed-loop system is unstable.

Care should be taken when interpreting the gain margin of nonminimum-phase systems using the Nyquist plot of $G(j\omega)H(j\omega)$, since in this case, the $(-1, j0)$ point may have to be encircled for stability, and thus a negative gain margin may correspond to a stable system.

Phase Margin

The gain margin is only a one-dimensional representation of the relative stability of a closed-loop system. As the name implies, gain margin gives indication only on the stability of a system with respect to the loop gain. In principle, a system with a large gain margin should be relatively more stable than one with a smaller gain margin. Unfortunately, gain margin alone is inadequate to indicate relative stability when system parameters other than the loop gain are subject to variation. For instance, the two sys-

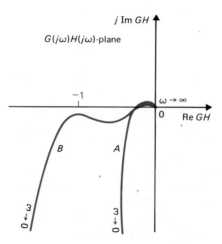

FIGURE 10-51 Nyquist plots showing systems with the same gain margin but different degrees of relative stability.

tems represented by the $G(j\omega)H(j\omega)$ plots in Fig. 10-51 apparently have the same gain margin. However, locus A actually corresponds to a more stable system than locus B. The reason is that with any change in the system parameter that affects the phase of $G(j\omega)H(j\omega)$, it is easier for locus B to pass through or even enclose the $(-1, j0)$ point. Furthermore, we can show that system B has a much larger M_p than system A.

To include the effect of phase shift on stability, we introduce the phase margin, which requires the following definitions:

gain crossover

GAIN-CROSSOVER POINT. The gain-crossover point is a point on the $G(j\omega)H(j\omega)$ plot at which the magnitude of $G(j\omega)H(j\omega)$ is equal to 1.

GAIN-CROSSOVER FREQUENCY. The gain-crossover frequency ω_g is the frequency of $G(j\omega)H(j\omega)$ at the gain-crossover point. Or

$$\left| G(j\omega_g)H(j\omega_g) \right| = 1 \qquad (10\text{-}143)$$

The definition of phase margin is stated as:

phase margin

Phase margin is defined as the angle in degrees through which the $G(j\omega)H(j\omega)$ plot must be rotated about the origin in order that the gain-crossover point on the locus passes through the $(-1, j0)$ point.

Figure 10-52 shows the Nyquist plot of a typical minimum-phase $G(j\omega)H(j\omega)$ plot, and the phase margin is shown as the angle between the phasor that passes through the gain-crossover point and the negative real axis of the $G(j\omega)H(j\omega)$-plane. In contrast to the gain margin, which gives a measure of the effect of the loop gain on the stability of the closed-loop system, the phase margin indicates the effect on stability due to changes of system parameters, which theoretically alter the phase of $G(j\omega)H(j\omega)$ by an equal amount at all frequencies.

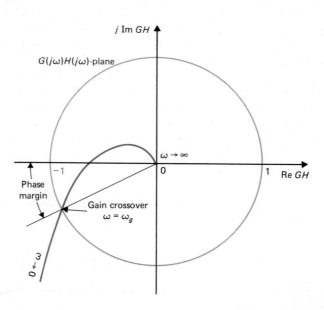

FIGURE 10-52 Phase margin defined in the $G(j\omega)H(j\omega)$-plane.

When the system is of the minimum-phase type, the analytical expression of the phase margin can be expressed as

phase margin

$$\text{phase margin} = \Phi.\text{M.} = \angle G(j\omega_g)H(j\omega_g) - 180° \qquad (10\text{-}144)$$

where ω_g is the gain-crossover frequency. For a minimum-phase system, if the system is unstable, the gain-crossover point usually would not be found in the third quadrant of the $G(j\omega)H(j\omega)$-plane, as is the case shown in Fig. 10-52, and the phase margin would be negative.

It is important to point out that although the gain-crossover point, phase-crossover point, and gain and phase margins are defined analytically, in practice, it is more convenient to obtain these quantities graphically using the Bode plots or using the FREQRP program from the ACSP software accompanying this text.

10-12
STABILITY ANALYSIS WITH THE BODE PLOT

The Bode plot of a transfer function $G(s)$ described in Appendix A is a very useful graphical tool for control-systems analysis and design. Although we introduced the Nyquist stability criterion using the Nyquist plot of $G(s)H(s)$ in the polar coordinates, in practice, the Bode plot is more convenient to apply when the system is of the minimum-phase type. The logarithmic scale on the magnitude coordinate makes the Bode plot attractive when changing gains. The separate phase plot also makes it convenient to study the effects of adding a pure phase shift. Since the Bode plot corresponds to only the positive portion of the $j\omega$-axis from $\omega = \infty$ to $\omega = 0$ in the s-plane, its use in stability studies is limited to the determination of the gain- and phase-crossover points and the corresponding gain and phase margins. The angle Φ_{11} of Eq. (10-61) cannot be determined from the Bode plot unless the $\angle[1 + G(j\omega)H(j\omega)]$-versus-$\omega$ curve is plotted. Thus, the Bode plot is most useful only when the system is of the minimum-phase type, so that the *enclosure* criterion can be used for stability. Care must be taken in interpreting the Bode diagram for closed-loop stability information when the transfer function $G(s)H(s)$ is of the nonminimum-phase type.

EXAMPLE 10-11   As an illustrative example, consider that the open-loop transfer function of a control system with unity feedback is given by

$$G(s) = \frac{K}{s(1 + 0.02s)(1 + 0.2s)} \qquad (10\text{-}145)$$

The Bode plot of $G(j\omega)$ is shown in Fig. 10-53 using the computer program FREQRP. The

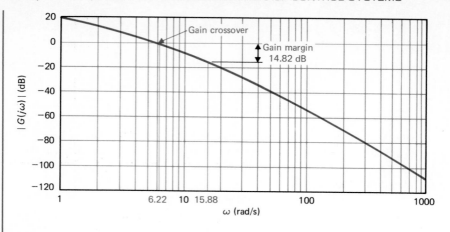

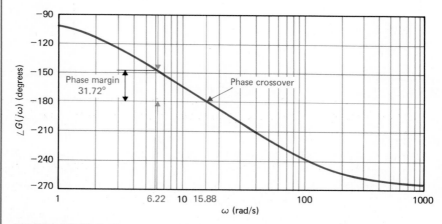

FIGURE 10-53 Bode plot of $G(s) = 10/[s(1 + 0.2s)(1 + 0.02s)]$.

gain- and phase-crossover points are determined as shown in Fig. 10-53. The phase-crossover frequency is 15.88 rad/s, and the magnitude of $G(j\omega)$ at this frequency is -14.82 dB. This means that if the loop gain of the system is increased by 14.82 dB, the magnitude curve of $G(j\omega)$ will cross the 0-dB axis at the phase-crossover frequency. This condition corresponds to the Nyquist plot of $G(j\omega)$ passing through the $(-1, j0)$ point, and the system is marginally unstable. Therefore, by definition, the gain margin of the system is 14.82 dB.

phase margin To determine the phase margin, we note that the gain-crossover frequency is at $\omega = 6.22$ rad/s. The phase of $G(j\omega)$ at this frequency is -142.28 degrees. The phase margin is the angle the phase curve must be shifted so that it will pass through the $-180°$ axis at the gain-crossover frequency. In this case,

$$\Phi.M. = 180° - 148.28° = 31.72° \tag{10-146}$$

Figure 10-54 shows the gain- and phase-margin analysis using the Nyquist plot of $G(j\omega)$. Apparently, the gain and phase margins of this system are all positive, and the system is stable.

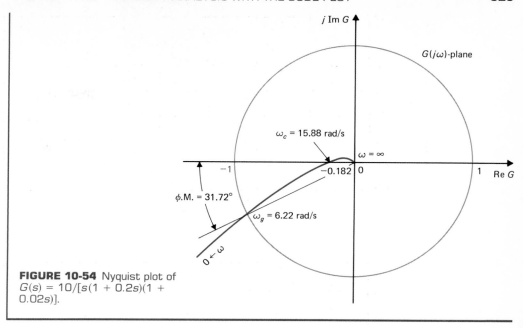

FIGURE 10-54 Nyquist plot of $G(s) = 10/[s(1 + 0.2s)(1 + 0.02s)]$.

The procedure of determining the gain margin and the phase margin of a minimum-phase system from the Nyquist plot or the Bode plot of $G(j\omega)H(j\omega)$ is outlined as follows:

gain margin

1. The gain margin is measured at the phase-crossover frequency ω_c:

$$\text{G.M.} = -|G(j\omega_c)H(j\omega_c)| \text{ dB} \qquad (10\text{-}147)$$

phase margin

2. The phase margin is measured at the gain-crossover frequency ω_g:

$$\Phi.\text{M.} = -180° + \angle G(j\omega_g)H(j\omega_g) \qquad (10\text{-}148)$$

As shown in Fig. 10-53, the gain margin is positive and the system is stable when the intersect on the magnitude curve of $G(j\omega)$ at the phase-crossover frequency is below the 0-dB axis. If the intersect is above the 0-dB axis, the gain margin is negative, and the system is unstable. Similarly, the phase margin is positive and the system is stable when the intersect on the phase curve of $G(j\omega)$ at the gain-crossover frequency is above the $-180°$ axis; otherwise, the phase margin is negative, and the system is unstable. Figure 10-55 illustrates the conclusions just given. It should be emphasized again that the analysis is valid only for minimum-phase transfer functions.

Bode Plot of Systems with Time Delay

The stability analysis of a system with a pure time delay discussed in Section 10-10 can be conducted easily with the Bode plot.

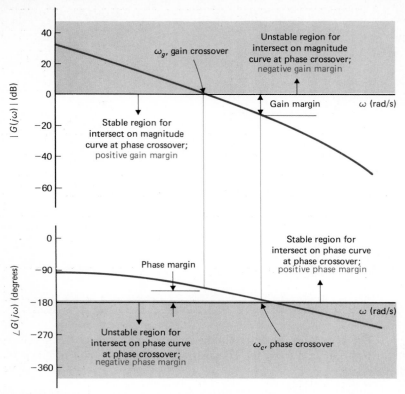

FIGURE 10-55 Determination of gain margin and phase margin on the Bode plot.

EXAMPLE 10-12
time-delay example

As an illustrative example, consider that the loop transfer function of a closed-loop system is

$$G(s)H(s) = \frac{Ke^{T_d s}}{s(s+1)(s+2)} \qquad (10\text{-}149)$$

Figure 10-56 shows the Bode plot of $G(j\omega)H(j\omega)$ with $K = 1$ and $T_d = 0$. The gain-crossover frequency is 0.446 rad/s, and the phase margin is 53.4°. The effect of the pure time-delay term, $e^{-jT_d\omega}$, is to add the phase shift of $-T_d\omega$ radians to the phase curve while not affecting the magnitude curve. The adverse effect of the time delay on stability is apparent, since the negative phase shift caused by the time delay increases rapidly with the increase in ω. To find the critical value of the time delay for stability with $K = 1$, we read from Fig. 10-56 that the phase margin with $T_d = 0$ is 53.4° at $\omega_g = 0.446$ rad/s. We set

$$T_d\omega_g = 53.4°\pi/180° = 0.932 \text{ radians} \qquad (10\text{-}150)$$

Solving for T_d from the last equation, we get $T_d = 2.09$ seconds.

As an extension to the above example, we can set T_d arbitrarily at 1 second, and find the critical value of K for stability. Figure 10-56 shows the Bode plot for this new time delay. With K still equal to 1, the magnitude curve is unchanged. The phase curve droops with the increase in ω, and the phase-crossover frequency is now at 0.66 rad/s. The gain margin is 4.5 dB. Thus,

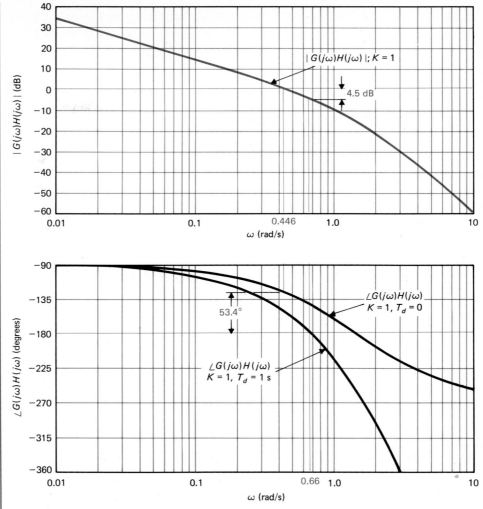

FIGURE 10-56 Bode plot of $G(s)H(s) = Ke^{-T_ds}/[s(s + 1)(s + 2)]$.

the critical value of K for stability when $T_d = 1$ is 1.67, which is determined from the fact that K can be increased by 4.5 dB before instability sets in, or $20 \log_{10} 1.67 = 4.5$ dB.

10-13

RELATIVE STABILITY FROM THE SLOPE
OF THE MAGNITUDE CURVE OF THE BODE PLOT

In addition to gain and phase margins, the relative stability of a closed-loop system can also be observed qualitatively by examining the slope of the magnitude curve of the Bode plot of $G(j\omega)H(j\omega)$ at the gain crossover. For example, in Fig. 10-53, if the loop gain of the system is decreased from the nominal value, the magnitude curve is shifted downward, while the phase curve is unchanged. This causes the gain-crossover frequency to be lower, and the slope of the magnitude curve at this frequency is reduced or less negative; the corresponding phase margin is increased. On the other hand, if the loop gain is increased, the gain-crossover frequency is increased, and the slope of the magnitude curve is more negative. This corresponds to a smaller phase margin, and the system is less stable. The reason behind these stability evaluations is quite simple. For a minimum-phase transfer function, the relation between its magnitude and phase is unique. Since the negative slope of the magnitude curve is a result of having more poles than zeros in the transfer function, the corresponding phase is also negative. In general, the steeper the slope of the magnitude curve, the more negative the phase. Thus, if the gain crossover is at a point where the slope of the magnitude curve is steep, it is likely that the phase margin will be small or negative.

EXAMPLE 10-13   The example given in Fig. 10-53 is uncomplicated, and the slopes of the magnitude and phase curves are monotonically decreasing as a function of ω. Let us consider a conditionally stable system for the purpose of illustrating relative stability. Consider that a control system with unity feedback has the open-loop transfer function

$$G(s) = \frac{K(1 + 0.2s)(1 + 0.025s)}{s^3(1 + 0.01s)(1 + 0.005s)} \qquad (10\text{-}151)$$

The Bode plot of $G(j\omega)$ is shown in Fig. 10-57 for $K = 1$. The gain-crossover frequency is 1 rad/s, and the phase margin is negative ($-78°$). The closed-loop system is unstable even for this small value of K. There are two phase-crossover points: one at $\omega = 25.8$ rad/s and the other at $\omega = 77.7$ rad/s. The phase characteristics between these two frequencies indicate that if the gain crossover lies in this range, the system would be stable. From the magnitude curve, the range of K for stable operation is found to be between 69 and 85.5 dB. For values of K above and below this range, the phase lag of $G(j\omega)$ exceeds $-180°$ and the system is unstable. This system serves as a good example of the relation between relative stability and the slope of the magnitude curve at the gain crossover. As observed from Fig. 10-57, at both very low and very high frequencies, the slope of the magnitude curve is -60 dB/decade; if the gain crossover falls in either one of these two regions, the phase margin is negative, and the system is unstable. In the two sections of the magnitude curve that have a slope of -40 dB/decade, the system is stable only if the gain crossover falls in about half of these regions, but even then the phase margin is small. If the gain crossover falls in the region in which the magnitude curve has a slope of -20 dB/decade, the system is stable. Figure 10-58 shows the Nyquist plot of $G(j\omega)$. It is interesting to compare the results on stability derived from the Bode plot and the Nyquist plot. The root-locus diagram of the system is shown in Fig. 10-59. Remember that when constructing the root loci, the transfer function in Eq. (10-151) should be conditioned by factoring out all the time constants in the numerator and the denominator. Thus, for this case, there is a factor of 100 in front of K. Figure 10-59 gives a clear picture on the stability condition of the system with respect to K. The number of crossings of the root loci on the $j\omega$-axis of the s-plane should equal the crossings of the phase curve of $G(j\omega)$ of the $-180°$ axis of the Bode plot or the crossings of the Nyquist plot of $G(j\omega)$ with the negative real axis. The reader should check the gain margins obtained from the Bode plot and the coordinates of the crossover points on the negative real axis of the Nyquist plot with the values of K at the $j\omega$-axis crossings on the root loci.

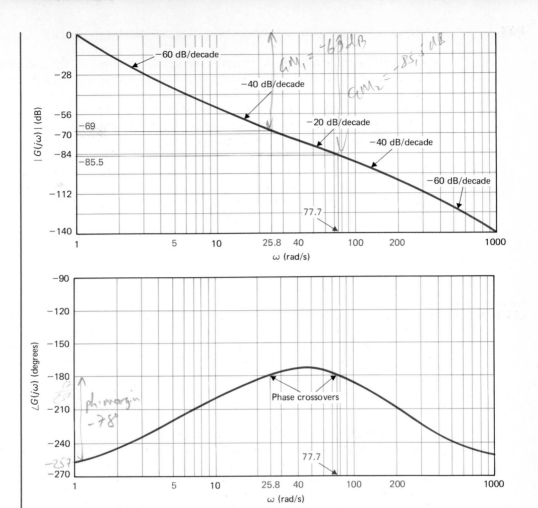

FIGURE 10-57 Bode plot of $G(s) = [K(1 + 0.2s)(1 + 0.025s)]/[s^3(1 + 0.01s)(1 + 0.005s)]$. $K = 1$.

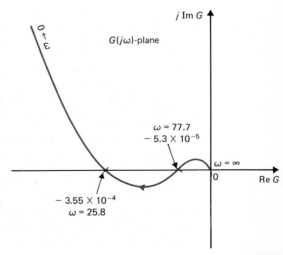

FIGURE 10-58 Nyquist plot of $G(s) = [K(1 + 0.2s)(1 + 0.025s)]/[s^3(1 + 0.01s)(1 + 0.005s)]$. $K = 1$.

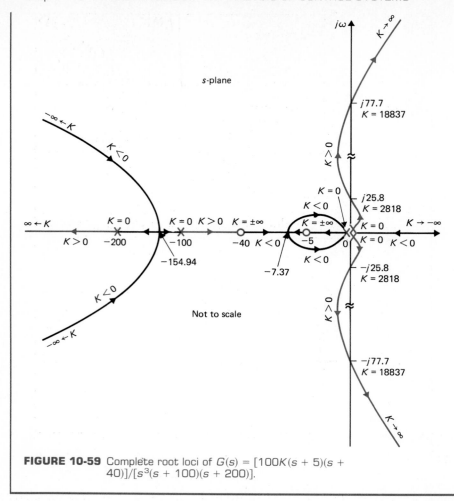

FIGURE 10-59 Complete root loci of $G(s) = [100K(s + 5)(s + 40)]/[s^3(s + 100)(s + 200)]$.

10-14
STABILITY ANALYSIS WITH THE GAIN-PHASE PLOT

One of the shortcomings of the Bode-plot method is that it does not directly give graphical solutions to closed-loop system parameters such as M_p, ω_p, and BW. These closed-loop system parameters are determined from the constant-M loci in the polar coordinates, which are discussed in the next section. For design purposes, it is more convenient to present the constant-M loci in the magnitude-versus-phase coordinates. This is done by displaying the $G(j\omega)$ curve in $|G(j\omega)|$ (dB) versus $\angle G(j\omega)$. For the transfer function in Eq. (10-145), the magnitude-versus-phase plot is shown in Fig. 10-60, which is constructed by use of the data from the Bode plot of Fig. 10-53. The gain and phase crossovers and the gain and phase margins are clearly indicated on the magni-

gain margin tude-versus-phase plot of $G(j\omega)$. The phase-crossover point is where the locus intersects the $-180°$ axis, and the gain crossover is where the locus intersects the 0-dB axis. Thus, the gain margin is simply the distance in dB measured from the phase

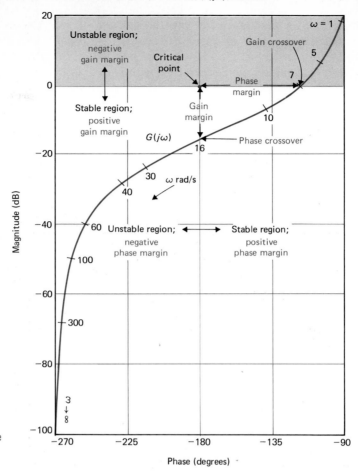

FIGURE 10-60 Gain-phase plot of $G(s) = 10/[s(1 + 0.2s)(1 + 0.02s)]$.

phase margin   crossover to the critical point point at 0 dB and $-180°$, and the phase margin is the horizontal distance in degrees measured from the gain crossover to the critical point. The regions in which the gain- and phase-crossover points should be located for stability are also indicated.

advantage   Another advantage of the magnitude-versus-phase plot is that when the loop gain of $G(jω)$ changes, the locus is simply shifted up and down along the vertical axis in dB. When a constant phase is added to $G(jω)$, the locus is shifted horizontally without distortion to the locus.

10-15

CONSTANT-*M* LOCI IN THE *G(jω)*-PLANE

It has been pointed out that analytically, the resonance peak M_p of high-order systems is difficult to obtain, and graphically the Bode plot provides information on the closed-loop system only in the form of gain margin and phase margin. It is necessary to develop a graphical method for the determination of M_p, $ω_p$, and BW using the open-loop transfer function $G(s)$. For the purpose of analysis, we can always obtain the magnification curve of $|M(j\omega)|$-versus-ω by use of a computer program such as FREQRP. However, the graphical method will allow not only a clear presentation of the closed-loop system performance with respect to the open-loop transfer function, but also the design of a system with a specified value for M_p.

Consider that the closed-loop transfer function of a control system with unity feedback is given by

$$M(s) = \frac{C(s)}{R(s)} = \frac{G(s)}{1 + G(s)} \tag{10-152}$$

For sinusoidal steady state, $G(s) = G(j\omega)$, and $G(j\omega)$ is

$$G(j\omega) = \text{Re } G(j\omega) + j \text{ Im } G(j\omega)$$
$$= x + jy \tag{10-153}$$

where x is the real component of $G(j\omega)$, and y is the imaginary component of $G(j\omega)$. Then, the magnitude of the closed-loop transfer function is

$$|M(j\omega)| = \left| \frac{G(j\omega)}{1 + G(j\omega)} \right| = \frac{\sqrt{x^2 + y^2}}{\sqrt{(1 + x)^2 + y^2}} \tag{10-154}$$

For simplicity of notation, let $M = |M(j\omega)|$; then Eq. 10-154 leads to

$$M\sqrt{(1 + x)^2 + y^2} = \sqrt{x^2 + y^2} \tag{10-155}$$

Squaring both sides of the last equation gives

$$M^2[(1 + x)^2 + y^2] = x^2 + y^2 \tag{10-156}$$

Rearranging the last equation yields

$$(1 - M^2)x^2 + (1 - M^2)y^2 - 2M^2x = M^2 \tag{10-157}$$

This equation is conditioned by dividing through by $(1 - M^2)$ and adding the term $[M^2/(1 - M^2)]^2$ on both sides. We have

$$x^2 + y^2 - \frac{2M^2}{1 - M^2}x + \left(\frac{M^2}{1 - M^2}\right)^2 = \frac{M^2}{1 - M^2} + \left(\frac{M^2}{1 - M^2}\right)^2 \tag{10-158}$$

which is finally simplified to

$$\left(x - \frac{M^2}{1 - M^2}\right)^2 + y^2 = \left(\frac{M}{1 - M^2}\right)^2 \tag{10-159}$$

For a given M, Eq. (10-159) represents a circle with the center at

$$x = \frac{M^2}{1 - M^2} \qquad y = 0$$

The radius of the circle is $r = |M/(1 - M^2)|$. Equation (10-159) is invalid for $M = 1$. For $M = 1$, Eq. (10-156) gives

$$x = -\tfrac{1}{2}$$

which is the equation of a straight line parallel to the j Im $G(j\omega)$-axis and passing through the $(-\tfrac{1}{2}, j0)$ point in the $G(j\omega)$-plane.

constant-M loci

When M takes on different values, Eq. (10-159) describes in the $G(j\omega)$-plane a family of circles that are called the **constant-*M* loci**, or the **constant-*M* circles**. The coordinates of the centers and the radii of the constant-M loci for various values of M are given in Table 10-2, and some of the loci are shown in Fig. 10-61.

Note that when M becomes infinite, the circle degenerates into a point at the critical point $(-1, j0)$. This agrees with the well-known fact that when the Nyquist plot of $G(j\omega)$ passes through the $(-1, j0)$ point, the system is marginally stable, and M_p is infinite. Figure 10-61 shows that the constant-M loci in the $G(j\omega)$-plane are symmetrical with respect to the $M = 1$ line and the real axis. The circles to the left of the

TABLE 10-2 Constant-*M* Circles

M	Center $x = \dfrac{M^2}{1 - M^2}$; $y = 0$	Radius $r = \left\| \dfrac{M}{1 - M^2} \right\|$
0.3	0.01	0.33
0.5	0.33	0.67
0.7	0.96	1.37
1.0	∞	∞
1.1	−5.76	5.24
1.2	−3.27	2.73
1.3	−2.45	1.88
1.4	−2.04	1.46
1.5	−1.80	1.20
1.6	−1.64	1.03
1.7	−1.53	0.90
1.8	−1.46	0.80
1.9	−1.38	0.73
2.0	−1.33	0.67
2.5	−1.19	0.48
3.0	−1.13	0.38
4.0	−1.07	0.27
5.0	−1.04	0.21
6.0	−1.03	0.17

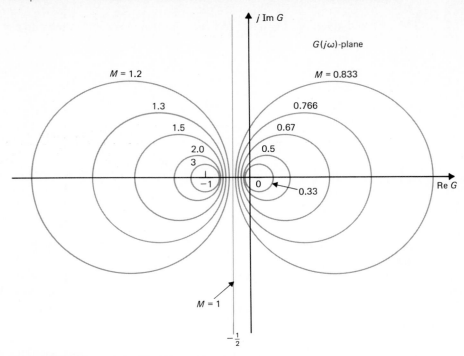

FIGURE 10-61 Constant-M circles in the polar coordinates.

$M = 1$ locus correspond to values of M greater than 1, and those to the right of the $M = 1$ line are for M less than 1.

Graphically, the intersections of the $G(j\omega)$ plot and the constant-M loci give the value of M at the frequency denoted on the $G(j\omega)$ curve. If it is desired to keep the value of M_p less than a certain value, the $G(j\omega)$ curve must not intersect the corresponding M circle at any point, and at the same time must not enclose the $(-1, j0)$ point. The constant-M circle with the smallest radius that is tangent to the $G(j\omega)$ curve gives the value of M_p, and the resonant frequency ω_p is read off at the tangent point on the $G(j\omega)$ curve.

Figure 10-62(a) illustrates the Nyquist plot of $G(j\omega)$ for a unity-feedback control system, together with several constant-M loci. For a given loop gain $K = K_1$, the intersects between the $G(j\omega)$ curve and constant-M loci give the points on the $|M(j\omega)|$-versus-ω curve. The peak resonance M_p is found by locating the smallest circle that is tangent to the $G(j\omega)$ plot. The resonant frequency is found at the point of tangency, and is designated as ω_{p1}. If the loop gain is increased to K_2, and if the system is still stable, a constant-M circle with a smaller radius that corresponds to a larger M is found tangent to the $G(j\omega)$ curve, and thus the peak resonance M_p is larger. The resonant frequency is shown to be ω_{p2}, which is closer to the phase-crossover frequency ω_c than ω_{p1}. When K is increased to K_3, so that the $G(j\omega)$ curve passes through the $(-1, j0)$ point, the system is at the margin of stability, M_p is infinite, and $\omega_{p3} = \omega_c$. In all cases, the bandwidth of the closed-loop system is found at the intersect of the $G(j\omega)$ curve and the $M = 0.707$ locus. For values of K beyond K_3, the system is unstable, and the

peak resonance

bandwidth

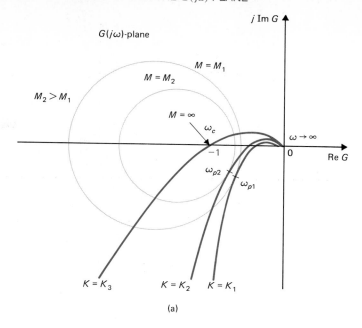

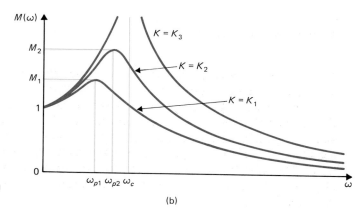

FIGURE 10-62 (a) Polar plots
of *G*(*s*) and constant-*M* loci.
(b) Corresponding magnification
curves.

magnification curve and M_p no longer have any meaning. When enough points of intersections between the $G(j\omega)$ curve and the constant-M loci are obtained, the magnification curves are plotted, as shown in Fig. 10-62(b).

10-16

CONSTANT PHASE LOCI IN THE *G(jω)*-PLANE

The loci of constant phase of the closed-loop system may also be determined in the
$G(j\omega)$-plane by a method similar to that used to secure the constant-M loci. In principle, we need both the magnitude and the phase of the closed-loop frequency response

to analyze the performance of the system. However, information on M_p, ω_p, and BW are all derived from the magnitude curve, so that the constant-phase loci are less used in practice.

With reference to Eqs. (10-152) and (10-153), the phase of the closed-loop system is

$$\phi_m(j\omega) = \angle M(j\omega) = \tan^{-1}\left(\frac{x}{y}\right) - \tan^{-1}\left(\frac{y}{1+x}\right) \qquad (10\text{-}160)$$

Taking the tangent on both sides of Eq. (10-160), and letting $\phi_m = \phi_m(j\omega)$, we have

$$\tan \phi_m = \frac{y}{x^2 + x + y} \qquad (10\text{-}161)$$

Let $N = \tan \phi_m$; then Eq. (10-161) becomes

$$x^2 + x + y^2 - \frac{y}{N} = 0 \qquad (10\text{-}162)$$

Adding the term $(1/4 + 1/4N^2)$ to both sides of the last equation, we get

$$x^2 + x + \frac{1}{4} + y^2 - \frac{y}{N} + \frac{1}{4N^2} = \frac{1}{4} + \frac{1}{4N^2} \qquad (10\text{-}163)$$

which is regrouped to give

$$\left(x + \frac{1}{2}\right)^2 + \left(y - \frac{1}{2N}\right)^2 = \frac{1}{4} + \frac{1}{4N^2} \qquad (10\text{-}164)$$

When N assumes various values, this equation represents a family of circles with centers at $(x, y) = (-1/2, 1/2N)$. The radii are given by

$$r = \left(\frac{N^2 + 1}{4N^2}\right)^{1/2} \qquad (10\text{-}165)$$

constant-N loci

These circles are known at the **constant-N loci** or the **constant-N circles** in the $G(j\omega)$-plane. The centers and the radii of the constant-N circles for various values of N are tabulated in Table 10-3, and the loci are shown in Fig. 10-63.

TABLE 10-3   Constant-N Circles

$\phi_m = 180°n$ $n = 0, 1, 2, \ldots$	$N = \tan \phi_m$	Center $x = -\dfrac{1}{2}$; $y = \dfrac{1}{2N}$	Radius $r = \left(\dfrac{N^2 + 1}{4N^2}\right)^{1/2}$
−90	−∞	0	0.500
−60	−1.732	−0.289	0.577
−45	−1.000	−0.500	0.707
−30	−0.577	−0.866	1.000
−15	−0.268	−1.866	1.931
0	0	∞	∞
15	0.268	1.866	1.931
30	0.577	0.866	1.000
45	1.000	0.500	0.707
60	1.732	0.289	0.577
90	∞	0	0.500

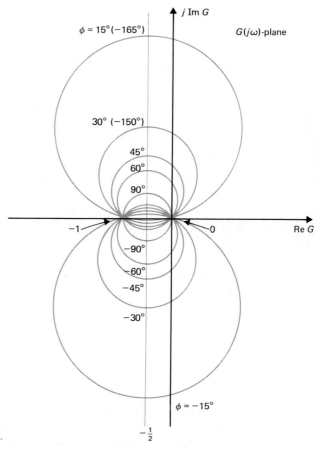

FIGURE 10-63 Constant-N circles in the polar coordinates.

10-17

CONSTANT-*M* AND CONSTANT-*N* LOCI IN THE MAGNITUDE-VERSUS-PHASE PLANE: THE NICHOLS CHART

Nichols chart

A major disadvantage in working with the polar coordinates for the $G(j\omega)$ plot is that the curve no longer retains its original shape when a simple modification such as the change of the loop gain is made to the system. Frequently, in design problems, not only the loop gain must be altered, but series or feedback controllers are added to the original system that require the complete reconstruction of the modified $G(j\omega)$. For design work involving M_p and BW, it is more convenient to work with the magnitude-versus-phase plot of $G(j\omega)$. In the magnitude-versus-phase plot, the entire $G(j\omega)$ curve is shifted up or down vertically without distortion when the loop gain is altered. When the phase properties of $G(j\omega)$ are changed independently, without affecting the gain, the magnitude-versus-phase plot is affected only in the horizontal axis, along the phase axis. The constant-*M* loci in the polar coordinates are transferred to the magnitude-versus-phase coordinates, and the loci are called the **Nichols chart**. A typical Nichols chart of selected constant-*M* loci is shown in Fig. 10-64. The constant-*N* loci can be transferred in a similar manner if needed. Once the $G(j\omega)$ locus is constructed in the Nichols chart, the intersects between the constant-*M* loci and the $G(j\omega)$ trajectory give the value of $|M(j\omega)|$ at the corresponding frequencies of $G(j\omega)$. The resonance peak M_p is found by locating the smallest of the constant-*M* locus ($M \geq 1$) that is tangent to the $G(j\omega)$ plot from above. The resonant frequency is the frequency of $G(j\omega)$ at the

bandwidth

point of tangency. *The bandwidth of the closed-loop system is the frequency at which the $G(j\omega)$ curve intersects the $M = 0.707$ locus.*

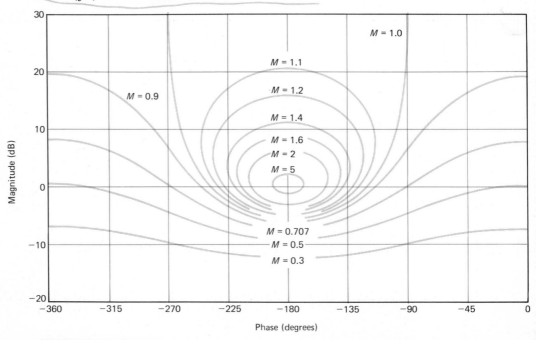

FIGURE 10-64 Nichols chart.

The following example illustrates the relationship among the analysis methods using the Bode plot, the magnitude-versus-phase plot, and the Nichols chart.

EXAMPLE 10-14

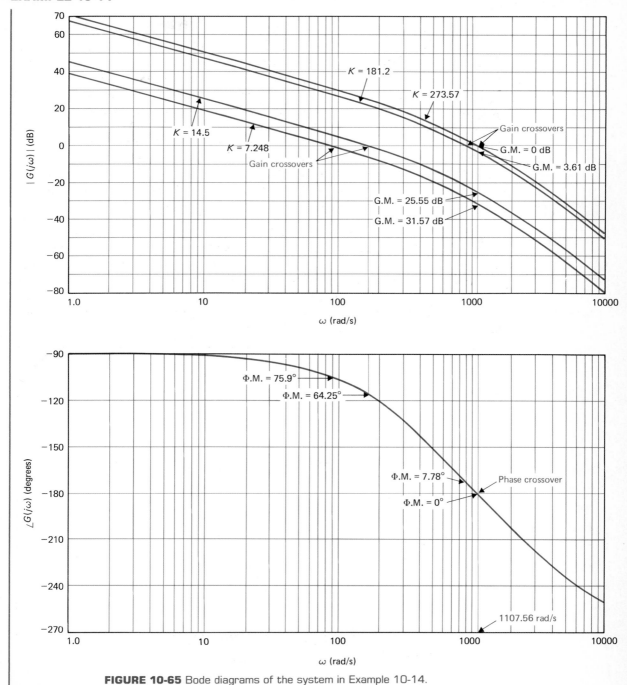

FIGURE 10-65 Bode diagrams of the system in Example 10-14.

*attitude-
control
system*

Consider the attitude-control system discussed in Section 7-6. When the inductance of the dc motor is not neglected, the system is of the third order, and the open-loop transfer function is given by Eq. (7-115), and is

$$G(s) = \frac{1.5 \times 10^7 K}{s(s + 400.26)(s + 3008)} \qquad (10\text{-}166)$$

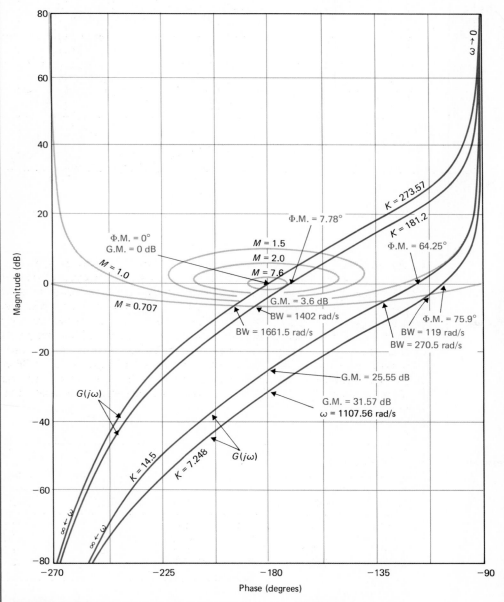

FIGURE 10-66 Gain-phase plots of the system in Example 10-14.

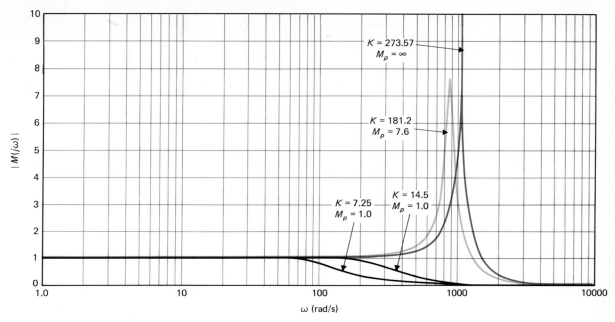

FIGURE 10-67 Closed-loop frequency responses of the system in Example 10-14.

M_p, ω_p, BW

The Bode plots for $G(s)$ are shown in Fig. 10-65 for $K = 7.248$, 14.5, 181.2, and 273.57. The gain and phase margins of the closed-loop systems are all determined from the Bode plots. The gain-phase plots of $G(j\omega)$ corresponding to the Bode plots are shown in Fig. 10-66. These gain-phase plots, together with the Nichols chart, give information on the peak resonance M_p, resonant frequency ω_p, and the bandwidth BW. The gain and phase margins are also clearly indicated on the gain-phase plots. Figure 10-67 shows the closed-loop frequency responses. The Bode and the gain-phase plots, with the Nichols chart, as well as the closed-loop frequency-response plots, can all be obtained with the program FREQRP in the ACSP computer program package. Table 10-4 summarizes all the results of the frequency-domain analysis for the four different values of K together with the time-domain maximum overshoots determined in Chapter 7.

summary

TABLE 10-4

K	Maximum Overshoot (%)	M_p	ω_p (rad/s)	Gain Margin (dB)	Phase Margin (degrees)	BW rad/s
7.25	0	1.0	1.0	31.57	75.90	119.0
14.5	4.3	1.0	43.33	25.55	64.25	270.5
181.2	15.2	7.6	900.00	3.61	7.78	1402.0
273.57	100.0	∞	1000.00	0	0	1661.5

10-18

NICHOLS-CHART APPLICATION
FOR NONUNITY-FEEDBACK SYSTEMS

nonunity-feedback control

The constant-M and -N loci and the Nichols-chart analysis discussed in the preceding sections are limited to closed-loop systems with unity feedback, whose transfer function is given by Eq. (10-152). When a system has nonunity feedback, $H(s) \neq 1$, the closed-loop transfer function of the single-loop system is

$$M(s) = \frac{C(s)}{R(s)} = \frac{G(s)}{1 + G(s)H(s)} \qquad (10\text{-}167)$$

Nichols chart

The constant-M loci and the Nichols chart cannot be applied directly to obtain the closed-loop frequency response. We shall show that with proper handling, the constant-M loci can still be applied to a nonunity-feedback systems. Let us consider the function

$$P(s) = H(s)M(s) = \frac{G(s)H(s)}{1 + G(s)H(s)} \qquad (10\text{-}168)$$

Apparently, Eq. (10-168) is of the same form as Eq. (10-152). The frequency response of $P(j\omega)$ can be determined by plotting the function $G(j\omega)H(j\omega)$ in the gain-phase coordinates along with the Nichols chart. Once this is done, the frequency-response information for $M(j\omega)$ is obtained as follows.

$$|M(j\omega)| = \frac{|P(j\omega)|}{|H(j\omega)|} \qquad (10\text{-}169)$$

$$\phi_m(j\omega) = \angle M(j\omega) = \angle P(j\omega) - \angle H(j\omega) \qquad (10\text{-}170)$$

10-19

SENSITIVITY STUDIES IN THE FREQUENCY DOMAIN

The frequency-domain study of linear control systems has the advantage in that higher-order systems can be handled more easily than in the time-domain. Furthermore, the sensitivity of the system with respect to parameter variations can be easily interpreted using the frequency-domain plots. We shall show how the Nyquist plot and the Nichols chart can be utilized for the analysis and design of control systems based on sensitivity considerations.

Consider that a linear control system with unity feedback has the transfer function

$$M(s) = \frac{C(s)}{R(s)} = \frac{G(s)}{1 + G(s)} \qquad (10\text{-}171)$$

sensitivity

The sensitivity of $M(s)$ with respect to $G(s)$ is defined as

$$S_G^M(s) = \frac{dM(s)/M(s)}{dG(s)/G(s)} = \frac{dM(s)}{dG(s)} \frac{G(s)}{M(s)} \qquad (10\text{-}172)$$

It should be pointed out that the sensitivity function defined in Eq. (10-172) is equivalent to considering that the variable parameter is the gain of $G(s)$. Substituting Eq. (10-171) into Eq. (10-172) and simplifying, we have

$$S_G^M(s) = \frac{1}{1 + G(s)} = \frac{G(s)^{-1}}{1 + G(s)^{-1}} \qquad (10\text{-}173)$$

Clearly, the sensitivity function $S_G^M(s)$ is a function of the complex variable s. Figure 10-68 shows the magnitude plot of $S_G^M(j\omega)$ when $G(s)$ is the transfer function given in Eq. (10-145) with $K = 10$. It is interesting to note that the sensitivity of the closed-loop system is inferior at frequencies greater than 4.8 rad/s to the open-loop system whose sensitivity to the variation of K is always unity. In general, it is desirable to formulate a design criterion on sensitivity in the following manner:

*design
criterion*

$$|S_G^M(j\omega)| = \frac{1}{|1 + G(j\omega)|} = \frac{|G(j\omega)^{-1}|}{|1 + G(j\omega)^{-1}|} \le k \qquad (10\text{-}174)$$

where k is a positive real number. This sensitivity criterion is in addition to the regular performance criteria on the steady-state error and the relative stability.

Equation (10-174) is analogous to the magnitude of the closed-loop transfer function, $|M(j\omega)|$ given in Eq. (10-154), with $G(j\omega)$ replaced by $G(j\omega)^{-1}$. Thus, the sensitivity function of Eq. (10-174) can be determined by plotting $G(j\omega)^{-1}$ in the polar coordinates along with the constant-M circles or the Nichols chart. Figure 10-69 shows the polar plots of $G(j\omega)$ of Eq. (10-145) and $G(j\omega)^{-1}$ with $K = 10$. Notice that $G(j\omega)$ is tangent to the $M = 1.84$ circle from below, which means that the peak resonance of the closed-loop system is 1.84. The $G(j\omega)^{-1}$ curve is tangent to the $M = 2.18$ curve from above, which according to Fig. 10-68 is the maximum value of $|S_G^M(j\omega)|$.

Equation (10-174) shows that for low sensitivity, the loop gain of $G(j\omega)$ must be high, but it is known that in general, high gain could cause instability. Thus, the designer is again challenged by the task of designing a system with both high degree of stability and low sensitivity.

Figure 10-70 shows the gain-phase plots of $G(j\omega)$ and $G(j\omega)^{-1}$ in the Nichols chart. These results again substantiate those determined from the polar plots. The Bode diagrams of $G(j\omega)$ and $G(j\omega)^{-1}$ are shown in Fig. 10-71. These are useful for compensation design, since the vertical coordinate is in decibels. The design of robust control systems (low sensitivity) with the frequency-domain methods is discussed in Chapter 11.

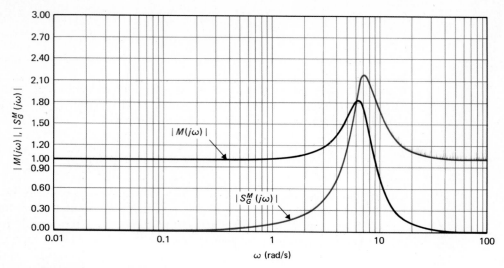

FIGURE 10-68 $|M(j\omega)|$ and $|S_G^M(j\omega)|$ versus ω for
$G(s) = 10/[s(1 + 0.02s)(1 + 0.002s)]$.

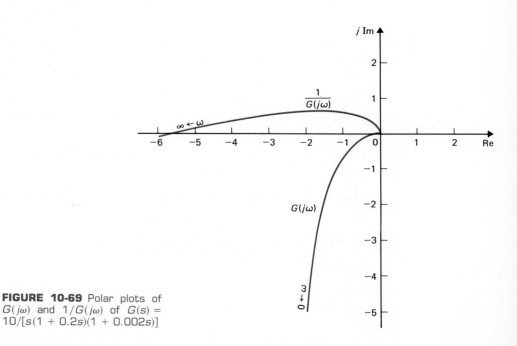

FIGURE 10-69 Polar plots of
$G(j\omega)$ and $1/G(j\omega)$ of $G(s) =$
$10/[s(1 + 0.2s)(1 + 0.002s)]$

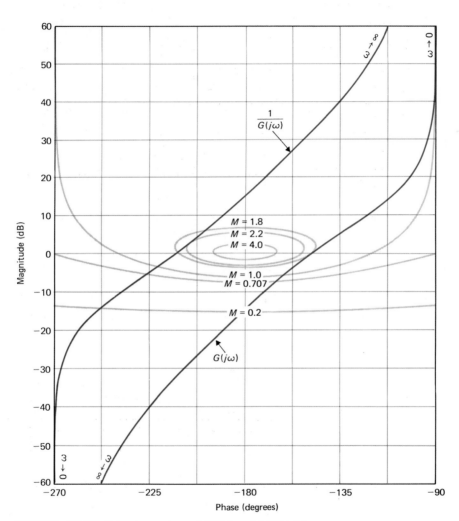

FIGURE   10-70 Gain-phase plots of $G(j\omega)$ and $1/G(j\omega)$ for
$G(s) = 10/[s(1 + 0.02s)(1 + 0.2s)]$.

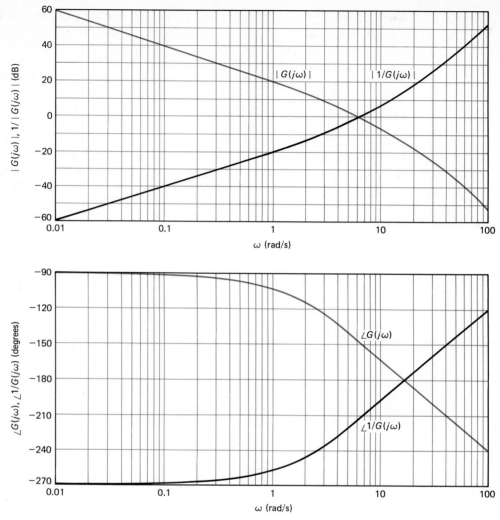

FIGURE   10-71 Bode plots of $G(j\omega)$ and $1/G(j\omega)$ of
$G(s) = 10/[s(1 + 0.02s)(1 + 0.2s)]$.

10-20

FREQUENCY-DOMAIN ANALYSIS
OF DISCRETE-DATA CONTROL SYSTEMS

All the frequency-domain methods discussed in the preceding sections can be extended for the analysis of discrete-data control systems. Consider the discrete-data system shown in Fig. 10-72. The closed-loop transfer function of the closed-loop system is

$$\frac{C(z)}{R(z)} = \frac{G_{h0}\, G(z)}{1 + G_{h0}\, G(z)} \tag{10-175}$$

where $G_{h0}G(z)$ is the z-transform of $G_{h0}(s)G(s)$. Just as in the case of continuous-data systems, the absolute and relative stability conditions of the closed-loop discrete-data system can be investigated by making the frequency-domain plots of $G_{h0}G(z)$. Since the positive $j\omega$-axis of the s-plane corresponds to real frequency, the frequency-domain plots of $G_{h0}G(z)$ are obtained by setting $z = e^{j\omega T}$ and then letting ω vary from 0 to ∞. This is also equivalent to mapping the points on the unit circle, $|z| = 1$, in the z-plane onto the $G_{h0}G(e^{j\omega T})$-plane. Since the unit circle repeats for every sampling frequency $\omega_s (= 2\pi/T)$, as shown in Fig. 10-73, when ω is varied along the $j\omega$-axis, the frequency-domain plot of $G(e^{j\omega T})$ repeats for $\omega = n\omega_s$ to $(n + 1)\omega_s$, $n = 0, 1, 2, \ldots$. Thus, it is necessary to plot $G_{h0}G(e^{j\omega T})$ only for the range of $\omega = 0$ to $\omega = \omega_s$. In fact, since the unit circle in the z-plane is symmetrical about the real axis, the plot of $G_{h0}G(e^{j\omega T})$ in the polar coordinates for $\omega = 0$ to $\omega = \omega_s$ will also be symmetrical about the real axis, so that only the portion that corresponds to $\omega = 0$ to $\omega = \omega_s/2$ needs to be plotted.

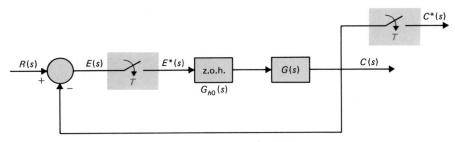

FIGURE 10-72 Closed-loop discrete-data control system.

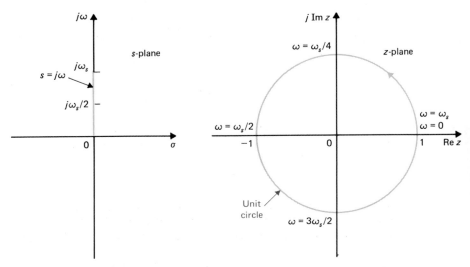

FIGURE 10-73 Relation between the $j\omega$-axis in the s-plane and the unit circle in the z-plane.

EXAMPLE 10-15   As an illustrative example, let the transfer function of the process in the system in Fig. 10-72 be

$$G(s) = \frac{1.57}{s(s + 1)} \tag{10-176}$$

and the sampling frequency is 4 rad/s. Let us first consider that the system does not have a zero-order hold, so that

$$G_{h0}G(z) = G(z) = \frac{1.243z}{(z - 1)(z - 0.208)} \tag{10-177}$$

The frequency response of $G_{h0}G(z)$ is obtained by substituting $z = e^{j\omega T}$ in Eq. (10-177). The polar plot of $G_{h0}G(e^{j\omega T})$ for $\omega = 0$ to $\omega_s/2$ is shown in Fig. 10-74. The mirror image of the locus shown, with the mirror placed on the real axis, represents the plot for $\omega = \omega_s/2$ to ω_s. The Bode plot of $G_{h0}G(e^{j\omega T})$ consists of the graphs of $|G_{h0}G(e^{j\omega T})|$ in dB versus ω, and $\angle G_{h0}G(e^{j\omega T})$ in degrees versus ω, as shown in Fig. 10-75 for three decades of frequency with the plots ended at $\omega = \omega_s/2 = 2$ rad. For the sake of comparison, the open-loop transfer function of the system with a zero-order hold is obtained:

$$G_{h0}G(z) = \frac{1.2215z + 0.7306}{(z - 1)(z - 0.208)} \tag{10-178}$$

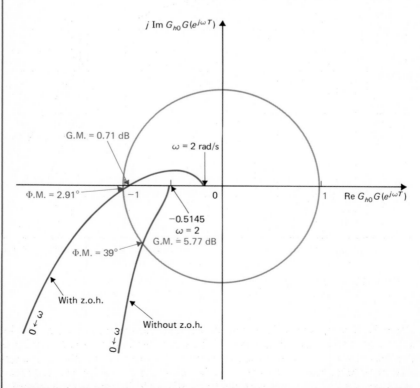

FIGURE 10-74 Frequency-domain plot of $G_{h0}G(z)$ of the system in Fig. 10-72, with $G(s) = 1.57/[s(s + 1)]$, $T = 1.57$ s, and with and without z.o.h.

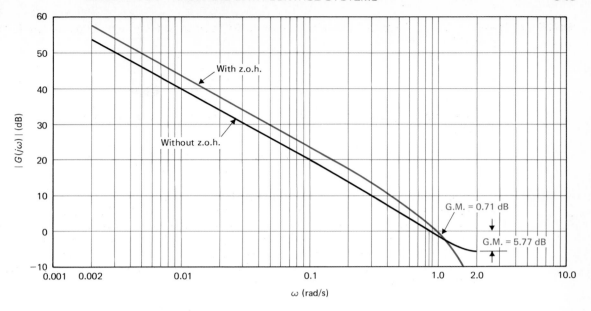

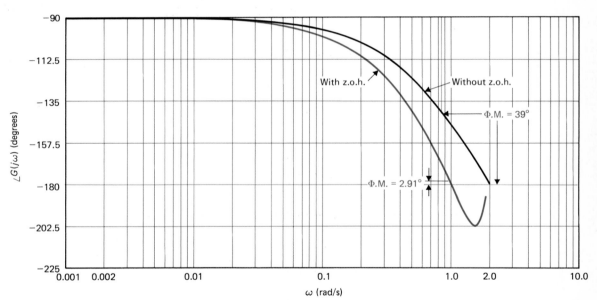

FIGURE 10-75 Bode plot of $G_{h0}G(z)$ of the system in Fig. 10-60, with $G(s) = 1.57/[s(s + 1)]$, $T = 1.57$ s, and with and without z.o.h.

The polar plot and the Bode plot of the last equation are shown in Figs. 10-74 and 10-75, respectively. Notice that the polar plot of the system with the zero-order hold intersects the negative real axis at a point that is closer to the $(-1, j0)$ point than that of the system without the zero-order hold. Thus, the system with the zero-order hold is less stable. Similarly, the phase of the Bode plot of the system with the zero-order hold is more negative than that of the system without the zero-order hold. The gain margin, phase margin, and peak resonance of the two systems are summarized as follows.

	Gain Margin (dB)	Phase Margin (degrees)	M_p
Without z.o.h.	5.77	39.0	1.58
With z.o.h.	0.71	2.91	22.64

The conclusion from this illustrative example is that once z is replaced by $e^{j\omega T}$ in the z-domain transfer function, all the frequency-domain analysis techniques available for continuous-data systems can be applied to discrete-data systems.

10-21
SUMMARY

This chapter is devoted to the frequency-domain analysis of linear control systems. The chapter begins by describing typical relationships between the open-loop and closed-loop frequency responses of linear systems. Performance specifications such as the peak resonance M_p, resonant frequency ω_p, and bandwidth BW are defined in the frequency domain. The relationships between these specifications and the parameters of a second-order prototype system are derived analytically. The effects of adding simple poles and zeros to an open-loop transfer function on M_p and BW are discussed.

The Nyquist criterion for the stability analysis of linear control systems is thoroughly developed. The stability of a single-loop control system can be investigated by studying the behavior of the Nyquist plot of the loop transfer function $G(s)H(s)$ for $\omega = 0$ to $\omega = \infty$ with respect to the critical point. The general Nyquist criterion is given in terms of the angular condition given in Eq. (10-62). If $G(s)H(s)$ is a minimum-phase transfer function, the condition of stability is simplified so that the Nyquist plot must not enclose the critical point. Application of the Nyquist criterion to multiloop systems is also outlined.

The relationship between the root loci and the Nyquist plot is pointed out in Section 10-6. The discussion should add more perspective on both subjects.

Relative stability is defined in terms of the gain margin and the phase margin. These quantities are defined in the polar coordinates as well as on the Bode diagram. The gain-phase plot allows the Nichols chart to be constructed for closed-loop analysis. The values of M_p and BW can be easily found by plotting the $G(j\omega)$ locus on the Nichols chart.

Sensitivity function $S_G^M(j\omega)$ is defined as a measure of the variation of $M(j\omega)$ due to variations in $G(j\omega)$. It is shown that the frequency-response plots of $G(j\omega)$ and $G(j\omega)^{-1}$ can be readily used for sensitivity studies.

Finally, the frequency response of discrete-data systems are investigated. Practically all the frequency-domain specifications defined for analog systems can be extended to discrete-data systems. The frequency-domain plots of discrete-data systems can be made by replacing the z-transform variable z by $e^{j\omega T}$, where T is the sampling period.

REVIEW QUESTIONS

1. Explain why it is important to conduct frequency-domain analyses of linear control systems.

2. By applying a sinusoidal signal of frequency ω_o to a linear system, the steady-state output of the system must also be of the same frequency.   **(T)**   **(F)**

3. Define the peak resonance M_p of a closed-loop control system.

4. Define the bandwidth BW of a closed-loop control system.

5. For a prototype second-order system, the value of M_p depends solely on the damping ratio ζ. **(T)**   **(F)**

6. Adding a zero to the open-loop transfer function will always increase the bandwidth of the closed-loop system. **(T)**   **(F)**

7. The general effect of adding a pole to the open-loop transfer function is to make the closed-loop system less stable, while decreasing the bandwidth. **(T)**   **(F)**

8. Define the following quantities:

Z_0 = number of zeros of $G(s)H(s)$ that are in the right-half s-plane
P_0 = number of poles of $G(s)H(s)$ that are in the right-half s-plane
Z_{-1} = number of zeros of $1 + G(s)H(s)$ that are in the right-half s-plane
P_{-1} = number of poles of $1 + G(s)H(s)$ that are in the right-half s-plane
P_ω = number of poles of $G(s)H(s)$ that are on the $j\omega$-axis.

Give the conditions on these parameters for the system to be **(a)** open-loop stable, and **(b)** closed-loop stable.

9. Consider that the Nyquist plot of $G(s)H(s)$ is drawn for $\omega = 0$ to $\omega = \infty$. Let Φ_{11} be the total angle traversed by the phasor drawn from the $(-1, j0)$ point to the Nyquist plot as ω varies from ∞ to 0. Give the condition between Φ_{11} and the parameters listed in Review Question 8 for closed-loop stability.

10. What condition must be satisfied by the function $G(s)H(s)$ so that the Nyquist criterion is simplified to investigating whether the $(-1, j0)$ is *enclosed* by the Nyquist plot or not?

11. Give all the properties of a minimum-phase transfer function.

12. Give the definitions of gain margin and phase margin.

13. For a minimum-phase loop transfer function $G(s)H(s)$, if the phase margin is negative, then the closed-loop system is always unstable. **(T)**   **(F)**

14. The gain margin is measured at the phase-crossover frequency. **(T)**   **(F)**

15. The phase margin is measured at the gain-crossover frequency. **(T)**   **(F)**

16. A system with a pure time delay is usually less stable than one without a time delay. **(T)**   **(F)**

17. How does the slope of the magnitude curve of the Bode plot of $G(j\omega)H(j\omega)$ at the gain crossover tell about the relative stability?

18. What are the constant-M circles in the $G(j\omega)$-plane?

19. What is a Nichols chart?

REFERENCES

1. H. NYQUIST, "Regeneration Theory," *Bell System. Tech. J.*, Vol. 11, pp. 126–147, Jan. 1932.

2. R. W. BROCKETT and J. L. WILLEMS, "Frequency Domain Stability Criteria—Part I," *IEEE Trans. Automatic Control*, Vol. AC-10, pp. 255–261, July 1965.

3. R. W. BROCKETT and J. L. WILLEMS, "Frequency Domain Stability Criteria—Part II," *IEEE Trans. Automatic Control*, Vol. AC-10, pp. 407–413, Oct. 1965.

4. T. R. NATESAN, "A Supplement to the Note on the Generalized Nyquist Criterion," *IEEE Trans. Automatic Control*, Vol. AC-12, pp. 215–216, April 1967.

5. A. GELB, "Graphical Evaluation of the Sensitivity Function Using the Nichols Chart," *IRE Trans. Automatic Control,* Vol. AC-7, pp. 57–58, July 1962.

6. K. S. YEUNG, "A Reformulation of Nyquist's Criterion," *IEEE Trans. Educ.,* Vol E-28, pp. 58–60, Feb. 1985.

7. K. S. YEUNG and H. M. LAI, "A Reformulation of Nyquist Criterion for Discrete Systems," *IEEE Trans. Educ.,* Vol. 31, No. 1, Feb. 1988.

PROBLEMS

The following problems can all be solved by either using a computer program or with computer-aided methods. Some of the simple problems should be solved analytically for a better understanding of the fundamental principles.

10-1.   The open-loop transfer function of a unity-feedback control system is

$$G(s) = \frac{K}{s(s + 6.54)}$$

Analytically, find the peak resonance M_p, resonant frequency ω_p, and the bandwidth BW of the closed-loop system for the following values of K: **(a)** $K = 5$. **(b)** $K = 21.39$. **(c)** $K = 100$.

10-2.   Use a computer program such as the FREQRP of the ACSP software package to solve the following parts. Do not attempt to obtain the solutions analytically. The open-loop transfer functions of the unity-feedback control systems are given in the following. Find the peak resonance M_p, resonant frequency ω_p, and the bandwidth BW of the closed-loop systems.

(a) $G(s) = \dfrac{5}{s(1 + 0.5s)(1 + 0.1s)}$ **(b)** $G(s) = \dfrac{20}{s(1 + 0.5s)(1 + 0.1s)}$

(c) $G(s) = \dfrac{500}{(s + 1.5)(s + 5)(s + 10)}$ **(d)** $G(s) = \dfrac{12(1 + s)}{s(1 + 0.5s)(1 + 0.1s)}$

(e) $G(s) = \dfrac{1}{2s(1 + s + s^2)}$ **(f)** $G(s) = \dfrac{100e^{-s}}{s(s^2 + 10s + 100)}$

10-3.   The specifications on a second-order control system with the closed-loop transfer function

$$M(s) = \frac{C(s)}{R(s)} = \frac{\omega_n^2}{s^2 + 2\zeta\omega_n s + \omega_n^2}$$

are that the maximum overshoot must not exceed 10 percent, and the rise time must be less than 0.1 second. Find the corresponding limiting values of M_p and BW analytically.

10-4.   The closed-loop frequency response $|M(j\omega)|$-versus-frequency of a second-order prototype system is shown in Fig. 10P-4. Sketch the corresponding unit-step response of the system; indicate the values of the maximum overshoot, peak time, and the steady-state error due to a unit-step input.

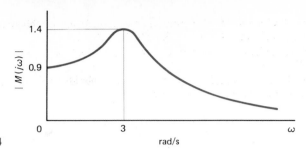

FIGURE 10P-4

10-5.   The open-loop transfer function of a unity-feedback control system is

$$G(s) = \frac{1 + Ts}{2s(s^2 + s + 1)}$$

Find the values of BW and M_p of the closed-loop system for $T = 0.05, 1, 2, 3, 4,$ and 5. Use the computer for solutions.

10-6.   The open-loop transfer function of a unity-feedback control system is

$$G(s) = \frac{1}{2s(s^2 + s + 1)(1 + Ts)}$$

Find the values of BW and M_p of the closed-loop system for $T = 0, 0.5, 1, 2, 3, 4,$ and 5. Use the computer for solutions.

10-7.   The loop transfer functions $G(s)H(s)$ of single-loop feedback control systems follow. Sketch the Nyquist plot of $G(j\omega)H(j\omega)$ for $\omega = 0$ to $\omega = \infty$. Determine the stability of the closed-loop system. If the system is unstable, find the number of poles of the closed-loop transfer function that are in the right-half s-plane. Solve for the intersect of $G(j\omega)H(j\omega)$ on the negative real axis of the GH-plane analytically. You may construct the Nyquist plot of $G(j\omega)H(j\omega)$ using the FREQRP program or any other suitable computer program.

(a) $G(s)H(s) = \dfrac{20}{s(1 + 0.1s)(1 + 0.5s)}$

(b) $G(s)H(s) = \dfrac{10}{s(1 + 0.1s)(1 + 0.5s)}$

(c) $G(s)H(s) = \dfrac{100(1 + s)}{s(1 + 0.1s)(1 + 0.2s)(1 + 0.5s)}$

(d) $G(s)H(s) = \dfrac{10}{s^2(1 + 0.2s)(1 + 0.5s)}$

(e) $G(s)H(s) = \dfrac{5(s - 2)}{s(s + 1)(s - 1)}$

(f) $G(s)H(s) = \dfrac{50}{s(s + 5)(s - 1)}$

(g) $G(s)H(s) = \dfrac{3(s + 2)}{s(s^3 + 3s + 1)}$

(h) $G(s)H(s) = \dfrac{0.1}{s(s + 1)(s^2 + s + 1)}$

(i) $G(s)H(s) = \dfrac{100}{s(s + 1)(s^2 + 2)}$

(j) $G(s)H(s) = \dfrac{s^2 - 5s + 2}{s(s^3 + 2s^2 + 2s + 10)}$

(k) $G(s)H(s) = \dfrac{-0.1(s^2 - 1)(s + 2)}{s(s^2 + s + 1)}$

(l) $G(s)H(s) = \dfrac{10(s + 10)}{s(s + 1)(s + 100)}$

10-8.   The loop transfer functions $G(s)H(s)$ of the single-loop feedback control systems follow. Apply the Nyquist criterion and determine the values of K for the system to be stable. Sketch the Nyquist plot of $G(j\omega)H(j\omega)$ with $K = 1$ for $\omega = 0$ to $\omega = \infty$.

(a) $G(s)H(s) = \dfrac{K}{s(s + 2)(s + 10)}$

(b) $G(s)H(s) = \dfrac{K(s + 1)}{s(s + 2)(s + 5)(s + 10)}$

(c) $G(s)H(s) = \dfrac{K}{s^2(s + 2)(s + 5)}$

(d) $G(s)H(s) = \dfrac{K(s - 2)}{s(s^2 - 1)}$

(e) $G(s)H(s) = \dfrac{K}{s(s + 5)(s - 1)}$

(f) $G(s)H(s) = \dfrac{K(s + 2)}{s(s^3 + 3s + 1)}$

(g) $G(s)H(s) = \dfrac{K(s^2 - 5s + 2)}{s(s^3 + 2s^2 + 2s + 10)}$

(h) $G(s)H(s) = \dfrac{K(s^2 - 1)(s + 2)}{s(s^2 + s + 1)}$

(i) $G(s)H(s) = \dfrac{K(s^2 - 5s + 1)}{s(s + 1)(s^2 + 4)}$

10-9.   The open-loop transfer function of a unity-feedback control system is

$$G(s) = \frac{K}{(s + 5)^n}$$

Determine by means of the Nyquist criterion, the range of $K(-\infty < K < \infty)$ for the closed-loop system to be stable. Sketch the Nyquist plot of $G(j\omega)$ for $\omega = 0$ to $\omega = \infty$. **(a)** $n = 2$. **(b)** $n = 3$. **(c)** $n = 4$.

10-10.   Figure 10P-10 shows the Nyquist plots of the loop transfer function $G(j\omega)H(j\omega)$ for $\omega = 0$ to $\omega = \infty$ for single-loop feedback control systems. The number of poles of $G(s)H(s)$ that are on the $j\omega$-axis and in the right-half s-plane are indicated for each case. Determine the stability of the closed-loop system by applying the Nyquist criterion. For the unstable systems, give the number of zeros of $1 + G(s)H(s)$ that are in the right-half s-plane.

10-11.   It is mentioned in the text that when the function $G(s)H(s)$ has more zeros than poles, it is necessary to plot the Nyquist plot of $1/G(s)H(s)$ in order to apply the simplified Nyquist criterion. Determine the stability of the systems described by the function $1/G(j\omega)H(j\omega)$ shown in Fig. 10P-11. For each case, the values of P_ω and P_{-1} for the

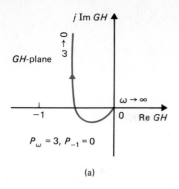

$P_\omega = 3, P_{-1} = 0$

(a)

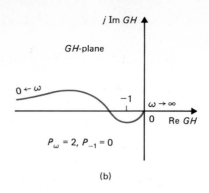

$P_\omega = 2, P_{-1} = 0$

(b)

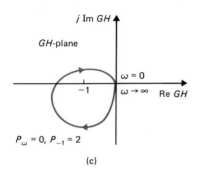

$P_\omega = 0, P_{-1} = 2$

(c)

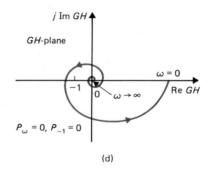

$P_\omega = 0, P_{-1} = 0$

(d)

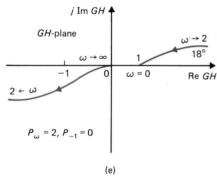

$P_\omega = 2, P_{-1} = 0$

(e)

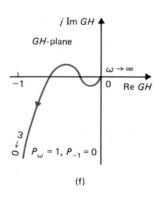

$P_\omega = 1, P_{-1} = 0$

(f)

FIGURE 10P-10

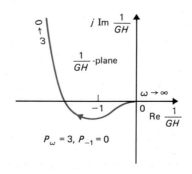

$P_\omega = 3, P_{-1} = 0$

(a)

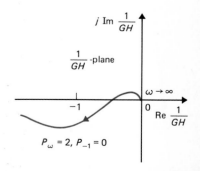

$P_\omega = 2, P_{-1} = 0$

(b)

FIGURE 10P-11

function $1/G(s)H(s)$ are given, where P_ω refers to the number of poles of $1/G(s)H(s)$ that are on the $j\omega$-axis, and P_{-1} refers to the number of poles of $1 + 1/G(s)H(s)$ that are in the right-half s-plane.

10-12. Figure 10P-12 shows the Nyquist plots of the loop transfer function $G(j\omega)H(j\omega)$ for $\omega = 0$ to $\omega = \infty$ for single-loop feedback control systems. The gain K appears as a multiplying factor in $G(s)H(s)$. The number of poles of $G(s)H(s)$ that are on the $j\omega$-axis and in the right-half s-plane are indicated in each case. Determine the range(s) of K for closed-loop system stability.

10-13. The characteristic equations of linear control systems follow. Apply the Nyquist criterion to determine the values of K for system stability. Check the answers by means of the Routh–Hurwitz criterion.
  (a) $s^3 + 4Ks^2 + (K + 5)s + 10 = 0$
  (b) $s^3 + K(s^3 + 2s^2 + 1) = 0$
  (c) $s(s + 1)(s^2 + 4) + K(s^2 + 1) = 0$
  (d) $s^3 + 2s^2 + 20s + 10K = 0$

10-14. The open-loop transfer function of a unity-feedback control system with a PD controller is

$$G(s) = \frac{10(K_P + K_D s)}{s^2}$$

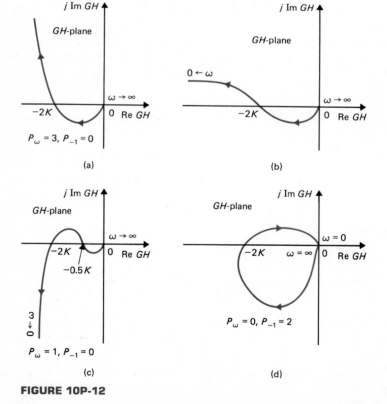

(a)   **(b)**

(c)   **(d)**

FIGURE 10P-12

Select the value of K_P so that the parabolic-error constant K_a is 100. Find the equivalent open-loop transfer function $G_{eq}(s)$ for stability analysis with K_D as a gain factor. Sketch the Nyquist plot of $G_{eq}(s)$ for $\omega = 0$ to $\omega = \infty$. Determine the range of K_D for stability by the Nyquist criterion.

10-15. The block diagram of a feedback control system is shown in Fig. 10P-15.
   (a) Apply the Nyquist criterion to determine the range of K for stability.
   (b) Check the answer obtained in part (a) with the Routh–Hurwitz criterion.

10-16. The open-loop transfer function of the liquid-level control system shown in Fig. 6P-13 is

$$G(s) = \frac{K_a K_i n K_l N}{s(R_a J s + K_i K_b)(As + K_o)}$$

The following system parameters are given: $K_a = 50$, $K_i = 10$, $K_l = 50$, $J = 0.006$, $K_b = 0.0706$, $n = 0.01$, and $R_a = 10$. The values of A, N, and K_o are variable.
   (a) For $A = 50$ and $K_o = 100$, sketch the Nyquist plot of $G(s)$ for $\omega = 0$ to ∞ with N as the variable parameter. Find the maximum integer value of N so that the closed-loop system is stable.
   (b) Let $N = 10$ and $K_o = 100$. Sketch the Nyquist plot of an equivalent transfer function $G_{eq}(s)$ that has A as a multiplying factor. Find the critical value of A for stability.
   (c) For $A = 50$ and $N = 10$, sketch the Nyquist plot of an equivalent transfer function $G_{eq}(s)$ that has K_o as a multiplying factor. Find the critical value of K_o for stability.

10-17. The block diagram of a dc-motor-control system is shown in Fig. 10P-17. Determine the range of K for stability using the Nyquist criterion when K_t has the following values: **(a)** $K_t = 0$. **(b)** $K_t = 0.01$. **(c)** $K_t = 0.1$.

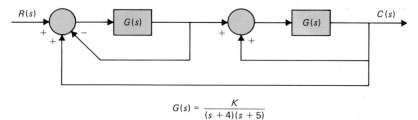

$$G(s) = \frac{K}{(s+4)(s+5)}$$

FIGURE 10P-15

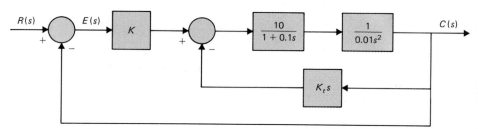

FIGURE 10P-17

10-18.   For the system shown in Fig. 10P-17, let $K = 10$. Find the range of K_t for stability with the Nyquist criterion.

10-19.   The steel-rolling system shown in Fig. 4P-26 has the open-loop transfer function

$$G(s) = \frac{100Ke^{-T_d s}}{s(s^2 + 10s + 100)}$$

(a)   When $K = 1$, determine the maximum time delay T_d in seconds for the closed-loop system to be stable.

(b)   When the time delay T_d is 1 second, find the maximum value of K for system stability.

10-20.   The system schematic shown in Fig. 10P-20 is devised to control the concentration of a chemical solution by mixing water and concentrated solution in appropriate proportions. The transfer function of the system components between the amplifier output $e_a(V)$ and the valve position x (in.) is

$$\frac{X(s)}{E_a(s)} = \frac{K}{s^2 + 10s + 100}$$

When the sensor is viewing pure water, the amplifier output voltage e_a is zero; when it is viewing concentrated solution, $e_a = 10$ V; 0.1 in. of the valve motion changes the output concentration from zero to maximum. The valve ports can be assumed to be shaped so that the output concentration varies linearly with the valve position. The output tube has a cross-sectional area of 0.1 in.^{2}, and the rate of flow is 10^3 in./s regardless of the valve position. To make sure that the sensor views a homogeneous solution, it is desirable to place it at some distance D in. from the valve.

(a)   Derive the loop transfer function of the system.

(b)   When $K = 10$, find the maximum distance D (in.) so that the system is stable. Use the Nyquist stability criterion.

(c)   Let $D = 10$ in. Find the maximum value of K for system stability.

10-21.   The approximation of high-order systems by low-order systems discussed in Section 7-9 is based on a frequency-domain criterion. The transfer functions of the high-order open-loop transfer functions and the corresponding low-order approximating transfer functions determined in the illustrative examples in Chapter 7 follow. Compute and compare the values of M_p, ω_p, and BW of the high-order systems with those of the

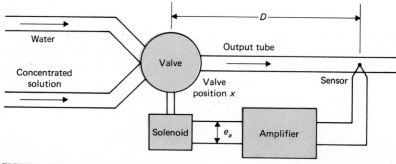

FIGURE 10P-20

low-order systems, so that the accuracy of the approximation can be judged. Use computer solutions.

(a) $G(s) = \dfrac{8}{s(s^2 + 6s + 12)}$ $G_L(s) = \dfrac{2.31}{s(s + 2.936)}$ (Example 7-3)

(b) $G(s) = \dfrac{0.909}{s(1 + 0.5455s + 0.0455s^2)}$ $G_L(s) = \dfrac{0.995}{s(1 + 0.4975s)}$ (Example 7-4)

(c) $G(s) = \dfrac{0.5}{s(1 + 0.75s + 0.25s^2)}$ $G_L(s) = \dfrac{0.707}{s(1 + 0.3536s)}$ (Example 7-4)

(d) $G(s) = \dfrac{90.3}{s(1 + 0.00283s + 8.3056 \times 10^7 s^2)}$ $G_L(s) = \dfrac{92.943}{s(1 + 0.002594s)}$

(Example 7-5)

(e) $G(s) = \dfrac{180.6}{s(1 + 0.00283s + 8.3056 \times 10^7 s^2)}$ $G_L(s) = \dfrac{189.54}{s(1 + 0.002644s)}$

(Example 7-5)

(f) $G(s) = \dfrac{1245.52}{s(1 + 0.00283s + 8.3056 \times 10^7 s^2)}$ $G_L(s) = \dfrac{2617.56}{s(1 + 0.0053s)}$

(Example 7-5)

10-22.   The open-loop transfer function of a unity feedback control system is

$$G(s) = \frac{1000}{s(s^2 + 105s + 600)}$$

(a) Find the the values of M_p, ω_p, and BW of the closed-loop system.
(b) Find the parameters of the second-order system with the open-loop transfer function

$$G_L(s) = \frac{\omega_n^2}{s(s + 2\zeta\omega_n)}$$

that will given the same value for M_p and ω_p as the third-order system. Compare the values of BW of the two systems.

10-23.   Repeat Problem 10-22 for the transfer function $G(s)$ given in Problem 10-21(f). (Although the approach used here may render results that are better or at least as good as those obtained by the method outlined in Seciton 7-9, the match here is restricted only to a prototype second-order system as the approximating low-order system.)

10-24.   Sketch or plot the Bode diagrams of the open-loop transfer functions given in Problem 10-2. Find the gain margin, gain-crossover frequency, phase margin, and the phase-crossover frequency for each system.

10-25.   The open-loop transfer functions of unity-feedback control systems are given in the following. Plot the Bode diagram of $G(j\omega)/K$ and do the following: (1) Find the value of K so that the gain margin of the system is 20 dB. (2) Find the value of K so that the phase margin of the system is 45°.

(a) $G(s) = \dfrac{K}{s(1 + 0.1s)(1 + 0.5s)}$

(b) $G(s) = \dfrac{K(s + 1)}{s(1 + 0.1s)(1 + 0.2s)(1 + 0.5s)}$

(c) $G(s) = \dfrac{K}{(s + 5)^3}$

(d) $G(s) = \dfrac{K}{(s + 5)^4}$

(e) $G(s) = \dfrac{Ke^{-s}}{s(1 + 0.1s + 0.01s^2)}$

(f) $G(s) = \dfrac{K(1 - s^2)(1 + 0.5s)}{s(s^2 + s + 1)}$

10-26. The open-loop transfer functions of unity-feedback control systems follow. Plot $G(j\omega)/K$ in the gain-phase coordinates of the Nichols chart, and do the following: (1) Find the value of K so that the gain margin of the system is 10 dB. (2) Find the value of K so that the phase margin of the system is 45°. (3) Find the value of K so that $M_p = 1.2$.

(a) $G(s) = \dfrac{10K}{s(1 + 0.1s)(1 + 0.5s)}$

(b) $G(s) = \dfrac{5K(s + 1)}{s(1 + 0.1s)(1 + 0.2s)(1 + 0.5s)}$

(c) $G(s) = \dfrac{Ke^{-s}}{s(1 + 0.1s + 0.01s^2)}$

10-27. The Bode diagram of the open-loop transfer function of a unity-feedback control system is obtained experimentally, and is shown in Fig. 10P-27 when the loop gain is set at its nominal value.

(a)   Find the gain margin and the phase margin of the system from the diagram as best you can read. Find the gain and the phase-crossover frequencies.

(b)   Repeat part (a) if the loop gain is doubled from its nominal value.

(c)   Repeat part (a) if the loop gain is 10 times its nominal value.

(d)   Find out how much the loop gain must be changed from its nominal value if the gain margin is to be 40 dB.

(e)   Find out how much the loop gain must be changed from its nominal value if the phase margin is to be 45°.

(f)   Find the steady-state error of the system if the reference input to the system is a unit-step function.

(g)   The open-loop system now has a pure time delay of T_d seconds, so that the open-loop transfer function is multiplied by $e^{-T_d s}$. Find the gain margin and the phase margin for $T_d = 0.05$. The loop gain is at nominal.

(h)   With the gain at nominal, find the maximum time delay T_d the system can tolerate without going into stability.

10-28. The open-loop transfer function of a unity-feedback control system is

$$G(s) = \frac{K(1 + 0.2s)(1 + 0.1s)}{s^2(1 + s)(1 + 0.01s)^2}$$

(a)   Construct the Bode and Nyquist plots of $G(j\omega)/K$ and determine the range of K for system stability.

(b)   Construct the root-locus diagram of the system for $K > 0$. Determine the values of K and ω at the points where the root loci cross the $j\omega$-axis, using the information found from the Bode diagram.

10-29. The open-loop transfer function of the dc-motor-control system described in Fig. 4P-17 is

$$G(s) = \frac{6.087 \times 10^8 K}{s(s^3 + 423.42s^2 + 2.6667 \times 10^6 s + 4.2342 \times 10^8)}$$

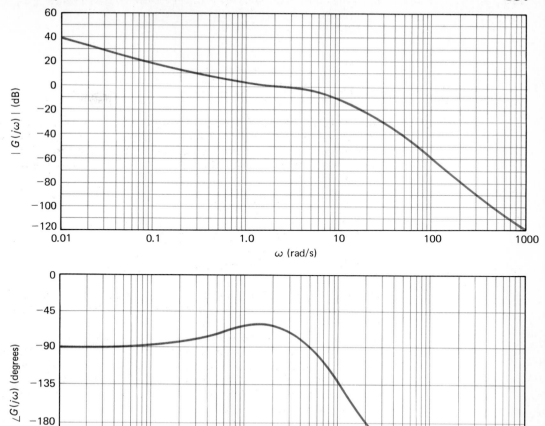

FIGURE 10P-27

Plot the Bode diagram of $G(j\omega)$ with $K = 1$, and determine the gain margin and phase margin of the system. Find the critical value of K for stability.

10-30.   The transfer function between the output position $\Theta_L(s)$ and the motor current $I_a(s)$ of the robot arm modeled in Fig. 4P-21 is

$$G_p(s) = \frac{\Theta_L(s)}{I_a(s)} = \frac{K_i(Bs + K)}{\Delta_o}$$

where

$$\Delta_o = s\{J_L J_m s^3 + [J_L(B_m + B) + J_m(B_L + B)]s^2$$
$$+ [B_L B_m + (B_L + B_m)B + (J_m + J_L)K]s + K(B_L + B_m)\}$$

The arm is controlled by a closed-loop system, as shown in Fig. 10P-30. The system parameters are $K_a = 65$, $K = 100$, $K_i = 0.4$, $B = 0.2$, $J_m = 0.2$, $B_L = 0.01$, $J_L = 0.6$, and $B_m = 0.25$.

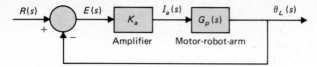

FIGURE 10P-30

(a) Derive the open-loop transfer function $G(s) = \Theta_L(s)/E(s)$.
(b) Draw the Bode diagram of $G(j\omega)$; find the gain and phase margins of the system.
(c) Draw the closed-loop frequency response, $|M(j\omega)|$ versus ω. Find M_p, ω_p, and BW.

10-31. The gain-phase plot of the open-loop transfer function $G(j\omega)/K$ of a unity-feedback control system is shown in Fig. 10P-31. Find the following performance characteristics of the system.

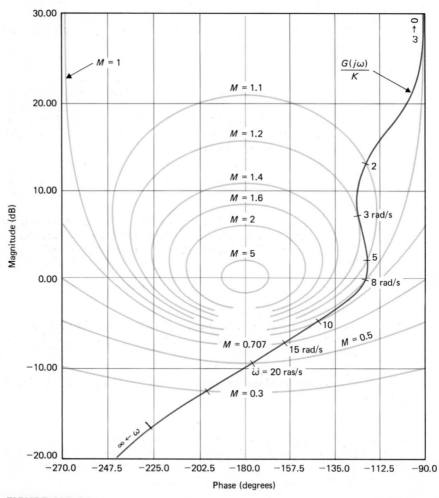

FIGURE 10P-31

(a) Gain-crossover frequency (rad/s) when $K = 1$.
(b) Phase-crossover frequency (rad/s) when $K = 1$.
(c) Gain margin (dB) when $K = 1$.
(d) Phase margin (degrees) when $K = 1$.
(e) Resonance peak M_p when $K = 1$.
(f) Resonant frequency ω_p when $K = 1$.
(g) Bandwidth of the closed-loop system when $K = 1$.
(h) The value of K so that the gain margin is 20 dB.
(i) The value of K so that the system is marginally stable. Find the frequency of sustained oscillation in rad/s.
(j) Steady-state error when the reference input is a unit-step function.

10-32. For the gain-phase plot of $G(j\omega)/K$ shown in Fig. 10P-31, the open-loop system now has a pure time delay T_d, so that the transfer function becomes $G(s)e^{-T_d s}$.
(a) With $K = 1$, find T_d so that the phase margin is 40°.
(b) With $K = 1$, find the maximum value of T_d so that the system will remain stable.

10-33. The block diagram of a furnace-control system is shown in Fig. 10P-33. The transfer function of the process is

$$G_p(s) = \frac{1}{(1 + 10s)(1 + 25s)}$$

(a) Plot the Bode diagram of $G(s) = C(s)/E(s)$ and find the gain-crossover and phase-crossover frequencies. Find the gain margin and the phase margin.
(b) Approximate the time delay by

$$e^{-2s} \cong \frac{1}{1 + 2s + 2s^2}$$

and repeat part (a). Comment on the accuracy of the approximation. What is the maximum frequency below which the polynomial approximation is accurate?
(c) Repeat part (b) for approximating the time delay term by

$$e^{-2s} \cong \frac{1 - s}{1 + s}$$

10-34. Plot the $|S_G^M(j\omega)|$-versus-ω plot for the system described in Problem 10-30 for $K = 1$. Find the frequency at which the sensitivity is maximum and the value of the maximum sensitivity.

More computer problems for this chapter can be found in the *User's Manual of the ACSP Software*.

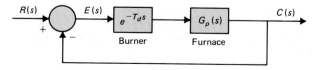

FIGURE 10P-33

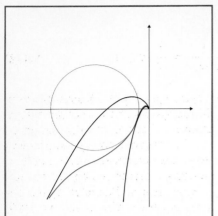

11

FREQUENCY-DOMAIN DESIGN OF CONTROL SYSTEMS

INTRODUCTION

In this chapter, the design of linear control systems will be carried out in the frequency domain, utilizing Bode plots and the magnitude-versus-phase plots. We realize from the time-domain and s-domain design techniques discussed in Chapter 9 that there is generally a lack of well-defined analytical design methods for higher-order systems. Although the root-locus and root-contour methods give indications on the effects of the various types of controllers, they do not provide precise solutions to the controller parameters unless a large number of root loci are plotted.

The frequency-domain design, on the other hand, possesses a wealth of semigraphical methods that allow the design of linear control systems to be carried out in a systematic way.

basic philosophy To illustrate the basic philosophy of design in the frequency domain, consider the unity-feedback control system that is characterized by the following process transfer function:

$$G_P(s) = \frac{K}{s(1 + s)(1 + 0.0125s)} \tag{11-1}$$

steady-state error Let us consider that the steady-state error of the system to a unit-ramp input must not exceed 0.01. The minimum value of K needed to satisfy this error requirement is deter-

664

mined as follows:

$$\text{steady-state error} = e_{ss} = \lim_{s \to 0} \frac{1}{sG_p(s)} = \frac{1}{K} \le 0.01 \qquad (11\text{-}2)$$

Thus, K must be greater than or equal to 100. However, applying the Routh–Hurwitz criterion to the characteristic equation of the closed-loop system shows that the system is unstable for all values of K greater than 81. This means that a controller should be applied to the system so that the steady-state error and the relative stability requirement can be satisfied simultaneously. Putting it another way, the controller must be able to keep the value of $sG(s)$ as $s \to 0$ at 100, while maintaining a prescribed degree of relative stability. The design principle in the frequency domain is best illustrated by the polar plot of $G_p(s)$ shown in Fig. 11-1, although in practice, we prefer to use the Bode plot for design purposes because it is simpler to construct. As shown in Fig. 11-1, when $K = 100$, the polar plot of $G_p(s)$ encloses the $(-1, j0)$ point, and the closed-loop system is unstable. Let us assume that we wish to realize a resonance peak of $M_p = 1.25$. This means that the polar plot of $G_p(s)$ must be tangent to the constant-M circle for $M = 1.25$ from below. If K is the only parameter that can be adjusted to achieve this design objective, the desired value of K is 1.2, as shown in Fig. 11-1. Apparently, we cannot set K to 1.2, since the ramp-error constant would only be 1.2, and the steady-state error requirement would not be satisfied.

Since the steady-state performance of the system is governed by the low-frequency characteristics of the transfer function, and the transient performance of the

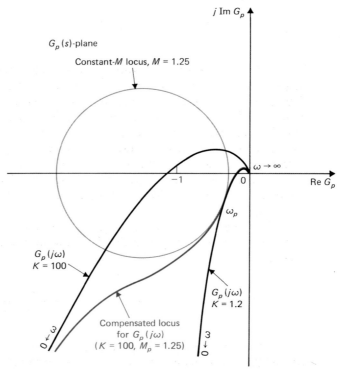

FIGURE 11-1 Nyquist plot of $G_p(s) = K/[s(1 + s)(1 + 0.0125s)]$.

system is governed by the high-frequency characteristics, Fig. 11-1 shows that to simultaneously satisfy the transient and the steady-state requirements, the frequency locus of $G_p(s)$ must be reshaped so that the high-frequency portion of the locus follows the $K = 1.2$ curve, and the low-frequency portion follows the $K = 100$ curve. The significance of this reshaping of the frequency locus is that the compensated locus shown in Fig. 11-1 will be tangent to the $M = 1.25$ circle at a relatively high frequency, while the zero-frequency gain is maintained at 100 to satisfy the steady-state requirement. When we inspect the plots in Fig. 11-1, we see that there are two alternative approaches in arriving at the compensated locus:

compensation approaches

> 1. Starting with the $K = 100$ locus and reshaping the locus in the region near the resonant frequency ω_p, while keeping the low-frequency region of $G_p(s)$ relatively unaltered.
> 2. Starting with the $K = 1.2$ locus and reshaping the low-frequency portion of $G_p(s)$ to obtain a ramp-error constant of $K_v = 100$, while keeping the locus near $\omega = \omega_p$ relatively unchanged.

In the first approach, the high-frequency portion of $G_p(s)$ is rotated in the counterclockwise direction, which means that more phase is added to the system in the positive direction in the proper frequency range. This scheme is basically referred to as **phase-lead compensation**, and controllers used for this purpose are often of the high-pass-filter type. The second approach apparently involves the shifting of the low-frequency part of the $K = 1.2$ trajectory in the clockwise direction, or alternatively, reducing the magnitude of $G_p(s)$ with $K = 100$ at the high-frequency range. This scheme is often referred to as the **phase-lag compensation**, since negative phase or phase lag is introduced to the system in the low-frequency range. The type of controllers used for the phase-lag compensation is often of the low-pass filter type.

phase-lead compensation

phase-lag compensation

Figures 11-2 and 11-3 further illustrate the philosophy of design in the frequency domain using the Bode diagram. In this case, the relative stability of the system is more conveniently represented by the gain margin and the phase margin. In Fig. 11-2, the Bode plot of $G_p(j\omega)$ shows that when $K = 100$, the gain and phase margins are both negative, and the system is unstable. By using the first approach, the phase-lead compensation, as described earlier, more positive phase is added to $G_p(j\omega)$ at the high frequencies to improve the phase margin. It should be noted that positive phase in the high-pass filter is accompanied by an additional gain at the high frequencies that tends to push up the gain-crossover frequency. In general, depending on the characteristics of the uncompensated system, if the design is carried out properly, it is possible to obtain a net improvement in relative stability using this approach. The Bode diagram in Fig. 11-3 serves to illustrate the principle of phase-lag compensation. If, instead of adding more positive phase to $G_p(j\omega)$ in the high-frequency range, as in Fig. 11-2, we attenuate the magnitude of $G_p(j\omega)$, with $K = 100$, in the high-frequency range by means of a low-pass filter, a similar stabilization effect can be achieved. The Bode diagram of Fig. 11-3 shows that if the attenuation starts at a sufficiently low-frequency range, the effect on the phase of $G_p(j\omega)$ due to the phase-lag compensation is insignificant at the new gain-crossover frequency. Thus, the net effect of the compensation scheme is the improvement of the relative stability of the system.

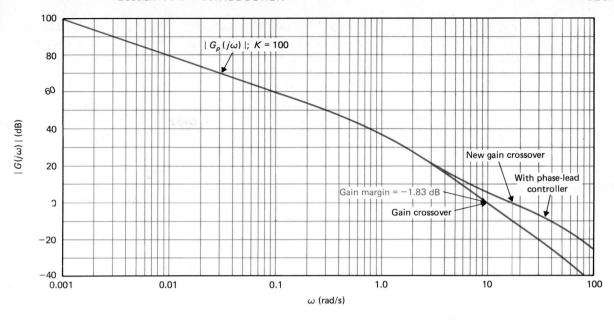

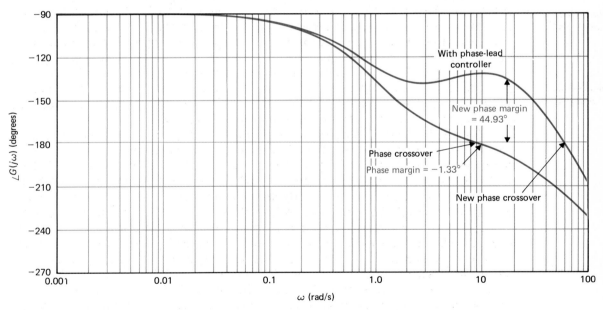

FIGURE 11-2 Bode diagram of $G_p(s) = 100/[s(1 + s)(1 + 0.0125s)]$ with phase-lead compensation.

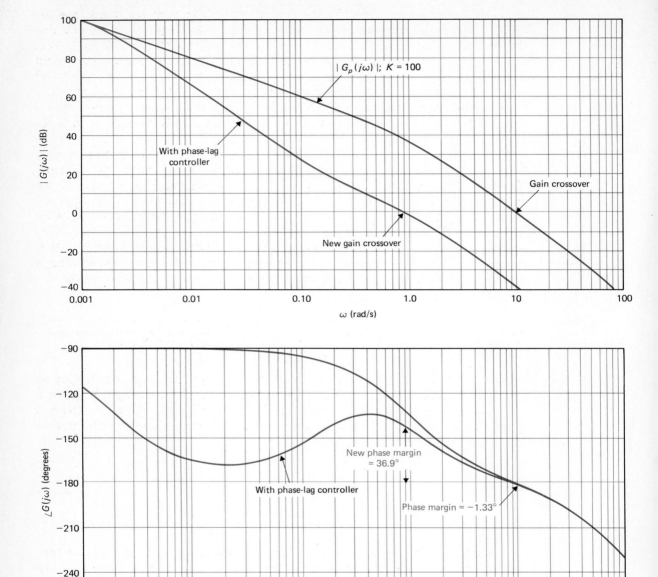

FIGURE 11-3 Bode diagram of $G_p(s) = 100/[s(1 + s)(1 + 0.0125s)]$ with phase-lag compensation.

lead-lag controller

The examples given are simply for the purpose of illustrating the principle of design of control systems in the frequency domain using controllers with phase-lead and phase-lag characteristics. In general, it may not be possible to satisfy all design criteria by simply using a phase-lead or a phase-lag controller. Situations often arise that both the low-frequency and the high-frequency loci of the process may need to be reshaped by using a controller with lead-lag or lag-lead characteristics. Or, the frequency locus of the controlled process must be reshaped in a certain midfrequency range so that a so-called "notch" controller such as the bridged-T network (see Section 11-8) is required.

It should be pointed out that the design of linear control systems in the frequency domain is not an exact science. For a given set of performance specifications, a number of controller schemes and parameter values may satisfy. Some degree of trial and error and fine tuning is inevitable. Therefore, a computer program for frequency-domain analysis, such as the FREQRP of the ACSP software package, should be invaluable for the design of linear control systems in the frequency domain. The designer can carry out a large number of computer runs with different controller parameters within a short period of time. FREQCAD in ACSP is a computer-aided design program for the design of various types of controllers for linear control systems in the frequency domain.

FREQCAD

11-2

DESIGN WITH THE PD CONTROLLER

PD controller

It was demonstrated in Section 9-2 that the PD controller may be effective in improving the relative stability of linear control systems. The design was carried out in the time domain or the s-plane using the root-locus method. We can gain a different perspective by carrying out the design of the PD controller in the frequency domain. The transfer function of the PD controller is

$$G_c(s) = K_P + K_D s = K_P\left(1 + \frac{K_D}{K_P}s\right) \tag{11-3}$$

design principle

The Bode plot of Eq. (11.3) is shown in Fig. 11-4 with $K_P = 1$. In general, the proportional-control gain K_P can be combined with the series gain of the system, so that the zero-frequency gain of the PD controller can be regarded as unity. The high-pass filter characteristics of the PD controller are clearly shown by the Bode plot in Fig. 11-4. The phase-lead properties are utilized to improve the phase margin of a control system. Unfortunately, the magnitude characteristics of the controller push the gain-crossover frequency to a higher-value. *Thus, the design principle of the PD controller involves the placing of the corner frequency of the controller, $\omega = K_P/K_D$, such that an effective improvement of the phase is realized at the new gain-crossover frequency.* As illustrated in Section 9-2, the value of K_P/K_D should be neither too large nor too small. The following example illustrates the essence of the design of the PD controller.

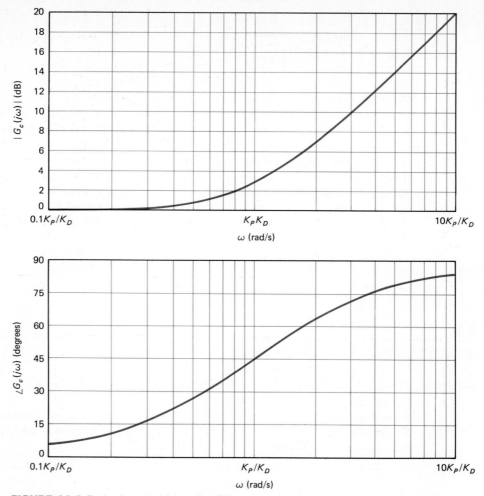

FIGURE 11-4 Bode diagram of $1 + K_D s / K_P$.

EXAMPLE 11-1   Consider the third-order process given in Eq. (9-26):

$$G_p(s) = \frac{K}{s(1 + 0.1s)(1 + 0.2s)} \quad (11\text{-}4)$$

where $K = 100$. The Bode diagram of $G_p(j\omega)$ is shown in Fig. 11-5. The gain margin of the uncompensated system is -16.48 dB and the phase margin is -40.31 degrees. With the PD controller, the open-loop transfer function is

$$G(s) = G_c(s)G_p(s) = \frac{100(1 + K_D s)}{s(1 + 0.1s)(1 + 0.2s)} \quad (11\text{-}5)$$

where K_P is set to 1. Figure 11-5 shows the Bode diagrams of $G(j\omega)$ for $K_D = 0.05$, 0.5, 1.0, and 2.0. It is interesting to note that as the value of K_D increases, the gain-crossover frequency is

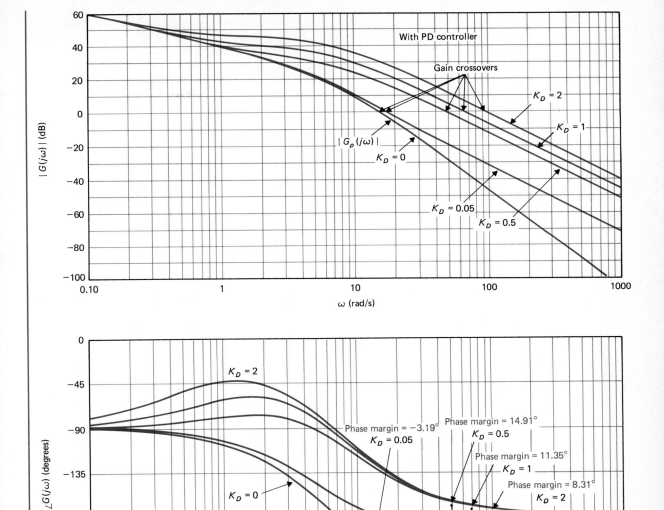

FIGURE 11-5 Bode diagrams of $G_p(s) = 100/[s(1 + 0.1s)(1 + 0.2s)]$ with PD compensations.

increased, so that although the maximum value of the phase of $G(j\omega)$ increases with the increase of K_D, the phase margin is actually decreased. The performance characteristics of the compensated system are tabulated in the following for various values of K_D. The maximum phase margin that can be obtained for the system with the PD controller is approximately $18°$, and this occurs when $K_D = 0.2$, which means that the controller zero cancels the open-loop pole at $s = -5$. Thus, this verifies the findings in Section 9-2 that the PD compensation is not very effective for the process given in Eq. (11-4) when the value of K is large.

phase-margin variations

K_D	Phase Margin (degrees)	Gain C.O. (rad/s)	BW (rad/s)
0	−40.31	15.97	20.00
0.05	−3.19	17.84	26.46
0.15	17.14	26.67	41.97
0.20	17.98	30.87	48.42
0.50	14.91	49.41	76.94
1.00	11.35	70.28	115.05
2.0	8.31	99.69	159.56
5.0	5.39	158.61	248.43
10.0	3.83	224.00	349.51

11-3

DESIGN WITH THE PI CONTROLLER

PI controller

The PI controller is described by the transfer function:

$$G_c(s) = K_P + \frac{K_I}{s} = \frac{K_I\left(1 + \dfrac{K_P}{K_I}s\right)}{s} \qquad (11\text{-}6)$$

From Section 9-2, we learned that an effective design of the PI controller is to place the zero at $s = -K_I/K_P$, very close to the origin in the s-plane, and the values of K_P and K_I should both be relatively small. Figure 11-6 shows the Bode diagram of $G_c(j\omega)$ given in Eq. (11-6). Notice that if K_P is less than 1, the magnitude of $G_c(j\omega)$ at $\omega = \infty$ represents an attenuation of $20\log_{10} K_P$ dB. This is the attenuation that may be utilized to improve the stability of the control system. The phase of $G_c(j\omega)$ is always negative, which is detrimental to stability. Thus, we should place the corner frequency of the PI controller, $\omega = K_I/K_P$, as far to the left as the bandwidth requirement allows, so that the phase-lag properties of $G_c(j\omega)$ will not degrade the achieved phase margin of the system.

The frequency-domain design procedure for the PI compensation to realize a given phase margin is as follows:

1.   The Bode plot of the open-loop transfer function $G_p(s)$ of the uncompensated system is made with the open-loop gain set according to the steady-state performance requirement.

2.   The phase margin and the gain margin of the uncompensated system are determined from the Bode plot. For a certain specified phase margin, the new gain-crossover frequency ω_c' corresponding to this phase margin is found on the Bode plot. The magnitude plot of the compensated transfer function must pass through the 0-db axis at this new gain-crossover frequency in order to realize the desired phase margin.

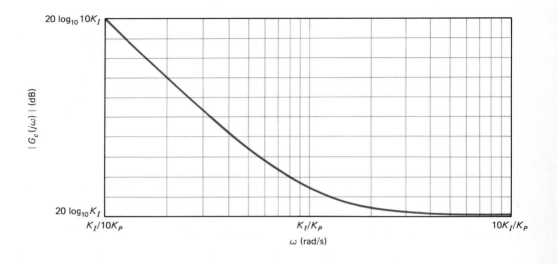

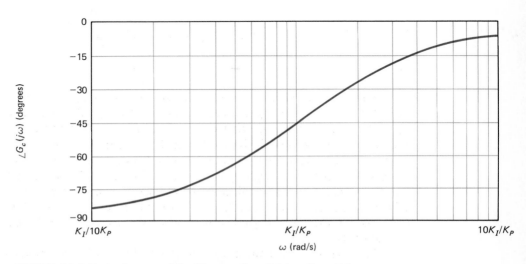

FIGURE 11-6 Bode diagram of the PI controller, $G_c(s) = K_P + K_I/s$.

3. To bring the magnitude curve of the uncompensated transfer function down to 0 dB at the new gain-crossover frequency ω_c' the PI controller must provide the amount of attenuation equal to the gain of the magnitude curve at the new gain-crossover frequency. In other words, set

$$|G_p(j\omega_c')|_{dB} = -20\log_{10}K_P \text{ dB} \qquad K_P < 1 \qquad (11\text{-}7)$$

from which

$$K_P = 10^{-|G_p(j\omega_c')|_{dB}/20} \qquad K_P < 1 \qquad (11\text{-}8)$$

Once the value of K_P is determined, it is necessary only to select the proper value of K_I to complete the design. Up to this point, we have assumed that although the gain-crossover frequency is altered by attenuating the magnitude of $G_p(j\omega)$ at ω_c', the original phase is not affected by the PI controller. This is not possible, however, since, as shown in Fig. 11-6, the attenuation property of the PI controller is accompanied with a phase lag. It is apparent that if the corner frequency $\omega = K_I/K_P$ is placed far below the new gain-crossover frequency ω_c', the phase lag of the PI controller will have a negligible effect on the phase of the compensated system near ω_c'. On the other hand, the value of K_I/K_P should not be too small or the bandwidth of the system will be too low, causing the system to be too sluggish. Usually, as a general guideline, it is recommended that K_I/K_P be placed at a frequency that is at least one decade below ω_c'. We set

$$\frac{K_I}{K_P} = \frac{\omega_c'}{10} \text{ rad/s} \qquad (11\text{-}9)$$

Thus,

$$K_I = \frac{\omega_c'}{10}K_P \text{ rad/s} \qquad (11\text{-}10)$$

4. The Bode plot of the compensated system is investigated to see if the performance specifications are met.
5. If all the design specifications are met, the values of K_I and K_P are substituted in Eq. (11-6) to give the desired transfer function of the PI controller.

    If the controlled process $G_p(s)$ is type 1, the value of K_I may be selected based on the ramp-error-constant requirement, and then there would only be one parameter, K_P, to determine. By computing the phase margins, gain margins, M_p, and bandwidths of the closed-loop system with a range of values of K_P, the best value for K_P can be easily selected.

EXAMPLE 11-2   Consider the following third-order process:

example on PI control

$$G_p(s) = \frac{100}{s(1 + 0.1s)(1 + 0.2s)} \qquad (11\text{-}11)$$

design specifications

The specifications of the system are as follows:

1. The ramp-error constant K_v is infinite.
2. The phase margin of the system should be approximately 45°.

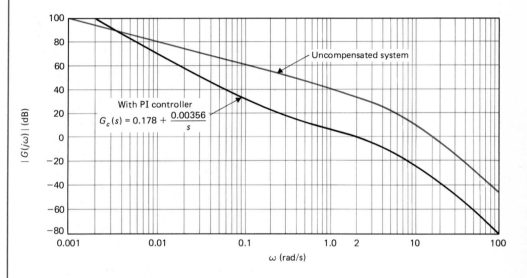

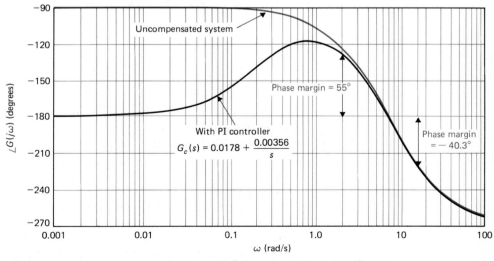

FIGURE 11-7 Bode diagram of $G_p(s) = 100/[s(1 + 0.01s)(1 + 0.2s)]$ with PI compensation.

To satisfy condition **1**, a PI controller is needed. The Bode diagram of $G_p(j\omega)$ is shown in Fig. 11-7. As seen, the closed-loop system is unstable, with the phase margin $= -40.3°$. From Fig. 11-7, the desired 45° phase margin can be obtained if the new gain-crossover frequency ω_c' is at approximately 2.8 rad/s. This means that the PI controller must reduce the magnitude of $G_p(j\omega)$ to 0 dB at $\omega = 2.8$ rad/s, while it does not appreciably affect the phase curve in the vicinity of this frequency. Actually, since a small negative phase is still accompanied by the PI controller at ω_c', even after the proper selection of the values of K_I and K_P, it is a safe measure to choose ω_c' somewhat less than 2.8 rad/s, say, 2 rad/s. From the magnitude plot of $G_p(j\omega)$, the value of $|G_p(j\omega)|_{dB}$ at $\omega_c' = 2$ rad/s is 35 dB. Thus, the PI controller must provide an attenuation of -35 dB at $\omega_c' = 2$ rad/s in order to bring the magnitude curve down to 0 dB at this frequency. By using Eq. (11-8), the value of K_P is determined as

$$K_P = 10^{-35/20} = 0.0178 \tag{11-12}$$

The reader should compare this value of K_P with that found in Section 9-2, Eq. (9-32), using the damping-ratio specification ($K_P = 0.0163$).

In order that the phase lag of the PI controller does not appreciably affect the phase of the compensated system at $\omega_c' = 2$ rad/s, we choose the corner frequency K_I/K_P to be at one decade below ω_c'. Thus,

$$K_I = \frac{\omega_c'}{10}K_P = \frac{2}{10}0.0178 = 0.00356 \text{ rad/s} \tag{11-13}$$

PI controller   The transfer function of the PI controller is

$$G_c(s) = 0.0178 + \frac{0.00356}{s} \tag{11-14}$$

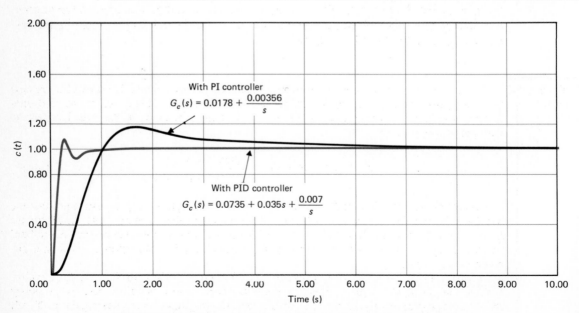

FIGURE 11-8 Unit-step responses of the systems in Example 11-2.

The open-loop transfer function of the compensated system is

$$G(s) = G_c(s)G_p(s) = \frac{0.356(1 + 5s)}{s^2(1 + 0.1s)(1 + 0.2s)} \tag{11-15}$$

*system
performance*

The phase margin of the compensated system is actually 55°. In fact, we can show that the design is very insensitive to the value of K_I as long as it is small. With $K_I = 0.00356$, the phase margin is greater than 45° for K_P from 0.0072 to 0.028. Figure 11-8 shows the unit-step response of the system. The maximum overshoot is 17.2 percent, but the rise and settling times are slow, which is typical for the PI control.

11-4

DESIGN WITH THE PID CONTROLLER

PID controller

The PID controller combines the advantages of the derivative and integral controls. We can make use of the controller equation established in Eq. (9-21), i.e.,

$$G_c(s) = K_P + K_D s + \frac{K_I}{s} = (1 + K_{D1}s)\left(K_{P2} + \frac{K_{I2}}{s}\right) \tag{11-16}$$

The relationships between K_P, K_D, and K_I, and K_{D1}, K_{P2}, and K_{I2} are given in Eqs. (9-22) through (9-24).

Since the PI portion is separated from the PD portion in Eq. (11-16), we can again carry out the design by first adding one portion and then the other to the uncompensated system. The following example will illustrate the design procedure in the frequency domain for the PID controller.

EXAMPLE 11-3
*PID control
example*

The controlled process and performance specifications described in Example 11-2 are now to be compensated by a PID controller. We shall first design the PD portion of the controller that has the transfer function $G_c(s) = 1 + K_{D1}s$. The transfer function of the open-loop system with the PD controller is

$$G(s) = G_c(s)G_p(s) = \frac{100(1 + K_{D1}s)}{s(1 + 0.1s)(1 + 0.2s)} \tag{11-17}$$

The effects of various values of K_{D1} have been studied in Example 11-1. The results are that the best the PD controller can do is to achieve a phase margin in the range of 14.91° to 17.14° for $K_{D1} = 0.5$ to 0.15. Let us select $K_{D1} = 0.5$, which happens to be the value determined in Section 9-2 using the s-plane method. Certainly, this value of K_{D1} does not represent an optimal solution by any measure.

Figure 11-9 shows the Bode diagram of $G(j\omega)$ in Eq. (11-17) with $K_{D1} = 0.5$. To realize a phase margin of 45°, without affecting the phase curve, the new gain-crossover frequency should be at approximately 13 rad/s, at which point the magnitude curve has a gain of 20 dB. Thus, the PI controller must provide an attenuation of −20 dB. Let us add a safety factor of

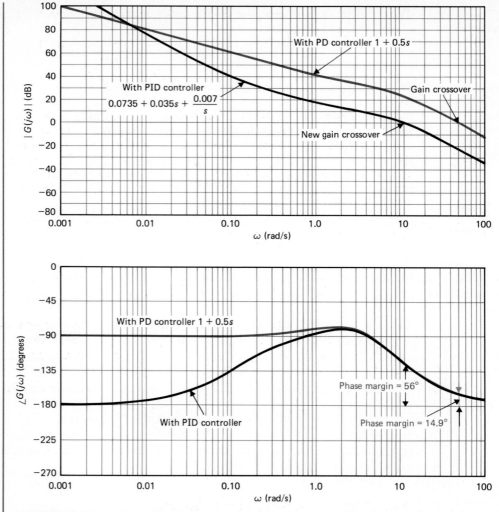

FIGURE 11-9 Bode diagrams of the system in Example 11-3 with PD and PID compensations.

−3 dB to this attenuation, since the phase curve will inevitably be degraded by the phase of the PI controller. Using Eq. (11-8), the value of K_{P2} is determined from

$$K_{P2} = 10^{-23/20} = 0.07 \tag{11-18}$$

which happens to be the same as the value of K_{P2} obtained in Section 9-2. Also, in Section 9-2, the value of K_{I2}/K_{P2} is chosen to be 0.1. In the frequency-domain design, the guideline on the selection of K_{I2}/K_{P2} is that the corner frequency should be sufficiently below the new gain-crossover frequency so that the phase lag accompanied by the PI controller will not significantly affect the phase of the compensated system. From this, we have $K_{I2} = 0.007$. By using Eqs.

PID controller parameters

(9-22) through (9-24), the parameters of the PID controller are determined to be $K_P = 0.0735$ $K_D = 0.035$, and $K_I = 0.007$

The open-loop transfer function of the PID-compensated system is

$$G(s) = \frac{0.7(1 + 0.5s)(1 + 10s)}{s^2(1 + 0.1s)(1 + 0.2s)} \tag{11-19}$$

performance summary

The Bode diagram of the compensated $G(j\omega)$ is shown in Fig. 11-9. The phase margin of the PID-compensated system is actually 56°. This is because the corner frequency K_{I2}/K_{P2} (0.1 rad/s) is placed a little too far below the new gain-crossover frequency of 13 rad/s, so that the effect of the phase lag of the PID controller is almost zero. The unit-step response of the compensated system is shown in Fig. 11-8. The maximum overshoot in this case is only 7.6 percent, and both the rise time and the settling time are vastly improved over the system with the PI controller.

11-5

DESIGN WITH THE PHASE-LEAD CONTROLLER

The phase-lead controller was applied in Section 9-3 for the compensation of a control system, and the design was carried out using the *s*-plane technique. In this section, the

phase-lead controller

phase-lead controller is to be designed using frequency-domain techniques.

The transfer function of the phase-lead controller is

$$G_c(s) = \frac{1 + aTs}{1 + Ts} \qquad a > 1 \tag{11-20}$$

Bode Plot of the Phase-Lead Controller

The Bode diagram of the phase-lead controller of Eq. (11-20) has two corner frequencies at $\omega = 1/aT$ and $\omega = 1/T$, as shown in Fig. 11-10. The maximum value of the phase, ϕ_m, and the frequency at which it occurs, ω_m, are derived as follows. Since ω_m is the geometric mean of the two corner frequencies, we write

$$\log_{10} \omega_m = \frac{1}{2}\left(\log_{10} \frac{1}{aT} + \log_{10} \frac{1}{T}\right) \tag{11-21}$$

Thus,

$$\omega_m = \frac{1}{\sqrt{a}\,T} \tag{11-22}$$

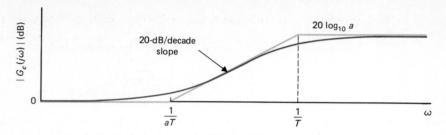

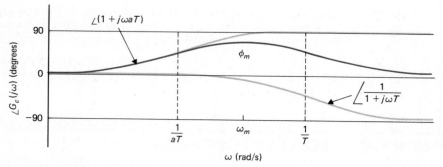

FIGURE 11-10 Bode diagram of phase-lead controller,
$G_c(s) = (1 + aTs)/(1 + Ts)$; $a > 1$.

To determine the maximum phase ϕ_m, the phase of $G_c(j\omega)$ is written

$$\angle G_c(j\omega) = \phi(j\omega) = \tan^{-1}\omega aT - \tan^{-1}\omega T \qquad (11\text{-}23)$$

from which we have

$$\tan \phi(j\omega) = \frac{\omega aT - \omega T}{1 + (\omega aT)(\omega T)} \qquad (11\text{-}24)$$

Substituting Eq. (11-22) into Eq.(11-24), we have

$$\tan \phi_m = \frac{a - 1}{2\sqrt{a}} \qquad (11\text{-}25)$$

or

$$\sin \phi_m = \frac{a - 1}{a + 1} \qquad (11\text{-}26)$$

Thus, by knowing ϕ_m, the value of a is determined from

$$a = \frac{1 + \sin \phi_m}{1 - \sin \phi_m}$$ (11-27)

Design of Phase-Lead Compensation with Bode Plots

Design of linear control systems in the frequency domain is more conveniently carried out using the Bode diagram. The reason is simply because the effect of the compensation is easily obtained by adding its magnitude and phase curves, respectively, to that of the original process. The general outline of phase-lead controller design in the frequency domain follows. It is assumed that the design specifications simply include steady-state-error and phase-margin requirements.

design
procedure

1. The Bode diagram of the uncompensated process $G_p(s)$ is constructed with the gain constant K set according to the steady-state-error requirement.
2. The phase margin and the gain margin of the uncompensated system are read from the Bode diagram, and the additional amount of phase lead needed to realize the desired phase margin is determined. From the additional phase lead required, the desired value of ϕ_m is estimated accordingly, and the value of a is calculated from Eq. (11-27).
3. Once a is determined, it is necessary only to determine the value of T, and the design is in principle completed. This is accomplished by placing the corner frequencies of the phase-lead controller, $1/aT$ and $1/T$, such that ϕ_m is located at the new gain-crossover frequency ω_c', so that the phase margin of the compensated system is benefited by ϕ_m. It is known that the high-frequency gain of the phase-lead controller of Eq. (11-20) is $20 \log_{10} a$ dB. Thus, to have the new gain crossover at ω_m, which is the geometric mean of $1/aT$ and $1/T$, we need to place ω_m at the frequency where the magnitude of the uncompensated $G_p(j\omega)$ is $-10 \log_{10} a$ dB.
4. The Bode diagram of the open-loop transfer function of the compensated system is investigated to check that all performance specifications are met; if not, a new value of ϕ_m must be chosen and the steps repeated.
5. If the design specifications are all satisfied, the transfer function of the phase-lead controller is established from the values of a and T.

The following numerical example illustrates the steps involved in the phase-lead compensation of a control system.

EXAMPLE 11-4

phase-lead design example

sun-seeker control

Consider the sun-seeker control system described in Example 9-5. The block diagram of the system is shown in Fig. 9-18. The open-loop transfer function of the uncompensated system is, from Eq. (9-40),

$$G_p(s) = \frac{\Theta_o(s)}{\alpha(s)} = \frac{2500K}{s(s + 25)} \qquad (11\text{-}28)$$

The performance specifications of the system are as follows:

specifications

1. The steady-state value of $\alpha(t)$ due to a unit-ramp-function input should be less than or equal to 0.01 rad/rad/s of the final steady-state-output velocity. In other words. the steady-state error due to a ramp input should be less than or equal to 1 percent.
2. The phase margin of the system should be greater than 45°.

Let us attempt to meet the above specifications with a series single-stage phase-lead controller described by Eq. (11-20). The following steps are carried out in the design of the phase-lead compensation:

1. Applying the final-value theorem to $\alpha(t)$, we have

design steps

$$\lim_{t \to \infty} \alpha(t) = \lim_{s \to 0} s\alpha(s) = \lim_{s \to 0} s\frac{\Theta_r(s)}{1 + G_p(s)} \qquad (11\text{-}29)$$

Since $\Theta_r(s) = 1/s^2$, by using Eq. (11-28), the last equation becomes

$$\lim_{t \to \infty} \alpha(t) = 0.01/K \qquad (11\text{-}30)$$

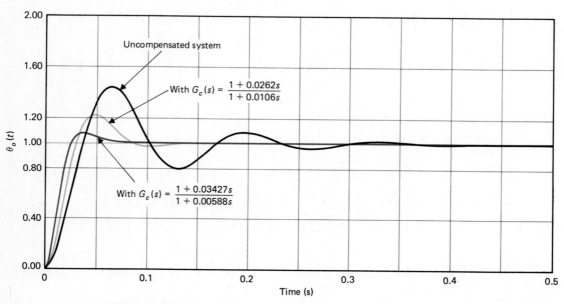

FIGURE 11-11 Step responses of the sun-seeker system in Example 11-4 with phase-lead compensations.

Thus, if $K = 1$, we have the steady-state error equal to 0.01, which is the maximum allowed value. However, for this value of K, the damping ratio of the closed-loop system is merely 25 percent, which corresponds to a maximum overshoot of 44.4 percent. Figure 11-11 shows the unit-step response of the closed-loop system with $K = 1$.

2. The Bode diagram of $G_p(s)$ with $K = 1$ is constructed as shown in Fig. 11-12.

3. The phase margin of the uncompensated system, read at the gain-crossover frequency, $\omega_c = 47$ rad/s, is 28°. Since the minimum desired phase margin is 45°, at least 17 more degrees of phase lead should be added to the open-loop system at the gain-crossover frequency.

4. The phase-lead controller of Eq. (11-20) must provide the additional 17° at the gain-crossover frequency of the compensated system. However, by applying the phase-lead controller, the magnitude curve of the Bode diagram is also affected in such a way that the gain-crossover frequency is shifted to a higher frequency. Although it is a simple matter to adjust the corner frequencies, $1/aT$ and $1/T$, of the

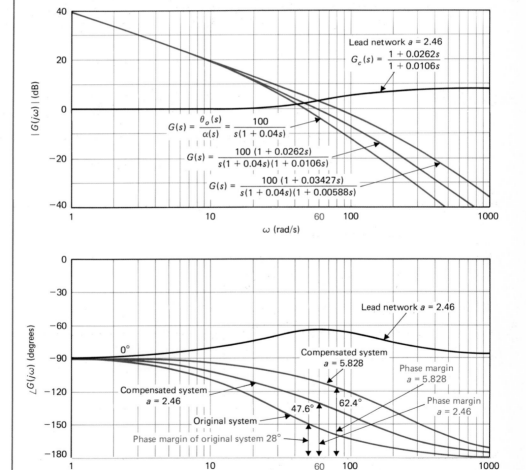

FIGURE 11-12 Bode diagrams of phase-lead compensated and uncompensated systems in Example 11-4.

controller, so that the maximum phase of the controller, ϕ_m, falls exactly at the new gain-crossover frequency, the original phase curve at this point is no longer 28°, and could be considerably less, as the phase of most control processes decreases with the increase in frequency. In fact, if the phase of the uncompensated process decreases rapidly with increasing frequency near the gain-crossover frequency, the single-stage phase-lead controller is no longer effective.

In view of the difficulty estimating the necessary amount of phase lead, it is essential to include some safety margin to account for the inevitable phase dropoff. Therefore, in the present case, instead of selecting a ϕ_m of a mere 17°, we let ϕ_m be 25°. Using Eq. (11-27), we have

$$a = \frac{1 + \sin 25°}{1 - \sin 25°} = 2.46 \tag{11-31}$$

5. To determine the proper location of the two corner frequencies, $1/aT$ and $1/T$, of the controller, it is known from Eq. (11-22) that the maximum phase lead ϕ_m occurs at the geometrical mean of the two corner frequencies. To achieve the maximum phase margin with the value of a determined, ϕ_m should occur at the new gain-crossover frequency ω_c', which is not known. The following steps are taken to ensure that ω_m occurs at ω_c'.

(a) The high-frequency gain of the phase-lead controller is

$$20 \log_{10} a = 20 \log_{10} 2.46 = 7.82 \text{ dB} \tag{11-32}$$

(b) The geometric mean ω_m of the two corner frequencies $1/aT$ and $1/T$ should be located at the frequency at which the magnitude of the uncompensated process transfer function $G_p(j\omega)$ in dB is equal to the negative value in dB of one-half of this gain. This way, the magnitude curve of the compensated transfer function will pass through the 0-dB axis at $\omega = \omega_m$. Thus, ω_m should be located at the frequency where

$$|G_p(j\omega)|_{dB} = -10 \log_{10} 2.46 = -3.91 \text{ dB} \tag{11-33}$$

From Fig. 11-12, this frequency is found to be $\omega_m = 60$ rad/s. Now using Eq. (11-22), we have

$$\frac{1}{T} = \sqrt{a}\, \omega_m = \sqrt{2.46} \times 60 = 94 \text{ rad/s} \tag{11-34}$$

Then, $1/aT = 94/2.46 = 38.2$ rad/s.

phase-lead controller

The transfer function of the phase-lead controller is

$$G_c(s) = \frac{U(s)}{E_o(s)} = \frac{1}{a}\frac{1 + aTs}{1 + Ts} = \frac{1}{2.46}\frac{1 + 0.0262s}{1 + 0.0106s} \tag{11-35}$$

Since it is assumed that the amplifier gains are increased by a factor of a, the open-loop transfer function of the compensated system is

$$G(s) = G_c(s)G_p(s) = \frac{6150(s + 38.2)}{s(s + 25)(s + 94)} \tag{11-36}$$

performance
summary

Figure 11-12 shows that the phase margin of the compensated system is 47.6°.

In Fig. 11-13, the magnitude and phase of the original and the compensated systems are plotted on the Nichols chart. These plots are obtained by taking the data directly from the Bode diagram of Fig. 11-12. From the Nichols chart, the resonance peak M_p of the uncompensated system is found to be 2.07. The value of M_p with the phase-lead controller is 1.25. One important observation is that the resonant frequency of the system is increased from 46.67 to approximately 53.5 rad/s, and the bandwidth is increased from 74.3 to 98.2 rad/s. These are natural consequences of the phase-lead compensation.

The unit-step response of the phase-lead compensated system is shown in Fig. 11-11. Notice that the response of the compensated system is better damped with a maximum overshoot of 24.5 percent, as against a maximum overshoot of 44.4 percent for the uncompensated system. The rise time of the system is also reduced by the phase-lead compensation, due to the increase of the bandwidth. On the other hand, excessive bandwidth may be objectionable in certain systems where noise and disturbance signals at high frequencies are to be rejected.

In the present design problem, we realized that a specification of 45° for the phase margin yielded a maximum overshoot of 24.5 percent in the step response. To make further improve-

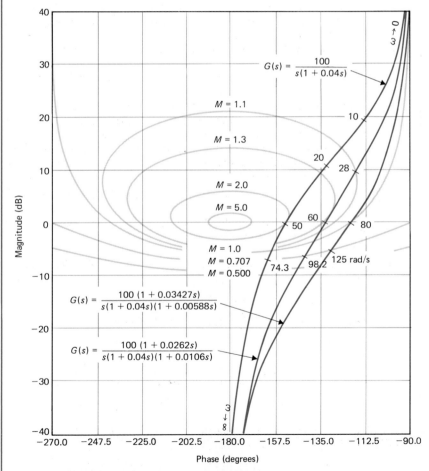

FIGURE 11-13 Plots of $G(s)$ in the Nichols chart for the system in Example 11-4.

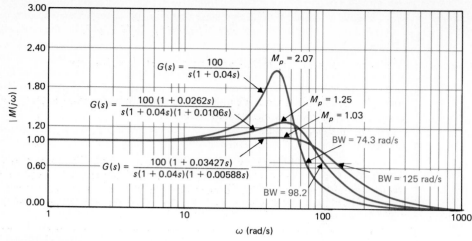

FIGURE 11-14 Closed-loop frequency responses of the sun-seeker system in Example 11-4.

another phase-lead controller

ment on the maximum overshoot, we can select the value of a to be 5.828. The resulting phase-lead controller has the transfer function

$$G_c(s) = \frac{1}{5.828} \frac{1 + 0.03427s}{1 + 0.00588s} \qquad (11\text{-}37)$$

The selection of $a = 5.828$ is made so that the results of the present frequency-domain design can be compared with those of the time-domain design carried out in Example 9-5. Since the time-domain design and the frequency-domain design use different specifications that are difficult to correlate exactly, especially for high-order systems, in general, we cannot obtain unique solutions for a given design problem when these different design methods are used.

performance summary

The Bode diagram of the compensated transfer function with $a = 5.828$ is shown in Fig. 11-12. The phase margin is improved to 62.4°, but the gain-crossover frequency is increased to 80 rad/s. The unit-step response of the new system is plotted in Fig. 11-11. In this case, the maximum overshoot is reduced to 7.3 percent, and the rise time is shorter still. From the Nichols chart, the value of M_p is now 1.03, and the bandwidth is increased to 125 rad/s.

By using the magnitude-versus-phase plots and the Nichols chart in Fig. 11-13, the closed-loop frequency responses of the sun-seeker system, before and after compensation, are plotted in Fig. 11-14.

EXAMPLE 11-5

high-order system with high gain

The control system described in Example 9-6 is again used to illustrate that under certain conditions, the single-stage phase-lead controller in Eq. (11-20) may not be effective for the purpose of improving the stability of control systems.

Let the open-loop transfer function of a control system with unity feedback be

$$G_p(s) = \frac{K}{s(1 + 0.1s)(1 + 0.2s)} \qquad (11\text{-}38)$$

specifications   The system must satisfy the following performance specifications:

1. $K_v = 100$; or the magnitude of the steady-state error of the system due to a unit-ramp function input is 0.01.
2. Phase margin $\geq 40°$.

From the steady-state requirement, we set $K = 100$. The Bode diagram of $G_p(s)$ when $K = 100$ is shown in Fig. 11-15. As observed from this Bode plot, the phase margin of the system is approximately $-40°$, which means that the system is unstable. In fact, the system is unstable for all values of K greater than 15. The rapid decrease of phase of $G_p(j\omega)$ at the gain-

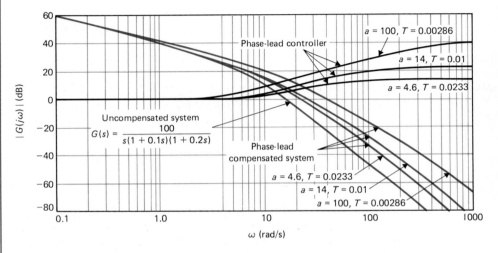

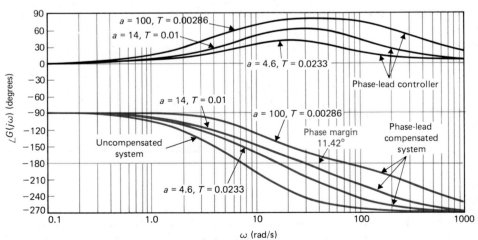

FIGURE 11-15 Bode plots of $G(s) = 100/[s(1 + 0.1s)(1 + 0.2s)]$ with phase-lead compensations.

crossover frequency, $\omega_c = 17$ rad/s, implies that single-stage phase-lead compensation may be ineffective for this system. To illustrate the point, the phase-lead controller of Eq. (11-20), with $a = 4.6$, 14, and 100, respectively, is used to compensate the system. Figure 11-15 illustrates the Bode diagrams of the compensated systems with the values of T chosen according to the procedure outlined for phase-lead compensation. It is clear from Fig. 11-15 that as more phase lead is being added to the open-loop system, the gain-crossover frequency is also pushed to a higher value. Therefore, for the present system, in which the uncompensated system is very unstable at the outset, it is impossible to realize a phase margin of 40° while simultaneously achieving a K_v of 100, with the phase-lead controller of Eq. (11-20). We can show with the computer program FREQCAD that the maximum phase margin that the single-stage phase-lead controller can achieve is 12.77°. In order to realize a phase margin of 45° or more, it would be necessary to use at least two stages of the phase-lead controller described by Eq. (11-20), as demonstrated in Example 11-6.

FREQCAD

Effects and Limitations of Phase-Lead Compensation

effects of phase-lead control

From the results of the last two illustrative examples, we can summarize that when the phase-lead compensation is effective, its effects on system performance are as follows:

1. The phase of the open-loop transfer function in the vicinity of the gain-crossover frequency is increased. This improves the phase margin of the closed-loop system.
2. The slope of the magnitude curve of the open-loop transfer function is reduced at the gain-crossover frequency. This usually corresponds to an improvement in the relative stability of the system in the form of improved gain and phase margins.
3. The bandwidth of the closed-loop system is increased.
4. The overshoot of the step response is reduced, and the rise time is reduced.
5. The steady-state error of the system is not affected.

We have shown that the stability of the second-order system in Example 11-4 is effectively improved by single-stage phase-lead compensation, whereas the improvement on the system in Example 11-5 is limited. In general, successful application of single-stage phase-lead compensation to improve the stability of a control system is hinged upon the following conditions:

conditions for phase-lead control

1. Bandwidth considerations: If the original system is unstable, the additional phase lead required to obtain a certain desired phase margin may be excessive. This requires a relatively large value of a in Eq. (11-20), which, as a result, will give rise to a large bandwidth for the compensated system, and the transmission of noise entering the system at the input may become objectionable. However, if the noise enters near the output of the system, as shown in Fig. 9-2(g), then the increased bandwidth may generally be beneficial to noise rejection. If a large value of a is required, two or more phase-lead controllers can

be connected in cascade to achieve the large phase lead. The larger bandwidth also has the advantage of robustness; i.e., the system is insensitive to parameter variations and noise rejection as described before.

2. If the original system is unstable, or stable but with a low stability margin, the phase curve of the Bode diagram of the open-loop transfer function has a steep negative slope near the gain-crossover frequency. Under this condition, the single-stage phase-lead compensation usually becomes ineffective because the additional phase lead at the new gain crossover is added to a much smaller phase angle than that at the old gain crossover. The desired phase margin can be realized only by using a very large value of a. However, the resulting system may still be unsatisfactory because a portion of the phase curve may still be below the 180° axis, and the system becomes conditionally stable.

conditionally
stable

In general, the following situations may also cause the phase of the open-loop transfer function to change rapidly near the gain-crossover frequency:

1. The open-loop transfer function has two or more poles that are close to each other and are close to the gain-crossover frequency.
2. The open-loop transfer function has one or more pairs of complex-conjugate poles near the gain-crossover frequency.

EXAMPLE 11-6

two-stage,
three-stage
phase-lead
control

In this example, the system considered in Example 11-5 will be compensated by two or more stages of the phase-lead controller of Eq. (11-20). In general, the stages of the controller need not be identical.

The design approach may be that of achieving a portion of the desired phase-margin improvement by each controller stage.

Consider first that the system is to achieve a 45° phase margin by using a two-stage phase-lead controller. Since the additional phase lead needed to achieve a phase margin of 45° is 85°, we can first add a single-stage phase-lead controller that will improve the phase margin initially to approximately 2.5°. The second stage will make up the final 42.5° of phase margin. Apparently, the design will not be unique.

Since the phase curve of the Bode diagram of the uncompensated system at the gain-crossover frequency has a large negative slope, the value of $a = 5.24$ given by Eq. (11-27) for $\phi_m = 42.5°$ will not yield a phase margin of 2.5°. In fact, when we set $a = 12.5$ and $T = 0.008$, the single-stage phase-lead controller improves the phase margin to 2.92°. Figure 11-16 shows the Bode diagram of $G(j\omega) = G_c(j\omega)G_p(j\omega)$ when

single-stage
controller

$$G_c(s) = \frac{1 + 0.1s}{1 + 0.008s} \tag{11-39}$$

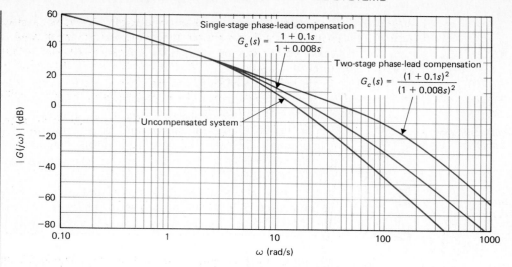

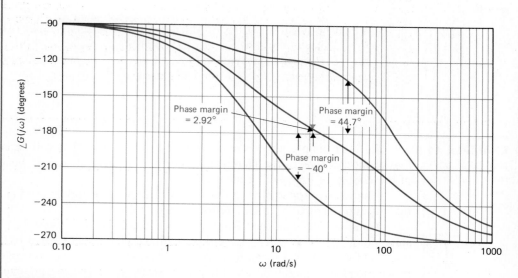

FIGURE 11-16 Bode diagrams of compensated and uncompensated systems in Example 11-6.

Notice that the single-stage phase-lead controller not only improves the phase margin, but also reduces the slope of $\angle G(j\omega)$ at the gain-crossover frequency. By adding another stage of the phase-lead controller with the transfer function in Eq. (11-39), the open-loop transfer function of the compensated system becomes

$$G(s) = G_c(s)G_p(s) = \frac{100(1 + 0.1s)}{s(1 + 0.2s)(1 + 0.008s)^2} \tag{11-40}$$

after the pole and zero at -10 cancel each other. Figure 11-16 shows the Bode diagram of the compensated system. The final compensated system has a phase margin of 44.7°. This design is carried out with the computer program FREQCAD of the ACSP software package. The program utilizes an iterative procedure in arriving at the values of a and T of a multiple-stage phase-lead controller with identical transfer functions for each stage.

FREQCAD

If the desired phase margin is set at 60°, the two-stage phase-lead controller would have the parameters $a = 22.3$ and $T = 0.00444$.

two-stage controller

For illustration purposes, a three-stage phase-lead controller is designed with the aid of FREQCAD to achieve a phase margin of 90°. The results are $a = 10$ and $T = 0.006$. The transfer function of the controller is

FREQCAD

three-stage controller

$$G_c(s) = \frac{(1 + 0.06s)^3}{(1 + 0.006)^3} \qquad (11\text{-}41)$$

The unit-step responses of the system with the two-stage phase-lead controller, $a = 22.3$ and $T = 0.00444$, and the three-stage controller are shown in Fig. 11-17. The magnitudes of the closed-loop frequency response and the sensitivity function $S_G^M(j\omega)$ are plotted in Fig. 11-18. Notice that the three-stage phase-lead compensated system has a high bandwidth, so that the sensitivity of the closed-loop system is less than unity for frequencies up to 100 rad/s. On the other hand, we can show that for the two-stage phase-lead compensated system, the magnitude of $S_G^M(j\omega)$ is less than unity only up to 50 rad/s. Thus, the compensated system with the three-stage phase-lead controller is shown to be more robust.

sensitivity

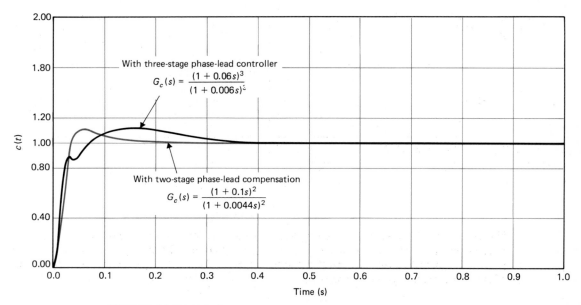

FIGURE 11-17 Unit-step responses of the system in Example 11-6 with two- and three-stage phase-lead compensations.

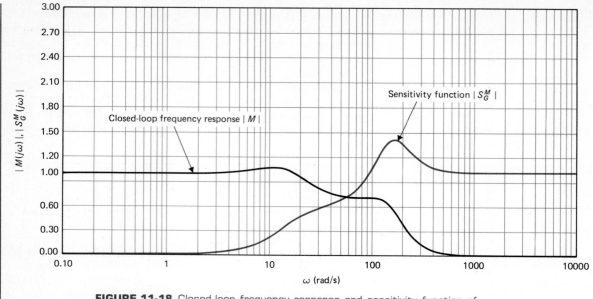

FIGURE 11-18 Closed-loop frequency response and sensitivity function of the system in Example 11-6 to illustrate the effects of phase-lead compensation.

11-6

DESIGN WITH THE PHASE-LAG CONTROLLER

In Section 9-4, we demonstrated that the single-stage phase-lag controller generally has a wider range of effectiveness than the single-stage phase-lead controller. The transfer function of the single-stage phase-lag controller given in Eq. (9-60) is

$$G_c(s) = \frac{1 + aTs}{1 + Ts} \qquad a < 1 \tag{11-42}$$

Bode Diagram of the Phase-lag Controller

Bode diagram   The Bode diagram of the transfer function in Eq. (11-42) is shown in Fig. 11-19. The magnitude curve has corner frequencies at $\omega = 1/aT$ and $1/T$. Since the transfer functions of the phase-lead and phase-lag controllers are identical in form, except for the value of a, it can be shown that the maximum phase lag ϕ_m of the phase curve of Fig. 11-19 satisfies

$$\sin \phi_m = \frac{a - 1}{a + 1} \tag{11-43}$$

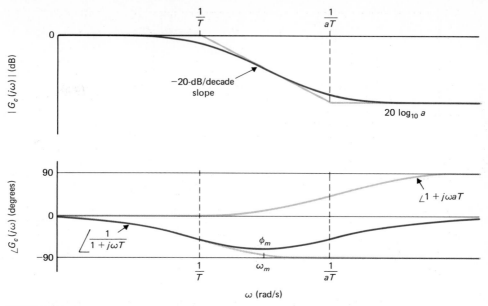

FIGURE 11-19 Bode diagram of the phase-lag controller,
$G_c(s) = (1 + aTs)/(1 + Ts)$; $a < 1$.

Figure 11-19 also shows that when ω approaches infinity, the phase-lag controller has an attentuation of $20 \log_{10} a$, which is the property that is utilized for compensation.

Design of Phase-Lag Compensation with Bode Plots

Unlike the phase-lead compensation that utilizes the maximum phase lead of the controller, phase-lag compensation utilizes the attenuation of the controller at high frequencies. For phase-lead compensation, the objective of the controller is to increase the phase of the open-loop system in the vicinity of the gain-crossover frequency while attempting to keep the magnitude curve of the Bode plot relatively unchanged near that frequency. In phase-lag compensation, the objective is to move the gain-crossover frequency to a lower frequency while keeping the phase curve of the Bode plot relatively unchanged at the gain-crossover frequency.

design procedure   The design procedure for phase-lag compensation using the Bode diagram is as follows:

1. The Bode diagram of the open-transfer function of the uncompensated system is drawn. The open-loop gain of the system is set according to the steady-state performance requirement.
2. The phase and gain margins of the uncompensated system are determined from the Bode diagram.
3. Assuming that the phase margin is to be increased, the frequency at which the desired phase margin is obtained is located on the Bode diagram. This frequency is also known as the new gain-crossover fre-

quency ω'_c, where the compensated magnitude curve passes through the 0-dB axis.

4.   To bring the magnitude curve down to 0 dB at the new gain-crossover frequency ω'_c, the phase-lag controller must provide the amount of attenuation equal to the value of the magnitude curve at ω'_c. In other words, let the open-loop transfer function of the uncompensated system be $G_p(s)$; then

$$\left| G_p(j\omega'_c) \right| = -20 \log_{10} a \text{ dB} \qquad a < 1 \qquad (11\text{-}44)$$

from which we have

$$a = 10^{-|G_p(j\omega'_c)|/20} \qquad a < 1 \qquad (11\text{-}45)$$

Once the value of a is determined, it is necessary only to select the proper value of T to complete the design. From the phase characteristics of the phase-lag controller shown in Fig. 11-19, it is observed that if the upper corner frequency $1/aT$ is placed far below the new gain-crossover frequency ω'_c, the phase lag of the controller will not appreciably affect the phase of the compensated system near ω'_c. On the other hand, the value of $1/aT$ should not be too low because the bandwidth of the system will be too low, causing the system to be too sluggish and less robust. Usually, as a general guideline, the frequency $1/aT$ should be approximately one decade below the new gain-crossover frequency ω'_c; that is,

$$\frac{1}{aT} = \frac{\omega'_c}{10} \text{ rad/s} \qquad (11\text{-}46)$$

Then,

$$\frac{1}{T} = \frac{\omega'_c}{10} a \text{ rad/s} \qquad (11\text{-}47)$$

5.   The Bode plot of the compensated system is investigated to see if performance specifications are met.

6.   If all the design specifications are met, the values of a and T are substituted in Eq. (11-42) to give the desired transfer function of the phase-lag controller.

EXAMPLE 11-7
sun-seeker
system

In this example, the sun-seeker system described in Example 11-4 is to be compensated by a phase-lag controller. The open-loop transfer function of the sun-seeker system is given in Eq. (11-28). The steady-state error requirement sets $K = 1$, and the phase margin is to be greater than 45°.

The Bode diagram of the uncompensated system is shown in Fig. 11-20. As can be seen from the figure, the phase margin is only 28°. From Fig. 11-20, it is observed that the desired

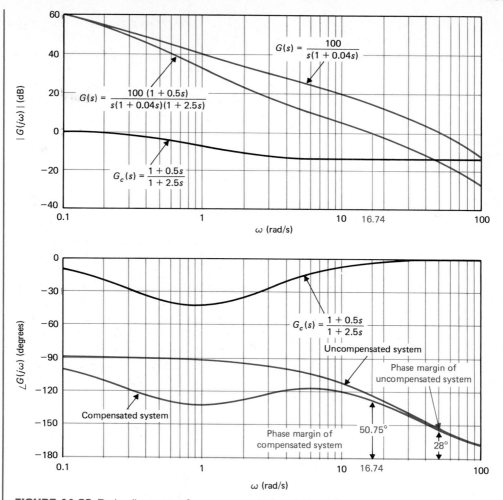

FIGURE 11-20 Bode diagrams of uncompensated and phase-lag compensated sun-seeker systems in Example 11-7.

$45°$ phase margin can be obtained if the gain-crossover frequency ω_c' is at 25 rad/s. This means that the phase-lag controller must reduce the magnitude of $G_c(j\omega)$ to 0 dB at $\omega = 25$ rad/s while it does not appreciably affect the phase curve in the vicinity of this frequency. Since the phase-lag controller still contributes a small negative phase when the upper corner frequency $1/aT$ is placed at $1/10$ of the value of ω_c' it is a safe measure to choose ω_c' at somewhat less than 25 rad/s, say, 20 rad/s.

From the Bode diagram, the value of $|G_p(j\omega)|_{dB}$ at $\omega_c' = 20$ rad/s is 14 dB. This means that the phase-lag controller must provide an attenuation of -14 dB at this frequency in order to bring the magnitude curve down to 0 dB at $\omega_c' = 20$ rad/s. Thus, using Eq. (11-45), we have

$$a = 10^{-|G_c(j\omega_c')|/20} = 10^{-0.7} = 0.2 \qquad (11\text{-}48)$$

In order that the phase lag of the controller does not appreciably affect the phase at the new gain-crossover frequency, we choose the value of $1/aT$ to be at one decade below $\omega_c' = 20$ rad/s.

Thus,

$$\frac{1}{aT} = \frac{\omega_c'}{10} = \frac{20}{10} = 2 \text{ rad/s} \tag{11-49}$$

Then, $1/T = 0.4$ rad/s. The transfer function of the phase-lag controller is

*phase-lag
controller*

$$G_c(s) = \frac{1 + 0.5s}{1 + 2.5s} \tag{11-50}$$

and the open-loop transfer function of the compensated system is

$$G(s) = G_c(s)G_p(s) = \frac{500(s + 2)}{s(s + 0.4)(s + 25)} \tag{11-51}$$

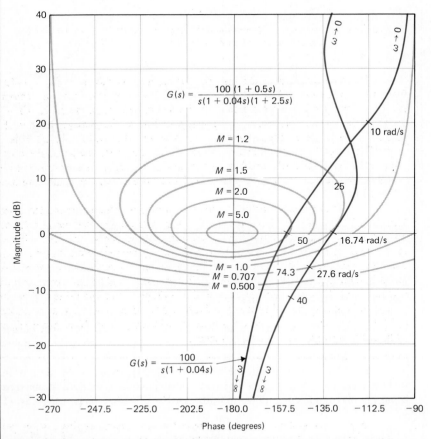

FIGURE 11-21 $G(s)$ plots of uncompensated and phase-lag-compensated sun-seeker systems in Example 11-7.

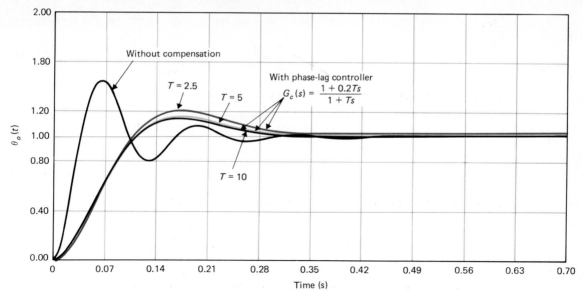

FIGURE 11-22 Unit-step responses of uncompensated and phase-lag-compensated sun-seeker systems in Example 11-7.

The Bode diagram of the last transfer function is shown in Fig. 11-20. We see that the magnitude curve beyond $\omega = 0.4$ rad/s is attenuated by the phase-lag controller whereas the low-frequency portion is not affected. In the meantime, the phase curve is not much affected by the phase-lag characteristic near the new gain-crossover frequency, which is at 16.74 rad/s. The phase margin of the compensated system is determined from Fig. 11-20 to be 50.75°.

The magnitude-versus-phase curves of the uncompensated and the compensated systems are plotted on the Nichols chart, as shown in Fig. 11-21. It is seen that the resonant peak, M_p, of the compensated system is 1.2. The bandwidth of the system is reduced from 75.3 to 27.6 rad/s.

performance evaluation

The unit-step responses of the uncompensated and the compensated systems are shown in Fig. 11-22. The effects of the phase-lag compensation are that the overshoot is reduced from 44.4 to 22 percent, but the rise time is increased considerably. The latter effect is apparently due to the reduction of the bandwidth by the phase-lag compensation. Figure 11-22 also shows the unit-step responses when the value of T of the phase-lag controller is increased to 5 and then to 10. It is seen that increasing the value of T gives only slight improvements on the overshoot. In Chapter 9, we have shown using the root-locus diagram that the value of T in the phase-lag compensation is not critical when it is relatively large. When $T = 5$, it is equivalent to setting $1/aT$ at $1/20$ of the new gain-crossover frequency. Similarly, $T = 10$ corresponds to placing $1/aT$ at $1/40$ of the new gain-crossover frequency.

EXAMPLE 11-8

design with phase-lag controller

Consider the system given in Example 11-6 for which the single-stage phase-lead compensation is ineffective. We shall show that the single-stage phase-lag controller is more versatile in compensating this system.

specifications

The open-loop transfer function of the uncompensated system and the performance specifications are as follows.

$$G_p(s) = \frac{K}{s(1 + 0.1s)(1 + 0.2s)} \tag{11-52}$$

1. $K_v = 100$.
2. Phase margin $\geq 45°$.

design steps

The design procedure with a phase-lag controller is as follows:

1. The Bode plot of $G_p(j\omega)$ is made as shown in Fig. 11-23 for $K = 100$.
2. The phase margin at the gain-crossover frequency, $\omega_c = 16$ rad/s, is $-40.3°$, and the closed-loop system is unstable.

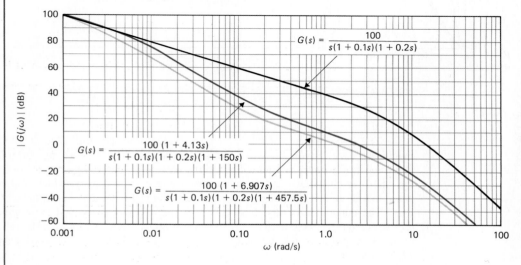

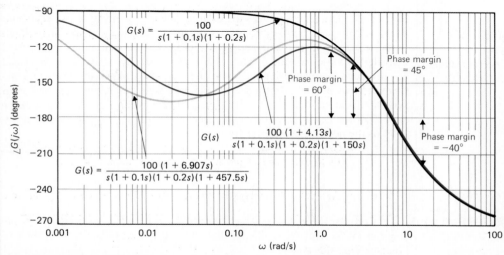

FIGURE 11-23 Bode diagrams of uncompensated and phase-lag compensated systems in Example 11-8.

3. To realize a phase margin of 45°, the gain-crossover frequency should be moved to $\omega_c' = 2.6$ rad/s. This means that the phase-lag controller must reduce the magnitude of $G_p(j\omega)$ to 0 dB while it does not significantly affect the phase curve at this new gain-crossover frequency. Since a small negative phase will eventually be introduced by the phase-lag controller at ω_c', it is proper to choose the new gain-crossover frequency somewhat less than 2.6 rad/s, say, at 2.42 rad/s.

4. From the Bode diagram, $|G_p(j\omega)|_{dB}$ at $\omega_c' = 2.42$ rad/s is 31.2 dB, which means that the controller must introduce -31.2 dB of attentution at this frequency in order to bring the $|G(j\omega)|_{dB}$ curve down to 0 dB. Thus, from Eq. (11-45),

$$a = 10^{-|G_p(j\omega_c')|/20} = 10^{-31.2/20} = 0.0275 \qquad (11\text{-}53)$$

5. Place the upper corner frequency of the controller, $1/aT$, at one decade below the new gain-crossover frequency, that is,

$$\frac{1}{aT} = \frac{\omega_c'}{10} = \frac{2.42}{10} = 0.242 \text{ rad/s} \qquad (11\text{-}54)$$

phase-lag controller

Thus, $T = 150$. The transfer function of the phase-lag controller is

$$G_c(s) = \frac{1 + aTs}{1 + Ts} = \frac{1 + 4.13s}{1 + 150s} \qquad (11\text{-}55)$$

The open-loop transfer function of the compensated system is

$$G(s) = G_c(s)G_p(s) = \frac{100(1 + 4.13s)}{s(1 + 0.1s)(1 + 0.2s)(1 + 150s)} \qquad (11\text{-}56)$$

6. The Bode diagram of the compensated system is constructed in Fig. 11-23. The phase margin of the compensated system is approximately 45°. The magnitude-versus-phase curves of the compensated and the uncompensated systems are plotted on the Nichols chart, as shown in Fig. 11-24. These curves show that the uncompensated system is unstable, and the compensated system has the following performance data in the frequency domain:

performance data

> resonant peak $M_p = 1.3$
> phase margin $= 45°$
> bandwidth $= 4.36$ rad/s

The unit-step response of the compensated system is shown in Fig. 11-25. The maximum overshoot of the system is approximately 24%.

controller

If we change the phase margin requirement to 60°, the phase-lag controller turns out to be

$$G_c(s) = \frac{1 + 6.907s}{1 + 457.5s} \qquad (11\text{-}57)$$

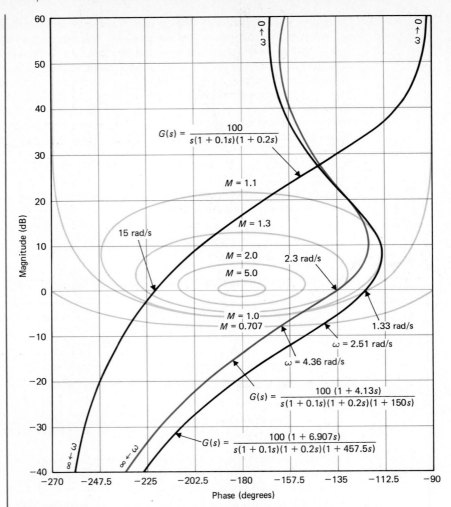

FIGURE 11-24 $G(s)$ plots of uncompensated and phase-lag-compensated systems in the Nichols chart for Example 11-8.

The Bode diagram of the compensated system is shown in Fig. 11-23, and the magnitude-versus-phase plot is shown in Fig. 11-24. The performance data in the frequency domain are

performance data

> resonant peak $M_p = 1.1$
> phase margin $= 59.5°$
> bandwidth $= 2.51$ rad/s

The unit-step response of the compensated system in Fig. 11-25 shows that the maximum overshoot is now reduced to 11.8 percent, except that the response takes a long time to reach the final steady-state value. At the end of 7 s, the output is still about 4 percent above the final value of 1.0. This is due to the low system bandwidth caused by the phase-lag control. Figure 11-26 shows the closed-loop frequency response and the sensitivity function $|S_G^M(j\omega)|$ for the system

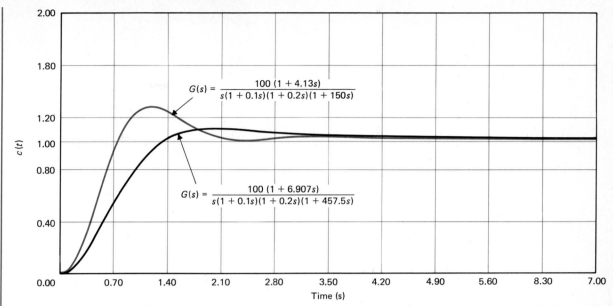

FIGURE 11-25 Unit-step responses of the phase-lag compensated system in Example 11-8.

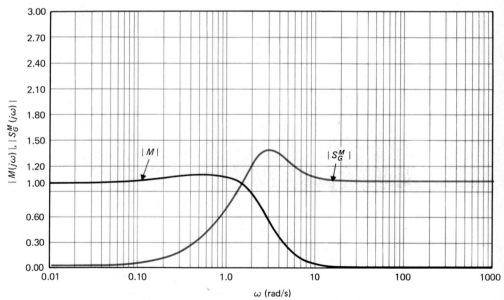

FIGURE 11-26 Closed-loop frequency response and sensitivity function of the phase-lag compensated system in Example 11-8.
$G_c(s) = (1 + 6.907s)/(1 + 457.6s)$.

with $G_c(s)$ given in Eq. (11-57). Notice that the sensitivity function is less than unity for frequencies up to only 1.5 rad/s.

Effects and Limitations of Phase-Lag Compensation

From the results of the preceding illustrative examples, the effects and limitations of phase-lag compensation on the performance of linear control systems can be summarized as follows on page 703:

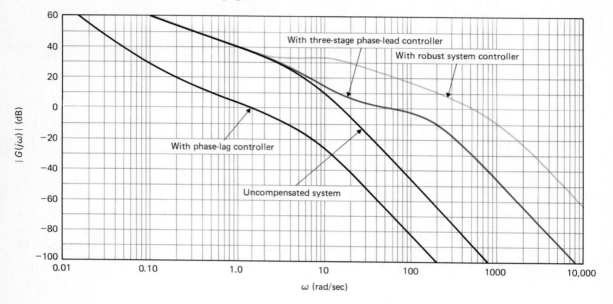

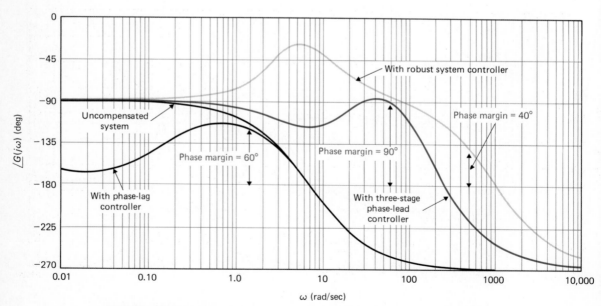

FIGURE 11-27 Bode diagrams of $G_p(s)$ in Eq. (11-52) with phase-lag, three-stage lead, and robust controllers.

*effects and
limitations*

1. For a given loop gain, K, the magnitude of the open-loop transfer function is attenuated near and above the gain-crossover frequency, thus allowing improvement in the relative stability of the system.
2. The gain-crossover frequency is decreased, and thus the bandwidth of the closed-loop system is decreased.
3. The rise time and the settling time of the system are usually slower, since the bandwidth is usually decreased.
4. The system is more sensitive to parameter variations in the sense that the sensitivity function is greater than unity for all frequencies approximately greater than the bandwidth of the system.

As a comparison of robustness of systems, Fig. 11-27 shows the Bode diagrams of the system described by the process in Eq. (11-38) with $K = 100$, and with

(a) phase-lag controller of Eq. (11-57),
(b) three-stage phase-lead controller of Eq. (11-41),

robustness

(c) robust system controller of Eq. (9-141).

Notice that the robust system controller gives a system with the highest bandwidth. The sensitivity function plot shown in Fig. 11-28 indicates that the sensitivity of the system is less than unity for frequencies up to 420 rad/s. At low frequencies, the sensitivity function is nearly zero.

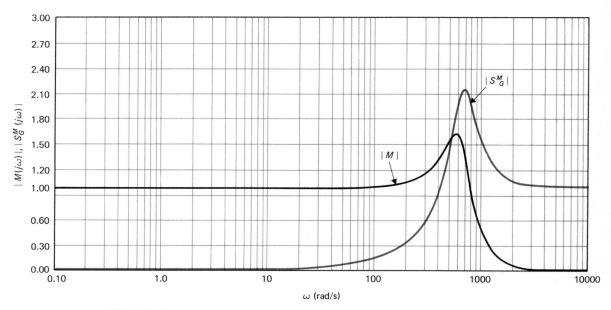

FIGURE 11-28 Closed-loop frequency response and sensitivity function of the system with the robust controller in Eq. (9-141).

11-7

DESIGN WITH LEAD-LAG (LAG-LEAD) CONTROLLERS

The lag-lead (lead-lag) controller discussed in Section 9-5 can also be designed using the frequency-domain methods. When designed properly, lag-lead compensation combines the advantages of the phase-lead and the phase-lag controllers. The following example repeats the design problem considered in Examples 11-5 and 11-8 by means of a lag-lead controller.

EXAMPLE 11-9   The open-loop transfer function of the system considered in Example 11-8 is

$$G_p(s) = \frac{K}{s(1 + 0.1s)(1 + 0.2s)} \tag{11-58}$$

The performance specifications are

$$K_v = 100$$
$$\text{phase margin} \geq 45°$$

These requirements have been satisfied by the phase-lag controller designed in Example 11-8, except that the designed system has a unit-step response that has large rise and settling times. The purpose of this example is to design a lag-lead controller so that the rise time is reduced while maintaining the phase margin and K_v values.

Let the series controller be represented by the transfer function

controller transfer function

$$G_c(s) = \frac{(1 + aT_1s)(1 + bT_2s)}{(1 + T_1s)(1 + T_2s)} \tag{11-59}$$

where $a > 1$, and $b < 1$.

It was demonstrated in Example 11-5 that a single-stage phase-lead controller is inadequate in achieving a phase margin of 45° when $K = 100$. Let us first select a to be 14, and $T_1 = 0.01$, for the phase-lead portion of the controller, according to the method outlined in Section 11-5. The Bode diagram of the phase-lead-compensated system is shown in Fig. 11-29. The phase margin is only 2.72°. Now starting with the phase-lead compensated system, we notice that a phase margin of 45° is realized if the gain-crossover frequency is moved to 6 rad/s. With the safety factor, let us set the new gain-crossover frequency at 5 rad/s. The magnitude of $G(j\omega)$ at $\omega = 5$ rad/s is 23.7 dB. Thus, the value of b is found from

$$20 \log_{10} b = -23.7 \text{ dB} \tag{11-60}$$

Thus, $b = 0.0653$. Setting $1/bT_2$ to be at 0.1 times the new gain-crossover frequency of 5 rad/s, we get $T_2 = 30.6$. Thus, the transfer function of the lag-lead controller is

lag-lead controller

$$G_c(s) = \frac{(1 + 0.14s)(1 + 1.998s)}{(1 + 0.01s)(1 + 30.6s)} \tag{11-61}$$

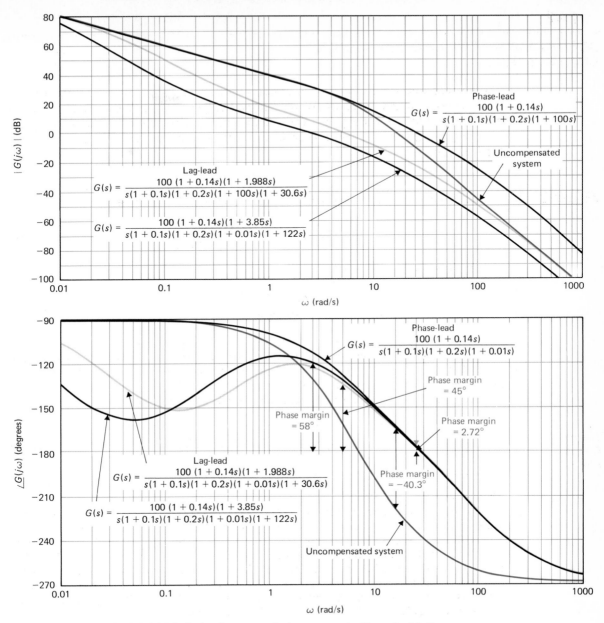

FIGURE 11-29 Bode diagrams of the system in Example 11-9, compensated with a lag-lead controller.

Figure 11-29 shows that the lag-lead compensated system has a phase margin of approximately 45°.

If we increase the phase-margin requirement to 60°, the new gain-crossover frequency should be moved to 2.6 rad/s. The required attenuation from the phase-lag portion of the controller is -30 dB. This corresponds to $b = 0.0316$. Setting $1/bT$ at 0.26 rad/s, the value of T is

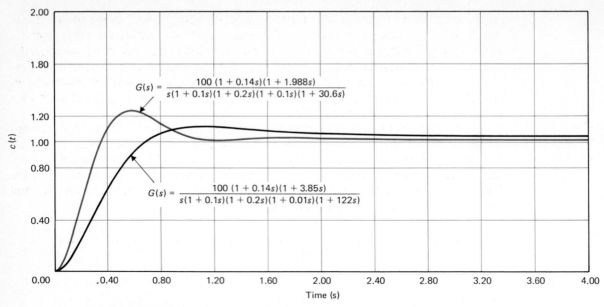

FIGURE 11-30 Unit-step responses of the system in Example 11-9 with lag-lead controllers.

found to be 122. Thus, the transfer function of the lag-lead controller is

lag-lead controller

$$G_c(s) = \frac{(1 + 0.14s)(1 + 3.85s)}{(1 + 0.01s)(1 + 122s)} \qquad (11\text{-}62)$$

performance summary

Figure 11-29 shows that the phase margin is actually approximately 58°. Figure 11-30 shows the unit-step responses of the system compensated with the two lag-lead controllers. The maximum overshoots are comparable to those of the two phase-lag-compensated systems in Example 11-8, shown in Fig. 11-25, but the rise and settling times are much shorter.

11-8

DESIGN WITH BRIDGED-T (NOTCH) CONTROLLER

bridged-T networks

In Section 9-6, the bridged-T controllers are used for pole-zero-cancellation compensation of linear control systems. The reason for the association with pole-zero cancellation is that in the *s*-domain, it is easier to grasp the controller design by canceling the undesirable poles of the process transfer function.

We can gain more perspective of the design of control systems with the bridged-T networks of Fig. 9-35 by working in the frequency domain. Figure 11-31 illustrates the Bode diagram of the transfer function of a typical bridged-T controller. Notice that the magnitude plot of the bridged-T controller typically has a "notch" at the resonant frequency ω_n. The phase plot is negative below and positive above the resonant frequency, while passing through zero degrees at the resonant frequency. The attenuation of the magnitude curve and the positive-phase characteristics can be used effectively to improve the stability of a linear control system. Because of the "notch" characteristic in the amplitude curve, the bridged-T controller or network is also referred to in the industry as a notch controller or network.

notch
controller

The notch controller has advantages over the phase-lag and the phase-lead controllers in certain conditions, since the magnitude and phase characteristics do not affect the high- and low-frequency properties of the system. Without using the pole-zero-cancellation principle, the design of the bridged-T controller for compensation in the frequency domain involves the determination of the amount of attenuation required and the resonant frequency of the controller.

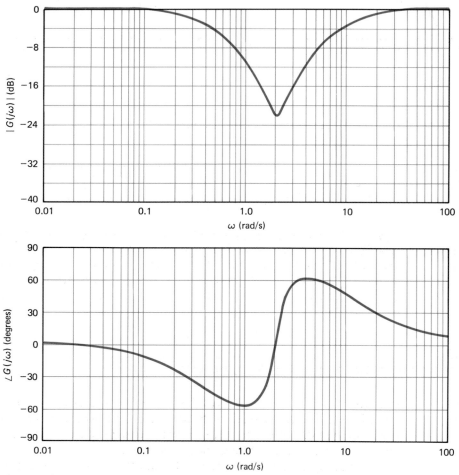

FIGURE 11-31 Bode plot of a bridged-T network with the transfer function
$G_p(s) = (s^2 + 0.8s + 4)/(s + 0.384)(s + 10.42)$

We write the transfer function of the notch controller as

$$G_c(s) = \frac{s^2 + 2\zeta_z \omega_n s + \omega_n^2}{s^2 + 2\zeta_p \omega_n s + \omega_n^2} \tag{11-63}$$

where we have utilized the fact that $\omega_{nz} = \omega_{np} = \omega_n$. The attenuation provided by the magnitude of $G_c(j\omega)$ at the resonant frequency ω_n is

design
equations

$$|G_c(j\omega_n)| = \frac{\zeta_z}{\zeta_p} \tag{11-64}$$

Using Eq. (9-103), we have

$$\zeta_z = \left[\frac{|G_c(j\omega_n)|}{2[1 - |G_c(j\omega_n)|]} \right]^{1/2} \tag{11-65}$$

Thus, given the amount of maximum attenuation required, the value of ζ_z is found using Eq. (11-65). The resonant frequency ω_{nz} ($= \omega_{np}$) is known once the location of the "notch" or the maximum attenuation is determined.

In practice, Eqs. (11-64) and (11-65) are useful when the characteristics of $G_c(j\omega)$ are determined experimentally, and the transfer function $G_c(s)$ is not available. Otherwise, if the expression of $G_c(s)$ is known, the values of ζ_z and ω_n can always be determined by aiming at cancelling the complex resonant poles of $G_c(s)$.

EXAMPLE 11-10   As an illustrative example on the design of the bridged-T controller in the frequency domain, consider the design problem given in Example 9-11. The transfer function of the controlled process is

$$G_p(s) = \frac{20,000(s + 100,000)}{(s + 16.69)(s^2 + 95.3s + 1,198,606.6)} \tag{11-66}$$

As shown in Example 9-11, the uncompensated system is unstable. We wish to achieve a phase margin of approximately 60° with cascade compensation.

We can set the zeros of the bridged-T controller to cancel the complex poles of $G_p(s)$. Thus, from Eq. (11-64), we get

$$\zeta_z = \frac{95.3}{2 \times 1095} = 0.0435 \tag{11-67}$$

bridged-T controller

Using Eq. (9-103), ζ_p is found to be 11.54. This is equivalent to introducing an attenuation of 0.00377 or -48.47 dB at $\omega_n = 1095$ rad/s. The transfer function of the bridged-T controller is

$$G_{c1}(s) = \frac{s^2 + 2\zeta_z\omega_n s + \omega_n^2}{s^2 + 2\zeta_p\omega_n s + \omega_n^2} = \frac{s^2 + 95.3s + 1{,}198{,}606.6}{s^2 + 25267.7s + 1{,}198{,}606.6}$$

$$= \frac{s^2 + 95.3s + 1{,}198{,}606.6}{(s + 47.53)(s + 25{,}215)} \qquad (11\text{-}68)$$

Figure 11-32 shows the Bode plots of the uncompensated system and the bridged-T controller. In this case, the phase margin of the compensated system is only 12.6°. Thus, we need another stage of controller to further improve the relative stability of the system.

Let us first consider a single-stage phase-lead controller described by Eq. (11-20). To arrive at a phase margin of 60°, the phase-lead controller must provide a phase lead of 47.4°. Let the maximum phase of the phase-lead controller be 53°. Then, using the design method outlined in Section 11-5, we have $a = 9$, $T = 0.000667$, and $aT = 0.006$. The transfer function of the phase-lead controller is

phase-lead controller

$$G_{c2}(s) = \frac{1 + aTs}{1 + Ts} = \frac{1 + 0.006s}{1 + 0.000667s} \qquad (11\text{-}69)$$

The open-loop transfer function of the compensated system with bridged-T and phase-lead controllers is

$$G(s) = G_{c1}(s)G_{c2}(s)G_p(s) = \frac{180{,}000(s + 10^5)(s + 166.67)}{(s + 16.67)(s + 47.53)(s + 1500)(s + 25{,}215)} \qquad (11\text{-}70)$$

Next, consider that the phase-lag controller described by Eq. (11-42) is used to improve the system already compensated by the bridged-T controller. From Fig. 11-32, we see that in order to realize a phase margin of 60°, the bridged-T compensated system should have an attenuation of -28 dB at the high frequencies. By using the design method outlined in Section 11-6, the value of a is found to be 0.04. The value of $1/aT$ is selected to be 4.5, then $T = 5.556$. The transfer function of the phase-lag controller is

phase-lag controller

$$G_{c3}(s) = \frac{1 + aTs}{1 + Ts} = \frac{1 + 0.222s}{1 + 5.556s} \qquad (11\text{-}71)$$

The open-loop transfer function of the system compensated with the bridged-T and phase-lag controllers is

$$G(s) = G_{c1}(s)G_{c3}(s) = \frac{800(s + 10^5)(s + 4.5)}{(s + 16.69)(s + 47.53)(s + 0.18)(s + 25{,}215)} \qquad (11\text{-}72)$$

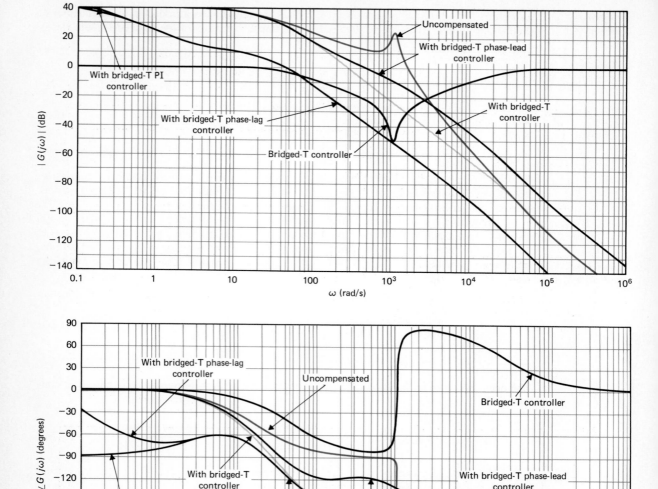

FIGURE 11-32 Bode diagrams of the system designed in Example 11-10.

Figure 11-32 shows the Bode plots of the system with the bridged-T phase-lead compensation and the bridged-T phase-lag compensation. The objective of improving the phase margin to 60° has been accomplished.

Figure 11-33 shows the unit-step responses of the system compensated by the bridged-T phase-lead controller. The maximum overshoot is 11.2 percent, and the rise time is very short. Figure 11-34 shows the unit-step response of the system compensated with the bridged-T phase-

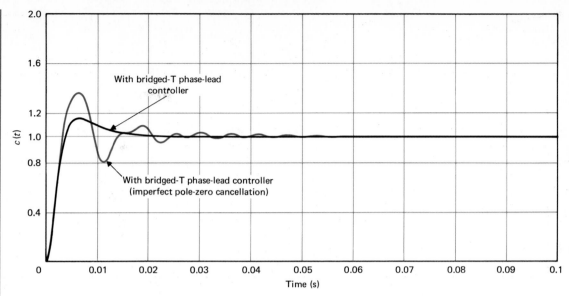

FIGURE 11-33 Step responses of the system in Example 11-10 showing the effect of imperfect pole-zero cancellation with the bridged-T-phase-lead controller.

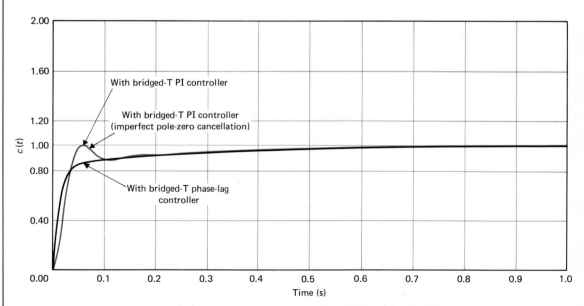

FIGURE 11-34 Step responses of the system in Example 11-10 showing the effects of imperfect pole-zero cancellation with the bridged-T-PI controller and bridged-T-phase-lag controller.

lag controller. The overshoot is very small, but the rise and settling times are increased. Since the system with either compensation is of type 0, the unit-step responses in Figs. 11-33 and 11-34 have a steady-state error of 1 percent. In order to eliminate this error, let us consider a PI controller, described by Eq. (11-6). The procedure of designing the PI controller is very similar to that of the phase-lag controller. As described in Section 11-3, Eq. (11-7), for an attenuation of -28 dB, $K_P = 0.04$. The value of K_I is selected to be 0.16. Thus, the transfer function of the PI controller is

PI controller

$$G_{c4}(s) = K_P + \frac{K_I}{s} = \frac{0.16(1 + 0.25s)}{s} \tag{11-73}$$

The open-loop transfer function of the system with the bridged-T PI controller is

$$G(s) = G_{c1}(s)G_{c4}(s) = \frac{800(s + 10^5)(s + 4)}{s(s + 16.69)(s + 47.53)(s + 25215)} \tag{11-74}$$

The Bode diagram of the bridged-T PI compensated system is shown in Fig. 11-32. The magnitude and phase curves are very close to those of the phase-lag-compensated system except that the magnitude curve has a slope of -20 dB/decade and the phase curve approaches $-90°$ as ω approaches 0. Figure 11-34 shows that the unit-step response of the bridged-T PI-compensated system is very similar to that of the bridged-T phase-lag-compensated system, except that the steady-state error is now zero since the system is type 1.

To investigate the effects of imperfect pole-zero cancellation and the utilization of Eqs. (11-64) and (11-65), let us consider the situation that the transfer function $G_p(s)$ is not known, and the Bode plot of $G_p(j\omega)$ is obtained experimentally. From the magnitude plot of $G_p(j\omega)$, the amount of attenuation that must be provided by the bridged-T controller at the resonant frequency ω_n of the complex poles of $G_p(s)$ is estimated to be -45 dB (0.00562), and ω_n is observed to be 1000 rad/s. From Eq. (11-65),

bridged-T controller

$$\zeta_z = \left[\frac{0.00562}{2(1 - 0.00562)}\right]^{1/2} = 0.0532 \tag{11-75}$$

Equation (11-64) gives

$$\zeta_p = \frac{\zeta_z}{|G_c(j\omega_n)|} = \frac{0.0532}{0.00562} = 9.46 \tag{11-76}$$

The transfer function of the bridged-T controller is

$$G_c(s) = \frac{s^2 + 2\zeta_z\omega_n s + \omega_n^2}{s^2 + 2\zeta_p\omega_n s + \omega_n^2} = \frac{s^2 + 106.4s + 10^6}{s^2 + 18,920s + 10^6} \tag{11-77}$$

The open-loop transfer function of the compensated system is effectively

$$G(s) = G_c(s)G_p(s) = \frac{800(s + 10^5)(s + 4)(s^2 + 106.4s + 10^6)}{s(s + 16.69)(s^2 + 95.3s + 1,198,606.6)(s^2 + 18,920s + 10^6)} \tag{11-78}$$

imperfect
pole-zero
cancellation

where no pole-zero cancellation takes place, and the PI controller in Eq. (11-73) has been added. The unit-step response of the system output is plotted in Fig. 11-34, along with the other responses with pole-zero cancellation. The effects due to the imperfect cancellation when the attenuation is selected to be −45 dB are negligible.

Next, let us assign the attenuation to be −40 dB, and ω_n is 1000 rad/s. Using Eqs. (11-64) and (11-65), we have $\zeta_z = 0.00505$ and $\zeta_p = 0.505$. The open-loop transfer function of the bridged-T PI-compensated system is now

$$G(s) = \frac{800(s + 10^5)(s + 4)(s^2 + 10.1s + 10^6)}{s(s + 16.69)(s^2 + 95.3s + 1,198,606.6)(s^2 + 1010s + 10^6)} \qquad (11\text{-}79)$$

The unit-step response of the system is shown in Fig. 11-34. Since the estimated attenuation of −40 dB is substantially different from the actual value of −48.47 dB required to cancel the complex poles of $G_p(s)$, the system response deviates noticeably from that with perfect cancellation or near cancellation. Actually, the resonance mode of uncancelled poles causes oscillations with the frequency of 1095 rad/s in the output response, but they are difficult to see from Fig. 11-34.

As further illustration of the effects of imperfect pole-zero cancellation, let us consider that the system with the bridged-T phase-lead controller is designed with an estimated attenuation of −45 dB. The transfer function of the open-loop system is

$$G(s) = \frac{82,237.76(s + 100,000)(s + 85)(s^2 + 106.4s + 10^6)}{(s + 16.69)(s + 350)(s^2 + 95.3s + 1,198,606.6)(s^2 + 18,920s + 10^6)} \qquad (11\text{-}80)$$

Figure 11-33 shows that not only the overshoot of the step response is higher than that of the system with perfect pole-zero cancellation, but the resonance mode of the uncancelled pole shows up as oscillations in the output response.

11-9

SUMMARY

This chapter discusses the design of linear control systems utilizing the frequency-domain techniques. The wealth of the graphical tools available in the frequency domain greatly facilitates the design of linear control systems. Since gain margin and phase margin are readily read off the Bode diagrams of open-loop transfer functions, it is very convenient to use these parameters as design criteria in the frequency domain. In addition, the effects of the controllers are easily implemented on the Bode diagram.

The design of the PD, PI, and PID controllers is carried out in the frequency domain using the Bode diagram. Design procedures are also outlined for the single-stage and multiple-stage phase-lead and the phase-lag controllers. Design with lead-lag or lag-lead and the bridged-T controllers is also covered. The computer programs FREQRP and FREQCAD are useful for the design of the various types of controllers discussed in this text.

It should be pointed out that, in general, a great variety of controllers can be applied to the compensation of control systems. One should not be limited to the types of controllers discussed in this chapter.

REVIEW QUESTIONS

1. How does the PD controller affect the bandwidth of a control system?
2. Once the value of K_D is fixed, increasing the value of K_P of the PD controller will increase the phase margin monotonically.   (T)   (F)
3. How does the PI controller affect the bandwidth of a control system?
4. How does the phase-lead controller affect the bandwidth of a control system?
5. The maximum phase that is available from a single-stage phase-lead controller is $90°$.   (T)   (F)
6. The design objective of the phase-lead controller is to place the maximum phase lead at the new gain-crossover frequency.   (T)   (F)
7. The design objective of the phase-lead controller is to place the maximum phase lead at the frequency where the magnitude of the uncompensated $G_p(j\omega)$ is $-10 \log_{10} a$, where a is the gain of the phase-lead controller.   (T)   (F)
8. A system compensated with a PD controller is usually more robust than the same system compensated with a PI controller.   (T)   (F)
9. How does the phase-lag controller affect the bandwidth of a control system?
10. The principle of design of the phase-lag controller is to utilize the zero-frequency attenuation property of the controller.   (T)   (F)
11. The corner frequencies of the phase-lag controller should not be too low or the bandwidth of the system will be too low.   (T)   (F)
12. The phase-lead controller may not be effective if the negative slope of the uncompensated process transfer function is too steep near the gain-crossover frequency.   (T)   (F)
13. How does the sensitivity function relate to the bandwidth of a system?
14. A system compensated with a phase-lead controller is usually less robust than the same system compensated with a phase-lag controller.   (T)   (F)

PROBLEMS

11-1. The control system described in Problem 9-2 has the open-loop transfer of $G(s) = G_c(s)G_p(s)$, where $G_c(s) = K_P + K_D s$, and

$$G_p(s) = \frac{1000}{s(s + 10)}$$

Set the value of K_P so that the ramp-error constant is 1000. Vary the value of K_D from 0.2 to 1.0 in increments of 0.2, and determine the values of phase margin, gain margin, M_p, and bandwidth of the closed-loop system. Find the value of K_D when the phase margin is maximum.

11-2. The liquid-level control system with a PD controller described in Problem 9-10 has the open-loop transfer function $G(s) = G_c(s)G_p(s)$, where $G_c(s) = K_P + K_D s$, and

$$G_p(s) = \frac{200}{s(s + 1)(s + 10)}$$

(a)   Set K_P so that the ramp-error constant is 1. Vary K_D from 0 to 0.5 and find the value of K_D that gives the maximum phase margin. Record the gain margin, M_p, and bandwidth.

(b)   Plot the sensitivity functions $|S_G^M(j\omega)|$ of the uncompensated system and the compensated system with the values of K_D and K_P determined in part (a). How does the PD control affect the sensitivity?

11-3.   The control system considered in Problem 9-3 has the open-loop transfer function $G(s) = G_c(s)G_p(s)$, where

$$G_c(s) = K_P + \frac{K_I}{s} \quad \text{and} \quad G_p(s) = \frac{100}{s^2 + 10s + 100}$$

(a)   Set K_I so that the ramp-error constant is 10. Find the value of K_P so that the phase margin is minimum. Record the values of the phase margin, gain margin, M_p, and BW. How does this optimal K_P relate to the root contours constructed in Problem 9-3(c)?

(b)   Plot the sensitivity function $|S_G^M(j\omega)|$ of the uncompensated system and the compensated system with the value of K_I and K_P selected in part (a). Comment on the effect of the PI control on sensitivity.

11-4.   One of the advantages of the frequency-domain analysis and design methods is that systems with pure time delays can be treated without approximations.

(a)   The automobile-engine-control system treated in Problem 9-7 has the open-loop transfer function $G(s) = G_c(s)G_p(s)$, where

$$G_p(s) = \frac{e^{-0.2s}}{1 + 0.2s}$$

and the controller transfer function is PI, $G_c(s) = K_P + K_I/s$. Set the value of K_I so that the ramp-error constant K_v is 2. Find the value of K_P so that the phase margin is a maximum. How does this "optimal" K_P compare with the value of K_P found in Problem 9-7(a)? Record the values of K_P beyond which the system is unstable. Compute and plot the unit-step response of the output by approximating $G_p(s)$ as

$$G_p(s) \cong \frac{125}{(s + 5)(s^2 + 5s + 25)}$$

(b)   In order to improve the system response, the PID controller $G_c(s) = K_P + K_D s + K_I/s$ is applied. For $K_v = 2$, and using the value of K_P arrived at in part (a), find the values of the phase margin, gain margin, M_p, and BW for various values of K_D. Find the best value of K_D so that the combined effects of phase margin, gain margin, M_p, and BW correspond to the best step response in the time domain. Compute and plot the unit-step response to verify your design. Use the second-order power-series approximation for the time delay given in part (a). Find the value of K_D beyond which the system is unstable.

11-5.   Consider the controlled process:

$$G_p(s) = \frac{100}{s^2 + 10s + 100}$$

The ramp-error constant is to be 100. Design a controller so that the maximum overshoot of the step response is less than 2 percent, and the rise time is less than 0.02 second. Carry out the design in the frequency domain.

11-6.   The transfer function of the process of a unity-feedback control system is

$$G_p(s) = \frac{6}{s(1 + 0.2s)(1 + 0.5s)}$$

(a)   Construct the Bode diagram of $G_p(j\omega)$ and determine the values of the phase margin, gain margin, M_p, and BW of the closed-loop system.

(b)   Design a single-stage series phase-lead controller with the transfer function

$$G_c(s) = \frac{1 + aTs}{1 + Ts}$$

so that the phase margin is greater than 30°. Determine the maximum phase margin that can be achieved by the single-stage phase-lead controller.

(c)   Using the system designed in part (b) as a basis, design a two-stage phase-lead controller so that the phase margin is 50°.

(d)   Compute and plot the unit-step responses of the output of the systems in parts (a), (b), and (c).

(e)   Plot the sensitivity functions $|S_G^M(j\omega)|$ of the uncompensated system and of the systems designed in parts (b) and (c) and compare.

11-7.   Figure 11P-7 shows the block diagram of a position-control system with a series controller $G_c(s)$.

(a)   Determine the minimum value of the amplifier gain K so that the steady-state value of the output $c(t)$ due to a unit-step torque disturbance is ≤ 0.01 (1 percent).

(b)   Show that the uncompensated system is unstable with the minimum value of K determined in part (a). Construct the Bode diagram for the open-loop transfer function $G(s) = C(s)/E(s)$, and find the values of the phase margin and gain margin.

(c)   Design a single-stage phase-lead controller with the transfer function $G_c(s) = (1 + aTs)/(1 + Ts)$ so that the phase margin is 30°. Find the phase margin, gain margin, M_p, and BW of the compensated system.

(d)   Design a two-stage phase-lead controller, using the system arrived at in part (c) as a basis, so that the phase margin is 55°. Find the phase margin, gain margin, M_p, and BW of the compensated system.

11-8.   The transfer function of the process of the inventory-control system described in Fig. 9P-9 is $G_p(s) = 4/s^2$. Design a single-stage phase-lead controller $G_c(s)$ so that the phase margin of the system is 80°. Show the Bode diagram of the design.

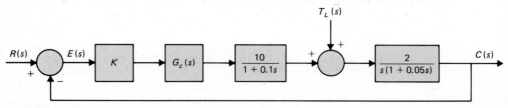

FIGURE 11P-7

11-9.   For the phase-lock-loop motor-speed control system described in Fig. 9P-18, the process transfer function is

$$G_p(s) = \frac{68.76}{s(1 + 0.05s)}$$

(a)   Plot the Bode diagram for $G_p(j\omega)$, and determine the phase margin, gain margin, M_p, and BW of the system when $G_c(s) = 1$.

(b)   Design a phase-lead controller, $G_c(s) = (1 + aTs)/(1 + Ts)$, so that the phase margin is 60°. Determine the gain margin, M_p, and BW of the compensated system.

11-10.   The computer-tape-drive control system described in Fig. 9P-26 has the open-loop transfer function

$$G(s) = \frac{1333.33K_p(s + K_I/K_p)(s + 300)}{s(s + 0.96443)(s^2 + 317.186s + 60,388.23)}$$

where a PI controller has been incorporated. The value of K_I is to be selected so that the ramp-error constant K_v is 100. Vary the value of K_p and calculate the values of phase margin, gain margin, M_p, and BW for each K_p. Find the value of K_p when the phase margin is a maximum. How does this value of K_p compare with the value selected in Problem 9-26? Discuss the results.

11-11.   The transfer function of the process of a unity-feedback control system is

$$G_p(s) = \frac{60}{s(1 + 0.2s)(1 + 0.5s)}$$

Due to the high gain, the uncompensated system is unstable.

(a)   Design a two-stage phase-lead controller with

$$G_c(s) = \frac{1 + aT_1s}{1 + T_1s}\frac{1 + bT_2s}{1 + T_2s} \qquad a > 1, b > 1 \qquad (1)$$

so that the phase margin of the compensated system is greater than 60°. Conduct the design by first determining the values of a and T_1 to realize a maximum phase margin that can be achieved with a single-stage phase-lead controller. The second stage of the controller is then designed to realize the balance of the 60° phase margin. Determine the gain margin, M_p, and BW of the compensated system. Compute and plot the unit-step response of the compensated system. Comment on the results.

(b)   Design a single-stage phase-lag controller with

$$G_c(s) = \frac{1 + aTs}{1 + Ts} \qquad a < 1$$

so that the phase margin of the compensated system is greater than 60°. Determine the gain margin, M_p, and BW of the compensated system. Compute and plot the unit-step response of the compensated system. Comment on the result.

(c)   Design a lag-lead controller with $G_c(s)$ as in Eq. (1) in part (a). Design the phase-lag portion first by setting the phase margin at 40°. The resulting system is then compensated by the phase-lead portion to achieve a total of 60° of phase margin. Determine the gain margin, M_p, and BW of the compensated system. Compute and plot the unit-step response of the compensated system.

11-12. The block diagram of the steel-rolling system described in Problem 4P-26 is shown in Fig. 11P-12. The transfer frunction of the process is

$$G_p(s) = \frac{5e^{-0.1s}}{s(1 + 0.1s)(1 + 0.5s)}$$

Design a series controller so that the phase margin of the compensated system is 60°. Determine the gain margin, M_p, and BW of the compensated system. Compute and plot the unit-step responses of the compensated and the uncompensated systems. Approximate the time delay by

$$e^{-0.1s} \cong \frac{1 - 0.5s}{1 + 0.05s}$$

11-13. Human beings breathe in order to provide for gas exchange for the entire body. A respiratory control system is needed to ensure that the body's needs for this gas exchange are adequately met. The criterion of control is adequate ventilation, which ensures satisfactory levels of both oxygen and carbon dioxide in the arterial blood. Respiration is controlled by neural impulses that originate within the lower brain and are transmitted to the chest cavity and diaphragm to govern the rate and tidal volume. One source of signals is the chemoreceptors located near the respiratory center, which are sensitive to carbon dioxide and oxygen concentrations. Figure 11P-13 shows the block diagram of a simplified model of the human respiratory control system. The objective is to control the effective ventilation of the lungs so that a satisfactory balance of concentrations of carbon dioxide and oxygen is maintained in the blood circulated at the chemoreceptor.

(a)   Plot the Bode diagram of the open-loop transfer function $G(s)$ when $G_c(s) = 1$. Find the phase margin and gain margin.

(b)   Design a PI controller, $G_s(s) = K_p + K_I/s$, so that the following specifications are satisfied:

   ramp-error constant $K_v = 1$
   phase margin is maximized

   Plot the Bode diagram to verify the design.

11-14. Repeat parts (a) and (b) of Problem 11-7. Design a single-stage phase-lag controller with the transfer function

$$G_c(s) = \frac{1 + aTs}{1 + Ts} \qquad a < 1$$

so that the phase margin is 60°. Plot the Bode diagram of the compensated system and determine the values of the gain margin, M_p, and BW.

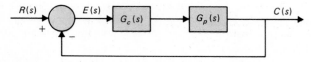

FIGURE 11P-12

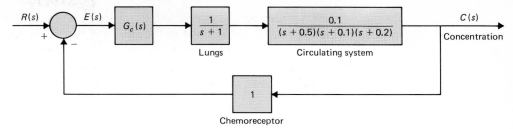

FIGURE 11P-13

11-15. Figure 11P-15 shows the block diagram of a motor-control system that has a flexible shaft between the motor and the load, such as that shown in Fig. 4P-10. The transfer function between the motor torque and the motor displacement is

$$\frac{\Theta_m(s)}{T_m(s)} = \frac{J_L s^2 + B_L s + K_L}{s[J_m J_L s^3 + (B_m J_L + B_L J_m)s^2 + (K_L J_m + K_L J_L + B_m B_L)s + B_m K_L]}$$

where $J_L = 0.01$, $B_L = 0.1$, $K_L = 10$, $J_m = 0.01$, $B_m = 0.1$, and $K = 100$.

(a) Plot the Bode diagram of the uncompensated system, $G_c(s) = 1$. Find the phase margin, gain margin, M_p, and BW of the system. Compute and plot the unit-step response of $\theta_m(t)$.

(b) Design a bridge-T controller to cancel the complex poles of $\Theta_m(s)/T_m(s)$. Plot the Bode diagram of $G(s) = \Theta_m(s)/E(s)$. Find the phase margin, gain margin, M_p, and BW of the compensated system. Compute and plot the unit-step response of $\theta_m(t)$.

11-16. The transfer function of the process of a unity-feedback control system is

$$G_p(s) = \frac{500(s + 10)}{s(s^2 + 10s + 1000)}$$

(a) Construct the Bode diagram of $G_p(j\omega)$ and determine the phase margin, gain margin, M_p, and BW of the closed-loop system. Compute and plot the unit-step response of the output.

(b) Design a series bridged-T controller with the transfer function

$$G_p(s) = \frac{s^2 + 2\zeta_z \omega_n s + \omega_n^2}{s^2 + 2\zeta_p \omega_n s + \omega_n^2}$$

so that its zeros cancel the complex poles of $G_p(s)$. Plot the Bode diagram of the open-loop transfer function $G(s) = G_c(s)G_p(s)$, and determine the phase margin, gain margin, M_p, and BW of the compensated system. Compute and plot the unit-step response of the output.

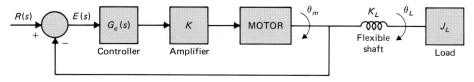

FIGURE 11P-15

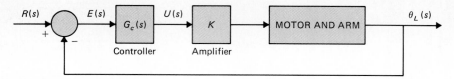

FIGURE 11P-17

11-17. The block diagram shown in Fig. 11P-17 represents a robot-arm control system such as the one described in Problem 4-21. The process transfer function is

$$G_p(s) = \frac{\Theta_L(s)}{U(s)} = \frac{1444.33(s + 15)}{s(s^3 + 2.6s^2 + 667.12s + 216.67)}$$

(a) Plot the Bode diagram of $G(s) = \Theta_L(s)/E(s)$ when $G_c(s) = 1$. Find the phase margin, gain margin, M_p, and BW of the uncompensated system. Compute and plot the unit-step response of the system.

(b) Design a bridged-T controller so that its zeros cancel the two complex poles of $G_p(s)$. The compensated system should have a phase margin of at least 70°. If the bridged-T controller alone cannot meet the phase-margin requirement, add one or more stages of the phase-lead controller. Plot the Bode diagram of the compensated system, and determine the phase margin, gain margin, M_p, and BW of the compensated system. Compute and plot the unit-step response of $\theta_L(t)$.

(c) Pretend that the transfer function $G_p(s)$ is not known, but the Bode diagram of $G_p(j\omega)$ is obtained experimentally. The resonant frequency of the second-order pole of $G_p(s)$ is observed to be at 25.8 rad/s. Design the bridged-T controller based on an estimate of the attenuation needed. Do not calculate the attenuation based on the values of ζ_z and ζ_p obtained in part (b). Repeat the phase-lead-controller designed as in part (b) for a phase margin of 70° if necessary. Determine the gain margin, M_p, and BW of the compensated system. Compute and plot the unit-step response of $\theta_L(t)$. Compare the step response with that obtained in part (b).

FREQUENCY-DOMAIN PLOTS

Let $G(s)$ be the open-loop transfer function of a linear feedback control system. The frequency-domain analysis of the closed-loop system can be conducted from the frequency-domain plots of $G(s)$ with s replaced by $j\omega$.

The function $G(j\omega)$ is generally a complex function of the frequency ω, and can be written as

$$G(j\omega) = |G(j\omega)|\angle G(j\omega) \tag{A-1}$$

where $|G(j\omega)|$ denotes the magnitude of $G(j\omega)$, and $\angle G(j\omega)$ is the phase of $G(j\omega)$.

The following forms of frequency domain plots of $G(j\omega)$, or of $G(j\omega)H(j\omega)$ in case $H(s)$ is nonunity, versus ω are often used in the analysis and design of linear control systems in the frequency domain.

1. *Polar plot.* A plot of the magnitude versus phase in the polar coordinates as ω is varied from zero to infinity.
2. *Bode plot.* A plot of the magnitude in decibels versus ω (or $\log_{10} \omega$) in semilog (or rectangular) coordinates.
3. *Magnitude-versus-phase-plot.* A plot of the magnitude in decibels versus the phase on rectangular coordinates with ω as a variable parameter on the curve.

Computer-Aided Construction of the Frequency-Domain Plots

computer solutions

FREQRP

The data for the plotting of the frequency-domain plots are usually quite time consuming to generate if the computation is to be carried out manually, especially if the function is of high order. In practice, a digital computer should be used to do the computation as well as the plotting of the final graph. The program FREQRP of the ACSP software package or any other program available commercially can be used for the construction of all three plots listed.

From an analytical standpoint, the engineer should still be familiar with the properties of the frequency-domain plots, so that proper interpretations can be made on these plots made by the computer.

A-1
POLAR PLOTS

The polar plot of a function of the complex variable s, $G(s)$, is a plot of the magnitude of $G(j\omega)$ versus the phase of $G(j\omega)$ on polar coordinates as ω is varied from zero to infinity. From the mathematical viewpoint, the process can be regarded as the mapping of the positive half of the imaginary axis of the s-plane onto the $G(j\omega)$ plane. A simple example of this mapping is shown in Fig. A-1. For any frequency $\omega = \omega_1$, the magnitude and phase of $G(j\omega_1)$ are represented by a phasor that has the corresponding magnitude and phase angle in the $G(j\omega)$-plane. In measuring the phase, counterclockwise is referred to as positive, and clockwise is negative.

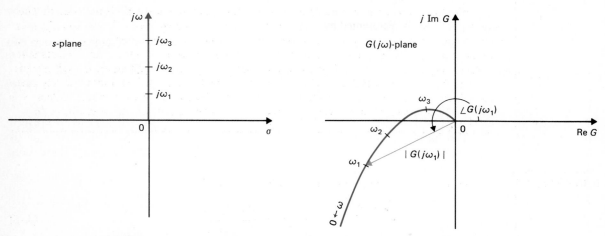

FIGURE A-1 Polar plot shown as a mapping of the positive half of the $j\omega$-axis in the s-plane onto the $G(j\omega)$-plane.

To illustrate the construction of the polar plot of a function $G(s)$, consider the function

$$G(s) = \frac{1}{1 + Ts} \qquad \text{(A-2)}$$

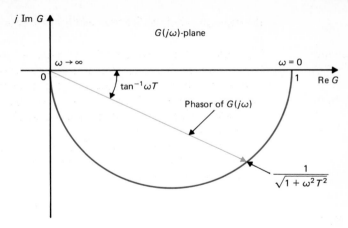

FIGURE A-2 Polar plot of
$G(j\omega) = 1/(1 + j\omega T)$.

where T is a positive constant. Setting $s = j\omega$, we have

$$G(j\omega) = \frac{1}{1 + j\omega T} \tag{A-3}$$

In terms of magnitude and phase, Eq. (A-3) is written

$$G(j\omega) = \frac{1}{\sqrt{1 + \omega^2 T^2}} \angle -\tan^{-1} \omega T \tag{A-4}$$

When ω is zero, the magnitude of $G(j\omega)$ is unity, and the phase of $G(j\omega)$ is at 0°. Thus, at $\omega = 0$, $G(j\omega)$ is represented by a phasor of unit length directed in the 0° direction. As ω increases, the magnitude of $G(j\omega)$ decreases, and the phase becomes more negative. As ω increases, the length of the phasor in the polar coordinates decreases, and the phasor rotates in the clockwise (negative) direction. When ω approaches infinity, the magnitude of $G(j\omega)$ becomes zero, and the phase reaches −90°. This is represented by a phasor with an infinitesimally small length directed along the −90° axis in the $G(j\omega)$-plane. By substituting other finite values of ω into Eq. (A-4), the exact plot of $G(j\omega)$ turns out to be a semicircle, as shown in Fig. A-2.

EXAMPLE A-2   As a second illustrative example, consider the function

$$G(j\omega) = \frac{1 + j\omega T_2}{1 + j\omega T_1} \tag{A-5}$$

where T_1 and T_2 are positive real constants. Equation (A-5) is written

$$G(j\omega) = \sqrt{\frac{1 + \omega^2 T_2^2}{1 + \omega^2 T_1^2}} \angle (\tan^{-1} \omega T_2 - \tan^{-1} \omega T_1) \tag{A-6}$$

The polar plot of $G(j\omega)$, in this case, depends on the relative magnitudes of T_1 and T_2. If T_2 is greater than T_1, the magnitude of $G(j\omega)$ is always greater than unity as ω is varied from zero to

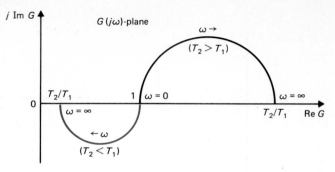

FIGURE A-3 Polar plots of
$G(j\omega) = (1 + j\omega T_2)/(1 + j\omega T_1)$.

infinity, and the phase of $G(j\omega)$ is always positive. If T_2 is less than T_1, the magnitude of $G(j\omega)$ is always less than unity, and the phase is always negative. The polar plots of $G(j\omega)$ of Eq. (A-6) that correspond to the two above-mentioned conditions are shown in Fig. A-3.

In many control-systems applications, such as the Nyquist stability criterion, an exact plot of the frequency-response plot is not essential. Often, a rough sketch of the polar plot of the transfer function $G(j\omega)H(j\omega)$ is quite adequate for stability analysis in the frequency domain. The general shape of the polar plot of a function $G(j\omega)$ can be determined from the following information.

1. The behavior of the magnitude and phase of $G(j\omega)$ at $\omega = 0$ and $\omega = \infty$.
2. The points of intersections of the polar plot with the real and imaginary axes, and the values of ω at these intersections.

EXAMPLE A-3

Consider that it is desired to make a rough sketch of the polar plot of the transfer function

$$G(s) = \frac{10}{s(s + 1)} \tag{A-7}$$

By susbstituting $s = j\omega$ in Eq. (A-7), the magnitude and phase of $G(j\omega)$ at $\omega = 0$ and $\omega = \infty$ are computed as follows:

$$\lim_{\omega \to 0} |G(j\omega)| = \lim_{\omega \to 0} \frac{10}{\omega} = \infty \tag{A-8}$$

$$\lim_{\omega \to 0} \angle G(j\omega) = \lim_{\omega \to 0} \angle 10/j\omega = -90° \tag{A-9}$$

$$\lim_{\omega \to \infty} |G(j\omega)| = \lim_{\omega \to \infty} \frac{10}{\omega^2} = 0 \tag{A-10}$$

$$\lim_{\omega \to \infty} \angle G(j\omega) = \lim_{\omega \to \infty} \angle 10/(j\omega)^2 = -180° \tag{A-11}$$

Thus, the properties of the polar plot of $G(j\omega)$ at $\omega = 0$ and $\omega = \infty$ are ascertained. Next, we determine the intersections, if any, of the polar plot with the two axes of the $G(j\omega)$-plane. If

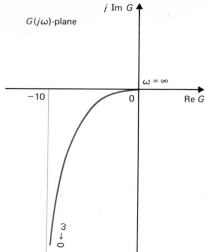

FIGURE A-4 Polar plot of
$G(s) = 10/[s(s + 1)]$

the polar plot of $G(j\omega)$ intersects the real axis, at the point of intersection, the imaginary part of $G(j\omega)$ is zero; that is,

$$\text{Im } [G(j\omega)] = 0 \qquad\qquad (A\text{-}12)$$

To express $G(j\omega)$ as the sum of its real and imaginary parts, we must rationalize $G(j\omega)$ by multiplying its numerator and denominator by the complex conjugate of its denominator. Therefore, $G(j\omega)$ is written

$$G(j\omega) = \frac{10(-j\omega)(-j\omega + 1)}{j\omega(j\omega + 1)(-j\omega)(-j\omega + 1)} = \frac{-10\omega^2}{\omega^4 + \omega^2} - j\frac{10\omega}{\omega^4 + \omega^2}$$

$$= \text{Re } [G(j\omega)] + j \text{ Im } [G(j\omega)] \qquad\qquad (A\text{-}13)$$

When we set $\text{Im } [G(j\omega)]$ to zero, we get $\omega = \infty$, meaning that the $G(j\omega)$ plot intersects only with the real axis of the $G(j\omega)$-plane at the origin.

Similarly, the intersection of $G(j\omega)$ with the imaginary axis is found by setting $\text{Re } [G(j\omega)]$ of Eq. (A-13) to zero. The only real solution for ω is also $\omega = \infty$, which corresponds to the origin of the $G(j\omega)$-plane. The conclusion is that the polar plot of $G(j\omega)$ does not intersect any one of the two axes at any finite nonzero frequency. Under certain circumstances, we are interested in the properties of the $G(j\omega)$ plot at infinity, which corresponds to $\omega = 0$ in this case. From Eq. (A-13), we see that $\text{Im } [G(j\omega)] = \infty$ and $\text{Re } [G(j\omega)] = -10$ at $\omega = 0$. Based on this information, as well as knowledge on the angles of $G(j\omega)$ at $\omega = 0$ and $\omega = \infty$, the polar plot of $G(j\omega)$ is easily sketched, as shown in Fig. A-4.

EXAMPLE A-4   Given the transfer function

$$G(s) = \frac{10}{s(s + 1)(s + 2)} \qquad\qquad (A\text{-}14)$$

it is desired to make a rough sketch of the polar plot of $G(j\omega)$. The following calculations are made for the properties of the magnitude and phase of $G(j\omega)$ at $\omega = 0$ and $\omega = \infty$:

$$\lim_{\omega \to 0} |G(j\omega)| = \lim_{\omega \to 0} \frac{5}{\omega} = \infty \tag{A-15}$$

$$\lim_{\omega \to 0} \angle G(j\omega) = \lim_{\omega \to 0} \angle 5/j\omega = -90° \tag{A-16}$$

$$\lim_{\omega \to \infty} |G(j\omega)| = \lim_{\omega \to \infty} \frac{10}{\omega^3} = 0 \tag{A-17}$$

$$\lim_{\omega \to \infty} \angle G(j\omega) = \lim_{\omega \to \infty} \angle 10/(j\omega)^3 = -270° \tag{A-18}$$

To find the intersections of the $G(j\omega)$ plot on the real and imaginary axes of the $G(j\omega)$-plane, we rationalize $G(j\omega)$ to give

$$G(j\omega) = \frac{10(-j\omega)(-j\omega + 1)(-j\omega + 2)}{j\omega(j\omega + 1)(j\omega + 2)(-j\omega)(-j\omega + 1)(-j\omega + 2)} \tag{A-19}$$

After simplification, the last equation is written

$$G(j\omega) = \text{Re}\,[G(j\omega)] + j\,\text{Im}\,[G(j\omega)] = \frac{-30}{9\omega^2 + (2 - \omega^2)^2} - \frac{j10(2 - \omega^2)}{9\omega^3 + \omega(2 - \omega^2)^2} \tag{A-20}$$

Setting Re $[G(j\omega)]$ to zero, we have $\omega = \infty$, and $G(j\infty) = 0$, which means that the $G(j\omega)$ plot intersects the imaginary axis only at the origin. Setting Im $[G(j\omega)]$ to zero, we have $\omega = \pm\sqrt{2}$ rad/s. This gives the point of intersection on the real axis at

$$G(\pm j\sqrt{2}) = -5/3 \tag{A-21}$$

The result of $\omega = -\sqrt{2}$ rad/s has no physical meaning, as the frequency is negative; it simply represents a mapping point on the negative $j\omega$-axis of the s-plane. In general, if $G(s)$ is a rational function of s (a quotient of two polynomials of s), the polar plot of $G(j\omega)$ for negative

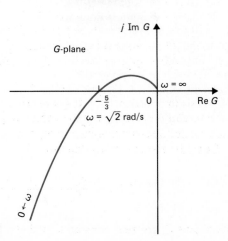

FIGURE A-5 Polar plot of
$G(s) = 10/[s(s + 1)(s + 2)]$

values of ω is the mirror image of that for positive ω, **with the mirror placed on the real axis of the $G(j\omega)$-plane. From Eq. (A-20), we also see that Re $[G(j0)] = \infty$ and Im $[G(j0)] = \infty$. With this information, it is now possible to make a sketch of the polar plot for the transfer function in Eq. (A-14), as shown in Fig. A-5.**

Although the method of obtaining the rough sketch of the polar plot of a transfer function as described is quite straightforward, in general, for complex transfer functions that may have multiple crossings on the real and imaginary axes of the transfer-function plane, the algebraic manipulation may again be quite involved. Furthermore, the polar plot is basically a tool for analysis; it is somewhat awkward for design purposes. We shall show in the next section that approximate information on the polar plot can always be obtained from the Bode plot, which can be sketched without any calculations. Thus, for more complex transfer functions, other than using the digital computer, sketches of the polar plots are preferably obtained with the help of the Bode plots.

A-2

BODE PLOT (CORNER PLOT OR ASYMPTOTIC PLOT)

The Bode plot of the function $G(j\omega)$ is composed of two plots, one with the magnitude of $G(j\omega)$ plotted in decibels (dB) versus $\log_{10} \omega$ or ω, and the other with the phase of $G(j\omega)$ in degrees as a function of $\log_{10} \omega$ or ω. The Bode plot is also known as the **corner plot** or the **asymptotic plot** of $G(j\omega)$. These names stem from the fact that the Bode plot can be constructed by using straight-line approximations that are asymptotic to the actual plot.

In simple terms, the Bode plot has the following features:

1. Since the magnitude of $G(j\omega)$ in the Bode plot is expressed in dB, product and division factors in $G(j\omega)$ become additions and subtractions, respectively. The phase relations are also added and subtracted from each other algebraically.
2. The magnitude plot of the Bode plots of $G(j\omega)$ can be approximated by straight-line segments, which allow the simple sketching of the Bode plot without detailed computation.

Since the straight-line approximation of the Bode plot is relatively easy to construct, the data necessary for the other frequency-domain plots, such as the polar plot and the magnitude-versus-phase plot, can be easily generated from the Bode plot.

Consider the following function:

$$G(s) = \frac{K(s + z_1)(s + z_2) \cdots (s + z_m)}{s^j(s + p_1)(s + p_2) \cdots (s + p_n)} e^{-T_d s} \qquad \text{(A-22)}$$

where K and T_d are real constants, and the z's and the p's may be real or complex (in conjugate pairs) numbers. In Chapter 8, Eq. (A-22) is the preferred form for root-locus construction. For the Bode plot, $G(s)$ must be first written in the following form.

$$G(s) = \frac{K_1(1 + T_1s)(1 + T_2s) \cdots (1 + T_ms)}{s^j(1 + T_as)(1 + T_bs) \cdots (1 + T_ns)} e^{-T_ds} \qquad \text{(A-23)}$$

where K_1 is a real constant, the T's may be real or complex (in conjugate pairs) numbers, and T_d is the real time delay. Since practically all the terms in Eq. (A-23) are of the same form, without loss of generality, we can use the following transfer function to illustrate the contruction of the Bode diagram.

$$G(s) = \frac{K(1 + T_1s)(1 + T_2s)}{s(1 + T_as)(1 + 2\zeta s/\omega_n + s^2/\omega_n^2)} e^{-T_ds} \qquad \text{(A-24)}$$

where $K, T_d, T_1, T_2, T_a, \zeta$, and ω_n are real constants. It is assumed that the second-order polynomial in the denominator has complex-conjugate zeros.

The magnitude of $G(j\omega)$ in dB is obtained by multiplying the logarithm to the base 10 of $|G(j\omega)|$ by 20; we have

$$|G(j\omega)|_{\text{dB}} = 20 \log_{10} |G(j\omega)| = 20 \log_{10} |K| + 20 \log_{10} |1 + j\omega T_1|$$
$$+ 20 \log_{10} |1 + j\omega T_2| - 20 \log_{10} |j\omega| \qquad \text{(A-25)}$$
$$- 20 \log_{10} |1 + j\omega T_a| - 20 \log_{10} |1 + j2\zeta\omega - \omega^2/\omega_n^2|$$

The phase of $G(j\omega)$ is written

$$\angle G(j\omega) = \angle K + \angle(1 + j\omega T_1) + \angle(1 + j\omega T_2) - \angle j\omega$$
$$- \angle(1 + j\omega T_a) - \angle(1 + 2\zeta\omega/\omega_n - \omega^2/\omega_n^2) - \omega T_d \text{ rad} \qquad \text{(A-26)}$$

In general, the function $G(j\omega)$ may be of higher order than that of Eq. (A-24) and have many more factored terms. However, Eqs. (A-25) and (A-26) indicate that additional terms in $G(j\omega)$ would simply produce more similar terms in the magnitude and phase expressions, so that the basic method of construction of the Bode plot would be the same. We have also indicated that, in general, $G(j\omega)$ can contain just five simple types of factors:

types of factors

1. Constant factor: K
2. Poles or zeros at the origin of order p: $(j\omega)^{\pm p}$
3. Poles or zeros at $s = -1/T$ of order q: $(1 + j\omega T)^{\pm q}$
4. Complex poles or zeros of order r: $(1 + j2\zeta\omega/\omega_n - \omega^2/\omega_n^2)^{\pm r}$
5. Pure time delay $e^{-j\omega T_d}$ where T_d, p, q, and r are positive integers.

Equations (A-25) and (A-26) verify one of the unique characteristics of the Bode plot in that each of the five types of factors listed can be considered as a separate plot; the individual plots are then added or subtracted accordingly to yield the total magnitude in dB and the phase plot of $G(j\omega)$. The curves can be plotted on semilog graph paper or linear rectangular-coordinate graph paper, depending on whether ω or $\log_{10} \omega$ is used as the abscissa.

We shall now investigate sketching the Bode plot of different types of factors.

Real Constant *K*

Since

$$K_{dB} = 20 \log_{10} K = \text{constant} \qquad (A\text{-}27)$$

and

$$\angle K = \begin{cases} 0° & K > 0 \\ 180° & K < 0 \end{cases} \qquad (A\text{-}28)$$

the Bode plot of the real constant *K* is shown in Fig. A-6 in semilog coordinates.

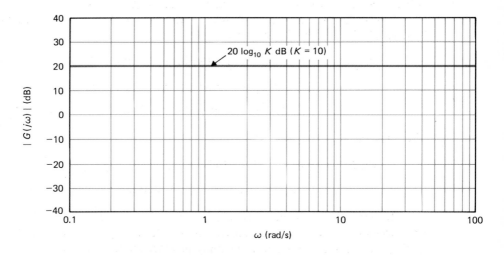

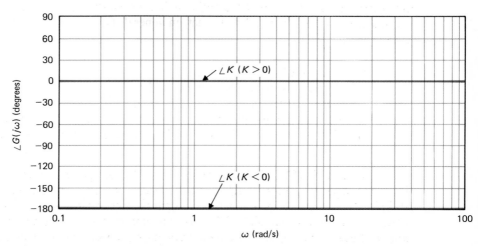

FIGURE A-6 Bode plot of constant *K*.

Poles and Zeros at the Origin, $(j\omega)^{\pm p}$

The magnitude of $(j\omega)^{\pm p}$ in dB is given by

$$20 \log_{10} |(j\omega)^{\pm p}| = \pm 20p \log_{10} \omega \;\; \text{dB} \qquad \text{(A-29)}$$

for $\omega \geq 0$. The last expression for a given p represents a straight line in either semilog or rectangular coordinates. The slopes of these lines are determined by taking the derivative of Eq. (A-29) with respect to $\log_{10} \omega$; that is,

$$\frac{d}{d \log_{10} \omega}(\pm 20p \log_{10} \omega) = \pm 20p \;\; \text{dB/decade} \qquad \text{(A-30)}$$

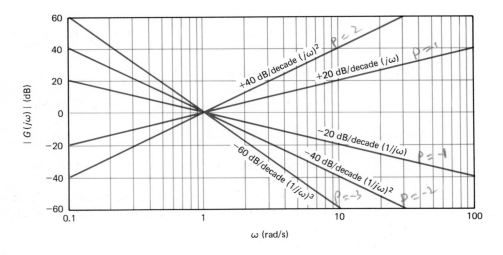

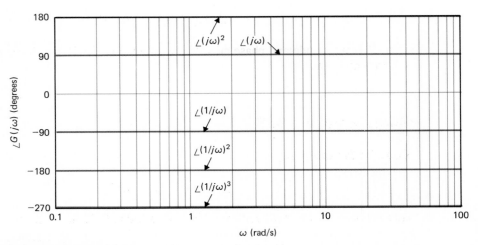

FIGURE A-7 Bode plots of $(j\omega)^{\pm p}$.

These lines pass through the 0-dB axis at $\omega = 1$. Thus, a unit change in $\log_{10} \omega$ corresponds to a change of $\pm 20p$ dB in magnitude. Furthermore, a unit change in $\log_{10} \omega$ in the rectangular coordinates is equivalent to one **decade** of variation in ω, that is, from 1 to 10, 10 to 100, and so on, in the semilog coordinates. Thus, the slopes of the straight lines described by Eq. (A-29) are said to be $\pm 20p$ dB/decade of frequency.

decade

Instead of using decades, sometimes **octaves** are used to represent the separation of two frequencies. The frequencies ω_1 and ω_2 are separated by an octave if $\omega_2/\omega_1 = 2$. The number of decades between any two frequencies ω_1 and ω_2 is given by

octaves

$$\text{number of decades} = \frac{\log_{10}(\omega_2/\omega_1)}{\log_{10} 10} = \log_{10}\left(\frac{\omega_2}{\omega_1}\right) \qquad \text{(A-31)}$$

Similarly, the number of octaves between ω_2 and ω_1 is

$$\text{number of octaves} = \frac{\log_{10}(\omega_2/\omega_1)}{\log_{10} 2} = \frac{1}{0.301}\log_{10}\left(\frac{\omega_2}{\omega_1}\right) \qquad \text{(A-32)}$$

Thus, the relation between octaves and decades is

$$\text{number of octaves} = 1/0.301 \text{ decades} = 3.32 \text{ decades} \qquad \text{(A-33)}$$

Substituting Eq. (A-33) into Eq. (A-30), we have

$$\pm 20p \text{ dB/decade} = \pm 20p \times 0.301 \cong 6p \text{ dB/octave} \qquad \text{(A-34)}$$

For the function $G(s) = 1/s$, which has a simple pole at $s = 0$, the magnitude of $G(j\omega)$ is a straight line with a slope of -20 dB/decade, and passes through the 0-dB axis at $\omega = 1$ rad/s.

The phase of $(j\omega)^{\pm p}$ is written

$$\angle(j\omega)^{\pm p} = \pm p \times 90° \qquad \text{(A-35)}$$

The magnitude and phase curves of the function $(j\omega)^{\pm p}$ are shown in Fig. A-7 for several values of P.

Simple Zero, $1 + j\omega T$

Consider the function

$$G(j\omega) = 1 + j\omega T \qquad \text{(A-36)}$$

where T is a positive real constant. The magnitude of $G(j\omega)$ in dB is

$$|G(j\omega)|_{\text{dB}} = 20\log_{10}|G(j\omega)| = 20\log_{10}\sqrt{1 + \omega^2 T^2} \qquad \text{(A-37)}$$

To obtain asymptotic approximations of $|G(j\omega)|_{dB}$, we consider both very large and very small values of ω. At very low frequencies, $\omega T \ll 1$, Eq. (A-37) is approximated by

$$|G(j\omega)|_{dB} \cong 20 \log_{10} 1 = 0 \text{ dB} \qquad (A\text{-}38)$$

since $\omega^2 T^2$ is neglected when compared with 1.

At very high frequencies, $\omega T \gg 1$, we can approximate $1 = \omega^2 T^2$ by $\omega^2 T^2$; then Eq. (A-37) becomes

$$|G(j\omega)|_{dB} \cong 20 \log_{10} \sqrt{\omega^2 T^2} = 20 \log_{10} \omega T \qquad (A\text{-}39)$$

Equation (A-38) represents a straight line with a slope of $+20$ dB/decade of frequency. The intersect of these two lines is found by equating Eq. (A-38) to Eq. (A-39), which gives

$$\omega = 1/T \qquad (A\text{-}40)$$

This frequency is also the intersect of the high-frequency approximate plot and the low-frequency approximate plot, which is the 0-dB axis. The frequency given in Eq. (A-40) is also known as the **corner frequency** of the Bode plot of Eq. (A-36), since the asymptotic plot forms the shape of a corner at this frequency, as shown in Fig. A-8. The actual $|G(j\omega)|_{dB}$ plot of Eq. (A-36) is a smooth curve, and deviates only slightly from the straight-line approximation. The actual values and the straight-line approximation of $|1 + j\omega T|_{dB}$ as functions of ωT are tabulated in Table A-1. The error between the actual magnitude curve and the straight-line asymptotes is symmetrical with respect to the corner frequency $\omega = 1/T$. It is useful to remember that the error is 3 dB at the corner frequency, and 1 dB at 1 octave above ($\omega = 2/T$) and 1 octave below ($\omega = 1/2T$) the corner frequency. At 1 decade above and below the corner frequency, the error is dropped to approximately 0.3 dB. Based on these facts, the procedure of making a sketch of $|1 + j\omega T|_{dB}$ is as follows:

corner frequency

1. Locate the corner frequency $\omega = 1/T$ on the frequency axis.
2. Draw the 20-dB/decade (or 6-dB/octave) line and the horizontal line at 0 dB, with the two lines intersecting at $\omega = 1/T$.
3. If necessary, the actual magnitude curve is obtained by adding the errors to the asymptotic plot at the strategic frequencies. Usually, a smooth curve can be sketched simply by locating the 3-dB point at the corner frequency and the 1-dB points at 1 octave above and below the corner frequency.

The phase of $G(j\omega) = 1 + j\omega T$ is

$$\angle G(j\omega) = \tan^{-1} \omega T \qquad (A\text{-}41)$$

Similar to the magnitude curve, a straight-line approximation can be made for the phase curve. Since the phase of $G(j\omega)$ varies from $0°$ to $90°$, we can draw a line from $0°$ at 1 decade below the corner frequency to $+90°$ at 1 decade above the corner frequency. As

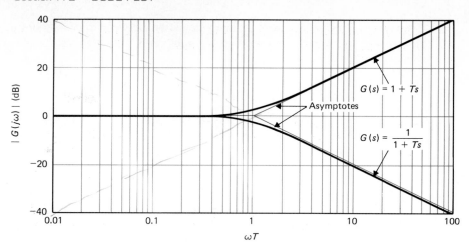

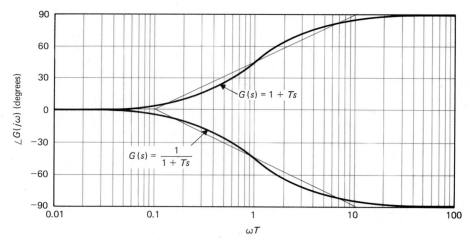

FIGURE A-8 Bode plots of $G(s) = 1 + Ts$ and $G(s) = 1/(1 + Ts)$.

TABLE A-1

ωT	$\log_{10} \omega T$	$|1 + j\omega T|$	$|1 + j\omega T|_{dB}$	STRAIGHT-LINE APPROXIMATION $|1 + j\omega T|_{dB}$	ERROR (dB)	$\angle(1 + j\omega T)$ (DEGREES)
0.01	-2	1	0.00043	0	0.00043	0.5
0.1	-1	1.04	0.043	0	0.043	5.7
0.5	-0.3	1.12	1	0	1	26.6
0.76	-0.12	1.26	2	0	2	37.4
1.0	0	1.41	3	0	3	45.0
1.31	0.117	1.65	4.3	2.3	2	52.7
2.0	0.3	2.23	7.0	6.0	1	63.4
10.0	1.0	10.4	20.043	20.0	0.043	84.3
100.0	2.0	100.005	40.00043	40.0	0.00043	89.4

shown in Fig. A-8, the maximum deviation between the straight-line approximation and the actual curve is less than 6°. Table A-1 gives the values of $\angle(1 + j\omega T)$ versus ωT.

Simple Pole, $1/(1 + j\omega T)$

For the function

$$G(j\omega) = \frac{1}{1 + j\omega T} \tag{A-42}$$

the magnitude, $|G(j\omega)|$ in dB, is given by the negative of the right side of Eq. (A-37), and the phase $\angle G(j\omega T)$ is the negative of the angle in Eq. (A-41). Therefore, it is simple to extend all the analysis for the case of the simple zero to the Bode plot of Eq. (A-42). The asymptotic approximations of $|G(j\omega)|_{dB}$ at low and high frequencies are

$$\omega T \ll 1 \qquad |G(j\omega)|_{dB} \cong 0 \text{ dB} \tag{A-43}$$

$$\omega T \gg 1 \qquad |G(j\omega)|_{dB} \cong -20 \log_{10} \omega T \tag{A-44}$$

Thus, the corner frequency of the Bode plot of Eq. (A-42) is still at $\omega = 1/T$, except that at high frequencies, the slope of the straight-line approximation is -20 dB/decade. The phase of $G(j\omega)$ is 0 degrees at $\omega = 0$, and $-90°$ when $\omega = \infty$. The magnitude in dB and phase of the Bode plot of Eq. (A-42) are shown in Fig. A-8. The data in Table A-1 are still useful for the simple-pole case if appropriate sign changes are made to the numbers. For instance, the numbers in the $|1 + j\omega T|_{dB}$, the straight-line approximation of $|1 + j\omega T|_{dB}$, the error (dB), and the $\angle(1 + j\omega T)$ columns should all be negative. At the corner frequency, the error between the straight-line approximation and the actual magnitude curve is -3 dB.

Quadratic Poles and Zeros

Now consider the second-order transfer function

$$G(s) = \frac{\omega_n^2}{s^2 + 2\zeta\omega_n s + \omega_n^2} = \frac{1}{1 + (2\zeta/\omega_n)s + (1/\omega_n^2)s^2} \tag{A-45}$$

We are interested only in the case when $\zeta \leq 1$, since otherwise $G(s)$ would have two unequal real poles, and the Bode plot can be obtained by considering $G(s)$ as the product of two transfer functions with simple poles.

By letting $s = j\omega$, Eq. (A-45) becomes

$$G(j\omega) = \frac{1}{[1 - (\omega/\omega_n)^2] + j2\zeta(\omega/\omega_n)} \tag{A-46}$$

The magnitude of $G(j\omega)$ in dB is

$$20 \log_{10} |G(j\omega)| = -20 \log_{10} \sqrt{[1 - (\omega/\omega_n)^2]^2 + 4\zeta^2(\omega/\omega_n)^2} \tag{A-47}$$

At very low frequencies, $\omega/\omega_n \ll 1$; Eq. (A-47) can be approximated as

$$|G(j\omega)|_{dB} = 20 \log_{10} |G(j\omega)| \cong -20 \log_{10} 1 = 0 \text{ dB} \qquad \text{(A-48)}$$

Thus, the low-frequency asymptote of the magnitude plot of Eq. (A-45) is a straight line that lies on the 0-dB axis. At very high frequencies, $\omega/\omega_n \gg 1$; the magnitude in dB of $G(j\omega)$ in Eq. (A-45) becomes

$$|G(j\omega)|_{dB} \cong -20 \log_{10} \sqrt{(\omega/\omega_n)^4} = -40 \log_{10} (\omega/\omega_n) \text{ dB} \qquad \text{(A-49)}$$

This equation represents a straight line with a slope of -40 dB/decade in the Bode-plot coordinates. The intersection of the two asymptotes is found by equating Eq. (A-48), to Eq. (A-49), yielding the corner frequency at $\omega = \omega_n$. The actual magnitude curve of $G(j\omega)$ in this case may differ strikingly from the asymptotic curve. The reason for this is that the amplitude and phase curves of the second-order $G(j\omega)$ depend not only on the corner frequency ω_n, but also on the damping ratio ζ, which does not enter the asymptotic curve. The actual and the asymptotic curves of $|G(j\omega)|_{dB}$ are shown in Fig. A-9 for several values of ζ. The errors between the two sets of curves are shown

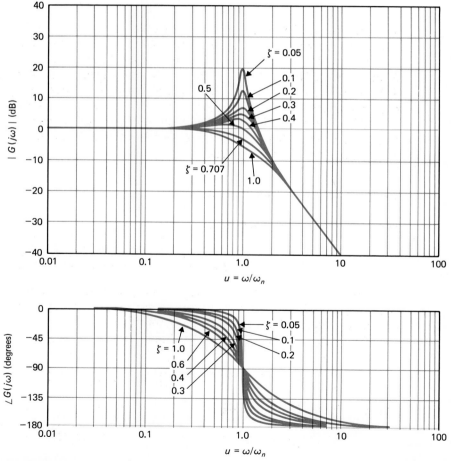

FIGURE A-9 Bode plots of $G(s) = 1/[1 + 2\zeta(s/\omega_n) + (s/\omega_n)^2]$.

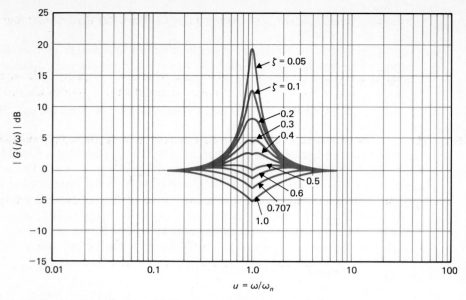

FIGURE A-10 Errors in magnitude curves of Bode plots of
$G(s) = 1/[1 + 2\zeta(s/\omega_n) + (s/\omega_n)^2]$.

in Fig. A-10 for the same set of values of ζ. The standard procedure of constructing the second-order $|G(j\omega)|_{dB}$ is to first locate the corner frequency ω_n, and then sketch the asymptotic lines: 0-dB line to the left of ω_n, and -40-dB/decade line to the right of ω_n. The actual curve is obtained by making corrections to the asymptotes by using either the data from the error curves of Fig. A-10 or the curves in Fig. A-9 for the corresponding ζ.

The phase of $G(j\omega)$ is given by

$$\angle G(j\omega) = -\tan^{-1}\left\{\frac{2\zeta\omega}{\omega_n}\left[1 - \left(\frac{\omega}{\omega_n}\right)^2\right]\right\} \qquad \text{(A-50)}$$

and is plotted as shown in Fig. A-9 for various values of ζ.

The analysis of the Bode plot of the second-order transfer function of Eq. (A-45) can be applied to the second-order transfer function with two complex zeros. For

$$G(s) = 1 + \frac{2\zeta}{\omega_n}s + \frac{1}{\omega_n^2}s^2 \qquad \text{(A-51)}$$

The magnitude and phase curves are obtained by inverting those in Fig. A-9. The errors between the actual and the asymptotic curves in Fig. A-10 are also inverted.

Pure Time Delay, $e^{-j\omega T_d}$

The magnitude of the pure time delay term is equal to unity for all values of ω. The phase of the pure time-delay term is

$$\angle e^{-j\omega T_d} = -\omega T_d \qquad \text{(A-52)}$$

which decreases linearly as a function of ω. Thus, for the transfer function

$$G(j\omega) = G_1(j\omega)e^{-j\omega T_d} \qquad \text{(A-53)}$$

the magnitude plot $|G(j\omega)|_{dB}$ is identical to that of $|G_1(j\omega)|_{dB}$. The phase plot $\angle G(j\omega)$ is obtained by subtracting ωT_d radians from the phase curve of $G_1(j\omega)$ at various ω.

EXAMPLE A-5

As an illustrative example on the construction of the Bode plot, consider the function

$$G(s) = \frac{10(s + 10)}{s(s + 2)(s + 5)} \qquad \text{(A-54)}$$

The first step is to express $G(s)$ in the form of Eq. (A-23) and set $s = j\omega$; we have

$$G(j\omega) = \frac{10(1 + j0.1\omega)}{j\omega(1 + j0.5\omega)(1 + j0.2\omega)} \qquad \text{(A-55)}$$

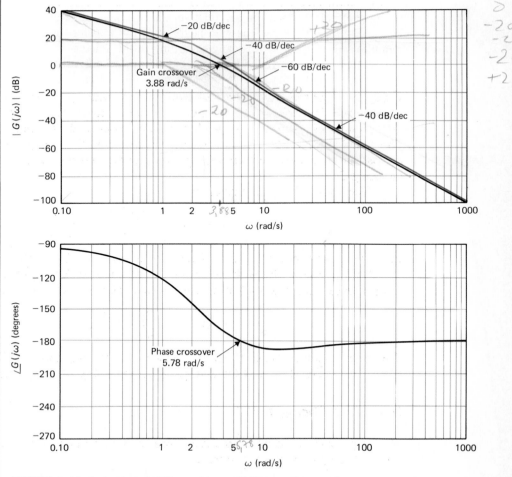

FIGURE A-11 Bode plot of $G(s) = 10(s + 10)/[s(s + 2)(s + 5)]$.

This equation shows that $G(j\omega)$ has corner frequencies at $\omega = 2, 5,$ and 10 rad/s. The pole at $s = 0$ gives a magnitude curve that is a straight line with a slope of -20 dB/decade, passing through the $\omega = 1$ rad/s point on the 0-dB axis. The complete Bode plot of the magnitude and phase of $G(j\omega)$ is obtained by adding the component curves together, point by point, as shown in Fig. A-11. The actual curves can be obtained by a computer program and are shown in Fig. A-11.

A-3

MAGNITUDE-VERSUS-PHASE PLOT

The magnitude-versus-phase plot of $G(j\omega)$ is a plot of the magnitude of $G(j\omega)$ in dB versus its phase in degrees, with ω as a parameter on the curve. One of the most important applications of this type of plot is that when $G(j\omega)$ is the open-loop transfer function of a control system, the plot can be superposed on the Nichols chart (see Chapter 10) to give information on the relative stability and the frequency response of the closed-loop system. When the gain factor K of the transfer function varies, the plot is simply raised or lowered vertically according to the value of K in dB. However, in the construction of the plot, the property of adding the curves of the individual components of the transfer function in the Bode plot does not carry over to this case. Thus, it is best to make the magnitude-versus-phase plot by computer or transfer the data from the Bode plot.

EXAMPLE A-6   As an illustrative example, the polar plot and the magnitude-versus-phase plot of Eq. (A-54) are shown in Figs. A-12 and A-13, respectively, The Bode plot of the function is already

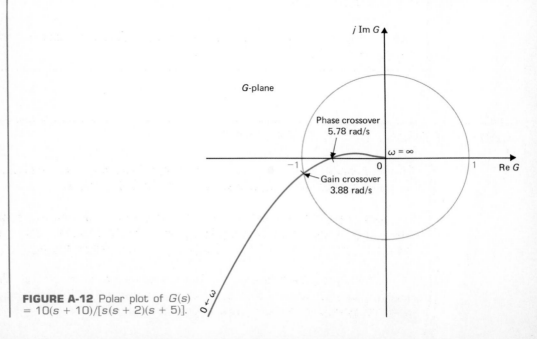

FIGURE A-12 Polar plot of $G(s)$ = $10(s + 10)/[s(s + 2)(s + 5)]$.

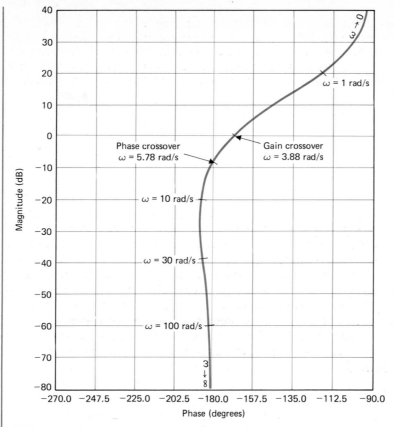

FIGURE A-13 Magnitude-versus-phase plot of $G(s) = 10(s + 10)/[s(s + 2)(s + 5)]$.

shown in Fig. A-11. The relationships among these three plots are easily identified by comparing the curves in Figs. A-11, A-12, and A-13.

A-4
GAIN AND PHASE CROSSOVER POINTS

Gain and phase crossover points on the frequency-domain plots are important for analysis and design of control systems. These are defined as follows.

GAIN-CROSSOVER POINT.   The gain-crossover point on the frequency-domain plot of $G(j\omega)$ is a point at which $|G(j\omega)| = 1$ or $|G(j\omega)|_{dB} = 0$ dB. The frequency at the gain-crossover point is called the **gain-crossover frequency** ω_g.

PHASE-CROSSOVER POINT.   The phase-crossover point on the frequency-domain plot of $G(j\omega)$ is a point at which $\angle G(j\omega) = 180°$. The frequency at the phase-crossover point is called the **phase-crossover frequency** ω_c.

The gain and phase crossovers are interpreted with respect to the three types of plots.

POLAR PLOT.   The gain-crossover point (or points) is where $|G(j\omega)| = 1$. The phase-crossover point (or points) is where $\angle G(j\omega) = 180°$ (see Fig. A-12).

BODE PLOT.   The gain-crossover point (or points) is where the magnitude curve $|G(j\omega)|_{dB}$ crosses the 0-dB axis. The phase-crossover point (or points) is where the phase curve crosses the 180° axis (see Fig. A-11).

MAGNITUDE-VERSUS-PHASE PLOT.   The gain-crossover point (or points) is where the $G(j\omega)$ curve crosses the 0-dB axis. The phase-crossover point (or points) is where the $G(j\omega)$ curve crosses the 180° axis (see Fig. A-13).

A-5
MINIMUM-PHASE AND NONMINIMUM-PHASE FUNCTIONS

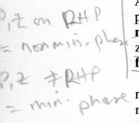

A majority of the transfer functions encountered in linear control systems do not have poles or zeros in the right-half s-plane. This class of transfer function is called a **minimum-phase transfer functions**. When a transfer function has either poles or zeros, or both, in the right-half s-plane, it is called a **nonminimum-phase transfer function**.

Minimum-phase transfer functions have an important property in that their magnitude and phase characteristics are uniquely related. In other words, given a minimum-phase function $G(s)$, then knowing its magnitude characteristics, $|G(j\omega)|$, completely defines the phase characteristics, $\angle G(j\omega)$. Conversely, given $\angle G(j\omega)$, $|G(j\omega)|$ is completely defined.

Nonminimum-phase transfer functions do not have the unique magnitude–phase relationships. For instance, given the function

$$G(j\omega) = \frac{1}{1 - j\omega T} \tag{A-56}$$

the magnitude of $G(j\omega)$ is the same whether T is positive (nonminimum phase) or negative (minimum phase). However, the phase of $G(j\omega)$ is different for positive and negative T.

Additional properties of the minimum-phase transfer functions are as follows:

1. For a minimum-phase transfer function $F(s)$ with m zeros and n poles, excluding the poles at $s = 0$, when $s = j\omega$, and as ω varies from ∞ to 0, the total phase variation of $F(j\omega)$ is $(n - m)\pi/2$.
2. The value of a minimum-phase transfer function cannot become zero or infinity at any finite nonzero frequency.
3. A nonminimum-phase transfer function will always have a more positive phase shift as ω is varied from ∞ to 0.

EXAMPLE A-7   As an illustrative example on the properties of the nonminimum-phase transfer function, consider that the zero of the transfer function of Eq. (A-54) is in the right-half s-plane; that is,

$$G(s) = \frac{10(s - 10)}{s(s + 2)(s + 5)} \qquad (A-57)$$

(handwritten: $w \to \infty, G(\omega) = \frac{10s}{s^3} = \frac{10}{s^2}$ $= -180°$)

The magnitude plot of the Bode diagram of $G(j\omega)$ is identical to that of the minimum-phase function in Eq. (A-54), and is shown in Fig. A-11. The phase curve of the Bode plot of $G(j\omega)$ of Eq. (A-57) is shown in Fig. A-14(a), and the polar plot is shown in Fig. A-14(b). Notice that the nonminimum-phase function has a net phase shift of 270° (from $-180°$ to $+90°$) as ω varies from ∞ to 0, whereas the minimum-phase function in Eq. (A-54) has a net phase change of only 90° (from $-180°$ to $-90°$) over the same frequency range.

Care should be taken when using the Bode diagram for the analysis and design of systems with nonminimum-phase transfer functions. For stability studies, the polar plot, when used along with the Nyquist criterion discussed in Chapter 10, is more convenient for nonminimum-phase systems.

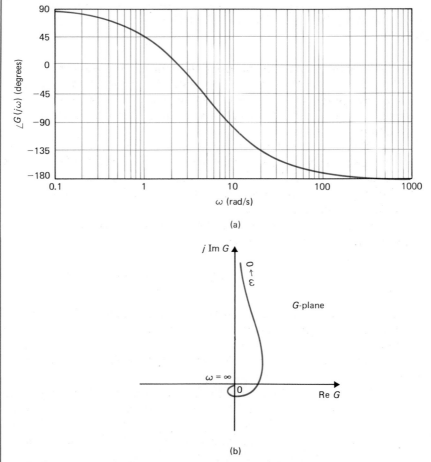

FIGURE A-14 (a) Phase curve of the Bode plot. (b) Polar plot. $G(s) = 10(s - 10)/[s(s + 2)(s + 5)]$.

LAPLACE TRANSFORM $F(s)$	TIME FUNCTION $f(t)$
1	Unit-impulse function $\delta(t)$
$\dfrac{1}{s}$	Unit-step function $u_s(t)$
$\dfrac{1}{s^2}$	Unit-ramp function t
$\dfrac{n!}{s^{n+1}}$	$t^n \quad$ (n = positive integer)
$\dfrac{1}{s+\alpha}$	$e^{-\alpha t}$
$\dfrac{1}{(s+\alpha)^2}$	$te^{-\alpha t}$
$\dfrac{n!}{(s+\alpha)^{n+1}}$	$t^n e^{-\alpha t} \quad$ (n = positive integer)
$\dfrac{1}{(s+\alpha)(s+\beta)}$	$\dfrac{1}{\beta-\alpha}(e^{-\alpha t} - e^{-\beta t}) \quad (\alpha \neq \beta)$
$\dfrac{s}{(s+\alpha)(s+\beta)}$	$\dfrac{1}{\beta-\alpha}(\beta e^{-\beta t} - \alpha e^{-\alpha t}) \quad (\alpha \neq \beta)$
$\dfrac{1}{s(s+\alpha)}$	$\dfrac{1}{\alpha}(1 - e^{-\alpha t})$
$\dfrac{1}{s(s+\alpha)^2}$	$\dfrac{1}{\alpha^2}(1 - e^{-\alpha t} - \alpha t e^{-\alpha t})$
$\dfrac{1}{s^2(s+\alpha)}$	$\dfrac{1}{\alpha^2}(\alpha t - 1 + e^{-\alpha t})$
$\dfrac{1}{s^2(s+\alpha)^2}$	$\dfrac{1}{\alpha^2}\left[t - \dfrac{1}{\alpha} + \left(t + \dfrac{2}{\alpha}\right)e^{-\alpha t}\right]$

LAPLACE TRANSFORM $F(s)$	TIME FUNCTION $f(t)$
$\dfrac{s}{(s + \alpha)^2}$	$(1 - \alpha t)e^{-\alpha t}$
$\dfrac{\omega_n^2}{s^2 + \omega_n^2}$	$\sin \omega_n t$
$\dfrac{s}{s^2 + \omega_n^2}$	$\cos \omega_n t$
$\dfrac{\omega_n^2}{s(s^2 + \omega_n^2)}$	$1 - \cos \omega_n t$
$\dfrac{\omega_n^2(s + \alpha)}{s^2 + \omega_n^2}$	$\omega_n\sqrt{\alpha^2 + \omega_n^2}\,\sin(\omega_n t + \theta)$ where $\theta = \tan^{-1}(\omega_n/\alpha)$
$\dfrac{\omega_n}{(s + \alpha)(s^2 + \omega_n^2)}$	$\dfrac{\omega_n}{\alpha^2 + \omega_n^2}e^{-\alpha t} + \dfrac{1}{\sqrt{\alpha^2 + \omega_n^2}}\sin(\omega_n t - \theta)$ where $\theta = \tan^{-1}(\omega_n/\alpha)$
$\dfrac{\omega_n^2}{s^2 + 2\zeta\omega_n s + \omega_n^2}$	$\dfrac{\omega_n}{\sqrt{1 - \zeta^2}}e^{-\zeta\omega_n t}\sin\omega_n\sqrt{1 - \zeta^2}\,t \quad (\zeta < 1)$
$\dfrac{\omega_n^2}{s(s^2 + 2\zeta\omega_n s + \omega_n^2)}$	$1 - \dfrac{1}{\sqrt{1 - \zeta^2}}e^{-\zeta\omega_n t}\sin(\omega_n\sqrt{1 - \zeta^2}\,t + \theta)$ where $\theta = \cos^{-1}\zeta \quad (\zeta < 1)$
$\dfrac{s\omega_n^2}{s^2 + 2\zeta\omega_n s + \omega_n^2}$	$\dfrac{-\omega_n^2}{\sqrt{1 - \zeta^2}}e^{-\zeta\omega_n t}\sin(\omega_n\sqrt{1 - \zeta^2}\,t - \theta)$ where $\theta = \cos^{-1}\zeta \quad (\zeta < 1)$
$\dfrac{\omega_n^2(s + \alpha)}{s^2 + 2\zeta\omega_n s + \omega_n^2}$	$\omega_n\sqrt{\dfrac{\alpha^2 - 2\alpha\zeta\omega_n + \omega_n^2}{1 - \zeta^2}}e^{-\zeta\omega_n t}\sin(\omega_n\sqrt{1 - \zeta^2}\,t + \theta)$ where $\theta = \tan^{-1}\dfrac{\omega_n\sqrt{1 - \zeta^2}}{\alpha - \zeta\omega_n} \quad (\zeta < 1)$
$\dfrac{\omega_n^2}{s^2(s^2 + 2\zeta\omega_n s + \omega_n^2)}$	$t - \dfrac{2\zeta}{\omega_n} + \dfrac{1}{\omega_n^2\sqrt{1 - \zeta^2}}e^{-\zeta\omega_n t}\sin(\omega_n\sqrt{1 - \zeta^2}\,t + \theta)$ where $\theta = \cos^{-1}(2\zeta^2 - 1) \quad (\zeta < 1)$

Laplace-Transform Table—Cont.

ANSWERS AND HINTS FOR SELECTED PROBLEMS

The answers given in the following are for certain selected problems. These answers are given for checking purposes, and thus are often for only certain parts of a problem. Hints on how the problems should be solved are also given in selected situations.

CHAPTER 2

2-1. **(a)** Poles: $s = 0, 0, -1, -10$; zeros: $s = -2, \infty, \infty, \infty$.
   (c) Poles: $s = 0, -1 + j, -1 - j$; zero: $s = -2$

2-3. **(a)** $G(s) = \dfrac{1 - e^{-s}}{s(1 + e^{-s})}$ **(b)** $G(s) = \dfrac{2(1 - e^{-s/2})}{s^2(1 + e^{-s/2})}$

2-5. **(a)** $f(t) = \frac{1}{6}e^{-4t} + \frac{1}{3}e^{-t} - \frac{1}{2}e^{-2t}$
   (b) $x_1(t) = 0.5 + e^{-t} - 0.5e^{-2t}$; $x_2(t) = -e^{-t} + e^{-2t}$; $t \geq 0$

2-7. **(a)** $\begin{bmatrix} 15 & -9 \\ 2 & 0 \end{bmatrix}$ **(c)** $\begin{bmatrix} \dfrac{s+2}{s+1} & 3 & \dfrac{1}{s} - 10 \\[2mm] \dfrac{2}{s} + s & \dfrac{1}{s} & \dfrac{s-2}{s-3} \end{bmatrix}$

2-9. **(a)** $\mathbf{A} = \begin{bmatrix} 1 & 1 & -1 \\ -1 & 3 & -1 \\ 3 & -5 & -2 \end{bmatrix}$ $\mathbf{A}^{-1} = \begin{bmatrix} 0.9167 & -0.5830 & -0.1667 \\ 0.4167 & -0.0833 & -0.1667 \\ 0.3333 & -0.6667 & -0.3333 \end{bmatrix}$ $\mathbf{B} = \begin{bmatrix} 1 \\ 1 \\ 0 \end{bmatrix}$

2-11. **(a)** $\mathbf{A}^{-1} = \dfrac{-1}{52}\begin{bmatrix} -1 & -5 \\ -10 & 2 \end{bmatrix}$ **(d)** $\mathbf{A}^{-1} = \begin{bmatrix} 1.3 & 0.5 & -0.3 \\ 1 & 0 & 0 \\ -0.2 & 0 & 0.2 \end{bmatrix}$

2-13. **(a)** $F^*(s) = \displaystyle\sum_{k=0}^{\infty} kTe^{-kT(s+3)} = \dfrac{Te^{-T(s+3)}}{(1 - e^{-T(s+3)})^2}$

   (d) $F^*(s) = \dfrac{T^2 e^{-T(s+2)}(1 + e^{-T(s+2)})}{(1 - e^{-T(s+2)})^3}$

2-15. $F(z) = z/(z + 1)$

2-17. **(a)** $f(k) = 12.5[1 - (0.2)^k]; \quad k = 0, 1, 2, \ldots$
   (c) $f(k) = 0.541[1 - (-0.85)^k]; \quad k = 0, 1, 2, \ldots$

2-19. **(a)** $x(k) = 6.25 - 8.333(0.8)^k + 2.0833(0.2)^k; \quad k = 0, 1, 2, \ldots$

CHAPTER 3

3-1. **(a)** $\dfrac{C(s)}{R(s)} = \dfrac{3s + 1}{s^3 + 2s^2 + 5s + 6}$ **(c)** $\dfrac{C(s)}{R(s)} = \dfrac{s(s + 2)}{s^4 + 10s^3 + 2s^2 + s + 2}$

3-3. **(d)** Steady-state speed = 6.66 ft/s

3-7. **(a)** $\dfrac{Y_5}{Y_1} = \dfrac{G_1 G_2 G_3 + G_3 G_4}{\Delta}$ $\dfrac{Y_2}{Y_1} = \dfrac{1 + G_3 H_2}{\Delta}$ $\dfrac{Y_5}{Y_2} = \dfrac{G_1 G_2 G_3 + G_3 G_4}{1 + G_3 H_2}$

   $\Delta = 1 + G_1 H_1 + G_3 H_2 + G_3 G_4 H_3 + G_1 G_2 G_3 H_3 + G_1 G_3 H_1 H_2$

3-9. $k = -1$

3-11. **(a)** The three loops are not in touch.
   (b) The three loops are in touch. $\Delta = 1 + G_1 H_1 + G_2 H_2 + G_3 H_3 + G_1 G_3 H_1 H_3$

3-15. **(b)** $1 + G_1(s)G_2(s)H_1(s) = 0$

3-17. **(b)** $K = 6.67$

3-21. **(b)** Characteristic equation: $s^2 + 7s + 25 = 0$

3-23. **(c)** Characteristic equation: $s^4 + 5s^3 + 3s^2 + 10s + 1 = 0$
   Roots: $-10.263, -4.8, -0.04869 + j1.424, -0.04869 - j1.424$
   (d) $\dfrac{C(s)}{R(s)} = \dfrac{2}{s^4 + 5s^3 + 3s^2 + 10s + 1}$

3-25. $G(z) = (1 - e^{-2T})/(z - e^{-2T})$

3-27. **(a)** $G(z) = Tz/(z - 1)$ **(b)** $Y(z) = Tz/(z - 1)$
   (c) $Y(z) = T/(z - 1)$

CHAPTER 4

4-1. **(c)** Force equations:

$$\frac{dy_1}{dt} = \frac{dy_2}{dt} + \frac{1}{B_1}f \qquad \frac{d^2y_2}{dt^2} = -\frac{B_1 + B_2}{M}\frac{dy_2}{dt} + \frac{B_1}{M}\frac{dy_1}{dt} - \frac{K}{M}y_2$$

4-3. **(a)** Torque equation:

$$\frac{d^2\theta}{dt^2} = -\frac{B}{J}\frac{d\theta}{dt} + \frac{1}{J}T(t)$$

State equations: $x_1 = \theta$; $x_2 = d\theta/dt$

$$\frac{dx_1}{dt} = x_2 \qquad \frac{dx_2}{dt} = -\frac{B}{J}x_2 + \frac{1}{J}T$$

Transfer function: $\Theta(s)/T(s) = 1/s(Js + B)$

4-6. **(b)** State variables: $x_1 = y_1 - y_2$; $x_2 = dy_2/dt$

State equations:

$$\frac{dx_1}{dt} = -\frac{K_h}{B_h}x_1 + \frac{1}{B_h}f(t) \qquad \frac{dx_2}{dt} = -\frac{B_t}{M}x_2 + \frac{1}{M}f(t)$$

4-7. **(a)** Optimal gear ratio:

$$n^* = -\frac{J_m T_L}{2J_L T_m} + \frac{\sqrt{J_m^2 T_L^2 + 4J_m J_L T_m^2}}{2J_L T_m}$$

(b) $T_L = 0$: Optimal gear ratio: $n^* = \sqrt{J_m/J_L}$

4-9. **(c)** State equations:

$$\frac{dx_1}{dt} = x_3 - x_2 \qquad \frac{dx_2}{dt} = \frac{K_1 + K_2}{M}x_1 \qquad \frac{dx_3}{dt} = \frac{-r(K_1 + K_2)}{J_m}x_1 + \frac{1}{J_m}T_m$$

(d) Transfer function:

$$\frac{Y(s)}{T_m(s)} = \frac{r(K_1 + K_2)}{s^2[J_m Ms^2 + (K_1 + K_2)(J_m + rM)]}$$

(e) Characteristic equation:

$$s^2[J_m Ms^2 + (K_1 + K_2)(J_m + rM)] = 0$$

4-12. **(b)**

$$\left.\frac{\Omega_m(s)}{\Omega_r(s)}\right|_{T_L=0} = \frac{K_1 K_i}{(R_a + L_a s)(B + Js) + K_i K_b + K_1 K_i K_b H_e(s)} \cong \frac{1}{K_b H_e(s)}$$

4-15. **(a)** The amplifier is not saturated: The amplifier is saturated:

Slope of speed–torque curve: Slope of speed–torque curve:

$$k = \frac{dT_m}{d\Omega_m} = \frac{-K_i K_b}{R_a + K_1 K_2} \qquad\qquad k = -\frac{K_i K_b}{R_a}$$

(b) $K_b = 0.00667$ V/rpm, $K_i = 0.011$ oz-in./A, $R_a = 5.0087\ \Omega$,
$K_1 = 100$ V/V, $K_2 = 0.85$ V/A

4-17. **(d)** Characteristic equation:

$$s^4 + 423.42s^3 + 2.6667 \times 10^6 s^2 + 4.2342 \times 10^8 s + 6.087 \times 10^8 K = 0$$

4-20. **(e)** $G_c(s) = (1 + R_2 Cs)/R_1 Cs$

Open-loop transfer function:

$$\frac{\Omega_m(s)}{E(s)} = \frac{K(1 + R_2 Cs)}{R_1 Cs(K_b K_i + R_a J_L s)}$$

Closed-loop transfer function:

$$\frac{\Omega_m(s)}{F_r(s)} = \frac{K_\phi K(1 + R_2 Cs)}{R_1 Cs(K_b K_i + R_a J_L s) + K_\phi K K_e N(1 + R_2 Cs)}$$

(f) For $\omega_m = 200$ rpm, $N = 1$. For $\omega_m = 1800$ rpm, $N = 9$.

4-26. **(b)** Closed-loop transfer function:

$$\frac{C(s)}{R(s)} = \frac{K K_i n G_c(s) e^{-T_D s}}{s(R_a + L_a s)[(J_m + J_L)s + B_m] + K_b K_i s + K G_c(s) K_i n e^{-T_D s}}$$

CHAPTER 5

5-1. **(a)** State equations:

$$\frac{de_1}{dt} = -\frac{1}{C_1}\left(\frac{1}{R_1} + \frac{1}{R_2}\right)e_1 + \frac{1}{R_1 C_1}e_2 + \frac{1}{R_1 C_1}e$$

$$\frac{di_L}{dt} = -\frac{R_3}{L}i_L + \frac{1}{L}e_2 \qquad \frac{de_2}{dt} = \frac{1}{R_2 C_2}e_1 - \frac{1}{R_2 C_2}e_2 - \frac{1}{C_2}i_L$$

5-3. Hint: multiply both sides of $(s\mathbf{I} - \mathbf{A})^{-1}$ by $(s\mathbf{I} - \mathbf{A})$.

5-4. **(a)** Characteristic equation: $\Delta(s) = s^2 + s + 2 = 0$
Eigenvalues: $s = -0.5 + j1.323, -0.5 - j1.323$
State transition matrix:

$$\phi(t) = \begin{bmatrix} \cos 1.323t + 0.378 \sin 1.323t & 0.756 \sin 1.323t \\ -1.512 \sin 1.323t & -1.069 \sin (1.323t - 69.3°) \end{bmatrix} e^{-0.5t}$$

(e) Characteristic equation: $\Delta(s) = s^2 + 4 = 0$. Eigenvalues: $s = j2, -j2$.
State transition matrix:

$$\phi(t) = \begin{bmatrix} \cos 2t & \sin 2t \\ -\sin 2t & \cos 2t \end{bmatrix}$$

5-5. **(c)** $\begin{bmatrix} 0 \\ 0.333(1 - e^{-3t}) \end{bmatrix}$ **(e)** $\begin{bmatrix} 2 \\ 0.5 \sin 2t \end{bmatrix}$

5-7. **(b)** $\mathbf{Q} = \begin{bmatrix} 0 & 0 & 1 \\ -1 & 1 & 1 \\ 0 & 1 & 1 \end{bmatrix}$ **(d)** $\mathbf{Q} = \begin{bmatrix} 1 & -1 & 1 \\ -1 & 2 & -2 \\ 1 & -3 & 4 \end{bmatrix}$

5-9. **(c)** (1) $\Delta(s) = s(s^2 + 2s + 1) = 0$, $s = -1, -1$

(2) $$\mathbf{X}(s) = \frac{1}{\Delta(s)} \begin{bmatrix} 1 \\ s \\ s^2 \end{bmatrix} U(s)$$

5-12. **(b)** (1) $\phi(t) = \begin{bmatrix} e^{-t} & 0 \\ 0 & e^{-2t} \end{bmatrix}$

5-13.   **(c)**   Characteristic equation: $\Delta(s) = s^2 + 80.65s + 322.58 = 0$

5-15.   **(a)**   Open-loop transfer function:
$$G(s) = \frac{5(K_1 + K_2s)}{s[s(s + 4)(s + 5) + 10]}$$
   (c)   Final value of $c(t) = 1$

5-18.   **(d)**   Characteristic equation: $\Delta(s) = s^3 + 3s^2 + 3s + 1 = 0$

5-21.   **(a)**   Characteristic equation: $\Delta(s) = s^2 - 2\sigma + (\sigma^2 + \omega^2) = 0$

5-25.   **(b)**   Hint: the system is of the fourth order, so there should be only four integrators. To achieve this, one of the integrators associated with the s^2 term should be shared by two states.

5-29.   **(c)**
$$\Omega(s) = \frac{-(s + 20)}{\Delta_c(s)}T_D(s) + \frac{30e^{-0.2s}}{\Delta_c(s)(s + 5)}U(s)$$

$$\Delta_c(s) = (s + 2)(s + 20) + 0.1e^{-0.2s}$$

5-32.   **(b)**   Characteristic equation: $s^3 + 12s^2 + 70s + 200 = 0$
   (c) $g_1 = -80$; final value of $c(t) = 5$
   (e) $g_1 = 0$

5-34.   **(d)**   Characteristic equation: $s^2 + 1037s + 20{,}131.2 = 0$

5-40.   The system is completely controllable for $b_2 \neq 0$.

5-42.   **(b)**   Eigenvalues: $0, -1, -11.767$
   (d)   Only one of the three cases is observable.

5-45.   **(b)**   Characteristic equation: $J_vs^2(J_Gs^2 + K_ps + K_I + K_N) = 0$

5-48. $C(z)/U(z) = (z - 3)/(2z + 15)$; characteristic equation: $2z + 15 = 0$

CHAPTER 6

6-1.   **(c)**   Poles at $-1, 0.5 \pm j1.937$; unstable
   (e)   Poles at $-0.4656, 1.233 \pm j0.7926$; unstable

6-2.   **(d)**   No roots in the RHP.   **(f)**   Two roots in the RHP.

6-3.   **(a)**   Stable for $0 < K < 1.49$. Frequency of oscillation: 0.316 rad/s.
   (d)   Stable for $K > 2$. Frequency of oscillation: 2.236 rad/s.

6-5.   **(a)**   Stable for $1.2166 \times 10^6 < K < 1.7535 \times 10^8$

6-9. $K_t > 0.081$

6-12. **(b)** All three roots are on the $s = -1$ line.

6-14. **(a)** $N_{max} = 1$

6-17. **(d)** One root outside the unit circle.

6-19. For stability, $-0.2 < K < 0.07$.

6-20. **(c)** For stability, $0 < K < 3.9535$.

CHAPTER 7

7-2. **(b)** Type 0   **(b)** Type 3

7-3. **(c)** $K_p = \infty, K_v = K, K_a = 0$ **(e)** $K_p = \infty, K_v = 1, K_a = 0$

7-5. **(b)** $e_s(t) = 0.00558 \sin (5t - 63.66°)u_s(t)$

7-7. **(c)** Minimum $e_{ss} = 1/12$

7-9. **(a)** $K > 0$ and $K_t > 0.02$

7-12. $\omega_n = 10.82$ rad/s, $K_t = 0.0206$

7-17. $\omega_n = 125.45$ rad/s, $K_t = 0.2188$

7-21. **(a)** For stability, $0 > K > -1.5$.

7-23. **(e)** $K_p = 10$

7-25. **(b)** Characteristic equation:
$$s^5 + 5000s^4 + 1.582 \times 10^6 s^3 + 5.05 \times 10^9 s^2$$
$$+ 5.724 \times 10^{11} s + 9 \times 10^{11} K = 0$$
(c) Critical value of K for stability: 3179.78

7-30. **(a)**
$$L(s) = \frac{0.995}{s^2 + 0.8955s + 0.995}$$

7-31. **(b)**
$$L(s) = \frac{18,248}{s^2 + 135.1s + 18,248}$$

7-33. **(c)** Third-order approximation:
$$L(s) = \frac{1.9876}{s^3 + 2.9886s^2 + 1.8932s + 1.9876}$$

7-35. **(a)** $K_p^* = \infty, K_v^* = 10/K_t, K_a^* = 0$ **(c)** For stability, $0.5 < K_t < 20$.

CHAPTER 8

8-1. **(c)** Asymptotes: $K > 0$: $180°$; $K < 0$: $0°$

8-2. **(a)** $K > 0$: $\theta_1 = 135°$; $K < 0$; $\theta_1 = -45°$

8-4. **(c)** Breakaway-point equation:
$3s^6 + 54s^5 + 347.5s^4 + 925s^3 + 867.2s^2 - 781.25s - 1953 = 0$
Breakaway points: -2.5, 1.09

8-7. **(a)** $K = 13.07$

8-8. **(a)** At $s = -37.86$, $K = -0.2317$.

8-20. **(c)** For one nonzero breakaway point, $\alpha = 9$.

8-21. **(c)** $\alpha < 0.333$

8-25. **(d)** The system is stable for $-3 < K < 1$.

CHAPTER 9

9-1. $a = 90$, $K = 18,000$

9-3. **(b)** $K_p = 1.5765$

9-7. **(b)** $K_I = 2$. To get K_I, K_P, and K_D, divide the characteristic equation by the second-order equation that corresponds to the desired ζ and ω_n.

9-8. **(b)** $K_P = 290.18$

9-9. **(b)** Hint: select a relatively large value of K_D and a small value of K_P so that the closed-loop poles are real.

9-11. **(a)** $K_P > 0$, $K_D > 0.1K_P$

9-13. Hint: select a small T, then select a to get the desired overshoot.

9-16. Hint: the value of a is determined to yield the desired ζ; T should be very large.

9-17. **(a)** Hint: select relatively small T and large a.

9-19. **(a)** $a = 80$

9-21. Nominal $K = 5000$. Maximum overshoot varies between 6.7 to 2.4 percent when K varies between 4000 and 6000.

9-24. $K = 5.93$

9-26. $K_I = 14.56$

9-29. $g_1 = 11$

9-32. $g_1 = -64.84$

9-34.   **(a)** $$X_1(s) = \frac{-s^2 - 11s + 20}{s(s^3 + 12s^2 + 22s + 20)}$$

9-36. $$G_c(z) = \frac{40z - 39 - e^{-100T}}{z - e^{-100T}}$$

CHAPTER 10

10-1.   **(a)** $M_P = 1$, $\omega_n = 2.24$ rad/s, $\zeta = 1.46$

10-2.   **(c)** $M_p = 2.28$, $\omega_n = 6$ rad/s, BW $= 8.82$ rad/s

10-4. $t_{max} = 0.95$ second

10-5.   When $T = 3$ seconds, BW $= 1.96$ rad/s and $M_p = 1.29$.

10-7.   **(a)** $Z_{-1} = 2$ **(f)** $\Phi_{11} = 90°$, $Z_{-1} = 2$ **(k)** $\Phi_{11} = -90°$, $Z_{-1} = 0$

10-8.   **(e)**   Unstable for all K.   **(h)** $0 > K > -\frac{1}{3}$

10-10.   **(c)** $Z_{-1} = 0$; stable.

10-12.   **(c)** $0 < K < 0.5$, stable; $0.5 < K < 1$, unstable; $K > 1$, stable, $K < 0$, unstable.

10-15.   **(a)** $|K| < 14.14$

10-19.   **(b)**   Critical $K = 1.407$

10-21.   **(c)**   For $G(s)$, $M_p = 1$, $\omega_p = 0$, and BW $= 0.87$ rad/s. For $G_L(s)$, $M_p = 1$, $\omega_p = 0$, and BW $= 0.91$ rad/s.

10-27.   **(c)**   Gain C.O. frequency $= 19.1$ rad/s, phase margin $= 4.07°$, phase C.O. frequency $= 20.31$ rad/s, and gain margin $= 1.13$ dB

10-34. $|S_G^M|_{max} = 17.15$, $\omega_{max} = 5.75$ rad/s

CHAPTER 11

11-1.   When $K_D = 0.6$, Φ.M. $= 89.36°$, G.M. $= \infty$, $M_p = 1.02$, and BW $= 607$ rad/s.

11-2.   **(a)**   The phase margin is minimum when $K_D = 0.09$.

11-4.   **(a)**   Maximum phase margin occurs at $K_P = 0.9$.

11-7. **(b)** For $K = 10$, $\Phi.M. = -9.65°$ and G.M. $= -5.19$ dB.
(c) and **(d)** Use FREQCAD for design.

11-10. The phase margin is maximum when $K_P = 7$.

11-12. Suggest a phase-lag controller. Required high-frequency attenuation is -17.5 dB. Use FREQCAD.

11-13. **(b)** $K_I = 0.01$. Vary K_P from 0.01 to 0.2 to find K_P for maximum phase margin.

11-15. **(b)** $G_c(s) = \dfrac{s^2 + 15.06s + 2025.6}{s^2 + 284.5s + 2025.6}$

INDEX

A

Absolute stability, 279
Acceleration, 128
AC control systems, 12, 154
ACSP, 32, 47, 58, 108, 257, 285, 669, 691
Actuating signal, 2
Aircraft attitude control, 338, 469
Analog-to-digital converter, 14
Analytic function, 18
Approximation of high-order systems, 357–69
Asymptotic plot (*see* Bode plot)
Asymptotic stability, 283, 294
Attitude control, 11, 338, 640
Auxiliary equation, 290, 417

B

Back emf, 165, 170–71, 521
Back emf constant, 170, 172
Backlash, 141
Backward-rectangular integration, 539
Bandwidth, 6, 351, 566, 634, 638
    second-order systems, 566, 570–72
    third-order systems, 575
Belt and pulley, 135

BIBO stability, 280, 294
Bilinear transformation, 295
Block diagram, 2–4, 71–76, 83, 231
    control systems, 2–4, 72
    DC motor system, 171
    digital process, 101
    multivariable system, 75
Bode diagram, 285, 623, 721, 727
    bridged-T controller, 707
    phase-lag controller, 693
    phase-lead controller, 679
    time-delay systems, 625, 736
Bode plot (*see* Bode diagram)
Bridged-T compensation, 503–8, 706, 712
Bridged-T controller (*see* bridged-T compensation)
Bridged-T network, 503–4
Bridged-T phase-lag controller, 507
Bridged-T PI controller, 508
Brushless DC motor (*see* DC motors)
Brushless DC motor controller, 169

C

Cascade compensation, 461
Cascade decomposition (*see* decomposition of transfer functions)
Causal system, 23

Cause-and-effect relationship, 6, 77, 170
Characteristic equation, 70, 229–30, 257, 279, 281–82, 285, 330, 341, 346, 400, 528, 577
  discrete-data system, 248–49, 297
  second-order system, 328
Closed-loop control system, 4, 5, 74
  frequency response, 564
Closed-loop stability, 587, 591
CLRSP, 341
Compensation schemes, 461
Complex convolution (*see* convolution)
Complex multiplication theorem, 27
Complex shifting, 27
Complex shifting theorem, 27, 102
Complex s-plane (*see* s-plane)
Complex variables, 16, 17
  functions of, 17
Computer solutions,
  matrix multiplication, 47
  partial-fraction expansion, 32, 58
  sampled-data systems, 108
Conditional frequency, 329
Constant conditional frequency loci, 330, 375
Constant damping factor loci, 330, 375
Constant damping ratio loci, 330, 376
Constant M circles (*see* constant M loci)
Constant M loci, 631–35
Constant natural undamped frequency loci, 330, 375
Constant N loci, 635–37
Constant-phase loci (*see* constant N loci)
Control system:
  basic elements, 2
  block diagram, 72
  closed-loop, 5
  continuous-data, 12
  digital, 13, 14
  discrete-data, 13
  frequency-response, 562–663
  nonlinear, 11
  open-loop, 4
  sampled-data, 14
  transient response, 307, 325, 327, 341
  types, 10–14, 315, 318–19, 532
Controllability, 236–41
  definition, 238
  general concept, 238
  state, 238–39
  tests, 239
Controlled process, 4
Controlled variable, 2
Controller, 4
Convolution:
  complex, 27, 73

Laplace transform, 27
z-transform, 55
Convolution integral, 27, 320
Corner frequency, 732
Corner plot (*see* Bode plot)
Coulomb friction (*see also* friction), 313, 344
Coulomb friction coefficient, 131, 139
Critical damping, 328
Critical point, 582
Critical trajectory, 615

D

Damped frequency (*see* conditional frequency)
Damping factor, 328
Damping ratio, 328, 340, 469, 569
  relative, 356
Data hold (*see* zero-order hold)
DC control system, 12, 153, 338
DC motors, 164–75
  brushless, 167–69
  classifications, 165
  iron-core, 166
  mathematical modeling, 170–73
  moving-coil, 166
  operational principles, 164
  separately excited, 165
  state diagram, 171
  surface-wound, 165–6
  torque-speed curve, 174–75
DC motor control system, 72, 518
Dead zone (*see also* backlash), 311
Decade, 731
Decomposition of transfer functions, 95, 231–36
  cascade, 234, 252
  direct, 221, 232–33, 252
  discrete-data systems, 252–54
  parallel, 234, 252
Delay time, 327, 335
Demodulator, 155
Derivative control, 464
Design criteria, 460
Diagonalization of a matrix, 235
Difference equations, 2, 20, 48, 250, 252
Differential equations, 2, 19, 69, 96, 217, 227, 229, 231
  linear ordinary, 20, 33
  nonlinear, 20
Digital autopilot system, 14
Digital control systems (*see* discrete-data control systems)

Digital controller, 14, 108
Digital-to-analog converter, 14, 100
Direct decomposition (*see* decomposition of) transfer functions)
Discrete-data system, 11, 13, 50, 100–108, 293, 369
  frequency-domain analysis, 646
  stability of, 293–98
Discrete state diagram, 250, 252–53, 255
Discrete state equations (*see also* state equations)
  solutions of, 245–48
Discrete state transition equation, 245–46
Displacement, 128
Distance, 128
Disturbance, 9, 205, 463, 509
Dominant poles (*see* dominant roots)
Dominant roots, 347, 355, 357
Dynamic equations, 204, 226, 231, 238

E

Eigenvalues, 230, 237, 257
Electric networks, 124
Electrical time constant, 340
Encirclement, 580
Enclosure, 580, 592, 623
Encoders, 150–63
  absolute, 160
  incremental, 161–63
Error coefficients, 321
Error constant:
  parabolic, 318, 384
  ramp, 317, 345, 383, 466
  step, 316, 344, 382
Error series, 320
Error transfer function, 320

F

Feedback:
  effect on external disturbance, 9
  effect on overall gain, 7
  effect on sensitivity, 8
  effect on signal-to-noise ratio, 9
  effect on stability, 7
  effects of, 6–10
  negative, 7
Feedback compensation, 461
Feedback control systems, 5
  types, 10–14

Feedforward compensation, 461, 508, 512–3, 515, 517
Final-value theorem:
  Laplace transform, 26, 314
  z-transform, 54, 381
Fixed-configuration design, 461
Force, 128
Forward compensation, 462, 508
Forward-rectangular integration, 539
Free-body diagram, 142, 144
FREQCAD, 669, 691
FREQRP, 285, 569, 588, 604, 632, 669, 722
Frequency-domain analysis, 562–663
  discrete-data systems, 646–50
  sensitivity studies, 642
Frequency-domain design, 664–720
Frequency-domain plots, 721–41
Frequency-domain specifications, 566
Friction,
  Coulomb, 131, 134, 170, 312
  rotational motion, 134
  static, 130, 134
  translational motion, 129–31, 142
  viscous, 130, 134

G

Gain-crossover frequency, 622, 739
Gain-crossover point, 622, 739
Gain margin, 618–20
Gain-phase plot, 630
Gear train, 137–40
General gain formula (*see* signal flow graph)
Generalized error coefficients (*see* error coefficients)

H

Horsepower, 137
Hurwitz criterion (*see* Routh-Hurwitz criterion)
Hurwitz determinant, 286–8

I

Ideal sampler, (*see* sampler)
Idle-speed control system, 2, 4, 70
Impedance, 6

Impulse response, 68
  zero-order hold, 105
Incremental encoder, 160–63
Inertia, 131–33, 136, 139
Inexact pole-zero cancellation, 501, 503
Initial conditions, 69, 282
Initial state, 99, 124
Initial-value theorem:
  Laplace transform, 26
  z-transform, 54
Input node, 80
Input vector, 71, 205
Integral control, 535
Inverse Laplace transform, 24–32
Inverse z-transformation, 56–59
  inversion formula, 58
  partial-fraction expansion, 56

J

Jordan canonical form, 236
Jury's stability tests, 297

K

Kinetic energy, 136

L

Lag-lead controller (see lead-lag controller)
Lag-lead network (see lead-lag network)
Laplace operator, 23
Laplace transform, 22–35, 51, 68, 211, 227
  application to solving differential equa-
    tions, 33
  definitions, 22
  inverse, 24–32
  one-sided, 23
  table, 742–43
  theorems, 24–28
Large space telescope, 311
Lead-lag controller, 496, 704
  digital implementation of, 542
Lead-lag network, 499
Lead screw, 135
Levers, 137, 140–41
Linearization, 176–181
Linear systems, 11

LINSYS, 47, 239, 257, 529
Low-pass filter, 565

M

Magnetic-ball suspension system, 255
Magnitude-versus-phase plot, 638, 721, 738
Marginal stability (*see* stability)
Mason, S. J., 77
Mason's gain formula (*see* signal flow
    graph)
Mass, 128, 142
Matrix:
  addition, 41
  adjoint of, 40, 230
  algebra, 41–47
  associative law, 42
  cofactor, 38
  column, 37
  commutative law, 42
  computer-aided solutions, 47
  definition, 37
  determinant of, 37, 38
  diagonal, 37
  distributive law, 44
  equality of, 41
  identity (*see* unity matrix)
  inverse, 45
  multiplication, 36, 42
  nonsingular, 39
  null, 38
  order of, 37
  product of, 36
  rank of, 46
  row, 37
  singular, 39
  square, 37
  subtraction, 41
  symmetric, 38
  transpose of, 40
  unity, 38
MATRIXM, 47
Maximum overshoot, 326, 334, 352, 371,
    566
Mechanical energy, 136
Mechanical power, 136
Mechanical systems, 127–50
  equations of, 142–50
  modeling, 127–41
Mechanical time constant, 340
Microprocessor, 3, 13, 311
Minimum-phase transfer functions, 591,
    598, 620, 623, 740

Minor-loop feedback control, 521, 523
Modulated control system, 11, 12
Multivariable system, 2, 70, 75, 226–7

N

Natural undamped frequency, 329, 336, 340
Nichols chart, 638
  application to nonunity-feedback
    systems, 642
Nonlinear control systems, 11
Nonlinear differential equations, 20
Nonlinear system, 21, 176, 178
  linearization, 176–81
Nonminimum-phase transfer function, 598,
  740
Notch controller, 707
Nyquist criterion, 284, 577–617
  application, 595–606
  multiloop system, 611
  noncausal systems, 591
  relation with root loci, 592
  simplified, 588
  time delays, 613
Nyquist path, 585, 588, 596

O

Observability, 237, 241, 258
  definition, 242
  tests, 242
Observer, 237, 461
Octave, 731
Open-loop control system, 4
  elements of, 4
Open-loop stability, 587
Output equations, 98, 204–5, 218, 244
Output node, 214
Output regulation, 532
Output variables, 2, 124
Output vector, 71, 205
Overshoot (*see* maximum overshoot)

P

Pade approximation, 183, 617
Parabolic error constant (*see* error constant)
Parabolic input, 309, 318

Parallel decomposition (*see* decomposition
  of transfer functions)
Parameter plane, 469
Partial fraction expansion:
  computer solution, 32, 58
  Laplace transform, 28–32
  z-transform, 56–58
PD controller, 464–70, 669–72
Peak resonance, 566, 634
Percent maximum overshoot, 326, 334
Periodic strips, 373, 374
PFE, 32, 58
Phase-crossover frequency, 620, 739
Phase-crossover point, 618, 739
Phase-lag compensation (*see also* phase-lag
  controller), 666
Phase-lag controller, 488, 498, 510, 692,
  709
  Bode plot, 693
  digital implementation, 542
  limitations, 702
  pole-zero configuration, 489
  root-contour design, 493
  root-locus design, 491
  transfer function, 488
Phase-lag network, 489
Phase-lead compensation (*see also* phase-
  lead controller), 666
Phase-lead controller, 479, 486, 497, 679,
  709
  Bode plot, 679
  digital implementation, 542
  limitations, 487, 688
  pole-zero configuration, 481
  transfer function, 481
  two-state, 487
Phase-lead network, 480
Phase margin, 621–23
Phase-variable canonical form, 219, 222–26,
  233, 528
PI control, 471, 477, 507, 672–77, 712
PID controller, 463, 474, 677–79
  digital implementation, 538
Polar plot, 721–27
Poles, definition, 18
  simple, 18
Pole-placement design, 237, 356, 527
Pole-zero cancellation control, 500, 506,
  713
POLYROOT, 284
Position-control system, 11, 338
Potential energy, 136
Potentiometer, 150–55
  block diagram, 152
Principle of the argument, 580, 586

Printwheel control system, 3, 5, 6, 13
Proportional control, 463
Prototype second-order system, 327,
335–37, 566
Pulse transfer function, 102

Q

Quantization error, 312
Quantizer, 312

R

Rack and pinion, 135
Ramp error constant (*see* error constant)
Ramp input, 309, 317, 344
Rate feedback (*see* Tachometer feedback)
RC network, 523
Real convolution, 27, 55
Real multiplication, 27
Regulator system, 5, 526
Relative damping, 356, 494
Relative stability, 279, 461, 606, 618, 627
Resonant frequency, 566
Rise time, 327, 335–6, 352, 474, 493, 497,
566
RL network, 208, 211, 215
RLC network, 19, 85, 125
Robust controller, 516, 520, 703
Robust system, 423, 479, 514
Root contours, 399, 440, 468, 476, 483,
485, 524
ROOTLOCI, 285, 398
Root loci, 182, 285, 343, 349, 372,
398–451, 472, 601, 604
angle of arrival, 414
angle of departure, 414
asymptotes, 409
breakaway points, 417
calculation of K, 426
complementary, 399
complete, 399
constructuon (*see* properties)
discrete-data systems, 447–51
effects of adding poles, 433
effects of adding zeros, 435
intersection with imaginary axis, 416
on real axis, 412
properties, 399, 427
relation with Nyquist plot, 592
rules of construction (*see* properties)
sensitivity, 423

Root sensitivity, 423
Rotational motion, 127, 135
Routh-Hurwitz criterion, 182, 284,
285–93, 295, 416
limitations of, 293
special cases, 289
Routh's tabulation (array), 287

S

s-plane, 17, 284, 329
Sample-and-hold (*see* zero-order hold)
Sampled-data control systems (*see also* dis-
crete-data control systems), 254
Sampler, 14, 50,
fictitious, 102
ideal, 50, 100
Sampling frequency, 101, 648
Sampling period, 50, 100, 108, 370, 373
Saturation nonlinearity, 178, 311
SDCS, 108, 370
Second-order system, 298, 327
prototype, 327, 352
Sensitivity, 6, 8, 515, 642, 691
root, 423
Sensors (*see also* encoders), 72
Series compensation, 461
Series-feedback compensation, 461
Settling time, 327, 337, 493
Signal flow graphs:
algebra, 82
basic elements, 77
basic properties, 79
branches, 77
construction, 84
definitions, 77
feedback control system, 83
forward path, 81
forward path gain, 82
gain formula, 87–99, 126, 143, 145, 147
input node, 80
loop, 81
loop gain, 82
nodes, 77
nontouching loops, 88–89
output node, 80, 92
path, 81
path gain, 82
Signal-to-noise ratio, 9
Single-value function, 17, 581
Singularity, 18
Sinusoidal signals, 310, 324
Sinusoidal steady state, 563
Speed-control system, 505

Spring,
  linear, 129, 142
  stiffness, 129
  torsional, 133–34, 140
Spring constant, 129
  torsional, 133, 140
Stability, 6–8, 279–98
  absolute, 279
  asymptotic, 282, 294
  Bode plot, 623
  bounded-input bounded-output (BIBO),
    280–82
  closed-loop, 587, 591
  continuous-data systems, 280–93
  discrete-data systems, 293–98
  gain-phase plot, 630
  marginal, 283, 330, 347
  multiloop systems, 611
  open-loop, 587
  relative, 279, 461, 606, 618
  time delays, 613
  zero-input, 282, 294
State diagram, 94–99, 126, 144–50, 171,
    203, 214, 231, 233, 241
  discrete-data systems, 250–53
  sampled-data systems, 254
State equations, 21, 47, 98, 124, 127,
    143–50, 171, 178, 204, 211, 213,
    217–22, 226, 244
  discrete-data systems, 243–50
  matrix representation, 47, 204–6
  nonlinear, 48
  vector-matrix form, 47
State feedback, 237, 461, 525
  integral control, 531
State transition, 95
State transition equation, 211–7, 231
  discrete (*see* discrete state-transition
    equations)
State transition matrix, 206–11, 257
  properties, 209–11
State variables, 22, 97, 124, 143–50
  definition, 97, 124
State vector, 47, 205
Static friction (*see* friction)
Steady-state error, 308, 310, 346, 471, 664
  definition, 313
  discrete-data systems, 380–84
  due to friction, 312
  due to nonlinear systems elements, 311
  summary of, 319
Steady-state response, 307
Step error constant (*see* error constant)
Step input, 309, 316, 323
Sun-seeker control system, 482, 523, 682
Suppressed-carrier modulation, 155

Synchro control transformer, 155–56
Synchros, 155–59
Synchro transmitter, 155

T

Tachometers, 159–60, 521
Tachometer-feedback control, 521, 526
  effects, 522
Taylor series, 320
Test signals (*see* typical test signals)
Third-order system, 346
Time-delay, 181–83, 613
  approximation, 182
  Bode diagram, 736
Time-domain analysis, 307–97
Time-domain design, 460–543
Time-invariant systems, 11
Time-varying systems, 11
Timing belts, 137, 140
Torque, 132–34
Torque constant, 170, 172
Torsional spring, 133–34
Total response, 280
Transfer function, 67–71, 97, 143, 145,
    147, 150, 226, 231, 257, 349
  closed-loop discrete-data systems, 106
  closed-loop systems, 74
  decomposition (*see* decomposition of
    transfer functions)
  discrete-data systems, 100–108, 249
  forward path, 74
  minimum-phase, 591
  multivariable systems, 70, 75
  single-input single-output systems, 68
  zero-order hold, (*see* zero-order hold)
Transfer function matrix, 71, 227–8, 248
Transient response, 307, 325, 327, 341
Translational motion, 127–8, 135, 143
Transportation lag (*see* time delay)
Trapezoidal integration, 539
Turboprop control, 70
Two-degree-of-freedom controller, 463,
    509
Types of control systems, 10–14, 315,
    318–9, 532
Typical test signals, 308

U

Undershoot, 334
Unit circle, 294
Unit-ramp input, 323

Unit-step function, 23, 52, 309, 323
Unit-step response, 326, 333–34, 341, 348, 352
Units:
  angular displacement, 133
  angular velocity, 133
  distance, 128
  force, 128
  friction, 129–31
  inertia, 133
  linear spring, 129
  mass, 128
  mechanical energy, 136
  mechanical power, 136
  spring constant, 134
  torque, 133
  torque constants, 172

V

Velocity, 128
Velocity control system, 11, 313
Viscous friction (*see* friction)
Viscous frictional coefficient, 130, 137, 139, 170

W

w-transformation, 295

Z

Zeros, 19
  definition, 19
Zero-input response, 280, 292
Zero-input stability, 282, 294
Zero-order hold, 100, 182, 243, 254, 648
  impulse response, 105
  state diagram, 254
  transfer function, 105
Zero-state response, 280
z-plane, 294, 374
z-transfer function, 103
z-transform, 50–60
  application to solution of linear difference equations, 59
  definition, 51
  discrete-data systems, 247, 370
  inverse, 56–58
  inversion formula, 58
  table, 53
  theorems, 53–56

Laplace-Transform Table

LAPLACE TRANSFORM $F(s)$	TIME FUNCTION $f(t)$
1	Unit-impulse function $\delta(t)$
$\dfrac{1}{s}$	Unit-step function $u_s(t)$
$\dfrac{1}{s^2}$	Unit-ramp function t
$\dfrac{n!}{s^{n+1}}$	t^n (n = positive integer)
$\dfrac{1}{s + \alpha}$	$e^{-\alpha t}$
$\dfrac{1}{(s + \alpha)^2}$	$te^{-\alpha t}$
$\dfrac{n!}{(s + \alpha)^{n+1}}$	$t^n e^{-\alpha t}$ (n = positive integer)
$\dfrac{1}{(s + \alpha)(s + \beta)}$	$\dfrac{1}{\beta - \alpha}(e^{-\alpha t} - e^{-\beta t})$ $(\alpha \neq \beta)$
$\dfrac{s}{(s + \alpha)(s + \beta)}$	$\dfrac{1}{\beta - \alpha}(\beta e^{-\beta t} - \alpha e^{-\alpha t})$ $(\alpha \neq \beta)$
$\dfrac{1}{s(s + \alpha)}$	$\dfrac{1}{\alpha}(1 - e^{-\alpha t})$
$\dfrac{1}{s(s + \alpha)^2}$	$\dfrac{1}{\alpha^2}(1 - e^{-\alpha t} - \alpha t e^{-\alpha t})$
$\dfrac{1}{s^2(s + \alpha)}$	$\dfrac{1}{\alpha^2}(\alpha t - 1 + e^{-\alpha t})$
$\dfrac{1}{s^2(s + \alpha)^2}$	$\dfrac{1}{\alpha^2}\left[t - \dfrac{1}{\alpha} + \left(t + \dfrac{2}{\alpha}\right)e^{-\alpha t}\right]$

LAPLACE TRANSFORM $F(s)$	TIME FUNCTION $f(t)$
$\dfrac{s}{(s+\alpha)^2}$	$(1-\alpha t)e^{-\alpha t}$
$\dfrac{\omega_n^2}{s^2+\omega_n^2}$	$\sin \omega_n t$
$\dfrac{s}{s^2+\omega_n^2}$	$\cos \omega_n t$
$\dfrac{\omega_n^2}{s(s^2+\omega_n^2)}$	$1-\cos \omega_n t$
$\dfrac{\omega_n^2(s+\alpha)}{s^2+\omega_n^2}$	$\omega_n\sqrt{\alpha^2+\omega_n^2}\,\sin(\omega_n t+\theta)$ where $\theta=\tan^{-1}(\omega_n/\alpha)$
$\dfrac{\omega_n}{(s+\alpha)(s^2+\omega_n^2)}$	$\dfrac{\omega_n}{\alpha^2+\omega_n^2}e^{-\alpha t}+\dfrac{1}{\sqrt{\alpha^2+\omega_n^2}}\sin(\omega_n t-\theta)$ where $\theta=\tan^{-1}(\omega_n/\alpha)$
$\dfrac{\omega_n^2}{s^2+2\zeta\omega_n s+\omega_n^2}$	$\dfrac{\omega_n}{\sqrt{1-\zeta^2}}e^{-\zeta\omega_n t}\sin \omega_n\sqrt{1-\zeta^2}\,t \qquad (\zeta<1)$
$\dfrac{\omega_n^2}{s(s^2+2\zeta\omega_n s+\omega_n^2)}$	$1-\dfrac{1}{\sqrt{1-\zeta^2}}e^{-\zeta\omega_n t}\sin(\omega_n\sqrt{1-\zeta^2}\,t+\theta)$ where $\theta=\cos^{-1}\zeta \qquad (\zeta<1)$
$\dfrac{s\omega_n^2}{s^2+2\zeta\omega_n s+\omega_n^2}$	$\dfrac{-\omega_n^2}{\sqrt{1-\zeta^2}}e^{-\zeta\omega_n t}\sin(\omega_n\sqrt{1-\zeta^2}\,t-\theta)$ where $\theta=\cos^{-1}\zeta \qquad (\zeta<1)$
$\dfrac{\omega_n^2(s+\alpha)}{s^2+2\zeta\omega_n s+\omega_n^2}$	$\omega_n\sqrt{\dfrac{\alpha^2-2\alpha\zeta\omega_n+\omega_n^2}{1-\zeta^2}}e^{-\zeta\omega_n t}\sin(\omega_n\sqrt{1-\zeta^2}\,t+\theta)$ where $\theta=\tan^{-1}\dfrac{\omega_n\sqrt{1-\zeta^2}}{\alpha-\zeta\omega_n} \qquad (\zeta<1)$
$\dfrac{\omega_n^2}{s^2(s^2+2\zeta\omega_n s+\omega_n^2)}$	$t-\dfrac{2\zeta}{\omega_n}+\dfrac{1}{\omega_n^2\sqrt{1-\zeta^2}}e^{-\zeta\omega_n t}\sin(\omega_n\sqrt{1-\zeta^2}\,t+\theta)$ where $\theta=\cos^{-1}(2\zeta^2-1) \qquad (\zeta<1)$